OXFORD SERIES ON MATERIALS MODELLING

Series Editors

Adrian P. Sutton, FRS
Department of Physics, Imperial College London

Robert E. Rudd
Lawrence Livermore National Laboratory

OXFORD SERIES ON MATERIALS MODELLING

Materials modelling is one of the fastest growing areas in the science and engineering of materials, both in academe and in industry. It is a very wide field covering materials phenomena and processes that span ten orders of magnitude in length and more than twenty in time. A broad range of models and computational techniques has been developed to model separately atomistic, microstructural and continuum processes. A new field of multi-scale modelling has also emerged in which two or more length scales are modelled sequentially or concurrently. The aim of this series is to provide a pedagogical set of texts spanning the atomistic and microstructural scales of materials modelling, written by acknowledged experts. Each book will assume at most a rudimentary knowledge of the field it covers and it will bring the reader to the frontiers of current research. It is hoped that the series will be useful for teaching materials modelling at the postgraduate level.

APS, London
RER, Livermore, California

1. M.W. Finnis: *Interatomic forces in condensed matter*
2. K. Bhattacharya: *Microstructure of martensite—Why it forms and how it gives rise to the shape-memory effects*
3. V.V. Bulatov, W. Cai: *Computer simulations of dislocations*
4. A.S. Argon: *Strengthening mechanisms in crystal plasticity*
5. L.P. Kubin: *Dislocations, mesoscale simulations and plastic flow*
6. A.P. Sutton: *Physics of elasticity and crystal defects*
7. D. Steigmann: *A Course on Plasticity Theory*
8. J.A. Moriarty: *Theory and application of quantum-based interatomic potentials in metals and alloys*

Forthcoming:
D.N. Theodorou, V. Mavrantzas: *Multiscale modelling of polymers*

Theory and Application of Quantum-Based Interatomic Potentials in Metals and Alloys

John A. Moriarty

OXFORD
UNIVERSITY PRESS

Great Clarendon Street, Oxford, OX2 6DP,
United Kingdom

Oxford University Press is a department of the University of Oxford.
It furthers the University's objective of excellence in research, scholarship,
and education by publishing worldwide. Oxford is a registered trade mark of
Oxford University Press in the UK and in certain other countries

© John A. Moriarty 2023

The moral rights of the author have been asserted

All rights reserved. No part of this publication may be reproduced, stored in
a retrieval system, or transmitted, in any form or by any means, without the
prior permission in writing of Oxford University Press, or as expressly permitted
by law, by licence or under terms agreed with the appropriate reprographics
rights organization. Enquiries concerning reproduction outside the scope of the
above should be sent to the Rights Department, Oxford University Press, at the
address above

You must not circulate this work in any other form
and you must impose this same condition on any acquirer

Published in the United States of America by Oxford University Press
198 Madison Avenue, New York, NY 10016, United States of America

British Library Cataloguing in Publication Data
Data available

Library of Congress Control Number: 2023930923

ISBN 978–0–19–882217–2

DOI: 10.1093/oso/9780198822172.001.0001

Printed and bound by
CPI Group (UK) Ltd, Croydon, CR0 4YY

Links to third party websites are provided by Oxford in good faith and
for information only. Oxford disclaims any responsibility for the materials
contained in any third party website referenced in this work.

Contents

Preface		xi
1	**Introduction**	**1**
1.1	Why quantum-based interatomic potentials	1
1.2	Basic concepts and nomenclature	5
	1.2.1 Total-energy functional, multi-ion potentials and radial vs. angular forces	5
	1.2.2 Environmental dependence, transferability and physical convergence	8
	1.2.3 Extension to alloys and intermetallic compounds	9
	1.2.4 Units, length scales and reference physical data	10
1.3	Failure of pure pair potentials in metals	11
1.4	Guiding principles in metals for quantum-based potentials	15
	1.4.1 Linear embedding in a uniform electron gas: volume-dependent potentials for bulk simple metals	16
	1.4.2 Extension of volume-dependent potentials to bulk transition metals	17
	1.4.3 Nonlinear embedding and free surfaces: glue and bond-order potentials	20
	1.4.4 Absence of bond charges in the electron density	23
	1.4.5 Finite ion-ion interaction range and order-N scaling	24
	1.4.6 Quantum-mechanical pillar of structural phase stability	26
	1.4.7 Machine learning and statistical interatomic potentials	29
2	**Fundamental Principles in Metals Physics**	**35**
2.1	Born-Oppenheimer or adiabatic approximation	35
2.2	Density functional theory	37
	2.2.1 Exchange and correlation functions in the local density approximation	39
	2.2.2 Exchange and correlation beyond the LDA	40
2.3	Small-core approximation and the valence binding energy in metals	41
2.4	Guidance from the DFT electronic structure: simple metals vs. d-band metals	46
2.5	Weak pseudopotentials and perturbation theory for simple metals	54
	2.5.1 Plane waves and nearly free-electron valence energy bands	58
	2.5.2 First-order electron density and second-order valence band energies	60
2.6	Localized d-states for the narrow d bands in transition-series metals	62

	2.6.1	Equivalence of resonance and tight-binding descriptions of the d bands	63
	2.6.2	Canonical d bands and their simplifying features	68
	2.6.3	Density of states moments in a tight-binding representation	71
2.7	Generalized pseudopotential theory for d-band metals		74
	2.7.1	Hybrid nearly free-electron tight-binding energy bands	77
	2.7.2	Pseudo-Green's functions in a mixed plane-wave, d-state basis	79
	2.7.3	Transition-metal ion with a d resonance in a free-electron gas	81
	2.7.4	Valence band-structure energy for bulk transition-series metals	85
	2.7.5	Valence electron density for bulk transition-series metals	88

3 Interatomic Potentials in Simple Metals — 91

3.1	Simple-metal cohesive-energy functional in DFT		92
	3.1.1	Reciprocal-space representation	93
	3.1.2	Real-space representation	100
3.2	Self-consistent electron screening		105
3.3	Evaluation of the energy-wavenumber characteristic and volume term		110
3.4	First-principles pair potentials for simple metals		114
	3.4.1	Impact of pseudopotential nonlocality and exchange and correlation	114
	3.4.2	Energy dependence and the optimization of nonlocal pseudopotentials	118
	3.4.3	Trends with valence, volume and atomic number	123
3.5	Long-range Friedel oscillations and materials application		125
	3.5.1	Exact methods for static lattice properties	125
	3.5.2	Damping and smooth truncation techniques for atomistic simulations	126
3.6	Higher-order corrections		130

4 Interatomic Potentials in Metals with Empty or Filled d Bands — 135

4.1	Inclusion of sp-d hybridization and d-state overlap in the GPT cohesive-energy functional		136
	4.1.1	Valence band-structure energy	137
	4.1.2	Valence electron density and self-consistent screening	144
	4.1.3	Volume term, energy-wavenumber characteristic and overlap potential	150
	4.1.4	Evaluation of the real-space volume term and pair potential	153
4.2	Zero-order pseudoatoms and optimized d basis states		157
4.3	Modified FDB-GPT treatment for the special case of the noble metals		166
4.4	Alternate resonant model potential approach		169
4.5	Trends in first-principles GPT pair potentials with atomic number and volume		172

5 Interatomic Potentials in Transition Metals — 176

- 5.1 GPT multi-ion potentials for metals with partially filled d bands — 178
  - 5.1.1 Essential new elements of the valence band-structure energy — 179
    - 5.1.1.1 Formal multi-ion d-state potential series — 180
    - 5.1.1.2 Tight-binding moments and multi-ion series convergence — 183
  - 5.1.2 Valence electron density and self-consistent screening — 188
    - 5.1.2.1 Zero-order pseudoatoms for transition metals — 188
    - 5.1.2.2 Oscillatory screening and orthogonalization-hole components — 194
  - 5.1.3 Completion of the cohesive-energy functional — 197
  - 5.1.4 First-principles pair and multi-ion potentials — 201
  - 5.1.5 Multi-ion screening of sp-d hybridization — 205
- 5.2 Simplified MGPT potentials for robust atomistic simulations — 208
  - 5.2.1 Baseline analytic MGPT for canonical d bands — 209
  - 5.2.2 Matrix MGPT for large-scale MD, noncanonical d bands and more — 216
  - 5.2.3 Optimized canonical d-band MGPT potentials for central bcc metals — 219
  - 5.2.4 Re-inclusion of sp-d hybridization and extension to late-series metals — 224
  - 5.2.5 Development of five- and six-ion potentials for mid-period metals — 227
- 5.3 Bond-order potentials for transition metals — 234
  - 5.3.1 Localized d-state moments expansion for the bond energy — 234
  - 5.3.2 Simplified analytic bond-order potentials — 242
  - 5.3.3 Parameterization of bond integrals and the repulsive energy — 246
- 5.4 Inclusion of magnetism in bond-order and MGPT potentials — 249

6 Structural Phase Stability and High-Pressure Phase Transitions — 253

- 6.1 Useful basic concepts and computational tools — 253
  - 6.1.1 Separation of cohesion and structure — 253
  - 6.1.2 Transformation paths connecting multiple structures — 257
  - 6.1.3 Simplified calculation of the total enthalpy difference between two structures at finite pressure — 260
- 6.2 QBIP-predicted structures and structural energies of the elements — 262
  - 6.2.1 Nontransition metals — 262
  - 6.2.2 Transition metals — 267
- 6.3 High-pressure phase stability and pressure-induced phase transitions — 271
  - 6.3.1 sp-d electron transfer across the Periodic Table and systematic trends — 272
  - 6.3.2 Successful GPT and MGPT predictions of new high-pressure phases — 275

7 Elastic Moduli and Phonons — 282

- 7.1 Quasiharmonic lattice dynamics for QBIP applications — 282
  - 7.1.1 Dynamical matrix and the calculation of normal-mode phonon frequencies — 284
  - 7.1.2 Tangential and radial force-constant functions for pair potentials — 287
  - 7.1.3 Tangential and radial force-constant functions for multi-ion potentials — 289
  - 7.1.4 Important caveats and alternate approaches — 294
- 7.2 Calculated quasiharmonic phonon spectra for elemental metals — 296
  - 7.2.1 Nontransition metals — 296
  - 7.2.2 Transition metals — 301
- 7.3 Elastic moduli for QBIP applications — 308
  - 7.3.1 Elastic constants from stress-strain in linear elasticity — 309
  - 7.3.2 Elastic constants and the long-wavelength limit of QHLD — 314
  - 7.3.3 EOS calculation of the bulk modulus and local environment corrections — 322
- 7.4 Thermodynamic properties in the QHLD limit — 326
- 7.5 Temperature-induced solid-solid phase transitions — 331

8 High-Temperature Properties, Melting and Phase Diagrams — 336

- 8.1 Important QBIP computational tools at high temperature — 336
  - 8.1.1 Molecular dynamics simulation with fast algorithms — 337
  - 8.1.2 Reversible-scaling MD for ion-thermal free energies — 342
    - 8.1.2.1 RSMD applied to a stable solid phase — 344
    - 8.1.2.2 RSMD applied to the liquid — 347
    - 8.1.2.3 RSMD applied to a metastable solid phase — 348
  - 8.1.3 Variational perturbation theory for liquid metals — 349
  - 8.1.4 Two-phase melting simulations and other dynamic methods — 352
- 8.2 Equation of state and high-temperature thermodynamic properties — 355
  - 8.2.1 Electron-thermal free energy in metals — 356
  - 8.2.2 Shock physics and the Hugoniot — 359
  - 8.2.3 Thermodynamic derivatives — 361
  - 8.2.4 Thermoelasticity and sound velocity — 363
- 8.3 Melting and the pressure-temperature phase diagram — 370
- 8.4 Rapid solidification and polymorphism in transition metals — 375

9 Defects and Mechanical Properties — 382

- 9.1 Point defect formation and migration energies — 382
- 9.2 Salient elastic and deformation properties of bcc transition metals — 388
  - 9.2.1 Shear elastic moduli and their pressure dependence — 389
  - 9.2.2 Ideal shear strength — 390
  - 9.2.3 Generalized stacking-fault energy surfaces — 394

9.3	Screw dislocation atomic structure and mobility in bcc transition metals		398
	9.3.1	Green's function simulation method for dislocation calculations	398
	9.3.2	Equilibrium dislocation core structures	401
	9.3.3	Movement under shear stress: kink-pair formation and the Peierls stress	405
	9.3.4	Kink-pair activation enthalpy and high-temperature mobility	409
9.4	Multiscale modeling of single-crystal plasticity in bcc transition metals		413
9.5	Grain-boundary atomic structure in bcc transition metals		419
9.6	Defect properties in fcc transition metals		422

10 Alloys and Intermetallic Compounds 425

10.1	General constrains with composition as an independent environmental variable		425
10.2	Nontransition-metal binary alloys and compounds		429
10.3	Transition-metal aluminides and their phase diagrams		436
	10.3.1	First-principles GPT interatomic potentials	436
	10.3.2	Basic trends in cohesion and structure	440
	10.3.3	Al-Co and Al-Ni binary phase diagrams	445
	10.3.4	Extension to ternary phase diagrams and quasicrystals	448
10.4	The special case of Ca-Mg		451
10.5	BOP treatment of transition-metal aluminides: TiAl		453
10.6	Treating pure transition-metal alloys with the MGPT		456

11 Local Volume Effects on Defects and Free Surfaces 460

11.1	Local-density representations of the GPT and their application		460
	11.1.1	Electron-density modulation	460
	11.1.2	Local density-of-states modulation	468
11.2	First-principles forces and stresses: the aGPT		471
	11.2.1	Force equations and the stress tensor	472
	11.2.2	Initial applications of the aGPT	476

12 Extension to f-Band Actinide Metals and p-Band Simple Metals 480

12.1	Localized p and f basis states in the GPT		481
	12.1.1	Canonical p bands for strong-PP simple metals	481
	12.1.2	Canonical f bands for lanthanide and actinide metals	484
12.2	MGPT representations of the early actinides U and Pu		487
	12.2.1	Weak electron correlation: a canonical f-band treatment of uranium	487
	12.2.2	Strong electron correlation: a novel canonical d-band treatment of δ-Pu	490

13 Interatomic Potentials with Electron Temperature — 493

    13.1 Some perspective on the importance of T_{el} in transition-metal melting — 494
    13.2 Extending the first-principles GPT to finite electron temperature — 496
    13.3 Temperature-dependent MGPT potentials and the simulation of melt for Mo — 502

Appendix A1 Units, Conversion Factors and Useful Physical Data — 508

Appendix A2 Additional Elements of Generalized Pseudopotential Theory — 519

    A2.1 Analytic forms for the LDA correlation energy ε_c — 519
    A2.2 Free-atom ionization energy $E_{\text{bind}}^{\text{atom}}$ and preparation energy E_{prep} — 520
    A2.3 Exchange-correlation correction terms $\delta\mu_{\text{xc}}^{\star}(n_i, n_j)$ and $\delta\varepsilon_{\text{xc}}^{\star}(n_i, n_j)$ — 521
    A2.4 The orthogonalization hole for an AHS pseudopotential — 521
    A2.5 Band-structure energies for fcc metals at high-symmetry BZ points — 523
    A2.6 Solution of the pseudo-Green's function equations for a d-band metal — 525
    A2.7 Approximate calculation of the d-state overlap kinetic energy $\delta v_{\text{ke}}^{ij}$ — 526
    A2.8 Self-consistent GPT pseudoatoms and the zero of energy for metals and alloys — 528

Glossary of Acronyms and Abbreviations — 531
Bibliography — 536
Subject Index — 564

Preface

Historically, the interatomic potentials used in most atomistic simulations have been simple empirical constructions, typically chosen in fixed analytic form with arbitrary parameters that are fitted to experimental or theoretical data. We know, however, that predictive power at atomic length scales comes from quantum mechanics, as amply demonstrated by the enormous success of density functional theory (DFT) over the past fifty years. At the same time, quantum simulations based on DFT are confined to small systems that are often no more than a few hundred atoms with time scales of a few picoseconds. In metals and alloys especially, a viable path forward to the vastly larger length and time scales offered by empirical potentials, while retaining the predictive power of DFT quantum mechanics, is to coarse-grain the underlying electronic structure and systematically derive quantum-based interatomic potentials (QBIPs) from first-principles considerations. This is possible because the valence energy bands in metals and alloys are amenable to simplified quantum treatments, leading to robust expansion of the total energy in terms of weak interatomic matrix elements that define the potentials.

The idea for a focused book on quantum-based interatomic potentials in metals and alloys first arose in the early 1990s in discussions with Roger Taylor, an early pioneer in the field in the 1970s and 1980s. Roger and his collaborators were not only developing first-principles pair potentials for simple sp-bonded metals based on pseudopotential perturbation theory, they were also leaders in using such potentials in atomistic simulations, including molecular dynamics (MD). During the same time period, we were developing generalized pseudopotential theory (GPT) within DFT quantum mechanics to allow calculation of QBIPs for d-band metals as well, including the essential d-bonding contributions to angular-force multi-ion potentials for transition metals. While Roger and I had a clear vision for a QBIP book, the reality of our busy non-academic careers soon cooled our passion for the book project, and it was reluctantly pushed to the back burner.

Although Roger subsequently moved on to other pursuits, my interest in the development and application of QBIPs has only continued to grow in the last thirty years. In the 1990s, we developed simplified *model* GPT (MGPT) potentials for mid-period transition metals, potentials that for the first time allowed large-scale MD simulations on such materials with full quantum-mechanical realism. In my Metals and Alloys Group at Lawrence Livermore National Laboratory (LLNL), the development and application of GPT and MGPT QBIPs opened up many new research areas with numerous collaborators both inside and outside of LLNL. Starting from our traditional work on equation of state and thermodynamic properties, this research extended to high-pressure phase transitions, melting and full phase diagrams; to defects, mechanical properties and multiscale modeling; to the treatment of surfaces, alloys and actinide metals; and most recently to high-temperature polymorphism and electron-temperature-dependent potentials. Current frontiers include the extension of MGPT across the entire transition-metals series. At the same time in the wider academic community, the general renaissance in tight-binding (TB) theory that began in the 1970s also led in the 1990s to significant QBIP-related research on semiconductors, transition metals and intermetallic compounds. Especially noteworthy in this regard are the advanced TB bond-order potentials (BOPs) of Pettifor and collaborators directed at transition metals, whose development and application continues with an eye toward large-scale atomistic simulation. Most recently, possible new pathways to more accurate and far-reaching QBIPs in metals and alloys have also emerged though the use of machine learning and high-performance computing.

In retirement, I have finally found the time to complete the book on QBIPs that I had tried to write thirty years ago. Fortunately, the present book is a much richer and more useful product today than would have been the case at that early time. It captures the substantial progress that has been made in the last three decades and puts the remaining challenges more clearly in focus. It is hoped that the book will be useful to students and researchers engaged in materials modeling on metals and alloys in a wide variety of disciplines. These disciplines include physics, chemistry, materials science, geophysics, metallurgy, materials engineering, mechanical engineering and related fields. In particular, the book has been designed to be as useful to interatomic-potential users as it is to interatomic-potential developers. In this regard, the book spans the entire QBIP process from foundation in first-principles theory, to the development of state-of-the-art potentials for real materials, to application of the potentials to materials modeling and simulation, to access to potential data and computer codes, as well as open-source computing resources such as LAMMPS.

Most major universities today have one or more postgraduate courses on materials modeling and/or atomistic simulation. This book could usefully serve as a supplemental text for such a course. The prerequisites for the book are postgraduate-level courses in quantum mechanics and condensed-matter theory or equivalent technical training. Readers wanting refreshers on these subjects are recommended to the early chapters of the previous Oxford books by Finnis (*Interatomic Forces in Condensed Matter*, 2003) and Pettifor (*Bonding and Structure of Molecules and Solids*, 1995). I have tried to design the book so that it would be equally useful to newcomers in the field and to experienced researchers. For the former audience, the book follows a logical progression starting from fundamental theory and proceeding all the way through potential development and materials physics application. For the latter audience, each chapter is reasonably self-contained, with numerous inter-chapter references where necessary, so knowledgeable readers can also skip to any specific topic of interest.

The specific content of the book is as follows. Chapter 1 provides an introduction to QBIPs and outlines the guiding physics principles in their development, while also giving an overview and roadmap of what is to follow in subsequent chapters. Chapter 2 elaborates the essential fundamental metals physics that provides the necessary theoretical starting points to develop QBIPs in both simple and transition metals. These theoretical results are then used in Chapter 3, 4 and 5 to develop real potentials for real materials in the respective cases of sp-bonded simple metals, pre- and post-transition metals with nearby empty and filled d bands, and d-bonded transition metals with partially filled d bands. The developed potentials are next applied to important materials physics problems in Chapters 6, 7, 8, 9 and 10. These applications are to structural phase stability and high-pressure phase transitions in Chapter 6; to elastic moduli and phonons in Chapter 7; to high-temperature properties, melting and phase diagrams in Chapter 8; to defects and mechanical properties in Chapter 9; and to alloys and intermetallic compounds in Chapter 10. The final three chapters discuss important ongoing extensions of the theory and its applications. Chapter 11 treats local-volume corrections to certain defect energies and the corresponding extension from the homogeneous bulk to inhomogeneous free surfaces. Chapter 12 discusses the extension of the d-band transition-metal theory to f-band actinide metals and also to the special case of non-ideal p-band simple metals. Finally, in Chapter 13 we address the extension of the usual zero-electron-temperature quantum mechanics to finite electron temperature and the development of QBIPs in transition metals with electron temperature.

Finally, I would like to acknowledge the large number of scientific colleagues and collaborators who have impacted either directly or indirectly the work described in this book. First on that list are my early-career academic and scientific mentors, Walt Harrison, John Wood and Volker Heine, who taught me the importance of quantum mechanics in understanding materials behavior and how to apply quantum techniques productively in that pursuit. Next, are the large number of talented and knowledgeable theorists I have had the pleasure to engage, and in many cases collaborate with, on QBIP-related research during my scientific career, including students, postdocs and senior researchers. These colleagues include (in alphabetical order) Jeff Althoff, Neil Ashcroft, Jim Belak, Anatoly Belonoshko, Lorin Benedict, Dave Boercker, Vasily Bulatov, Leonid Burakovsky, Anders Carlson, Murray Daw, Mike Duesbery, Mike Finnis, Stephen Foiles, Mike Gillan, Jim Glosli, Carl Greef, Jürgen Hafner, Justin Haskins, John Hirth, Randy Hood, John Klepeis, Ladislas Kubin, Alex Landa, Ben Liu, Andy McMahan, Tomas Oppelstrup, Daniel Orlikowski, Mehul Patel, Tony Paxton, David Pettifor, Rob Phillips, Satish Rao, Marvin Ross, Rob Rudd, Guy Skinner, Per Söderlind, Fred Streitz, Ellad Tadmor, Meijie Tang, Roger Taylor, Vasek Vitek, Mike Widom, John Wills, Chris Woodward, Christine Wu, Wei Xu, Lin Yang, Sid Yip, David Young and Jing Zhu. Additionally, I have had the good fortune to work with many equally talented and knowledgeable experimentalists, whose accurately measured data have provided important validation of QBIP-predicted materials properties. These researchers include Geoff Campbell, Jonathon Crowhurst, Hyunchae Cynn, Roger Gathers, Rob Hixson, Neil Holmes, Wilfried Holzapfel, Wayne King, Art Mitchell, Bill Nellis, Jeff Nguyen, John Shaner, Joe Wong, Choong-Shik Yoo and Joe Zaug. Special thanks to Guy Skinner, Lin Yang and David Young for reading and providing constructive comments on portions of the original manuscript.

1
Introduction

1.1 Why quantum-based interatomic potentials

Atomistic computer simulations are often at the heart of modern attempts to predict and understand the physical properties of real materials, including the vast domain of metals and alloys that is the central focus of this book. True predictive power in these materials emanates directly from quantum mechanics, which controls such fundamental properties as structural phase stability, elastic moduli and lattice vibrations, point and line defect energetics, phase boundaries and melting, and, more generally, the thermodynamic behavior of both the solid and the liquid. A quantum-mechanical starting point can also be important to the predictive multiscale modeling of higher length-scale phenomena in metals such as plasticity and mechanical strength, fracture, and rapid solidification and grain formation, all phenomena that occur across multiple length and time scales. Here a key challenge is to bridge the gap between quantum mechanics and the large-scale atomistic simulation of the thousands or even millions of atoms needed to describe and quantify the relevant atomic processes, such as defect formation, motion and interaction, which in turn control higher-scale behavior. For example, modeling single-crystal plasticity via micron-scale dislocation-dynamics (DD) simulations (Tang, 2005; Bulatov and Cai, 2006), which track the detailed evolution of the dislocation microstructure, requires accurate atomistic input information on the mobility and interaction of individual dislocations (Yang et al., 2010).

There currently exists a wide spectrum of available atomic-scale computational methods in condensed-matter and materials physics, ranging all the way from essentially exact quantum-mechanical techniques to classical descriptions based on empirical interatomic force laws. A representative sample of these methods relevant to the present discussion on metals and alloys is indicated schematically in Fig. 1.1 with respect to the approximate number of atoms that can be simulated in each case via high performance computing (HPC) on modern supercomputers. All of these methods may be grouped into one of two general classes or categories, separated by a material-dependent length-scale gap. On one side of this gap (the left-hand side or LHS of Fig. 1.1) are electronic methods based on direct quantum-mechanical treatments of a material's valence electrons in the presence of compensating ions. These methods include rigorous many-body approaches based on correlated electron theory, such as quantum Monte Carlo

2 Introduction

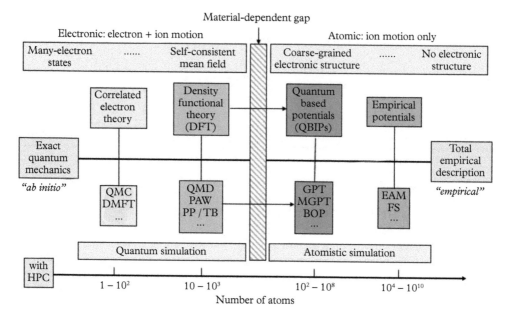

Fig. 1.1 *Representative sample of the wide spectrum of electronic and atomic simulation methods used in condensed-matter and materials physics and the material-dependent length-scale gap separating them. Indicated are the estimated numbers of atoms that could be usefully simulated in each category via high-performance computing (HPC) on modern supercomputers.*

(QMC), as well as more widely tractable self-consistent mean-field approaches based on density-functional theory (DFT) (Hohenberg and Kohn, 1964; Kohn and Sham, 1965). Numerous first-principles DFT electronic-structure methods have now been developed and implemented (Martin, 2004). Especially noteworthy in the present context are DFT methods utilizing simplifying plane-wave electron-ion pseudopotentials (PPs), projector-augmented waves (PAWs), or localized atomic orbitals in tight-binding (TB) approaches. Efficient PP, PAW and TB techniques also permit calculation of reliable interatomic forces for use in quantum molecular dynamics (QMD) simulations. In a QMD simulation one treats electron and ion motion on an equal footing, simultaneously solving DFT equations on the fly for the electronic states of the system and the forces that move the individual ions. For materials with weak electron correlation effects, which include *sp*-bonded simple metals, *d*-bonded transition metals, and early-series *f*-bonded actinide metals, all DFT approaches can provide a reliable description of the energetics of the system. But DFT-based QMD simulation methods come at the price of being severely limited in terms of the maximum number of atoms that can be treated at a reasonable cost (typically 250 – 500) and the corresponding short duration of the simulations (typically 1 – 10 ps).

On the other side of the material-dependent length-scale gap (the right-hand side or RHS of Fig. 1.1) are atomistic methods, which treat only ion motion and allow dramatically larger numbers of atoms to be simulated (up to $10^8 - 10^{10}$) and dramatically longer simulations (up to $1 - 10$ μs) to be performed. This is accomplished through explicit interatomic potentials that provide the needed interatomic forces for atomistic simulations such as molecular dynamics (MD). Highly simplified empirical potentials usually provide maximum computational efficiency in this regard and are widely used. Such potentials are established by assuming convenient analytic functional forms with free parameters that are then fit to either experimental or theoretical data. Popular examples of empirical potentials in metals are the short-ranged central-force forms of Daw and Baskes (1983, 1984) used in the embedded-atom method (EAM) and of Finnis and Sinclair (1984) (FS) in their N-body potential. Both the EAM and FS potential forms contain a nonlinear many-body embedding term inspired by quantum-mechanical considerations, allowing for both bulk and surface calculations. However, the simple fixed forms adopted do not carry forward predictive DFT quantum mechanics in any reliable way, and this is true of empirical potentials in general. This shortcoming is particularly apparent in directionally bonded transition metals, where an accurate treatment of quantum-mechanical angular forces arising from partially filled d-state energy bands is necessary to explain structural phase stability and other fundamental properties.

Alternatively, one may attempt to model across the electronic-atomic gap in Fig. 1.1 to derive explicit quantum-based interatomic potentials (QBIPs) by systematically coarse-graining the DFT electronic structure, while retaining the essential quantum-mechanical content of the electron density and total energy. In simple *sp*-bonded metals and alloys, for example, this has been done entirely from first principles within DFT-based PP perturbation theory (Dagens et al., 1975; McMahan and Moriarty, 1983; Hafner, 1987a). The result is a derived, rather than assumed, many-body total-energy functional with self-consistent pair potentials that can reliably predict fundamental properties of bulk materials, including structural phase stability. In the more challenging d-bonded transition metals, localized-d-state TB expansions can capture the unique physics of the narrow, partially filled d bands, and make possible the development of predictive multi-ion interatomic potentials and angular forces in these materials. This has been accomplished within the first-principles DFT-based generalized pseudopotential theory (GPT) of Moriarty (1985, 1988a), and in subsequent simplified *model* GPT (MGPT) potentials (Moriarty, 1990a, 1994; Yang et al., 2001, 2010; Moriarty et al., 2002a, 2006, 2008, 2012; Haskins et al., 2012; Moriarty and Haskins, 2014; Haskins and Moriarty, 2018), as well as in the TB bond-order potentials (BOPs) of Pettifor and co-workers (Pettifor, 1989, 1995; Horsfield et al., 1996a, 1996b; Aoki, 2003; Aoki et al., 2007; Mrovec et al., 2004, 2007, 2011; Drautz and Pettifor, 2006, 2011; Drautz et al., 2015). The development of fast MGPT algorithms have further allowed large-scale MD simulations in mid-period transition metals with quantum-mechanical realism, extending to more than 30 million atoms (Streitz et al., 2006a). The first-principles GPT also provides additional *sp-d* hybridization contributions to the multi-ion potentials, contributions that are especially important to early and late transition-series metals, and we are now adding back neglected *sp-d* hybridization to MGPT transition-metal potentials in simplified form (see Chapter 5). In addition, the fundamental physics content of GPT, MGPT and BOP is transferable from d states to f states for application to f-bonded actinide metals, and this has been accomplished in the case of MGPT multi-ion potentials (see Chapter 12).

In the past fifty years we have witnessed an enormous worldwide research and development effort on DFT electronic-structure methods, with many mature technologies and accessible computer codes now available to implement the various approaches. We view QBIP research and development in metals and alloys as fully complementary to that effort, but collectively QBIP activity has so far proceeded at a much less vigorous and sustained pace, especially relative to its potential importance. Consequently, QBIP development is still at a relatively early stage, with many practical challenges remaining and with new ideas continuing to emerge and impact the landscape. Because QBIPs are based on sound quantum-mechanical principles, they share in common the essential property of being systematically improvable. Often this is a matter of modifying or eliminating nonessential secondary approximations that were initially introduced for reasons of simplicity or computational convenience. Systematic improvement may also be achieved by the introduction of new pathways such as the HPC statistical concept of machine learning (ML) to infuse complex quantum-mechanical content into an existing or generalized QBIP framework.

In this book, we will review the underlying principles essential to QBIP development in metals and alloys, the substantial progress that has already been made, and the remaining challenges and promising future research directions. Toward this end, it is perhaps useful to list at the outset some of the important long-term, high-impact objectives or goals that we foresee as achievable with the mature development of reliable and efficient QBIPs in metals and alloys:

i) to extend the effective reach of DFT simulations by up to five or six orders of magnitude in both space and time within acceptable limits of accuracy;

ii) to add quantum-level understanding and insight into complex physical processes such as phase transitions and dynamic materials response, including finite size and time effects in the modeling;

iii) to facilitate and enhance predictive multiscale modeling of mechanical properties and physical processes such as solidification and grain formation, including more rigorous and certain links to higher length- and time-scale modeling;

iv) to aid in the search for new intermetallic alloys with tailored thermodynamic and mechanical properties; and

v) to aid in the search for new high-pressure and high-temperature structural phases of elemental metals and intermetallic alloys, including reliable characterization of high-temperature mechanical and thermodynamic stability.

1.2 Basic concepts and nomenclature

1.2.1 Total-energy functional, multi-ion potentials and radial vs. angular forces

The central theoretical quantities in DFT, namely the electron density $n(\mathbf{r})$ and the total energy of the system as a functional of the electron density, $E_{\text{tot}}[n(\mathbf{r})]$, are also major focal points in QBIP development. For an elemental metal with N identical ions, the total-energy functional may also be considered to be a function of the N ion positions $\mathbf{R} \equiv \mathbf{R}_1, \mathbf{R}_2, \ldots \mathbf{R}_N$ in the system. In a DFT electronic-structure calculation or QMD simulation, the dependence of E_{tot} on the positions of the ions is usually implicit and hidden from view within a reciprocal-space formalism, with the solution method depending heavily on the symmetry of the system and typically leading to order-N^3 scaling of E_{tot} with system size. Within a real-space QBIP representation of the total energy, on the other hand, one seeks to make the dependence of E_{tot} on the ion positions explicit and independent of symmetry, with order-N scaling and the ability to treat large systems. In either DFT or QBIP applications, $E_{\text{tot}}(\mathbf{R}_1, \mathbf{R}_2, \ldots \mathbf{R}_N)$ becomes the total potential energy U_{tot} of the system when used in a classical simulation method like MD:

$$U_{\text{tot}}(\mathbf{R}) = E_{\text{tot}}(\mathbf{R}_1, \mathbf{R}_2, \ldots \mathbf{R}_N). \tag{1.1}$$

The corresponding force $\mathbf{F}_i$ on the ith ion,

$$\mathbf{F}_i(\mathbf{R}) = -\nabla_i U_{\text{tot}}(\mathbf{R}), \tag{1.2}$$

can then be used to move the ion via an appropriate classical equation of motion. In the general context of atomistic simulations and interatomic potentials, $U_{\text{tot}}(\mathbf{R})$ is also often referred to as the potential-energy surface (PES).

In a QBIP representation of the total energy, one may also anticipate some other general features of $E_{\text{tot}}(\mathbf{R}_1, \mathbf{R}_2, \ldots \mathbf{R}_N)$. Consider, for example, an sp-bonded simple metal, where one envisages the mobile atomic species to consist of positively charged ions of charge $+Ze$ immersed in a compensating sea of ZN itinerant valence electrons that screen the ions and modify their interactions. Here Z is the number of valence s and p electrons per atom, as determined by the column number of the element in the Periodic Table (e.g., $Z = 1$ for Na, $= 2$ for Mg, and $= 3$ for Al). In this case, the immediate thing one can anticipate about E_{tot} is that it consists of a pairwise additive sum of repulsive Coulomb potentials between bare ions plus a compensating set of indirect ion-electron-ion interactions that screen the ions by the valence electrons and produce cohesion in the condensed metallic state. Thus, within DFT quantum mechanics, one can write

$$E_{\text{tot}}(\mathbf{R}_1, \mathbf{R}_2, \ldots \mathbf{R}_N) = \frac{1}{2} {\sum_{i,j}}' v_2^{\text{ion}}(R_{ij}) + E_{\text{ion-el}}[n(\mathbf{r})], \tag{1.3}$$

where $v_2^{\text{ion}}(R_{ij}) = (Ze)^2/R_{ij}$ with $R_{ij} = |\mathbf{R}_i - \mathbf{R}_j|$, and the prime on the sum excludes the $i = j$ self-interaction term. The term $E_{\text{ion-el}}[n(\mathbf{r})]$ in Eq. (1.3) denotes the remaining ion-electron-ion interaction energy, which is an intrinsically many-body property of the metal, only a part of which can be represented by a pairwise potential and added to v_2^{ion} to form a screened two-ion pair potential.

A similar conceptual picture actually holds more generally for transition and actinide metals as well, with appropriate caveats. In an elemental d-bonded transition metal, one has, in addition to Z loosely bound s and p valence electrons per atom, some number Z_d of more tightly bound d valence electrons per atom, whose localized electron density effectively neutralizes the corresponding nuclear charge $+Z_d e$ on each ion in the metal. The d electrons contribute to the overall valence electron screening through sp-d hybridization or mixing of the electronic states of the system, making Z and Z_d non-integer variable numbers that must be determined self-consistently, with only $Z + Z_d$ fixed by the column number of the element in the Periodic Table. Thus Eq. (1.3) continues to apply with all remaining complications engendered by the d electrons being absorbed into $E_{\text{ion-el}}$. Similarly, in an elemental f-bonded actinide metal, Z_d is replaced by Z_f, corresponding to a variable number of tightly bound f electrons, and Z then represents all remaining loosely bound non-f valence electrons, with $Z + Z_f$ fixed.

To express $E_{\text{ion-el}}[n(\mathbf{r})]$ in Eq. (1.3) theoretically and develop QBIPs for metals from first-principles considerations, one must deal directly with the quantum mechanics of the electron-ion interaction. This subject will be treated in detail in Chapters 2–5 of this book for bulk simple and transition metals, with extensions to alloys and intermetallic compounds in Chapter 10, to free surfaces in Chapter 11, and to actinide metals in Chapter 12. Here we first discuss a few of the key general concepts that will arise.

In both DFT electronic-structure and QBIP approaches, one seeks to obtain the electron density and total energy by means of an efficient wave-function basis set. For QBIPs, one also needs a basis set that produces weak interatomic matrix elements coupling one ion with another, so that $n(\mathbf{r})$ and $E_{\text{ion-el}}$ can be developed as useful expansions in these matrix elements. In metals and alloys, this can be accomplished with plane waves and weak pseudopotentials for broad nearly free-electron (NFE) sp bands, and with localized atomic d or f states and short-ranged TB bond integrals for narrow d or f bands. It is convenient here to introduce QBIPs into Eq. (1.3) somewhat more generally through an envisaged expansion of $E_{\text{ion-el}}$ about an appropriate zero-order reference system with known total energy E_0^{ref}. Specifically, we consider a many-body cluster expansion in an unspecified basis set, but corresponding to a series of successively higher-order real-space interactions between ions, such that in an elemental bulk metal E_{tot} takes the form

$$E_{\text{tot}}(\mathbf{R}_1, \mathbf{R}_2, \ldots \mathbf{R}_N) = E_0^{\text{ref}} + \frac{1}{2}{\sum_{i,j}}' v_2(ij) + \frac{1}{6}{\sum_{i,j,k}}' v_3(ijk) + \frac{1}{24}{\sum_{i,j,k,l}}' v_4(ijkl) + \cdots. \tag{1.4}$$

1.2.1 Total-energy functional, multi-ion potentials and radial vs. angular forces

In this expansion, we suppose that $v_2(ij) \equiv v_2(R_{ij})$ is a screened two-ion *pair* potential between ions i and j that has absorbed v_2^{ion} in Eq. (1.3) and remains a continuous function of the radial separation distance R_{ij}. Likewise,

$$v_3(ijk) \equiv v_3(R_{ij}, R_{jk}, R_{ki}) \tag{1.5}$$

is a continuous three-ion *triplet* potential of the three separation distances R_{ij}, R_{jk} and R_{ki} defining a triangle among ions i, j and k, as shown in Fig. 1.2, while

$$v_4(ijkl) \equiv v_4(R_{ij}, R_{jk}, R_{kl}, R_{li}, R_{ki}, R_{lj}) \tag{1.6}$$

is a continuous four-ion *quadruplet* potential of the six separation distances connecting ions i, j, k and l, as also shown in Fig. 1.2. As in Eq. (1.3), the prime on each sum in Eq. (1.4) excludes all self-interaction terms where two indices are equal and the fraction multiplying each sum compensates for the multiple counting inherent in the otherwise unrestricted summations over all N ions. Higher-order potentials in this expansion would be similarly defined, but obviously the complexity of these potentials increases very rapidly beyond the level of pair interactions, so that there is a high premium on rapid convergence and an early termination of the series. It is also to be noted that v_2 will have only radial derivatives and hence corresponds to purely central-force or *radial-force* interactions between ions, while all higher many-body or *multi-ion* potentials will have nonradial derivatives with respect to any ion pair and hence correspond to noncentral-force or *angular-force* interactions. It should be further noted in Eq. (1.4) that the total-energy functional E_{tot} varies linearly with the near-neighbor coordination number Z_{cn} for each ion. This is a good approximation for bulk conditions, where Z_{cn} is large and changes in it are small when ion positions are rearranged. But more generally,

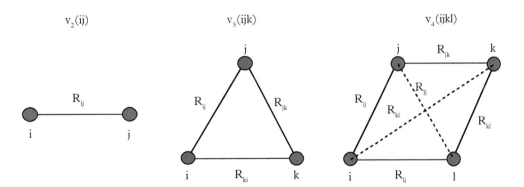

Fig. 1.2 *Atomic configuration geometry corresponding to the multi-ion potentials v_2, v_3 and v_4 in Eqs. (1.4)–(1.6).*

the dependence of E_{tot} on Z_{cn} is nonlinear, and thus an expansion like Eq. (1.4) must necessarily be altered as one goes from the bulk to the free surface.

1.2.2 Environmental dependence, transferability and physical convergence

The reference energy E_0^{ref} in Eq. (1.4) implicitly includes any one-ion intra-atomic contribution to the total energy, but more generally it will also include some collective many-body contribution to the total energy as well. In any case, the reference system itself is not unique and is a matter of physical choice. Consequently, the interatomic potentials v_2, v_3, v_4, etc. and the convergence of this series depend on the reference system chosen and are not unique properties of the metal. Rather, the pair and multi-ion potentials in Eq. (1.4) are dependent on the environment in which they are derived. Nonetheless, such *environmentally dependent* potentials may be perfectly well defined and calculable from first principles if our envisaged series is consistent with the physics of the system we are trying to describe and thus derivable from quantum mechanics. The environmental dependence of given interatomic potentials can be either implicit and hidden from view or explicit in terms of specific physical variables that characterize the state of the system and are directly accessible. Empirical potentials typically have an implicit environmental dependence on the particular data that was used to fit the potential. Quantum-based interatomic potentials, on the other hand, can have an explicit environmental dependence that arises from the particular quantum-mechanical formulation used. For example, using plane waves in whole or in part in the basis set leads to an explicit dependence of the total energy and the potentials on the volume of the system V, or equivalently the atomic volume $\Omega = V/N$. This is the case in PP perturbation theory, as well as in GPT and MGPT potentials, and can actually be a decided advantage for applications that extend over significant volume ranges such as materials properties at high pressure. Another, somewhat less common, example of an explicit environmental variable is electron temperature T_{el}, which, in addition to the ordinary ion temperature T_{ion} that appears in MD simulations, can impact $3d$ and $4d$ transition metals at high temperatures near melt and above. While standard practice is to construct QBIPs at $T_{\text{el}} = 0$, temperature-dependent MGPT potentials with $T = T_{\text{ion}} = T_{\text{el}}$ have been developed and successfully applied to transition-metal melting (Moriarty et al., 2012). This extension, which puts QBIPs at the same level of T_{el} treatment as QMD simulations, will be discussed in Chapter 13.

To be truly useful, however, one needs not only well-defined potentials and a rapidly convergent total-energy expansion, but potentials that are also *transferable* to a wide variety of situations. That is, one wants the same interatomic potentials to apply as the local environment of any individual ion is continuously changed. If one represents the elemental metal by an arbitrarily large box of atoms with N independent ion positions, then *full* transferability means that a given functional $E_{\text{tot}}(\mathbf{R}_1, \mathbf{R}_2, \ldots \mathbf{R}_N)$, and its associated interatomic potentials, can accurately represent the total energy of the system for any arrangement of the ion positions. Clearly, neither DFT itself nor any DFT-based QBIP representation has full transferability, since neither can treat arbitrarily small ion-ion separation distances or arbitrarily sparse near-neighbor environments, where electron correlation effects become large. One must therefore define various sub-classes of transferability with specific restrictions. Two useful sub-classes for QBIPs in metals are *bulk* and *surface* transferability. For bulk transferability, one adds the restrictions of a minimum ion-ion separation distance and a dense near-neighbor environment, so as to include all metallic atomic structures, either ordered or disordered, and structures with the presence of point and line defects. For surface transferability, one appropriately modifies the near-neighbor environment requirement, so as to include cleaved free surfaces, macroscopic voids and cracks in the bulk, and nanostructures.

In practice, both rapid convergence of the total-energy expansion and high transferability of its defined potentials are sought. Ultimately, the importance of the multi-ion potentials v_3, v_4, etc. in an expansion like Eq. (1.4) will depend on the reference system chosen, the type of metal (e.g., simple or transition), and the macroscopic environment (e.g., bulk or surface) under consideration. In bulk *sp*-bonded simple metals (e.g., Na, Mg or Al), pair potentials will normally suffice with an advantageous choice of a many-body reference system. The expansion (1.4) will necessarily require modification at surfaces, but multi-ion potentials are usually not needed for either the bulk or surface. In *d*-bonded transition metals (e.g., Ta, Mo or Fe), on the other hand, multi-ion potentials will generally be important for both the bulk and the surface.

As to the convergence of a total-energy expansion like (1.4), one is normally interested in its effective *physical* convergence as opposed to its strict *mathematical* convergence. Physical convergence is the point at which the relevant physics has been captured and $E_{\text{tot}}(\mathbf{R})$ can reliably predict materials properties to some reasonable level of accuracy (say, to 5–10% or so). In this situation, forces obtained via Eq. (1.2) normally need to be calculated by differentiating the total-energy expansion term by term to control numerical precision. If, on the other hand, forces via Eq. (1.2) are obtained through a total-energy sum rule, such as the Hellman-Feynman theorem (Finnis, 2003), then first converging the total-energy expansion to high numerical precision (say, to 1% or better) is necessary to obtain reliable forces.

1.2.3 Extension to alloys and intermetallic compounds

All the same concepts apply to alloys and intermetallic compounds as well with one major complication, namely, that one must distinguish between potentials for like and unlike atomic species. In an AB binary system, therefore, one must allow for v_2^{AA}, v_2^{AB} and v_2^{BB} pair potentials; v_3^{AAA}, v_3^{AAB}, v_3^{ABB} and v_3^{BBB} triplet potentials; etc. In addition, one can expect that such potentials will have an environmental dependence on the concentrations of the different alloy components. On the other hand, one's objective in making an expansion like Eq. (1.4) may sometimes be more limited in the case of alloys.

For example, one might be interested in how the total energy changes with atomic composition for a fixed geometrical arrangement of lattice sites. Thus, the reference system in that case could be the completely disordered or random alloy for a given underlying crystal structure, and the interatomic potentials would represent the changes in total energy with the partial ordering of two or more atomic species. Such potentials would take on values only for discrete near-neighbor distances and are sometimes referred to as "ordering potentials" of an alloy (Ducastelle, 1991). More generally, however, the alloy potentials $v_n^{\alpha\beta\cdots}$ that we consider in this book will be fully continuous functions of position, as envisaged above for elemental metals, with the same issues of bulk and surface transferability as well as the issues of concentration-dependent and volume-dependent potentials.

1.2.4 Units, length scales and reference physical data

In applying interatomic potentials to physical problems, a variety of length, energy, mass and time scales are in common usage. In this book we will acknowledge that fact and establish only a minimal number of conventions. In translating quantum mechanical equations into computational forms, the introduction of atomic units (a.u.) is convenient, with lengths expressed in Bohr radii and energies in either Rydbergs (Ry) or Hartrees. We will confine our use of atomic units here to Rydberg a.u., which are realized by setting $\hbar^2 = 2m = e^2/2 = 1$ in the DFT equations, where m is the mass of the electron. In many applications we will use electron volts (eV) instead of Rydbergs for energy, and in either case we will use gigapascals (GPa) for pressure and elastic moduli. Useful conversion factors among these and all other needed units are given in Appendix A1.

It is also convenient here to define a dimensionless length scale for interatomic distances, $r/R_{\rm WS}$, where $R_{\rm WS}$ is the familiar Wigner-Seitz radius,

$$R_{\rm WS} = \left(\frac{3\Omega}{4\pi}\right)^{1/3}. \tag{1.7}$$

This quantity is alternately referred to as the atomic sphere radius. In units of $r/R_{\rm WS}$, neighbor distances in any given crystal structure are fixed numbers independent of the material or the atomic volume Ω under consideration. For example, in the fcc and ideal hcp structures, the nearest-neighbor distance in these units is 1.809, while in the bcc structure the first- and second-neighbor distances are 1.759 and 2.031, respectively.

An additional useful concept is that of a reference atomic volume Ω_0. This quantity is defined here as the equilibrium or zero-pressure atomic volume of a metal at a chosen temperature. For example, Ω_0 could be the predicted zero-pressure volume at $T = 0$ for a given metal, if one desires to establish a theoretical value for this quantity. It is usually more convenient, however, to choose Ω_0 to be the observed equilibrium volume at room temperature (or other experimental temperature) and ambient pressure, where most experimental data on materials properties have been measured. This latter definition will be the default choice of reference atomic volume in this book unless otherwise specified. Useful data tables of observed values of Ω_0 and other basic physical data for elemental metals are also given in Appendix A1.

1.3 Failure of pure pair potentials in metals

Perhaps the most basic reference system one could consider in contemplating the total-energy expansion (1.4) is a collection of N isolated free atoms. Then $E_0^{\text{ref}}/N = E_{\text{tot}}^{\text{atom}}$, the total energy of the free atom, and the *cohesive energy* of the metal per atom is just

$$E_{\text{coh}}(\mathbf{R}) = \frac{1}{N} E_{\text{tot}}(\mathbf{R}_1, \mathbf{R}_2, \ldots \mathbf{R}_N) - E_{\text{tot}}^{\text{atom}}$$

$$= \frac{1}{2N} \sum_{i,j}{}' v_2(R_{ij}) + \cdots . \qquad (1.8)$$

In this expansion, it is implicitly assumed that the condensed phases of the material can be represented by a system of rather weakly interacting free atoms. For the ideal case of simple insulators such as rare-gas solids (e.g., Ar), with tightly bound closed-shell s and p valence electrons, this is entirely justified. This physical picture is clearly less appropriate for the loosely bound open-shell s and p valence electrons in metals; nonetheless, it is quite instructive to see at the outset exactly what practical difficulties are encountered with such a starting point. For simplicity, let us suppose that the multi-ion potentials beyond v_2 in Eq. (1.8) can be neglected, or at least averaged over and combined with v_2, and attempt to describe E_{coh} solely in terms of a pair potential. Much of the early history of empirical interatomic potentials in metals prior to the mid 1980s is closely related to this point of view (Gehlen, 1972; Torrens, 1972; Johnson, 1973; Lee, 1981). If the envisaged pair potential is dominated by its short-range behavior, as is usually assumed, then to explain the basic cohesive properties of a given metal $v_2(r)$ must have a deep negative minimum in the vicinity of the first nearest-neighbor distance d_{nn}. At very short distances, on the other hand, one expects v_2 to be large and positive to reflect the inability of one atom to penetrate the closed-shell inner core of another (so-called "hard-core" repulsion). Simple analytic pair potentials consistent with these requirements are, for example, the classic Morse potential (Morse, 1929):

$$v_2^{\text{M}}(r) = v_0 \{\exp[-2\alpha(r - r_0)] - 2\exp[-\alpha(r - r_0)]\}, \qquad (1.9)$$

with free parameters v_0, r_0 and α, as well as the classic Lennard-Jones (LJ) potential (Lennard-Jones, 1924):

$$v_2^{\text{LJ}}(r) = v_0[(r_0/r)^{12} - 2(r_0/r)^6], \qquad (1.10)$$

with free parameters v_0 and r_0. For simple insulators, the second r^{-6} term in Eq. (1.10) represents the real van der Waals attraction that binds rare-gas solids, as is well known. Less well known is the interesting fact that the LJ potential form also arises more generally in metals from localized p-state interactions, as discussed in Chapter 12.

Regardless of its mathematical form, the free parameters in any such empirical pair potential are determined by the elementary cohesive properties of the metal, principally the cohesive energy, the pressure and the bulk modulus. The cohesive energy of the perfect crystal with equilibrium lattice sites $\mathbf{R}_i^0$ reduces to

$$E_{\text{coh}}^0 = \frac{1}{2N} {\sum_{i,j}}' v_2(R_{ij}^0) = \frac{1}{2} \sum_{i \neq 0} v_2(R_i^0), \qquad (1.11)$$

where in writing the second equality we have assumed for simplicity that all N sites in the crystal are equivalent. Additionally, one requires that the solid be in mechanical equilibrium, so that the total internal pressure must vanish:

$$P_{\text{tot}}^0 = -\frac{dE_{\text{coh}}}{d\Omega}\bigg|_{\Omega=\Omega_0} = -\frac{1}{6\Omega_0} \sum_{i \neq 0} R_i^0 \frac{dv_2(R_i^0)}{dr} \equiv P_{\text{vir}}^0 = 0. \qquad (1.12)$$

Here we have used the chain-rule relationship $\Omega dv_2/d\Omega = (r/3)\, dv_2/dr$ for any volume Ω to differentiate $v_2(r)$ and equate P_{tot}^0 to the equilibrium *virial* pressure P_{vir}^0. Using Eq. (1.12), the corresponding equilibrium bulk modulus is then just

$$B_{\text{tot}}^0 = -\Omega \frac{dP_{\text{tot}}}{d\Omega}\bigg|_{\Omega=\Omega_0} = \frac{1}{18\Omega_0} \sum_{i \neq 0} (R_i^0)^2 \frac{d^2 v_2(R_i^0)}{dr^2}. \qquad (1.13)$$

The constraints (1.11)–(1.13) are clearly adequate to determine the free parameters in the Morse and LJ potentials. For example, if the range of the v_2^{M} and v_2^{LJ} potentials were restricted to nearest-neighbor interactions, then $r_0 = d_{\text{nn}}$; $v_0 = |2E_{\text{coh}}^0/N_{\text{nn}}|$, where N_{nn} is the number of nearest neighbors; and in Eq. (1.9) for v_2^{M}, $\alpha = d_{\text{nn}}^{-1}\sqrt{9B_{\text{tot}}^0 \Omega_0/(2|E_{\text{coh}}^0|)}$. In real materials, the actual range of $v_2^{\text{M}}(r)$ and $v_2^{\text{LJ}}(r)$ extends beyond nearest neighbors, which acts to lower the magnitude of v_0 and raise r_0 beyond the first neighbor distance, as illustrated in Fig. 1.3 for fcc aluminum. More generally, the shape and range of the pair potential $v_2(r)$ in Eq. (1.8) is fully determined by cohesion alone. In this regard, Carlsson et al. (1980) showed that one could systematically invert first-principles electronic-structure calculations of E_{coh} as a function of volume for metals into a completely parameter-free pair potential of just this type.

While a pair potential is adequate to describe metallic cohesion, such a representation of the total energy suffers from at least four fundamental problems in metals. First, in using interatomic potentials in solid or liquid metals one is mostly interested in how the total energy changes as the atomic species are rearranged within the condensed state. In metals this energy scale is very small relative to the cohesive energy, to which the major details of the pair potential in Eq. (1.8) are tied. Typically, such structural energies are in the range of 0.001–0.1 eV (or about 0.1–10 mRy) as compared with metallic cohesive

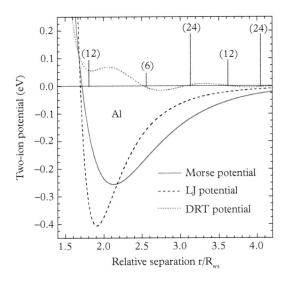

Fig. 1.3 *Empirical Morse and LJ cohesive-energy pair potentials in fcc aluminum. For comparison, also plotted is the first-principles structural-energy pair potential, a QBIP, obtained from PP perturbation theory by Dagens et al. (1975). Also shown are the locations and number of neighbors in the first five nearest-neighbor shells of the fcc crystal structure.*

energies of 1–10 eV (or about 0.1–1 Ry). Moreover, since cohesive properties calculated from Eq. (1.8) depend on a particular crystal structure, a pure pair potential v_2 will necessarily be *structure dependent* on at least this same 0.001–0.1 eV energy scale. Thus, the transferability of v_2 to any other arbitrary ion configurations is neither assured nor, in fact, very likely, and the calculation of structural properties with such a pair potential is always in serious doubt. The strong contrast between cohesive and structural energy scales is also shown in Fig. 1.3, where the *cohesive-energy* v_2^M and v_2^{LJ} pair potentials for Al are compared with the *structural-energy* pair potential of Dagens et al. (1975) arising from a first-principles PP perturbation expansion of the total energy (see Chapter 3).

A second more specific problem with a pure pair-potential representation of the cohesive energy in metals is that it requires the well-known Cauchy elastic-constant relations to be satisfied. In general, the second-order elastic constants C_{ij} can be expressed in terms of first and second strain derivatives of the total energy, as discussed in Chapter 7. In a pure pair-potential representation, these derivatives become the first and second radial derivatives of $v_2(r)$. As with the bulk modulus (1.13), the first derivative terms vanish by virtue of the zero-pressure condition (1.12). In cubic systems, one is then left with the rigid symmetry requirement $C_{12} = C_{44}$ or equivalently the Cauchy-ratio condition

$$\sigma_C \equiv C_{12}/C_{44} = 1. \tag{1.14}$$

This condition is seldom satisfied in real metallic systems, neither in simple metals (e.g., $\sigma_C = 2.2$ in fcc Al and $= 1.7$ in bcc K) nor in transition-series metals (e.g., $\sigma_C = 1.6$

in fcc Cu and $= 2.8$ in bcc V, while $= 0.35$ in bcc Cr). Consequently, one cannot expect either elastic properties or lattice vibrational properties to be described within this framework.

A third problem with a pure pair potential description in metals concerns the vacancy-formation energy $E_{\text{vac}}^{\text{f}}$. This is the energy required to remove a single atom from the interior of the metal and place it on the surface. Removing an atom from the bulk breaks one complete set of near-neighbor bonds, while placing it on the surface restores on average one-half of these bonds. Thus, ignoring any small relaxation of the lattice about the vacancy created, the vacancy-formation energy is from Eq. (1.11) equal in magnitude to the cohesive energy itself:

$$E_{\text{vac}}^{\text{f}} = -\sum_{i \neq 0} v_2(R_i^0) + \frac{1}{2} \sum_{i \neq 0} v_2(R_i^0) = -E_{\text{coh}}^0 . \qquad (1.15)$$

Measured vacancy-formation energies in metals, however, are only a small fraction of the cohesive energy. Typically, in both simple metals and transition-series d-band metals, $|E_{\text{vac}}^{\text{f}}/E_{\text{coh}}^0| \approx 0.2 - 0.4$. Consequently, one can also not expect to describe defect energies in metals with pure pair potentials of this type.

Finally, a fourth problem with pure pair potentials in metals concerns the melt temperature T_{m} and the relationship of the corresponding melt energy $k_{\text{B}} T_{\text{m}}$ to E_{coh}^0. In rare-gas solids $|E_{\text{coh}}^0/k_{\text{B}} T_{\text{m}}| \approx 10 - 12$, consistent with what one would predict using a pure pair potential. For example, $|E_{\text{coh}}^0/k_{\text{B}} T_{\text{m}}| = 12.8$ for an LJ potential in any fcc material (Ladd and Woodcock, 1978). In metals, however, melt energies are much smaller relative to the cohesive energy and are also significantly material dependent, with $|E_{\text{coh}}^0/k_{\text{B}} T_{\text{m}}| \approx 20 - 40$. Thus, a pure cohesive-energy pair potential will overestimate the melt temperature of a metal by a factor of 2–3 and is not adequate to describe the high-temperature thermodynamic properties of either the solid or the liquid.

Historically, the stop-gap empirical solution to these difficulties in the 1960s and 1970s was to acknowledge that there must be additional terms beyond a pair potential in the metal total energy, but rather than attempt an explicit treatment of such terms, one simply abandoned using the constraints (1.11)–(1.13) on the pair potential. In place of cohesive properties, the pair potential could be fit instead to the elastic constants and/or to point-defect energies to address mechanical properties in metals. For example, in the context of the above Morse or LJ potentials, one might fit the parameters v_0 and r_0 to $E_{\text{vac}}^{\text{f}}$ and the Cauchy pressure $P_{\text{C}} = C_{12} - C_{44}$ rather than to E_{coh}^0 and P_{tot}^0. This produces a qualitatively similar pair potential, but one whose depth is reduced to $\sim 2E_{\text{vac}}^{\text{f}}/N_{\text{nn}}$. This idea was extended to more general short-ranged potential forms represented by a series of spline polynomials, with the parameters fit to additional elastic-constant, phonon, and defect-energy data. Empirical pair potentials of this type were pioneered by Johnson (1964, 1973) and are often referred to as "Johnson potentials." While such a pair potential is certainly more closely related to the structural energy scale and properties one desires to describe, at the same time, it is not consistent with Eq. (1.8) for the cohesive energy. More seriously, such potentials can display considerable nonuniqueness even when fitted to essentially the same experimental data (Taylor, 1981), and thus Johnson potentials typically have no reliable transferability beyond the data fit.

The major conclusion to be drawn here is that having at least some many-body content in the total- or cohesive-energy functional is essential for a metal. Within the framework of a free-atom reference system, the difficulties associated with a pure pair potential description of the cohesive energy can only be overcome mathematically by adding triplet, quadruplet and possibly higher potential terms to the expansion (1.7). One may anticipate, however, that at best such a series will be slowly convergent, especially for the loosely bound s and p valence electrons. Thus, as a practical matter, the free-atom starting point is not very attractive in metals, and alternative reference systems that acknowledge the condensed metallic state for the valence s and p electrons need to be considered.

1.4 Guiding principles in metals for quantum-based potentials

The essential many-body nature of the total- or cohesive-energy functional in metals is an idea that is fully embraced in all QBIPs, as well as in most modern empirical interatomic potentials such as EAM and FS. In QBIPs, of course, this is a property that derives directly from the quantum mechanics itself, beginning from Eq. (1.3). Depending on the particular quantum formulation employed, the many-body content may appear in different specific forms in the cohesive-energy functional $E_{\mathrm{coh}}(\mathbf{R})$, but these forms generally follow the simple schematic structure

$$E_{\mathrm{coh}}(\mathbf{R}) = \boxed{\begin{array}{c}\text{Collective, non-directional}\\ \text{many-body contribution}\end{array}} + \frac{1}{2N}\sum_{i,j}{}' v_2(R_{ij})$$

$$+ \boxed{\begin{array}{c}\text{Directional multi-ion}\\ \text{contributions: } v_3, v_4, \ldots\end{array}}, \qquad (1.16)$$

or else straightforward extensions of this structure. In Eq. (1.16) the many-body energy content appears first and foremost in a collective, nondirectional contribution that is essential for nontransition metals with only radial forces. In the context of the expansion (1.4), this contribution arises naturally from reference systems based on embedding the ions in a uniform electron gas representing the valence s and p electrons. Any additional many-body content associated with directional bonding in transition or actinide metals appears in explicit multi-ion potentials or an equivalent representation. Discussion of the specific content of Eq. (1.16) for both simple metals and transition metals in the bulk, as well as at free surfaces, will lead us though the major guiding principles for QBIP development that will be more fully explored in subsequent chapters in this book.

1.4.1 Linear embedding in a uniform electron gas: volume-dependent potentials for bulk simple metals

Let us first consider an *sp*-bonded simple metal in the limit in which the Z valence electrons in zero order are spread out uniformly into a free-electron gas. Specifically, we choose a reference system consisting of N noninteracting ions placed in a compensating uniform electron gas of density $n_{\text{unif}} = Z/\Omega$. The reference energy E_0^{ref} now depends on both the properties of a single ion embedded in the electron gas and on the specific state variable Ω for the metal. Thus, to the extent that any additional multi-ion potentials beyond v_2 can be neglected, one expects the cohesive-energy functional to take the form

$$E_{\text{coh}}(\mathbf{R}, \Omega) = E_{\text{vol}}(\Omega) + \frac{1}{2N} {\sum_{i,j}}' v_2(R_{ij}, \Omega), \tag{1.17}$$

where $E_{\text{vol}}(\Omega) = E_0^{\text{ref}}(\Omega) - E_{\text{tot}}^{\text{atom}}$ is a volume term and a collective property of the metal that does not depend on the actual positions of the ions. In addition, since the reference system now depends on volume, one may anticipate that the pair potential v_2 will also depend on volume, as explicitly indicated in Eq. (1.17). For real materials, the functions $E_{\text{vol}}(\Omega)$ and $v_2(r, \Omega)$ can be calculated directly from PP perturbation theory, as first recognized by Harrison (1966) and a number of others, whose work is covered in the early review by Heine and Weaire (1970). Harrison's approach in particular emphasized representing the electron-ion interaction in the metal rigorously by a weak nonlocal PP. Then using a plane-wave basis set to calculate the electron density and self-consistent screening of the ions within first-order perturbation theory, the cohesive energy is obtained self-consistently within second-order perturbation theory. Advanced formulations of this approach based on DFT quantum mechanics were subsequently obtained by Rasolt and Taylor (1975) and Dagens et al. (1975) in terms of nonlocal model PPs or model potentials (MPs), by Moriarty (1977, 1982) in terms of the simple-metal limit of the GPT, and by Hafner (1987) in terms of Harrison's original nonlocal PP. These formulations are discussed in Chapter 3.

In applying Eq. (1.17) to calculate materials properties, the pair potential $v_2(r, \Omega)$ exhibits full bulk transferability at a given volume Ω, allowing reliable treatment of a wide range of structural, thermodynamic, defect and mechanical properties. In particular, all four of the objections raised in Sec. 1.3 for a pure pair-potential representation of E_{coh} are removed. First, most of the metallic cohesion is now contained in the volume term E_{vol}, so that at a given Ω, $v_2(r, \Omega)$ is not only structure independent, but its magnitude is on the desired scale of structural energies rather than the cohesive energy, as was illustrated for aluminum in Fig. 1.3. Second, the equilibrium virial pressure P_{vir}^0 is no longer required to vanish as in Eq. (1.12). Instead, P_{vir} is now a positive outward pressure that is balanced in equilibrium by the negative inward pressure coming from the volume term $E_{\text{vol}}(\Omega)$ plus the volume dependence of the pair potential $v_2(r, \Omega)$, $P_{\text{vol}} \equiv -\partial E_{\text{coh}}(\mathbf{R}, \Omega)/\partial \Omega$, so that $P_{\text{tot}}^0 = P_{\text{vol}}^0 + P_{\text{vir}}^0 = 0$. As a consequence, the Cauchy relation $C_{12} = C_{44}$ is replaced with the positive Cauchy pressure condition (see Sec. 7.3.2 of Chapter 7):

$$P_C = C_{12} - C_{44} = 2P_{\text{vir}}^0 > 0, \quad (1.18)$$

as is observed in cubic simple metals. Third, the vacancy-formation energy is reduced in magnitude from $|E_{\text{coh}}^0|$ to the smaller observed range, with the unrelaxed value of E_{vac}^f in Eq. (1.15) replaced by (see Sec. 9.1 of Chapter 9):

$$\begin{aligned} E_{\text{vac}}^f &= -(E_{\text{coh}}^0 - E_{\text{vol}}^0) + \Omega_0 P_{\text{vir}}^0 \\ &= -\frac{1}{2}\sum_{i\neq 0} v_2(R_i^0, \Omega_0) - \frac{1}{6}\sum_{i\neq 0} R_i^0 \frac{\partial v_2(R_i^0, \Omega_0)}{\partial r}, \end{aligned} \quad (1.19)$$

which only depends on the smaller-magnitude QBIP v_2 in Eq. (1.17). Fourth, calculated melt temperatures also now depend only on the smaller-magnitude v_2 and are reduced to the observed range, allowing reliable calculation of melt curves as a function of pressure (see Chapter 8).

The cohesive-energy functional (1.17) can also be extended from ideal *sp*-bonded simple metals (e.g., Na, Mg and Al) to include pre-transition metals influenced by nearby *empty d* bands above the Fermi level (e.g., Ca) and post-transition elements influenced by nearby *filled d* bands below the Fermi level (e.g., Cu and Zn). The primary additional physics addressed is that of *sp-d* hybridization between the broad NFE *sp* energy bands and the narrow empty or filled *d* bands. This hybridization acts to lower the energies of electronic states below the *d* bands but raise energies of states above *d* bands, significantly modifying the cohesive energy and materials properties, including structural phase stability. By adding localized atomic-like *d* states to the basis set to account for the narrow *d* bands, the *sp-d* hybridization can be treated as an additional perturbation and readily absorbed into the theory. Thus, Eq. (1.17) remains valid but with generalized forms for E_{vol} and v_2. Detailed quantum treatments of such empty-*d*-band (EDB) and filled-*d*-band (FDB) metals were first developed from perturbation theory by Harrison (1969), from a mixed-basis pseudo-Green's-function formalism by Moriarty (1972a), and from resonant model potentials (RMPs) by Dagens (1976, 1977a, 1977b). Full elaboration of the theory and the modified forms of E_{vol} and v_2 obtained within the first-principles GPT of Moriarty (1977, 1982) are discussed in Chapter 4 and compared there with the RMP treatment of Dagens.

1.4.2 Extension of volume-dependent potentials to bulk transition metals

In bulk transition metals, more extensive generalizations are needed, beginning with the zero-order reference system. Unlike the inert tightly bound inner-core states, the valence *d* states of a single transition-metal ion placed in a free-electron gas will exhibit a sharp *d* resonance at an energy E_d corresponding to the center of the d bands in the bulk metal. Most of the d-electron density will remain localized near the ion center, but at the same time, the extended wave-function tails of the resonant d states allow electron transfer with the uniform gas. In equilibrium, the effective number of d electrons retained by the ion is established by the $\ell = 2$ phase shift δ_2 of the electron-ion potential (see Chapter 2):

$$Z_d = \frac{10}{\pi}\delta_2(\varepsilon_\text{F}), \qquad (1.20)$$

where ε_F is the Fermi energy determined by the ZN valence s and p electrons of the free-electron gas:

$$\varepsilon_\text{F} = \frac{\hbar^2}{2m}\left(\frac{3\pi^2 Z}{\Omega}\right)^{2/3}. \qquad (1.21)$$

In a dilute transition-metal alloy, both Z and ε_F would be fixed by the host sp electron gas, but in a bulk transition metal, one must calculate Z and Z_d self-consistently subject to the constraint that $Z+Z_d$ is a constant determined by the Periodic Table. The resulting reference system of N self-consistent, noninteracting resonant ions immersed in a free-electron gas is then an excellent starting point for developing an interatomic-potential expansion for the bulk metal.

In general, deriving an interatomic-potential expansion from quantum mechanics for transition metals is greatly facilitated by using a real-space TB description of the narrow d bands in these materials and capitalizing on the weakness and the short-ranged extent of the localized d-state matrix elements coupling different ion sites. By retaining a mixed basis set of both plane waves and localized atomic-like d states within a pseudo-Green's-function representation, as in the GPT treatment of pre- and post-transition elements discussed above, one can rigorously synthesize the d-resonant zero-order reference system, a nonlocal PP perturbation treatment of the valence s and p electrons, and a non-orthogonal d-state TB treatment of the valence d electrons together with the inclusion of sp-d hybridization, all within DFT quantum mechanics. This synthesis is fully embodied in the first-principles GPT interatomic potentials for transition metals (Moriarty, 1988a). The GPT transition-metal cohesive-energy functional generalizes Eq. (1.16) to allow for angular-force multi-ion potentials arising from the electrons in partially filled d bands:

$$E_\text{coh}(\mathbf{R},\Omega) = E_\text{vol}(\Omega) + \frac{1}{2N}\sum_{i,j}{}' v_2(ij,\Omega) + \frac{1}{6N}\sum_{i,j,k}{}' v_3(ijk,\Omega)$$
$$+ \frac{1}{24N}\sum_{i,j,k,l}{}' v_4(ijkl,\Omega) + \cdots. \qquad (1.22)$$

Equation (1.22) also retains the two key formal features of Eq. (1.17). First, the pair and multi-ion potentials v_n remain structure independent and possess full bulk transferability. Second, most of the cohesive energy itself is still contained in the

1.4.2 Extension of volume-dependent potentials to bulk transition metals

volume term E_{vol}, so that the potentials v_n remain relatively small in magnitude and on the order of structural energies, as desired. In addition, as in Eq. (1.17), the PP sp interactions normally only need to be extended to the pair potential v_2, while the additional TB d-state and hybridization sp-d interactions contribute to all the potentials, both pair and multi-ion.

For atomistic simulations, however, there is significant complexity in the first-principles form of Eq. (1.22) that must be addressed. This complexity arises from the nonanalytic, multi-dimensional nature of the angular-force potentials v_3, v_4, etc. and the long-range character of the sp-d hybridization. Unlike the two-ion pair potential v_2, the multi-ion potentials v_3, v_4, etc. can't be efficiently stored for later use, and one has to re-calculate the latter potentials at each usage. This complication has led to the development of simplified MGPT potentials (Moriarty, 1990a, 1990b, 1994), in which additional approximations, valid for mid-period transition metals with nearly half-filled d bands, have been introduced. The most important of these approximations are: (i) the neglect of sp-d hybridization contributions beyond E_{vol} and d-state nonorthogonality contributions beyond v_2; (ii) the introduction of parameterized canonical d bands (see Chapter 2) to express the remaining two-center TB d-state matrix elements, or *bond integrals*, $h_{ij}^{dd'}(R_{ij})$ analytically; and (iii) the termination of the cohesive-energy expansion (1.22) at four-ion interactions, with the forces calculated term by term. Quantum mechanically, the GPT expansion (1.22) is linear in the moments of the d-band density of states (DOS), and the MGPT termination at four-ion interactions retains all contributions from the second, third, and fourth low-order TB moments. The MGPT simplifications allow efficient evaluation and use of the angular-force potentials v_3 and v_4, and have enabled large-scale MD simulations in mid-period transition metals (Moriarty et al., 2002a, 2006).

The detailed character of MGPT multi-ion potentials in mid-period transition metals is illustrated in Fig. 1.4 for the three representative bcc elements V, Mo and Ta at their observed equilibrium atomic volumes. The two-ion pair potential v_2 is seen to be attractive at near-neighbor separations in each case, but with material-dependent well shapes and shallow well depths. In each of three examples, the well depth of v_2 is less than 4% of the cohesive energy. The corresponding three-ion triplet potential v_3 is all or mostly repulsive in each case, partially balancing the attractive contribution of v_2, and is also small in magnitude in each case. The four-ion quadruplet potential v_4, on the other hand, is always oscillatory in nature, contributing very little to the cohesive energy, but having a large impact on structural phase stability. In particular, note that v_4 is strongly favorable to the {110} planar 70.5° and 109.5° near-neighbor bond angles in the bcc structure, but equally unfavorable to the {100} planar 90° near-neighbor bond angles in the fcc structure. In the cases of V and Ta, v_4 contributes less than 2% to the bcc cohesive energy but 55% to the fcc-bcc energy difference, while in Mo, with more near half d-band filling, the respective contributions of v_4 are 6% to the cohesive energy and 88% to the fcc-bcc energy difference. The theoretical foundations underlying the GPT and MGPT are discussed in Chapter 2, while the full details of E_{vol} and v_n for transition metals are elaborated in Chapter 5, together with a discussion of recent work aimed at extending MGPT capabilities beyond the mid-period elements.

20 Introduction

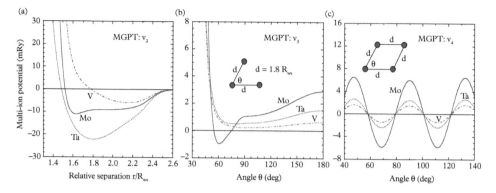

Fig. 1.4 *Multi-ion MGPT potentials in the mid-period transition metals V, Mo and Ta at their respective equilibrium atomic volumes (Moriarty et al., 2006). The specific potentials shown are V6.1, Mo5.2 and Ta4, as defined and discussed in Chapter 5. (a) Two-ion potential v_2; (b) three-ion potential v_3; (c) four-ion potential v_4.*

1.4.3 Nonlinear embedding and free surfaces: glue and bond-order potentials

While volume-dependent interatomic potentials have been highly successful in treating a wide range of structural, thermodynamic and mechanical properties of bulk metals, such an approach encounters difficulties at free surfaces, including voids, cracks and cleaved surfaces of the perfect crystal. At a free surface, the near-neighbor environment of an individual ion or atom is clearly very nonuniform and the definition of an atomic volume becomes uncertain. Even if an expansion like Eq. (1.17) or (1.22) were applicable at a surface, it is not obvious how the volume term is to be treated or at what atomic volume the interatomic potentials are to be calculated. In the 1980s, this led to the development of a seemingly very different type of approach in metals that makes no reference to atomic volume. In this approach, one begins with the following *an satz* for the general form of the cohesive-energy functional:

$$E_{\text{coh}}(\mathbf{R}) = \frac{1}{N}\sum_i F\left(\sum_j{}' g(R_{ij})\right) + \frac{1}{2N}\sum_{i,j}{}' v_2(R_{ij}), \qquad (1.23)$$

where $F(x)$ is a nonlinear function of x, with $g(r)$ a short-ranged function of the ion-ion separation distance. The nonlinearity of $F(x)$ ensures that Eq. (1.23) satisfies the many-body requirement for the cohesive-energy functional, as expressed in Eq. (1.16), with only nondirectional radial-force interactions. The functions F, g and v_2, however, are subject to different interpretations in different schemes and are easily seen to be nonunique quantities. In particular, any linear component of $F(x)$ can be equally well

1.4.3 Nonlinear embedding and free surfaces: glue and bond-order potentials

included with v_2, so that the corresponding energy contributions are interchangeable between the first and second terms in Eq. (1.23). Nonetheless, to the extent that transferable functions F, g and v_2 can be found, Eq. (1.23) may be applied without ambiguity to either the surface or the bulk. At the same time, Eq. (1.23) cannot be directly derived from quantum mechanics, so neither bulk nor surface transferability of F, g and v_2 are ensured by theory, and they can only be established by trial and error.

Schemes that fall within the framework of Eq. (1.23) are sometimes collectively referred to as "glue" potentials (Ercolessi et al., 1988, 1994), or also as pair functionals (Carlsson, 1990). Notable specific examples include the EAM potential (Daw et al., 1983, 1984, 1993), the effective medium theory (EMT) of Jacobsen et al. (1987), and the N-body FS potential (Finnis and Sinclair, 1984; Ackland and Thetford, 1987). The EAM and EMT have been inspired in part by the central DFT idea that the total energy is a functional of the electron density. In these approaches, the function $g(R_{ij})$ is viewed as an average atomic-like electron density $\bar{n}_{ij}(R_{ij})$ at the site i arising from site j, while $F(n_b)$ is the energy required to embed an atom in the metal, approximated at the site i as a background medium of electron density

$$n_b = {\sum_{j}}' \bar{n}_{ij}(R_{ij}). \tag{1.24}$$

In the EAM, the embedding function $F(n_b)$ and the electron-density function $\bar{n}_{ij}(R_{ij})$ are treated as empirical quantities, while in the EMT, $F(n_b)$ is calculated within DFT as the energy required to embed an atom in a free-electron gas of density n_b. Both of these views of nonlinear potentials are most compatible with the physics of simple metals and series-end transition metals, where directional bonding is absent. In this regard, the main initial application targets of both the EAM and EMT were the fcc metals on the RHS of the Periodic Table (e.g., Al, Cu and Ni).

In bulk metals, the expected ranges of validity for the simple-metal volume-dependent cohesive energy (1.17) and the EAM and EMT versions of the pair-functional cohesive energy (1.23) are clearly the same, suggesting a possible link between the two functionals. Moriarty and Phillips (1991) first demonstrated such a link by making an approximate transformation of Eq. (1.17) to the form (1.23), thus paving the way for volume-dependent GPT potentials to be used to calculate reliable free-surface energies. In this transformation, the embedding function comes out quite naturally as just the volume term E_{vol}. Specifically, $F(n_b) = E_{\text{vol}}(\Omega)$ with the simple identification $n_b = Z/\Omega - n_a$, where n_a is the average on-site component of the local valence electron density. This local-density GPT approach is discussed in Chapter 11 together with a second approach based on the local DOS that allows use of the transition-metal MGPT volume-dependent potentials in Eq. (1.22) at surfaces. In addition, these local representations of volume-dependent potentials can also be used to examine local volume corrections to defect energies such as the vacancy formation energy. Such corrections are small and negligible in transition metals, but can be significant in high sp-electron-density simple metals like Mg and Al, and in post-transition metals like Cu. Moreover,

Skinner et al. (2019) have now generalized the local-density GPT to create an *adaptive GPT* (aGPT) that provides accurate forces and stresses in addition to energies, and is also discussed in Chapter 11.

In contrast to the original intent of the EAM and EMT schemes, the N-body FS potential is actually aimed at mid-period transition metals, and arises from somewhat different reasoning. In this case, one invokes the well-known argument that the transition-metal cohesive energy should be proportional to the d-band width and hence to the square-root of the second moment μ_2 of the d-band DOS. In an orthogonal d-state TB representation of the electronic structure, one can write $\mu_2 = \sum_i \mu_2^i/N$, where the ith local DOS moment μ_2^i is just a lattice sum of product d-state bond-integral matrix elements $\sum_j \sum_{d,d'} h_{ij}^{dd'} h_{ji}^{d'd}$ coupling sites i and j (see Chapter 2). Thus, one chooses $F(x) = x^{1/2}$ in Eq. (1.23) and interprets $g(R_{ij})$ as a quantity proportional to $\sum_{d,d'} h_{ij}^{dd'}(R_{ij}) h_{ji}^{d'd}(R_{ij})$. In practice, however, the functions g and v_2 are both treated empirically in the FS potential. In addition, the interatomic interactions contained in the first term in Eq. (1.23) remain radial, so that the many-body content of F does not include angular forces and the directional bonding contained in v_3, v_4, etc. of the volume-dependent GPT/MGPT expansion (1.22).

The lack of any directional bonding in the FS and EAM potentials was subsequently addressed in the 1990s by attempting to add angular-dependent, higher-moment TB contributions to the cohesive-energy functional. Carlsson (1991) proposed a simple TB model for mid-period transition metals that included both a square-root second-moment contribution, as in the FS potential, and an additional contribution linear in the fourth moment μ_4^i of the local d-band DOS. Similarly, Foiles (1993) added a fourth-moment μ_4^i contribution to the standard form of the EAM potential for application to mid-period transition metals. Expressed as real-space interactions, the μ_4^i contribution in both cases adds two-ion, three-ion and four-ion d-state potential terms to $E_{\text{coh}}(\mathbf{R})$, terms that qualitatively have a one-to-one correspondence with d-state potential contributions in the GPT/MGPT cohesive-energy functional (1.22).

In contrast to such quantum-based TB model building in transition metals, others have attempted more empirical extensions of the EAM or EMT potentials to include angular dependence. Baskes and co-workers (Baskes, 1992; Baskes and Johnson, 1994; Lee and Baskes, 2000) have developed a modified EAM or MEAM potential for metals by adding additional angular-dependent components to the background electron density n_b entering the embedding function $F(n_b)$. Mishin et al. (2005, 2006) have developed an angular-dependent potential (ADP) by adding ad hoc bond-angle potential terms to the EAM form. While the MEAM and ADP extensions of EAM have enjoyed success as empirical potentials, providing as they do many extra fitting parameters, the models themselves are not directly supported by quantum mechanics, as will be further discussed for the MEAM potential in Sec. 1.4.4. It should also be noted that Frederiksen et al. (2004) have added empirical angular terms involving many free parameters to the EMT to create a modified EMT (MEMT) potential for Mo, and successfully tested their potential against the MEAM form. Empirical potentials with angular terms such as MEAM, ADP and MEMT are now sometimes referred to collectively as cluster functionals (Tadmor and Miller, 2011).

The most systematic and complete real-space TB potential model that includes angular dependence for transition metals is the BOP d-band local-moment expansion of Pettifor and co-workers (Pettifor, 1995; Aoki et al., 2007; Drautz et al., 2015). The nonlinear BOP expansion includes local DOS moment contributions to the dominant d-band energy, or so-called d-bond energy E_{bond}, exactly to all orders needed for mathematical convergence within the orthogonal-d-state, two-center bond-integral approximations:

$$E_{\text{bond}}(\mathbf{R}) = \frac{1}{N} {\sum_{i,j}}' \sum_{d,d'} h_{ij}^{dd'} \Theta_{ji}^{d'd}, \qquad (1.25)$$

where $\Theta_{ji}^{d'd}$ is the chemical bond order of the $d'd$ bond between ions j and i (see Chapter 5). Although $E_{\text{bond}}(\mathbf{R})$ does not break out individual multi-ion potentials as in the GPT/MGPT expansion (1.22), all d-state directional-bonding contributions are implicitly included in Eq. (1.25). In the standard BOP scheme for transition metals (e.g., Mrovec et al., 2004, 2007, 2011), the attractive d-bond energy E_{bond} is balanced by an empirical repulsive energy E_{rep} such that

$$E_{\text{coh}}(\mathbf{R}) = E_{\text{bond}}(\mathbf{R}) + E_{\text{rep}}(\mathbf{R}), \qquad (1.26)$$

where E_{rep} is modeled in an ad hoc pair-functional form discussed in Chapter 5. In this scheme, the valence s and p electrons are not treated quantum mechanically as in the GPT and MGPT, but rather their assumed secondary effect on E_{coh} is simply absorbed into E_{rep}. Consequently, there is also no attempt in the transition-metal BOP to treat sp-d hybridization or a self-consistent d-band occupation number Z_d. Optionally, the effect of d-state nonorthogonality can be treated, however, through an effective environmental dependence of the bond integrals $h_{ij}^{dd'}(R_{ij})$ (Nguyen-Manh et al., 2000).

Forces in the standard transition-metal BOP are calculated via the Hellman-Feynman theorem, requiring good numerical convergence of the bond-energy expansion (1.25). In practice, this requires that nine or more local DOS moments be retained in the expansion, making the evaluation of E_{bond} and the corresponding forces expensive. While it has been possible to perform molecular statics (MS) calculations on defects in this manner, large-scale MD simulations with the transition-metal BOP have so far remained beyond reach. The transition-metal BOP is further elaborated and compared with the MGPT in Chapter 5, including the discussion of recent efforts to simplify E_{bond} for simulations.

1.4.4 Absence of bond charges in the electron density

It is important to recognize that the principal origin of the multi-ion angular forces in transition metals comes directly from the total energy through the partial filling of the

d bands, and *not* through any large covalent bonding contribution to the electron density, as is the case in semiconductors like silicon. To a good approximation, the electron density $n(\mathbf{r})$ in nontransition and transition metals alike can be expressed as a superposition of overlapping, spherically symmetric densities n_i centered on individual ion sites:

$$n(\mathbf{r}) = \sum_i n_i(\mathbf{r} - \mathbf{R}_i). \tag{1.27}$$

In simple metals, this result is a well-known consequence of the fact that the first-order linear screening of the ions can be recast in terms of neutral overlapping pseudoatoms (Ziman, 1964; see Chapter 11), with the valence electron density exactly in the form of Eq. (1.27). In transition metals, the use of a zero-order resonant-ion reference system in the GPT allows one to construct analogous self-consistent pseudoatoms from a generalized first-order screening treatment, as is also discussed in Chapter 11.

The basic experimental confirmation of Eq. (1.27) in metals comes from the X-ray diffraction of their observed crystal structures. Any measurable deviation of the electron density from Eq. (1.27) will produce so-called forbidden diffraction lines corresponding to extra bond-centered accumulations of electron density, or so-called bond charges. Such forbidden diffraction lines are observed in Si and other covalently bonded semiconductors (Colella and Merlini, 1966), but they are not seen in metals.

Thus, multi-ion angular-force contributions to the total energy occur for somewhat opposite reasons in semiconductors and transition metals. In semiconductors, the valence energy bands are completely filled, but their observed diamond crystal structure requires that TB wavefunction basis functions be constructed from overlapping and directionally oriented sp^3 hybrid orbitals, which form saturated covalent bonds that lead to bond charges in the electron density. In transition metals, on the other hand, the TB d basis functions are well localized and structure-neutral atomic-like orbitals, with only small near-neighbor overlap, and it is the partial filling of the resulting TB d bands that controls directional bonding and multi-ion angular forces.

The physical distinction between the origin of angular forces in semiconductors and transition metals has implications for empirical interatomic potential models such as the MEAM. While the MEAM's addition of angular components to the electron density makes good physical sense in the case of covalently bonded semiconductors, an original application target of the method (Baskes, 1992), the application of the MEAM electron-density model to simple and transition metals (Baskes, 1992; Baskes and Johnson, 1994; Lee and Baskes, 2000), as well as to actinide metals (Baskes, 2000), is physically misguided for the pure bulk material. This underscores the importance of quantum-based theory to the development of physically correct interatomic potentials in metals.

1.4.5 Finite ion-ion interaction range and order-N scaling

An important practical issue in the application of all metal QBIPs is the effective range of interaction between ions, which directly impacts one's ability to attain the order-N scaling needed for large-scale MD simulations. For empirical potentials, an arbitrarily short ion-ion interaction range is often imposed in metals as a matter of simplicity and convenience through a sharp cutoff of the potential functions at a chosen radius R_{cut}. In typical empirical potentials $R_{\text{cut}} < 2.25 R_{\text{WS}}$, which corresponds to the direct interaction of a given ion with only 6–14 of its near neighbors. This short interaction range clearly serves to economize large-scale MD simulations and maximize the number of atoms that can be so treated, with order-N scaling in the simulations easily achieved. Unfortunately, such a short ion-ion interaction range in metals is not well supported by quantum considerations.

In quantum mechanics, the ion-ion interaction range extends, in principle, across the entire system, and is only made finite in metals though the physical phenomenon of screening. Screening in metals can have many specific components, but is dominated by two main effects. The first effect is the direct screening of an individual ion by the highly mobile valence s and p electrons. This effect is handled very elegantly in PP perturbation theory and leads to the well-known Friedel oscillations in the long-range behavior of the two-ion pair potential in simple metals (Harrison, 1966):

$$v_2(r, \Omega) \sim \frac{9\pi Z^2 \tilde{w}^2}{\varepsilon_F} \frac{\cos(2k_F r)}{(2k_F r)^3}, \tag{1.28}$$

where $\tilde{w} = w(2k_F)$ is the back-scattering form factor of the electron-ion pseudopotential (see Chapter 3), and k_F is the Fermi wavenumber of the uniform electron gas:

$$k_F = \left(\frac{3\pi^2 Z}{\Omega}\right)^{1/3}. \tag{1.29}$$

Such Friedel oscillations in simple metals, as well as in pre- and post-transition metals, are real with observable consequences such as Kohn anomalies in the phonon spectra (Kohn, 1959; Woll and Kohn, 1962). More generally, the long-range nature of $v_2(r, \Omega)$ in these metals strongly impacts fundamental properties such as structural phase stability, as well as defect properties such as stacking fault energies. This results in a very long ion-ion interaction range, with an effective cutoff radius $R_{\text{cut}} \geq 8.25 R_{\text{WS}}$, corresponding to the interaction of a given ion with no fewer than 500–600 neighbors. For general MS and MD applications, an accurate numerical treatment of the Friedel oscillations is somewhat difficult, and historically this has been a significant barrier to the widespread usage of QBIPs in these materials. There are, however, a number of viable options available today that significantly lessen this barrier, as we will discuss in Chapter 3.

The second main component of screening in metals arises from the collective effect of multi-ion interactions, which can also act to screen the ion-ion interactions. This effect is largely one of destructive interference, and an effect that is more difficult to model quantitatively, with a fully rigorous theoretical treatment still an open challenge in QBIP research. Multi-ion screening is a secondary effect in nontransition metals, adding at most some modest damping to the Friedel oscillations. In transition metals, however, multi-ion screening heavily damps both the sp Friedel oscillations and the corresponding long-range oscillations that arise from sp-d hybridization, as numerical simulations discussed in Chapter 5 confirm. This damping is maximized for mid-period transition metals with nearly half-filled d bands, a fact that has motivated the neglect of *sp-d* hybridization in the standard MGPT treatment of multi-ion potentials (Moriarty, 1990a). As one moves away from half d-band filling toward the beginning or end of the transition-metal series, however, the multi-ion damping generally diminishes and the importance of *sp-d* hybridization increases. There can also be effective multi-ion damping of the nonorthogonal d-state interactions, as has been modeled in the BOP (Nguyen-Manh et al., 2000).

At the same time, the direct d-state interactions that establish the multi-ion potentials in mid-period transition metals necessarily extend the effective ion-ion interaction range through the multiple connected couplings of the short-ranged bond integrals: $\sum_{j,k\cdots} \sum_{d',d'',d'''\cdots} h_{ij}^{dd'} h_{jk}^{d'd''} h_{kl}^{d''d'''} \cdots$, etc. In principle, if the bond integral $h_{ij}^{dd'}(R_{ij})$ has a range R_{bond}, then the M-ion potential (or M^{th}-moment total-energy contribution) can require a cutoff as large as $R_{\text{cut}} = (M-1)R_{\text{bond}}$. In practice, however, destructive interference can lessen the effective cutoff for such multi-ion interactions considerably, often by a factor of two or more. For example, including d-state interactions in the MGPT through four-ion potentials, with a bond-integral range of $R_{\text{bond}} = 2.75R_{\text{WS}}$, reduces the cutoff radius from a theoretical maximum of $R_{\text{cut}} = 8.25R_{\text{WS}}$ to an effective value of just $\sim 4.25R_{\text{WS}}$. This corresponds to a given ion directly interacting with only about 60–80 near neighbors.

To achieve order-N scaling in a QBIP-based MD simulation requires, at a minimum, that the effective ion-ion interaction sphere of radius R_{cut} be contained entirely within the simulation cell. For example, in the case of transition-metal MGPT potentials with $R_{\text{cut}} = 4.25R_{\text{WS}}$, an order-$N$ scaling simulation cell requires at least ~ 250 atoms. Similarly, for a simple-metal GPT potential with $R_{\text{cut}} = 8.25R_{\text{WS}}$, an order-$N$ scaling simulation cell requires at least ~ 1000 atoms. In practical MGPT-MD and GPT-MD simulations, these expectations for order-N scaling are indeed met, as shown in Fig. 1.5 for liquid tantalum and liquid magnesium examples. These 2016 results demonstrated typical single-processor serial-simulation times in the range $0.5-2.0 \times 10^{-4}$ seconds per atom per time step. With parallel processing on HPC machines, one can therefore indeed perform large-scale MD simulations on $10^6 - 10^8$ atoms, as was first demonstrated in the case of MGPT by Streitz et al. (2006a). For MGPT potentials, these capabilities have been made possible by the development of fast algorithms to evaluate multi-ion energies and forces on the fly during an MD simulation, as discussed in Chapter 8. The most recent and efficient of these algorithms, due to Oppelstrup (2015), is now encoded in a USER-MGPT package (Oppelstrup and Moriarty, 2018) included in the open-source molecular dynamics code LAMMPS (Plimpton, 1995).

1.4.6 Quantum-mechanical pillar of structural phase stability

The vast array of crystal structures observed in metals and alloys is perhaps the most direct and striking manifestation of quantum mechanics in these materials. In the context

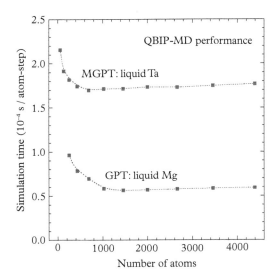

Fig. 1.5 *Normalized single-processor simulation time for typical bulk thermodynamic MGPT-MD simulations on liquid Ta performed in a canonical (NVT) ensemble, and corresponding GPT-MD simulations on liquid Mg, both as a function of the number of atoms N treated (Moriarty, 2016). The simulation conditions here are as follows: for liquid Ta, $\Omega = 121.6$ a.u., $T = 6000$ K and $P = 22$ GPa; for liquid Mg, $\Omega = 113.3$ a.u., $T = 3000$ K and $P = 35$ GPa.*

of QBIPs, the prediction of structural phase stability is consequently an important fundamental test of the physical convergence of the bulk total-energy expansions such as Eq. (1.17) for simple metals and Eqs. (1.22) and (1.25) for transition metals, as well as a test of the transferability of the associated interatomic potentials. For simple metals, and both pre- and post-transition metals, the use of first-principles DFT-based functionals $E_{\text{vol}}(\Omega)$ and $v_2(r, \Omega)$ in the total-energy expansion (1.17) has proven to be remarkably successful in describing the main key features of structural phase stability in prototype elemental materials. These features include: the observed ground-state structure; the small total-energy differences between alternate crystal structures; total-energy variations along deformation paths connecting different structures; and the prediction of pressure-induced solid-solid phase transitions. The latter predictions, in fact, historically have resulted in the identification of several new high-pressure phases that were subsequently confirmed by experiment. In addition, when the same first-principles QBIP ingredients are used in a corresponding quasiharmonic lattice dynamics (QHLD) treatment of the low-temperature solid as well as in MD simulations of the high-temperature solid and the liquid, temperature-induced solid-solid phase transitions, melting and temperature-pressure phase diagrams can also be well described. These applications and others are discussed in Chapters 6, 7 and 8.

Structural phase stability plays an especially important guiding role to QBIP development in the case of transition metals. In these materials, the central question is: exactly how far does one need to go in a multi-ion-potential expansion like Eq. (1.22) or a d-band moments expansion like Eq. (1.25) to achieve good physical convergence of the total energy? Fortunately, consideration of structural phase stability in elemental transition metals provides an illuminating answer to that question. Among the nonmagnetic $3d$, $4d$ and $5d$ transition-metal elements, one has an observed hcp → bcc → hcp → fcc sequence of crystal structures across each series. This sequence is fully explained by DFT quantum mechanics in terms of the filling of the d bands, as one moves across a given series. This physics is also captured by the TB d-bond energy E_{bond} in the BOP, as first shown by Aoki (1993). Specifically, Aoki calculated bcc-fcc and hcp-fcc d-bond energy differences as a function of d-band filling and the number of DOS moments retained for canonical d bands. His central result in this regard is displayed in Fig. 1.6. This result shows that retaining the first four d-band moments explains the large negative bcc-fcc and bcc-hcp energies in mid-period transition metals. This is fully consistent with the MGPT treatment of mid-period bcc metals at the level of four-ion multi-ion potentials and the large contribution of v_4 to the bcc-fcc energy, as pointed out in connection with Fig. 1.4. Moreover, Aoki's results further show that one must go to at least a six-moment treatment to adequately resolve the smaller fcc-hcp energy difference and explain the appearance of the hcp metals to the left and right of the mid-period bcc metals. At the same time, a fully converged eighteen-moment description adds little significant detail beyond the six-moment description. Thus, one can conclude that good physical convergence of E_{bond} in the BOP model of transition metals occurs at the point of retaining the first six d-band moments. At the same time, however, one should note that abrupt truncation at the six-moment level would require modifications in the current BOP methodology to allow a term-by-term calculation of forces instead of the Hellman-Feynman treatment that is actually used. In contrast, extending MGPT to hcp transition metals should only require adding corresponding five- and six-body multi-ion potentials, v_5 and v_6, in the total-energy expansion (1.22), with a continued use of a term-by-term calculation of forces. Current progress on this challenge is discussed in Chapter 5.

Available QBIP applications on phase stability, high-pressure phase transitions, melting, pressure-temperature phase diagrams and other related topics, including applications for mid-period transition metals within the current MGPT and BOP frameworks, are discussed in Chapters 6, 7 and 8. It should also be noted that the importance of angular-force contributions to atomic structure in transition metals extends beyond bulk phase stability and phase transitions to fundamental defect issues such as self-interstitial-atom (SIA) formation and migration, and grain-boundary (GB) atomic structure, as well as to cleaved-surface structural issues such as surface reconstruction. Available MGPT and BOP applications on SIAs and GBs will be discussed in Chapter 9 and MGPT applications on surface reconstruction in Chapter 11.

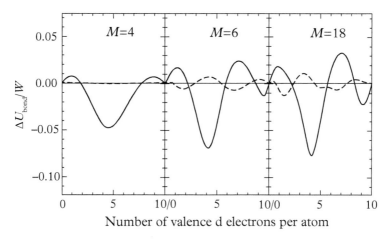

Fig. 1.6 *Calculated bcc-fcc (solid lines) and hcp-fcc (dashed lines) bond-energy differences, $\Delta U_{bond}/W$, as a function of d-band filling and the number of DOS moments retained in a BOP treatment of canonical d bands of width W. Left panel: first four moments retained; center panel: first six moments retained; right panel: numerically converged result with first 18 moments retained. Note that here, $U_{bond} \equiv E_{bond}$ in Eq. (1.25).*
From Aoki (1993), with publisher (APS) permission.

1.4.7 Machine learning and statistical interatomic potentials

With the advent of modern supercomputers and fast parallel processing has come the idea of using machine learning to help "train" the correct behavior of interatomic potentials through the input of large quantities of first-principles DFT and QMD training data. Beginning in the mid-1990s, machine learning was first used in an attempt to improve empirical EAM potentials for simple metals. Working with the general pair-functional cohesive-energy form (1.23), Eroclessi and Adams (1994) developed a so-called force-matching method for an aluminum prototype, supplementing experimental and DFT input data on cohesion, elastic constants, and unrelaxed defect and surface energies with large quantities of small-cell (≤ 150 atoms) QMD training data on forces at finite temperature in various environments. In this force-matching method, the unknown functions F, g and v_2 in Eq. (1.23) are each expressed as a series of cubic splines connected by spline knots that together with weighing factors for the input data and a cutoff radius R_{cut} act as fitting parameters. In total, the Al potential was thereby parameterized as a function of 40 such parameters, which were determined by minimizing an objective function of the errors in calculated quantities compared to their input values. The final force-matched EAM Al potential so obtained displayed generally improved finite-temperature properties, including an excellent melting temperature within 1% of experiment. This improvement required an increased cutoff radius of

$R_\text{cut} = 3.51 R_\text{WS}$ to accommodate the QMD data, while at the same time reproducing input QMD forces with only an average accuracy of about 20%.

Force matching with the fixed pair-functional form (1.23) provides no guarantees about the transferability of the final potential, however, even for simple metals. For example, in a very similar procedure applied to magnesium by Liu et al. (1996), the force-matched EAM potential obtained yielded a relatively poor melt temperature, some 20% below experiment. This occurred in spite of the fact that the Mg potential reproduced the input QMD forces with the same average accuracy of about 20%, and the potential had a similar and even slightly expanded range, with $R_\text{cut} = 3.78 R_\text{WS}$. In addition, the force-matched EAM potential also produced about 20% agreement with high-temperature solid and liquid forces obtained from first-principles GPT Mg potentials, potentials that in contrast, independently yield a Mg melt temperature within 4% of experiment (Moriarty and Althoff, 1995).

In more recent years, significant progress has been made on moving past fixed potential forms in applying machine learning to interatomic-potential development through the use of artificial neural networks and Gaussian process regression. For applications to metals and semiconductors, the generalized neural-network potential (NNP) method of Behler and Parrinello (2007) (BP) and the Gaussian approximation potential (GAP) method of Bartók et al. (2010) have been two of the leading approaches used to develop machine-learning potentials (MLPs). In both methods, one begins by replacing the fixed pair-functional cohesive energy (1.22) with a more general cohesive-energy functional of the form

$$E_\text{coh}(\mathbf{R}) = \frac{1}{N} \sum_i \varepsilon_i(\{R_{ij}\}). \tag{1.30}$$

Here one assumes only that $E_\text{coh}(\mathbf{R})$ is a smooth and continuous function that can be expressed as a sum of single-site atomic potential energies ε_i of finite spatial range, with all interatomic interactions of the central atom i with its neighbors confined to a central sphere of radius R_cut in the case of GAPs or to multiple localized cutoff spheres in the case of NNPs. The general idea is to expand each retained energy component $\varepsilon_i(\{R_{ij}\})$ in a suitable basis set that serve as local environment descriptors within the defined localized region surrounding atom i, and then, using extensive input DFT training data on energies and forces for the same system, determine all unknown coefficients and other variable parameters statistically through ML and mathematical regression techniques. The final NNP or GAP representation is the average defined by the RHS of Eq. (1.30).

In the BP NNP approach, the basis set consists of a chosen collection of localized radial and angular symmetry functions of the respective forms

$$G_i^\text{rad} = \sum_{j \neq i} e^{-\eta_R (R_{ij} - R_s)} f_\text{cut}(R_{ij}), \tag{1.31}$$

with predetermined parameters η_R and R_s, and

$$G_i^{\text{ang}} = 2^{1-\zeta} \sum_{j,k \neq i} (1 + \lambda \cos \theta_{ijk})^\zeta e^{-\eta_\theta (R_{ij}^2 + R_{jk}^2 + R_{ik}^2)} f_{\text{cut}}(R_{ij}) f_{\text{cut}}(R_{jk}) f_{\text{cut}}(R_{ik}), \quad (1.32)$$

where $\cos \theta_{ijk} = \mathbf{R}_{ij} \cdot \mathbf{R}_{ik}/(R_{ij} R_{ik})$, with predetermined parameters η_θ, λ and ζ. One may define as many independent symmetry functions as needed, which can then be labeled G_i^μ for the μth function, each with its own radial cutoff function of the form

$$f_{\text{cut}}(R_{ij}) = \begin{cases} 0.5 \left[\cos(\pi R_{ij}/R_c) + 1\right] & \text{for } R_{ij} \leq R_c \\ 0 & \text{for } R_{ij} > R_c \end{cases}, \quad (1.33)$$

where R_c is an additional predetermined parameter. With a four-layer neural network, including an input layer 0, two hidden layers 1 and 2, and an output layer 3, then the energy ε_i in Eq. (1.30) is obtained in the form

$$\varepsilon_i = f_a^3 \left\{ b_1^3 + \sum_{k=1}^{N_2} w_{k1}^{23} f_a^2 \left[b_k^2 + \sum_{j=1}^{N_1} w_{jk}^{12} f_a^1 \left(b_j^1 + \sum_{\mu=1}^{N_{\text{sym}}} w_{\mu j}^{01} G_i^\mu \right) \right] \right\}, \quad (1.34)$$

where N_{sym} is the number of symmetry functions used, N_1 and N_2 are the number of nodes in layers 1 and 2, f_a^n are chosen activation functions for layers $n = 1, 2, 3$, and the remaining quantities are variable biases b_1^3, b_k^2, b_j^1 and weights $w_{k1}^{23}, w_{jk}^{12}, w_{\mu j}^{01}$ that are to be optimized via the DFT training data.

In a recent application of the BP NNP formalism directed at the metallurgically important properties of hcp Mg, Stricker et al. (2020) used 27 radial and 5 angular input symmetry functions and DFT training data spanning some 12,000 atomic configurations of 443 bulk, surface and defect atomic structures. The resulting NNP so developed was found to be consistent with a number of basic aspects of Mg plasticity and fracture predicted by DFT, including the dislocation core structure, the atomistic decohesion energy and crack-tip behavior.

To implement the GAP method, on the other hand, 4D bispectrum mathematics is introduced to represent the atomic environment within the 3D cutoff spheres in terms of K-component bispectrum vectors $\mathbf{b}_i = \{b_1^i, \ldots, b_K^i\}$. This bispectrum representation is then combined with a basis set of Gaussian functions G to yield an expansion of the form

$$\varepsilon_i(\mathbf{b}_i) = \sum_n \alpha_n G(\mathbf{b}_i, \mathbf{b}_n), \quad (1.35)$$

where the expansion coefficients α_n are to be determined from the input DFT training data. Using training data derived from atomic configurations in which atoms were randomly displaced from their equilibrium positions in various small unit cells,

Bartók et al. (2010) successfully applied their GAP method to describe the elastic moduli, phonons, surface energies, and a few finite-temperature properties of covalent semiconductors, as well as the phonons of bcc iron.

Bartók et al. (2013) later added an important refinement to the GAP method in the form of a smooth overlap of atomic positions (SOAP) descriptor for a more accurate treatment of multi-body interactions. With the refined GAP method applied to defects in the bcc transition metal tungsten, Szlachta et al. (2014) studied the accuracy and transferability of GAP as a function of the input DFT training data for a chosen cutoff radius of $R_{\text{cut}} = 3.21 R_{\text{WS}}$. They showed that quantum accuracy of a few percent could be achieved across a range of basic properties, including elastic constants, phonons, vacancy and surface energies, and dislocation structure and mobility, with input training data from some 10,000 atomic configurations. At this level of treatment, however, the GAP method becomes very expensive and out of the reach of MD applications.

In an attempt to simplify the GAP approach for metals and make it readily applicable to MD simulations, Thompson et al. (2015) developed an alternative spectral neighbor analysis potential (SNAP) method. In this method, one assumes a computationally efficient linear relationship between ε_i and $\mathbf{b}_i$, such that Eq. (1.35) is replaced with

$$\varepsilon_i(\mathbf{b}_i) = \alpha_0 + \sum_{k=1}^{K} \alpha_k b_k^i. \tag{1.36}$$

This simplification was then combined with an applications strategy of using far fewer atomic configurations in the DFT training data establishing the expansion coefficients α_k and a minimum cutoff radius R_{cut}. Thompson et al. (2015) applied their SNAP model to tantalum using $R_{\text{cut}} = 2.88 R_{\text{WS}}$ with some success, demonstrating both the feasibility of MD simulations with their approach and the ability to calculate an accurate bcc screw dislocation migration barrier. At the same time, however, many basic bcc Ta properties, such as the C' elastic constant, (110) and (112) unstable stacking fault energies, self-interstitial formation energies, and the melt temperature were only calculated to accuracies of 20% or less. The linear SNAP approach has also now been applied to bcc Mo by Chen et al. (2017) and to Ni-Mo compounds and fcc Ni and Cu metals by Li et al. (2018).

In addition to SNAP, several other recent approaches to linearized MLPs have been proposed. These include the moment tensor potentials (MTP) of Shapeev (2016), the atomic cluster expansion (ACE) potentials of Drautz (2019), the group theoretical invariants (GTI) potentials of Seko et al. (2019) and the permutation-invariant polynomial potentials of van der Oord et al. (2020). These methods share in common the expansion of the site energy ε_i in Eq. (1.30) into a many-body series of body-ordered potential contributions. Qualitatively, this treatment mirrors the many-body cluster expansion of the total energy envisaged in Eq. (1.1) that gives rise to the corresponding multi-ion GPT/MGPT potentials in Eq. (1.22). The MTP and ACE linear MLP methods have also demonstrated enhanced computational efficiency over GAP,

1.4.7 Machine learning and statistical interatomic potentials

allowing the possibility of MD simulations. In the case of fcc copper, SNAP, ACE and GTI calculations of the elastic constants, surface energies, point-defect energies and (except for GTI) the melting temperature have been compared with DFT and experiment by Lysogorskiy et al. (2021).

In addition, a somewhat intermediate Gaussian potential approach to bcc transition metals was recently considered by Byggmästar et al. (2020), who retained the full nonlinear GAP formalism of Bartók et al. (2010, 2013), but in contrast to Szlachta et al. (2014), used only about 4,000 atomic configurations in their DFT training sets for each of the five metals V, Nb, Ta, Mo and W. A cutoff radius of $R_{cut} \cong 3.2 R_{WS}$ was maintained in each metal, close to that used by Szlachta et al. for W. Good results were obtained over a range of basic physical properties, including cohesion, elastic constants, phonons, point defects, surface energies and melt properties. The GAP vacancy and SIA energies and volumes obtained by Byggmästar et al. are compared against MGPT, BOP and DFT results for V, Ta and Mo in Table 9.2 of Chapter 9.

Machine-learning potentials such as NNP, GAP, SNAP and ACE are all developed using mathematical regression techniques requiring large DFT databases of calculated physical properties to provide useful content. While clearly distinct in character, these MLPs do share in common two basic features with QBIPs such as GPT, MGPT and BOP. These features are first, the absence of arbitrary fixed forms in representing the potentials, and second, the inevitable practical trade-off between accuracy and cost in developing and applying the potentials. While MLPs do contain general many-body content, they are otherwise empirical potentials, whose mathematical parameters have no clear physical meaning and can change in a discontinuous fashion with small changes in the underlying DFT database or the assumed radial cutoffs. Thus, at any given level of treatment, it is unknown what many-body physics is actually included in MLP potentials, and hence there is no systematic path forward to improve the physics content of these potentials. In contrast, QBIPs are derived directly from quantum mechanics, so at any level of treatment the many-body physics content of the potentials is known, and the important possibility of systematic physics improvement is fully retained.

An interesting and important question moving forward is: to what extent can ML be used to advantage in improving QIBPs such as GPT, MGPT and BOP? In the case of GPT and MGPT potentials for d-band metals, three main areas come to mind where ML could have immediate positive impact. The first is in the full optimization of the localized d basis states for a given material at a given density or pressure. This is especially so for metals near the beginning and end of the transition-metal series, where the GPT treatment is particularly sensitive to the choice made. An initial attempt at such optimization is discussed in Sec. 4.2 of Chapter 4 for the prototype cases of Ca, treated as an EDB metal, and Zn, treated as a FDB metal. A second area where ML could be useful is in helping to model and restore neglected sp-d hybridization contributions to MGPT potentials for transition metals. A current attempt to develop such a model for a Ni prototype is discussed in Sec. 5.2.4 of Chapter 5, but extending this model back across the Periodic Table to more central transition metals will be computationally challenging and will require machine-learning guidance. Finally, a third area where ML could have near-term impact is in developing advanced MGPT multi-ion potentials with the inclusion of five- and six-ion interactions, as discussed in Chapter 5, where an important practical question is: how far do such interactions have to extend in real space to capture the essential physics?

A closely related issue for both MLPs and QBIPs is the importance and ability to treat long-range interactions in metals. As we have discussed above, long-range interactions in simple metals in the form of Friedel oscillations in the pair potential are well known and their effect is observable through Kohn anomalies in the phonon spectrum. These interactions also impact other important properties such as stacking faults in fcc and hcp metals. While first-principles QBIPs like GPT and DRT can treat such properties accurately, they are generally more problematic for DFT calculations and consequently for MLPs. In this regard, a recent study by Ruffino et al. (2020) of the four stable basal planar stacking faults in hcp Mg compared the relative ability of accurate GPT pair potentials and high-precision DFT calculations to obtain reliable, converged fault energies. To maintain high numerical fidelity in these studies, they were performed with the aid of a generalized Ising model carried out to sixth-neighbor interactions. While the calculated GPT fault energies all showed good convergence and internal consistency, the DFT energies did not, demonstrating that there are definite limitations to what direct DFT calculations can do. This issue, of course, is not restricted to Mg, but extends to other simple metals such as fcc Al, to EDB metals such as fcc Ca and Sr, to FDB metals such as hcp Zn and Cd, as well as to series end transition metals such as fcc Ni and Cu. In all such metals, smooth long-range radial cutoffs of the GPT pair potential at $R_{\text{cut}} = 8.25 R_{\text{WS}}$ or beyond are required and can be implemented as described in Sec. 3.5.2 of Chapter 3.

For central transition metals, the standard MGPT intermediate-range radial cutoff of $R_{\text{cut}} = 4.25 R_{\text{WS}}$ is fully adequate to capture the physics of the localized d-state multi-ion interactions through four-ion contributions that dominate physical propeties, as discussed above. This cutoff is more or less consistent with the $R_{\text{cut}} \cong 3.2 R_{\text{WS}}$ MLP values used by Szlachta et al. (2014) and Byggmästar et al. (2020) in their transition-metal studies. Still to be determined in the case of the MGPT, however, is the impact of sp-d hybridization and five- and six-ion interactions on long-range interactions and the calculation of physical properties.

2
Fundamental Principles in Metals Physics

In this chapter we summarize some of the important principles of metals physics that we will require in later chapters to develop interatomic potentials from quantum mechanics. For simplicity, we discuss these principles in the context of elemental metals; extension of the results to alloys and intermetallic compounds is reasonably straightforward and will be deferred until required in Chapter 10. We also confine our attention to the nonmagnetic and nonrelativistic limits. Magnetic effects in transition metals are considered in Chapter 5, while relativistic effects, apart from the spin-orbit interaction, can be incorporated into the nonrelativistic treatment in a straightforward way where required. In addition, most of the material we introduce here is rather well established condensed-matter theory that does not need elaborate justification. We concentrate, therefore, on reviewing and highlighting the central results; additional useful details are given in Appendix A2.

2.1 Born-Oppenheimer or adiabatic approximation

The cornerstone of most quantum-mechanical treatments of solids and liquids is the separation of nuclear and electronic motion, and the central results were first derived by Born and Oppenheimer (1927) in a classic paper. This separation of motion is possible because of the great disparity in mass between electrons and nuclei. To the electrons, which travel at speeds 50–100 times greater than the nuclei, the nuclei appear to be almost stationary. The nuclei, on the other hand, see essentially an average potential created by the swiftly moving electrons. At the outset, it is convenient to consider all of the electrons, both core and valence, in the presence of compensating nuclei. In Sec. 2.3 we will distinguish between bound inner-core electrons that move rigidly with the nuclei and itinerant valence electrons that do not and thus recover the conceptual picture of a metal that we introduced in Chapter 1.

Let H be the total many-body Hamiltonian of an elemental metal with N nuclei and $Z_a N$ electrons, where Z_a is the atomic number. Next divide H into an electronic and a nuclear component, $H = H_e + H_n$, such that the electronic Hamiltonian H_e is that for the electrons in the presence of infinitely massive nuclei:

$$H_e = -\frac{\hbar^2}{2m}\sum_e \nabla_e^2 + V_{ee} + V_{en}. \tag{2.1}$$

Here m is the mass of the electron and the sum in the kinetic-energy operator runs over all electron coordinates, while the final two terms are the Coulomb potential energies of all electron-electron and electron-nucleus interactions. Then H_n is just the remaining kinetic-energy operator for the nuclei plus the Coulomb potential energy of all nucleus-nucleus interactions:

$$H_n = -\frac{\hbar^2}{2M}\sum_n \nabla_n^2 + V_{nn}, \tag{2.2}$$

where M is the mass of each nucleus. As an approximate trial eigenfunction of H, we try a product wavefunction of the form $\psi = \psi_e \psi_n$ where ψ_e is the eigenfunction of H_e:

$$H_e \psi_e = E_e \psi_e. \tag{2.3}$$

Then one has the following equation to solve:

$$H\psi = (H_e + H_n)\psi_e\psi_n = E_n\psi_e\psi_n. \tag{2.4}$$

Since all of the electron coordinates in the total Hamiltonian H are contained in H_e, ψ_n will depend only on the nuclear coordinates and hence will commute with H_e. That is, $H_e\psi_e\psi_n = \psi_n H_e \psi_e = E_e \psi_n \psi_e$. The eigenfunction ψ_e, on the other hand, will depend on both the electron and the nuclear coordinates, as is clear from Eq. (2.1). If, however, one ignores the latter dependence in Eq. (2.4) and assumes that ψ_e similarly commutes with H_n, one is led to the familiar Born-Oppenheimer or adiabatic approximation:

$$\left[-\frac{\hbar^2}{2M}\sum_n \nabla_n^2 + E_{tot}(\mathbf{R}_1, \mathbf{R}_2, \ldots \mathbf{R}_N)\right]\psi_n = E_n\psi_n, \tag{2.5}$$

where $E_{tot}(\mathbf{R}_1, \mathbf{R}_2, \ldots \mathbf{R}_N) = V_{nn} + E_e$ is the total-energy functional that we introduced in Chapter 1. The physical content of Eq. (2.5) is that the nuclei move on an average potential-energy surface $U_{tot}(\mathbf{R}) = E_{tot}(\mathbf{R}_1, \mathbf{R}_2, \ldots \mathbf{R}_N)$ created by the electrons. Born and Oppenheimer showed that the leading corrections to Eqs. (2.3) and (2.5) are of order $(m/M)^{1/4}$ and involve the coupling of ψ_e and ψ_n. In the solid, this is physically the electron-phonon interaction, which is normally treated in perturbation theory based on solutions of Eqs. (2.3) and (2.5). In addition, except for the very lightest elements hydrogen and helium, the eigenfunctions ψ_n are sufficiently well localized on the nuclei that Eq. (2.5) can be replaced by the corresponding Newtonian mechanics with the forces derived from $U_{tot}(\mathbf{R})$, as we have assumed in Chapter 1.

2.2 Density functional theory

The major formal task before us is to develop adequate means of solving Eq. (2.3) and thereby actually find E_{tot} for the case of a metal. This equation is, however, a full many-electron Schrödinger equation, so that further approximation is essential in any real condensed-matter system. The most tractable approach to this problem involves reducing Eq. (2.3) to a set of one-electron self-consistent-field equations that can be readily solved. The traditional way this has been done is to write ψ_e as a product (or sum of products) of $Z_a N$ one-electron orbitals ψ_α and then variationally minimize the quantity $\langle \psi_e | H_e | \psi_e \rangle / \langle \psi_e | \psi_e \rangle$. This leads to a one-electron Schrödinger equation for the orbitals ψ_α of the form

$$\left(-\frac{\hbar^2}{2m}\nabla^2 + V(\mathbf{r})\right)\psi_\alpha(\mathbf{r}) = E_\alpha \psi_\alpha(\mathbf{r}), \qquad (2.6)$$

where $V(\mathbf{r})$ is a self-consistent electron potential that depends on all of the occupied states α, including the one being calculated. The historically oldest approaches of this type are based on the Hartree and Hartree-Fock (HF) approximations (Kittel, 1963). In the Hartree approximation ψ_e is taken as a single product wavefunction, leading to a potential $V(\mathbf{r})$ that contains only direct Coulomb contributions from the nuclei and filled one-electron orbitals. The absence of the expected anti-symmetry in ψ_e results in a complete neglect of exchange and correlation contributions to $V(\mathbf{r})$, although the most important consequence of anti-symmetry, namely the Pauli principle, is added separately by requiring that only two electrons (one spin up and one spin down) can occupy any given orbital ψ_α. In real materials, the Hartree approximation is qualitatively reasonable, but is not quantitatively accurate due the neglect of exchange and correlation in $V(\mathbf{r})$. In the HF approximation, on the other hand, one begins instead with ψ_e in the form of a Slater determinant made up of the same one-electron orbitals ψ_α. This satisfies the anti-symmetry requirement for ψ_e and leads to a potential $V(\mathbf{r})$ that includes, by definition, a full exchange contribution but no correlation. The resulting exchange term in $V(\mathbf{r})$, however, comes in the form of a complicated nonlocal operator acting on ψ_e, making the HF method difficult to apply to condensed-matter systems. More seriously, the HF approximation is well known to lead to unphysical results in the case of metals. For example, in the limiting case of a uniform or homogeneous electron gas, the HF density of one-electron energy states exactly vanishes at the Fermi level.

The unsuitability of the HF approximation for metals historically resulted in great emphasis being placed on developing a modified Hartree method. It was proposed by Slater (1951) that one add an exchange term to the Hartree potential proportional to $n^{1/3}$, where n is the local electron density

$$n(\mathbf{r}) = \sum_\alpha \psi_\alpha^\star(\mathbf{r})\psi_\alpha(\mathbf{r}), \qquad (2.7)$$

with the sum over all occupied states α. The $n^{1/3}$ functional dependence comes from the form of the HF exchange energy for a homogeneous electron gas (Kittel, 1963):

$$\varepsilon_x(n) = -\frac{3}{2}e^2\left(\frac{3n}{8\pi}\right)^{1/3}. \tag{2.8}$$

Slater originally proposed choosing the coefficient such that the added exchange potential was exactly $2\varepsilon_x(n)$. Later Gaspar (1954) suggested using an exchange potential that was 2/3 of this value. In subsequent years, Slater (1974) advocated taking the coefficient as a variable parameter in the so-called $X\alpha$ method. Modified Hartree schemes of this type, often collectively referred to as Hartree-Fock-Slater methods, led in the 1960s to the first self-consistent band-structure calculations on real materials and produced many new and useful results. Nonetheless, such schemes were still theoretically unsatisfying in that they seemed to have no real rigorous basis.

The sought-after justification of self-consistent-field methods, and their subsequent consolidation into a single unified scheme, was provided by the density functional theory (DFT) of the inhomogeneous electron gas set forth by Hohenberg and Kohn (1964) and Kohn and Sham (1965). Hohenberg and Kohn reformulated the entire problem, isolating the electron density $n(\mathbf{r})$ as the fundamental variable of the system and treating exchange and correlation on an equal footing. They then proved a simple but powerful theorem that states that the total energy E_{tot} is unique functional of $n(\mathbf{r})$. For arbitrary $n(\mathbf{r})$, the complete form of this functional is not known, but if $n(\mathbf{r})$ is sufficiently slowly varying, the exchange and correlation contribution to E_{tot} is well approximated by

$$E_{xc}^{LDA}[n] = \int n(\mathbf{r})\varepsilon_{xc}(n(\mathbf{r}))d\mathbf{r}, \tag{2.9}$$

where $\varepsilon_{xc}(n)$ is the total exchange-correlation (xc) energy of the homogeneous electron gas of density n. This is the so-called local density approximation (LDA). Kohn and Sham (1965) used this result together with the stationary property of total energy, namely $\delta E_{tot}[n]/\delta n = 0$, to show that there is a one-to-one correspondence between solving the exact many-electron Schrödinger equation, Eq. (2.3), and the one-electron Schrödinger equation, Eq. (2.6), so long as the self-consistent electron potential $V(\mathbf{r})$ in the latter equation is taken in the form

$$V(\mathbf{r}) = -\sum_i \frac{Z_a e^2}{|\mathbf{r} - \mathbf{R}_i|} + \int \frac{e^2 n(\mathbf{r}')}{|\mathbf{r} - \mathbf{r}'|}d\mathbf{r}' + \mu_{xc}(n(\mathbf{r})), \tag{2.10}$$

where μ_{xc} is the effective xc potential given by

$$\mu_{xc}(n) = \frac{d}{dn}[n\varepsilon_{xc}(n)]. \tag{2.11}$$

Once the Schrödinger equation (2.6) is actually solved with the electron potential (2.10), the total energy may be expressed in the form

$$E_{\text{tot}} = \frac{1}{2}\sum_{i,j}{}' \frac{(Z_a e)^2}{|\mathbf{R}_i - \mathbf{R}_j|} + \sum_{\alpha} E_{\alpha} - \frac{1}{2} \iint \frac{e^2 n(\mathbf{r}) n(\mathbf{r}')}{|\mathbf{r} - \mathbf{r}'|} d\mathbf{r} d\mathbf{r}' + \int^n (\mathbf{r})[\varepsilon_{\text{xc}}(n(\mathbf{r}))$$
$$- \mu_{\text{xc}}(n(\mathbf{r}))] d\mathbf{r}. \tag{2.12}$$

The first term here is the usual Coulomb potential energy between charged nuclei, while the second term is a sum over the occupied one-electron energy levels of Eq. (2.6). The third term in Eq. (2.12) subtracts off the electron-electron interaction energy, which is double counted in the second term, and the final term adds an additional correction designed to yield the correct xc energy. The stationary property of E_{tot} provides the added bonus that any small error in the calculated electron density, say Δn, will only give rise to an error of order $(\Delta n)^2$ in E_{tot}.

For a given form of ε_{xc}, the essential LDA equations in DFT are then Eqs. (2.6), (2.7) and (2.10)–(2.12). In addition to the assumed regime of nearly uniform $n(\mathbf{r})$, Kohn and Sham argued that these equations should also be valid in the limit of very high electron density, since then the kinetic energy dominates E_{xc} and there should still be little net error in using Eq. (2.10). Moreover, because the high-density regime applies to the inner-core regions near the nuclei, while the nearly uniform-density regime applies to the outer-core and interstitial regions between nuclei, Kohn and Sham further argued that the LDA equations could, in fact, be applied to the whole material.

2.2.1 Exchange and correlation functions in the local density approximation

To actually use the LDA equations, of course, one needs a specific form for $\varepsilon_{\text{xc}}(n)$ from the theory of the homogeneous electron gas. By definition, the total xc energy is an additive sum of the HF exchange energy ε_x, as given by Eq. (2.8), and the remaining correlation energy ε_c, so that $\varepsilon_{\text{xc}} = \varepsilon_x + \varepsilon_c$. There is not an exact analytic result for $\varepsilon_c(n)$ valid at all densities, but over the years useful approximation formulas have been developed. Most of the older forms developed before the mid-1970s are based on interpolations between high and low electron-density limits. A popular example of these forms is one due to Hedin and Lundqvist (HL, 1971), which is based on the correlation energy results of Singwi et al. (1970). In subsequent years, accurate numerical results for $\varepsilon_c(n)$ were obtained over a wide range of densities by Ceperley and Alder (1980) using first-principles QMC simulations. These results in turn have been fitted to analytic forms by Vosko, Wilk and Nusiar (VWN, 1980) and by Perdew and Zunger (PZ, 1981). The HL, VWN and PZ analytic forms for ε_c are given in Sec. A2.1 of Appendix A2. These schemes for ε_c are compared with each other and with the larger magnitude exchange energy ε_x in Fig. 2.1(a), as a function of the electron radius

$$r_s \equiv \left(\frac{4\pi n}{3}\right)^{-1/3}. \tag{2.13}$$

In the range $2 < r_s < 6$ a.u. relevant to the equilibrium valence electron densities of metals (see Table A1.1 in Appendix A1), $\varepsilon_c(n)$ is slowly varying with negligible difference between the VMN and PZ schemes, and with the older HL scheme overestimating the magnitude of the negative correlation energy by a near-constant 7 mRy.

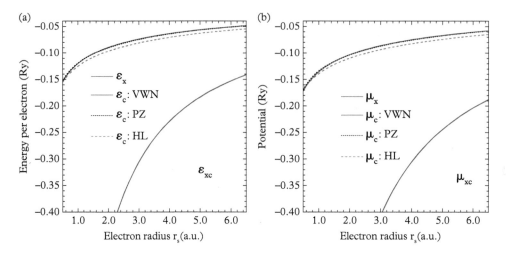

Fig. 2.1 *Exchange-correlation functions used in the LDA of DFT. (a) Total xc energy: $\varepsilon_{xc} = \varepsilon_x + \varepsilon_c$; (b) self-consistent xc potential: $\mu_{xc} = \mu_x + \mu_c$.*

Corresponding results for $\mu_{xc} = \mu_x + \mu_c$ can be worked out from ε_{xc} using Eq. (2.11). The exchange potential μ_x is given by

$$\mu_x(n) = -2e^2 \left(\frac{3n}{8\pi}\right)^{1/3}. \tag{2.14}$$

Thus $\mu_x = (4/3)\varepsilon_x$, which is precisely the exchange formula proposed by Gaspar (1954). For this reason, Eq. (2.14) is sometimes referred to as the Kohn-Sham-Gaspar exchange potential. The HL, VMN and PZ analytic schemes for $\mu_c(n)$ are compared with each other and with $\mu_x(n)$ in Fig. 2.1(b). At metallic densities, it is clear that $|\mu_x| \gg |\mu_c|$ and that $\mu_x(n)$ is a much more rapidly varying function of electron density than $\mu_c(n)$, with the HL correlation potential lowered by the same near-constant 7 mRy below the VMN and PZ potentials. These factors explain much of the early success of the Hartree-Fock-Slater methods, which neglected μ_c entirely.

2.2.2 Exchange and correlation beyond the LDA

In DFT electronic structure calculations, the most common approach to go beyond the LDA in the treatment of exchange and correlation is to use the so-called generalized gradient approximation (GGA). The GGA, which developed out of the early work of Langreth and co-workers (1980, 1983), takes into account the gradient in the density in addition to the density itself in constructing the xc energy functional:

$$E_{\text{xc}}^{\text{GGA}}[n, \nabla n] = \int n(\mathbf{r}) f_{\text{xc}}(n(\mathbf{r}), \nabla n(\mathbf{r})) d\mathbf{r}, \quad (2.15)$$

where $f_{\text{xc}}(n, \nabla n)$ is an unknown function that replaces $\varepsilon_{\text{xc}}(n)$ in Eq. (2.9). Over the years, many forms of f_{xc} have been proposed and used in various GGA methods. In recent years, one of the most popular and successful GGA methods is that due to Perdew, Burke and Ernzerhof (PBE, 1996). In the PBE approach, the function f_{xc} is expressed as an enhancement factor F_{xc} multiplying the exchange energy ε_x in the LDA:

$$f_{\text{xc}}^{\text{PBE}}[n, \nabla n] = \varepsilon_x(n) F_{\text{xc}}(r_s, s), \quad (2.16)$$

with r_s given by Eq. (2.13) and $s = |\nabla n|/2k_F n$. At metallic densities, F_{xc} is in the small range 1.2–1.4 for $s = 0$, consistent with Fig. 2.1(a), but rises in value up to 1.6 for $s = 3$.

In the context of QBIPs, the GGA is probably most significant for transition and actinide metals because of its impact on non-NFE d and f states. Enhanced treatment of exchange and correlation for NFE s and p valence states, however, is handled more conveniently in reciprocal space via the traditional many-body perturbation theory of the interacting electron gas. For a perturbation in density $\delta n(q)$, the induced Coulomb plus xc potential is given by

$$\delta V(q) = \frac{4\pi e^2}{q^2}[1 - G(q)]\delta n(q), \quad (2.17)$$

where the density-dependent, local function $G(q)$ represents the effect of exchange and correlation on the potential. In the LDA limit, one finds $G(q) = -(q^2/4\pi e^2) d\mu_{\text{xc}}(n)/dn$. Notable treatments of $G(q)$ beyond the LDA are found in the work of Geldart and Taylor (GT, 1970) and Ishimaru and Utsumi (IU, 1981). The LDA, GT and IU treatments of $G(q)$ are compared in Fig. 2.2 for an electron density corresponding to aluminum at equilibrium. Note that at long wavelengths where $q < 2k_F$, the three treatments agree closely, but in the short-wavelength limit where $q \to \infty$, the GT and IU functions properly tend to a constant, while the LDA function diverges.

2.3 Small-core approximation and the valence binding energy in metals

We now take advantage of the fact that the electrons governed by the DFT equations of the last section can be separated into two broad groups: localized core electrons, whose one-electron orbitals $\psi_\alpha = \phi_c$ are concentrated very close to the nuclei and essentially move rigidly with them, and the remaining valence electrons, whose orbitals $\psi_\alpha = \psi_k$ are concentrated outside of the core regions between the nuclei and are the cohesive glue that binds the material together. In most simple metals the distinction between core and

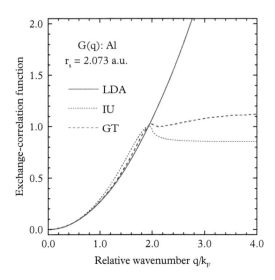

Fig. 2.2 *Exchange-correlation function $G(q)$ entering Eq. (2.17) for an electron density corresponding to aluminum at equilibrium, as calculated from LDA, IU and GT treatments.*

valence states is sharp, with the core comprising the last complete rare-gas configuration [e.g., $(1s)^2(2s)^2(2p)^6$ in Na, Mg and Al], leaving Z valence electrons per atom. In the small-core approximation, one assumes that there is negligible overlap between core orbitals ϕ_c centered on different sites, such that the ϕ_c are amenable to a purely atomic-like treatment. The core states satisfy the Schrödinger equation

$$\left(-\frac{\hbar^2}{2m}\nabla^2 + V(\mathbf{r})\right)\phi_c(\mathbf{r}) = E_c\phi_c(\mathbf{r}). \tag{2.18}$$

In practice, the core orbitals ϕ_c are essentially the same as in the free atom, although the core energy levels in the metal, E_c, will be shifted away their free-atom values. The valence states, which remain orthogonal to the core states, then satisfy the Schrödinger equation

$$\left(-\frac{\hbar^2}{2m}\nabla^2 + V(\mathbf{r})\right)\psi_\mathbf{k}(\mathbf{r}) = E(\mathbf{k})\psi_\mathbf{k}(\mathbf{r}). \tag{2.19}$$

In the periodic solid, the index $\mathbf{k}$ corresponds to a particular wavevector in reciprocal space. In that case the valence orbitals $\psi_\mathbf{k}$ are itinerate Bloch wavefunctions and the energies $E(\mathbf{k})$ are the familiar valence energy bands. If $\mathbf{k}$ is restricted to the first Brillouin zone (BZ) in reciprocal space, then one must also specify a band index n, in which case $\psi_\mathbf{k} \to \psi_{n\mathbf{k}}$ and $E(\mathbf{k}) \to E_n(\mathbf{k})$.

In the remaining DFT equations, one may then effectively combine the core electrons with the nucleus to form an ion of net charge $+Ze$ and treat the electron potential due

2.3 Small-core approximation and the valence binding energy in metals

to ions as a combined external potential. To do this, we first write the electron density in the form

$$n(\mathbf{r}) = n_{\text{val}}(\mathbf{r}) + \sum_i n_{\text{core}}(\mathbf{r} - \mathbf{R}_i), \quad (2.20)$$

where n_{val} is the valence electron density

$$n_{\text{val}}(\mathbf{r}) = 2 \sum_k \psi_k^*(\mathbf{r}) \psi_k(\mathbf{r}), \quad (2.21)$$

with the sum over $\mathbf{k}$ restricted to occupied states, and $n_{\text{core}}(\mathbf{r} - \mathbf{R}_i)$ is the localized core electron density on the site i, such that

$$n_{\text{core}}(\mathbf{r} - \mathbf{R}_i) = 2 \sum_c \phi_c^*(\mathbf{r} - \mathbf{R}_i) \phi_c(\mathbf{r} - \mathbf{R}_i). \quad (2.22)$$

Here the spin factor of 2 has been taken out explicitly in both n_{val} and n_{core}, and the index c in Eq. (2.22) runs over the remaining core quantum numbers on one site. Within the LDA, the self-consistent potential $V(\mathbf{r})$ in Eqs. (2.18) and (2.19) can then be written

$$V(\mathbf{r}) = V_{\text{ion}}(\mathbf{r}) + V_{\text{val}}(\mathbf{r}) + \mu_{\text{xc}}(n_{\text{val}}(\mathbf{r})), \quad (2.23)$$

where V_{ion} is the total ionic contribution to the electron potential,

$$V_{\text{ion}}(\mathbf{r}) = \sum_i \left[-\frac{Z_a e^2}{|\mathbf{r} - \mathbf{R}_i|} + v_{\text{core}}(\mathbf{r} - \mathbf{R}_i) + \mu_{\text{xc}}^*(n_{\text{core}}(\mathbf{r} - \mathbf{R}_i)) \right], \quad (2.24)$$

with v_{core} the Coulomb potential arising from n_{core}, i.e.,

$$v_{\text{core}}(\mathbf{r}) = \int \frac{e^2 n_{\text{core}}(\mathbf{r}')}{|\mathbf{r} - \mathbf{r}'|} d\mathbf{r}', \quad (2.25)$$

and μ_{xc}^* defined such that for $n_i = n_{\text{core}}(\mathbf{r} - \mathbf{R}_i)$

$$\mu_{\text{xc}}^*(n_i) \equiv \mu_{\text{xc}}(n_{\text{val}} + n_i) - \mu_{\text{xc}}(n_{\text{val}}), \quad (2.26)$$

and where V_{val} is the Coulomb potential arising from n_{val}

$$V_{\text{val}}(\mathbf{r}) = \int \frac{e^2 n_{\text{val}}(\mathbf{r}')}{|\mathbf{r} - \mathbf{r}'|} d\mathbf{r}'. \quad (2.27)$$

The effective core xc potential $\mu_{\text{xc}}^*(n_i)$ is localized about the site i but retains a residual dependence on the valence electron density n_{val} because $\mu_{\text{xc}}(n)$ is a *nonlinear* function of n. In practice, it is usually sufficient to replace n_{val} in Eq. (2.26) with its leading component, namely the uniform density n_{unif} (Moriarty, 1974, 1977). Then V_{ion} is volume dependent, but otherwise is a fixed external potential in the solution of the Schrödinger equation (2.19) for the valence electrons.

The LDA total energy given by Eq. (2.12) can be similarly simplified. In the metal, the total energy per atom may be decomposed into a valence-binding energy for the Z valence electrons, a large core energy for the remaining $Z_a - Z$ electrons, and a residual valence-core overlap energy associated with the nonlinear nature of μ_{xc} and ε_{xc}:

$$E_{tot} = N(E_{bind} + E_{core} + E_{val\text{-}core}) . \tag{2.28}$$

The valence binding energy E_{bind} is given by an expression analogous to Eq. (2.12) for E_{tot} with n replaced by n_{val} and Z_a replaced by Z:

$$E_{bind}(Z) = \frac{1}{N} \left\{ \frac{1}{2} \sum_{i,j}{}' \frac{(Ze)^2}{|\mathbf{R}_i - \mathbf{R}_j|} + 2 \sum_{\mathbf{k}} E(\mathbf{k}) - \frac{1}{2} n_{val} V_{val} \right.$$

$$\left. + n_{val}[\varepsilon_{xc}(n_{val}) - \mu_{xc}(n_{val})] \right\} . \tag{2.29}$$

Here we have introduced a shorthand notation for the terms involving n_{val} in which the required integration over the volume of the system is implied. That is,

$$n_{val} V_{val} \equiv \int n_{val}(\mathbf{r}) V_{val}(\mathbf{r}) d\mathbf{r} , \tag{2.30}$$

and similarly for the other terms. Corresponding expressions for the core energy E_{core} and the valence-core overlap energy $E_{val\text{-}core}$ are given by Moriarty (1988a), but within the small-core approximation, E_{core} and $E_{val\text{-}core}$ are the same in the metal and in the free atom and hence cancel out in the cohesive energy

$$E_{coh} = \frac{1}{N} E_{tot} - E_{tot}^{atom} = E_{bind}(Z) - E_{bind}^{atom}(Z) , \tag{2.31}$$

where $E_{bind}^{atom}(Z)$ is the valence binding energy in the free atom. Physically, $E_{bind}^{atom}(Z)$ is the ionization energy required to remove the Z valence electrons from the free atom and, if desired, can also be calculated entirely within DFT (Moriarty, 1988a), as discussed in Sec. A2.2 of Appendix A2. For fixed Z, of course, this term contributes just a constant to the cohesive energy E_{coh}.

Equations (2.18)–(2.31) thus represent the appropriate starting LDA self-consistent-field equations for a simple metal within the small-core approximation. The primary situation where the small-core approximation breaks down is for the d electrons of transition-series metals. Of course, if one simply chooses to include the d electrons as part of the valence electrons, Eqs. (2.18)–(2.31) still apply without modification. However, even in the case of transition metals with partially filled d bands, the d electrons retain a sufficiently localized character that it is advantageous to generalize the above description to accommodate them with the remaining inner-core electrons as far as possible. To do this, one normally attempts to describe the d states in terms of localized basis orbitals ϕ_d, analogous to the inner-core orbitals ϕ_c, but unlike the core states, the orbitals ϕ_d are *not* assumed to be eigenstates of the metal Hamiltonian, as in Eq. (2.18). In the electron density one can include a contribution to n_{core} from occupied orbitals ϕ_d at each site i, corresponding to Z_d electrons per atom, so that

$$n_{core}(\mathbf{r}-\mathbf{R}_i) = 2\sum_c \phi_c^\star(\mathbf{r}-\mathbf{R}_i)\phi_c(\mathbf{r}-\mathbf{R}_i) + \frac{Z_d}{5}\sum_d \phi_d^\star(\mathbf{r}-\mathbf{R}_i)\phi_d(\mathbf{r}-\mathbf{R}_i), \quad (2.32)$$

provided, of course, that one subtracts out this density at each site in defining n_{val}:

$$n_{val}(\mathbf{r}) = 2\sum_\mathbf{k} \psi_\mathbf{k}^\star(\mathbf{r})\psi_\mathbf{k}(\mathbf{r}) - \frac{Z_d}{5}\sum_{i,d} \phi_d^\star(\mathbf{r}-\mathbf{R}_i)\phi_d(\mathbf{r}-\mathbf{R}_i). \quad (2.33)$$

Equations (2.20) and (2.23)–(2.27) then still apply without alteration. In the total energy, one may still focus on an effective binding energy associated with the valence electrons by (i) adding an appropriate d-state contribution to the core energy E_{core}; (ii) subtracting an energy NZ_dE_d from the total band-structure energy to form the net valence band-structure energy

$$E_{band}^{val} = 2\sum_\mathbf{k} E(\mathbf{k}) - NZ_dE_d, \quad (2.34)$$

with the d-state energy E_d defined as the expectation value

$$E_d = \frac{1}{5N}\sum_{i,d}\int \phi_d^\star(\mathbf{r}-\mathbf{R}_i)\left(-\frac{\hbar^2}{2m}\nabla^2 + V(\mathbf{r})\right)\phi_d(\mathbf{r}-\mathbf{R}_i)d\mathbf{r}; \quad (2.35)$$

and (iii) keeping careful track of all extra Coulomb and xc corrections that arise from the overlap of the orbitals ϕ_d centered on neighboring sites. Retaining all such two-center overlap corrections, Eq. (2.29) is generalized to the form (Moriarty, 1988a)

$$\begin{aligned}E_{bind}(Z,Z_d) = &\frac{1}{2N}{\sum_{i,j}}'\frac{(Ze)^2}{|\mathbf{R}_i-\mathbf{R}_j|} + E_{band}^{val} - \frac{1}{2}n_{val}V_{val} + n_{val}[\varepsilon_{xc}(n_{val}) - \mu_{xc}(n_{val})]\\
&+ \frac{1}{2N}{\sum_{i,j}}'\left[\frac{(Z_a-Z)^2e^2}{|\mathbf{R}_i-\mathbf{R}_j|} - 2n_i\frac{(Z_a-Z)e^2}{|\mathbf{r}-\mathbf{R}_j|} + n_iv_j\right]\\
&+ \frac{1}{N}{\sum_{i,j}}'\left[\frac{(Z_a-Z)Ze^2}{|\mathbf{R}_i-\mathbf{R}_j|} - n_i\frac{Ze^2}{|\mathbf{r}-\mathbf{R}_j|}\right]\\
&+ \frac{1}{2N}{\sum_{i,j}}'\left[n_i\mu_{xc}^\star(n_j) + \delta\varepsilon_{xc}^\star(n_i,n_j) - n_{val}\delta\mu_{xc}^\star(n_i,n_j)\right],\end{aligned} \quad (2.36)$$

where $v_j \equiv v_{core}(\mathbf{r}-\mathbf{R}_j)$ is the Coulomb potential arising from n_j and an integration over volume is implied in all terms involving n_{val} and n_i. The five terms on the first line of

Eq. (2.36) are analogous to those in the simple-metal result (2.29), except that now the valence band-structure energy $E_{\text{band}}^{\text{val}}$ contains a large extra contribution to the cohesion from the d electrons. The two groups of terms on the second line of Eq. (2.36) represent the exact Coulomb corrections due to overlapping densities n_i, while the final group of terms on the third line represents the corresponding contribution from exchange and correlation. The latter contribution arises from systematic expansions of $\mu_{\text{xc}}(n)$ and $\varepsilon_{\text{xc}}(n)$ in terms of the same overlapping densities, neglecting only three-ion and higher overlap contributions, as discussed by Moriarty (1988a). The definitions of the nonlinear correction terms $\delta\mu_{\text{xc}}^{\star}(n_i, n_j)$ and $\delta\varepsilon_{\text{xc}}^{\star}(n_i, n_j)$ are given in Sec. A2.3 of Appendix A2. Note that since the Coulomb and xc corrections in Eq. (2.36) consist of pairwise additive terms, they will contribute directly to the total two-ion pair potential in a d-band metal.

One final generalization must be incorporated to cancel core and valence-core overlap energies between the metal and the free atom and obtain the cohesive energy in terms of valence binding energies as in Eq. (2.31). As discussed in Chapter 1, the values of Z and Z_d in the metal must be determined in a self-consistent manner. Consequently, they will normally differ from their corresponding values in the free atom, say Z_0 and Z_d^0. To achieve the desired cancellations, one must first transfer the required number of outer s and d electrons in the free atom to match Z and Z_d. This costs an atomic preparation energy E_{prep}, which can also be calculated within DFT (Moriarty, 1988a), as discussed in Sec. A2.2 of Appendix A2. Then the cohesive energy is just

$$E_{\text{coh}} = \frac{1}{N}E_{\text{tot}}(Z, Z_d) - E_{\text{tot}}^{\text{atom}}(Z_0, Z_d^0) = E_{\text{bind}}(Z, Z_d) - E_{\text{bind}}^{\text{atom}}(Z, Z_d) + E_{\text{prep}}. \quad (2.37)$$

The last two promoted-atom terms in the second equality of Eq. (2.37) are now individually volume dependent in general (because Z and Z_d are volume dependent), but the sum of these terms still contributes just a constant to the volume term E_{vol} in the cohesive-energy functional (1.22). Equations (2.32)–(2.37) together with Eqs. (2.20) and (2.23)–(2.27) then represent the appropriate starting LDA self-consistent-field equations for a d-band transition-series metal.

2.4 Guidance from the DFT electronic structure: simple metals vs. *d*-band metals

Within DFT quantum mechanics, the rigorous physical quantities are the electron density $n(\mathbf{r})$ and the total energy E_{tot}. Although these are indeed the quantities of greatest interest to us in the context of developing QBIPs for metals from first principles, valuable insight needed to develop useful treatments of n_{val} starting from Eqs. (2.21) and (2.33) and E_{bind} starting from Eqs. (2.29) and (2.36) are to be found in the electronic structure of the valence energy bands $E(\mathbf{k})$ obtained from Eq. (2.19). In this section, we briefly survey relevant calculated DFT energy bands with the purpose of contrasting simple and d-band metals and identifying features of their respective electronic structures that will allow simplification. In this regard, a useful compilation of DFT band-structure data for metals with atomic numbers $Z_a \leq 49$ was published by Moruzzi et al. (1978), and we will appeal to these results below. These data are based on the self-consistent solutions of the LDA equations with HL exchange and correlation at the observed equilibrium volumes, as obtained using the Green's function or Korringa-Kohn-Rostoker (KKR) band-structure method (Korringa, 1947; Kohn and Rostoker, 1954; Martin, 2004) and treating each metal in an equivalent manner.

The simple and d-band elemental metals of primary interest to us in this book are summarized in the abbreviated Periodic Table shown in Fig. 2.3. We exclude here the nonmetals and the heavy f-electron lanthanide and actinide metals, which are considered in Chapter 12, but for completeness include Si and Ge, which although semiconductors at ambient temperature and pressure, become metals both in the liquid state and at moderate pressures. The *nominal* simple metals are those with completely empty or filled d shells in the atomic state. Of the elements shown in Fig. 2.3, these are the group-IA alkali metals (Li, Na, K, Rb and Cs), the group-IIA alkaline-earth metals (Be, Mg, Ca, Sr and Ba), the group-IB noble metals (Cu, Ag and Au), the divalent group-IIB metals (Zn, Cd and Hg), the trivalent group-IIIA metals (Al, Ga, In and Ta), and the tetravalent group-IV elements (Si, Ge, Sn and Pb). In addition, the sixth-period elements to the right of Pb, Bi in group V and sometimes Po in group VI, are considered simple metals as well. The *nominal d*-band metals in Fig. 2.3 are then the remaining $3d$, $4d$ and $5d$ transition elements with incomplete d shells in the atomic state.

3 Li	4 Be												
11 Na	12 Mg											13 Al	14 Si
19 K	20 Ca	21 Sc	22 Ti	23 V	24 Cr	25 Mn	26 Fe	27 Co	28 Ni	29 Cu	30 Zn	31 Ga	32 Ge
37 Rb	38 Sr	39 Y	40 Zr	41 Nb	42 Mo	43 Tc	44 Ru	45 Rh	46 Pd	47 Ag	48 Cd	49 In	50 Sn
55 Cs	56 Ba	57 La	72 Hf	73 Ta	74 W	75 Re	76 Os	77 Ir	78 Pt	79 Au	80 Hg	81 Tl	82 Pb

Fig. 2.3 *Simple and d-band metals in the Periodic Table.*

In the condensed state, however, the distinction between simple and d-band metals is less sharp, with the elements directly to the left and right of the transition metals displaying considerable d character in their electronic structures. The practical working definition of a good simple metal is one whose electronic structure is substantially s and p in character and NFE-like in its actual behavior. Thus, the valence energy bands of a good simple metal should be well characterized to a first approximation by the free-electron parabola

$$E_0(\mathbf{k}) = \frac{\hbar^2 k^2}{2m} \equiv \varepsilon_\mathbf{k}, \qquad (2.38)$$

when the wavevector **k** is expressed in an extended-zone representation. By this criterion, the prototype simple metals are the third-period elements Na, Mg and Al. The calculated LDA valence band structure of fcc Al is shown in Fig. 2.4(a). Apart from the definitional folding of the energy bands back into the first Brillouin zone, the free-electron character of these bands is altered only by the appearance of small band gaps at the BZ boundaries. The magnitude of such a gap, say W_{gap}, is a direct measure of the effective strength of the electron-ion interaction and can be expressed in terms of an appropriate plane-wave PP matrix element, as will be discussed in Sec. 2.5. The important thing to notice here is that all such band gaps are small compared to the occupied valence-band width. Since the latter is well approximated by the free-electron Fermi energy ε_F, as given by Eq. (1.21), one has the condition

$$W_{\text{gap}}/\varepsilon_F \ll 1, \qquad (2.39)$$

which can then form the mathematical basis for NFE perturbation expansions of the valence electron density and binding energy in terms of the electron-ion PP.

The degree of free-electron character of a simple metal can be seen more readily in terms of the corresponding density of electronic states, or simply the density of states (DOS), the total number of **k** states per unit energy, as given by

$$\rho(E) = N\frac{2\Omega}{(2\pi)^3} \int_S \frac{1}{|\nabla_{\mathbf{k}} E(\mathbf{k})|} dS, \qquad (2.40)$$

where the integral is over a surface S of constant energy E and the gradient is with respect to the wavevector **k**. For the free-electron energy bands $E(\mathbf{k}) = E_0(\mathbf{k})$, the DOS for $E > 0$ is readily determined to be

$$\rho_0(E) = N\left(\frac{2m}{\hbar^2}\right)^{3/2} \frac{\Omega}{2\pi^2} E^{1/2}. \qquad (2.41)$$

The $E^{1/2}$ behavior of $\rho_0(E)$ is reasonably well obeyed in good simple metals apart from small sharp oscillations due to the band gaps, as is evident from Fig. 2.4(b) for Al. A mathematically smoother quantity, and one also more closely related to the binding or cohesive energy, is the *integrated* DOS,

$$N(E) = \int_0^E \rho(E) dE, \qquad (2.42)$$

which for the free-electron case is just

$$N_0(E) = N\left(\frac{2m}{\hbar^2}\right)^{3/2} \frac{\Omega}{3\pi^2} E^{3/2}. \qquad (2.43)$$

Unlike $\rho(E)$, $N(E)$ for a NFE simple metal is a quite smooth function of energy, as can be seen in Fig. 2.4(b) for the case of Al.

2.4 Guidance from the DFT electronic structure: simple metals vs. d-band metals 49

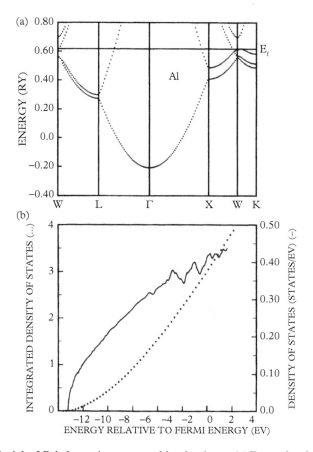

Fig. 2.4 *First-principles LDA electronic structure of fcc aluminum. (a) Energy bands $E_n(\mathbf{k})$ along symmetry directions; (b) DOS (solid curve) and the integrated DOS (dotted curve).*
From Moruzzi et al. (1978), with publisher (Elsevier) permission.

Metals other than Na, Mg and Al with substantially NFE band structures are the heavy alkalis K, Rb and Cs, the trivalent metals Ga, In and Tl, and under high pressure the group-IV elements Si, Ge and Sn. Special cases are the light second-period metals Li and Be, and the heavy sixth-period metals Pb and Bi. Both Li and Be have considerably stronger electron-ion interactions with more localized p character than the other alkalis and alkaline-earths due to the absence of p electrons in their cores. This is manifest in much larger band gaps and a departure from a NFE DOS over a considerable energy range. In the case of Li, this energy range is largely above the Fermi energy, as shown in Fig. 2.5(a), so that a simple-metal treatment of the valence electron density and binding energy still remains reasonable. In the case of Be, however, the DOS is driven far beneath $\rho_0(E)$ both at and below the Fermi level, as shown in Fig. 2.5(b), so that any

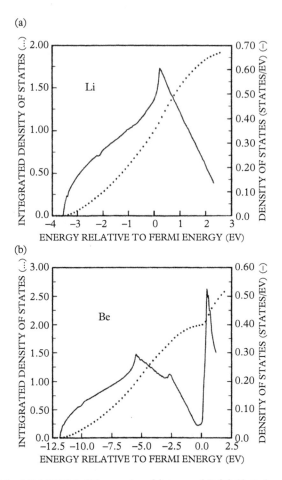

Fig. 2.5 *First-principles LDA DOS (solid curves) and integrated DOS (dotted curves) of second-period simple metals. (a) Bcc lithium; (b) (hypothetical) fcc beryllium.*
From Moruzzi et al. (1978), with publisher (Elsevier) permission.

NFE treatment of this metal is expected to be significantly less accurate. In Chapter 12, we will consider a possible alternate treatment of Li and Be as p-band metals.

The other special cases Pb and Bi also have effectively large PPs, associated in part with their complex relativistic electronic structures and in part with their large valences, $Z = 4$ for Pb and $Z = 5$ for Bi. As discussed in Chapter 3, these metals consequently do not meet the small band-gap criterion given by Eq. (2.39) for an NFE treatment with second-order PP perturbation theory. Nonetheless, these metals have been traditionally treated as simple metals, usually with some ad hoc modification.

All of these purely sp-bonded materials should be contrasted with the noble metals, which each possess a relatively narrow set of occupied d bands in the midst of their NFE

2.4 Guidance from the DFT electronic structure: simple metals vs. d-band metals

s and p bands, as revealed in the DOS shown in Fig. 2.6 for the case of fcc Cu. There can be strong sp-d mixing or hybridization between these energy bands, introducing a gap near the d-band center E_d. Furthermore, the influence of the hybridization can extend to states well above and below the d bands themselves, which in turn acts to unfill the d bands slightly, affecting the valence electron density and binding energy. In the noble metals, the sp-d hybridization significantly perturbs the electronic structure in the vicinity of the Fermi energy E_F and leads to a substantial contribution to the cohesive energy. Consequently, these materials must be treated as d band metals. The quantitative strength of such d-electron effects decreases in magnitude as one moves to the right in the Periodic Table away from the noble metals and the d bands narrow and fall farther below the Fermi energy. In the group II-B elements (Zn, Cd and Hg), the d bands are positioned at the bottom of the NFE s and p bands, but their influence still remains too strong to warrant treating these materials as simple metals. On the other side of the transition elements, the heavy alkaline-earth metals (Ca, Sr and Ba) are similarly influenced by sp-d hybridization with the unoccupied d bands above the Fermi level, leading to a substantial perturbation of their electronic structure at and below E_F, as shown in Fig. 2.7(a) for the case of fcc Ca. These elements too must be treated as d band metals. Simplifications are possible in treating the valence electron density and binding energy of such series-end d-band metals when the effective strength of the sp-d hybridization W_{hyb} is weak such that

$$W_{\text{hyb}}/|\varepsilon_F - E_d| \ll 1, \tag{2.44}$$

allowing useful expansions to be developed, as discussed in Sec. 2.7 and Chapter 4. This is definitely the case in the heavy alkaline-earth and group-IIB metals and approximately so in the noble metals.

The electronic structure of the pure transition elements may be thought of as continuously evolving between that of the alkaline earths on the left and that of the noble metals on the right. The general qualitative features of the d bands remain similar for all transition metals of the same crystal structure, except for a smooth increase in their width as one approaches the center of each series and the fact that now the Fermi level comes in the midst of these bands, as illustrated in Fig. 2.7(b) for (hypothetical) fcc Mn. In the central transition metals, however, the physics associated with the partial occupation of the d bands becomes the central focus as opposed to the sp-d hybridization. Important quantities in seeking a simplified treatment of the binding or cohesive energy are then the width W_d, position E_d and occupation Z_d of the d bands, as an elementary model due to Friedel (1969) nicely demonstrates. In the Friedel model, one ignores the NFE s and p bands entirely and crudely approximates the remaining d-band density of states in the very simple rectangular form $\rho_d(E) = 10N/W_d$ for $E_d - W_d/2 < E < E_d + W_d/2$ and $\rho_d(E) = 0$ otherwise. Substituting this result in Eq. (2.39), it follows immediately that the integrated DOS is linear in energy across the width of the d bands:

$$N_d(E) = N[10(E - E_d)/W_d + 5]. \tag{2.45}$$

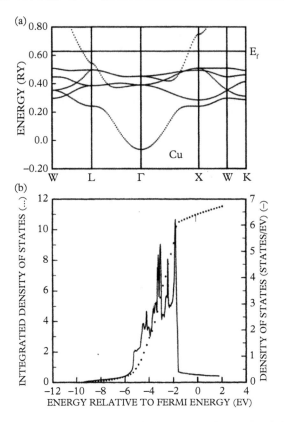

Fig. 2.6 *First-principles LDA electronic structure of fcc copper. (a) Energy bands $E_n(\mathbf{k})$ along symmetry directions; (b) DOS (solid curve) and the integrated DOS (dotted curve).*
From Moruzzi et al. (1978), with publisher (Elsevier) permission.

As shown in Fig. 2.6(b) for Cu and Fig. 2.7(b) for Mn, $N_d(E)$ for the true DOS indeed tends to be linear within the d-band region. Using Eq. (2.45), it then follows that the d-band occupation number Z_d and the Fermi energy E_F can be directly related through E_d and W_d:

$$Z_d = \frac{1}{N} N_d(E_F) = 10(E_F - E_d)/W_d + 5. \tag{2.46}$$

or $E_F - E_d = (Z_d/5 - 1)W_d/2$. Finally, if the cohesive energy E_{coh} is approximated as just the one-site component of the valence band-structure energy defined in Eq. (2.34), one obtains

$$E_{\text{coh}} \approx \frac{1}{N} E_{\text{band}}^{\text{val}} = \frac{1}{N} \int_0^{E_F} (E - E_d)\rho_d(E) dE = \frac{5}{4}\left[(Z_d/5 - 1)^2 - 1\right] W_d. \tag{2.47}$$

2.4 Guidance from the DFT electronic structure: simple metals vs. d-band metals

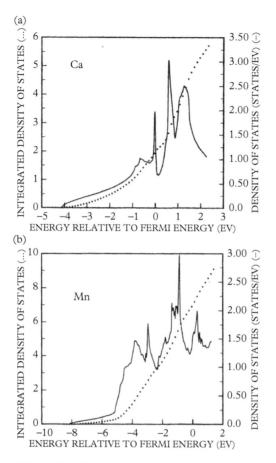

Fig. 2.7 *First-principles LDA DOS (solid curves) and integrated DOS (dotted curves) of fourth-period d-band metals. (a) Fcc calcium; (b) (hypothetical) fcc manganese.*
From Moruzzi et al. (1978), with publisher (Elsevier) permission.

This result contains the observed parabolic variation of the transition-metal cohesive energy with Z_d (Moruzzi et al., 1978), as well as the expected proportionality of E_{coh} to W_d. Thus, quite reasonable first approximations to Z_d and E_{coh} can be derived from only the barest information about ρ_d.

More generally in transition metals, integrated properties of the d-band DOS such as E_{coh} can be systematically related to energy *moments* of ρ_d, with the nth moment defined as

$$M_n \equiv \frac{1}{N} \int_0^\infty (E - E_d)^n \rho_d(E) dE. \tag{2.48}$$

For example, the d-band width squared is proportional to the second moment M_2, as can be readily seen from the simple Friedel model, where one obtains from Eq. (2.48)

$$M_2 = \frac{5}{6} W_d^2, \qquad (2.49)$$

and hence $E_{\text{coh}} \propto M_2^{1/2}$. The higher moments $n > 2$, on the other hand, contain information on the *shape* of $\rho_d(E)$ in addition to its width and can contribute importantly to the structural components of E_{coh}. The moments themselves can be developed in terms of short-ranged TB matrix elements linking d states on neighboring ion sites in the metal, as will be discussed in Sec. 2.6. The detailed relationship between E_{coh} and the moments M_n for transition metals will be considered in Chapter 5.

2.5 Weak pseudopotentials and perturbation theory for simple metals

We now wish to consider specific quantum-mechanical approaches that can capitalize on the simplifications identified above for simple and d-band metals. In all cases, our starting point is the one-electron Schrödinger equation (2.19), which it is now convenient to rewrite in standard bra and ket notation:

$$(T + V) |\psi_\mathbf{k}\rangle = E(\mathbf{k}) |\psi_\mathbf{k}\rangle, \qquad (2.50)$$

where $\langle \mathbf{r} | \psi_\mathbf{k} \rangle \equiv \psi_\mathbf{k}(\mathbf{r})$, and where we have defined T as the kinetic energy operator, such that

$$\langle \mathbf{r} | T | \mathbf{r} \rangle = -\frac{\hbar^2}{2m} \nabla^2. \qquad (2.51)$$

The NFE nature of simple-metal band structures can be reconciled with the deep potential wells of $V(\mathbf{r})$ and corresponding oscillatory valence wavefunctions $\psi_\mathbf{k}(\mathbf{r})$ in the vicinity of the ion cores by introducing appropriate pseudopotentials. Specifically, within the small-core approximation, it is possible to transform the true Schrödinger equation (2.50) to an exactly equivalent pseudo-Schrödinger equation of the form

$$(T + W) |\phi_\mathbf{k}\rangle = E(\mathbf{k}) |\phi_\mathbf{k}\rangle, \qquad (2.52)$$

where W is a relatively weak electron-ion PP with no bound core states and $\langle \mathbf{r} | \phi_\mathbf{k} \rangle \equiv \phi_\mathbf{k}(\mathbf{r})$ is a smooth (nodeless) pseudowavefunction. The schematic relationship between $\psi_\mathbf{k}(\mathbf{r})$ and $\phi_\mathbf{k}(\mathbf{r})$, as well as between V and W, is illustrated in Fig. 2.8. Physically, the effective weakness of W for the s and p valence electrons is a consequence of both the Pauli principle, which effectively excludes these electrons from the inner-core regions of space where V is large, and the fact that the potential V is smooth and slowly varying in the interstitial regions where the s and p electrons reside.

2.5 Weak pseudopotentials and perturbation theory for simple metals

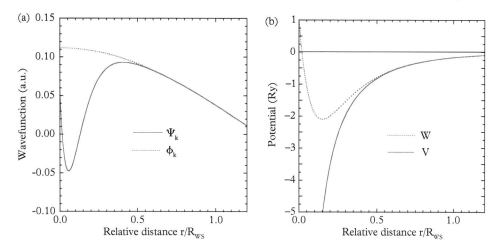

Fig. 2.8 *Simple schematic representations of the wavefunctions and potentials defining the pseudopotential transformation in metals. (a) Valence wavefunction ψ_k in Eq. (2.50) compared with the pseudowavefunction ϕ_k in Eq. (2.52); (b) full potential V in Eq. (2.50) compared with the pseudopotential W in Eq. (2.52).*

Equation (2.52) was first derived by Phillips and Kleinman (1959) on the basis of the orthogonalized-plane-wave (OPW) band-structure method (Herring, 1940). A much more general form of the pseudopotential W entering (2.52) as an exact transformation of Eq. (2.50), however, was subsequently obtained by Austin, Heine and Sham (AHS, 1962). For our purposes, the latter may be written as the *nonlocal* operator

$$W = V + \sum_{i,c} |\phi_c^i\rangle \langle \phi_c^i | O_k , \qquad (2.53)$$

where $\langle \mathbf{r} | \phi_c^i \rangle \equiv \phi_c(\mathbf{r} - \mathbf{R}_i)$ a core orbital centered on site i and O_k is an arbitrary operator that may or may not depend on the state $\mathbf{k}$ and energy $E(\mathbf{k})$ under consideration in the pseudo-Schrödinger equation. The validity of Eq. (2.53) for W is readily verified by inserting this form in Eq. (2.52), operating on the left with $\langle \psi_k |$, and noting that $\langle \psi_k | \phi_c^i \rangle = 0$. The arbitrariness in W can be used to advantage by seeking an *optimized* form with some desired property such as the PP producing the smoothest pseudo-wavefunction $\phi_k(\mathbf{r})$. In practice, useful PPs of the AHS form exist in a rather narrow range and are never very far removed from the simple choice $O_k = -V$ in Eq. (2.53), which yields the compact and interesting baseline result

$$W = (1 - P_c)V, \qquad (2.54)$$

where P_c is the core-state projection operator

$$P_c = \sum_{i,c} |\phi_c^i\rangle \langle\phi_c^i|. \qquad (2.55)$$

Note that if the core orbitals $|\phi_c^i\rangle$ about any site i were a complete set of states, then one would have $P_c = 1$ and $W = 0$ in the core regions with this form of PP. In practice, the core states are only approximately complete in the core regions, and the cancellation between the attractive potential V and the repulsive term $-P_c V$ in Eq. (2.54) is substantial if not exact. Outside the core regions, on the other hand, the core orbitals vanish and one effectively has $P_c = 0$ and $W = V$. Note also that the AHS PP (2.54) is an example of a *non-Hermitian* operator, where $\langle\phi_{k'}|W|\phi_k\rangle \neq \langle\phi_k|W|\phi_{k'}\rangle^*$. One can further show that this PP has the plane-wave matrix elements

$$\langle k'|W|k\rangle = \langle k'|V|k\rangle + \sum_{i,c}[\varepsilon_k - E_c]\langle k'|\phi_c^i\rangle\langle\phi_c^i|k\rangle. \qquad (2.56)$$

The final term in Eq. (2.56) is derived from Eq. (2.54) by exactly replacing $-\langle k'|P_c V|k\rangle$ with $-\langle k'|P_c(H-T)|k\rangle$ and then operating to the left with H to yield E_c and operating on the right with T to yield ε_k. For simple metals, $E(k) \simeq \varepsilon_k$, so that the PP (2.54) is effectively energy dependent, i.e., W depends on the energy of the state being calculated.

Shortly after the formal development of pseudopotential theory in the early 1960s as a rigorous transformation of the Schrödinger equation, it was realized that a more intuitive formulation was also possible in which one viewed the strong ionic-potential field V_{ion} [i.e., Eq. (2.24)] in V as simply being replaced by a weak PP field made up of an individual ionic component from each site:

$$W_{ion} = \sum_i w_{ion}^i. \qquad (2.57)$$

The arbitrariness of the operator O_k in the AHS formulation effectively ensures one that the explicit appearance of core orbitals in the PP is not an essential feature. Thus, one may discard the core entirely and choose w_{ion} in an angular-momentum representation that preserves both the scattering properties of the ion at valence-band energies and the nonlocal operator character of Eq. (2.53). Specifically, one can write

$$w_{ion} = \sum_\ell w_\ell P_\ell, \qquad (2.58)$$

where $\langle r|w_\ell|r'\rangle = w_\ell(r,E)\delta(\mathbf{r}-\mathbf{r}')$, with E the energy of the state under consideration, and where P_ℓ is the angular-momentum projection operator

$$P_\ell = \sum_m |Y_{\ell m}\rangle\langle Y_{\ell m}|. \qquad (2.59)$$

where $\langle \theta, \phi | Y_{\ell m} \rangle \equiv Y_{\ell m}(\theta, \phi)$ is the usual spherical harmonic angular function. The ℓ components of an individual ionic PP, w_{ion}, can then be modeled in the form

$$w_\ell(r, E) = \begin{cases} -A_\ell(r, E) & r < R_\ell \\ -Ze^2/r & r > R_\ell \end{cases}, \quad (2.60)$$

where A_ℓ and R_ℓ are arbitrary and may be chosen to satisfy any desired optimization criteria. Model PPs or model potentials (MPs) of this form were first introduced by Heine and Abarenkov (1964) and have since been highly developed and widely used, with important early contributions made by Shaw and Harrison (1967), Shaw (1968), Rasolt and Taylor (1975) and Hamann, Schlüter and Chiang (1979).

Once W_{ion} has been established, either implicitly by choosing O_k in Eq. (2.53) or explicitly through the A_ℓ and R_ℓ in Eq. (2.60), the full pseudopotential W is formed by adding the appropriate Coulomb and xc potentials due to the valence electrons. Thus, in DFT quantum mechanics within the LDA, $W = W_{\text{ion}} + V_{\text{val}} + \mu_{\text{xc}}$ enters the pseudo-Schrödinger equation (2.52) exactly as the full potential V from Eq. (2.23) enters the true Schrödinger equation (2.50). The valence-binding energy expressions (2.29) and (2.36) continue to hold so long as one calculates the correct electron density $n(\mathbf{r})$ from $\phi_k(\mathbf{r})$. In this regard, the relationship between $\phi_k(\mathbf{r})$ and $\psi_k(\mathbf{r})$ must be established and this depends on the properties of the pseudopotential W. In practice, $W = V$ outside the core regions, so that one is ensured that $\phi_k(\mathbf{r})$ and $\psi_k(\mathbf{r})$ are at least proportional there. If the constant of proportionality is unity, the PP is said to be *norm conserving*. It can be shown that a norm-conserving PP must be a Hermitian operator as well as independent of energy. MPs are norm conserving if the $A_\ell(r, E)$ in Eq. (2.60) are independent of the energy E. If this is the case, then the pseudo-electron density

$$n_{\text{val}}^{\text{ps}}(\mathbf{r}) = 2 \sum_k \phi_k^*(\mathbf{r}) \phi_k(\mathbf{r}), \quad (2.61)$$

is equal to the true valence electron density $n_{\text{val}}(\mathbf{r})$ except inside the core regions. Since the cores have been discarded in setting up the MP, however, the difference between $n_{\text{val}}(\mathbf{r})$ and $n_{\text{val}}^{\text{ps}}(\mathbf{r})$ in the core regions cannot be determined from the $\phi_k(\mathbf{r})$ alone, so that this difference is either neglected, or in the energy-dependent optimized model potential (OMP) method of Shaw (1968), is modeled in terms of a depletion hole in the electron density around each ion site. For the core-based AHS PPs, on the other hand, there is a unique relationship between ψ_k and ϕ_k over all space of the form (Harrison, 1966):

$$|\psi_k\rangle = C_k(1 - P_c)|\phi_k\rangle, \quad (2.62)$$

where C_k is a normalization constant. Given the ϕ_k in this case, one can construct the full valence electron density as

$$n_{\text{val}}(\mathbf{r}) = n_{\text{val}}^{\text{ps}}(\mathbf{r}) + \delta n_{\text{oh}}^c(\mathbf{r}), \quad (2.63)$$

where δn_{oh}^c is a so-called orthogonalization-hole correction, which can be derived from Eq. (2.62), as discussed in Sec. A2.4 of Appendix A2. In general, $\delta n_{\text{oh}}^c(\mathbf{r})$ extends into both the core and valence regions of space, but it carries no net charge in the metal, so that

$$\int \delta n_{\text{oh}}^{c}(\mathbf{r})d\mathbf{r} = 0. \tag{2.64}$$

The nonlocality of both AHS PPs and MPs is an important and usually unavoidable complication for quantitatively accurate calculations. In real space, it implies that W is not a simple function of a local position $\mathbf{r}$ but depends on a second nonlocal coordinate $\mathbf{r}'$ as well: $W(\mathbf{r},\mathbf{r}') \equiv \langle \mathbf{r} | W | \mathbf{r}' \rangle$. Thus, for example,

$$\langle \phi_{\mathbf{k}'} | W | \phi_{\mathbf{k}} \rangle = \iint \phi_{\mathbf{k}'}^{*}(\mathbf{r}) W(\mathbf{r},\mathbf{r}') \phi_{\mathbf{k}}(\mathbf{r}') d\mathbf{r} d\mathbf{r}'. \tag{2.65}$$

It is often instructive, nonetheless, to consider the special case of a *local* PP to gain insight, since this greatly simplifies the mathematical description while retaining most of the qualitative features of the full nonlocal result. Formally, a local PP has the property $W(\mathbf{r},\mathbf{r}') = W(\mathbf{r})\delta(\mathbf{r}-\mathbf{r}')$, where $W(\mathbf{r})$ behaves like an ordinary function of position. The nonlocal MP described by Eqs. (2.58)–(2.60) becomes a local MP in the limit that A_ℓ and R_ℓ are the same for each ℓ. For example, if we take $A_\ell = 0$ and $R_\ell = R_c$, a fixed core radius, then w_{ion} immediately reduces to

$$w_{\text{ion}}(r) = \begin{cases} 0 & r < R_c \\ -Ze^2/r & r > R_c \end{cases}, \tag{2.66}$$

a form that is known as the Ashcroft (1966) empty-core PP or MP.

2.5.1 Plane waves and nearly free-electron valence energy bands

For simple metals, the weakness of the PP makes $|\phi_{\mathbf{k}}\rangle$ amenable to an expansion in plane waves. In the perfect crystal one can write

$$|\phi_{\mathbf{k}}\rangle = \sum_{\mathbf{G}} a_{\mathbf{k}-\mathbf{G}} |\mathbf{k}-\mathbf{G}\rangle, \tag{2.67}$$

where $\mathbf{G}$ is a reciprocal lattice vector for the crystal structure under consideration, and here the plane waves $|\mathbf{k}\rangle$ are normalized in the volume of the system, $N\Omega$:

$$\langle \mathbf{r} | \mathbf{k} \rangle = (N\Omega)^{-1/2} \exp(i\mathbf{k} \cdot \mathbf{r}). \tag{2.68}$$

2.5.1 Plane waves and nearly free-electron valence energy bands

Equation (2.67) satisfies Bloch's theorem, so that substituting this expansion in Eq. (2.52) and operating on the left with $\langle \phi_{\mathbf{k}} |$ immediately leads to an appropriate *secular equation* for the band structure $E(\mathbf{k})$:

$$|[\varepsilon_{\mathbf{k}-\mathbf{G}} - E(\mathbf{k})]\delta_{\mathbf{GG}'} + \langle \mathbf{k} - \mathbf{G}'| W |\mathbf{k} - \mathbf{G}\rangle| = 0, \qquad (2.69)$$

where the rows and columns of the determinant span all possible values of $\mathbf{G}$ and $\mathbf{G}'$. Whenever the full PP of the metal can be written as a sum of individual atomic PPs,

$$W(\mathbf{r}, \mathbf{r}') = \sum_{i} w(\mathbf{r} - \mathbf{R}_i, \mathbf{r}' - \mathbf{R}_i), \qquad (2.70)$$

then the plane-wave matrix elements of W can be factored into structural and atomic components to produce a major simplification of the problem. For a general plane-wave matrix element, one has

$$\langle \mathbf{k} + \mathbf{q}| W |\mathbf{k}\rangle = \frac{1}{N\Omega} \iint \exp[-i(\mathbf{k}+\mathbf{q}) \cdot \mathbf{r}] \sum_{i} w(\mathbf{r} - \mathbf{R}_i, \mathbf{r}' - \mathbf{R}_i) \exp(i\mathbf{k} \cdot \mathbf{r}') d\mathbf{r} d\mathbf{r}'$$

$$= \left(\frac{1}{N} \sum_{i} \exp(-i\mathbf{q} \cdot \mathbf{R}_i) \right) \left(\frac{1}{\Omega} \iint \exp[-i(\mathbf{k}+\mathbf{q}) \cdot \mathbf{r}] w(\mathbf{r}, \mathbf{r}') \right.$$

$$\left. \exp(i\mathbf{k} \cdot \mathbf{r}') d\mathbf{r} d\mathbf{r}' \right) \equiv S(q) \langle \mathbf{k}+\mathbf{q}|w|\mathbf{k}\rangle . \qquad (2.71)$$

To obtain the second line of Eq. (2.71), we have replaced, without approximation, $\mathbf{r}$ and $\mathbf{r}'$ with $\mathbf{r} + \mathbf{R}_i$ and $\mathbf{r}' + \mathbf{R}_i$, respectively, inside the double integral. Here $S(\mathbf{q})$ is a geometrical *structure factor* depending only on the ion positions and $\langle \mathbf{k}+\mathbf{q}|w|\mathbf{k}\rangle$ is an atomic *form factor* depending only on w and atomic volume Ω. For a perfect crystal, this separation is possible because the periodicity of the lattice ensures one that an atomic PP can always be defined, although it may depend on structure. In this case,

$$\langle \mathbf{k} - \mathbf{G}'| W |\mathbf{k} - \mathbf{G}\rangle = S(\mathbf{G} - \mathbf{G}') \langle \mathbf{k} - \mathbf{G}'|w|\mathbf{k} - \mathbf{G}\rangle \qquad (2.72)$$

enters Eq. (2.69) for the band structure. More generally, the RHS of Eq. (2.71) may be considered to be the leading term in an expansion in which w is independent of structure, so that Eq. (2.71) precisely separates volume and structure contributions to lowest order. Also note that in Eqs. (2.71) and (2.72) we have adopted the convention that for plane-wave matrix elements of atomic or one-ion quantities, such as the form factor, the normalization volume is Ω rather than $N\Omega$.

The full particulars of the band structure arising from Eq. (2.69) will not concern us further in this book, but it instructive nonetheless to see how the energy band gaps seen in Fig. 2.4(a) for fcc Al arise from this approach. Consider such a simple metal in a crystal structure with one atom per primitive cell (e.g., fcc), so that $S(\mathbf{G} - \mathbf{G}') = 1$

for all values of **G** and **G**′. For simplicity take the PP to be local, so that the form factor reduces to the Fourier transform of w:

$$\langle \mathbf{k} - \mathbf{G}' | w | \mathbf{k} - \mathbf{G} \rangle = \frac{1}{\Omega} \int w(\mathbf{r}) \exp[-i(\mathbf{G} - \mathbf{G}') \cdot \mathbf{r}] d\mathbf{r} \equiv w(\mathbf{G} - \mathbf{G}') . \quad (2.73)$$

In the vicinity of a given Brillouin-zone boundary or Bragg diffraction plane, only a few reciprocal lattice vectors will contribute significantly to the band structure in Eq. (2.69). Consider the case when this number is two: $\mathbf{G} = 0$ and $\mathbf{G} = \mathbf{G}_0$, where the Bragg plane in question bisects $\mathbf{G}_0$. Then the LHS of Eq. (2.69) reduces to a simple 2×2 determinant and the resulting equation is readily solved for $E(\mathbf{k})$:

$$E(\mathbf{k}) = \frac{1}{2}(\varepsilon_\mathbf{k} + \varepsilon_{\mathbf{k}-\mathbf{G}_0}) + w(0) \pm \frac{1}{2}\left[(\varepsilon_\mathbf{k} - \varepsilon_{\mathbf{k}-\mathbf{G}_0})^2 + 4|w(\mathbf{G}_0)|^2\right]^{1/2}, \quad (2.74)$$

noting only that $w(-\mathbf{G}_0) = w^\star(\mathbf{G}_0)$. The energy bands (2.74) are plotted in Fig. 2.9 for conditions simulating Al with **k** in the [100] direction and should be compared with the full KKR result in Fig. 2.4(a). At the BZ boundary, where $\mathbf{k} = \mathbf{G}_0/2$,

$$E(\mathbf{G}_0/2) = \varepsilon_{\mathbf{G}_0/2} + w(0) \pm |w(\mathbf{G}_0)| , \quad (2.75)$$

so that the magnitude of the band gap is just $2|w(\mathbf{G}_0)|$. Combining this result with the small band-gap criterion given by Eq. (2.39), one sees that a specific condition for the weakness of the PP in a simple metal is that the form factor in the vicinity of reciprocal lattice vectors be small compared to the Fermi energy: $2|w(\mathbf{G}_0)|/\varepsilon_F \ll 1$ for all $\mathbf{G}_0$.

2.5.2 First-order electron density and second-order valence band energies

The alternative to seeking exact solutions for $|\phi_\mathbf{k}\rangle$ and $E(\mathbf{k})$ via Eqs. (2.67)–(2.69) is to capitalize on the weakness of w and use perturbation theory based on a zero-order state of a single plane wave $|\mathbf{k}\rangle$ with energy $\varepsilon_\mathbf{k}$. That is, one expands $|\phi_\mathbf{k}\rangle$ in the form

$$|\phi_\mathbf{k}\rangle = |\mathbf{k}\rangle + \sum_\mathbf{q}{}' a_{\mathbf{k}+\mathbf{q}} |\mathbf{k}+\mathbf{q}\rangle , \quad (2.76)$$

where the prime on the summation denotes the exclusion of the term $\mathbf{q} = 0$. Application of standard first-order perturbation theory then yields for the expansion coefficients $a_{\mathbf{k}+\mathbf{q}}$ the well-known result (Harrison, 1966)

$$a_{\mathbf{k}+\mathbf{q}} = \frac{\langle \mathbf{k}+\mathbf{q}| W |\mathbf{k}\rangle}{\varepsilon_\mathbf{k} - \varepsilon_{\mathbf{k}+\mathbf{q}}} = S(\mathbf{q})\frac{\langle \mathbf{k}+\mathbf{q}| w |\mathbf{k}\rangle}{\varepsilon_\mathbf{k} - \varepsilon_{\mathbf{k}+\mathbf{q}}} . \quad (2.77)$$

Using Eqs. (2.76) and (2.77) in Eq. (2.61) for the pseudo-electron density, one obtains to first order in w the result

2.5.2 First-order electron density and second-order valence band energies 61

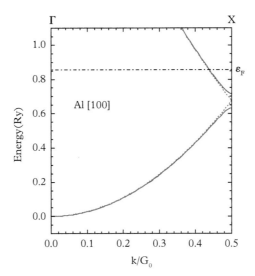

Fig. 2.9 *NFE energy bands for fcc aluminum in the [100] direction from Γ to X in the Brillouin zone (solid lines), as calculated from Eq. (2.74) at the observed equilibrium volume. Shown for comparison are the corresponding free-electron energy bands (dashed lines) and Fermi energy ε_F (dot-dashed line).*

$$n_{\text{val}}^{\text{ps}}(\mathbf{r}) = 2\sum_{\mathbf{k}} \langle \mathbf{r}|\mathbf{k}\rangle \langle \mathbf{k}|\mathbf{r}\rangle + 2\sum_{\mathbf{k}}\sum_{\mathbf{q}}{}' (a_{\mathbf{k}+\mathbf{q}} \langle \mathbf{r}|\mathbf{k}+\mathbf{q}\rangle \langle \mathbf{k}|\mathbf{r}\rangle + \text{c.c.})$$

$$= \frac{Z}{\Omega} + \frac{4}{N\Omega} \sum_{\mathbf{k}}\sum_{\mathbf{q}}{}' S(\mathbf{q}) \frac{\langle \mathbf{k}+\mathbf{q}|\, w\, |\mathbf{k}\rangle}{\varepsilon_{\mathbf{k}} - \varepsilon_{\mathbf{k}+\mathbf{q}}} \exp(i\mathbf{q}\cdot\mathbf{r})$$

$$= n_{\text{unif}} + \delta n_{\text{scr}}(\mathbf{r}). \tag{2.78}$$

In deriving the first term on the second line of Eq. (2.78), we have noted that the number of doubly occupied **k** states is just ZN, so that $ZN/N\Omega = Z/\Omega = n_{\text{unif}}$. In obtaining the remaining term on the second line, the so-called screening electron density $\delta n_{\text{scr}}(\mathbf{r})$, we have first noted that the sums over **k** and **q** for the complex conjugate (c.c.) term on the first line can be replaced by sums over $-\mathbf{k}$ and $-\mathbf{q}$, respectively, and have then used $a^{\star}_{-\mathbf{k}-\mathbf{q}} = a_{\mathbf{k}+\mathbf{q}}$. In Eq. (2.78) the screening electron density δn_{scr} is a functional of the pseudopotential w. In turn, w depends on δn_{scr} through the full electron potential V, as in Eq. (2.23). Combining Eqs. (2.23) and (2.78) thus provides the basis for a self-consistent determination of w. This procedure becomes the *self-consistent screening* calculation in PP perturbation theory and is fully discussed in Chapter 3.

In a similar manner, the valence band energies $E(\mathbf{k})$ obtained from second-order perturbation theory are given by the expression (Harrison, 1966)

$$E(\mathbf{k}) = \varepsilon_\mathbf{k} + \langle \mathbf{k}|\, W\, |\mathbf{k}\rangle + {\sum_\mathbf{q}}' \frac{\langle \mathbf{k}|\, W\, |\mathbf{k+q}\rangle \langle \mathbf{k+q}|\, W\, |\mathbf{k}\rangle}{\varepsilon_\mathbf{k} - \varepsilon_\mathbf{k+q}}$$

$$= \varepsilon_\mathbf{k} + \langle \mathbf{k}|\, w\, |\mathbf{k}\rangle + {\sum_\mathbf{q}}' |S(\mathbf{q})|^2 \frac{\langle \mathbf{k}|\, w\, |\mathbf{k+q}\rangle \langle \mathbf{k+q}|\, w\, |\mathbf{k}\rangle}{\varepsilon_\mathbf{k} - \varepsilon_\mathbf{k+q}}. \tag{2.79}$$

The combination of Eqs. (2.78) and (2.79) then provide the necessary ingredients to evaluate the valence binding energy (2.29) for a simple metal. In this regard, two important and special features of these results need to be emphasized. First, the application of perturbation theory does *not* require Bloch's theorem or lattice periodicity, as in Eq. (2.69). Thus Eqs. (2.78) and (2.79) apply equally to the ordered or disordered metal, with all information on the ion positions carried in the structure factor $S(\mathbf{q})$. This aspect is crucial to the development of highly transferable, structure-independent interatomic potentials from PP perturbation theory. Second, note that Eqs. (2.78) and (2.79) are results obtained with *nondegenerate* perturbation theory. If one tries to apply Eq. (2.79) to the calculation of the band structure of the periodic solid, it will be seen that $E(\mathbf{k})$ diverges on the Bragg planes, where the free-electron energy bands $\varepsilon_\mathbf{k}$ and $\varepsilon_\mathbf{k+q}$ are degenerate. These divergences are readily removable by applying *degenerate* perturbation theory, although this turns out to be unnecessary for the electron density and the valence binding energy. Specifically, it can be shown that such divergences at most contribute negligible higher-order corrections to n_{val} and E_{bind} (Harrison, 1966).

Full completion of the total-energy calculation in simple metals to obtain the binding energy (2.29) in a reciprocal-space representation and the corresponding cohesive-energy functional (1.17) in a real-space representation, including explicit forms for $E_{\text{vol}}(\Omega)$ and $v_2(r,\Omega)$, is considered in Chapter 3.

2.6 Localized *d*-states for the narrow *d* bands in transition-series metals

The transformation from the true Schrödinger equation to the pseudo-Schrödinger equation (2.52) remains valid for *d*-band metals, since this transformation is based only on the small-core approximation for the inner-core electrons. Just as for simple metals, the pseudopotential W is effectively weak for the *s* and *p* valence electrons of transition-series elements. This is not true, however, for the more localized *d* electrons, which still feel the full potential V, as is evident from the general AHS form of the PP, Eq. (2.53). This means that an expansion of $|\phi_\mathbf{k}\rangle$ in plane waves can no longer be used to obtain the electron density and total energy via the perturbation expressions (2.78) and (2.79). Instead, one must either replace or supplement the plane-wave basis with appropriate localized atomic-like *d*-state orbitals centered on each ion site, for example,

$$\langle \mathbf{r}|\phi_d\rangle \equiv \phi_d(\mathbf{r}) = \frac{P_d(r)}{r} Y_{2m}(\theta,\phi), \tag{2.80}$$

where P_d is a normalized radial wavefunction and Y_{2m} is the usual spherical harmonic for $\ell = 2$ with $m = 0, \pm 1$ or ± 2. Such localized orbitals can more readily describe the electron density and total energy associated with the narrow d bands in transition-series metals. In the transition-metal BOP, for example, these orbitals (or linear combinations of them) form the basis set for a TB description of the d bands and the d-bond energy E_{bond}, as given by Eq. (1.25). In the GPT, on the other hand, a *mixed basis* set of plane waves $|\mathbf{k}\rangle$ and localized d states $|\phi_d\rangle$ on each ion site is used to encompass both the NFE treatment of the sp bands developed above as well as TB and resonance descriptions of the d bands, and the additional sp-d hybridization between bands. It is useful, therefore, to consider first the description and some of the properties of isolated d bands in terms of the latter TB and resonance concepts. We proceed with that task in the present section before returning to the full GPT description of d-band metals in Sec. 2.7.

2.6.1 Equivalence of resonance and tight-binding descriptions of the *d* bands

Pure d bands in transition-series metals can be mathematically described in two distinct, yet essentially equivalent, ways. In an angular-momentum representation, such as the KKR band-structure method used to obtain the DFT results displayed in Figs. 2.4–2.7, the d bands are closely linked to the narrow d resonance exhibited by an individual atomic potential $v(\mathbf{r})$. For energies in the vicinity of the resonance, the $\ell = 2$ scattering phase shift of each potential, $\delta_2(E)$, varies rapidly between a small positive value below the resonance to a value approaching π above the resonance. For a sharp resonance over this energy range, the phase shift δ_2 has the approximate form

$$\delta_2(E) \cong \tan^{-1}\left(\frac{W_{\text{res}}/2}{E_{\text{res}} - E}\right), \quad (2.81)$$

where E_{res} and W_{res} are the position and width of the resonance, respectively, such that $\delta_2(E_{\text{res}}) = \pi/2$. The full structure of the d bands in the metal arises out of the multiple scattering among the individual atomic resonances on all sites, but their properties may be expressed approximately in terms of the geometry of the lattice and the *intra-atomic* parameters W_{res} and E_{res}. In particular, the width of the d bands is directly proportional to the width of the resonance, $W_d \propto W_{\text{res}}$, and the center of the d bands approximates the position of the resonance, $E_d \cong E_{\text{res}}$.

From the TB perspective, on the other hand, the d bands are instead linked to the properties of individual atomic orbitals such as Eq. (2.80). Then the d-band center approximates the average one-ion *intra-atomic* expectation value,

$$E_d \cong \frac{1}{N}\sum_i \left(\langle\phi_d^i|\, T + v_i\,|\phi_d^i\rangle = \int \phi_d^\star(\mathbf{r} - \mathbf{R}_i)[T + v(\mathbf{r} - \mathbf{R}_i)]\phi_d(\mathbf{r} - \mathbf{R}_i)d\mathbf{r}\right), \quad (2.82)$$

while the structure and width of the d bands arise out of *interatomic* matrix elements linking two or more sites. These are principally the two-centered integrals connecting the potential v_i centered on a site i with a d-state orbital centered on a neighboring site j:

$$h_{ij}^{dd'} \cong \langle \phi_d^i | v_i | \phi_{d'}^j \rangle = \int \phi_d^*(\mathbf{r} - \mathbf{R}_i) v(\mathbf{r} - \mathbf{R}_i) \phi_{d'}(\mathbf{r} - \mathbf{R}_j) d\mathbf{r}. \tag{2.83}$$

Such a quantity is frequently referred to either as a *hopping* matrix element or as a *transfer* or *bond* integral in TB theory.

Let us now examine these two points of view further and try to understand their basic equivalence. In the KKR method, one first approximates the electron potential $V(\mathbf{r})$ for the periodic solid in so-called muffin-tin form. In the muffin-tin approximation, one takes $V(\mathbf{r}) = \sum_i v(\mathbf{r} - \mathbf{R}_i)$, where each ion-centered atomic component $v(\mathbf{r})$ is assumed to be spherically symmetric within a sphere of radius R_s and constant in the interstitial region outside the sphere such that

$$v(\mathbf{r}) = \begin{cases} v(r) & r < R_s \\ V_0 & r > R_s \end{cases}. \tag{2.84}$$

Here the atomic potentials $v(r)$ are assumed to be nonoverlapping, so that R_s is normally chosen as the radius of the largest sphere that can be inscribed in the atomic polyhedron of the lattice in question. Next, the Schrödinger equation is written as an integral equation for $|\psi_\mathbf{k}\rangle$ in terms of a lattice Green's function and then expanded in terms of energy-dependent partial waves about one particular site:

$$|\psi_\mathbf{k}\rangle = \sum_{\ell m} a_{\ell m}(\mathbf{k}) |\phi_{\ell m}\rangle, \tag{2.85}$$

where the orbital $\langle \mathbf{r} | \phi_{\ell m} \rangle$ has the form

$$\phi_{\ell m}(\mathbf{r}, E) = Y_{\ell m}(\theta, \phi) \begin{cases} R_\ell(r, E) & r < R_s \\ A_\ell[j_\ell(\kappa r) - \tan \delta_\ell(E) n_\ell(\kappa r)] & r > R_s \end{cases}, \tag{2.86}$$

and is a regular solution of the local one-site Schrödinger equation $(T+v)|\phi_{\ell m}\rangle = E|\phi_{\ell m}\rangle$ for $E = E(\mathbf{k})$. In Eq. (2.86), j_ℓ and n_ℓ are the familiar spherical Bessel and Neumann functions, respectively; δ_ℓ is the ℓth phase shift of the muffin-tin potential $v(r)$; and $\kappa \equiv [(2m/\hbar^2)(E - V_0)]^{1/2}$. This leads to the well-known KKR secular equation for the band structure, of the form

$$|\kappa \cot \delta_\ell(E) \delta_{\ell \ell'} \delta_{mm'} + A_{\ell m, \ell' m'}(\mathbf{k}, \kappa)| = 0, \tag{2.87}$$

where the row and columns of the determinant span all values of angular momenta. Here $A_{\ell m, \ell' m'}$ are geometrical *structure constants* that depend on angular momenta ℓ and ℓ', the wavevector $\mathbf{k}$ and energy E, and the crystal structure in question, but are completely independent of the potential $v(r)$. All information regarding the potential is contained

2.6.1 Equivalence of resonance and tight-binding descriptions of the d bands

in the phase shifts δ_ℓ, which in turn can be written in terms of the *logarithmic derivatives* D_ℓ of the radial wavefunction R_ℓ:

$$\cot \delta_\ell(E) = \frac{\kappa R_s n'_\ell(\kappa R_s) - n_\ell(\kappa R_s) D_\ell(E)}{\kappa R_s j'_\ell(\kappa R_s) - j_\ell(\kappa R_s) D_\ell(E)}, \quad (2.88)$$

where $j'_\ell(z) \equiv dj_\ell(z)/dz$, $n'_\ell(z) \equiv dn_\ell(z)/dz$ and $D_\ell(E) \equiv R_s \partial [\ln R_\ell(R_s, E)]/\partial r$.

In the KKR approach, energy bands of pure ℓ symmetry are neatly separated into the diagonal $\ell = \ell'$ blocks, while any hybridization between the bands is contained in the off-diagonal $\ell \neq \ell'$ blocks. Thus, the unhybridized d bands are determined by the 5×5 secular determinant for $\ell = 2$:

$$|\kappa \cot \delta_2(E)\delta_{mm'} + A_{2m,2m'}(\mathbf{k}, \kappa)| = 0. \quad (2.89)$$

Using Eq. (2.81) for the phase shift $\delta_2(E)$ with $E = E(\mathbf{k})$ and setting $E_{\text{res}} = E_d$, this result can be written in the more suggestive form

$$|[E_d - E(\mathbf{k})]\delta_{mm'} + (W_{\text{res}}/2\kappa)A_{2m,2m'}(\mathbf{k}, \kappa)| = 0, \quad (2.90)$$

from which it is immediately evident that the width of the d bands is proportional to W_{res}.

In the TB method applied to a periodic solid, on the other hand, one normally begins by writing $|\psi_\mathbf{k}\rangle$ as an appropriate Bloch sum of localized d-state orbitals centered on each ion site:

$$|\psi_\mathbf{k}\rangle = \frac{1}{N^{1/2}} \sum_{i,d} a_d(\mathbf{k}) \exp(i\mathbf{k} \cdot \mathbf{R}_i) |\phi_d^i\rangle. \quad (2.91)$$

Using this expansion in the Schrödinger equation (2.50) leads directly to a 5×5 secular determinant for the band structure of the form

$$\left| [E_d - E(\mathbf{k})]\delta_{dd'} + \frac{1}{N} \sum_{i,j}{'} \exp[-i\mathbf{k} \cdot (\mathbf{R}_i - \mathbf{R}_j)] h_{ij}^{dd'}(E) \right| = 0, \quad (2.92)$$

where the prime on the summation excludes the $i = j$ term, and we have defined

$$E_d \equiv \frac{1}{N} \sum_i \langle \phi_d^i | H | \phi_d^i \rangle = \frac{1}{N} \sum_i \langle \phi_d^i | T + v_i + \sum_{k \neq i} v_k | \phi_d^i \rangle \quad (2.93)$$

and for $i \neq j$

$$h_{ij}^{dd'}(E) \equiv \langle \phi_d^i | H - E | \phi_{d'}^j \rangle = \langle \phi_d^i | T + \sum_k v_k - E | \phi_{d'}^j \rangle$$

$$= \langle \phi_d^i | v_i | \phi_{d'}^j \rangle + \langle \phi_d^i | T + v_j - E | \phi_{d'}^j \rangle + \sum_{k \neq i,j} \langle \phi_d^i | v_k | \phi_{d'}^j \rangle. \quad (2.94)$$

Here we have assumed for simplicity and comparison with the KKR method that $V = \sum_k v_k$. Equations (2.93) and (2.94) then directly generalize Eqs. (2.82) and (2.83), respectively, to include two-center contributions to E_d and contributions to $h_{ij}^{dd'}$ from d-state *nonorthogonality* integrals

$$\langle \phi_d^i | \phi_{d'}^j \rangle = \int \phi_d^\star(\mathbf{r} - \mathbf{R}_i) \phi_{d'}(\mathbf{r} - \mathbf{R}_j) d\mathbf{r} \tag{2.95}$$

and three-center potential integrals

$$\langle \phi_d^i | v_k | \phi_{d'}^j \rangle = \int \phi_d^\star(\mathbf{r} - \mathbf{R}_i) v(\mathbf{r} - \mathbf{R}_k) \phi_{d'}(\mathbf{r} - \mathbf{R}_j) d\mathbf{r}. \tag{2.96}$$

In the most basic form of TB theory, however, one neglects d-state nonorthogonality, as well as the extra two-center contributions to E_d and three-center contributions to $h_{ij}^{dd'}$. In addition, one also implicitly assumes that $|\phi_d^j\rangle$ is an exact eigenfunction of the one-site Hamiltonian, $(T + v_j)|\phi_d^j\rangle = E_d |\phi_d^j\rangle$, so that the second term on the second line of Eq. (2.94) vanishes, since $(E_d - E)\langle \phi_d^i | \phi_{d'}^j \rangle = 0$ for $i \neq j$. Then one has $h_{ij}^{dd'} = \langle \phi_d^i | v_i | \phi_{d'}^j \rangle$, independent of the energy E, and the secular equation (2.92) can be reduced to the form

$$\left| [E_d - E(\mathbf{k})]\delta_{dd'} + \sum_{j \neq 0} \exp[i\mathbf{k} \cdot \mathbf{R}_j)] \langle \phi_d^{i=0} | v_{i=0} | \phi_{d'}^j \rangle \right| = 0, \tag{2.97}$$

where we have further assumed for simplicity that all lattice sites are equivalent. Note that both Eqs. (2.92) and (2.97) span the five possible d states on any given site. If atomic-like d states of the form (2.80) are used, the determinant (2.97) may be indexed with the quantum number m, as in the KKR approach.

Comparison of the KKR result (2.90) with the TB result (2.97) suggests that if the same d bands are being described, then there ought to be some direct relationship between the *intra-atomic* width of the resonance W_{res} and the *interatomic* bond integral $\langle \phi_d^i | v_i | \phi_{d'}^j \rangle$. Physically, the origin of this relationship lies in the subtle fact that the principal contribution to the latter two-centered integral comes from deep within the site i and *not* from the interstitial region between the sites i and j. To see how a specific relationship between W_{res} and $\langle \phi_d^i | v_i | \phi_{d'}^j \rangle$ can be derived, we first focus on a pure resonant orbital of energy $E = E_{\text{res}}$ in Eq. (2.86). For $r > R_s$, the n_2 term in Eq. (2.86) completely dominates, and from standard scattering theory (e.g., Landau and Lifshitz, 1965) one can rewrite the resonant orbital exactly in the form

$$\phi_{2m}(r, E_{\text{res}}) = Y_{2m}(\theta, \phi) \begin{cases} R_2(r, E_{\text{res}}) & r < R_s \\ -(\kappa_{\text{res}} W_{\text{res}}/2)^{1/2} n_2(\kappa_{\text{res}} r) & r > R_s \end{cases}, \tag{2.98}$$

provided only that one chooses a normalization of R_2 such that $\int_0^{R_s} [rR_2(r, E_{\text{res}})]^2 dr = 1$. The *virtually bound* state described by Eq. (2.98) has a simple physical interpretation.

2.6.1 Equivalence of resonance and tight-binding descriptions of the d bands

If an electron of energy $E = E_{\text{res}}$ were initially confined to $r < R_s$ at some time $t = 0$, then for $t > 0$ it would slowly "leak" away from that site with a current amplitude proportional to W_{res}. In addition, with this normalization one can further derive via scattering theory (e.g., Schiff, 1955) the following expression for W_{res} in terms of the potential $v(r)$:

$$(W_{\text{res}})^{1/2} = -(2\kappa_{\text{res}})^{1/2} \int_0^{R_s} j_2(\kappa_{\text{res}} r) v(r) R_2(r, E_{\text{res}}) r^2 dr, \qquad (2.99)$$

which is again exact provided only that one chooses the zero of energy such that $V_0 = 0$. Following Heine (1967), this result can be usefully simplified in the limit $\kappa_{\text{res}} R_s \ll 1$, which is the physically appropriate regime for transition-series metals. Using the small argument expansion of $j_2(\kappa_{\text{res}} r)$ and keeping only the leading $(\kappa_{\text{res}} r)^2$ term, one finds

$$W_{\text{res}} \cong (2\kappa_{\text{res}}^5/225) M^2 \qquad (2.100)$$

with

$$M \equiv -\int_0^{R_s} v(r) R_2(r, E_{\text{res}}) r^4 dr. \qquad (2.101)$$

To connect Eqs. (2.99)–(2.101) to elementary TB theory, we next develop the two-centered bond integral $\langle \phi_d^{i=0} | v_{i=0} | \phi_{d'}^j \rangle$ in terms of the same resonant state (2.98), assuming only that $v_{i=0}$ has the muffin-tin form (2.84) with $V_0 = 0$. Specifically, we consider the $i = 0$ site and, following Löwdin (1956), expand $\phi_{d'}(\mathbf{r} - \mathbf{R}_j) \equiv \phi_{2m'}(\mathbf{r} - \mathbf{R}_j; E_{\text{res}})$ in spherical harmonics about that site. This procedure yields a result of the general form

$$\langle \phi_d^{i=0} | v_{i=0} | \phi_{d'}^j \rangle = \delta_{mm'} \int_0^{R_s} \alpha_2^m(R, r) v(r) R_2(r, E_{\text{res}}) r^2 dr, \qquad (2.102)$$

where $\alpha_2^m(R, r)$ is an expansion coefficient and $R \equiv R_{ij} = |\mathbf{R}_j|$ is the separation distance between the two sites. For $R > R_s$ α_2^m can be developed in terms of the tail function $\phi_{2m'} \propto n_2$ from site j:

$$\alpha_2^m(R, r) = -(\kappa_{\text{res}} W_{\text{res}})^{1/2} \int_{|R-r|}^{|R+r|} n_2(\kappa_{\text{res}} s) f_2^m(R, r, s) ds$$

$$= -(\kappa_{\text{res}} W_{\text{res}})^{1/2} [b_1^m(R) r^2 + b_2^m(R) r^4 + \cdots]. \qquad (2.103)$$

Here the function f_2^m depends only on d symmetry and may be written entirely in terms of $\ell = 2$ Legendre polynomials. The coefficients b_1^m, b_2^m, etc. in the second line of

Eq. (2.103) arise from a Taylor expansion of $n_2(\kappa_{res}s)$ about the point $s = R$ inside the integral on the first line. Keeping only the r^2 term in the second line of Eq. (2.103) and using Eqs. (2.100) and (2.101), one can derive the explicit final results (Moriarty, 1975)

$$dd\sigma(R) = \frac{W_{res}}{2}\left[5n_0(\kappa_{res}R) - 30\frac{n_1(\kappa_{res}R)}{\kappa_{res}R} + 90\frac{n_2(\kappa_{res}R)}{(\kappa_{res}R)^2}\right] \quad (2.104a)$$

$$dd\pi(R) = \frac{W_{res}}{2}\left[15\frac{n_1(\kappa_{res}R)}{\kappa_{res}R} - 60\frac{n_2(\kappa_{res}R)}{(\kappa_{res}R)^2}\right] \quad (2.104b)$$

$$dd\delta(R) = \frac{W_{res}}{2}\frac{15n_2(\kappa_{res}R)}{(\kappa_{res}R)^2}, \quad (2.104c)$$

where, $dd\sigma$, $dd\pi$ and $dd\delta$ are the $m = 0$, ± 1 and ± 2 components of $\langle \phi_d^{i=0}|v_{i=0}|\phi_{d'}^j\rangle$, and where n_0 and n_1 are, respectively, the $\ell = 0$ and $\ell = 1$ spherical Neumann functions. In this result, one factor of $W_{res}^{1/2}$ has come directly from the amplitude of the tail of the resonant state (2.98) and the other from the integral $M \propto W_{res}^{1/2}$ in Eq. (2.100). While obtained here in an approximate manner, Eqs. (2.104a)–(2.104c) turn out to be exact results due to the fact that $\alpha_2^m(R, r)$ is directly proportional to $j_2(\kappa_{res}r)$ for a pure $\ell = 2$ resonant state. Pettifor (1972), for example, has obtained equivalent results from a direct manipulation of the KKR band-structure equations.

2.6.2 Canonical d bands and their simplifying features

In spite of the exact nature of Eqs. (2.104a)–(2.104c), these results are of limited use because they correspond to extremely long-range interactions between sites and cannot be summed over a crystal lattice in real space. This is a consequence of the fact that $\phi_{2m}(r, E_{res})$ is not a true localized state, as normally envisaged in TB theory, but only a resonance. As shown by Moriarty (1975), this shortcoming is easily overcome, however, by simply damping the $n_\ell(\kappa_{res}r)$ tails in an appropriate manner. Such damping preserves the proportionality of $dd\sigma$, $dd\pi$ and $dd\delta$ to W_{res}, but replaces the spherical Neumann functions by rapidly decaying functions. A particularly important special case occurs when the damping corresponds to adding a constant barrier potential of height E_{res} to the muffin-tin potential for $r > R_s$. This is mathematically equivalent to choosing $V_0 = E_{res}$ and taking the limit $\kappa_{res} \to 0$ in Eqs. (2.104a)–(2.104c). Using the small-argument expansion for the Neumann functions,

$$n_\ell(\kappa_{res}R) = -\frac{1 \cdot 1 \cdot 3 \cdot 5 \ldots (2\ell - 1)}{(\kappa_{res}R)^{\ell+1}} + \cdots, \quad (2.105)$$

2.6.2 Canonical d bands and their simplifying features

together with Eq. (2.100), then yields without approximation the simple results

$$dd\sigma(R) = -\frac{6}{5}\frac{M^2}{R^5} \tag{2.106a}$$

$$dd\pi(R) = \frac{4}{5}\frac{M^2}{R^5} \tag{2.106b}$$

$$dd\delta(R) = -\frac{1}{5}\frac{M^2}{R^5}. \tag{2.106c}$$

In this limit, the two-center bond integrals have the fixed constant ratios of $-6:4:-1$ and a common R^{-5} dependence on interatomic distance. This distinctive behavior is dictated by d symmetry alone and is a characteristic property of so-called *canonical d* bands, which in practice are quite representative of the true d bands of all transition metals.

More generally, the concept of canonical energy bands was formalized by Andersen (1973) directly from the KKR band-structure equations. Andersen noted that calculated energy bands are insensitive to the precise value of the muffin-tin constant V_0, especially if the interstitial volume is a small fraction of the total atomic volume, as is true in close-packed metallic structures. Thus, he suggested working in the following two limits: (i) $V_0 = E$, so that $\kappa \to 0$ in the KKR equations; and (ii) $R_s \to R_{\text{WS}}$, the full atomic-sphere or Wigner–Seitz radius, so that the interstitial volume exactly vanishes. The latter is now referred to as the atomic-sphere approximation or ASA. He further replaced the unbounded orbitals (2.86) with localized *muffin-tin orbitals* of the form

$$\chi_{\ell m}(\mathbf{r}, E) = Y_{\ell m}(\theta, \phi) \begin{cases} R_\ell(r, E) + p_\ell(E)(r/R_{\text{WS}})^\ell & r < R_{\text{WS}} \\ (R_{\text{WS}}/r)^{\ell+1} & r > R_{\text{WS}} \end{cases}, \tag{2.107}$$

where the terms proportional to r^ℓ and $r^{-(\ell+1)}$ derive from the small-argument expansions of j_ℓ and n_ℓ, respectively. Here the quantity

$$p_\ell(E) \equiv \frac{D_\ell(E) + \ell + 1}{D_\ell(E) - \ell} \tag{2.108}$$

ensures that the muffin-tin orbital and its first derivative are continuous at $r = R_{\text{WS}}$, where now the logarithmic derivative is also evaluated at the atomic-sphere radius:

$$D_\ell(E) \equiv R_{\text{WS}} \frac{d}{dr}[\ln R_\ell(R_{\text{WS}}, E)]. \tag{2.109}$$

Because of the r^ℓ term, however, the muffin-tin orbital (2.107) is no longer a solution of the local one-site Schrödinger equation. Therefore, in place of Eq. (2.85) a linear combination of muffin-tin orbitals is used in the expansion of $|\psi_\mathbf{k}\rangle$ together with the requirement that on any given site the tail contributions arising from the $r^{-(\ell+1)}$ term on all other sites exactly cancel the r^ℓ term on the given site. This leads to a secular equation for the band structure $E = E(\mathbf{k})$ of the form

$$|-2(2\ell+1)p_\ell(E)\delta_{\ell\ell'}\delta_{mm'} + S_{\ell m,\ell'm'}(\mathbf{k})| = 0, \qquad (2.110)$$

in place of Eq. (2.87). In Eq. (2.110) simplified geometrical structure constants $S_{\ell m,\ell'm'}$ now appear that are not only independent of the potential but also independent of energy and atomic volume:

$$S_{\ell m,\ell'm'}(\mathbf{k}) = c_{\ell m,\ell'm'} \sum_{j\neq 0} \exp(i\mathbf{k}\cdot\mathbf{R}_j)\left(\frac{R_{\mathrm{WS}}}{R_j}\right)^{\ell''+1} Y_{\ell''m''}(\theta_j,\phi_j), \qquad (2.111)$$

with $\ell'' \equiv \ell' + \ell$ and $m'' = m' - m$. Here the quantities $c_{\ell m,\ell'm'}$ are constants that bear simple relationships to the well-known Gaunt coefficients. Equations (2.110) and (2.111) are often referred to as the KKR-ASA equations. Andersen (1973, 1975) went on to parameterize the energy-dependent functions $p_\ell(E)$ by expanding each muffin-tin orbital about a fixed energy, producing a secular equation that is essentially equivalent to Eq. (2.110) but linear in energy. This resulted in the development of the efficient linear-muffin-tin-orbital (LMTO) method, also known as the LMTO-ASA method, for calculating energy-band structures and total energies within DFT (Andersen, 1975; Andersen and Jepsen, 1975; Skriver, 1984). Subsequently, it has been possible to go beyond the ASA and treat the full potential (FP) entering DFT within the LDA and GGA approximations in robust FP-LMTO approaches (Methfessel et al., 2000; Wills et al., 2000, 2010). In addition, a first-principles TB version of the LMTO method, with self-consistent screened bond integrals, has also been developed (Andersen and Jepsen, 1984; Andersen et al., 1998).

For our purposes, the most important of these results that impact QBIP development are already contained in the KKR-ASA equations. By definition, the canonical ℓ bands for a given potential are those obtained by diagonalizing the $\ell = \ell'$ sub-block of the full secular determinant (2.110). Thus canonical d bands arise from the 5×5 secular determinant

$$|-10p_2(E)\delta_{mm'} + S_{2m,2m'}(\mathbf{k})| = 0. \qquad (2.112)$$

Comparing Eq. (2.112) with the two-center, orthogonal-d-state TB result (2.97), it is evident that the center of the d bands, E_d, is established by the condition $p_2(E_d) = 0$, or from Eq. (2.108) with $\ell = 2$ the condition $D_2(E_d) = -3$. This provides a very specific boundary condition on the wavefunction at the d-band center and is useful guidance in optimizing the choice of d-basis states in the GPT and MGPT QBIP methods, as discussed in Sec. 2.7 and in Chapters 4 and 5. It is also evident from Eq. (2.111) for the structure constants $S_{2m,2m'}$ that the corresponding two-center bond integrals in elementary TB theory should vary with interatomic distance as R^{-5}, in agreement with Eqs. (2.106a)–(2.106c). In fact, had we started with a suitably normalized muffin-tin orbital instead the resonant orbital (2.98) in evaluating the bond integral $\langle \phi_d^{i=0}|v_{i=0}|\phi_{d'}^j\rangle$, we would have been directly led to this result. More generally, for energy bands of any angular momentum ℓ, the center of the bands E_ℓ is determined from the condition $D_\ell(E_\ell) = -(\ell+1)$, while the corresponding TB matrix elements vary with interatomic distance as $R^{-(2\ell+1)}$. In addition, the width of canonical ℓ bands can be well calculated from Wigner-Seitz boundary conditions applied at $r = R_{WS}$. That is, for $\ell = 2$ the d-band width is $W_d = E_{top} - E_{bot}$, where the top of the d bands E_{top} is determined from the antibonding or zero-wavefunction condition $p_2(E_{top}) = 1$, or equivalently $D_2(E_{top}) = -\infty$, and the bottom of the d bands E_{bot} is determined from the bonding or zero-derivative condition $p_2(E_{bot}) = -3/2$, or equivalently $D_2(E_{bot}) = 0$.

2.6.3 Density of states moments in a tight-binding representation

Whether or not canonical d bands are used, the magnitude of W_d, and more generally the d-band contribution to E_{coh}, can be directly related to two-center, orthogonal-d-state TB matrix elements through the DOS moments M_n defined in Eq. (2.48). It is useful, however, to first decompose $\rho_d(E)$ and M_n into atomic components on each ion site, $\rho_d^i(E)$ and μ_n^i, respectively, since the relationships we obtain below do not rely on lattice periodicity and may be applied to arbitrary atomic configurations. Specifically, we write

$$\rho_d(E) = 2\sum_i \rho_d^i(E) \tag{2.113}$$

and

$$M_n = \frac{2}{N}\sum_i \mu_n^i, \tag{2.114}$$

where the factor of 2 accommodates the spin degeneracy in each case. In the infinite periodic solid with equivalent lattice sites, of course, ρ_d^i and μ_n^i are the same on each site, so that $\rho_d = 2N\rho_d^i$ and $M_n = 2\mu_n^i$. More generally, $\rho_d^i(E)$ is the *local d-band density of states*, or simply the local DOS (LDOS), on the site i and may be expressed in terms of the one-electron Green's function G associated with the exact Schrödinger equation for the system in question, $H|\psi_\alpha\rangle = E_\alpha|\psi_\alpha\rangle$:

$$\begin{aligned}
\rho_d^i(E) &= -\frac{1}{\pi}\text{Im}\sum_d \langle\phi_d^i|G|\phi_d^i\rangle \\
&= -\frac{1}{\pi}\text{Im}\sum_d\sum_\alpha \frac{\langle\phi_d^i|\psi_\alpha\rangle\langle\psi_\alpha|\phi_d^i\rangle}{E + i0^+ - E_\alpha} \\
&= \sum_d\sum_\alpha \langle\phi_d^i|\psi_\alpha\rangle\langle\psi_\alpha|\phi_d^i\rangle \delta(E - E_\alpha).
\end{aligned} \tag{2.115}$$

Here it is assumed that the localized basis states $|\phi_d^i\rangle$ comprise a complete set of states with which to represent $|\psi_\alpha\rangle$. In the periodic solid, $|\psi_\alpha\rangle = |\psi_\mathbf{k}\rangle$ and $E_\alpha = E(\mathbf{k})$ for the

d bands, as assumed in Eq. (2.91). More generally, Eq. (2.115) may be used to develop μ_n^i in terms of d-state bond integrals:

$$\mu_n^i \equiv \int_0^\infty (E - E_d)^n \rho_d^i(E) dE$$

$$= \sum_d \sum_\alpha (E_\alpha - E_d)^n \langle \phi_d^i | \psi_\alpha \rangle \langle \psi_\alpha | \phi_d^i \rangle$$

$$= \sum_d \langle \phi_d^i | (H - E_d)^n | \phi_d^i \rangle. \tag{2.116}$$

In obtaining the third line of Eq. (2.116) from the second, we have used the fact that $(H - E_d)^n |\psi_\alpha\rangle = (E_\alpha - E_d)^n |\psi_\alpha\rangle$, which follows from the Schrödinger equation and the completeness relation $\sum_\alpha |\psi_\alpha\rangle \langle \psi_\alpha| = 1$. Finally, the matrix element in the third line of Eq. (2.116) may be factored into individual TB components by inserting the additional completeness relation for the d states,

$$\sum_j \sum_{d'} |\phi_{d'}^j\rangle \langle \phi_{d'}^j| = 1, \tag{2.117}$$

between successive factors of $(H - E_d)$. Thus for $n = 2$, one has

$$\mu_2^i = \sum_j \sum_{d,d'} \langle \phi_d^i | H - E_d | \phi_{d'}^j \rangle \langle \phi_{d'}^j | H - E_d | \phi_d^i \rangle$$

$$= \sum_j \sum_{d,d'} h_{ij}^{dd'} h_{ji}^{d'd} \equiv \sum_j \mathrm{Tr}\,[\mathbf{H}_{ij}\mathbf{H}_{ji}], \tag{2.118}$$

with $h_{ij}^{dd'} = h_{ij}^{dd'}(E_d)$ as given by Eq. (2.94) for $E = E_d$, and where $\mathbf{H}_{ij}$ is the 5×5 matrix made up of elements $h_{ij}^{dd'}$ and Tr denotes the trace of the matrix product which follows. Similarly, for $n = 3$

$$\mu_3^i = \sum_{j,k} \mathrm{Tr}\,[\mathbf{H}_{ij}\mathbf{H}_{jk}\mathbf{H}_{ki}], \tag{2.119}$$

for $n = 4$

$$\mu_4^i = \sum_{j,k,l} \mathrm{Tr}\,[\mathbf{H}_{ij}\mathbf{H}_{jk}\mathbf{H}_{kl}\mathbf{H}_{li}], \tag{2.120}$$

and so on for all higher moments.

In the basic two-center, orthogonal-d-state TB theory, several further simplifications are possible. One then has $h_{ii}^{dd'} = 0$, so that all contributions to μ_n^i where two site indices are equal in the summations automatically vanish. In addition, for $i \ne j$, $h_{ij}^{dd'}$ couples only the sites i and j, so that each moment μ_n^i has a simple diagrammatic representation involving the possible paths of n steps starting and ending on the site i. Evaluation of the

2.6.3 Density of states moments in a tight-binding representation

moments may be done in terms of the basic parameters $dd\sigma$, $dd\pi$ and $dd\delta$ evaluated at appropriate separation distances R_{ij}, R_{jk}, etc. For $n = 2$, this becomes entirely straightforward if one chooses the z axis to lie along the direction $\mathbf{R}_j - \mathbf{R}_i$, so that $\mathbf{H}_{ij}$ is diagonal with the d-basis states given by Eq. (2.80). One then easily finds

$$\mu_2^i = \sum_{j \neq i} \{[dd\sigma(R_{ij})]^2 + 2[dd\pi(R_{ij})]^2 + 2[dd\delta(R_{ij})]^2\}. \tag{2.121}$$

In a perfect crystal with equivalent lattice sites, where $M_2 = 2\mu_2^i$, one can readily use this result to estimate the d-band width W_d. For example, if the simple Friedel-model formula (2.49) is used, where $W_d^2 = (6/5)M_2$, the corresponding d-band width is just

$$W_d = \left(\frac{12}{5} \sum_{j \neq i} \{[dd\sigma(R_{ij})]^2 + 2[dd\pi(R_{ij})]^2 + 2[dd\delta(R_{ij})]^2\} \right)^{1/2}$$

$$= \left(\frac{12}{5} \sum_n N_n \{[dd\sigma(R_n)]^2 + 2[dd\pi(R_n)]^2 + 2[dd\delta(R_n)]^2\} \right)^{1/2}, \tag{2.122}$$

where in the second line the sum is over near-neighbor shells of radius R_n containing N_n neighbors. A similar, but somewhat more accurate formula was obtained by Moriarty (1975), using the TB recursion method of Haydock et al. (1972) and Haydock (1980); see also Finnis (2003):

$$W_d = 4 \left(\frac{1}{5} \sum_n N_n \{[dd\sigma(R_n)]^2 + 2[dd\pi(R_n)]^2 + 2[dd\delta(R_n)]^2\} \right)^{1/2}. \tag{2.123}$$

In the case of fcc canonical d bands, for example, with an assumed nearest-neighbor bond-integral range, so $N_{n=1} = 12$ and $R_{n=1} = 1.809 R_{\text{WS}}$, Eq. (2.123) yields the result

$$W_d = 10.4 M^2 / R_{n=1}^5 = 0.535 M^2 / R_{\text{WS}}^5, \tag{2.124}$$

using Eqs. (2.106a)–(2.106c) for $dd\sigma$, $dd\pi$ and $dd\delta$. This example shows that for canonical d bands, W_d varies with atomic volume as $\Omega^{-5/3}$, a result first obtained by Heine (1967) from resonance-theory arguments together with Eqs. (2.100) and (2.101).

To evaluate the higher moments μ_n^i for $n \geq 3$ in a systematic fashion, one needs a more efficient mathematical representation of the bond-integral matrix $\mathbf{H}_{ij}$ coupling sites i and j. This representation can be obtained by first taking appropriate linear combinations of the five m-indexed d-states (2.80) to express $\phi_d(\mathbf{r})$ alternatively in Cartesian coordinates:

$$\phi_d(\mathbf{r}) = \left(\frac{15}{4\pi} \right)^{1/2} \frac{P_d(r)}{r} \begin{cases} xy/r^2 \\ yz/r^2 \\ zx/r^2 \\ (x^2 - y^2)/(2r^2) \\ (3z^2 - r^2)/(2\sqrt{3}r^2) \end{cases}. \tag{2.125}$$

Then for any orientation of the vector $\mathbf{R}_j - \mathbf{R}_i$, each matrix element of $\mathbf{H}_{ij}$ can be evaluated directly in terms of $dd\sigma$, $dd\pi$ and $dd\delta$ and the direction cosines l, m and n of the vector:

$$\mathbf{R}_j - \mathbf{R}_i = l\hat{\mathbf{x}} + m\hat{\mathbf{y}} + n\hat{\mathbf{z}}. \tag{2.126}$$

Such real symmetric matrices were first elaborated by Slater and Koster (SK, 1954) for s, p and d states, and their possible interactions (e.g., pp', pd, dd'). For the dd' interactions considered here, the rows and columns of the SK matrix are labeled xy, yz, zx, $x^2 - y^2$ and $3z^2 - r^2$ in accord with the five d states defined in Eq. (2.125). For any arrangement of ion positions, the dd' SK matrix may be used repeatedly to perform the required matrix multiplications in Eqs. (2.119) and (2.120) to calculate the moments μ_3^i and μ_4^i, as well as to calculate any needed higher moments. The SK matrices are thus valuable tools for QBIP development, both for TB-moment methods like BOP and for corresponding multi-ion potential methods like GPT and MGPT. The full SK tables are available from a number of references, including Harrison (1980).

2.7 Generalized pseudopotential theory for *d*-band metals

While angular-momentum-based approaches such as the KKR and LMTO methods can efficiently solve the full band-structure problem in transition metals, these methods do not readily lend themselves to the simplified treatment of the electron density and total energy that we seek to develop QBIPs from first principles. The desired simplifications can be achieved, however, within the framework of a mixed basis of plane waves and localized d states, as is used in GPT. The groundwork for such a mixed-basis treatment of transition-series metals was laid in the 1960s beginning with the work of Hodges et al. (1966) and Mueller (1967), who first successfully parameterized the band structure of copper in a hybrid nearly free-electron tight-binding (H-NFE-TB) form. Then Heine (1967), working with a plane-wave representation of the KKR band-structure equations due to Ziman (1965), derived a very approximate model Hamiltonian of this form for transition metals which embraced elements of both the NFE pseudopotential theory of Sec. 2.5 and the d-resonance theory discussed in Sec. 2.6. The method was reformulated by Hubbard (1967), Hubbard et al. (1968, 1969), Jacobs (1968), Kanamori et al. (1969, 1970) and Pettifor (1969, 1970a, 1972), who all produced rigorous H-NFE-TB band-structure schemes. These schemes demonstrated very clearly that the entire transition-metal band structure could be cast in terms of weak matrix elements, which in turn could be related to fundamental parameters such as the position and width of the d resonance.

The H-NFE-TB band-structure schemes did not show, however, how the electron density and total energy could be systematically developed as expansions in the identified small parameters. A key contribution towards establishing that connection was provided by Harrison (1969), who was able to extend the simple-metal perturbation expressions

(2.78) and (2.79), essential for the electron density and the total energy, to the special limiting cases of empty and filled d bands. This was accomplished by introducing a weak d-state hybridization potential and establishing a second perturbation expansion in terms of matrix elements linking plane waves and localized d states. Building on these ideas, the desired theory applicable to all d-band metals was achieved with the development of the GPT by Moriarty (1972a). He utilized the Green's-function techniques of Anderson and McMillan (1967) for describing a d resonance directly in terms of localized d orbitals to generalize Harrison's results to partially filled d bands, including in the theory both two-center bond-integral and nonorthogonal TB d-state matrix elements, as well as sp-d hybridization matrix elements, in a multi-ion total-energy expansion. Subsequently, the GPT was fully integrated into DFT in a series of three papers (Moriarty, 1977, 1982, 1988a), culminating in a first-principles theory of multi-ion interatomic potentials for transition-series metals.

As is the case in simple-metal perturbation theory, the starting point for the GPT method is the exact pseudo-Schrödinger equation (2.52) for the valence-energy states, with the general AHS form of the pseudopotential W still given by Eq. (2.53). In a transition-series d-band metal, an appropriate basis set with which to represent the valence states $|\phi_{\mathbf{k}}\rangle$ is one consisting of both plane waves $|\mathbf{k}\rangle$ and five localized d states $|\phi_d\rangle$ centered on each ion site. In the GPT, the $\phi_d(\mathbf{r})$ basis functions are explicit, calculated orbitals and are defined to be exact eigenstates of a suitable atomic-like reference Hamiltonian:

$$(T + v_{\text{at}})|\phi_d\rangle = E_d^0 |\phi_d\rangle \ . \tag{2.127}$$

It is envisaged that at short range v_{at} approximates the full potential V, so that just as the pseudopotential W remains weak for the valence s and p electrons, a second effectively weak potential for the d electrons is then

$$\Delta \equiv \delta V - \langle \phi_d | \delta V | \phi_d \rangle \ , \tag{2.128}$$

where

$$\delta V \equiv v_{\text{at}} - V. \tag{2.129}$$

These central definitions are those first introduced by Harrison (1969), who denoted Δ as the *hybridization* potential. Note that both Δ and δV are centered on the same site as v_{at} and that Δ is independent of any constant in either v_{at} or δV. When a specific site i is referenced for v_{at}, one may add, if necessary, a superscript i in the above definitions to avoid ambiguity: $\Delta^i \equiv \delta V^i - \langle \phi_d^i | \delta V^i | \phi_d^i \rangle$ and $\delta V^i \equiv v_{\text{at}}^i - V$. In particular, for an optimum choice of v_{at}^i on site i, both δV^i and Δ^i are small within the interior of that site, where $v_{\text{at}}^i \approx V$, and only become large outside, as illustrated schematically in Fig. 2.10. At the same time, $\langle \mathbf{r} | \phi_d^i \rangle = \phi(\mathbf{r} - \mathbf{R}_i)$ is highly localized on the site i, as also shown in Fig. 2.10, so that the operator $\Delta^i |\phi_d^i\rangle$ is effectively weak in all matrix elements coupling $|\phi_d^i\rangle$ to either plane waves or localized d states on other sites.

In the GPT, the NFE *sp* bands are described in terms of free-electron energies ε_k and small plane-wave PP matrix elements $W_{kk'} = \langle k| W |k' \rangle$, while the TB *d* bands are characterized by a mean energy

$$E_d \equiv \langle \phi_d | T + V | \phi_d \rangle = E_d^0 - \langle \phi_d | \delta V | \phi_d \rangle \tag{2.130}$$

plus small and short-ranged *d*-state overlap matrix elements

$$S_{dd'}(R_{ij}) = \langle \phi_d^i | \phi_{d'}^j \rangle \tag{2.131}$$

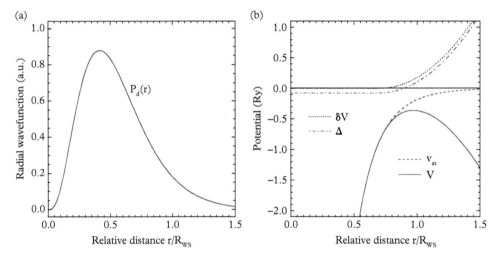

Fig. 2.10 *Schematic representations of the localized atomic-like d-state and associated potentials used in the GPT. (a) Radial wavefunction $P_d(r)$ defined in Eq. (2.80) and used in Eq. (2.127); (b) potentials v_{at}, δV and Δ defined in Eqs. (2.127)–(2.129).*

and

$$\Delta_{dd'}(R_{ij}) = \langle \phi_d^i | \Delta^j | \phi_{d'}^j \rangle = \int \phi_d^\star(\mathbf{r} - \mathbf{R}_i) \Delta(\mathbf{r} - \mathbf{R}_j) \phi_{d'}(\mathbf{r} - \mathbf{R}_j) d\mathbf{r}. \tag{2.132}$$

The quantity $S_{dd'}(R_{ij})$ is just the TB nonorthogonality integral (2.95), while for the same potential V, $\Delta_{dd'}(R_{ij})$ is the negative of the general TB bond integral $h_{ij}^{dd'}(E)$, Eq. (2.94), evaluated at $E = E_d$:

$$\Delta_{dd'}(R_{ij}) = -h_{ij}^{dd'}(E_d) = -\langle \phi_d^i | H - E_d | \phi_{d'}^j \rangle. \tag{2.133}$$

Equation (2.133) may be readily verified by first adding and subtracting v_{at} inside the final matrix element, and then using Eqs. (2.127) and (2.130). More generally, if we define $V' \equiv H - E$, then

$$V'_{dd'} = h_{ij}^{dd'}(E) = -(E - E_d)S_{dd'}(R_{ij}) - \Delta_{dd'}(R_{ij}). \tag{2.134}$$

The remaining sp-d hybridization between the NFE sp bands and the TB d bands is described by small plane-wave–d-state matrix elements

$$S_{kd} = \langle k | \phi_d \rangle = \Omega^{-1/2} \int \exp(-i\mathbf{k} \cdot \mathbf{r}) \phi_d(\mathbf{r}) d\mathbf{r} \tag{2.135}$$

and

$$\Delta_{kd} = \langle k | \Delta | \phi_d \rangle = \Omega^{-1/2} \int \exp(-i\mathbf{k} \cdot \mathbf{r}) \Delta(\mathbf{r}) \phi_d(\mathbf{r}) d\mathbf{r}. \tag{2.136}$$

These matrix elements combine more generally as

$$V'_{kd} = -(E - E_d)S_{kd} - \Delta_{kd}. \tag{2.137}$$

2.7.1 Hybrid nearly free-electron tight-binding energy bands

While our primary focus here is not on the details of the band structure, it is nevertheless instructive to formulate the appropriate secular determinant and see exactly how the above NFE, TB and hybridization components enter. Formally, one expands the pseudo-eigenstate $|\phi_k\rangle$ of Eq. (2.52) in an overcomplete set of plane waves and localized d states that combines the simple-metal expansion (2.67) with the TB expansion (2.91):

$$|\phi_k\rangle = \sum_G a_{k-G} |k - G\rangle + \frac{1}{N^{1/2}} \sum_{i,d} a_d(k) \exp(i\mathbf{k} \cdot \mathbf{R}_i) |\phi_d^i\rangle. \tag{2.138}$$

The resulting pseudo-Hamiltonian matrix and secular equation, however, are singular due to the mathematical overcompleteness of the basis set. To remove this singularity, one can follow Pettifor (1972) and make a set of three exact transformations of the pseudo-Hamiltonian matrix to effectively orthogonalize the basis functions and, at the same time, remove all energy dependence from the off-diagonal terms. For illustrative purposes, one can further simplify the exact result by neglecting in each entry of the transformed matrix all terms higher than first order in W or Δ, and also invoking the usual two-center, orthogonal-d-state TB approximations. Further, if all lattice sites are equivalent and one renormalizes the plane waves to the volume Ω, the secular determinant for the band structure $E(k)$ then reduces to the simple H-NFE-TB block structure

$$\begin{vmatrix} [\varepsilon_{k-G} - E(k)]\delta_{GG'} + \langle k - G' | w_0 | k - G \rangle & -\langle k - G' | \Delta | \phi_{d'} \rangle \\ -\langle \phi_d | \Delta | k - G \rangle & [E_d - E(k)]\delta_{dd'} - \sum_{j \neq 0} \exp[i\mathbf{k} \cdot \mathbf{R}_j] \langle \phi_d^{i=0} | \Delta | \phi_{d'}^j \rangle \end{vmatrix} = 0. \tag{2.139}$$

Here the rows and columns span all values of the reciprocal lattice vectors $\mathbf{G}$ and $\mathbf{G}'$ plus the five localized d states $|\phi_d\rangle$ and $|\phi_{d'}\rangle$. Apart from the normalization, the NFE block of Eq. (2.139) generalizes the simple-metal result (2.69), with the effective plane-wave PP matrix element now becoming

$$\langle \mathbf{k} - \mathbf{G}'| w_0 |\mathbf{k} - \mathbf{G}\rangle = \langle \mathbf{k} - \mathbf{G}'| w |\mathbf{k} - \mathbf{G}\rangle + \sum_d [(\varepsilon_{\mathbf{k}-\mathbf{G}} - E_d) \langle \mathbf{k} - \mathbf{G}'|\phi_d\rangle \langle \phi_d|\mathbf{k} - \mathbf{G}\rangle$$
$$+ \langle \mathbf{k} - \mathbf{G}'| \Delta |\phi_d\rangle \langle \phi_d|\mathbf{k} - \mathbf{G}\rangle + \langle \mathbf{k} - \mathbf{G}'|\phi_d\rangle \langle \phi_d| \Delta |\mathbf{k} - \mathbf{G}\rangle]. \quad (2.140)$$

The potential w_0 is the single-site component of the full potential W_0, which was denoted by Harrison (1969) as the *transition-metal* PP. These potentials also appear quite naturally in the electron-density and total-energy expressions for empty- and filled-d-band metals (see Chapter 4). The TB block of Eq. (2.139), on the other hand, appears as it does in the two-center, orthogonal d-state result, Eq. (2.97). The NFE and TB blocks are then coupled by sp-d hybridization blocks in the form of plane-wave–d-state matrix elements.

To see the influence of the sp-d hybridization more explicitly in Eq. (2.139), consider the simplest special case of a single free-electron band $E(\mathbf{k}) = \varepsilon_\mathbf{k}$ crossing and mixing with a single flat d band $E(\mathbf{k}) = E_d$. Then the LHS of Eq. (2.139) collapses to a simple 2×2 determinant and the corresponding secular equation is just

$$\begin{vmatrix} \varepsilon_\mathbf{k} - E(\mathbf{k}) & -\Delta_{kd} \\ -\Delta_{dk} & E_d - E(\mathbf{k}) \end{vmatrix} = 0 \text{ or } [\varepsilon_k - E(\mathbf{k})][E_d - E(\mathbf{k})] - |\Delta_{kd}|^2 = 0. \quad (2.141)$$

This equation has the solutions

$$E(\mathbf{k}) = \frac{1}{2} \left\{ (\varepsilon_\mathbf{k} + E_d) \pm \left[(\varepsilon_\mathbf{k} - E_d)^2 + 4|\Delta_{kd}|^2 \right]^{1/2} \right\}. \quad (2.142)$$

The resulting energy bands are plotted in Fig. 2.11 for conditions simulating fcc Cu in the [100] direction. In the vicinity of $\mathbf{k} = \mathbf{k}_0$, where $|\mathbf{k}_0| = \sqrt{2mE_d}/\hbar$, a hybridization gap has opened up separating the two bands with a magnitude of $\approx 2|\Delta_{\mathbf{k}_0 d}|$. These model energy bands should be compared with the full KKR band structure of Cu in Fig. 2.6(a). Note that along the very high symmetry [100] and [111] directions, only a single Cu d band hybridizes with the NFE sp band, so that our simple model qualitatively reproduces the behavior of the hybridizing bands. More generally, where high symmetry is lacking, all of the d bands will experience hybridization. In both fcc and bcc transition-series metals, only four plane waves and hence a 9×9 secular determinant in Eq. (2.139) are typically needed to produce an accurate band structure. In addition, at high-symmetry points in the BZ further simplification is possible. Useful analytic formulas for $E(\mathbf{k})$ at the Γ, X and L points in the BZ for fcc metals are given in Sec. A2.5 of Appendix A2.

Our simple hybridization model can also be used to derive an approximate measure of the hybridization strength W_{hyb} needed in Eq. (2.44) to establish the weak-hybridization criterion for treating series-end transition metals as having completely empty or filled d bands. As is clear from Fig. 2.11, hybridization raises the energies of electronic states above E_d and lowers the energies for states below E_d. In the weak hybridization limit, one can expand the band energies $E(\mathbf{k})$ in Eq. (2.142) in powers of $|\Delta_{kd}|^2/(\varepsilon_\mathbf{k} - E_d)$, so that the net hybridization contribution to the cohesive energy depends on properly summing such terms over occupied states up to the Fermi energy ε_F. This task is considered in Chapter 4, but one can anticipate that cancelling contributions from the hybridizing bands above and below E_d will make the behavior near $E(\mathbf{k}) = \varepsilon_F$ dominate the result. To qualify as good empty or filled d-band metals, this suggests a weak-hybridization criterion $|\Delta_{k_F d}|^2/|\varepsilon_F - E_d|\varepsilon_F \ll 1$, so that $W_{\text{hyb}} = |\Delta_{k_F d}|^2/\varepsilon_F$ in Eq. (2.44).

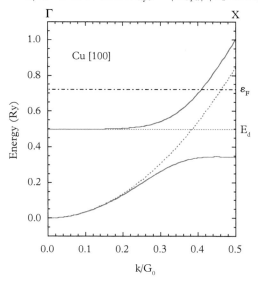

Fig. 2.11 *Hybridized energy bands for fcc copper in the [100] direction from Γ to X in the Brillouin zone (solid lines), as calculated from Eq. (2.142) at the observed equilibrium volume. Shown for comparison are the unhybridized free-electron band $\varepsilon_\mathbf{k}$ and flat d-band E_d (dashed lines), together with the free-electron Fermi energy ε_F (dot-dashed line).*

2.7.2 Pseudo-Green's functions in a mixed plane-wave, d-state basis

To proceed more generally and develop the valence electron density and binding energy for transition-series metals as systematic expansions in the weak matrix elements $W_{\mathbf{kk}'}$, $S_{dd'}$, $\Delta_{dd'}$, S_{kd} and Δ_{kd} it is advantageous to utilize Green's functions. For the present case of interest, however, one needs first to make two generalizations in the usual one-electron Green's function G associated with the true Schrödinger equation (2.50) and introduced in Eq. (2.115). Formally, the Green's function is defined by the relation

$$(E + i0^+ - H)G = 1, \qquad (2.143)$$

with $H = T + V$. Projected into any chosen complete, orthonormal set of basis states $|\beta\rangle$ Eq. (2.143) becomes

$$(E + i0^+)G_{\beta\beta'} - \sum_{\beta''} H_{\beta\beta''} G_{\beta''\beta'} = \delta_{\beta\beta'}, \qquad (2.144)$$

where $G_{\beta\beta'} = \langle\beta|G|\beta'\rangle$ and $H_{\beta\beta'} = \langle\beta|H|\beta'\rangle$. One may then attempt to solve this equation for the Green's function elements $G_{\beta\beta'}$ in terms of $H_{\beta\beta'}$ by standard matrix methods. With respect to any overcomplete mixed-basis set of states $|\alpha\rangle$, however, the matrices $H_{\alpha\alpha'}$ and $G_{\alpha\alpha'}$ become singular and this procedure cannot be used directly. Anderson and McMillan (1967) showed that in such a case Eq. (2.144) should be replaced by

$$(E + i0^+)\sum_{\alpha''} S_{\alpha\alpha''} G_{\alpha''\alpha'} - \sum_{\alpha''} H_{\alpha\alpha''} G_{\alpha''\alpha'} = S_{\alpha\alpha'}, \qquad (2.145)$$

where $S_{\alpha\alpha'} = \langle\alpha|\alpha'\rangle$. Now the singularities in the nonorthogonality matrix $S_{\alpha\alpha'}$ and the Hamiltonian matrix $H_{\alpha\alpha'}$ cancel out, leaving $G_{\alpha\alpha'}$ well behaved. In particular, properties such as the density of states $\rho(E)$ may be formally calculated from $G_{\alpha\alpha'}$ in the same manner as for complete, orthonormal basis sets:

$$\rho(E) = -\frac{2}{\pi}\text{Im}[\text{Tr } G] = -\frac{2}{\pi}\text{Im}\sum_{\alpha} G_{\alpha\alpha}. \qquad (2.146)$$

The second generalization needed here is to adapt the Anderson-McMillan results to the pseudo-Schrödinger equation (2.52). Moriarty (1972a) has shown that Eqs. (2.145) and (2.146) continue to hold if one makes the replacements $H \to H^{ps}$ and $G \to G^{ps}$, where H^{ps} is the pseudo-Hamiltonian $H^{ps} = T + W$ and G^{ps} is the corresponding pseudo-Green's function. The appropriate pseudo-Green's-function equations in a $|\mathbf{k}\rangle$, $|\phi_d\rangle$ basis can then be readily generated from Eq. (2.145). Using the definitions of $V'_{dd'}$ and $V'_{\mathbf{k}d}$ in Eqs. (2.134) and (2.137), respectively, one obtains the following four coupled equations:

$$(E - E_d)G_{dd'} = S_{dd'} + \sum_{\mathbf{k}} V'_{d\mathbf{k}} G_{\mathbf{k}d'} + \sum_{d''\neq d} V'_{dd''} G_{d''d'} \qquad (2.147)$$

$$(E - \varepsilon_\mathbf{k})G_{\mathbf{k}d} = S_{\mathbf{k}d} + \sum_{\mathbf{k}'} W_{\mathbf{k}\mathbf{k}'} G_{\mathbf{k}'d} + \sum_{d'} V'_{\mathbf{k}d'} G_{d'd} \qquad (2.148)$$

$$(E - \varepsilon_\mathbf{k})G_{\mathbf{k}\mathbf{k}'} = \delta_{\mathbf{k}\mathbf{k}'} + \sum_{\mathbf{k}''} W_{\mathbf{k}\mathbf{k}''} G_{\mathbf{k}''\mathbf{k}'} + \sum_{d} V'_{\mathbf{k}d} G_{d\mathbf{k}'} \qquad (2.149)$$

$$(E - E_d)G_{d\mathbf{k}} = S_{d\mathbf{k}} + \sum_{\mathbf{k}'} V'_{d\mathbf{k}'} G_{\mathbf{k}'\mathbf{k}} + \sum_{d'\neq d} V'_{dd'} G_{d'\mathbf{k}}. \qquad (2.150)$$

For notational simplicity in Eqs. (2.147)–(2.150), the superscript "ps" has been dropped from G^{ps}, $E + i0^+$ has been written as E, and in all sums over d states we have temporarily absorbed a sum over ion sites as well as individual quantum numbers m (or xy, yz, zx, etc.).

2.7.3 Transition-metal ion with a *d* resonance in a free-electron gas

The formal solution of the coupled equations (2.147)–(2.150) for a *d*-band metal, as obtained by Moriarty (1972a), is discussed and applied in Sec. 2.7.4. To gain some perspective and familiarity with these equations, it is instructive to consider first the important special case of a single *d*-resonant transition-metal ion in a free-electron gas of density Z/Ω. Not only is this an appropriate reference system for a dilute transition-metal alloy with a simple metal, but the reader will recall from Chapter 1 that the system of N non-interacting *d*-resonant ions represents the zero-order reference system for the bulk transition metal itself.

For the single transition-metal-ion case, *d*-state orthogonality on the ion site requires that $G_{dd'} = \delta_{dd'} G_{dd}$, and in this limit, the final terms in Eqs. (2.147) and (2.150) involving $V'_{dd'}$ for $d \neq d'$ vanish identically. In addition, for simplicity, we consider the limit of an arbitrarily weak pseudopotential such that $W_{\mathbf{k}\mathbf{k}'} = 0$. The single-ion pseudo-Green's-function equations then reduce to:

$$(E - E_d) G_{dd} = 1 + \sum_{\mathbf{k}} V'_{d\mathbf{k}} G_{\mathbf{k}d} \qquad (2.151)$$

$$(E - \varepsilon_{\mathbf{k}}) G_{\mathbf{k}d} = S_{\mathbf{k}d} + V'_{\mathbf{k}d} G_{dd} \qquad (2.152)$$

$$(E - \varepsilon_{\mathbf{k}}) G_{\mathbf{k}\mathbf{k}'} = \delta_{\mathbf{k}\mathbf{k}'} + \sum_{d} V'_{\mathbf{k}d} G_{d\mathbf{k}'} \qquad (2.153)$$

$$(E - E_d) G_{d\mathbf{k}} = S_{d\mathbf{k}} + \sum_{\mathbf{k}'} V'_{d\mathbf{k}'} G_{\mathbf{k}'\mathbf{k}}, \qquad (2.154)$$

where the sum in Eq. (2.153) has collapsed to one over *d* states on the single ion site in question. Equations (2.151)–(2.154) represent the nonmagnetic limit of the formalism first introduced by Anderson to treat the classic local magnetic-moment problem (Anderson, 1961; Kittel, 1963). The solution of these equations proceeds as follows. First use Eq. (2.152) in Eq. (2.151) and solve the resulting equation for G_{dd}. This yields immediately

$$G_{dd}(E) = \left[1 + \sum_{\mathbf{k}} \frac{V'_{d\mathbf{k}} S_{\mathbf{k}d}}{E - \varepsilon_{\mathbf{k}}} \right] [E - E_d - \Gamma_{dd}(E)]^{-1}, \qquad (2.155)$$

where $\Gamma_{dd}(E)$ is the *d*-state self-energy

$$\Gamma_{dd}(E) \equiv \sum_{\mathbf{k}} \frac{V'_{d\mathbf{k}} V'_{\mathbf{k}d}}{E - \varepsilon_{\mathbf{k}}}. \qquad (2.156)$$

Substituting Eq. (2.155) back into Eq. (2.152) will then give $G_{kd}(E)$. To obtain the remaining components, $G_{kk'}$ and G_{dk}, first use Eq. (2.153) in Eq. (2.154) to obtain

$$(E - E_d)G_{dk} = S_{dk} + \frac{V'_{dk}}{E - \varepsilon_k} + \sum_{d'}\sum_{k'} \frac{V'_{dk'}V'_{k'd'}}{E - \varepsilon_{k'}} G_{d'k}. \qquad (2.157)$$

Because of the spherical symmetry of the problem, only the $d' = d$ contribution to the sum over d' survives in the last term of Eq. (2.157). Using the definition (2.156) and solving Eq. (2.157) for $G_{dk}(E)$ one has

$$G_{dk}(E) = \left[S_{dk} + \frac{V'_{dk}}{E - \varepsilon_k}\right][E - E_d - \Gamma_{dd}(E)]^{-1}. \qquad (2.158)$$

Finally, inserting Eq. (2.158) back into Eq. (2.153) yields

$$G_{kk'}(E) = \frac{\delta_{kk'}}{E - \varepsilon_k} + \sum_d \left[\frac{V'_{kd}S_{dk'}}{E - \varepsilon_k} + \frac{V'_{kd}V'_{dk'}}{(E - \varepsilon_k)(E - \varepsilon_{k'})}\right][E - E_d - \Gamma_{dd}(E)]^{-1}. \qquad (2.159)$$

Having obtained the pseudo-Green's-function elements, physical properties may then be calculated directly in terms of these quantities. In particular, the density of states is found by using Eqs. (2.155) and (2.159) in Eq. (2.146):

$$\rho(E) = -\frac{2}{\pi}\mathrm{Im}\left[\sum_k G_{kk} + \sum_d G_{dd}\right]$$

$$= -\frac{2}{\pi}\mathrm{Im}\left\{\sum_k \frac{1}{E - \varepsilon_k} + \sum_d\left[1 + \sum_k\left(\frac{V'_{kd}S_{dk} + S_{kd}V'_{dk}}{E - \varepsilon_k} + \frac{V'_{kd}V'_{dk}}{(E - \varepsilon_k)^2}\right)\right]\right.$$

$$\left.[E - E_d - \Gamma_{dd}(E)]^{-1}\right\}. \qquad (2.160)$$

The first term on the second line of Eq. (2.160) is just the free-electron density of states

$$\rho_0(E) = -\frac{2}{\pi}\mathrm{Im}\sum_k \frac{1}{E - \varepsilon_k} = \left(\frac{2m}{\hbar^2}\right)^{3/2}\frac{N\Omega}{2\pi^2}E^{1/2}. \qquad (2.161)$$

The latter equality can be verified by first converting the sum over $\mathbf{k}$ to an integral, and then performing the simple contour integration to yield the RHS of Eq. (2.161). In addition, using the definition (2.156) of $\Gamma_{dd}(E)$ and noting from Eq. (2.134) that $dV'_{kd}/dE = -S_{kd}$, it immediately follows that

2.7.3 Transition-metal ion with a d resonance in a free-electron gas

$$-\frac{d}{dE}[\Gamma_{dd}(E)] = \sum_k \left(\frac{V'_{kd} S_{dk} + S_{kd} V'_{dk}}{E - \varepsilon_k} + \frac{V'_{kd} V'_{dk}}{(E - \varepsilon_k)^2} \right). \qquad (2.162)$$

Equation (2.160) for the density of states thus collapses to the relatively simple form

$$\rho(E) = \rho_0(E) - \frac{2}{\pi} \text{Im} \sum_d \frac{d}{dE} \{\ln[E - E_d - \Gamma_{dd}(E)]\}. \qquad (2.163)$$

To see the physical significance of the second term in Eq. (2.163), we can compare this result with that obtained via the standard treatment of this problem in an angular-momentum representation. That is, if the electron potential $V(\mathbf{r})$ here were in the form of a single muffin-tin, as in Eq. (2.84) with the constant-potential region extended to infinity, then the full solutions to the Schrödinger equation are just the energy-dependent partial waves $\phi_{\ell m}(\mathbf{r}, E)$ given by Eq. (2.86). If the Green's function is then developed in terms of these partial waves, the density of states is obtained in the well-known form (Anderson and McMillan, 1967)

$$\rho(E) = \rho_0(E) + \frac{2}{\pi} \sum_\ell (2\ell + 1) \frac{d\delta_\ell(E)}{dE} + \cdots, \qquad (2.164)$$

where $\cdots$ represents small additional oscillatory terms. In the context of Eq. (2.163), the $\ell \neq 2$ components of the second term in Eq. (2.164), as well as the additional oscillatory terms, have been suppressed in our treatment by setting $W_{kk'}$ to zero. The remaining $\ell = 2$ component of Eq. (2.164) is thus to be identified with the second term in (2.163), so the corresponding phase shift δ_2 is just

$$\delta_2(E) = -\frac{1}{5} \text{Im} \sum_d \ln[E - E_d - \Gamma_{dd}(E)]$$

$$= -\tan^{-1} \left(\frac{-\text{Im}[\Gamma_{dd}(E)]}{E - E_d - \text{Re}[\Gamma_{dd}(E)]} \right), \qquad (2.165)$$

where in the second line we have used the fact that by symmetry E_d and Γ_{dd} are the same for each of the five d states centered on the ion. Further noting from the expected form of $\delta_2(E)$ for a sharp resonance, as given by Eq. (2.81), one concludes that the position of the resonance E_{res} is

$$E_{\text{res}} = E_d + \text{Re}[\Gamma_{dd}(E_{\text{res}})], \qquad (2.166)$$

while the width of resonance W_{res} is given by

$$W_{\text{res}} = -2\text{Im}[\Gamma_{dd}(E_{\text{res}})]. \qquad (2.167)$$

In developing Eq. (2.165), we have thus succeeded in expressing the $\ell = 2$ resonance in terms of localized atomic d states, defined by Eq. (2.127). This result also goes beyond the angular-momentum-based treatment by applying to an arbitrary potential $V(\mathbf{r})$ and not just one of muffin-tin form. In addition, Eq. (2.165) allows for the full energy dependence of the resonance to be taken into account through the self-energy $\Gamma_{dd}(E)$, and thus generalizes the sharp-resonance approximation (2.82). This energy dependence is important in practice because $-\text{Im}\,[\Gamma_{dd}(E)]$ increases rapidly with energy, varying approximately as $E^{5/2}$ for small E. The calculated behaviors of $\Gamma_{dd}(E)$ and $\delta_2(E)$ are illustrated in Fig. 2.12 for the series-end, sharp-resonance case of copper.

In the full GPT for the bulk metal, the zero-order electron density, including the sp valence Z and the d-state occupation number Z_d, is based on applying the above results to a system of N noninteracting transition-metal ions placed in a compensating uniform electron gas of density Z/Ω. The total integrated density of states for this reference system,

$$N(E) = N_0(E) + N_d(E) = \left(\frac{2m}{\hbar^2}\right)^{3/2} \frac{N\Omega}{3\pi^2} E^{3/2} + \frac{10N}{\pi}\delta_2(E), \qquad (2.168)$$

when evaluated at the Fermi energy of the free-electron gas, $E = \varepsilon_F$, provides three interdependent self-consistency constraints on the system. First, the strong d resonance exhibited by each ion in the presence of the free-electron gas determines the number of valence d electrons retained by the ion, Z_d, by the condition

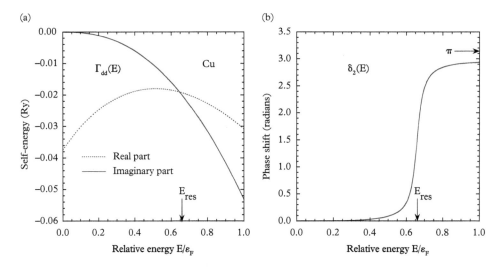

Fig. 2.12 *Resonant response of a Cu ion embedded self-consistently in a free-electron gas of density Z/Ω_0, with $Z = 1.65$, $\Omega_0 = 79.68$ a.u. and $Z_d = 9.35$.*
From Moriarty (1988a), with publisher (APS) permission.

$$N_d(\varepsilon_F) = NZ_d = \frac{10N}{\pi}\delta_2(\varepsilon_F),\tag{2.169}$$

where $\delta_2(\varepsilon_F)$ is to be calculated from Eq. (2.165). Second, the remaining NZ free electrons fix the Fermi energy ε_F of the free-electron gas:

$$N_0(\varepsilon_F) = NZ = \left(\frac{2m}{\hbar^2}\right)^{3/2}\frac{N\Omega}{3\pi^2}\varepsilon_F^{3/2}.\tag{2.170}$$

Third, for a pure transition metal Z effectively represents the number of non-d valence electrons per atom, so that for overall charge conservation one must additionally require

$$N(\varepsilon_F) = N(Z + Z_d) = N(Z_a - Z_c),\tag{2.171}$$

where Z_a is the atomic number and Z_c the number of inner-core electrons per atom for the element in question. Thus, for example, in the case of group-VIB elements (Cr, Mo, W), $Z + Z_d = 6$. Equations (2.164)–(2.166) then constitute three equations in three unknowns and must be iterated to yield self-consistent values of Z, Z_d and ε_F. The calculated equilibrium values of these quantities for $3d$ and $4d$ elemental transition metals are discussed in Chapter 5. The actual physical entity, self-consistent ion plus compensating free-electron gas, is modeled in an atomic-like construction that we call a zero-order pseudoatom (Moriarty, 1974). In practice, the zero-order pseudoatom self-consistently determines the inner-core orbitals $\phi_c(\mathbf{r})$ and the basis d-state orbitals $\phi_d(\mathbf{r})$, as defined by Eq. (2.127), as well as the values of Z, Z_d and ε_F. A full elaboration of the zero-order pseudoatom and its properties is given in Chapters 4 and 5.

2.7.4 Valence band-structure energy for bulk transition-series metals

The solution of the pseudo-Green's-function equations (2.147)–(2.150) for the bulk transition-metal problem of N interacting ions can be obtained in a similar manner, as discussed in Sec. A2.6 of Appendix A2, and builds upon the above results. In particular, the density-of-states expression (2.163) for a single transition-metal ion is generalized in the d-band metal to the form

$$\rho(E) = \rho_0(E) - \frac{2}{\pi}\text{Im}\left(\sum_{\mathbf{k}}\frac{\Sigma_{\mathbf{kk}}(E)}{(E-\varepsilon_\mathbf{k})^2} + \frac{d}{dE}\ln[D(E)]\right),\tag{2.172}$$

where we have defined the energy-dependent functionals

$$\Sigma_{\mathbf{kk}'}(E) \equiv W_{\mathbf{kk}'} + \sum_{\mathbf{k}''}\frac{W_{\mathbf{kk}''}W_{\mathbf{k}''\mathbf{k}'}}{E-\varepsilon_{\mathbf{k}''}}+\cdots\tag{2.173}$$

86 *Fundamental Principles in Metals Physics*

and the $5N \times 5N$ determinant

$$D(E) \equiv \det\left[-(V'_{dd'} + \Gamma_{dd'} + \Lambda_{dd'})\right], \quad (2.174)$$

with

$$\Gamma_{dd'}(R_{ij}, E) \equiv \sum_{\mathbf{k}} \frac{V'_{d\mathbf{k}} V'_{\mathbf{k}d'}}{E - \varepsilon_{\mathbf{k}}} \quad (2.175)$$

and

$$\Lambda_{dd'}(R_{ij}, E) \equiv \sum_{\mathbf{k},\mathbf{k}'} \frac{V'_{d\mathbf{k}} \Sigma_{\mathbf{k}\mathbf{k}'} V'_{\mathbf{k}'d'}}{(E - \varepsilon_{\mathbf{k}})(E - \varepsilon_{\mathbf{k}'})}. \quad (2.176)$$

The quantity $\Sigma_{\mathbf{k}\mathbf{k}'}$ involves an infinite series expansion in powers of the pseudopotential $W_{\mathbf{k}\mathbf{k}'}$ such that the first two terms in the density of states (2.172) are formally the same as would be obtained for a simple metal in a plane-wave basis. The quantity $D(E)$, on the other hand, is a large $5N \times 5N$ determinant that embraces the entire d-band complex, with the d states in Eqs. (2.172)–(2.176) again spanning both ion position and individual quantum numbers. Equation (2.175) is a generalization of the one-ion d-state self-energy (2.156), which is recovered in the $R_{ij} = 0$ limit.

Although formally exact, Eq. (2.172) cannot be used in this form to calculate the density of states numerically for a real material. The usefulness of this result comes rather through its powerful analytic properties, which we can exploit in developing a multi-ion expansion of the valence band-structure energy (2.34). Written in terms of the density of states $\rho(E)$, Eq. (2.34) takes the form

$$E_{\text{band}}^{\text{val}} = \int_0^{E_F} E\rho(E)dE - NZ_d E_d, \quad (2.177)$$

where the zero of energy has been chosen at the bottom of the valence energy bands. In the context of Eq. (2.172) for the density of states, considerable simplification will result by then re-expressing $E_{\text{band}}^{\text{val}}$ in terms of the integrated density of states $N(E)$. Using $\rho(E) = dN(E)/dE$, integrating Eq. (2.177) once by parts and adding the conservation of electrons condition $N(E_F) = N(Z + Z_d)$, one obtains without approximation

$$E_{\text{band}}^{\text{val}} = NZE_F + NZ_d(E_F - E_d) - \int_0^{E_F} N(E)dE. \quad (2.178)$$

The next step is to break $\rho(E)$ and $N(E)$ up into separate components. We write

$$\rho(E) = \rho_0(E) + \rho_d(E) + \delta\rho_{sp}(E) + \delta\rho_d(E) \quad (2.179)$$

2.7.4 Valence band-structure energy for bulk transition-series metals

and

$$N(E) = N_0(E) + N_d(E) + \delta N_{sp}(E) + \delta N_d(E), \qquad (2.180)$$

where $\rho_d(E)$ and $N_d(E)$ correspond to the one-ion values, as in Eq. (2.168). We then identify $\delta \rho_{sp}$ as the second term on the RHS of Eq. (2.172) involving the pseudopotential W, from which one can readily derive an explicit expression for δN_{sp}:

$$\delta N_{sp}(E) = \frac{2}{\pi} \text{Im} \left(\sum_k \frac{W_{kk}}{E - \varepsilon_k} + \frac{1}{2} \sum_{k,k'} \frac{W_{kk'} W_{k'k}}{(E - \varepsilon_k)(E - \varepsilon_{k'})} + \cdots \right). \qquad (2.181)$$

The correctness of Eq. (2.181) can be easily verified by differentiating once with respect to energy and comparing the result with the second term in Eq. (2.172). Likewise, we associate $\delta \rho_d$ and δN_d with the multi-ion contributions to the third term involving $D(E)$ in Eq. (2.172). To obtain an explicit expression for δN_d, we first note that the one-ion $R_{ij} = 0$ terms in $\ln[D(E)]$ can be extracted in the form

$$\ln[D(E)] = \sum_d \ln(E - E_d - \Gamma_{dd} - \Lambda_{dd}) + \ln[\det(T_{dd'})], \qquad (2.182)$$

where $T_{dd'}$ is the relative d-state coupling strength between different ion sites

$$T_{dd'}(R_{ij}, E) = \frac{-(V'_{dd'} + \Gamma_{dd'} + \Lambda_{dd'})}{E - E_d - \Gamma_{dd} - \Lambda_{dd}}, \qquad (2.183)$$

and $\det(T_{dd'})$ remains a $5N \times 5N$ determinant. To proceed further, it is necessary to separate out the zero-order volume-dependent contributions to E_d and Γ_{dd} occurring in the one-ion reference system equations from the additional higher-order structure-dependent contributions to these quantities arising in the bulk transition-metal equations. Specifically, we write

$$E_d = E_d^{\text{vol}} + E_d^{\text{struc}} \qquad (2.184)$$

and

$$\Gamma_{dd} = \Gamma_{dd}^{\text{vol}} + \Gamma_{dd}^{\text{struc}}. \qquad (2.185)$$

In this notation, $E_d = E_d^{\text{vol}}$ and $\Gamma_{dd} = \Gamma_{dd}^{\text{vol}}$ in the reference-system equations of Sec. 2.7.3, Eqs. (2.151)–(2.171). Thus, from Eqs. (2.165) and (2.168) we have

$$N_d(E) = -\frac{2}{\pi} \text{Im} \sum_d \ln(E - E_d^{\text{vol}} - \Gamma_{dd}^{\text{vol}}). \qquad (2.186)$$

Since $N_d(E) + \delta N_d(E) = \ln[D(E)]$, it follows from Eqs. (2.182) and (2.186) that

$$\delta N_d(E) = -\frac{2}{\pi} \operatorname{Im} \left(\sum_d \ln\left[1 - \frac{E_d^{\text{struc}} - \Gamma_{dd}^{\text{struc}} - \Lambda_{dd}}{E - E_d^{\text{vol}} - \Gamma_{dd}^{\text{vol}}}\right] + \ln[\|T_{dd'}\|] \right). \quad (2.187)$$

Note that the d-state sums in Eqs. (2.182), (2.186) and (2.187) are over both ion sites and individual quantum numbers.

Using the above-defined components of $N(E)$, some straightforward manipulation of Eq. (2.178) then yields, still *without approximation*, the useful result

$$E_{\text{band}}^{\text{val}} = \frac{3}{5} N Z \varepsilon_F + N E_{\text{vol}}^d - \int_0^{\varepsilon_F} [\delta N_{sp}(E) + \delta N_d(E)] \, dE - N Z_d E_d^{\text{struc}} + \delta E_{\text{band}}. \quad (2.188)$$

The first term in Eq. (2.188) is the familiar free-electron kinetic energy, which derives from the integration of $\rho_0(E)$, while the second term is the pure volume component of the d-state energy, which derives from the integration of $\rho_d(E)$:

$$E_{\text{vol}}^d = \frac{1}{N} \int_0^{\varepsilon_F} (E - E_d^{\text{vol}}) \rho_d(E) dE$$

$$= Z_d(\varepsilon_F - E_d^{\text{vol}}) - \frac{10}{\pi} \int_0^{\varepsilon_F} \delta_2(E) dE. \quad (2.189)$$

As is discussed in Chapter 5, the quantity E_{vol}^d already contains much of the increased cohesive energy in transition metals over simple metals. The remaining components of $E_{\text{band}}^{\text{val}}$ are smaller in magnitude and primarily structural in character, with the final term in Eq. (2.188) representing a correction for the true Fermi energy:

$$\delta E_{\text{band}} \equiv N(Z + Z_d)(E_F - \varepsilon_F) - \int_{\varepsilon_F}^{E_F} N(E) dE. \quad (2.190)$$

As shown by Moriarty (1988a), this quantity is of order $(E_F - \varepsilon_F)^2$ and small so long as ε_F is indeed a good approximation to E_F, which is usually the case in practice. The most important components of $E_{\text{band}}^{\text{val}}$ that contribute to real-space interatomic potentials are the energy integrals over δN_{sp} and δN_d. The contribution from δN_{sp} is formally the same as one obtains for simple metals, leading to sp pair potentials, while the contribution from δN_d is the additional consequence of the partially filled d bands, leading to d-state and sp-d hybridized pair and multi-ion potentials.

2.7.5 Valence electron density for bulk transition-series metals

The electron density can be similarly developed in terms of the solutions to the pseudo-Green's-function equations (2.147)–(2.150). For an elemental transition metal, the total

2.7.5 Valence electron density for bulk transition-series metals

electron density in the GPT consists of a valence density n_{val} plus a localized d-state density n_d and inner-core density n_c centered about each site:

$$n(\mathbf{r}) = n_{\text{val}}(\mathbf{r}) + \sum_i n_d(\mathbf{r} - \mathbf{R}_i) + \sum_i n_c(\mathbf{r} - \mathbf{R}_i). \quad (2.191)$$

The sum of n_c and n_d on a given site is the total core density defined in Eq. (2.32), so that n_{val} is formally defined by Eq. (2.33). As shown by Moriarty (1972a), the first two terms on the RHS of Eq. (2.191) can be expressed exactly in terms of the pseudo-Green's-function elements $G_{\mathbf{k}\mathbf{k}'}$ and $G_{d\mathbf{k}}$:

$$n_{\text{val}}(\mathbf{r}) + \sum_i n_d(\mathbf{r} - \mathbf{R}_i) = -\frac{2}{\pi} \text{Im} \int_0^{E_F} \left(\sum_{\mathbf{k},\mathbf{k}'} \langle \mathbf{r} | (1 - P_c) | \mathbf{k} \rangle G_{\mathbf{k}\mathbf{k}'} \langle \mathbf{k}' | (1 - P_c) | \mathbf{r} \rangle \right.$$

$$\left. + \sum_{d,\mathbf{k}} \langle \mathbf{r} | \phi_d \rangle G_{d\mathbf{k}} \langle \mathbf{k} | (1 - P_c) | \mathbf{r} \rangle \right) dE, \quad (2.192)$$

where P_c is the inner-core projection operator defined in Eq. (2.55) and the d-state sum in the final term is over both ion sites and individual quantum numbers. The several terms in Eq. (2.192) involving P_c either contribute to the inner-core orthogonalization-hole density δn_{oh}^c, or else they vanish as a consequence of the orthogonality of the localized d states to the inner-core states.

Extracting the known d-state density $\sum_i n_d(\mathbf{r} - \mathbf{R}_i)$, as given through Eq. (2.32), from the RHS of Eq. (2.192), one is left with a net expression for the valence electron density n_{val}. Analogous to the case of a simple metal, n_{val} can then be developed as a sum of a uniform density n_{unif}, a screening electron density δn_{scr} and an orthogonalization-hole density δn_{oh}, in addition to a possible Fermi-energy correction contribution, δn_{band}:

$$n_{\text{val}}(\mathbf{r}) = n_{\text{unif}} + \delta n_{\text{scr}}(\mathbf{r}) + \delta n_{\text{oh}}(\mathbf{r}) + \delta n_{\text{band}}(\mathbf{r}), \quad (2.193)$$

where

$$n_{\text{unif}} = -\frac{2}{\pi} \text{Im} \int_0^{\varepsilon_F} \sum_{\mathbf{k}} \frac{\langle \mathbf{r} | \mathbf{k} \rangle \langle \mathbf{k} | \mathbf{r} \rangle}{E - \varepsilon_{\mathbf{k}}} dE \quad (2.194)$$

and where δn_{oh} contains both inner-core and d-state components:

$$\delta n_{\text{oh}}(\mathbf{r}) = \delta n_{\text{oh}}^c(\mathbf{r}) + \delta n_{\text{oh}}^d(\mathbf{r}). \quad (2.195)$$

Separating δn_{scr} from δn_{oh}^d in Eq. (2.192) is not entirely unique, but the present preferred approach was first established by Moriarty (1982) for pre- and post-transition metals, and later extended to all transition-series metals (Moriarty, 1988a). Retaining contributions to δn_{scr} and δn_{oh}^d to first order in W and $(V')^2$, and separating volume and structure components of V', one obtains the results

$$\delta n_{\text{scr}}(\mathbf{r}) = -\frac{2}{\pi}\text{Im}\int_0^{\varepsilon_F}\left[\sum_{k,k'}{}'\frac{\langle\mathbf{r}|\mathbf{k}'\rangle W_{\mathbf{k}'\mathbf{k}}\langle\mathbf{k}|\mathbf{r}\rangle}{(E-\varepsilon_{\mathbf{k}})(E-\varepsilon_{\mathbf{k}'})}\right.$$

$$\left.+\sum_d \frac{\langle\mathbf{r}|\mathbf{k}'\rangle v'_{\mathbf{k}'d}v'_{d\mathbf{k}}\langle\mathbf{k}|\mathbf{r}\rangle}{(E-\varepsilon_{\mathbf{k}})(E-\varepsilon_{\mathbf{k}'})(E-E_d^{\text{vol}}-\Gamma_{dd}^{\text{vol}})}\right]dE \qquad (2.196)$$

and

$$\delta n_{\text{oh}}^d(\mathbf{r}) = -\frac{2}{\pi}\text{Im}\int_0^{\varepsilon_F}\sum_k\sum_d\left[\frac{\langle\mathbf{r}|\mathbf{k}\rangle v'_{\mathbf{k}d}v'_{d\mathbf{k}}\langle\mathbf{k}|\mathbf{r}\rangle - \langle\mathbf{r}|\phi_d\rangle v'_{d\mathbf{k}}v'_{\mathbf{k}d}\langle\phi_d|\mathbf{r}\rangle}{(E-\varepsilon_{\mathbf{k}})^2(E-E_d^{\text{vol}}-\Gamma_{dd}^{\text{vol}})}\right.$$

$$\left.+\frac{\langle\mathbf{r}|\phi_d\rangle v'_{d\mathbf{k}}(\langle\mathbf{k}|\mathbf{r}\rangle - S_{\mathbf{k}d}\langle\phi_d|\mathbf{r}\rangle)+\text{c.c.}}{(E-\varepsilon_{\mathbf{k}})(E-E_d^{\text{vol}}-\Gamma_{dd}^{\text{vol}})}\right]dE, \qquad (2.197)$$

where v' is the volume component of V', such that

$$v'_{\mathbf{k}d} = -(E-E_d^{\text{vol}})S_{\mathbf{k}d} - \Delta_{\mathbf{k}d}^{\text{vol}}. \qquad (2.198)$$

In the usual way, both δn_{scr} and δn_{oh}^d, in addition to δn_{oh}^c, are charge neutral, so that

$$\int \delta n_{\text{scr}}(\mathbf{r})d\mathbf{r} = \int \delta n_{\text{oh}}^d(\mathbf{r})d\mathbf{r} = 0, \qquad (2.199)$$

as can be verified from Eqs. (2.196) and (2.197) by inspection.

Finally, as shown by Moriarty (1988a), the correction term δn_{band} in Eq. (2.193) is on the order of $E_F - \varepsilon_F$ and hence is normally small. In addition, the conservation of electrons requires that δn_{band} is also charge neutral, so that

$$\int \delta n_{\text{band}}(\mathbf{r})d\mathbf{r} = 0. \qquad (2.200)$$

The simplest realistic approximation that will satisfy Eq. (2.200) is to take $E_F = \varepsilon_F$, which makes both δE_{band} in Eqs. (2.188) and (2.190) and δn_{band} in Eq. (2.193) vanish identically.

3
Interatomic Potentials in Simple Metals

In this chapter, we elaborate the theory of interatomic potentials in bulk simple metals that follows from the PP perturbation theory discussed in Sec. 2.5 of Chapter 2. Specifically, we seek to develop the cohesive-energy functional in the general real-space form anticipated in Chapter 1,

$$E_{\text{coh}}(\mathbf{R}, \Omega) = E_{\text{vol}}(\Omega) + \frac{1}{2N} \sum_{i,j}{}' v_2(R_{ij}, \Omega), \qquad (3.1)$$

and thereby obtain explicit first-principles expressions for both the volume term $E_{\text{vol}}(\Omega)$ and the two-ion pair potential $v_2(r, \Omega)$. In the general expansion leading to Eq. (3.1), there is a one-to-one correspondence between the order of the potentials and the order of PP perturbation theory. Terminating this expansion at the pair-potential level is consistent with a second-order perturbation-theory treatment of the total energy. For good simple metals with weak nonlocal PPs this is the essence of the method and will be the focal point of the present chapter. It is possible, of course, to carry perturbation theory to higher order and, in principle, thereby obtain three-ion and higher potentials, but this is extremely complicated in general and, at the very least, requires significant additional approximations including the use of a local PP. After developing the full theory of the volume term and the two-ion pair potential in Eq. (3.1), and testing the results on real materials, the role of higher-order multi-ion interactions in simple metals will be discussed in the final section of this chapter.

As we have discussed, the term simple metal in the present context refers to systems that can be regarded as a collection of well-defined ions of charge $+Ze$ immersed in a compensating sea of itinerant, nearly free valence electrons. Consistent with the small-core approximation, it is assumed that each ion consists of a nucleus surrounded by tightly bound closed shells of non-overlapping core electrons that move rigidly with the ion, while the s and p valence electrons exhibit NFE behavior, with their motion perturbed only by weak electron-ion PPs. Although the formalism developed in this chapter can be readily applied to any nontransition metal for which the valence Z can be assigned without ambiguity from the Periodic Table, only a subset of nontransition elements are

92 Interatomic Potentials in Simple Metals

actually good simple metals with both weak PPs and negligible d-electron hybridization and overlap effects arising from nearby d bands. It is primarily to such ideal simple metals that we address ourselves in this chapter, with the third-row elements Na, Mg and Al representing prototype cases. At the same time, by applying the simple-metal theory to all candidate elements we provide a baseline from which to improve the theory systematically where necessary, a subject that will be addressed in Chapters 4, 5 and 12.

3.1 Simple-metal cohesive-energy functional in DFT

Within the adiabatic, small-core and local-density approximations introduced in Chapter 2, the cohesive energy of a simple metal can be calculated as the difference between metal and free-atom binding energies for the valence electrons. From Eqs. (2.29) and (2.31) one thus has the starting equation

$$E_{\text{coh}} = \frac{1}{N}\left\{\frac{1}{2}{\sum_{i,j}}'\frac{(Ze)^2}{|\mathbf{R}_i - \mathbf{R}_j|} + 2\sum_{\mathbf{k}} E(\mathbf{k}) - \frac{1}{2}n_{\text{val}} V_{\text{val}} + n_{\text{val}}[\varepsilon_{\text{xc}}(n_{\text{val}}) - \mu_{\text{xc}}(n_{\text{val}})]\right\}$$
$$- E_{\text{bind}}^{\text{atom}}(Z). \tag{3.2}$$

Recall that the sum over band energies $E(\mathbf{k})$ is over occupied $\mathbf{k}$ states and the shorthand notation (2.30) is used for the implied integrations over all space in the three terms involving n_{val}. In Eq. (3.2) the free-atom binding energy $E_{\text{bind}}^{\text{atom}}$ acts only as a constant, and the ingredients that must be supplied by the PP perturbation theory are the valence electron density n_{val} and the band-structure energy involving $E(\mathbf{k})$.

In general, the valence electron density consists of the uniform free-electron density $n_{\text{unif}} = Z/\Omega$ plus small oscillatory and charge-neutral screening and orthogonalization-hole components, δn_{scr} and δn_{oh}:

$$n_{\text{val}}(\mathbf{r}) = n_{\text{unif}} + \delta n_{\text{scr}}(\mathbf{r}) + \delta n_{\text{oh}}(\mathbf{r}). \tag{3.3}$$

Within first-order PP perturbation theory, the screening electron density for a simple metal is from Eq. (2.78)

$$\delta n_{\text{scr}}(\mathbf{r}) = \frac{4}{N\Omega}\sum_{\mathbf{k}}{\sum_{\mathbf{q}}}' S(\mathbf{q})\frac{\langle \mathbf{k}+\mathbf{q}|w|\mathbf{k}\rangle}{\varepsilon_{\mathbf{k}} - \varepsilon_{\mathbf{k}+\mathbf{q}}}\exp(i\mathbf{q}\cdot\mathbf{r}), \tag{3.4}$$

where again $\mathbf{k}$ is summed over occupied states and the prime on the summation over $\mathbf{q}$ denotes the exclusion of the $\mathbf{q} = 0$ term. The final orthogonalization-hole contribution to n_{val} has the general form

$$\delta n_{\text{oh}}(\mathbf{r}) = (Z^*/Z - 1)n_{\text{unif}} + \sum_{i} n_{\text{oh}}(\mathbf{r} - \mathbf{R}_i), \qquad (3.5)$$

where Z^* is an effective valence ($Z^* \geq Z$) and n_{oh} is a localized hole (i.e., negative electron) density. For a simple metal, $n_{\text{oh}}(\mathbf{r} - \mathbf{R}_i)$ is confined to the inner-core region of the site i, but both Z^* and n_{oh} depend on the properties of the PP w. For nonlocal, energy-dependent PPs of the AHS form, Z^* and $n_{\text{oh}}(\mathbf{r}) = n_{\text{oh}}^c(\mathbf{r})$ are directly calculable quantities in terms of the inner core states, as is the case with GPT. For nonlocal, but energy-independent, norm-conserving MPs, on the other hand, $Z^* = Z$ and $n_{\text{oh}}(\mathbf{r})$ is undetermined, so one normally neglects δn_{oh} entirely. This is the case, for example, with the nonlocal MPs developed for simple metals by Rasolt and Taylor (1975) and Dagens, Rasolt and Taylor (1975), collectively referred to here as DRT.

In a similar manner from Eq. (2.96), the band energy $E(\mathbf{k})$ obtained within second-order PP perturbation theory is given by

$$E(\mathbf{k}) = \varepsilon_{\mathbf{k}} + \langle \mathbf{k} | w | \mathbf{k} \rangle + \sum_{\mathbf{q}}' |S(\mathbf{q})|^2 \frac{\langle \mathbf{k} | w | \mathbf{k} + \mathbf{q} \rangle \langle \mathbf{k} + \mathbf{q} | w | \mathbf{k} \rangle}{\varepsilon_{\mathbf{k}} - \varepsilon_{\mathbf{k}+\mathbf{q}}}. \qquad (3.6)$$

From Eqs. (3.2)–(3.6) it is clear that the simple-metal cohesive energy is a functional of the screened atomic PP w and all structure dependence of the result is explicit through the summations over ion positions in the first term involving the ion-ion Coulomb energy in Eq. (3.2), the second term involving n_{oh} in Eq. (3.5), and in the structure factor

$$S(\mathbf{q}) \equiv \frac{1}{N} \sum_{i} \exp(-i\mathbf{q} \cdot \mathbf{R}_i). \qquad (3.7)$$

To develop a quantum-based interatomic pair potential from these equations requires three general steps. First, we must combine Eqs. (3.2)–(3.6) and express E_{coh} in a reciprocal-space representation involving the PP form factor $\langle \mathbf{k} + \mathbf{q} | w | \mathbf{k} \rangle$ and $S(\mathbf{q})$. Second, we must then use Eq. (3.7) to transform this result to an exactly equivalent real-space representation of the general form (3.1). At this step, the functional dependence of both the volume term E_{vol} and the pair potential v_2 on the PP w is explicitly obtained. Finally, we must perform a self-consistent screening calculation and actually determine the mathematical forms of w and $\langle \mathbf{k} + \mathbf{q} | w | \mathbf{k} \rangle$, so that E_{vol} and v_2 can be entirely expressed from first principles in terms of the fundamental input quantities of the problem, i.e., Z, Ω and the bare-ion component of the PP w_{ion}. We proceed with the first two tasks in this section and return to the screening calculation in Sec. 3.2.

3.1.1 Reciprocal-space representation

For simplicity in manipulating Eqs. (3.2)–(3.6), we shall work in the limit of a vanishing orthogonalization-hole density, $\delta n_{\text{oh}} = 0$, with $Z^* = Z$ and $n_{\text{oh}} = 0$. This limit is exactly realized for a local PP and, as we have indicated, is also the usual approximation made in the case of a nonlocal norm-conserving PP. The more general AHS case is discussed in in Sec. A2.4 of Appendix A2, and we shall indicate the generalizations required at the end of our derivation. It is also convenient at this point to introduce a more explicit notation for sums over occupied $\mathbf{k}$ states, which in the present context are sums over occupied plane-wave states $|\mathbf{k}\rangle$. Specifically, for any function $f(\mathbf{k})$ we write

$$\sum_{\mathbf{k}} f(\mathbf{k}) \to \sum_{\mathbf{k}} f(\mathbf{k})\Theta^<(\mathbf{k}) = \frac{N\Omega}{(2\pi)^3} \int f(\mathbf{k})\Theta^<(\mathbf{k})d\mathbf{k}, \tag{3.8}$$

where we have defined the step function

$$\Theta^<(\mathbf{k}) \equiv \Theta(k_F - k) = \begin{cases} 1 & k \leq k_F \\ 0 & k > k_F \end{cases}, \tag{3.9}$$

with $k = |\mathbf{k}|$. Here k_F is the Fermi wavenumber such that $\varepsilon_F = \hbar^2 k_F^2/2m$.

Using Eqs. (3.6) and (3.8), the band-structure energy in Eq. (3.2) can be written as

$$\frac{2}{N}\sum_{\mathbf{k}} E(\mathbf{k}) = \frac{2}{N}\sum_{\mathbf{k}} \left(\varepsilon_{\mathbf{k}} + \langle \mathbf{k}|w|\mathbf{k}\rangle + \sum_{\mathbf{q}}' |S(\mathbf{q})|^2 \frac{\langle \mathbf{k}|w|\mathbf{k}+\mathbf{q}\rangle\langle\mathbf{k}+\mathbf{q}|w|\mathbf{k}\rangle}{\varepsilon_{\mathbf{k}} - \varepsilon_{\mathbf{k}+\mathbf{q}}} \right) \Theta^<(\mathbf{k}). \tag{3.10}$$

The first term on the RHS of Eq. (3.10) can be directly evaluated to yield the kinetic energy of the free-electron gas:

$$\frac{2}{N}\sum_{\mathbf{k}} \varepsilon_{\mathbf{k}}\Theta^<(\mathbf{k}) = \frac{2\Omega}{(2\pi)^3}\left[\int \varepsilon_{\mathbf{k}}\Theta^<(\mathbf{k})d\mathbf{k} = \int_0^{k_F}(\hbar^2 k^2/2m)(4\pi k^2)dk\right] = \frac{3}{5}Z\varepsilon_F, \tag{3.11}$$

noting that $k_F = (3\pi^2 Z/\Omega)^{1/3}$. The evaluation of the second term in Eq. (3.10) requires more care because it includes large electrostatic interactions involving n_{unif} and the Coulomb field from the ions, which must be balanced against similar terms in Eq. (3.2). Backing up one step and extracting these interactions explicitly from the full interaction $n_{\text{unif}}W$ implicit in $\langle \mathbf{k}|W|\mathbf{k}\rangle$, this term can be readily manipulated into the form

$$\frac{2}{N}\sum_{\mathbf{k}}\langle \mathbf{k}|w|\mathbf{k}\rangle \Theta^<(\mathbf{k}) = \frac{1}{N}n_{\text{unif}}\left[V_{\text{unif}} - \sum_i' \frac{Ze^2}{|\mathbf{r}-\mathbf{R}_i|} + \mu_{\text{xc}}(n_{\text{val}})\right]$$

$$+ \frac{2\Omega}{(2\pi)^3}\int \langle \mathbf{k}|w_{\text{core}}|\mathbf{k}\rangle \Theta^<(\mathbf{k})d\mathbf{k}. \tag{3.12}$$

where V_{unif} is the Coulomb potential arising from n_{unif}, the prime on the sum over ion sites denotes the exclusion of the $i=0$ term, w_{core} is the one-ion core component of W,

$$w_{\text{core}} = w_{\text{ion}} + Ze^2/r, \tag{3.13}$$

3.1.1 Reciprocal-space representation

and in the terms involving n_{unif} there is the usual implied integration over all space. In deriving Eq. (3.12), we have noted that with $\delta n_{\text{oh}} = 0$ the full PP has the form

$$W = V_{\text{val}} + \mu_{\text{xc}} + W_{\text{ion}} = V_{\text{unif}} + \delta V_{\text{scr}} + \mu_{\text{xc}} + W_{\text{ion}} \tag{3.14}$$

and that

$$n_{\text{unif}} \delta V_{\text{scr}} \equiv \int n_{\text{unif}} \delta V_{\text{scr}}(\mathbf{r}) d\mathbf{r} = 0. \tag{3.15}$$

Equation (3.15) follows from the fact that the screening field carries no net charge, so that

$$\int \delta n_{\text{scr}}(\mathbf{r}) d\mathbf{r} = \int \delta V_{\text{scr}}(\mathbf{r}) d\mathbf{r} = 0, \tag{3.16}$$

as can be verified from Eq. (3.4).

The electrostatic terms in Eq. (3.12) can be combined with the Coulomb-repulsion and electron-electron interaction terms in Eq. (3.2). To do this we evaluate the latter electron-electron term via Eq. (3.3) with $\delta n_{\text{oh}} = 0$:

$$-\frac{1}{2N} n_{\text{val}} V_{\text{val}} = -\frac{1}{2N} (n_{\text{unif}} + \delta n_{\text{scr}})(V_{\text{unif}} + \delta V_{\text{scr}}) = -\frac{1}{2N} n_{\text{unif}} V_{\text{unif}} - \frac{1}{2N} \delta n_{\text{scr}} \delta V_{\text{scr}} \tag{3.17}$$

again using Eq. (3.15) and noting that $\delta n_{\text{scr}} V_{\text{unif}} = n_{\text{unif}} \delta V_{\text{scr}}$. Then, adding the first term in Eq. (3.17) to the first term in Eq. (3.2) plus the first two terms in Eq. (3.12) yields the electrostatic energy per ion of N point ions of charge $+Ze^2$ in a compensating uniform electron gas:

$$E_{\text{es}}(Z, \Omega) = \frac{1}{N} \left\{ \frac{1}{2} {\sum_{i,j}}' \frac{(Ze)^2}{|\mathbf{R}_i - \mathbf{R}_j|} - n_{\text{unif}} {\sum_{i}}' \frac{Ze^2}{|\mathbf{r} - \mathbf{R}_i|} + \frac{1}{2} n_{\text{unif}} V_{\text{unif}} \right\}. \tag{3.18}$$

The electrostatic energy E_{es} depends on the valence Z, the atomic volume Ω and the structure of the metal. In practice, E_{es} can be evaluated most efficiently using Ewald techniques, in which the long-range Coulomb interactions are divided into rapidly convergent real- and reciprocal-space contributions. This leads to the following well-known computational form first obtained by Fuchs (1935) and discussed in many modern references (e.g., Harrison, 1966; Wallace, 1972; Hafner, 1987):

96 Interatomic Potentials in Simple Metals

$$E_{es}(Z,\Omega) = \frac{(Ze)^2}{2}\left\{\frac{1}{N}{\sum_{i,j}}'\frac{\text{efrc}\,(\eta|\mathbf{R}_i-\mathbf{R}_j|)}{|\mathbf{R}_i-\mathbf{R}_j|} + \frac{4\pi}{\Omega}{\sum_{\mathbf{q}}}'|S(\mathbf{q})|^2\frac{\exp(-q^2/4\eta^2)}{q^2}\right.$$
$$\left. - \left[\frac{2\eta}{\sqrt{\pi}} + \frac{\pi}{\eta^2\Omega}\right]\right\}. \tag{3.19}$$

Here η is a variable convergence parameter that may be chosen at will to make the two summations both converge rapidly, and erfc is the complementary error function. The simple choice $\eta = 2/R_{\text{WS}}$ will work well in this regard for essentially all structures and all volumes of interest. For metallic structures, E_{es} will have a value close to $-0.9(Ze)^2/R_{\text{WS}}$, the electrostatic energy of a single point ion embedded at the center of a compensating sphere of radius R_{WS}. It is often convenient, therefore, to extract this large structure-independent contribution explicitly and write

$$E_{es}(Z,\Omega) = -\frac{9}{10}(Ze)^2/R_{\text{WS}} + \delta E_{es}(Z,\Omega), \tag{3.20}$$

where δE_{es} is the small structure-dependent contribution to the electrostatic energy. For the perfect crystal structure, one can express the latter contribution as

$$\delta E_{es}(Z,\Omega) = \frac{1}{2}(1.8 - \alpha_{es})(Ze)^2/R_{\text{WS}}, \tag{3.21}$$

where α_{es} is the electrostatic Ewald or Madelung constant of the structure in question: e.g., $\alpha_{es} = 1.791747$ for the fcc structure, 1.791676 for the ideal hcp structure, and 1.791859 for the bcc structure. A useful table of α_{es} values for a wide range of observed crystal structures has been given by Hafner (1987a) in his Table C.1 on p. 332.

Next, we treat the explicit LDA exchange-correlation terms contained in E_{coh} and its components by first combining the contributions from Eqs. (3.2), (3.10) and (3.12) to yield:

$$\delta E_{\text{xc}} \equiv \frac{1}{N}[n_{\text{val}}\varepsilon_{\text{xc}}(n_{\text{val}}) - (n_{\text{val}} - n_{\text{unif}})\mu_{\text{xc}}(n_{\text{val}})]. \tag{3.22}$$

This result may then be developed in terms of the components of $n_{\text{val}} = n_{\text{unif}} + \delta n_{\text{scr}}$ by expanding $n_{\text{val}}\varepsilon_{\text{xc}}(n_{\text{val}})$ and $\mu_{\text{xc}}(n_{\text{val}})$ in Taylor series about n_{unif} in powers of δn_{scr}. Keeping contributions to second order in δn_{scr}, one thereby obtains

$$\delta E_{\text{xc}} = \frac{1}{N}\left[n_{\text{unif}}\varepsilon_{\text{xc}}(n_{\text{unif}}) - \frac{1}{2}(\delta n_{\text{scr}})^2\frac{d\mu_{\text{xc}}(n_{\text{unif}})}{dn}\right], \tag{3.23}$$

where we have used from Eq. (2.11) the basic DFT relationship $\mu_{\text{xc}}(n) = d[n\varepsilon_{\text{xc}}(n)]/dn$. The first term inside the brackets of Eq. (3.23) is just the xc energy of the uniform

3.1.1 Reciprocal-space representation

electron gas, while the second term accounts for the second-order contribution of the oscillatory screening field.

Finally, the second-order terms in Eqs. (3.17) and (3.23) involving δn_{scr} can be combined with the final second-order term in the band-structure energy (3.10) by expressing the screening density in a reciprocal-space representation. To do this, we note from Eq. (3.4) that δn_{scr} has the form

$$\delta n_{\text{scr}}(\mathbf{r}) = {\sum_{\mathbf{q}}}' S(\mathbf{q}) n_{\text{scr}}(q) \exp(i\mathbf{q} \cdot \mathbf{r}), \tag{3.24}$$

where

$$n_{\text{scr}}(q) = \frac{4}{N\Omega} \sum_{\mathbf{k}} \frac{\langle \mathbf{k} + \mathbf{q} | w | \mathbf{k} \rangle}{\varepsilon_{\mathbf{k}} - \varepsilon_{\mathbf{k+q}}} \Theta^{<}(\mathbf{k}) = \frac{4}{(2\pi)^3} \int \frac{\langle \mathbf{k} + \mathbf{q} | w | \mathbf{k} \rangle}{\varepsilon_{\mathbf{k}} - \varepsilon_{\mathbf{k+q}}} \Theta^{<}(\mathbf{k}) d\mathbf{k}. \tag{3.25}$$

Physically, $n_{\text{scr}}(q)$ is the Fourier transform of the screening density arising from one site and, by symmetry, only depends on the magnitude of $\mathbf{q}$. The corresponding Coulomb potential can also be expressed in terms of $n_{\text{scr}}(q)$ as

$$\delta V_{\text{scr}}(\mathbf{r}) = e^2 \int \frac{\delta n_{\text{scr}}(\mathbf{r}')}{|\mathbf{r} - \mathbf{r}'|} d\mathbf{r}' = {\sum_{\mathbf{q}}}' S(\mathbf{q}) \left(\frac{4\pi e^2}{q^2}\right) n_{\text{scr}}(q) \exp(i\mathbf{q} \cdot \mathbf{r}), \tag{3.26}$$

so that

$$-\frac{1}{2N} \delta n_{\text{scr}} \delta V_{\text{scr}} \equiv -\frac{1}{2N} \int \delta n_{\text{scr}}(\mathbf{r}) \delta V_{\text{scr}}(\mathbf{r}) d\mathbf{r} = -{\sum_{\mathbf{q}}}' |S(\mathbf{q})|^2 \left(\frac{2\pi e^2 \Omega}{q^2}\right) [n_{\text{scr}}(q)]^2. \tag{3.27}$$

In a similar manner, one obtains for the final xc term in Eq. (3.23)

$$-\frac{1}{2N} (\delta n_{\text{scr}})^2 \frac{d\mu_{\text{xc}}(n_{\text{unif}})}{dn} = {\sum_{\mathbf{q}}}' |S(\mathbf{q})|^2 \left(\frac{2\pi e^2 \Omega}{q^2}\right) G(q) [n_{\text{scr}}(q)]^2, \tag{3.28}$$

where $G(q)$ is the LDA xc function

$$G(q) = -\frac{q^2}{4\pi e^2} \frac{d\mu_{\text{xc}}(n_{\text{unif}})}{dn}. \tag{3.29}$$

Combining Eqs. (3.27) and (3.28) with the second-order term in Eq. (3.10) produces the structure-dependent band-structure energy

$$E_{\text{bs}} = {\sum_{\mathbf{q}}}' |S(\mathbf{q})|^2 F(q, \Omega), \tag{3.30}$$

where we have defined the volume-dependent, but structure-independent, function

$$F(q,\Omega) = \frac{2\Omega}{(2\pi)^3} \int \frac{|\langle \mathbf{k}+\mathbf{q}|\,w\,|\mathbf{k}\rangle|^2}{\varepsilon_\mathbf{k} - \varepsilon_{\mathbf{k}+\mathbf{q}}} \Theta^<(\mathbf{k})d\mathbf{k} - \frac{2\pi e^2 \Omega}{q^2}[1 - G(q)][n_{\rm scr}(q)]^2 \,. \quad (3.31)$$

The function $F(q,\Omega)$ is commonly referred to as the *energy-wavenumber characteristic*, a term coined by Harrison (1966). In deriving Eq. (3.31) we have assumed that $\langle \mathbf{k}|\,w\,|\mathbf{k}+\mathbf{q}\rangle = \langle \mathbf{k}+\mathbf{q}|\,w|\mathbf{k}\rangle^*$, as is true for both local and nonlocal norm-conserving PPs.

The cohesive-energy functional (3.2) can now be reassembled into volume and structure components. We write

$$E_{\rm coh} = E_{\rm vol}^q(\Omega) + {\sum_\mathbf{q}}' |S(\mathbf{q})|^2 F(q,\Omega) + \delta E_{\rm es}(Z,\Omega) \,, \quad (3.32)$$

where from Eqs. (3.2), (3.11), (3.12), (3.20) and (3.23) the volume component is

$$E_{\rm vol}^q(\Omega) = \frac{3}{5}Z\varepsilon_F + Z\varepsilon_{\rm xc}(n_{\rm unif}) - \frac{9}{10}(Ze)^2/R_{\rm WS} + \frac{2\Omega}{(2\pi)^3}\int \langle \mathbf{k}|\,w_{\rm core}\,|\mathbf{k}\rangle\,\Theta^<(\mathbf{k})d\mathbf{k} - E_{\rm bind}^{\rm atom} \,. \quad (3.33)$$

The superscript "q" in $E_{\rm vol}^q$ has been introduced to emphasize that this quantity is given in the reciprocal-space representation. The volume term $E_{\rm vol}^q$ may actually be expressed in several slightly different forms depending on how the PP contribution is represented. In the GPT, for example, it is convenient to define a PP $w_{\rm pa}$ associated with the zero-order pseudoatom method (Moriarty, 1974) used to implement the theory (see Chapter 4, as well as Sec. A2.8 of Appendix A2):

$$w_{\rm pa} \equiv v_{\rm unif} + w_{\rm ion} - V_0' = v_{\rm unif} - Ze^2/r + w_{\rm core} - V_0' \,, \quad (3.34)$$

where $v_{\rm unif}$ is the Coulomb potential arising from the uniform electron gas contained within one atomic sphere and V_0' is a zero-of-energy constant chosen to ensure that $\langle 0|\,w_{\rm pa}\,|0\rangle = 0$. It is then straightforward to show that

$$\langle \mathbf{k}|\,w_{\rm pa}\,|\mathbf{k}\rangle = -\frac{3}{10}(Ze)^2/R_{\rm WS} + \langle \mathbf{k}|\,w_{\rm core}\,|\mathbf{k}\rangle - V_0' \,, \quad (3.35)$$

so that $E_{\rm vol}^q$ may alternately be expressed as

$$E_{\rm vol}^q(\Omega) = E_{\rm fe}^0(\Omega) + \frac{2\Omega}{(2\pi)^3}\int \langle \mathbf{k}|\,w_{\rm pa}\,|\mathbf{k}\rangle\,\Theta^<(\mathbf{k})d\mathbf{k} - E_{\rm bind}^{\rm atom} \,, \quad (3.36)$$

with $E_{\rm fe}^0$ defined as the zero-order free-electron energy

$$E_{\rm fe}^0(\Omega) \equiv \frac{3}{5}Z\varepsilon_F + Z\varepsilon_{\rm xc}(n_{\rm unif}) - \frac{3}{5}(Ze)^2/R_{\rm WS} + ZV_0' \,. \quad (3.37)$$

The four terms on the RHS of Eq. (3.37) represent the effective free-electron binding energy of the metal one obtains in the limit of an arbitrarily weak PP where $w_{pa} \approx 0$. Thus, $E_{\mathrm{vol}}^q(\Omega_0)$ alone represents a respectable first estimate of the equilibrium cohesive energy of most simple metals, as first shown by Moriarty (1979).

All of the above-derived results are readily generalized to nonlocal AHS PPs where the orthogonalization-hole density δn_{oh} does not vanish, such as the PP given by Eq. (2.54) that is used in the simple-metal limit of the GPT (Moriarty, 1974, 1977, 1979, 1982). In this case, as discussed in Sec. A2.4 of Appendix A2, one has an effective valence

$$Z^\star = Z + \frac{2\Omega}{(2\pi)^3} \int \langle \mathbf{k}| p_c |\mathbf{k}\rangle \Theta^<(\mathbf{k}) d\mathbf{k} \tag{3.38}$$

and a single-site orthogonalization hole density

$$n_{\mathrm{oh}}(\mathbf{r}) = n_{\mathrm{oh}}^c(\mathbf{r}) = -\frac{2\Omega}{(2\pi)^3} \int [\langle \mathbf{r}| p_c |\mathbf{k}\rangle \langle \mathbf{k}|\mathbf{r}\rangle + \mathrm{c.c.} - \langle \mathbf{r}| p_c |\mathbf{k}\rangle \langle \mathbf{k}| p_c |\mathbf{r}\rangle] \Theta^<(\mathbf{k}) d\mathbf{k}, \tag{3.39}$$

where $p_c = \sum_c |\phi_c\rangle \langle \phi_c|$ is the one-site core-state projection operator. In addition, the following three general modifications are required in the components of the cohesive energy functional:

(i) In the electrostatic-energy expressions (3.18)–(3.21), the valence Z is replaced by the effective valence $Z^\star$, so that in Eqs. (3.21) and (3.32)

$$\delta E_{\mathrm{es}}(Z, \Omega) \to \delta E_{\mathrm{es}}(Z^\star, \Omega) = \frac{1}{2}(1.8 - \alpha_{\mathrm{es}})(Z^\star e)^2 / R_{\mathrm{WS}}. \tag{3.40}$$

(ii) An additional xc contribution involving n_{oh} is added to the energy-wavenumber characteristic (3.31), also appearing in Eq. (3.32):

$$F(q, \Omega) \to F(q, \Omega) - \frac{2\pi e^2 \Omega}{q^2} G(q)[n_{\mathrm{oh}}(q)]^2$$

$$= \frac{2\Omega}{(2\pi)^3} \int \frac{|\langle \mathbf{k}+\mathbf{q}| w |\mathbf{k}\rangle|^2}{\varepsilon_{\mathbf{k}} - \varepsilon_{\mathbf{k}+\mathbf{q}}} \Theta^<(\mathbf{k}) d\mathbf{k} - \frac{2\pi e^2 \Omega}{q^2}$$

$$\times \left\{[1 - G(q)][n_{\mathrm{scr}}(q)]^2 + G(q)[n_{\mathrm{oh}}(q)]^2\right\}, \tag{3.41}$$

where, in analogy with $n_{\mathrm{scr}}(q)$, $n_{\mathrm{oh}}(q)$ is the Fourier transform of the orthogonalization-hole density $n_{\mathrm{oh}}(\mathbf{r})$.

(iii) Two additional contributions are added to the volume term E_{vol}^q. The first of these is a self-energy correction δE_{oh} associated with the finite size of the orthogonalization hole, as given in Sec. A2.4 of Appendix A2. The second contribution derives from either the energy dependence or the non-Hermitian nature of the PP and depends on its specific form. For an AHS PP of the form (2.54), one has the relationship

$$\langle \mathbf{k}| w |\mathbf{k}+\mathbf{q}\rangle = \langle \mathbf{k}+\mathbf{q}| w |\mathbf{k}\rangle^* - (\varepsilon_\mathbf{k} - \varepsilon_{\mathbf{k}+\mathbf{q}}) \langle \mathbf{k}| p_c |\mathbf{k}+\mathbf{q}\rangle, \qquad (3.42)$$

The second term in Eq. (3.42) then produces an extra volume contribution when used in Eq. (3.10) for the band-structure energy. In this case, Eq. (3.36) is generalized to

$$E_{\text{vol}}^q(\Omega) \to E_{\text{vol}}^q(\Omega) + \frac{2\Omega}{(2\pi)^3} \int \langle \mathbf{k}| w_{\text{pa}} |\mathbf{k}\rangle \langle \mathbf{k}| p_c |\mathbf{k}\rangle \Theta^<(\mathbf{k}) d\mathbf{k} + \delta E_{\text{oh}}(\Omega)$$

$$= E_{\text{fe}}^0(\Omega) + \frac{2\Omega}{(2\pi)^3} \int \langle \mathbf{k}| w_{\text{pa}} |\mathbf{k}\rangle [1 + \langle \mathbf{k}| p_c |\mathbf{k}\rangle]\Theta^<(\mathbf{k}) d\mathbf{k} + \delta E_{\text{oh}}(\Omega) - E_{\text{bind}}^{\text{atom}}(Z).$$
$$(3.43)$$

Also in this regard, note that it is still the true valence Z (and not Z^*) that enters the four terms in Eq. (3.37) for E_{fe}^0. In summary, for a nonlocal AHS PP of the form (2.54), the cohesive energy functional (3.32) becomes

$$E_{\text{coh}} = E_{\text{vol}}^q(\Omega) + \sum_\mathbf{q}{}' |S(\mathbf{q})|^2 F(q,\Omega) + \delta E_{\text{es}}(Z^*,\Omega), \qquad (3.44)$$

with $E_{\text{vol}}^q(\Omega)$ given by Eq. (3.43), $F(q,\Omega)$ by Eq. (3.41) and $\delta E_{\text{es}}(Z^*,\Omega)$ by Eq. (3.40).

In addition to providing a gateway to derive first-principles interatomic potentials, the reciprocal-space representation of the cohesive energy is useful in its own right as an efficient means of calculation. This is particularly so when the structure factor $S(\mathbf{q})$ is easily specified such as in the perfect lattice. The reciprocal-space representation actually provides superior convergence properties to the real-space representation (3.1), primarily because $F(q,\Omega)$ decays more rapidly for large q than $v_2(r,\Omega)$ does for large r. This feature may be used to advantage in checking the accuracy of real-space calculations. Another satisfying feature of the reciprocal-space formalism developed above is that the functions E_{vol}^q, F and δE_{es} defining E_{coh} are reflective of the very different energy scales for cohesive and structural properties of metals. In particular, $E_{\text{vol}}^q(\Omega)$ alone gives a good approximation to the cohesion curve (cohesive energy vs. volume) with a minimum close to the observed equilibrium volume. This is demonstrated in Fig. 3.1 for the case of Mg on the basis of first-principles GPT calculations. The additional structural component $E_{\text{struc}}^q = E_{\text{coh}} - E_{\text{vol}}^q$ involving F and δE_{es} is consequently relatively small and on the scale of energy differences between different structures.

3.1.2 Real-space representation

We now proceed to transform the above results to an exactly equivalent real-space representation. We again initially work in the limit of a vanishing orthogonalization-hole density relevant to a local PP or a nonlocal norm-conserving PP and then indicate the changes required for a more general nonlocal AHS PP at the end of our derivation.

3.1.2 Real-space representation

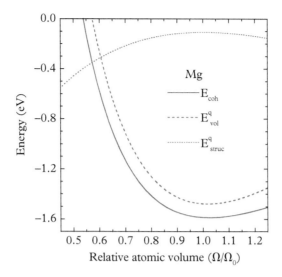

Fig. 3.1 The cohesive energy E_{coh} and its volume (E^q_{vol}) and structural (E^q_{struc}) components for Mg in a reciprocal-space representation based on first-principles GPT calculations parallel to the real-space treatment of Mg given by Althoff et al. (1993).

As was first pointed out by Jones (1973), a certain amount of care is required to get all of the correct contributions to the real-space volume term $E_{vol}(\Omega)$ and the pair potential $v_2(r, \Omega)$. We begin with Eqs. (3.30)–(3.33) and first add and subtract the quantity $0.9(Ze)^2/R_{WS}$ to recover the full electrostatic energy E_{es} in the cohesive energy:

$$E_{coh} = E^q_{vol} + E_{bs} + \delta E_{es} = E^q_{vol} + \frac{9}{10}(Ze)^2/R_{WS} + E_{bs} + E_{es}. \qquad (3.45)$$

Next, using Eq. (3.19) for E_{es} in the limit of large η together with Eq. (3.30) for E_{bs}, the sum of E_{bs} and E_{es} can be expressed directly in terms of the structure factor $S(\mathbf{q})$:

$$E_{bs} + E_{es} = \sum_{\mathbf{q}}{}' |S(\mathbf{q})|^2 \left[\frac{2\pi(Ze)^2}{q^2\Omega}\exp\left(-q^2/4\eta^2\right) + F(q,\Omega)\right] - \frac{(Ze)^2\eta}{\sqrt{\pi}}. \qquad (3.46)$$

We then manipulate Eq. (3.46) by (*i*) adding and subtracting the $\mathbf{q} = 0$ component to the sum over $\mathbf{q}$ in the first term; (*ii*) using Eq. (3.7) to express the structure-factor term as

$$|S(\mathbf{q})|^2 = \frac{1}{N^2}\sum_{i,j}\exp[-i\mathbf{q}\cdot(\mathbf{R}_i - \mathbf{R}_j)] = \frac{1}{N^2}\sum_{i,j}{}'\exp[-i\mathbf{q}\cdot(\mathbf{R}_i - \mathbf{R}_j)] + \frac{1}{N}, \qquad (3.47)$$

where we have separated $i \neq j$ and $i = j$ terms; and (iii) taking the limit $\eta \to \infty$. These three operations yield the result

$$E_{bs} + E_{es} = \frac{1}{N^2} {\sum_{i,j}}' \sum_q \left[\frac{2\pi(Ze)^2}{q^2 \Omega} + F(q, \Omega) \right] \exp[-i\mathbf{q} \cdot (\mathbf{R}_i - \mathbf{R}_j)] - \lim_{q \to 0} \left[\frac{2\pi(Ze)^2}{q^2 \Omega} \right.$$

$$\left. + F(q, \Omega) \right] + \lim_{\eta \to \infty} \left\{ \frac{1}{N} \sum_q \left[\frac{2\pi(Ze)^2}{q^2 \Omega} \exp\left(-q^2/4\eta^2\right) + F(q, \Omega) \right] - \frac{(Ze)^2 \eta}{\sqrt{\pi}} \right\}$$

(3.48)

In the second line of Eq. (3.48), the two terms involving η exactly cancel since

$$\frac{2\pi}{N\Omega} \sum_q \frac{1}{q^2} \exp\left(-q^2/4\eta^2\right) = \frac{1}{\pi} \int_0^\infty \exp\left(-q^2/4\eta^2\right) dq = \frac{\eta}{\sqrt{\pi}}. \tag{3.49}$$

The surviving pure volume terms in Eq. (3.48) can then be combined with the first two terms in Eq. (3.45) to give the real-space volume component of E_{coh}

$$E_{vol}(\Omega) = E_{vol}^q(\Omega) + \frac{9}{10}(Ze)^2/R_{WS} - \lim_{q \to 0} \left[\frac{2\pi(Ze)^2}{q^2 \Omega} + F(q, \Omega) \right] + \frac{1}{N} \sum_q F(q, \Omega). \tag{3.50}$$

The corresponding real-space structure component of E_{coh} is then the remaining term on the first line of Eq. (3.48). Comparing this contribution with Eq. (3.1), one immediate infers that the two-ion pair potential is given by

$$v_2(r, \Omega) = \frac{1}{N} \sum_q \left[\frac{4\pi(Ze)^2}{q^2 \Omega} + 2F(q, \Omega) \right] \exp(-i\mathbf{q} \cdot \mathbf{r}). \tag{3.51}$$

Equation (3.50) for E_{vol} and Eq. (3.51) for v_2 may be developed somewhat further and simplified to more useful forms. To do this, we first define a *normalized energy-wavenumber characteristic*

$$F_N(q, \Omega) \equiv -\frac{q^2 \Omega}{2\pi(Ze)^2} F(q, \Omega)$$

$$= \frac{q^2 \Omega^2}{4\pi(Ze)^2} \left\{ -\frac{4}{(2\pi)^3} \int \frac{|\langle \mathbf{k} + \mathbf{q}| w |\mathbf{k}\rangle|^2}{\varepsilon_\mathbf{k} - \varepsilon_{\mathbf{k}+\mathbf{q}}} \Theta^<(\mathbf{k}) d\mathbf{k} + \frac{4\pi e^2}{q^2} [1 - G(q)][n_{scr}(q)]^2 \right\}$$

(3.52)

where $F(q, \Omega)$ is given by Eq. (3.31). This quantity has the property $F_N(0, \Omega) = 1$, as may be readily confirmed by noting that at small q the leading term in $F_N(q, \Omega)$ is

$(\Omega/Z)^2 [n_{\text{scr}}(q)]^2$ and that $\lim_{q \to 0} n_{\text{scr}}(q) = Z/\Omega$. In a similar manner, it can be shown that the next leading term at small q is of order q^2, so that formally $F_N(q,\Omega)$ has the Taylor-series expansion

$$F_N(q,\Omega) = 1 + \frac{1}{2} q^2 \frac{\partial^2 F_N(0,\Omega)}{\partial q^2} + \cdots. \tag{3.53}$$

Using Eqs. (3.52) and (3.53) together with Eq. (3.33) for E_{vol}^q in Eq. (3.50) then gives the formal result

$$E_{\text{vol}}(\Omega) = \frac{3}{5} Z \varepsilon_F + Z \varepsilon_{\text{xc}}(n_{\text{unif}}) + \frac{2\Omega}{(2\pi)^3} \int \langle \mathbf{k} | w_{\text{core}} | \mathbf{k} \rangle \Theta^<(\mathbf{k}) d\mathbf{k}$$

$$+ \frac{\pi (Ze)^2}{\Omega} \frac{\partial^2 F_N(0,\Omega)}{\partial q^2} - \frac{(Ze)^2}{\pi} \int_0^\infty F_N(q,\Omega) dq - E_{\text{bind}}^{\text{atom}}(Z). \tag{3.54}$$

The small-q analytic properties of $F_N(q,\Omega)$, including an exact evaluation of the second derivative entering Eqs. (3.53) and (3.54), will be considered in Sec. 3.2, where the screening density $n_{\text{scr}}(q)$ is treated more explicitly.

To simplify Eq. (3.51) for the pair potential v_2, we note that the first term is an exact reciprocal-space representation of the ion-ion Coulomb potential, i.e.,

$$\frac{(Ze)^2}{r} = \frac{1}{N\Omega} \sum_{\mathbf{q}} \frac{4\pi (Ze)^2}{q^2} \exp(-i\mathbf{q} \cdot \mathbf{r}). \tag{3.55}$$

In a similar manner, for any spherically symmetric function $f(q)$ one has

$$\frac{1}{N\Omega} \sum_{\mathbf{q}} f(q) \exp(-i\mathbf{q} \cdot \mathbf{r}) = \frac{1}{(2\pi)^3} \int f(q) \exp(-i\mathbf{q} \cdot \mathbf{r}) d\mathbf{q} = \frac{1}{2\pi^2} \int_0^\infty q^2 f(q) \frac{\sin(qr)}{qr} dq. \tag{3.56}$$

Using Eqs. (3.55) and (3.56) together with the definition (3.52) of $F_N(q)$, Eq. (3.51) for v_2 reduces to the relatively simple form

$$v_2(r,\Omega) = \frac{(Ze)^2}{r} \left[1 - \frac{2}{\pi} \int_0^\infty F_N(q,\Omega) \frac{\sin(qr)}{q} dq \right]. \tag{3.57}$$

The interatomic pair potential in a simple metal is thus the sum of two contributions: the direct ion-ion Coulomb repulsion $(Ze)^2/r$ and an indirect ion-electron-ion screening interaction involving F_N. In other words, v_2 is a screened Coulomb potential between ions. From the form of Eq. (3.57), it is clear that as $r \to 0$ the screening contribution becomes small and the Coulomb repulsion dominates, as one would expect. In the limit $r \to \infty$, on the other hand, one can note that $(2/\pi)\sin(qr)/q$ tends to a delta function $\delta(q)$, so that the long-ranged Coulomb repulsion between ions is exactly cancelled by the screening contribution, as it physically should be. As we shall see later, the behavior of $v_2(r,\Omega)$ at intermediate values of r is generally oscillatory, with the long-range limit being a Friedel oscillation of the form given by Eq. (1.28).

Equations (3.52), (3.54) and (3.57) represent the principal real-space results for a local PP or a nonlocal norm-conserving PP. These equations can again be readily generalized to the case of a nonlocal AHS PP with a finite orthogonalization-hole density. To do this, one must first modify the definition of the normalized energy-wavenumber characteristic to ensure the desired normalization property $F_N(0,\Omega) = 1$. Using Eq. (3.41) and noting that now $\lim_{q \to 0} n_{\text{scr}}(q) = Z^*/\Omega$, as discussed in Sec. 3.2, one thus defines

$$F_N(q,\Omega) \equiv -\frac{q^2 \Omega}{2\pi(Z^*e)^2} F(q,\Omega)$$

$$= \frac{q^2 \Omega^2}{4\pi(Z^*e)^2} \left[-\frac{4}{(2\pi)^3} \int \frac{|\langle \mathbf{k+q}| w |\mathbf{k}\rangle|^2}{\varepsilon_\mathbf{k} - \varepsilon_\mathbf{k+q}} \Theta^<(\mathbf{k}) d\mathbf{k} \right.$$

$$\left. + \frac{4\pi e^2}{q^2}[1 - G(q)][n_{\text{scr}}(q)]^2 + \frac{4\pi e^2}{q^2} G(q)[n_{\text{oh}}(q)]^2 \right]. \quad (3.58)$$

One must then take care to replace Z with Z^* in the appropriate places for v_2 and E_{vol}. Equation (3.57) for the pair potential becomes

$$v_2(r,\Omega) = \frac{(Z^*e)^2}{r}\left[1 - \frac{2}{\pi}\int_0^\infty F_N(q,\Omega)\frac{\sin(qr)}{q}dq \right]. \quad (3.59)$$

For an AHS PP of the form (2.56) used in the simple-metal limit of the GPT, the volume term (3.54) becomes

$$E_{\text{vol}}(\Omega) = E_{\text{fe}}^0(\Omega) + \frac{2\Omega}{(2\pi)^3}\int \langle \mathbf{k}| w_{\text{pa}} |\mathbf{k}\rangle [1 + \langle \mathbf{k}| p_c |\mathbf{k}\rangle]\Theta^<(\mathbf{k})d\mathbf{k} + \frac{9}{10}\frac{(Z^*e)^2}{R_{\text{WS}}}$$

$$+ \frac{\pi(Z^*e)^2}{\Omega}\frac{\partial^2 F_N(0,\Omega)}{\partial q^2} - \frac{(Z^*e)^2}{\pi}\int_0^\infty F_N(q,\Omega)dq + \delta E_{\text{oh}}(\Omega) - E_{\text{bind}}^{\text{atom}}(Z),$$

$$(3.60)$$

with $E_{\text{fe}}^0(\Omega)$ still given by Eq. (3.37).

Near equilibrium ($\Omega = \Omega_0$) the real-space volume and structural components of the cohesive energy are typically similar in magnitude to their reciprocal-space counterparts. Under such conditions, the energy scale of v_2 is expected to be on the order of structural energies rather than the cohesive energy, as desired. It should be noted,

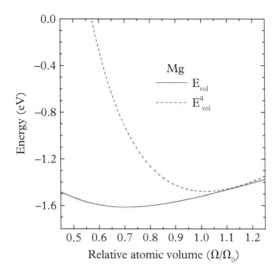

Fig. 3.2 *Comparison of real-space (E_{vol}) and reciprocal-space (E^q_{vol}) volume terms in the cohesive energy of Mg based on first-principles GPT calculations (Althoff et al., 1993).*

however, that under high compression, where the volume Ω is decreased well below Ω_0, the real-space structural energy, unlike the reciprocal-space structural energy, will grow rapidly in magnitude due to the strong positive Coulomb repulsion contribution to v_2. As a consequence, the corresponding volume term $E_{\text{vol}}(\Omega)$ will have a significantly different behavior than $E^q_{\text{vol}}(\Omega)$ as a function of volume. This is illustrated in Fig. 3.2 for the case of Mg. In particular, note that unlike $E^q_{\text{vol}}(\Omega)$, $E_{\text{vol}}(\Omega)$ alone does *not* give a good approximation to the cohesion curve.

3.2 Self-consistent electron screening

To evaluate Eq. (3.57) or Eq. (3.59) for the pair potential $v_2(r, \Omega)$, and Eq. (3.54) or Eq. (3.60) for the volume term $E_{\text{vol}}(\Omega)$, one must calculate the appropriate normalized energy-wavenumber characteristic $F_N(q, \Omega)$, starting from Eq. (3.52) or Eq. (3.58), respectively. This brings us to the issue of electron screening and the self-consistent determination of the screening density $n_{\text{scr}}(q)$ and the form factor $\langle \mathbf{k} + \mathbf{q} | w | \mathbf{k} \rangle$, which are the central ingredients in $F_N(q, \Omega)$. The most essential feature of electron screening within first-order PP perturbation theory is that each ion is screened independently of all the others. This is usually called *linear screening* and is embodied in the fact that both $\langle \mathbf{k} + \mathbf{q} | w | \mathbf{k} \rangle$ and $n_{\text{scr}}(q)$ are one-ion quantities independent of atomic structure. Equation (3.25) expresses $n_{\text{scr}}(q)$ directly as a functional of the form factor $\langle \mathbf{k} + \mathbf{q} | w | \mathbf{k} \rangle$. In turn the form factor depends on the screening density through the corresponding electron screening potential $v_{\text{scr}}(q)$. For a general nonlocal PP, one has $W = V_{\text{val}} + \mu_{\text{xc}} + W_{\text{ion}}$, where, corresponding to n_{val} in Eq. (3.3), the Coulomb potential V_{val} has uniform, screening and orthogonalization-hole components: $V_{val} = V_{unif} + \delta V_{scr} + \delta V_{oh}$. For the one-ion form factor one thus has

$$\langle \mathbf{k+q}| w |\mathbf{k}\rangle = v_{scr}(q) + v_{oh}(q) + \langle \mathbf{k+q}| w_{ion} |\mathbf{k}\rangle$$

$$= \frac{4\pi e^2}{q^2}[1 - G(q)][n_{scr}(q) + n_{oh}(q)] + \langle \mathbf{k+q}| w_{ion} |\mathbf{k}\rangle, \quad (3.61)$$

where the contribution of the uniform electron density has been absorbed as the $q = 0$ component of $n_{scr}(q)$, and the xc contributions to $v_{scr}(q)$ and $v_{oh}(q)$ are included through the function $G(q)$. Equations (3.25) and (3.61) can then be combined and solved for $n_{scr}(q)$ to establish a self-consistent screening field that can be used back in Eq. (3.61). To do this, we first substitute the second line of Eq. (3.61) directly into Eq. (3.25) to obtain the relation

$$n_{scr}(q) = -\left\{\frac{4\pi e^2}{q^2}[1 - G(q)][n_{scr}(q) + n_{oh}(q)] + \bar{w}_{ion}(q)\right\}\Pi_0(q), \quad (3.62)$$

where $\Pi_0(q)$ is the *polarizability* of the uniform electron gas in the Hartree or random phase approximation (RPA),

$$\Pi_0(q) \equiv -\frac{4}{(2\pi)^3}\int \frac{1}{\varepsilon_k - \varepsilon_{k+q}}\Theta^<(\mathbf{k})d\mathbf{k}, \quad (3.63)$$

and where we have retained the nonlocal character of w_{ion} by defining the useful average of $\langle \mathbf{k+q}| w_{ion} |\mathbf{k}\rangle$:

$$\bar{w}_{ion}(q) \equiv -\frac{4}{(2\pi)^3}\int \frac{\langle \mathbf{k+q}| w_{ion} |\mathbf{k}\rangle}{\varepsilon_k - \varepsilon_{k+q}}\Theta^<(\mathbf{k})d\mathbf{k}/\Pi_0(q). \quad (3.64)$$

One next solves Eq. (3.62) for the self-consistent screening density $n_{scr}(q)$ to obtain

$$n_{scr}(q) = -\left\{\bar{w}_{ion}(q) + \frac{4\pi e^2}{q^2}[1 - G(q)]n_{oh}(q)\right\}\Pi_0(q)/\varepsilon(q)$$

$$= -[\bar{w}_{ion}(q) + v_{oh}(q)]\Pi_0(q)/\varepsilon(q), \quad (3.65)$$

where we have additionally defined the *dielectric function* $\varepsilon(q)$ for the interacting electron gas:

$$\varepsilon(q) \equiv 1 + \frac{4\pi e^2}{q^2}[1 - G(q)]\Pi_0(q). \quad (3.66)$$

In the limit $G(q) \to 0$, $\varepsilon(q) \to \varepsilon_0(q)$, which is the familiar Hartree or RPA dielectric function of the noninteracting electron gas. Finally, one can use Eq. (3.65) in Eq. (3.61) to obtain the self-consistent form factor

$$\langle \mathbf{k}+\mathbf{q}| w |\mathbf{k}\rangle = [\bar{w}_{\text{ion}}(q) + v_{\text{oh}}(q)]/\varepsilon(q) + \langle \mathbf{k}+\mathbf{q}| w_{\text{ion}} |\mathbf{k}\rangle - \bar{w}_{\text{ion}}(q) \qquad (3.67)$$

Both $n_{\text{scr}}(q)$ as given by Eq. (3.65) and $\langle \mathbf{k}+\mathbf{q}| w |\mathbf{k}\rangle$ as given by Eq. (3.67) now depend on only the bare-ion PP w_{ion}, the orthogonalization-hole density n_{oh}, and the properties of the electron gas. These equations can then be used to evaluate $F_N(q, \Omega)$ from either Eq. (3.52) if $n_{\text{oh}} = 0$ or Eq. (3.58) for finite n_{oh}.

While we have indicated the importance of PP nonlocality to the physical accuracy of $E_{\text{vol}}(\Omega)$ and $v_2(r, \Omega)$, it is instructive to consider first the limit of a local PP to gain additional insight into the screening process and the determination of the important $q \to 0$ limits of the functions $n_{\text{scr}}(q)$, $\langle \mathbf{k}+\mathbf{q}| w |\mathbf{k}\rangle$ and $F_N(q, \Omega)$. For a local PP, one has three major simplifications: $n_{\text{oh}} = v_{\text{oh}} = 0$, $\langle \mathbf{k}+\mathbf{q}| w_{\text{ion}} |\mathbf{k}\rangle = w_{\text{ion}}(q)$ independent of $\mathbf{k}$, and $\bar{w}_{\text{ion}}(q) = w_{\text{ion}}(q)$. Equations (3.65) and (3.67) then immediately reduce to

$$n_{\text{scr}}(q) = -w_{\text{ion}}(q)\Pi_0(q)/\varepsilon(q) \qquad (3.68)$$

and

$$\langle \mathbf{k}+\mathbf{q}| w |\mathbf{k}\rangle \to w(q) = w_{\text{ion}}(q)/\varepsilon(q), \qquad (3.69)$$

respectively. Using these latter results, Eq. (3.52) for $F_N(q, \Omega)$ collapses to the relatively simple form for a local PP:

$$F_N(q, \Omega) = \left[-\frac{q^2}{4\pi e^2}\frac{\Omega}{Z}w_{\text{ion}}(q)\right]^2 \chi(q), \qquad (3.70)$$

where we have further defined a *susceptibility* $\chi(q)$ for the interacting electron gas as

$$\chi(q) \equiv \frac{4\pi e^2}{q^2}\Pi_0(q)/\varepsilon(q). \qquad (3.71)$$

The corresponding susceptibility of the noninteracting electron gas $\chi_0(q)$ is obtained by replacing $\varepsilon(q)$ with $\varepsilon_0(q)$ in Eq. (3.71).

The basic screening functions $\Pi_0(q)$, $\varepsilon_0(q)$, $\varepsilon(q)$, $\chi_0(q)$ and $\chi(q)$ entering the above expressions can be determined analytically by evaluating the integral in Eq. (3.63) for $\Pi_0(q)$. One finds

$$\Pi_0(q) = \frac{k_{\text{TF}}^2}{4\pi e^2}f_L(q/2k_F) = \frac{Z/\Omega}{(2/3)\varepsilon_F}f_L(q/2k_F), \qquad (3.72)$$

where k_{TF} is the Thomas-Fermi (TF) wavenumber $k_{\text{TF}}^2 \equiv 4k_F me^2/(\pi\hbar^2) = 6\pi Ze^2/(\varepsilon_F\Omega)$ and where f_L is the Lindhard function

$$f_L(x) \equiv \frac{1}{2} + \frac{(1-x^2)}{4x}\ln\left|\frac{1+x}{1-x}\right|, \qquad (3.73)$$

with $x = q/2k_F$. The Lindhard function contains a logarithmic singularity in its first derivative at $q = 2k_F$, which arises from the sharp cutoff of the integral in Eq. (3.63) at $k = k_F$. Although weak, this singularity is the origin of the long-range oscillatory behavior of the pair potential, culminating in the Friedel oscillations given by Eq. (1.28). The limiting behavior of $f_L(x)$ at small and large x is readily found to be

$$f_L(x) = \begin{cases} 1 - x^2/3 & x \to 0 \\ 1/(3x^2) & x \to \infty \end{cases}. \tag{3.74}$$

In the limit of small q where $f_L \to 1$, the RPA dielectric function $\varepsilon_0(q)$ approaches the Thomas-Fermi value

$$\varepsilon_0(q) \to 1 + k_{TF}^2/q^2 \equiv \varepsilon_{TF}(q), \tag{3.75}$$

while the corresponding susceptibility $\chi_0(q)$ approaches $\chi_{TF}(q) \equiv (k_{TF}^2/q^2)/\varepsilon_{TF}(q)$. At small q, the full susceptibility $\chi(q)$ has the limiting form

$$\chi(q) \to 1 - q^2/k_{TF}^2 + G(q) = 1 - \frac{q^2}{4\pi e^2}\left[\frac{(2/3)\varepsilon_F}{Z/\Omega} + \frac{d\mu_{xc}(n_{unif})}{dn}\right]. \tag{3.76}$$

The complete behavior of $\varepsilon_{TF}(q)$, $\varepsilon_0(q)$ and $\varepsilon(q)$ in the LDA is shown in Fig. 3.3(a) for electron-gas conditions corresponding to Al at equilibrium. The corresponding behavior of $\chi_{TF}(q)$, $\chi_0(q)$ and $\chi(q)$ in the LDA is illustrated in Fig. 3.3(b). Note that the inclusion of exchange and correlation in $\varepsilon(q)$ and $\chi(q)$ quantitatively lowers the magnitude of $\varepsilon(q)$, but raises the magnitude of $\chi(q)$, with the largest impact in the vicinity of $q = 2k_F$ and below. At large q, the leading xc contributions to $\varepsilon(q)$ and $\chi(q)$ are small, and in the case of $\chi(q)$ vary as q^{-6} and become negligibly small beyond about $q = 3k_F$.

The screening functions $\varepsilon_0(q)$ and $\chi_0(q)$ depend on the atomic volume Ω through the precise ratio Z/Ω of the uniform electron-gas density. This is a general constraint of electron-gas screening and remains true when exchange and correlation are included in $\varepsilon(q)$ and $\chi(q)$, and when arbitrary nonlocal PPs are considered. This constraint impacts the limiting behavior of both the screening density and the form factor as $q \to 0$. These limits can be readily deduced from Eqs. (3.68)–(3.70) for a local PP. To do this, we first expose the ionic Coulomb potential contained in w_{ion} entering Eq. (3.68) for $n_{scr}(q)$ and Eq. (3.69) for $w(q)$ by writing

$$w_{ion}(q) = -\frac{4\pi Z e^2}{q^2 \Omega} + w_{core}(q). \tag{3.77}$$

This result follows from the definition of w_{core} in Eq. (3.13). Assuming that $w_{core}(q)$ remains finite and that $G(q) \to 0$ as $q \to 0$, one immediately obtains for $n_{scr}(q)$ from Eqs. (3.66) and (3.68) the expected result

$$\lim_{q \to 0} n_{scr}(q) = Z/\Omega, \tag{3.78}$$

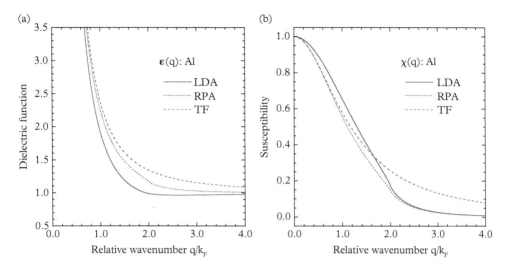

Fig. 3.3 *Screening functions for Al at its observed equilibrium volume ($\Omega_0 = 112.0$ a.u.) in the TF, RPA and LDA limits. (a) Dielectric functions $\varepsilon_{TF}(q)$, $\varepsilon_0(q)$ and $\varepsilon(q)$ in the LDA; (b) susceptibility functions $\chi_{TF}(q)$, $\chi_0(q)$ and $\chi(q)$ in the LDA.*

as was assumed in Eq. (3.52). This result remains true for a nonlocal norm-conserving PP as well since it is an intrinsic property of the electron-gas screening. For a nonlocal AHS PP, however, one must let $Z \to Z^*$ in Eq. (3.78) to account for the fact that the orthogonalization hole n_{oh} is screened by the extra uniform electron-gas density $(Z^*/Z - 1)n_{\text{unif}}$ contained in Eq. (3.5) for δn_{oh}.

In a similar manner, using Eqs. (3.66), (3.69) and (3.72) with $f_L(0) = 1$, one finds for the local PP form factor $w(q) \to -(2/3)\varepsilon_F$ as $q \to 0$. This again is an intrinsic property of electron-gas screening and can be generalized to an arbitrary nonlocal PP in the form

$$\lim_{q \to 0} \langle \mathbf{k}_F + \mathbf{q} | w | \mathbf{k}_F \rangle = -\frac{2}{3}\varepsilon_F, \qquad (3.79)$$

where the form factor is confined to the free-electron Fermi surface. This result was first proved for a nonlocal PP by Shaw and Harrison (1967) and can be verified numerically. In fact, for nonlocal PPs, where the form factor $\langle \mathbf{k} + \mathbf{q} | w | \mathbf{k} \rangle$ must be calculated numerically, the degree to which Eq. (3.79) is satisfied represents an excellent check on one's numerical methods.

Even for a nonlocal PP we shall continue to denote the Fermi-surface form factor as just $w(q)$, with the understanding that $w(q) \equiv \langle \mathbf{k}_F + \mathbf{q} | w | \mathbf{k}_F \rangle$ and the range of q is restricted from $q = 0$, the forward-scattering condition, to $q = 2k_F$, the back-scattering condition where $\mathbf{q} = -2\mathbf{k}_F$. Calculated self-consistent nonlocal GPT form factors for Na, Mg and Al are displayed in Fig. 3.4. In addition to its close connection with electron screening, the form factor $w(G_0)$ for reciprocal lattice vectors with magnitude $G_0 < 2k_F$

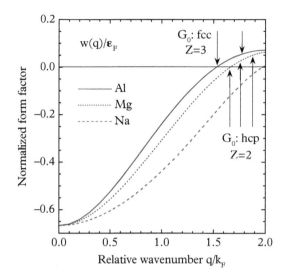

Fig. 3.4 *Normalized form factor $w(q)/\varepsilon_F$ on the respective free-electron Fermi surfaces of Na, Mg and Al, as obtained from first-principles GPT calculations with self-consistent screening at $\Omega = \Omega_0$. Also shown are the shortest reciprocal lattice vectors with magnitude $G_0 < 2k_F$ for the Mg hcp structure with $Z = 2$ and for the Al fcc structure with $Z = 3$.*

provides a useful approximate estimate of the size of corresponding BZ band gaps in the electronic structure. In particular, from Fig. 3.4 one sees that $|w(G_0)|/\varepsilon_F < 0.05$ for hcp Mg and fcc Al, so that the small band-gap criterion for perturbation theory, $2|w(G_0)|/\varepsilon_F \ll 1$, is indeed well satisfied. A similar test applied to bcc Na is also easily passed, but requires a small extrapolation of the form factor to $q = G_0 = 2.280$, the shortest reciprocal lattice vector for the $Z = 1$ bcc structure. It should also be noted that beyond the electronic structure and the calculation of total-energy properties, the form factor plays a central role in the calculation of transport properties, which are often dominated by electron scattering on the Fermi surface. While transport properties are not a main focus in this book, calculation of the liquid metal resistivity, which provides a straightforward and useful test of the form factor, has helped guide the development of the GPT for simple and other nontransition metals (Moriarty, 1974, 1977, 1982).

3.3 Evaluation of the energy-wavenumber characteristic and volume term

Returning to the full self-consistent evaluation of the components of the cohesive energy functional for simple metals, the explicit form of $F_N(q, \Omega)$ in Eq. (3.70) for a local PP can be extended to more general nonlocal PPs by using Eq. (3.65) for $n_{\text{scr}}(q)$. We first consider the case of a nonlocal norm-conserving PP and rewrite Eq. (3.65) with $n_{\text{oh}} = 0$ in terms of the susceptibility $\chi(q)$ as

3.3 Evaluation of the energy-wavenumber characteristic and volume term

$$n_{\text{scr}}(q) = -\frac{q^2}{4\pi e^2}\bar{w}_{\text{ion}}(q)\chi(q) \tag{3.80}$$

Then substituting this result into Eq. (3.52) for $F_N(q,\Omega)$ yields, after some manipulation,

$$F_N(q,\Omega) = \left[-\frac{q^2}{4\pi e^2}\frac{\Omega}{Z}\bar{w}_{\text{ion}}(q)\right]^2 \chi(q) + \frac{q^2}{4\pi e^2}\left(\frac{\Omega}{Z}\right)^2\left\{\overline{w_{\text{ion}}^2}(q) - [\bar{w}_{\text{ion}}(q)]^2\right\}\Pi_0(q), \tag{3.81}$$

where we have defined the additional average of $\langle \mathbf{k}+\mathbf{q}|\, w_{\text{ion}}\,|\mathbf{k}\rangle$

$$\overline{w_{\text{ion}}^2}(q) \equiv -\frac{4}{(2\pi)^3}\int\frac{|\langle\mathbf{k}+\mathbf{q}|\,w_{\text{ion}}\,|\mathbf{k}\rangle|^2}{\varepsilon_\mathbf{k}-\varepsilon_{\mathbf{k}+\mathbf{q}}}\Theta^<(\mathbf{k})d\mathbf{k}/\Pi_0(q). \tag{3.82}$$

One can also use Eq. (3.81) for $F_N(q,\Omega)$ to evaluate $\partial^2 F_N(0,\Omega)/\partial q^2$ analytically and obtain a more useful form for the volume term $E_{\text{vol}}(\Omega)$ given by Eq. (3.54). To do this, we first replace w_{ion} with w_{core} in Eq. (3.64) to define $\bar{w}_{\text{core}}(q)$ and also in Eq. (3.82) to define $\overline{w_{\text{core}}^2}(q)$. Noting, in analogy with Eq. (3.77) for a local PP, that

$$\bar{w}_{\text{ion}}(q) = -\frac{4\pi Z e^2}{q^2 \Omega} + \bar{w}_{\text{core}}(q) \tag{3.83}$$

Eq. (3.81) can then be rewritten as

$$F_N(q,\Omega) = \left[1-\frac{q^2}{4\pi e^2}\frac{\Omega}{Z}\bar{w}_{\text{core}}(q)\right]^2\chi(q) + \frac{q^2}{4\pi e^2}\left(\frac{\Omega}{Z}\right)^2\overline{\delta w_{\text{core}}^2}(q)\Pi_0(q), \tag{3.84}$$

where we have defined

$$\overline{\delta w_{\text{core}}^2}(q) \equiv \overline{w_{\text{core}}^2}(q) - [\bar{w}_{\text{core}}(q)]^2. \tag{3.85}$$

Using the small-q representation of $\chi(q)$ from Eq. (3.76) and assuming only that $\bar{w}_{\text{core}}(0)$ and $\overline{\delta w_{\text{core}}^2}(0)$ are finite, one arrives at the following limiting form for $F_N(q,\Omega)$ as $q\to 0$:

$$F_N(q,\Omega) \to 1 - \frac{q^2}{4\pi e^2}\left\{\frac{2\Omega}{Z}\left[\frac{1}{3}\varepsilon_F + \bar{w}_{\text{core}}(0)\right] + \frac{d\mu_{\text{xc}}(n_{\text{unif}})}{dn}\right\} + \frac{3q^2}{8\pi e^2\varepsilon_F}\frac{\Omega}{Z}\overline{\delta w_{\text{core}}^2}(0). \tag{3.86}$$

Differentiating Eq. (3.86) twice with respect to q, using the result in Eq. (3.54) for E_{vol}, and then manipulating the electron-gas terms gives the following computationally efficient form for a nonlocal norm-conserving PP:

$$E_{\text{vol}}(\Omega) = E_{\text{eg}}(\Omega) - \frac{1}{2}\Omega B_{\text{eg}}(\Omega) + \frac{2\Omega}{(2\pi)^3}\int [\langle \mathbf{k}|w_{\text{core}}|\mathbf{k}\rangle - \bar{w}_{\text{core}}(0)]\Theta^{<}(\mathbf{k})d\mathbf{k}$$

$$+ \frac{3}{4}\frac{Z}{\varepsilon_F}\overline{\delta w_{\text{core}}^2}(0) - \frac{(Ze)^2}{\pi}\int_0^\infty F_N(q,\Omega)dq - E_{\text{bind}}^{\text{atom}}(Z), \qquad (3.87)$$

where E_{eg} is the interacting electron-gas energy

$$E_{\text{eg}}(\Omega) \equiv \frac{3}{5}Z\varepsilon_F + Z\varepsilon_{\text{xc}}(n_{\text{unif}}) \qquad (3.88)$$

and $B_{\text{eg}} \equiv \Omega \partial^2 E_{\text{eg}}/\partial \Omega^2$ is the corresponding bulk modulus with

$$\Omega B_{\text{eg}}(\Omega) = \frac{2}{3}Z\varepsilon_F + Zn_{\text{unif}}\frac{d\mu_{\text{xc}}(n_{\text{unif}})}{dn}. \qquad (3.89)$$

The term $\overline{\delta w_{\text{core}}^2}(0)$ in Eq. (3.87) for $E_{\text{vol}}(\Omega)$ is generally small in magnitude, and for a DRT MP this term can be shown to be exactly zero.

It is also interesting to note that either Eq. (3.81) or (3.84) for $F_N(q,\Omega)$ can be expressed directly in terms of $n_{\text{scr}}(q)$ and n_{unif}. For example, combining Eqs. (3.80) and (3.83) to write

$$n_{\text{scr}}(q) = \left[n_{\text{unif}} - \frac{q^2}{4\pi e^2}\bar{w}_{\text{core}}(q)\right]\chi(q), \qquad (3.90)$$

Eq. (3.84) then becomes

$$F_N(q,\Omega) = [n_{\text{scr}}(q)/n_{\text{unif}}]^2 \chi^{-1}(q) + \frac{q^2}{4\pi e^2}\overline{\delta w_{\text{core}}^2}(q)\Pi_0(q)/n_{\text{unif}}^2. \qquad (3.91)$$

Figure 3.5 illustrates the full behavior of $F_N(q,\Omega)$ as a function of wavenumber q for a DRT MP with LDA screening in the case of Al at equilibrium. Note that F_N drops rapidly in magnitude from its small-q limit of 1 at $q = 0$ to a value less than 0.1 for $q > k_F$. Beyond that point, significant fine structure appears in $F_N(q,\Omega)$ near $q = 2k_F$, as shown in the inset of Fig. 3.5. As will be discussed in Sec. 3.4.1, this fine structure reflects both the nonlocality of the PP and the specific treatment of the electron screening, and can strongly impact the pair potential $v_2(r,\Omega)$ at near-neighbor distances.

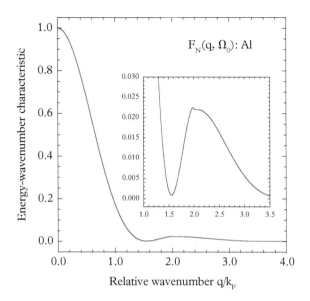

Fig. 3.5 *Normalized energy-wavenumber characteristic $F_N(q, \Omega_0)$ for Al calculated at its observed equilibrium volume ($\Omega_0 = 112.0$ a.u.), using a nonlocal DRT MP and LDA electron screening.*

The above results for a nonlocal norm-conserving PP can be readily generalized to a nonlocal AHS PP with nonzero orthogonalization-hole contributions to $F_N(q, \Omega)$. Using the general result (3.65) for the self-consistent screening electron density, Eq. (3.90) with a nonzero n_{oh} contribution included becomes

$$n_{scr}(q) = \left\{ n_{unif} - \frac{q^2}{4\pi e^2} \bar{w}_{core}(q) - [1 - G(q)] n_{oh}(q) \right\} \chi(q), \qquad (3.92)$$

while Eq. (3.91) for F_N generalizes to

$$F_N(q, \Omega) = [n_{scr}(q)/n^*_{unif}]^2 \chi^{-1}(q) + \frac{q^2}{4\pi e^2} \overline{\delta w^2_{core}}(q) \Pi_0(q)/(n^*_{unif})^2 + G(q) [n_{oh}(q)/n^*_{unif}]^2, \qquad (3.93)$$

where we have defined $n^*_{unif} \equiv (Z^*/Z) n_{unif} = Z^*/\Omega$. Determination of the small-q limit of $F_N(q, \Omega)$ starting from Eqs. (3.92) and (3.93) is basically similar to that accomplished in Eq. (3.86) for a nonlocal norm-conserving PP, except that now one must also take into account the small-q behavior of $n_{oh}(q)$, which can be expressed in the form

$$n_{oh}(q) \rightarrow -\frac{(Z^* - Z)}{\Omega} \left[1 - \gamma_{oh}(q/k_F)^2 \right]. \qquad (3.94)$$

Calculation of the constant γ_{oh} in Eq. (3.94) is discussed in Sec. A2.4 of Appendix A2. For the AHS PP used in the GPT, the volume term as given by Eq. (3.87) is then generalized to

$$E_{\text{vol}}(\Omega) = E_{\text{fe}}^0(\Omega) - \frac{1}{2}\Omega B_{\text{eg}}(\Omega) + \frac{2\Omega}{(2\pi)^3}\int \langle \mathbf{k}|w_{\text{pa}}|\mathbf{k}\rangle [1 + \langle \mathbf{k}|p_c|\mathbf{k}\rangle]\Theta^<(\mathbf{k})d\mathbf{k} - Z^*\bar{w}_{\text{core}}(0)$$
$$+ \frac{3}{4}\frac{Z}{\varepsilon_F}\overline{\delta w_{\text{core}}^2}(0) + \frac{9}{10}\frac{(Z^*e)^2}{R_{\text{WS}}} - \frac{(Z^*e)^2}{\pi}\int_0^\infty F_N(q,\Omega)dq + \delta E_{\text{oh}}^*(\Omega) - E_{\text{bind}}^{\text{atom}}(Z),$$

(3.95)

where we have defined

$$\delta E_{\text{oh}}^*(\Omega) \equiv \delta E_{\text{oh}}(\Omega) + [(Z^*)^2 - Z^2]\frac{\varepsilon_F}{3Z} - 4\pi Z^* e^2\frac{(Z^* - Z)}{\Omega k_F^2}\gamma_{\text{oh}}.$$

(3.96)

In spite of the more complicated form for $E_{\text{vol}}(\Omega)$ in Eq. (3.95), this result is still exact to second order in the AHS PP and can be numerically evaluated without difficulty.

3.4 First-principles pair potentials for simple metals

Armed with the theoretical results obtained in Secs. 3.2 and 3.3 for the self-consistent electron screening density n_{scr} and the normalized energy-wavenumber characteristic $F_N(q,\Omega)$, we are now in a position to analyze the remaining variable contributions to the two-ion pair potential $v_2(r,\Omega)$, as well as the volume term $E_{\text{vol}}(\Omega)$, in the cohesive-energy functional (3.1) for simple metals, and then to proceed to optimum treatments for real materials.

3.4.1 Impact of pseudopotential nonlocality and exchange and correlation

The principal variable components in the simple-metal formalism we have developed for $F_N(q,\Omega)$ and $v_2(r,\Omega)$ concern the PP w itself, or more precisely, its bare-ion component w_{ion} and the details of how this potential is screened by the valence electrons. In this section, we provide some additional insight as to specific importance of PP nonlocality and the choice of xc treatment in the electron screening. For simplicity and clarity, we will address these questions in the context of the energy-independent, norm-conserving DRT MP. The more complex questions of the importance of energy dependence and optimization of the PP will be discussed in Sec. 3.4.2.

For the norm-conserving nonlocal DRT MP, one has a bare-ion component

$$w_{\text{ion}} = \sum_\ell w_\ell P_\ell,$$

(3.97)

3.4.1 Impact of pseudopotential nonlocality and exchange and correlation

where w_ℓ is an energy-independent potential well for the ℓth angular momentum channel of the form

$$w_\ell(r) = \begin{cases} -A_\ell & r < R_\ell \\ -Ze^2/r & r > R_\ell \end{cases}. \tag{3.98}$$

Here both the well depths A_ℓ and the corresponding radii R_ℓ are treated as constants. In the DRT approach, the ℓ components A_ℓ and R_ℓ are explicitly modeled and optimized only for $\ell \leq 2$, subject to the typical constraints $A_0 = A_1$ and $R_1 = R_2$, while one takes $A_\ell = A_2$ and $R_\ell = R_2$ for $\ell > 2$. The bare-ion form factor $\langle \mathbf{k}+\mathbf{q}| w_{\text{ion}} |\mathbf{k}\rangle$ can then be numerically calculated, together with the needed averages $\bar{w}_{\text{ion}}(q)$ from Eq. (3.64) and $\overline{w^2_{\text{ion}}}(q)$ from Eq. (3.82). In turn $F_N(q,\Omega)$ can be evaluated from either Eq. (3.81) or Eq. (3.91).

For direct comparison to the nonlocal DRT MP, a companion local MP can be defined by the single potential well

$$w_{\text{ion}}(r) = \begin{cases} -A_0 & r < R_0 \\ -Ze^2/r & r > R_0 \end{cases} \tag{3.99}$$

to test the importance of nonlocality. The Fourier transform of this potential, $w_{\text{ion}}(q)$, which establishes the electron screening density $n_{\text{scr}}(q)$ through Eq. (3.68) and $F_N(q,\Omega)$ through Eq. (3.70), has the analytic representation

$$w_{\text{ion}}(q) = -\frac{4\pi Ze^2}{q^2 \Omega}\left[\left(1 - \frac{A_0 R_0}{Ze^2}\right)\cos(qR_0) + \frac{A_0 R_0}{Ze^2}\frac{\sin(qR_0)}{qR_0}\right]. \tag{3.100}$$

It is also possible to define an intermediate quasi-local treatment that falls between the use of a full nonlocal DRT MP and an approximate local MP to calculate $F_N(q,\Omega)$. In the quasi-local approximation, one neglects the second term involving $\delta w^2_{\text{core}}(q)$ in Eq. (3.91) for F_N such that

$$F_N^{\text{QL}}(q,\Omega) = [n_{\text{scr}}(q)/n_{\text{unif}}]^2 \chi^{-1}(q). \tag{3.101}$$

Historically, this form was first suggested by Manninen et al. (1981) as a means of directly calculating the pair potential v_2 from an independent DFT calculation of the screening density n_{scr}, without the intermediate step of defining a nonlocal MP. Equation (3.101) also retains the functional relationship between n_{scr} and F_N established by a local PP, as can be seen by combining Eqs. (3.68) and (3.70). In the present context, however, the idea is to retain the full nonlocal calculation of $n_{\text{scr}}(q)$, via either Eq. (3.80) or (3.90), in applying Eq. (3.101).

In Fig. 3.6 we compare nonlocal DRT and quasi-local and local MP calculations of $F_N(q,\Omega)$ and $v_2(r,\Omega)$ for the representative case of Al at its observed equilibrium volume Ω_0, maintaining an LDA treatment of exchange and correlation in the electron screening for each calculation. The visible impact of nonlocality on $F_N(q,\Omega_0)$ shown

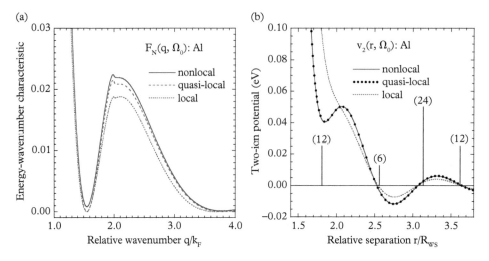

Fig. 3.6 *Nonlocal DRT, quasi-local and local MP calculations of (a) $F_N(q,\Omega_0)$ and (b) $v_2(r,\Omega_0)$ for Al at its observed equilibrium volume ($\Omega_0 = 112.0$ a.u.), using LDA electron screening for each calculation. Also shown in panel (b) are the positions and number of neighbors in the first four near-neighbor shells of the observed fcc crystal structure.*

in Fig. 3.6(a) is confined to the fine-structure portion of the curve between $q = 1.5k_F$ and $q = 3.5k_F$. The quasi-local approximation lowers the calculated peak in $F_N(q,\Omega_0)$ near $q = 2k_F$ only slightly, while the local approximation lowers the peak significantly. Much more dramatic, however, is the impact of the local approximation on the shape and magnitude of the pair potential $v_2(r,\Omega_0)$ at near-neighbor distances, as shown in Fig. 3.6(b). Here the absence of nonlocality removes the positive-energy nearest-neighbor minimum in the potential near $r = 1.8R_{WS}$, leaving only a slight inflection in curve. In complete contrast, the quasi-local approximation closely reproduces the entire nonlocal pair potential.

Similar MP results have been obtained by Walker and Taylor (1990) for several other prototype simple metals as well. We conclude, therefore, that the inclusion of nonlocality in the MP or PP is generally important to the reliable calculation of first-principles pair potentials for such metals, and that the most important contribution to the nonlocality is that contained in the screening electron density.

An independent argument in favor of unique shape of the nonlocal LDA pair potential for Al is provided by the work of Dharma-wardana and Aers (1983). These authors developed an MD simulation technique for extracting a pair potential from the liquid structure factor of molten simple metals and rare gas systems. Using the experimental structure factor for Al, they recovered a pair potential that very closely resembles the nonlocal LDA pair potential illustrated in Fig. 3.6(b). In particular, they were not able to force their method to produce a potential that was negative at the nearest-neighbor distance of Al. Hence, one can further conclude that the calculated shape of the nonlocal LDA pair potential for Al is experimentally verified.

3.4.1 Impact of pseudopotential nonlocality and exchange and correlation

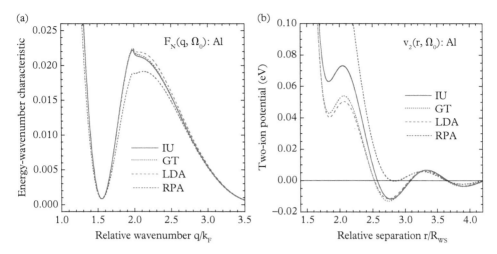

Fig. 3.7 *Nonlocal DRT MP calculations of (a) $F_N(q,\Omega_0)$ and (b) $v_2(r,\Omega_0)$ for Al at its observed equilibrium volume ($\Omega_0 = 112.0$ a.u.), using IU, GT, LDA and RPA treatments of the electron screening.*

One can perform an additional exercise to examine the impact of different treatments of exchange and correlation in the electron screening, as implemented through the function $G(q)$. In addition to the Hartree or RPA limit of no exchange and correlation [$G(q) = 0$] and the standard LDA treatment [Eq. (3.29) for $G(q)$], we consider here the treatments of Geldart and Taylor (1970) and Ishimaru and Utsumi (1981) that attempt to go beyond the LDA, as discussed in Sec. 2.2.2 of Chapter 2. In Fig. 3.7 we compare nonlocal DRT MP calculations of $F_N(q,\Omega)$ and $v_2(r,\Omega)$ for the case of Al at its observed equilibrium volume Ω_0, using RPA, LDA, GT and IU treatments of exchange and correlation. In the LDA, GT and IU treatments, the low-q behavior of $G(q)$ correctly conforms to the accurate VWN representation of the uniform-electron-gas exchange and correlation energy, as was discussed in Sec. 2.2.1 of Chapter 2.

For the energy-wavenumber characteristic $F_N(q,\Omega_0)$ plotted in Fig. 3.7(a), one sees that the GT and IU xc treatments produce almost identical results across the fine-structure peak near $q = 2k_F$, with the LDA peak only slightly displaced from those results. The RPA peak, on the other hand, is significantly lower. Consequently, one sees in Fig. 3.7(b) for $v_2(r,\Omega_0)$ a rather dramatic difference between the RPA pair potential and the IU, GT or LDA potentials. The latter three xc treatments fully preserve the shape of the pair potential, although clearly the height of the first minimum near $r = 1.8R_{WS}$ can be sensitive to the exact treatment used. In contrast, the absence of exchange and correlation in the RPA entirely eliminates the first minimum in the potential and also misses the depth of the second minimum near $r = 2.8R_{WS}$. We conclude that reliable first-principles pair potentials in simple metals indeed require a treatment of exchange and correlation in the electron screening at least at the LDA level.

118 *Interatomic Potentials in Simple Metals*

In terms of their mathematical representations, the additional GT and IU many-body treatments of exchange and correlation in the electron screening have the same pole structure as the LDA and thus are indeed expected to give similar pair potentials at metallic densities, as we have found for Al. Since the LDA and IU forms of $G(q)$ can also be expressed analytically, they are the simplest forms to use in practice. Thus, it is generally desirable to choose one or the other in calculating nonlocal pair potentials, at least for potentials derived from norm-conserving MPs or PPs. For pair potentials generated from nonlocal AHS PPs such as used in the GPT, on the other hand, there is an additional important technical distinction between the LDA and the GT or IU screening treatments. This distinction occurs because of the extra orthogonalization-hole contribution $G(q)[n_{\rm oh}(q)/n^*_{\rm unif}]^2$ that is present in the normalized energy-wavenumber characteristic (3.93). Since the LDA form of $G(q)$ continues to increase as q^2 at large q, $F_{\rm N}(q,\Omega)$ will acquire a quite long-ranged tail that can cause numerical convergence problems. In the GT or IU treatments, on the other hand, $G(q) \to$ constant at large q and such a tail is absent. Thus, for GPT pair potentials the IU treatment of screening is preferred over the LDA treatment, and will be used in all remaining comparisons with DRT and in the materials applications discussed below.

3.4.2 Energy dependence and the optimization of nonlocal pseudopotentials

The energy dependence and optimization of nonlocal PPs and MPs in the context of perturbation theory are closely related subjects. As was discussed in Chapter 2, the baseline AHS PP defined by Eqs. (2.54)–(2.56) is effectively energy dependent. This form is also closely related to the optimized PP proposed by Harrison (1966) for simple metals based on the criterion of the PP producing the smoothest pseudowavefunction $\phi_{\bf k}({\bf r})$, and hence, in principle, the most rapidly convergent plane-wave expansion. For the optimized Harrison PP, the plane-wave matrix element (2.56) is generalized to

$$\langle {\bf k}'|W|{\bf k}\rangle = \langle {\bf k}'|V|{\bf k}\rangle + \sum_{i,c}[\varepsilon_{\bf k} + \langle {\bf k}|W|{\bf k}\rangle - E_c]\langle {\bf k}'|\phi_c^i\rangle\langle \phi_c^i|{\bf k}\rangle. \quad (3.102)$$

This form was widely accepted at the time and, for example, has been used extensively by Hafner and coworkers (Hafner, 1987a). However, the inclusion of the additional $\langle {\bf k}|W|{\bf k}\rangle$ term in Eq. (3.102) complicates the needed form of the orthogonalization-hole density. This is largely unnecessary, as well as inconvenient, when one moves beyond simple metals to d-band metals. In the GPT (Moriarty, 1974, 1977, 1982), we have noted that the ${\bf k}$ dependence of $\langle {\bf k}|W|{\bf k}\rangle$ is weak for $k \leq k_{\rm F}$ and can be neglected, so that if one choses the zero of energy at the bottom of the NFE valence bands by requiring that $\langle 0|W|0\rangle = 0$, one can revert back to the simpler baseline form (2.56). The optimized bare-ion form factor in the simple-metal limit of the GPT is then given by

$$\langle {\bf k}+{\bf q}|w_{\rm ion}|{\bf k}\rangle = v_{\rm ion}(q) + \sum_c [\varepsilon_{\bf k} - E_c^{\rm vol}]\langle {\bf k}+{\bf q}|\phi_c\rangle\langle \phi_c|{\bf k}\rangle, \quad (3.103)$$

where E_c^{vol} is the dominant volume component of E_c, as obtained in practice from the zero-order pseudoatom, and $\langle \mathbf{k} + \mathbf{q} | w_{\text{ion}} | \mathbf{k} \rangle$ is fully structure independent as desired.

A hidden practical limitation of all AHS PPs is that the repulsive component of the PP involving the core projection operator P_c, which is intended to reduce the strength of the bare-ion potential in the core regions, is confined to angular momentum channels ℓ for which there actually are occupied core states. Thus, for example, in second-row elements (Li, Be), which have only 1s core states, one is effectively weakening only the $\ell = 0$ component of the bare-ion potential, leaving a strong $\ell = 1$ component. Similarly, for third-row elements (Na, Mg, Al, Si), with additional 2s and 2p core states, one is weakening only the $\ell = 0$ and $\ell = 1$ components of the bare-ion potential, leaving a potentially strong $\ell = 2$ component. Using the added basis-state approach discussed in Chapters 4 for d states, and also proposed in Chapter 12 for p states, the missing PP nonlocality can be remedied in necessary cases within the GPT by adding, respectively, localized d states or p states to the plane-wave basis set to account for empty, but strongly hybridizing, d or p bands above the Fermi level in these materials.

The flexibility of using both an energy-dependent and ℓ-dependent nonlocal MP to represent w_{ion}, in place of an AHS PP, allows one, in principle, to model the actual phase shifts $\delta_\ell(E)$ of the bare-ion potential v_{ion} accurately at all relevant energies E and angular momentum ℓ. Rasolt and Taylor (1973) further showed that so long as one does not have to consider energies too near a d-band resonance, such as occurs in the heavy alkaline-earth metals, the noble metals and the Group II-B metals, the energy dependence of w_{ion} can be made sufficiently weak over the occupied valence-band range that it often may be neglected. This is especially so in important energy-averaged quantities such as the screening electron density. These observations became the principal motivating factor behind the energy-independent, and hence norm-conserving, but strongly nonlocal DRT MPs, with modeled $\ell= 0$, 1 and 2 components, subsequently developed by Rasolt and Taylor (1975) and Dagens, Rasolt and Taylor (1975). These authors considered two general optimization criteria for determining the constant A_ℓ and R_ℓ parameters in Eq. (3.98) for $w_\ell(r)$: (*i*) the reproduction of the phase shifts of the full bare-ion potential; and (*ii*) the reproduction of the exact nonlinear screening electron density induced by a single isolated ion in a uniform electron gas, as obtained by an independent non-perturbative LDA calculation of the same. While these criteria produced very similar parameters, the latter approach, in fact, led to more accurate pair potentials $v_2(r, \Omega)$ in the context of second-order PP perturbation theory. In particular, DRT demonstrated for Li, Na, K and Al that the pair potentials obtained via criterion (*ii*) yield accurate phonon spectra for these metals. DRT parameters so obtained for eight simple metals from this and subsequent studies are summarized in Table 3.1. Of the metals listed in Table 3.1, only Be, with its non-NFE electronic structure as displayed in Fig. 2.5(b), has too strong of a PP to be accurately treated by second-order perturbation theory alone.

In Fig. 3.8 we compare optimized GPT and DRT pair potentials $v_2(r, \Omega_0)$ for the third-period metals Na, Mg and Al at their observed equilibrium volumes, based on an IU treatment of electron screening in each case. To complete the series, the corresponding GPT pair potential for Si is also shown. For Na the GPT and DRT potentials are nearly identical. For Mg and Al, the GPT and DRT potentials are substantially similar,

Table 3.1 *DRT MP parameters for eight simple metals, with A_ℓ in Rydbergs and R_ℓ in a.u. For $\ell > 2$, $A_\ell = A_2$ and $R_\ell = R_2$.*

Metal	A_0	R_0	A_1	R_1	A_2	R_2	Reference
Li	0.70	2.38	1.32	2.38	1.32	2.38	b
Na	0.255	1.87	0.255	1.93	0.50	1.93	a, c
K	0.32	2.65	0.32	3.08	1.80	3.08	b
Rb	0.32	2.93	0.32	3.70	1.65	3.70	d
Cs	0.36	3.76	0.36	5.10	0.83	5.10	d
Be	2.30	1.50	4.40	1.50	1.00	1.50	e
Mg	1.235	1.72	1.235	1.50	3.80	1.50	b
Al	2.22	1.45	2.22	1.31	6.70	1.31	b

[a] Rasolt and Taylor (1975).
[b] Dagens, Rasolt and Taylor (1975).
[c] Cohen et al. (1976).
[d] Taylor and MacDonald (1980).
[e] Duesbery and Taylor (1979).

with nearly equal nodes in the respective potentials, but with deeper minima and higher maxima for the DRT potentials and only a strong inflection in the GPT Al potential near the first-neighbor distance. The latter inflection nearly vanishes into a purely repulsive potential for GPT Si. The GPT-DRT differences in Mg and Al partly reflect the strong $\ell = 2$ nonlocal MP component present in the DRT potentials versus the complete absence of such a PP component in third-period GPT potentials obtained in the simple-metal limit. Near ambient pressure, these differences turn out to be most important in the case of Al, where the *sp-d* hybridization significantly impacts the pair potential. As also shown in Fig. 3.8(c), adding an effective $\ell = 2$ nonlocal component to the GPT PP by treating Al in the empty *d*-band limit of GPT (see Chapter 4) turns the near-neighbor inflection in the simple-metal pair potential into a positive-energy potential well that is quite similar to the well found in the DRT Al potential.

At the same time, the baseline simple-metal GPT treatment of Mg results in excellent calculated vibrational, structural, thermodynamic and mechanical properties, as shown by Althoff et al. (1993), Moriarty and Althoff (1995) and Greeff and Moriarty (1999). These results will be discussed further in Chapters 6, 7 and 8. They include an accurate hcp phonon spectrum and an accurate description of the observed high-pressure hcp → bcc structural phase transition at 50 GPa (Olijnyk and Holzapfel, 1985). We have also verified that the DRT MP for Mg gives a similarly good phonon spectrum, and as shown in Fig. 3.9(a), the calculated GPT and DRT Debye temperatures (see Chapter 7) agree well over a wide volume range and with experiment at $\Omega = \Omega_0$.

3.4.2 Energy dependence and the optimization of nonlocal pseudopotentials 121

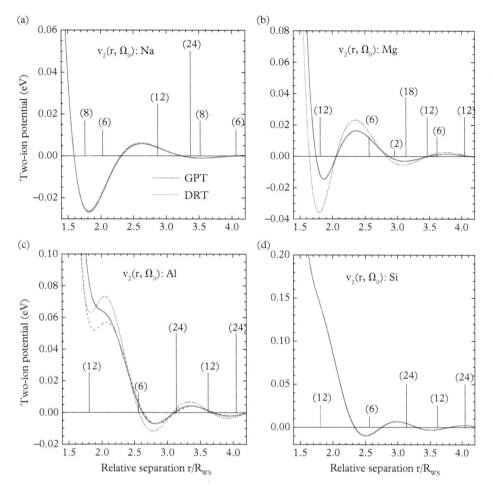

Fig. 3.8 *Optimized simple-metal GPT (solid lines) and DRT (dotted lines) pair potentials $v_2(r, \Omega_0)$ in third-period elements at their observed equilibrium volumes, using an IU treatment of electron screening for each calculation. Also shown for Al in panel (c) is the corresponding empty-d-band GPT pair potential (short dashed line). (a) Na at $\Omega_0 = 255.2$ a.u.; (b) Mg at $\Omega_0 = 156.8$ a.u.; (c) Al at $\Omega_0 = 112.0$ a.u.; (d) Si at $\Omega_0 = 135.1$ a.u. Also shown in each panel are the position and number of neighbors for near-neighbor shells of the observed or assumed crystal structures: (a) bcc, (b) hcp, (c) fcc and (d) fcc.*

In addition, as illustrated in Fig. 3.9(b), both GPT and DRT calculated structural energies are consistent with the observed equilibrium hcp structure of Mg, as well as with the location of the observed high-pressure hcp → bcc phase transition. The calculated DRT fcc-hcp energy difference for Mg, however, appears to be significantly overestimated, in contrast to the corresponding GPT result, which agrees well with DFT at $\Omega = \Omega_0$.

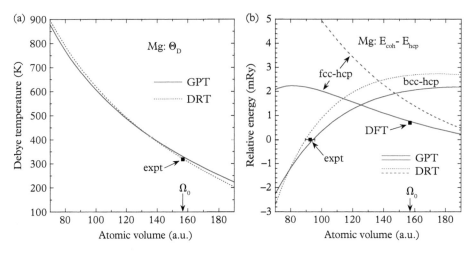

Fig. 3.9 *Vibrational and structural properties of Mg calculated over a wide volume range from GPT and DRT simple-metal pair potentials $v_2(r, \Omega)$. (a) Debye temperature $\Theta_D(\Omega)$, calculated from Eq. (7.48) as a measure of the average hcp phonon frequency, with the experimental value at the observed equilibrium volume ($\Omega_0 = 156.8$ a.u.) from Ashcroft and Mermin (1976); (b) bcc-hcp and fcc-hcp cohesive energy differences, predicting an hcp $\to$ bcc phase transition at the location observed by Olijnyk and Holzapfel (1985), and, in the case of GPT, matching the DFT fcc-hcp energy difference at $\Omega = \Omega_0$ of McMahan and Moriarty (1983).*

More generally, the simple-metal GPT treatments of Na and Mg are fully adequate except at extreme pressures above 100 GPa, where sp-d hybridization effects become significant and an empty d-band treatment is needed (Moriarty and McMahan, 1982; McMahan and Moriarty, 1983). In the cases of Al and Si, on the other hand, sp-d hybridization effects on physical properties are already significant at ambient pressure and an empty d-band GPT treatment, using the first-principles approach described in Chapter 4, is desirable there as well as at high pressure. For this reason, we will defer further comparisons between GPT and DRT treatments of Al until the application Chapters 6, 7 and 8.

It should also be noted that in comparison to the *energy-independent* DRT approach, one clear advantage of the *energy-dependent* GPT formulation is in the accuracy of the whole cohesive-energy functional $E_{\text{coh}}(\mathbf{R}, \Omega)$ including the volume term $E_{\text{vol}}(\Omega)$, allowing a simultaneous treatment of both cohesion and structure. Although the DRT parameters listed in Table 3.1 generally yield good-quality pair potentials $v_2(r, \Omega)$ as a function of volume, the same is not true for the respective volume terms. As was discovered by Walker and Taylor (1990), when the same parameters are used to evaluate $E_{\text{vol}}(\Omega)$ via Eq. (3.85), one is led in each case to a cohesion curve with an equilibrium volume that is 20–70% too small. No such difficulty is encountered in the GPT, however, and the striking difference is indeed mostly a consequence of very different calculated volume terms, as shown for Mg in Fig. 3.10. It remains an open question as to whether or not the situation would be improved if the DRT optimization procedure at $\Omega = \Omega_0$ were applied as a function of volume, allowing the DRT MP parameters to become volume dependent.

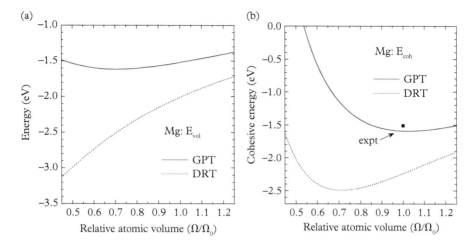

Fig. 3.10 *Calculated GPT and DRT real-space descriptions of cohesion in hcp Mg as a function of volume. (a) Volume term $E_{vol}(\Omega)$; (b) cohesive energy $E_{coh}(\boldsymbol{R},\Omega)$, with the experimental equilibrium value from Kittel (1976).*

Walker and Taylor (1990) opted instead for the much simpler ad hoc fix of maintaining the equilibrium-volume DRT parameters but adding a large empirical correction to $E_{vol}(\Omega)$ designed to minimize $E_{coh}(\boldsymbol{R},\Omega)$ at $\Omega = \Omega_0$. From a fundamental perspective, however, it appears likely that neglecting the energy dependence in the MP effectively pushes otherwise low-order contributions to the cohesive energy to higher order.

3.4.3 Trends with valence, volume and atomic number

Because of the strong dependence of the self-consistent screening on the uniform electron density $n_{unif} = Z/\Omega$, simple-metal pair potentials display some prominent characteristic trends with electron density. It is instructive to examine these trends in terms of the independent variables Z and Ω, as well as the atomic number Z_a, which strongly affects the equilibrium volume Ω_0. The behavior of $v_2(r,\Omega_0)$ in the third-row elements Na, Mg, Al and Si displayed in Fig. 3.8 is representative of the trend with increasing valence at equilibrium. In this case, Na, Mg and Al exemplify behavior at low, intermediate and high electron densities in simple metals, with the equilibrium density in Al being almost seven times that in Na and about two times that in Mg. For low electron density, the pair potential $v_2(r,\Omega)$ has a broad negative minimum in the vicinity of the nearest-neighbor distance, with the long-range oscillatory structure in the potential pushed out to larger distance r. For intermediate electron densities, the radial extend of the first minimum narrows, as does the entire long-range oscillatory structure, and the minimum may begin to move upward in energy, as seen in Fig. 3.8(b) for GPT Mg. Finally, at high electron density, the first minimum further narrows and moves to positive energy, as seen in Fig. 3.8(c) for DRT Al. This positive minimum soon becomes only a strong inflection in the pair-potential curve, as already occurs in simple-metal GPT Al at equilibrium, and eventually only purely repulsive potential as seen in the case of Si.

Similar trends can be seen in the pair potential of any individual element as a function of atomic volume. This is shown in Fig. 3.11(a) for Mg, where $v_2(r,\Omega)$ is plotted for five widely spaced volumes ranging from 26% expansion ($\Omega/\Omega_0 = 1.26$) to more than twofold compression ($\Omega/\Omega_0 = 0.44$). Under expansion to lower electron density, the first minimum in $v_2(r,\Omega)$ begins to broaden and lower in energy. Under compression, the first minimum narrows and moves to higher energy, becoming a shallow positive minimum above two-fold compression, similar to the minimum seen for Al in Fig. 3.8(c). Also note that in Fig. 3.11(a) we have plotted $v_2(r,\Omega)$ as a function of r rather than r/R_{WS}. This has been done to emphasize that at intermediate and high electron densities, the position of the first minimum r_{min} is nearly constant for a given material and only weakly dependent on atomic volume.

Finally, trends in the behavior of the pair potential with atomic number Z_a at equilibrium are illustrated in Fig. 3.11(b) for the low-density case of the alkali metals. Here there is first a clear discontinuity between the second-row element Li and the third- and higher-row elements. This discontinuity directly reflects the uniquely strong $\ell = 1$ component of the PP or MP for second-row elements. The remaining elements, however, exhibit more regular trends based on electron density. In particular, the observed equilibrium volume increases with atomic number in any given column of the Periodic Table, producing lower density as one moves down the column. As expected, the first minimum in $v_2(r,\Omega_0)$ broadens and deepens as one moves down the first column from Na, with $Z_a = 11$, to Cs, with $Z_a = 55$, where Ω_0 increases by a factor of three. Note, however, that in these low-density metals the position of the first minimum increases

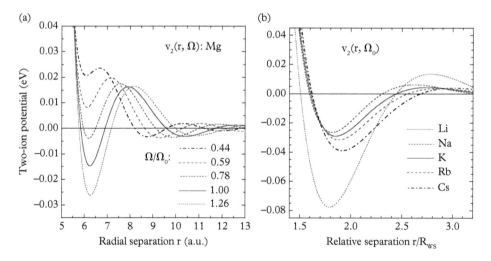

Fig. 3.11 *Trends in first-principles simple-metal pair potentials, using an IU treatment of electron screening for each calculation. (a) Variation with atomic volume for Mg, as obtained from optimized GPT potentials; (b) variation with atomic number for the alkali metals at their observed equilibrium volumes, as obtained from optimized DRT potentials.*

with atomic number even in units of r/R_{WS}. Thus, for the alkali metals we can conclude that $r_{\min} \propto R_{\mathrm{WS}}^{\alpha}$ with $\alpha > 1$.

3.5 Long-range Friedel oscillations and materials application

As we have discussed and seen in the above examples, an important characteristic feature of simple-metal pair potentials is the presence of long-range oscillations. Asymptotically these oscillations culminate in the classic Friedel form given by Eq. (1.28), with $v_2 \propto \cos(2k_F r)/(2k_F r)^3$. In this section, we consider different possible ways to overcome the numerical convergence difficulties that these oscillations present in computing materials properties.

3.5.1 Exact methods for static lattice properties

In constructing a simple-metal pair potential one has to Fourier transform the normalized energy-wavenumber characteristic $F_{\mathrm{N}}(q, \Omega)$, which depends directly on the Lindhard function $f_{\mathrm{L}}(q/2k_F)$ containing the logarithmic singularity that gives rise to the Friedel oscillations. As first pointed out by Basinski et al. (1970), these oscillations may be thought of as the leading term in an asymptotic series for the pair potential of the form

$$v_2^{\mathrm{as}}(r, \Omega) = A_3(\Omega)\frac{\cos(2k_F r)}{(2k_F r)^3} + A_4(\Omega)\frac{\sin(2k_F r)}{(2k_F r)^4} + A_5(\Omega)\frac{\cos(2k_F r)}{(2k_F r)^5} + \cdots. \quad (3.104)$$

For a large enough distance r, of course, the leading Friedel term will dominate, but this distance is usually well beyond $5R_{\mathrm{WS}}$ and involves hundreds of neighbors. In this regard, the number of neighbors of a given ion at a distance r increases roughly as r^2, so the sum over all ions of the Friedel term converges only slowly as $\cos(2k_F r)/(2k_F r)$. Special techniques are consequently needed to handle such long-range oscillations accurately. Duesbery and Taylor (1977) have shown that terms of the form $\cos(2k_F r)/(2k_F r)^m$ and $\sin(2k_F r)/(2k_F r)^m$ can be summed over all neighbors in an infinite lattice for $m > 1$ by use of a modified Ewald technique, similar to that used in obtaining Eq. (3.19) for the electrostatic energy. These authors have tabulated infinite sums for contributions to the cohesive energy and elastic moduli in bcc, fcc and ideal hcp lattices for valences $Z = 1$, 2, 3 and 4 and for values of $m = 3$, 4 and 5. They have also calculated selected phonon frequency contributions for bcc and fcc lattices this way.

By carefully fitting the long-range part of the potential to the asymptotic form (3.104) to establish the expansion coefficients A_3, A_4 and A_5 for a fixed atomic volume, one can construct a companion short-range potential

$$v_2^{\mathrm{sr}}(r, \Omega) = v_2(r, \Omega) - v_2^{\mathrm{as}}(r, \Omega). \quad (3.105)$$

Then one can determine the sum of $v_2^{as}(r,\Omega)$ over the entire lattice by using the appropriate tabulated values given by Duesbery and Taylor (1977). The corresponding sum for $v_2^{sr}(r,\Omega)$ can be performed numerically over the limited number of neighbors required, and thus one can determine the fully converged sum over the entire lattice for $v_2(r,\Omega)$. In addition, needed radial derivatives of $v_2^{as}(r,\Omega)$ and $v_2^{sr}(r,\Omega)$ for the elastic moduli and phonons can be obtained by differentiation of Eqs. (3.104) and (3.105).

This procedure, of course, is limited to performing infinite sums in perfect lattices, and is useful for calculating static lattice properties such as structural energy differences, elastic moduli and phonon frequencies. With special care, this procedure can be extended to the *unrelaxed* vacancy formation energy E_{vac}^u, as it too can be expressed from Eq. (1.18) in terms of infinite sums of the perfect lattice. In this case, however, the virial-pressure contribution in Eq. (1.18) for E_{vac}^u will produce a leading asymptotic term proportional to $\sin(2k_Fr)/(2k_Fr)^2$, which is only weakly convergent. Using DRT potentials with GT screening, Jacucci and Taylor (1979) have successfully calculated E_{vac}^u for bcc Li, Na and K in this manner, as have Jacucci et al. (1981) for fcc Al. To calculate relaxed vacancy formation energies in real space, however, requires special damping or smooth truncation approaches, which we discuss in Sec. 3.5.2.

3.5.2 Damping and smooth truncation techniques for atomistic simulations

For dynamic atomistic simulations of materials properties, such as MD, some form of damping or smooth truncation of the long-range oscillations in the pair potential is necessary in order to achieve convergence without incurring significant errors. Here we discuss three general methods of this kind that can be used to perform accurate atomistic simulations.

The first method, due to Duesbery et al. (1979), applies a similar modified Ewald technique to an MD simulation of a finite number of screened metal ions placed in a simulation cell with periodic boundary conditions applied. The essential trick in this approach is choose the variable Ewald convergence parameter η small enough to make the reciprocal-space contribution to the total energy negligibly small, but large enough that the real-space pair potential is significantly damped. When those conditions are achieved, the net effect is to replace each power law $(2k_Fr)^{-m}$ in Eq. (1.04) for $v_2^{as}(r,\Omega)$ by the function $2\mathcal{J}_m(2k_Fr,\eta)/\Gamma(m/2)$, where

$$\mathcal{J}_m(x,\eta) \equiv \int_\eta^\infty t^{m-1}\exp(-t^2x^2)dt \qquad (3.106)$$

and $\Gamma(z)$ is the Euler gamma function. For $\eta = 0$, $2\mathcal{J}_m(2k_Fr,\eta)/\Gamma(m/2) = (2k_Fr)^{-m}$, while for $\eta > 0$ $\mathcal{J}_m(2k_Fr,\eta)$ decays more rapidly than r^{-m} and the asymptotic potential is damped. Useful values of the convergence parameter η depend on the material being treated and the size of the simulation cell. For bcc alkali metals simulated in a cubic cell of 128 atoms, Duesbery et al. (1979) recommended values of η in the range 0.23–0.30 a^{-1}

3.5.2 Damping and smooth truncation techniques for atomistic simulations

for Li and K, and values in the range 0.35–0.40 a^{-1} for Na, where a is the bcc lattice parameter. Removing one atom from the center of the simulation cell to create a vacancy, Jacucci and Taylor (1979) successfully used this approach for Li, Na and K, first to confirm their values of E^u_{vac} obtained by the infinite perfect lattice method discussed above, and then to calculate the relaxed vacancy formation energy E_{vac} for each of these metals. For fcc Al, Jacucci et al. (1981) used a larger simulation cell containing 863 atoms plus a vacancy, together a value of 0.23 a^{-1} for η, with a the fcc lattice parameter, to calculate E_{vac}.

A second general method for damping the long-range oscillations in a simple-metal pair potential was developed by Pettifor and Ward (1984). Their approach is based on removing the logarithmic singularity that occurs in the Lindhard function $f_L(q/2k_F)$ and hence in the normalized energy-wavenumber characteristic $F_N(q, \Omega)$ determining the pair potential $v_2(r, \Omega)$. To do this, they replace f_L by a rational function f_{PW} made up of a ratio of polynomials chosen to fit $f_L(q/2k_F)$ very closely. Specifically, they take

$$f_{PW}(x) = P_1(x)/P_2(x), \tag{3.107}$$

where

$$P_1(x) \equiv 1 + bx^2 + (61/315 + 4b/15 - 3c/35)x^4 + (16/63 + b/3 - 8c/35)x^6 + cx^8 \tag{3.108}$$

and

$$P_2(x) \equiv 1 + (b + 1/3)x^2 + (13/35 + 3b/5 - 3c/35)x^4 + (3/7 + 3b/5 - 9c/35)x^6 \\ + (16/21 + b - 9c/7)x^8 + 3cx^{10}, \tag{3.109}$$

with $x = q/2k_F$, $b = -0.5395$ and $c = 0.3333$. The function $f_{PW}(q/2k_F)$ has the same analytic properties as the Lindhard function $f_L(q/2k_F)$ at small and large q. At $q = 2k_F$ the two functions are exactly equal with $f_{PW}(1) = f_L(1) = 0.5$.

In the Pettifor-Ward (PW) method one can evaluate the integral in Eq. (3.57) for $v_2(r, \Omega)$ analytically by contour methods using the known pole structure of f_{PW}, although this is only straightforward and useful for a local PP. In that case, replacing f_L by f_{PW} in the normal LDA treatment of screening, Pettifor and Ward showed that the pair potential for $r_s > 1.66$ could be written as the sum of three damped oscillatory terms:

$$v_2(r, \Omega) = \frac{(Ze)^2}{r} \sum_{n=1,3} A_n \cos(k_n r + \alpha_n) \exp(-\kappa_n r), \tag{3.110}$$

where A_n, k_n, α_n and κ_n are volume-dependent parameters determined by the poles in f_{PW} and by the value of the local PP at those poles. For a nonlocal PP, simply replacing f_L by f_{PW} in the screening treatment and then generating the pair potential by numerical integration in the usual way from Eq. (3.57) or Eq. (3.59), one automatically obtains a pair potential with damped long-range oscillations, as shown in Fig. 3.12 for Mg. As desired, the PW damping diminishes the minima and maxima in the pair potential more strongly with increasing distance, while accurately preserving the nodes in the potential. Percentagewise, however, the damping only becomes significant beyond about $5R_{WS}$, and even near $10R_{WS}$ the minima and maxima are reduced in magnitude by less than 50%. Consequently, PW damping can only be expected to improve the convergence of calculated properties by a modest amount, and the effective range of the damped pair potential is still quite long.

A third less rigorous but effective and easily implemented method for dealing with the long-range Friedel oscillations in the pair potential is to smoothly, but rapidly, truncate the potential at some optimum cutoff distance R_{cut}. This is possible without incurring large errors because calculated properties carried to long distance tend to oscillate in magnitude about their converged values in a regular way that mirrors the potential oscillations themselves. In this regard, the long-range nodes in the pair potential are determined by the condition $\cos(2k_F r) = 0$, suggesting possible useful values for R_{cut} given by the simple condition $2k_F R_{cut} = (2n+1)\pi/2$, where n a large integer. Using $k_F = (3\pi^2 Z/\Omega)^{1/3}$, this condition reduces to

$$R_{cut}/R_{WS} = (2n+1)\frac{\pi}{4}\left(\frac{4}{9\pi Z}\right)^{1/3}. \qquad (3.111)$$

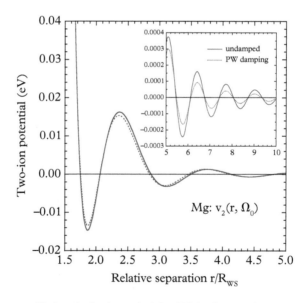

Fig. 3.12 *Long-range oscillations in the first principles GPT pair potential for Mg at equilibrium, $v_2(r, \Omega_0)$, as calculated in pure undamped form with IU screening, and as damped using the PW method discussed in the text.*

3.5.2 Damping and smooth truncation techniques for atomistic simulations

If one restricts one's attention to the regime $R_{\text{cut}} > 8R_{\text{WS}}$ that is typically required for good convergence, then the shortest values of $R_{\text{cut}}/R_{\text{WS}}$ satisfying Eq. (3.111) for $Z = 1$, 2, 3 and 4 are clustered in the narrow range 8.25 ± 0.25. It is not surprising, therefore, that for static lattice properties the simple universal choice of $R_{\text{cut}}/R_{\text{WS}} = 8.25$ often works quite well in practice over wide ranges of such applications in nontransition metals. As required to improve convergence, more distant useful cutoff distances can be found in a similar manner, such as, for example, $R_{\text{cut}}/R_{\text{WS}} = 10.30$, with the solutions to Eq. (3.111) for $Z = 1$, 2, 3 and 4 clustered in the range 10.30 ± 0.25.

More generally, and especially for dynamic simulations such as MD, one needs a smooth cutoff of the pair potential over a small but finite distance, say from $r/R_{\text{WS}} = 8.0$ to 8.5. This can be accomplished by multiplying the full potential $v_2(r, \Omega)$ by an appropriate cutoff function $f_{\text{cut}}(r)$, such as

$$f_{\text{cut}}(r) = \begin{cases} 1 & r < R_0 \\ \left[1 + \alpha(r/R_0 - 1)^2\right] \exp\left[-\alpha(r/R_0 - 1)^2\right] & r > R_0 \end{cases}, \quad (3.112)$$

with parameters $R_0 = 8.0 R_{\text{WS}}$ and $\alpha = 2.5 \times 10^3$. The cutoff function (3.112) has the very desirable properties of preserving the pair potential $v_2(r, \Omega)$ and its first three radial derivatives at $r = R_0$, while rendering $f_{\text{cut}}(R_0 + 0.5 R_{\text{WS}})$ negligibly small. This functional form was actually first introduced in the MGPT for transition metals to smoothly terminate canonical d-state bond integrals, albeit with very different values of R_0 and α (Yang et al., 2001; see Chapter 5). Thus, with appropriate parameters, Eq. (3.112) can be used to advantage in transition and nontransition metals alike.

In simple metals, both the sharp cutoff at $8.25 R_{\text{WS}}$ and the corresponding smooth cutoff that $f_{\text{cut}}(r)$ offers can be used effectively to achieve good convergence in simulations of materials properties with damped as well as undamped pair potentials. This effectiveness is demonstrated in Table 3.2 for representative calculated static lattice properties of Mg. Using an undamped GPT pair potential, the sharp $8.25 R_{\text{WS}}$ cutoff and the smooth $8.0 - 8.5 R_{\text{WS}}$ cutoff [i.e., Eq. (3.112)] give very similar results, with an effectively converged value of the cohesive energy, the unrelaxed vacancy formation energy converged to within 0.05 eV, the structural energies converged to within 0.1 mRy, the shear elastic moduli converged to within about 2 GPa or so, and the phonon frequencies converged to within 0.05 THz. In this regard, elastic moduli are typically among the hardest quantities to converge and short-wavelength phonon frequencies among the easiest. Damping the GPT potential with the PW method first before applying the smooth $8.0 - 8.5 R_{\text{WS}}$ cutoff has the desirable effect of improving the convergence of the shear elastic moduli to within about 1 GPa. At the same time, however, PW damping has the undesirable physical effect of lowering the $E_{\text{fcc}} - E_{\text{hcp}}$ energy difference 40% below its actual converged value. This outcome reflects the fact that the PW damping modifies the pair potential at all distances, and not just in the asymptotic region, so the convergence of any particular property is never fully guaranteed.

We also show in Table 3.2 the effect of moving the smooth cutoff outward in distance to $10.05 - 10.55 R_{\text{WS}}$, where the cutoff range now corresponds to 10.30 ± 0.25. Then $E_{\text{vac}}^{\text{u}}$

Table 3.2 *Representative static lattice properties for ideal hcp Mg at its observed equilibrium volume ($\Omega_0 = 156.8$ a.u.), as calculated from a first-principles GPT pair potential with IU screening, using the long-range termination schemes discussed in the text. Here the cohesive energy (E_{coh}) and unrelaxed vacancy formation energy (E_{vac}^u) are given in eV, the structural energies ($E_{bcc} - E_{hcp}$ and $E_{fcc} - E_{hcp}$) are given in mRy, the shear elastic moduli (C_{44} and C_{66}) are given in GPa, and the transverse and longitudinal optic phonon frequencies (ν_{TO} and ν_{LO}) are given in THz. The real-space results are compared with available converged reciprocal-space results in the final column, obtained from the cohesive energy functional given by Eq. (3.44).*

Termination	Sharp	Smooth	Smooth	Smooth	Smooth	Q space
R_0/R_{WS}	—	8.0	8.0	10.05	10.05	—
R_{cut}/R_{WS}	8.25	8.5	8.5	10.55	10.55	—
Damping	None	None	PW	None	PW	—
$|E_{coh}|$	1.594	1.594	1.589	1.596	1.590	1.594
E_{vac}^u	0.450	0.441	0.438	0.415	0.425	0.410
$E_{bcc} - E_{hcp}$	2.03	2.03	2.04	2.13	2.08	2.09
$E_{fcc} - E_{hcp}$	0.76	0.75	0.48	0.84	0.52	0.82
C_{44}	18.2	19.0	17.8	17.9	16.9	—
C_{66}	23.2	23.0	22.5	23.8	22.6	—
ν_{TO}	3.73	3.74	3.72	3.74	3.72	—
ν_{LO}	7.58	7.59	7.58	7.60	7.59	—

is converged to within 0.02 eV, and with the undamped GPT potential, the structural energies are converged to with 0.05 mRy.

3.6 Higher-order corrections

Generally speaking, one expects the importance of higher-order PP terms in the total energy of a simple metal to depend directly on the effective strength of the form factor $\langle \mathbf{k} + \mathbf{q}| w |\mathbf{k}\rangle$ in reciprocal space, which controls the size of band gaps and the degree of non-NFE character of the electronic structure. Leading candidates for important higher-order PP effects are second-period simple metals, especially Be, with its intrinsically strong $\ell = 1$ PP component and non-NFE electronic structure [e.g., Fig. 2.5(b)], and also heavy polyvalent simple metals with $Z \geq 4$, such as Pb and Bi, with relatively short reciprocal lattice vectors for their observed structures. If the metal is represented by a full energy-dependent, nonlocal PP, with a form factor $w(q) \equiv \langle \mathbf{k}_F + \mathbf{q}| w |\mathbf{k}_F\rangle$ on the free-electron Fermi surface, as in the GPT, one can ask how well the small band gap criterion for a weak PP, namely, $2|w(G_0)|/\varepsilon_F \ll 1$, is satisfied for $G_0 < 2k_F$. In the case of hcp Be, two of the three shortest G_0 are close in magnitude to $2k_F$, where the form factor is positive and large, as shown in Fig. 3.13(a). Consequently, the corresponding band gaps in Be are as large as $0.3\varepsilon_F$, as illustrated in Fig. 3.13(b). In the cases of Pb and Bi, the form factor near $q = 2k_F$ is much smaller, but because $G_0/k_F \propto Z^{-1/3}$, $w(G_0)$ is still large in magnitude for the shortest reciprocal lattice vectors G_0, as shown in Fig. 3.13(a) for fcc Pb. This results in Pb and Bi band gaps similar to those in Be, and much larger than in prototype simple metals like Mg and Al, as is also illustrated in Fig. 3.13(b).

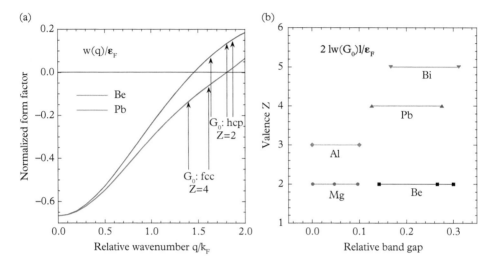

Fig. 3.13 *Measures of PP strength in selected simple metals, as obtained from self-consistent GPT calculations with IU screening at $\Omega = \Omega_0$. (a) Normalized form factor $w(q)/\varepsilon_F$ on the respective free-electron Fermi surfaces of Be and Pb. Also shown are the positions of the shortest reciprocal lattice vectors with magnitude $G_0 < 2k_F$ for the Be hcp structure with $Z = 2$ and for the Pb fcc structure with $Z = 4$. (b) Relative band gap $2\,|w(G_0)|/\varepsilon_F$ obtained for $G_0 < 2k_F$ (solid points) in hcp Mg and Be, and in fcc Al, Pb and (hypothetical) Bi.*

In a plane-wave representation, it is formally straightforward to extend the valence band-structure energy of a simple metal to include higher-order PP contributions, and the general features may be gleaned from the plane-wave parts of our treatment of bulk transition metals in Sec. 2.7.4 of Chapter 2. The density of valence-band states becomes

$$\rho(E) = \rho_0(E) - \frac{2}{\pi}\text{Im}\sum_k \frac{\Sigma_{kk}(E)}{(E - \varepsilon_k)^2}, \tag{3.113}$$

where

$$\Sigma_{kk}(E) = W_{kk} + \sum_{k'} \frac{W_{kk'}W_{k'k}}{E - \varepsilon_{k'}} + \sum_{k',k''} \frac{W_{kk'}W_{k'k''}W_{k''k}}{(E - \varepsilon_{k'})(E - \varepsilon_{k''})} + \cdots \tag{3.114}$$

In the usual way, the valence band-structure energy can be expressed in terms of the integrated density of states, $N(E) = N_0(E) + \delta N_{sp}(E)$, as

$$E_{\text{band}}^{\text{val}} = \frac{3}{5}NZ\varepsilon_F - \int_0^{\varepsilon_F} \delta N_{sp}(E)dE + \delta E_{\text{band}}, \qquad (3.115)$$

where

$$\delta N_{sp}(E) = \frac{2}{\pi}\text{Im}\left(\sum_{\mathbf{k}}\frac{W_{\mathbf{kk}}}{E-\varepsilon_{\mathbf{k}}} + \frac{1}{2}\sum_{\mathbf{k},\mathbf{k}'}\frac{W_{\mathbf{kk}'}W_{\mathbf{k}'\mathbf{k}}}{(E-\varepsilon_{\mathbf{k}})(E-\varepsilon_{\mathbf{k}'})}\right.$$
$$\left. + \frac{1}{3}\sum_{\mathbf{k},\mathbf{k}',\mathbf{k}''}\frac{W_{\mathbf{kk}'}W_{\mathbf{k}'\mathbf{k}''}W_{\mathbf{k}''\mathbf{k}}}{(E-\varepsilon_{\mathbf{k}})(E-\varepsilon_{\mathbf{k}'})(E-\varepsilon_{\mathbf{k}''})} + \cdots\right) \qquad (3.116)$$

and

$$\delta E_{band} = NZ(E_F - \varepsilon_F) - \int_{\varepsilon_F}^{E_F} N(E)dE. \qquad (3.117)$$

However, in addition to the explicit extra complexity introduced in Eq. (3.116) by third order and higher PP contributions, the implicit self-consistent screening in $W_{\mathbf{kk}'}$ itself should, in principle, now be carried to second order for a consistent third-order total energy and to order $n-1$ for a consistent order-n total energy. But even without second-order screening, the third order nonlocal PP terms in Eq. (3.116) produce coupled three-dimensional integrals in $\mathbf{k}$ space and consequently a reciprocal-space total-energy formalism that is effectively intractable without further approximation.

To proceed further with higher-order corrections in a plane-wave representation requires the introduction of a local PP. Historically, the treatment of third-order energy corrections with local PPs in simple metals was first explored in the late 1960s and early 1970s, most notably in the work of Lloyd and Sholl (1968) and Brovman et al. (1971). Lloyd and Sholl derived analytic closed-form results for both the second-order electron screening and for the third-order band-structure energy in reciprocal space, assuming a local PP but without including exchange and correlation in the screening. There were subsequently a number of attempts to use these results to calculate third-order contributions to structural energies and phonon spectra in simple metals, both with and without second-order screening, and with and without exchange and correlation. These applications produced rather mixed results. Third-order band-structure energies were typically found to be a third of corresponding second-order energies, with the largest impact on second-period elements and non-alkali metals. However, neither observed crystal structures nor structural-energy differences could be reliably predicted at either second or third order with the local PPs used. At the same time, and as might be expected from the discussion above, phonon frequencies were found to be more strongly affected by third-order contributions in Be and Pb than in other simple metals

(Bertoni et al., 1973, 1974), although again accurate phonon spectra were not obtained with the third-order corrections calculated in either metal.

Brovman et al. (1971) independently developed a third-order total-energy formalism based on a local PP, which is similar to that of Lloyd and Sholl (1968), but retains only first-order linear screening, and includes corrections for exchange and correlation. Using a carefully constructed empirical local PP for Mg, these authors were able to achieve an accurate description of the phonon spectrum, elastic moduli, equation of state and structural energies in this metal. The quality of the results obtained is comparable to that achieved for Mg with the second-order nonlocal GPT treatment we have discussed above. Thus, in a good simple metal like Mg, the Brovman et al. results would seem to speak to the inevitable slower convergence of a local PP perturbation expansion of the total energy than one with an energy-dependent, nonlocal PP such as that used in GPT.

An additional noteworthy example of using a third-order total-energy approach with first-order linear screening to address the properties of a simple metal is found in the work of Pélissier and Wetta (2001) on the challenging case of Bi. Bismuth has a complex and only partially understood pressure-temperature (PT) phase diagram, with an observed rhombohedral structure at ambient conditions, known as the Bi I phase, and as many as eight additional high-pressure solid phases below 15 GPa, including a high PT bcc phase, which is assumed to be the melting structure above about 4 GPa. Using a two-parameter local MP, these authors were able to explain the observed atomic volume, rhombohedral angle, phonon spectra and heat capacity of the Bi I phase, as well as two observed structure factors in the liquid above the ambient-pressure melt temperature of 545 K. Pélissier and Wetta also addressed the high-pressure equation of state in the Bi I and bcc phases and calculated a bcc melting curve. In a second companion paper (Wetta and Pélissier, 2001), they went on to consider additional dynamic material properties of Bi under shock loading.

Transforming the third-order band-structure energy obtained in reciprocal space into a real-space three-ion interatomic potential v_3 is also feasible for a local PP with first-order linear screening, and this was first accomplished by Hasegawa (1976), with applications to liquid simple metals. Hasegawa, in fact, showed that the third-order band-structure energy in reciprocal space produced both an additional contribution to the two-ion potential v_2 as well as an independent three-ion potential v_3. For both potential contributions, he found the near-neighbor interactions to be attractive in the low electron density metals Na and K. Vasiliu (2003) later found similar attractive short-range third-order interactions in the high electron density metals Be, Al and Pb.

Earlier, Harrison (1973) was able to derive the limiting asymptotic form for all n-ion interatomic potentials occurring in a simple metal directly from Eq. (3.116). Following the notation established in Chapter 1, his resulting n^{th}-ion asymptotic potential for large separation distances can be expressed in the form

$$v_n(ijk...n;\Omega) \sim \frac{4\varepsilon_F}{\pi}\left(\frac{-3\pi Z}{4\varepsilon_F k_F}\right)^n \sum_{paths} \frac{\tilde{w}_{ij}\tilde{w}_{jk}...\tilde{w}_{ni}}{R_{ij}R_{jk}...R_{ni}} \frac{\cos[k_F(R_{ij}+R_{jk}+...+R_{ni})]}{k_F(R_{ij}+R_{jk}+...+R_{ni})},$$

(3.118)

where $\tilde{w}_{ij} = w(q_{ij})$ is a value of the PP form factor determined by the real-space path connecting the ions. Here the path counting follows the geometry given in Fig. 1.2. For $n = 2$ there is a single path between two ions i and j, and Eq. (3.118) reduces to the familiar Friedel form given by Eq. (1.28) with $r = R_{ij}$. For $n = 3$ there is a single path shape (a triangle) for the three ions i, j and k, but two path directions to be counted, so Eq. (3.118) for the three-ion potential becomes

$$v_3(ijk;\Omega) \sim -\frac{27\pi^2 Z^3}{8\varepsilon_F^2} \frac{\tilde{w}_{ij}\tilde{w}_{jk}\tilde{w}_{ki}}{k_F^3 R_{ij}R_{jk}R_{ki}} \frac{\cos[k_F(R_{ij} + R_{jk} + R_{ki})]}{k_F(R_{ij} + R_{jk} + R_{ki})}. \tag{3.119}$$

However, as Hasegawa (1976) pointed out, when this result is extrapolated back to short interatomic distances, Eq. (3.119) qualitative disagrees with the full three-ion potentials in Na and K, producing a repulsive interaction instead of an attractive one. Thus, while the multi-ion potential forms produced from Eq. (3.118) are suggestive, one can't draw any reliable conclusions from them regarding their impact on physical properties.

Because of the difficulty posed by implementing a higher-order nonlocal PP treatment of simple metals, the most promising path forward beyond the second-order nonlocal PP theory developed in this chapter is to augment the plane-wave expansions of the electron density and total energy with appropriate localized basis orbitals that capture important missing physics. This includes localized d states to describe sp-d hybridization and d-band filling in transition-series metals, as is elaborated in Chapters 4 and 5, as well as localized p states for second-period metals to absorb the strong $\ell = 1$ component of the PP, and localized f states for lanthanide and actinide metals, as is discussed in Chapter 12.

4
Interatomic Potentials in Metals with Empty or Filled *d* Bands

In this chapter, we extend the theory of QBIPs for simple metals developed in Chapter 3 to treat pre- and post-transition metals whose electronic structure is strongly affected by nearby empty or filled d bands through sp-d hybridization. This hybridization is captured most completely in the GPT, where one proceeds from first principles in an enlarged basis set of plane waves and localized d states. In this basis, one can simultaneously expand the electron density and total energy in matrix elements of the d-state hybridization potential Δ, defined by Eq. (2.128), and the pseudopotential W. To second order in these matrix elements, the cohesive-energy functional retains the general real-space form of Eq. (3.1), but both the volume term $E_{\text{vol}}(\Omega)$ and the two-ion pair potential $v_2(r, \Omega)$ are substantially altered, including the addition of a short-range d-state overlap potential to v_2 that accounts for the spatial overlap of neighboring d states. Near ambient pressure, the elements most strongly affected by the sp-d hybridization and d-state overlap are the nominal empty-d-band heavy alkaline earths metals (Ca, Sr and Ba) and the nominal filled-d-band noble metals (Cu, Ag and Au) and group-IIB metals (Zn, Cd and Hg). Of these elements, Ca and Zn are ideal prototype materials, with their electronic structures well matched to the theory and with critical contributions from the sp-d hybridization and d-state overlap to structural phase stability and other basic properties. In some of the remaining materials, especially Ba and Cu, the d bands are sufficiently close to the Fermi level that one is already at the onset of transition-metal behavior (Moriarty, 1982, 1986b), so such cases are more challenging and require special attention. In all cases, machine learning can be used to advantage in optimizing the choice of d basis state and the quality of an empty-d-band or filled-d-band description of the metal. In addition, the heavy $5d$ elements beginning with Au and Hg require at least a semi-relativistic treatment of their core and d states in constructing Δ and W. In the case of Hg, for example, a semi-relativistic filled-d-band GPT treatment is both necessary and sufficient to explain the complex phase diagram of that material (Moriarty, 1988b).

At high pressure, the empty-d-band GPT treatment can also be applied effectively to the heavy alkali metals (K, Rb and Cs) as well as to the third-period simple metals (Na, Mg, Al and Si). In the latter metals, in fact, this treatment has been used to predict a series of high-pressure phase transitions (Moriarty and McMahan, 1982; McMahan and Moriarty, 1983), several of which have since been confirmed experimentally, as discussed in Chapter 6. In addition, in the empty-d-band GPT limit, the most important effects of the sp-d hybridization can be folded back into a generalized PP, which qualitatively adds the missing d-state component to the AHS PP, Eq. (2.56), for third-period metals. As indicated in Chapter 3, extending the empty-d-band GPT treatment of Al back to ambient pressure provides a refined description that compares more closely with the DRT model potential description, which directly includes an $\ell = 2$ component of the MP.

Similarly at high pressure, the filled d-band GPT treatment of post-transition metals can be usefully extended to the remaining group-III metals (Ga, In and Tl), group-IV metals (Sn and Pb) and group-V metals (Bi), although as indicated in Chapter 3, Pb and Bi are particularly challenging cases because of their strong PPs. In the filled d-band limit of the GPT, the effects of the sp-d hybridization and d-state overlap involve more than just a modified PP, but can be treated directly without any major complication of the theory.

4.1 Inclusion of sp-d hybridization and d-state overlap in the GPT cohesive-energy functional

The starting equation for the cohesive energy of an empty- or filled-d-band metal follows from Eqs. (2.36) and (2.37) applied to $Z_d = 0$ or $Z_d = 10$ transition-series metals:

$$\begin{aligned} E_{\text{coh}} = \frac{1}{N} \Bigg\{ & \frac{1}{2} \sum_{i,j}{}' \frac{(Ze)^2}{|\mathbf{R}_i - \mathbf{R}_j|} + E_{\text{band}}^{\text{val}} - \frac{1}{2} n_{\text{val}} V_{\text{val}} + n_{\text{val}} [\varepsilon_{\text{xc}}(n_{\text{val}}) - \mu_{\text{xc}}(n_{\text{val}})] \\ & + \frac{1}{2} \sum_{i,j}{}' \left[\frac{(Z_a - Z)^2 e^2}{|\mathbf{R}_i - \mathbf{R}_j|} - 2 n_i \frac{(Z_a - Z) e^2}{|\mathbf{r} - \mathbf{R}_j|} + n_i v_j \right] \\ & + \sum_{i,j}{}' \left[\frac{(Z_a - Z) Z e^2}{|\mathbf{R}_i - \mathbf{R}_j|} - n_i \frac{Z e^2}{|\mathbf{r} - \mathbf{R}_j|} \right] \\ & + \frac{1}{2} \sum_{i,j}{}' [n_i \mu_{\text{xc}}^*(n_j) + \delta \varepsilon_{\text{xc}}^*(n_i, n_j) - n_{\text{val}} \delta \mu_{\text{xc}}^*(n_i, n_j)] \Bigg\} - E_{\text{bind}}^{\text{atom}}(Z), \end{aligned} \quad (4.1)$$

where in this notation recall that in all terms involving n_{val} or n_i there is an implied integration over all space. Also, as for a simple metal, the valence Z here is a constant determined from the Periodic Table and is the same in the free atom and the metal, so $E_{\text{bind}}^{\text{atom}}(Z)$ is a constant as well. The top line in Eq. (4.1) is analogous to the binding energy of a simple metal, with the valence band-structure energy $E_{\text{band}}^{\text{val}}$ and electron density n_{val}

the main quantities to be determined. The second and third lines in Eq. (4.1) include electrostatic and xc corrections arising from overlapping occupied d states. Recall from Eq. (2.32) that the total core electron density $n_i \equiv n_{\text{core}}(\mathbf{r} - \mathbf{R}_i)$ includes contributions from both inner core electrons and from Z_d d electrons per atom occupying the d-state basis orbitals $\phi_d(\mathbf{r})$. For an empty-d-band (EDB) metal $Z_d = 0$, and the electrostatic and xc corrections in Eq. (4.1) formally vanish via the small-core approximation. For a filled-d-band (FDB) metal, on the other hand, $Z_d = 10$ and the electrostatic and xc correction terms directly contribute to a new d-state overlap potential v_{ol} that will add to the pair potential v_2. In both EDB and FDB metals, $E_{\text{band}}^{\text{val}}$ will include sp-d hybridization contributions to a modified volume term $E_{\text{vol}}^q(\Omega)$ and energy-wavenumber characteristic $F(q, \Omega)$, as well as d-state overlap contributions to $v_{\text{ol}}(r, \Omega)$. We can thus anticipate that the reciprocal-space representation of the cohesive-energy functional will be generalized to the form

$$E_{\text{coh}} = E_{\text{vol}}^q(\Omega) + \sum_{q}{}' |S(\mathbf{q})|^2 F(q, \Omega) + \delta E_{\text{es}}(Z^*, \Omega) + \frac{1}{2N} \sum_{i,j}{}' v_{\text{ol}}(R_{ij}, \Omega). \quad (4.2)$$

The valence electron density n_{val} is still in the general form given by Eq. (3.3), but will include sp-d hybridization contributions to a modified self-consistent screening electron density δn_{scr}, as well as localized d-state and hybridization contributions to a modified orthogonalization-hole density δn_{oh}. In the next section, we consider calculation of the modified valence band-structure energy $E_{\text{band}}^{\text{val}}$ for EDB and FDB metals within the GPT. Corresponding calculation of the modified screening and orthogonalization-hole densities are considered in Sec. 4.1.2, and the completion of the reciprocal-space cohesive-energy calculation is given in Sec. 4.1.3. Final evaluation of the real-space volume term and pair potential is then completed in Sec. 4.1.4.

4.1.1 Valence band-structure energy

To evaluate the valence band structure energy for an EDB or FDB metal in the GPT, it is convenient to work in terms of the valence plus d-state density of states $\rho(E)$, where $E_{\text{band}}^{\text{val}}$ is given by Eq. (2.177):

$$E_{\text{band}}^{\text{val}} = \int_0^{E_{\text{F}}} E\rho(E) dE - NZ_d E_d, \quad (4.3)$$

with the zero of energy chosen at the bottom of the valence plus d-state energy bands. The form needed here for $\rho(E)$ can be developed in two steps from the more general transition-metal result, Eq. (2.172). First, one can formally develop the $5N \times 5N$ d-band determinant $D(E)$ appearing in Eq. (2.172) in powers of the d-state inter-site coupling strength $V'_{dd'} + \Gamma_{dd'}$. Following Eq. (2.172) and the pseudo-Green's-function treatment of Moriarty (1972a), one has

$$\rho(E) = \rho_0(E) - \frac{2}{\pi}\text{Im}\left(\sum_k \frac{\Sigma_{kk}(E)}{(E-\varepsilon_k)^2} + \frac{d}{dE}\ln[D(E)]\right)$$

$$= \rho_0(E) - \frac{2}{\pi}\text{Im}\sum_k \frac{\Sigma_{kk}(E)}{(E-\varepsilon_k)^2} - \frac{2}{\pi}\text{Im}\left[\sum_d \frac{d}{dE}\ln(E-E_d-\Gamma_{dd}-\Lambda_{dd})\right.$$

$$\left. + \frac{d}{dE}\ln\left(1 - \frac{1}{2}{\sum_{d,d'}}' \frac{(V'_{dd'}+\Gamma_{dd'})(V'_{d'd}+\Gamma_{d'd})}{(E-E_d-\Gamma_{dd}-\Lambda_{dd})(E-E_{d'}-\Gamma_{d'd'}-\Lambda_{d'd'})} + \cdots\right)\right],$$

(4.4)

where $\Sigma_{kk'}, V'_{dd'}, \Gamma_{dd'}$ and $\Lambda_{dd'}$ are defined by Eqs. (2.173), (2.134), (2.175) and (2.176), respectively. Recall that in this notation, each d-state summation includes a sum over both ion sites and individual d quantum numbers on those sites, but the prime on the final d-state summations excludes only terms where d and d' refer to the same ion site. Also in this expansion, we have implicitly assumed a specific order of smallness for the sp-d hybridization and dd' overlap quantities. In comparison to the first-order PP $W_{kk'}$ contained in $\Sigma_{kk'}$, we have taken Γ_{dd}, $V'_{dd'}$ and $\Gamma_{dd'}$ to be first-order quantities, and Λ_{dd} to be second order.

Next one notes that for an EDB or FDB metal, $|E_F - E_d|$ is large in comparison to the effective sp-d hybridization strength embodied in Γ_{dd}. This allows one to expand out the terms in square brackets in Eq. (4.4) further in powers of $(E-E_d)^{-1}$ to obtain $\rho(E)$ in the useful form

$$\rho(E) = \rho_0(E) + \rho_d(E) + \frac{d}{dE}\delta N(E), \quad (4.5)$$

where we have redefined ρ_d for the EDB or FDB case as

$$\rho_d(E) \equiv -\frac{2}{\pi}\text{Im}\sum_d \frac{1}{E-E_d}, \quad (4.6)$$

and have further defined $\delta N(E) = \delta N_{sp}(E) + \delta N_d(E)$ as

$$\delta N(E) \equiv \frac{2}{\pi}\text{Im}\left[\sum_k \frac{W_{kk}}{E-\varepsilon_k} + \frac{1}{2}\sum_k \frac{W_{kk}^2}{(E-\varepsilon_k)^2} + {\sum_{k,k'}}' \frac{W_{kk'}W_{k'k}}{(\varepsilon_k-\varepsilon_{k'})(E-\varepsilon_k)}\right.$$

$$+ \sum_d \frac{\Gamma_{dd}+\Lambda_{dd}}{E-E_d} + \frac{1}{2}\sum_d \frac{\Gamma_{dd}^2}{(E-E_d)^2}$$

$$\left. + \frac{1}{2}{\sum_{d,d'}}' \frac{(V'_{dd'}+\Gamma_{dd'})(V'_{d'd}+\Gamma_{d'd})}{(E-E_d)(E-E_{d'})} + \cdots\right]. \quad (4.7)$$

4.1.1 Valence band-structure energy

In deriving Eq. (4.7), we have used Eq. (2.173) for $\Sigma_{kk}(E)$ together with the identity

$$\frac{1}{(E-\varepsilon_k)(E-\varepsilon_{k'})} = \frac{1}{(\varepsilon_k-\varepsilon_{k'})}\left(\frac{1}{E-\varepsilon_k} - \frac{1}{E-\varepsilon_{k'}}\right) \quad (4.8)$$

to obtain the terms involving W_{kk} and $W_{kk'}W_{k'k}$. These terms also comprise $\delta N_{sp}(E)$ in Eq. (2.181). Recalling that E represents $E + i0^+$, the analytic structure of Eqs. (4.6) and (4.7) is particularly simple, with the energy dependence of all terms consisting of low-order poles on the real axis at ε_k and E_d. This ensures that the terms in E_{band}^{val} can be directly evaluated by contour integration.

The special DOS expansion represented by Eqs. (4.5)–(4.7) requires replacing the true Fermi energy E_F in Eq. (4.3) for E_{band}^{val} by its free-electron value ε_F, as was done implicitly in the treatment of simple metals in Chapter 3. With E_F replaced by ε_F, we then insert Eq. (4.5) for $\rho(E)$ into Eq. (4.3) and integrate the $Ed(\delta N)/dE$ term by parts to obtain

$$E_{band}^{val} = \int_0^{\varepsilon_F} E\left[\rho_0(E) + \rho_d(E) + \frac{d}{dE}\delta N(E)\right] dE - NZ_d E_d$$

$$= \frac{3}{5}NZ\varepsilon_F - \int_0^{\varepsilon_F} \delta N(E)dE + \varepsilon_F \delta N(\varepsilon_F), \quad (4.9)$$

noting that the integral of $E\rho_d(E)$ is just $NZ_d E_d$. At the same time, the corresponding conservation of valence plus d electrons requires that

$$(Z+Z_d)N = \int_0^{\varepsilon_F} \rho(E)dE = \int_0^{\varepsilon_F} [\rho_0(E) + \rho_d(E)] dE + \delta N(\varepsilon_F). \quad (4.10)$$

Since the integral over $\rho_0(E)$ in Eq. (4.10) yields ZN and the integral over $\rho_d(E)$ yields $Z_d N$, one is left with the important constraint

$$\delta N(\varepsilon_F) = 0. \quad (4.11)$$

Thus, the final term on the second line of Eq. (4.9) must also vanish.

Next, we neglect the small higher-order difference between E_d and $E_{d'}$ in the second-order terms of Eq. (4.7), and combine the final two terms in that equation. Then using Eq. (4.7) so modified in Eq. (4.9), and imposing the constraint (4.11), the valence band-structure energy can be calculated to second order as

$$E_{\text{band}}^{\text{val}} = \frac{3}{5} N Z \varepsilon_F - \frac{2}{\pi} \text{Im} \int_0^{\varepsilon_F} \left[\sum_k \frac{W_{kk}}{E - \varepsilon_k} + \sum_{k,k'}{}' \frac{W_{kk'} W_{k'k}}{(\varepsilon_k - \varepsilon_{k'})(E - \varepsilon_k)} + \sum_d \frac{\Gamma_{dd} + \Lambda_{dd}}{E - E_d} \right.$$

$$\left. + \frac{1}{2} \sum_{d,d'} \frac{\Gamma_{dd'} \Gamma_{d'd}}{(E - E_d)^2} + \frac{1}{2} \sum_{d,d'}{}' \frac{V'_{dd'} V'_{d'd} + V'_{dd'} \Gamma_{d'd} + \Gamma_{dd'} V'_{d'd}}{(E - E_d)^2} \right] dE. \quad (4.12)$$

In this regard, care has been taken to remove second-order pole terms from Eq. (4.12) such as $W_{kk}^2/(E-\varepsilon_k)^2$, which result in only a surface contribution evaluated on the Fermi sphere because W_{kk}^2 is independent of energy. The sum of all such surface contributions formally must vanish because of the constraint (4.11).

Evaluation of the surviving PP terms in Eq. (4.12) is straightforward using standard contour integration:

$$-\frac{2}{\pi} \text{Im} \int_0^{\varepsilon_F} \left[\sum_k \frac{W_{kk}}{E - \varepsilon_k} + \sum_{k,k'}{}' \frac{W_{kk'} W_{k'k}}{(\varepsilon_k - \varepsilon_{k'})(E - \varepsilon_k)} \right] dE$$

$$= N \frac{2\Omega}{(2\pi)^3} \left[\int W_{kk} \Theta^<(\mathbf{k}) d\mathbf{k} + \sum_q{}' \int \frac{W_{k+qk} W_{kk+q}}{(\varepsilon_k - \varepsilon_{k+q})} \Theta^<(\mathbf{k}) d\mathbf{k} \right], \quad (4.13)$$

where in the last term on the second line we have replaced $\mathbf{k}'$ with $\mathbf{k+q}$. The two terms on the RHS of Eq. (4.13) are, of course, just the familiar first-order structure-independent and second-order structure-dependent PP contributions to $E_{\text{band}}^{\text{val}}$ one would have in the case of a simple metal. If the PP W is of the non-Hermitian AHS form (2.54), as is the case in the GPT, one has the property $W_{kk+q} = W_{k+qk}^* - (\varepsilon_k - \varepsilon_{k+q}) P_{kk+q}^c$, which is just the multi-site version of Eq. (3.42). Using this result in Eq. (4.13), one can readily derive the small modifications $W_{kk} \to W_{kk}(1 + P_{kk}^c)$, with $P_{kk}^c \equiv \langle \mathbf{k} | P_c | \mathbf{k} \rangle$, and $W_{k+qk} W_{kk+q} \to |W_{k+qk}|^2$ that one has for the GPT.

To evaluate the remaining d-state terms in Eq. (4.12), we first need to expose their full energy dependence by expressing Γ_{dd}, Λ_{dd}, $\Gamma_{dd'}$ and $V'_{dd'}$ in terms of the matrix elements $\Delta_{dd'}$, $S_{dd'}$, Δ_{kd} and S_{kd} associated with the hybridization potential Δ, using Eqs. (2.134) and (2.137). We can then write the first-order, structure-independent d-state term involving Γ_{dd} in Eq. (4.12) in the mathematically useful form

$$\frac{\Gamma_{dd}}{E - E_d} = \sum_k \left[\frac{(E - E_d) S_{dk} S_{kd} + S_{dk} \Delta_{kd} + \Delta_{dk} S_{kd}}{E - \varepsilon_k} + \frac{\Delta_{dk} \Delta_{kd}}{(E - \varepsilon_k)(E - E_d)} \right]. \quad (4.14)$$

The last contribution on the RHS of Eq. (4.14) can be reduced further to a sum of first-order poles by using the additional identity

$$\frac{1}{(E - \varepsilon_k)(E - E_d)} = \frac{1}{(\varepsilon_k - E_d)} \left(\frac{1}{E - \varepsilon_k} - \frac{1}{E - E_d} \right). \quad (4.15)$$

4.1.1 Valence band-structure energy

Then the corresponding contribution to $E_{\text{band}}^{\text{val}}$ is easily obtained by contour integration:

$$-\frac{2}{\pi}\text{Im}\int_0^{\varepsilon_F}\sum_d \frac{\Gamma_{dd}}{E-E_d}dE = N\frac{2\Omega}{(2\pi)^3}\left(\int\sum_d[(\varepsilon_{\mathbf{k}}-E_d)S_{\mathbf{k}d}S_{d\mathbf{k}}+\Delta_{\mathbf{k}d}S_{d\mathbf{k}}\right.$$

$$\left.+S_{\mathbf{k}d}\Delta_{d\mathbf{k}}]\Theta^<(\mathbf{k})d\mathbf{k}+\int\sum_d\frac{\Delta_{\mathbf{k}d}\Delta_{d\mathbf{k}}}{\varepsilon_{\mathbf{k}}-E_d}\Theta_>^<(\mathbf{k})d\mathbf{k}\right),\quad(4.16)$$

where we have defined the combined step function

$$\Theta_>^<(\mathbf{k})\equiv\Theta(\varepsilon_F-\varepsilon_{\mathbf{k}})-\Theta(\varepsilon_F-E_d).\quad(4.17)$$

This function arises directly from Eq. (4.15) as applied to Eq. (4.14). Thus, for EDB metals, where $E_d > \varepsilon_F$,

$$\Theta_>^<(\mathbf{k}) = \Theta^<(\mathbf{k}) = \begin{cases} 1 & k \leq k_F \\ 0 & k > k_F \end{cases},\quad(4.18)$$

and the second **k**-space integral in Eq. (4.16) is to be performed *inside* the free-electron Fermi sphere, just as the first **k**-space integral is in that equation. For FDB metals, on the other hand, where $\varepsilon_F > E_d$,

$$\Theta_>^<(\mathbf{k}) = -\Theta^>(\mathbf{k}) = -\begin{cases} 0 & k \leq k_F \\ 1 & k > k_F \end{cases},\quad(4.19)$$

and the second **k**-space integral in Eq. (4.16) is to be performed *outside* the free-electron Fermi sphere, with the inclusion of an overall minus sign in front of the integral. Note that in both cases the integration region in the second **k**-space integral is such as to avoid the singularity at $\varepsilon_{\mathbf{k}} = E_d$.

Processing the remaining second-order d-state quantities in Eq. (4.12) involving Λ_{dd}, $\Gamma_{dd'}$ and $V'_{dd'}$ is quite similar and we only briefly outline their evaluation. The Λ_{dd} term enters in the form

$$\frac{\Lambda_{dd}}{E-E_d} = \sum_{\mathbf{k},\mathbf{k}'}\left[\frac{(E-E_d)S_{d\mathbf{k}}W_{\mathbf{k}\mathbf{k}'}S_{\mathbf{k}'d}+S_{d\mathbf{k}}W_{\mathbf{k}\mathbf{k}'}\Delta_{\mathbf{k}'d}+\Delta_{d\mathbf{k}}W_{\mathbf{k}\mathbf{k}'}S_{\mathbf{k}'d}}{(E-\varepsilon_{\mathbf{k}})(E-\varepsilon_{\mathbf{k}'})}\right.$$

$$\left.+\frac{\Delta_{d\mathbf{k}}W_{\mathbf{k}\mathbf{k}'}\Delta_{\mathbf{k}'d}}{(E-\varepsilon_{\mathbf{k}})(E-\varepsilon_{\mathbf{k}'})(E-E_d)}\right].\quad(4.20)$$

Here, after the energy integration in Eq. (4.12), one obtains two structure-independent contributions from the $\mathbf{k} = \mathbf{k}'$ terms and four additional structure-independent contributions resulting from the non-Hermiticity of the PP. The remaining structure-dependent $\mathbf{k} \neq \mathbf{k}'$ terms can be readily evaluated using Eqs. (4.8) and (4.15) to produce the needed first-order poles.

The $\Gamma_{dd'}\Gamma_{d'd}$ hybridization term in Eq. (4.12) is evaluated in the form

$$\frac{\Gamma_{dd'}\Gamma_{d'd}}{(E-E_d)^2} = \sum_{\mathbf{k},\mathbf{k}'} \left[\frac{1}{E-\varepsilon_{\mathbf{k}}}\left((E-E_d)S_{d\mathbf{k}}S_{\mathbf{k}d'} + S_{d\mathbf{k}}\Delta_{\mathbf{k}d'} + \Delta_{d\mathbf{k}}S_{\mathbf{k}d'} + \frac{\Delta_{d\mathbf{k}}\Delta_{\mathbf{k}d'}}{E-E_d}\right)\right.$$
$$\left.\times \frac{1}{E-\varepsilon_{\mathbf{k}'}}\left((E-E_d)S_{d'\mathbf{k}'}S_{\mathbf{k}'d} + S_{d'\mathbf{k}'}\Delta_{\mathbf{k}'d} + \Delta_{d'\mathbf{k}'}S_{\mathbf{k}'d} + \frac{\Delta_{d'\mathbf{k}'}\Delta_{\mathbf{k}'d}}{E-E_d}\right)\right]. \tag{4.21}$$

In this case, the $\mathbf{k} = \mathbf{k}'$ terms yield ten further structure-independent contributions, and again the $\mathbf{k} \neq \mathbf{k}'$ structure-dependent terms are easily evaluated with the aid of Eqs. (4.8) and (4.15).

Finally, in the group of dd' overlap terms in Eq. (4.12) we note that $V'_{dd'}\Gamma_{d'd}$ and $\Gamma_{dd'}V'_{d'd}$ will produce complex conjugates of one another, so one only needs to consider

$$\frac{V'_{dd'}V'_{d'd} + V'_{dd'}\Gamma_{d'd}}{(E-E_d)^2} = S_{dd'}S_{d'd} + \frac{S_{dd'}\Delta_{d'd} + \Delta_{dd'}S_{d'd}}{E-E_d} + \frac{\Delta_{dd'}\Delta_{d'd}}{(E-E_d)^2}$$
$$- \left(S_{dd'} + \frac{\Delta_{dd'}}{E-E_d}\right) \times \sum_{\mathbf{k}} \frac{1}{E-\varepsilon_{\mathbf{k}}}\left((E-E_d)S_{d'\mathbf{k}}S_{\mathbf{k}d} + S_{d'\mathbf{k}}\Delta_{\mathbf{k}d}\right.$$
$$\left. + \Delta_{d'\mathbf{k}}S_{\mathbf{k}d} + \frac{\Delta_{d'\mathbf{k}}\Delta_{\mathbf{k}d}}{E-E_d}\right). \tag{4.22}$$

Of the first three terms on the RHS of Eq. (4.22), note that only the cross terms involving $S_{dd'}\Delta_{d'd}$ and its complex conjugate for a FDB metal will survive the contour integration in Eq. (4.12). The remaining terms are easily evaluated using Eq. (4.15).

The large number of individual sp-d hybridization and dd' overlap terms that are so generated from the energy integrations in Eq. (4.12) may be efficiently recombined into three major components of $E_{\text{band}}^{\text{val}}$:

$$E_{\text{band}}^{\text{val}} = N(E_{\text{fe}} + E_{\text{bs}} + E_{\text{ol}}). \tag{4.23}$$

Here the free-electron energy E_{fe} is the collection of explicitly structure-independent contributions from Eqs. (4.13), (4.16), (4.20) and (4.21) that will enter the final volume term E_{vol}^q in the cohesive-energy functional (4.2):

$$E_{\text{fe}} = \frac{3}{5}Z\varepsilon_{\text{F}} + \frac{2\Omega}{(2\pi)^3}\left[\int W_{\mathbf{kk}}^0(1+P_{\mathbf{kk}})\Theta^<(\mathbf{k})d\mathbf{k} + \int \sum_d \frac{\Delta_{\mathbf{k}d}\Delta_{d\mathbf{k}}}{\varepsilon_{\mathbf{k}}-E_d}\right.$$
$$\left.\times \left(1 + P_{\mathbf{kk}} - \frac{W_{\mathbf{kk}}^0}{\varepsilon_{\mathbf{k}}-E_d} - \sum_{d'}\frac{\Delta_{\mathbf{k}d'}\Delta_{d'\mathbf{k}}}{(\varepsilon_{\mathbf{k}}-E_d)^2}\right)\Theta^<(\mathbf{k})d\mathbf{k}\right], \tag{4.24}$$

where $W_{\mathbf{kk}}^0$ is analogous to the transition-metal PP defined by Harrison (1969),

$$W_{\mathbf{kk}}^0 \equiv W_{\mathbf{kk}} + \sum_d [(\varepsilon_{\mathbf{k}}-E_d)S_{\mathbf{k}d}S_{d\mathbf{k}} + \Delta_{\mathbf{k}d}S_{d\mathbf{k}} + S_{\mathbf{k}d}\Delta_{d\mathbf{k}}], \tag{4.25}$$

and $P = P_c + P_d$ is the inner-core plus d-state projection operator such that

$$P_{\mathrm{kk}} = \langle \mathbf{k}|\, P\, |\mathbf{k}\rangle = \langle \mathbf{k}|\, P_c\, |\mathbf{k}\rangle + \sum_d S_{kd} S_{dk}. \tag{4.26}$$

The band-structure energy E_{bs} in Eq. (4.23) includes second-order, structure-dependent PP and hybridization contributions from Eqs. (4.13), (4.20) and (4.21):

$$\begin{aligned}
E_{\mathrm{bs}} = \frac{2\Omega}{(2\pi)^3} \sum_{\mathbf{q}}{}' \Bigg[& \int \frac{\left|W^0_{\mathbf{k}+\mathbf{q}\mathbf{k}}\right|^2}{(\varepsilon_{\mathbf{k}} - \varepsilon_{\mathbf{k}+\mathbf{q}})} \Theta^<(\mathbf{k})d\mathbf{k} + \int \frac{1}{(\varepsilon_{\mathbf{k}} - \varepsilon_{\mathbf{k}+\mathbf{q}})} \\
& \times \left(\sum_d \frac{\Delta_{kd}\Delta_{dk+q} W^0_{\mathbf{k}+\mathbf{q}\mathbf{k}} + \mathrm{c.c.}}{\varepsilon_{\mathbf{k}} - E_d} + \left| \sum_d \frac{\Delta_{kd}\Delta_{dk+q}}{\varepsilon_{\mathbf{k}} - E_d} \right|^2 \right) \Theta^<_>(\mathbf{k})d\mathbf{k} \Bigg].
\end{aligned} \tag{4.27}$$

Equation (4.27) now includes the electronic-structure ingredients needed to establish the energy-wavenumber characteristic $F(q, \Omega)$ in Eq. (4.2). Lastly, the contributions to the overlap energy E_{ol} in Eq. (4.23) derive entirely from Eq. (4.22):

$$\begin{aligned}
E_{\mathrm{ol}} = \frac{1}{2} \sum_{d,d'}{}' \Bigg[& 2(S_{dd'}\Delta_{d'd} + \Delta_{dd'}S_{d'd})\Theta(\varepsilon_F - E_d) \\
& - \frac{2\Omega}{(2\pi)^3} \Bigg(\int \{ S_{dd'}[(\varepsilon_{\mathbf{k}} - E_d) S_{d'k} S_{kd} + S_{d'k}\Delta_{kd} \\
& + \Delta_{d'k} S_{kd}] + \Delta_{dd'} S_{d'k} S_{kd} \} \Theta^<(\mathbf{k})d\mathbf{k} \\
& + \int \Bigg\{ \frac{S_{dd'}\Delta_{d'k}\Delta_{kd} + \Delta_{dd'}(S_{d'k}\Delta_{kd} + \Delta_{d'k}S_{kd})}{\varepsilon_{\mathbf{k}} - E_d} \\
& + \frac{\Delta_{dd'}\Delta_{d'k}\Delta_{kd}}{(\varepsilon_{\mathbf{k}} - E_d)^2} \Bigg\} \Theta^<_>(\mathbf{k})d\mathbf{k} + \mathrm{c.c.} \Bigg) \Bigg].
\end{aligned} \tag{4.28}$$

The above results for E_{fe}, E_{bs} and E_{ol} are the same as those obtained by Moriarty (1972a), apart from the removal of the d-state energy $Z_d E_d$ from E_{fe}, which stems from the starting definition of $E^{\mathrm{val}}_{\mathrm{band}}$ in Eq. (4.3). The present FBD results for E_{fe} and E_{bs} are also substantially the same as those obtained earlier by Harrison (1969) via an extension of PP perturbation theory. The main differences concern the treatment of the PP terms, especially the diagonal matrix element W^0_{kk}, which Harrison included in his definition of the plane-wave energy $E_{\mathbf{k}} \equiv \varepsilon_{\mathbf{k}} + W^0_{\mathrm{kk}}$ and used in place of $\varepsilon_{\mathbf{k}}$. Harrison did not, however, consider the important practical complication of spatially overlapping d-states and the corresponding overlap energy E_{ol}, nor did not he specifically treat the EDB case.

There are several remaining steps to complete the development of the cohesive energy functional within the GPT for EDB and FDB metals. These steps require first separating the d-state energy E_d into its large volume component E_d^{vol} and a smaller first-order structural component E_d^{struc}, as in Eq. (2.184). We also make a corresponding separation of the sp-d hybridization matrix element Δ_{kd}:

$$\Delta_{kd} = \Delta_{kd}^{\text{vol}} + \Delta_{kd}^{\text{struc}}. \tag{4.29}$$

This will permit us to remove hidden structural components in the nominal first-order terms of E_{fe}, and then proceed to define final structure-independent functions in terms of E_d^{vol} and Δ_{kd}^{vol} for both the valence electron density and the cohesive energy. In this regard, we first address the impact of sp-d hybridization on the valence electron density and the self-consistent electron screening in the next section, before returning to the electrostatic energy, volume term and energy-wavenumber characteristic in Sec. 4.1.3. The practical details of defining the d basis states and hybridization potential so that a clean separation of E_d and Δ_{kd} into volume and structure components can be made will be discussed in Sec. 4.2.

4.1.2 Valence electron density and self-consistent screening

In this section, we consider the modifications in the electron screening density $\delta n_{\text{scr}}(\mathbf{r})$ and the orthogonalization-hole density $\delta n_{\text{oh}}(\mathbf{r})$ due to sp-d hybridization for EDB and FDB metals. Starting from Eq. (2.196) for $\delta n_{\text{scr}}(\mathbf{r})$, expanding in powers of $(E - E_d^{\text{vol}})^{-1}$, and using the identity (4.8), one has to first order

$$\delta n_{\text{scr}}(\mathbf{r}) = -\frac{4}{\pi}\text{Im}\int_0^{\varepsilon_F}\sum_{\mathbf{k},\mathbf{k}'}{}'\left[\frac{\langle \mathbf{r}|\mathbf{k}'\rangle W_{\mathbf{k}'\mathbf{k}}\langle \mathbf{k}|\mathbf{r}\rangle}{(\varepsilon_k - \varepsilon_{k'})(E - \varepsilon_k)} + \sum_d \frac{\langle \mathbf{r}|\mathbf{k}'\rangle v'_{\mathbf{k}'d}v'_{d\mathbf{k}}\langle \mathbf{k}|\mathbf{r}\rangle}{(\varepsilon_k - \varepsilon_{k'})(E - \varepsilon_k)(E - E_d^{\text{vol}})}\right]dE. \tag{4.30}$$

Recall here that v' is the volume component of V', such that $v'_{kd} = -(E - E_d^{\text{vol}})S_{kd} - \Delta_{kd}^{\text{vol}}$. Performing the contour integrations in Eq. (4.30) with $\mathbf{k}' = \mathbf{k} + \mathbf{q}$, combining PP and hybridization terms, and noting that $\langle \mathbf{r}|\mathbf{k}+\mathbf{q}\rangle\langle \mathbf{k}|\mathbf{r}\rangle = (N\Omega)^{-1}\exp(i\mathbf{q}\cdot\mathbf{r})$, then gives the net result

$$\delta n_{\text{scr}}(\mathbf{r}) = \sum_q{}'\frac{4}{(2\pi)^3}\left[\int \frac{W^0_{\mathbf{k}+\mathbf{q}\mathbf{k}}}{\varepsilon_k - \varepsilon_{k+q}}\Theta^<(\mathbf{k})d\mathbf{k}\right.$$
$$\left.+\int\sum_d \frac{\Delta_{k+qd}^{\text{vol}}\Delta_{dk}^{\text{vol}}}{(\varepsilon_k - \varepsilon_{k+q})(\varepsilon_k - E_d^{\text{vol}})}\Theta^<_>(\mathbf{k})d\mathbf{k}\right]\exp(i\mathbf{q}\cdot\mathbf{r}). \tag{4.31}$$

The matrix elements in Eq. (4.31) are still in a multi-ion format, but one can extract a structure factor $S(\mathbf{q})$ from both the PP term involving $W^0_{\mathbf{k}+\mathbf{q}\mathbf{k}}$ and the hybridization term involving $\Delta_{k+qd}^{\text{vol}}\Delta_{dk}^{\text{vol}}$. This allows one to put $\delta n_{\text{scr}}(\mathbf{r})$ in the desired form (3.24) that separates out the atomic structure contained in $S(\mathbf{q})$ from the Fourier transform of the single-ion screening density $n_{\text{scr}}(q)$. One then has

4.1.2 Valence electron density and self-consistent screening

$$n_{\text{scr}}(q) = \frac{4}{(2\pi)^3}\left[\int \frac{w_0(\mathbf{k},\mathbf{q})}{\varepsilon_\mathbf{k} - \varepsilon_{\mathbf{k}+\mathbf{q}}}\Theta^<(\mathbf{k})d\mathbf{k} + \int \frac{h_1(\mathbf{k},\mathbf{q})}{\varepsilon_\mathbf{k} - \varepsilon_{\mathbf{k}+\mathbf{q}}}\Theta^<_>(\mathbf{k})d\mathbf{k}\right], \quad (4.32)$$

where, following Moriarty (1982), we have defined a transition-metal PP form factor

$$w_0(\mathbf{k},\mathbf{q}) \equiv \langle \mathbf{k}+\mathbf{q}|w_0|\mathbf{k}\rangle$$
$$= v(q) + \sum_{\alpha=c,d} (\varepsilon_\mathbf{k} - E_\alpha^{\text{vol}})\langle \mathbf{k}+\mathbf{q}|\phi_\alpha\rangle\langle\phi_\alpha|\mathbf{k}\rangle$$
$$+ \sum_d \left(\langle \mathbf{k}+\mathbf{q}|\Delta_{\text{vol}}|\phi_d\rangle\langle\phi_\alpha|\mathbf{k}\rangle + \text{c.c.}\right), \quad (4.33)$$

with $v(q) = v_{\text{scr}}(q) + v_{\text{oh}}(q) + v_{\text{ion}}(q)$ the Fourier transform of the total one-ion potential, and where we have also defined a hybridization form factor

$$h_1(\mathbf{k},\mathbf{q}) \equiv \sum_d \frac{\langle \mathbf{k}+\mathbf{q}|\Delta_{\text{vol}}|\phi_d\rangle\langle\phi_d|\Delta_{\text{vol}}|\mathbf{k}\rangle}{\varepsilon_\mathbf{k} - E_d^{\text{vol}}}. \quad (4.34)$$

The sums over core and d states in Eqs. (4.33) and (4.34) are now only over quantum numbers on a single site.

Similarly, starting from Eq. (2.197) for the d-state component of the orthogonalization-hole density $\delta n_{\text{oh}}^d(\mathbf{r})$, and expanding in powers of $(E - E_d^{\text{vol}})^{-1}$, one has to first order

$$\delta n_{\text{oh}}^d(\mathbf{r}) = -\frac{2}{\pi}\text{Im}\int_0^{\varepsilon_F}\sum_\mathbf{k}\sum_d \left[\frac{\langle\mathbf{r}|\mathbf{k}\rangle v'_{kd}v'_{dk}\langle\mathbf{k}|\mathbf{r}\rangle - \langle\mathbf{r}|\phi_d\rangle v'_{dk}v'_{kd}\langle\phi_d|\mathbf{r}\rangle}{(E-\varepsilon_\mathbf{k})^2(E-E_d^{\text{vol}})}\right.$$
$$\left. + \frac{\langle\mathbf{r}|\phi_d\rangle v'_{dk}(\langle\mathbf{k}|\mathbf{r}\rangle - S_{kd}\langle\phi_d|\mathbf{r}\rangle) + \text{c.c.}}{(E-\varepsilon_\mathbf{k})(E-E_d^{\text{vol}})}\right]dE. \quad (4.35)$$

The individual contour integrals in Eq. (4.35) can be evaluated without difficulty, and the results combined and reordered to yield

$$\delta n_{\text{oh}}^d(\mathbf{r}) = \frac{2\Omega}{(2\pi)^3}\left(\int\sum_d [\langle\mathbf{r}|\mathbf{k}\rangle S_{kd}S_{dk}\langle\mathbf{k}|\mathbf{r}\rangle - (\langle\mathbf{r}|\phi_d\rangle S_{dk}\langle\mathbf{k}|\mathbf{r}\rangle + \text{c.c.})\right.$$
$$+ \langle\mathbf{r}|\phi_d\rangle S_{dk}S_{kd}\langle\phi_d|\mathbf{r}\rangle]\Theta^<(\mathbf{k})d\mathbf{k} + \int\sum_d\left[-\frac{\langle\mathbf{r}|\mathbf{k}\rangle \Delta_{kd}^{\text{vol}}\Delta_{dk}^{\text{vol}}\langle\mathbf{k}|\mathbf{r}\rangle}{(\varepsilon_\mathbf{k}-E_d^{\text{vol}})^2}\right.$$
$$+ \frac{\langle\mathbf{r}|\phi_d\rangle \Delta_{dk}^{\text{vol}}(S_{kd}\langle\phi_d|\mathbf{r}\rangle - \langle\mathbf{k}|\mathbf{r}\rangle) + \text{c.c.}}{\varepsilon_\mathbf{k}-E_d^{\text{vol}}}$$
$$\left.\left. + \frac{\langle\mathbf{r}|\phi_d\rangle \Delta_{dk}^{\text{vol}}\Delta_{kd}^{\text{vol}}\langle\phi_d|\mathbf{r}\rangle}{(\varepsilon_\mathbf{k}-E_d^{\text{vol}})^2}\right]\Theta^<_>(\mathbf{k})d\mathbf{k}\right). \quad (4.36)$$

The d-state component $\delta n_{\text{oh}}^d(\mathbf{r})$ can be combined with the GPT inner-core component $\delta n_{\text{oh}}^c(\mathbf{r})$, as given by Eqs. (3.5), (3.38) and (3.39), to form the total orthogonalization-hole density $\delta n_{\text{oh}}(\mathbf{r})$ for an EDB or FDB metal. One can then re-express $\delta n_{\text{oh}}(\mathbf{r})$ in the usual way in terms of an effective valence Z^* and a localized, single-site hole density $n_{\text{oh}}(\mathbf{r})$ centered on each ion site of the metal, as in Eq. (3.5):

$$Z^* = Z + \frac{2\Omega}{(2\pi)^3}\left(\int \langle \mathbf{k}|p|\mathbf{k}\rangle \Theta^<(\mathbf{k})d\mathbf{k} - \int h_2(\mathbf{k})\Theta^<_>(\mathbf{k})d\mathbf{k}\right), \qquad (4.37)$$

where we have defined

$$h_2(\mathbf{k}) \equiv \sum_d \frac{\langle \mathbf{k}|\Delta_{\text{vol}}|\phi_d\rangle \langle \phi_d|\Delta_{\text{vol}}|\mathbf{k}\rangle}{(\varepsilon_\mathbf{k} - E_d^{\text{vol}})^2}, \qquad (4.38)$$

and

$$n_{\text{oh}}(\mathbf{r}) = -\frac{2\Omega}{(2\pi)^3}\left(\int [\langle \mathbf{r}|p|\mathbf{k}\rangle\langle \mathbf{k}|\mathbf{r}\rangle + \text{c.c.} - \langle \mathbf{r}|p|\mathbf{k}\rangle\langle \mathbf{k}|p|\mathbf{r}\rangle]\Theta^<(\mathbf{k})d\mathbf{k}\right.$$
$$\left. - \int \tilde{h}_2(\mathbf{k},\mathbf{r})\Theta^<_>(\mathbf{k})d\mathbf{k}\right), \qquad (4.39)$$

where we have also defined

$$\tilde{h}_2(\mathbf{k},\mathbf{r}) \equiv \sum_d \left[\frac{\langle \mathbf{r}|\phi_d\rangle\langle \phi_d|\Delta_{\text{vol}}|\mathbf{k}\rangle(\langle \mathbf{k}|\phi_d\rangle\langle \phi_d|\mathbf{r}\rangle - \langle \mathbf{k}|\mathbf{r}\rangle) + \text{c.c.}}{\varepsilon_\mathbf{k} - E_d^{\text{vol}}}\right.$$
$$\left. + \frac{\langle \mathbf{r}|\phi_d\rangle\langle \phi_d|\Delta_{\text{vol}}|\mathbf{k}\rangle\langle \mathbf{k}|\Delta_{\text{vol}}|\phi_d\rangle\langle \phi_d|\mathbf{r}\rangle}{(\varepsilon_\mathbf{k} - E_d^{\text{vol}})^2}\right]. \qquad (4.40)$$

In Eqs. (4.37) and (4.39), $p = \sum_{\alpha=c,d}|\phi_\alpha\rangle\langle\phi_\alpha|$ is the single-site, inner-core plus d-state projection operator, so that $\langle \mathbf{k}|p|\mathbf{k}\rangle = \langle \mathbf{k}|P|\mathbf{k}\rangle$ within our convention of plane-wave normalization established in Chapter 2. Also note in Eqs. (4.38) and (4.40) that $h_2(\mathbf{k})$ and $\tilde{h}_2(\mathbf{k},\mathbf{r})$ are connected by the effective sum rule $h_2(\mathbf{k}) = \int \tilde{h}_2(\mathbf{k},\mathbf{r})d\mathbf{r}$.

Examples of the behavior of the localized, single-site orthogonalization-hole density $n_{\text{oh}}(\mathbf{r})$ in EDB and FDB metals are displayed in Fig. 4.1 for Ca and Zn prototypes. The sharp oscillations seen in the radial density $u_{\text{oh}}(r) = 4\pi r^2 n_{\text{oh}}(r)$ at small r directly reflect the orthogonalization to the inner core states contained in $n_{\text{oh}}(\mathbf{r})$, while the main negative-electron or hole density has a maximum depth near $r = 0.4 R_{\text{WS}}$ in both cases. In the EDB metal Ca, the net effect of sp-d hybridization on $u_{\text{oh}}(r)$ is to add an oscillation to positive density beyond $r = 0.7 R_{\text{WS}}$, while decreasing the depth of the main hole density near $0.4 R_{\text{WS}}$. Consequently, the magnitude of Z^* in Ca drops from a value of 2.371 in the simple-metal limit to 2.007 in the EDB limit. In contrast, for the FDB metal Zn, the effect of the sp-d hybridization on $u_{\text{oh}}(r)$ is to draw in the hole density to smaller r, while increasing slightly the depth of the main hole density near $0.4 R_{\text{WS}}$. In this case, the magnitude of Z^* in Zn increases modestly from 2.257 in the simple-metal limit to 2.280 in the FDB limit.

4.1.2 Valence electron density and self-consistent screening

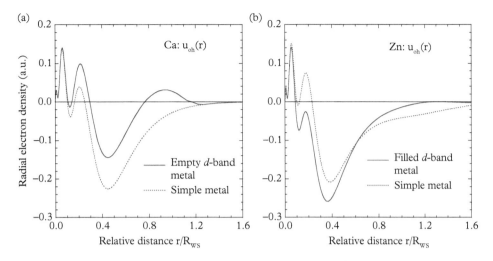

Fig. 4.1 *Single-site radial orthogonalization-hole density, $u_{oh}(r) = 4\pi r^2 n_{oh}(r)$, as calculated for prototype empty- and filled-d-band metals at $\Omega = \Omega_0$ from Eq. (4.39) using optimum d-basis states (see Sec. 4.2). In each case, the full result with sp-d hybridization included is compared with that obtained in the simple-metal limit where $\Delta_{kd}^{vol} \to 0$. (a) Ca as an EDB metal at $\Omega_0 = 294.5$ a.u.; (b) Zn as a FDB metal at $\Omega_0 = 102.7$ a.u.*

Finally, we return to Eqs. (4.32)–(4.34) to complete the full self-consistent screening calculation. To begin, we note that the first-order screening density δn_{scr} only enters the hybridization potential Δ through its small structure component Δ_{struc} and not through Δ_{vol}. Thus δn_{scr} is absent from h_1 in Eq. (4.32) and only enters the PP w_0. This makes the screening calculation for EDB and FDB metals completely analogous to that for simple metals in Chapter 3, beginning with Eq. (3.61). Breaking the form factor $w_0(\mathbf{k}, \mathbf{q})$ in Eq. (4.33) into valence electron and ionic components, one has in place of Eq. (3.61)

$$w_0(\mathbf{k}, \mathbf{q}) = v_{scr}(q) + v_{oh}(q) + w_0^{ion}(\mathbf{k}, \mathbf{q})$$
$$= \frac{4\pi e^2}{q^2}[1 - G(q)][n_{scr}(q) + n_{oh}(q)] + w_0^{ion}(\mathbf{k}, \mathbf{q}), \quad (4.41)$$

where $w_0^{ion}(\mathbf{k}, \mathbf{q})$ is given by the second line of Eq. (4.33) with $v(q)$ replaced by its ionic component $v_{ion}(q)$, and $n_{oh}(q)$ is the Fourier transform of Eq. (4.39). Substituting Eq. (4.41) back into Eq. (4.32) then gives

$$n_{scr}(q) = -\left\{\frac{4\pi e^2}{q^2}[1 - G(q)][n_{scr}(q) + n_{oh}(q)] + \bar{w}_0^{ion}(q) + \bar{h}_1(q)\Pi_{10}(q)\right\}\Pi_0(q), \quad (4.42)$$

where we have defined the smooth averages

$$\bar{w}_0^{\text{ion}}(q) \equiv -\frac{4}{(2\pi)^3} \int \frac{w_0^{\text{ion}}(\mathbf{k}, \mathbf{q})}{\varepsilon_\mathbf{k} - \varepsilon_{\mathbf{k}+\mathbf{q}}} \Theta^<(\mathbf{k}) d\mathbf{k} / \Pi_0(q) \qquad (4.43)$$

and

$$\bar{h}_1(q) \equiv -\frac{4}{(2\pi)^3} \int \frac{h_1(\mathbf{k}, \mathbf{q})}{\varepsilon_\mathbf{k} - \varepsilon_{\mathbf{k}+\mathbf{q}}} \Theta^<_>(\mathbf{k}) d\mathbf{k} / \Pi_1(q), \qquad (4.44)$$

with $\Pi_1(q)$ defined in analogy with $\Pi_0(q)$ in Eq. (3.63) as

$$\Pi_1(q) \equiv -\frac{4}{(2\pi)^3} \int \frac{1}{\varepsilon_\mathbf{k} - \varepsilon_{\mathbf{k}+\mathbf{q}}} \Theta^<_>(\mathbf{k}) d\mathbf{k}, \qquad (4.45)$$

such that $\Pi_{10}(q) = \Pi_1(q)/\Pi_0(q)$ in Eq. (4.42). For EDB metals, $\Pi_1 = \Pi_0$ and $\Pi_{10} = 1$. For FDB metals, it is implicitly assumed that the integrals over $\mathbf{k}$ in Eqs. (4.44) and (4.45) both extend above the free-electron Fermi surface up to some large but finite maximum sphere radius $|\mathbf{k}| = k_{\max}$, established by the convergence of the integral in Eq. (4.44). These integrals are be evaluated numerically in exactly the same way, so as to cancel any small errors that are produced in $\bar{h}_1(q)$ near the screening singularity at $q = 2k_\text{F}$. Then the remaining screening function $\Pi_{10}(q)$ in Eq. (4.42) can be evaluated analytically using Eq. (3.72) and the properties of $\Pi_1(q)$. One finds

$$\Pi_{10}(q) = \frac{k_{\max}}{k_\text{F}} \frac{f_\text{L}(q/2k_{\max})}{f_\text{L}(q/2k_\text{F})} - 1, \qquad (4.46)$$

where f_L is the Lindhard function. In the $q \to 0$ limit, note that $\Pi_{10}(0) = k_{\max}/k_\text{F} - 1$. Returning to Eq. (4.42) and solving for $n_{\text{scr}}(q)$ then yields the self-consistent result

$$\begin{aligned} n_{\text{scr}}(q) &= -\left\{ \bar{w}_0^{\text{ion}}(q) + \bar{h}_1(q)\Pi_{10}(q) + \frac{4\pi e^2}{q^2}[1 - G(q)]n_{\text{oh}}(q) \right\} \Pi_0(q)/\varepsilon(q) \\ &= -\left[\bar{w}_0^{\text{ion}}(q) + \bar{h}_1(q)\Pi_{10}(q) + v_{\text{oh}}(q) \right] \Pi_0(q)/\varepsilon(q). \end{aligned} \qquad (4.47)$$

The corresponding self-consistent screening potential is

$$v_{\text{scr}}(q) = \frac{4\pi e^2}{q^2}[1 - G(q)]n_{\text{scr}}(q) = \left[\bar{w}_0^{\text{ion}}(q) + \bar{h}_1(q)\Pi_{10}(q) + v_{\text{oh}}(q) \right] [1 - \varepsilon(q)]/\varepsilon(q). \qquad (4.48)$$

4.1.2 Valence electron density and self-consistent screening

Finally, one can use either Eq. (4.47) or (4.48) in Eq. (4.41) to obtain the total self-consistent form factor

$$w(\mathbf{k}, \mathbf{q}) \equiv w_0(\mathbf{k}, \mathbf{q}) + h_1(\mathbf{k}, \mathbf{q})$$
$$= [\bar{w}_0^{\text{ion}}(q) + \bar{h}_1(q)\Pi_{10}(q) + v_{\text{oh}}(q)]/\varepsilon(q) + w_0^{\text{ion}}(\mathbf{k}, \mathbf{q}) - \bar{w}_0^{\text{ion}}(q)$$
$$+ h_1(\mathbf{k}, \mathbf{q}) - \bar{h}_1(q)\Pi_{10}(q). \quad (4.49)$$

Physically, the total form factor given by Eq. (4.49) plays the same role for EDB and FDB metals as the form factor (3.67) does for simple metals, controlling the magnitude of zone-boundary band gaps in the electronic structure as well as Fermi-surface transport properties in the solid and liquid. The behavior of the form factor $w(q) = w(\mathbf{k}_F, \mathbf{q})$ on the free-electron Fermi surface is illustrated in Fig. 4.2 for Ca and Zn prototypes. First note that electron screening with sp-d hybridization included preserves the simple-metal small-q limit $w(q)/\varepsilon_F \to -2/3$ as $q \to 0$ for both EDB and FDB metals. In the case of Ca, the form factor naturally peaks at positive energy beyond $q = 1.5k_F$, and this behavior is strongly enhanced by the hybridization, with $w(q)$ driven sharply down to negative energy as $q \to 2k_F$. At the same time, the shortest fcc reciprocal lattice vectors, with a magnitude $G_0 = 1.759k_F$, produce only a small band gap, $2|w(G_0)| = 0.066\varepsilon_F$, either with or without the sp-d hybridization. In the case of Zn, hybridization impacts the form factor modestly, pushing $w(q)$ downward in energy for $q < 1.7k_F$, and pushing $w(q)$ upward in energy for $q > 1.7k_F$. For the shortest three hcp reciprocal lattice vectors, the small band-gap criterion, $2|w(G_0)|/\varepsilon_F \ll 1$, is also reasonably well obeyed in Zn with $2|w(G_0)|/\varepsilon_F \leq 0.166$, either with or without sp-d hybridization.

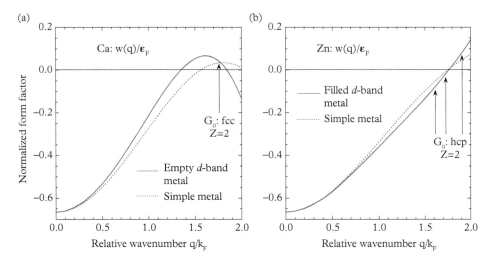

Fig. 4.2 Total form factor $w(q) = w(\mathbf{k}_F, \mathbf{q})$ evaluated on the free-electron Fermi surface, as obtained from Eq. (4.49) for empty- and filled-d-band metals at $\Omega = \Omega_0$ and compared with the corresponding results obtained from Eq. (3.67) in the simple-metal limit. (a) Ca as an EDB metal at $\Omega_0 = 294.5$ a.u.; (b) Zn as a FDB metal at $\Omega_0 = 102.7$ a.u.

4.1.3 Volume term, energy-wavenumber characteristic and overlap potential

In this section, we return to Eq. (4.1) to complete the development of the cohesive-energy functional for EDB and FDB metals. With the forms of $E_{\text{band}}^{\text{val}}$ and n_{val} established above, it remains to extract of an electrostatic energy $\delta E_{\text{es}}(Z^*, \Omega)$ from the first three terms in Eq. (4.1). This task is analogous to that for a simple metal, as described in Chapter 3, but is significantly more intricate for EDB and FDB metals because of the complexity of the orthogonalization-hole contribution δn_{oh} to n_{val}. The full procedure to complete this task is discussed in detail by Moriarty (1977, 1982). Here we only indicate a few of its salient features. The task requires the removal and direct treatment of the hidden second-order structure-dependent contributions arising from the nominal first-order PP and hybridization terms in the free-electron energy E_{fe}, as given by Eq. (4.24). The hybridization contributions enter that equation through both E_d^{struc} and $\Delta_{kd}^{\text{struc}}$ matrix elements of $\Delta_{\text{struc}} = \delta V_{\text{struc}} - \langle \phi_d | \delta V_{\text{struc}} | \phi_d \rangle$. As shown by Moriarty (1982), the total hidden second-order contributions reduce to the relatively simple real-space interaction

$$-\sum_i \int n_{\text{oh}}(\mathbf{r} - \mathbf{R}_i) \delta V_{\text{struc}}(\mathbf{r} - \mathbf{R}_i) d\mathbf{r} \equiv -\sum_i n_{\text{oh}}^i \delta V_{\text{struc}}^i$$

$$= \sum_{i,j}{'} n_{\text{oh}}^i v_{\text{ion}}^j + \delta n_{\text{oh}} \left(\delta V_{\text{val}} + [d\mu_{\text{xc}}(n_{\text{unif}})/dn] \delta n_{\text{val}} \right), \quad (4.50)$$

In turn, the individual elements on the RHS of the second line Eq. (4.50), together with other direct orthogonalization-hole contributions, are then re-absorbed into the theory in familiar ways: through Z^* in $\delta E_{\text{es}}(Z^*, \Omega)$, through a self-energy correction $\delta E_{\text{oh}}(\Omega)$ to $E_{\text{vol}}^q(\Omega)$, and through a $G(q)[n_{\text{oh}}(q)]^2$ contribution to $F(q, \Omega)$. In addition, one obtains a real-space orthogonalization-hole contribution $\delta v_{\text{oh}}^{ij}$ to the overlap potential $v_{\text{ol}}(R_{ij}, \Omega)$. Also note that to second order, the final xc term in Eq. (4.1) reduces to $n_{\text{unif}} \delta \mu_{\text{xc}}^*(n_i, n_j)$, which cancels the same contribution to the first W_{kk}^0 term in Eq. (4.24).

The remaining terms in Eq. (4.24) for E_{fe}, with Δ_{kd} replaced by Δ_{kd}^{vol}, then contribute directly to the final reciprocal-space volume term

$$E_{\text{vol}}^q(\Omega) = E_{\text{fe}}^0(\Omega) + \frac{2\Omega}{(2\pi)^3} \int w_0^{\text{pa}}(\mathbf{k})[1 + p(\mathbf{k})] \Theta^<(\mathbf{k}) d\mathbf{k}$$

$$+ \frac{2\Omega}{(2\pi)^3} \int h_1(\mathbf{k}) \left[1 + p(\mathbf{k}) - \frac{w_0^{\text{pa}}(\mathbf{k}) + h_1(\mathbf{k})}{\varepsilon_\mathbf{k} - E_d^{\text{vol}}} \right] \Theta_>^<(\mathbf{k}) d\mathbf{k}$$

$$+ \delta E_{\text{oh}}(\Omega) - E_{\text{bind}}^{\text{atom}}(\Omega), \quad (4.51)$$

where we have made the short-hand definitions $w_0^{\text{pa}}(\mathbf{k}) \equiv \langle \mathbf{k} | w_0^{\text{pa}} | \mathbf{k} \rangle$, $p(\mathbf{k}) \equiv \langle \mathbf{k} | p | \mathbf{k} \rangle$ and $h_1(\mathbf{k}) \equiv h_1(\mathbf{k}, 0)$, with

4.1.3 Volume term, energy-wavenumber characteristic and overlap potential

$$\langle \mathbf{k}|w_0^{\mathrm{pa}}|\mathbf{k}\rangle = \langle \mathbf{k}|v_{\mathrm{pa}}|\mathbf{k}\rangle - V_0' + \sum_{\alpha=c,d}(\varepsilon_{\mathbf{k}} - E_\alpha^{\mathrm{vol}})\langle \mathbf{k}|\phi_d\rangle\langle\phi_d|\mathbf{k}\rangle$$
$$+ \sum_d \left(\langle \mathbf{k}|\Delta_{\mathrm{vol}}|\phi_d\rangle\langle\phi_d|\mathbf{k}\rangle + \langle \mathbf{k}|\phi_d\rangle\langle\phi_d|\Delta_{\mathrm{vol}}|\mathbf{k}\rangle\right). \tag{4.52}$$

Here $v_{\mathrm{pa}} = v_{\mathrm{unif}} + v_{\mathrm{ion}}$ and V_0' is a constant chosen to make $\langle 0|w_0^{\mathrm{pa}}|0\rangle = 0$. Equation (4.51) for an EDB or FDB metal is analogous to Eq. (3.43) for a simple metal.

Extracting a structure factor $S(\mathbf{q})$ from each matrix element of the band-structure energy E_{bs} in Eq. (4.27), and then subtracting off the important screening and orthogonalization-hole contributions, the corresponding energy-wavenumber characteristic for an EDB or FDB metal takes the form

$$F(q,\Omega) = \frac{2\Omega}{(2\pi)^3}\left[\int \frac{[w_0(\mathbf{k},\mathbf{q})]^2}{\varepsilon_{\mathbf{k}} - \varepsilon_{\mathbf{k}+\mathbf{q}}}\Theta^<(\mathbf{k})d\mathbf{k} + \int \frac{2h_1(\mathbf{k},\mathbf{q})w_0(\mathbf{k},\mathbf{q}) + [h_1(\mathbf{k},\mathbf{q})]^2}{\varepsilon_{\mathbf{k}} - \varepsilon_{\mathbf{k}+\mathbf{q}}}\Theta_>^<(\mathbf{k})d\mathbf{k}\right]$$
$$- \frac{2\pi e^2 \Omega}{q^2}\left\{[1 - G(q)][n_{\mathrm{scr}}(q)]^2 + G(q)[n_{\mathrm{oh}}(q)]^2\right\}. \tag{4.53}$$

This result is analogous to Eq. (3.41) for a simple metal, with the explicit screening and orthogonalization-hole contributions formally entering in exactly the same way. One may further note that for EDB metals the terms on the first line of Eq. (4.53) can be combined and written in terms of the total form factor $w(\mathbf{k},\mathbf{q})$, as given by Eq. (4.49):

$$F(q,\Omega) = \frac{2\Omega}{(2\pi)^3}\int \frac{[w(\mathbf{k},\mathbf{q})]^2}{\varepsilon_{\mathbf{k}} - \varepsilon_{\mathbf{k}+\mathbf{q}}}\Theta^<(\mathbf{k})d\mathbf{k} - \frac{2\pi e^2 \Omega}{q^2}\left\{[1 - G(q)][n_{\mathrm{scr}}(q)]^2 + G(q)[n_{\mathrm{oh}}(q)]^2\right\}.$$
$$\tag{4.54}$$

This then makes the EDB form of $F(q,\Omega)$ the same as in Eq. (3.41) for simple metals.

The overlap potential includes the remaining real-space $i \neq j$ ion-site contributions from three sources: (i) the overlap energy E_{ol} given by Eq. (4.28); (ii) the electrostatic and xc overlap correction terms from the second and third lines of Eq. (4.1), collectively written here as $\delta v_{\mathrm{es-xc}}^{ij}$; and (iii) the additional correction terms $\delta v_{\mathrm{oh}}^{ij}$ and $\delta v_{\mathrm{ke}}^{ij}$:

$$v_{\mathrm{ol}}(R_{ij},\Omega) = \sum_{d,d'}\left[4S_{dd'}^{ij}\Delta_{d'd}^{ji}\Theta(\varepsilon_F - E_d^{\mathrm{vol}}) - \frac{4\Omega}{(2\pi)^3}\int\left\{S_{dd'}^{ij}[(\varepsilon_{\mathbf{k}} - E_d^{\mathrm{vol}})f_{d'd}^{ji}(\mathbf{k}) + 2g_{d'd}^{ji}(\mathbf{k})]\right.\right.$$
$$+ \Delta_{dd'}^{ij}f_{d'd}^{ji}(\mathbf{k})\right\}\Theta^<(\mathbf{k})d\mathbf{k} + \frac{4\Omega}{(2\pi)^3}\int\left\{S_{dd'}^{ij}h_{d'd}^{ji}(\mathbf{k})\right.$$
$$+ \left.\left.\frac{\Delta_{dd'}^{ij}[2g_{d'd}^{ji}(\mathbf{k}) + h_{d'd}^{ji}(\mathbf{k})]}{\varepsilon_{\mathbf{k}} - E_d^{\mathrm{vol}}}\right\}\Theta_>^<(\mathbf{k})d\mathbf{k}\right]$$
$$+ \delta v_{\mathrm{es-xc}}^{ij} + \delta v_{\mathrm{oh}}^{ij} + \delta v_{\mathrm{ke}}^{ij}. \tag{4.55}$$

152   *Interatomic Potentials in Metals with Empty or Filled d Bands*

In arriving at Eq. (4.55), we have explicitly exposed the site dependence of $S_{dd'}$ and $\Delta_{dd'}$ by writing $S^{ij}_{dd'} \equiv S_{dd'}(R_{ij}) = \langle \phi^i_d | \phi^j_{d'} \rangle$ and $\Delta^{ij}_{dd'} \equiv \Delta_{dd'}(R_{ij}) = \langle \phi^i_d | \Delta^j | \phi^j_{d'} \rangle$. We have further defined in Eq. (4.55) the hybridization-related quantities

$$f^{ji}_{d'd}(\mathbf{k}) \equiv \langle \phi^j_{d'} | \mathbf{k} \rangle \langle \mathbf{k} | \phi^i_d \rangle, \qquad (4.56)$$

$$g^{ji}_{d'd}(\mathbf{k}) \equiv \langle \phi^j_{d'} | \mathbf{k} \rangle \langle \mathbf{k} | \Delta_{\text{vol}} | \phi^i_d \rangle \qquad (4.57)$$

and

$$h^{ji}_{d'd}(\mathbf{k}) \equiv \frac{\langle \phi^j_{d'} | \Delta_{\text{vol}} | \mathbf{k} \rangle \langle \mathbf{k} | \Delta_{\text{vol}} | \phi^i_d \rangle}{\varepsilon_\mathbf{k} - E^{\text{vol}}_d}. \qquad (4.58)$$

The sums over d states in Eq. (4.55) are now over individual quantum numbers m (or xy, yz, zx, etc.) on the sites i and j. Also recall that the matrix elements $\Delta^{ij}_{dd'}$ are just the negative values of the TB bond integrals linking sites i and j, as expressed in Eq. (2.133). The first-principles evaluation of these integrals will be discussed in Sec. 4.2.

The overlap correction term $\delta v^{ij}_{\text{es-xc}}$ in Eq. (4.55) includes the electrostatic terms on the second line of Eq. (4.1) plus the first two xc terms on the third line of that equation. These terms are here rearranged in a computationally useful form and written in the shorthand notation of Moriarty (1977), which includes the implied integration over all space, as in Eq. (4.1):

$$\delta v^{ij}_{\text{es-xc}} = n^i_{\text{core}} \left[\left(\frac{Z_a - Z}{Z_a} \right) v^j_{\text{nuc}} + v^j_{\text{core}} + v^j_{\text{xc}} \right]$$

$$+ \left(\frac{Z_a + Z}{Z_a} \right) n^i_{\text{nuc}} \left[\left(\frac{Z_a - Z}{Z_a} \right) v^j_{\text{nuc}} + v^j_{\text{core}} \right] + \delta\varepsilon^*_{\text{xc}}(n^i_{\text{core}}, n^j_{\text{core}}). \qquad (4.59)$$

In this notation, $n^i_{\text{nuc}} \equiv -Z_a \delta(\mathbf{r} - \mathbf{R}_i)$ and $v^j_{\text{nuc}} \equiv -Z_a e^2 / |\mathbf{r} - \mathbf{R}_j|$, while $n^i_{\text{core}} \equiv n_{\text{core}}(\mathbf{r} - \mathbf{R}_i)$, v^j_{core} is the Coulomb potential arising from n^j_{core}, and $v^j_{\text{xc}} \equiv \mu^*_{\text{xc}}(n^j_{\text{core}})$. Here both μ^*_{xc} and $\delta\varepsilon^*_{\text{xc}}$ are evaluated with n_{val} replaced by n_{unif}.

In the same notation, the additional orthogonalization-hole overlap correction term $\delta v^{ij}_{\text{oh}}$ in Eq. (4.55), which derives from Eq. (4.50), is given by

$$\delta v^{ij}_{\text{oh}} = 2n^i_{\text{oh}} \left[\left(\frac{Z_a - Z}{Z_a} \right) v^j_{\text{nuc}} + v^j_{\text{core}} + v^j_{\text{xc}} \right] + n^i_{\text{oh}} \left[v^j_{\text{oh}} - \left(\frac{Z^* - Z}{Z_a} \right) v^j_{\text{nuc}} \right]$$

$$+ (n^i_{\text{oh}} + n^j_{\text{oh}}) \delta v^{ij}_{\text{xc}}, \qquad (4.60)$$

where $n^i_{\text{oh}} \equiv n_{\text{oh}}(\mathbf{r} - \mathbf{R}_i)$ and v^j_{oh} is the Coulomb potential arising from n^j_{oh}. In addition, $\delta v^{ij}_{\text{xc}} \equiv \delta\mu^*_{\text{xc}}(n^i_{\text{core}}, n^j_{\text{core}})$ with n_{val} replaced by n_{unif}.

The final overlap correction term $\delta v^{ij}_{\text{ke}}$ in Eq. (4.55) compensates for the fact that in the GPT formalism the overlap of localized d states on one site with the inner core of a neighboring site is only accounted for in $\delta v^{ij}_{\text{es-xc}}$ and not in the valence band-structure energy $E^{\text{val}}_{\text{band}}$. This short-ranged correction is positive and ensures that $v_{\text{ol}}(R_{ij}, \Omega) > 0$

as $R_{ij} \to 0$. An approximate procedure that can be used to evaluate $\delta v_{ke}(R_{ij}) \equiv \delta v_{ke}^{ij}$ is discussed in Sec. A2.7 of Appendix A2, with the form of $\delta v_{ke}(R_{ij})$ given by Eq. A2.50.

The character of the short-ranged overlap potential $v_{ol}(r, \Omega_0)$ for EDB and FDB metals is illustrated in Fig. 4.3, including the impact of the $\delta v_{oh}(r)$ and $\delta v_{ke}(r)$ correction terms. For EDB metals, $v_{ol}(r, \Omega_0)$ is generally quite small in magnitude and can be either attractive or repulsive at near-neighbor distances, as shown in Fig. 4.3(a) for Ca, where $|v_{ol}| < 0.2$ mRy for $r/R_{WS} > 1.5$. For FDB metals, on the other hand, $v_{ol}(r, \Omega_0)$ is much larger and repulsive at near-neighbor distances, with a significantly longer range, as shown in Fig. 4.3(b) for Zn, where $v_{ol} \simeq 4$ mRy for $r/R_{WS} = 1.75$. In both EDB and FDB metals, the correction terms $\delta v_{oh}(r)$ and $\delta v_{ke}(r)$ are generally positive at all near-neighbor distances, as is seen to be the case for Ca and Zn in Fig. 4.3. Also, as might be expected, the impact of $\delta v_{ke}(r)$ extends to a much longer distance in an FDB metal than in a EDB metal: for example, to $r/R_{WS} \simeq 2.2$ in Zn vs. to only $r/R_{WS} \simeq 1.6$ in Ca.

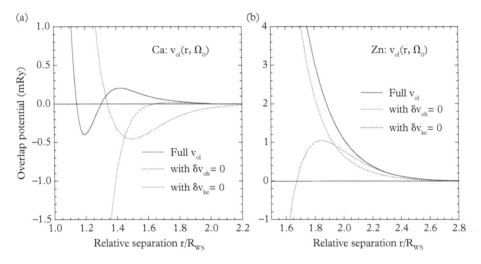

Fig. 4.3 *The overlap potential $v_{ol}(r, \Omega_0)$, as calculated for prototype empty- and filled-d-band metals from Eq. (4.55) using optimum d-basis states (see Sec. 4.2), and showing the impact of the δv_{oh} and δv_{ke} correction terms. (a) Ca as an EDB metal at $\Omega_0 = 294.5$ a.u.; (b) Zn as a FDB metal at $\Omega_0 = 102.7$ a.u.*

4.1.4 Evaluation of the real-space volume term and pair potential

In the previous section, we completed the formal evaluation of the four components of the reciprocal-space cohesive energy in Eq. (4.2). Since the overlap potential is already expressed in real space, what remains is to transform the first three terms of Eq. (4.2) into real-space as well, so the cohesive energy has the form (3.1). This transformation does not depend on the details of $E_{vol}^q(\Omega)$ or $F(q, \Omega)$ and thus is exactly the same as that obtained in Chapter 3 for a simple metal. Adding the overlap potential to Eq. (3.59) for the two-ion pair potential of a simple metal gives the corresponding result for an EDB or FDB metal:

$$v_2(r,\Omega) = \frac{(Z^*e)^2}{r}\left[1 - \frac{2}{\pi}\int_0^\infty F_N(q,\Omega)\frac{\sin(qr)}{q}dq\right] + v_{\text{ol}}(r,\Omega), \qquad (4.61)$$

where now, using Eq. (4.53), the normalized energy-wavenumber characteristic is given by

$$\begin{aligned}
F_N(q,\Omega) &\equiv -\frac{q^2\Omega}{2\pi(Z^*e)^2}F(q,\Omega) \\
&= \frac{q^2\Omega^2}{4\pi(Z^*e)^2}\left[-\frac{4}{(2\pi)^3}\left\{\int\frac{[w_0(\mathbf{k},\mathbf{q})]^2}{\varepsilon_\mathbf{k}-\varepsilon_{\mathbf{k+q}}}\Theta^<(\mathbf{k})d\mathbf{k} + \int\frac{2h_1(\mathbf{k},\mathbf{q})w_0(\mathbf{k},\mathbf{q})}{\varepsilon_\mathbf{k}-\varepsilon_{\mathbf{k+q}}}\Theta^\lessgtr(\mathbf{k})d\mathbf{k}\right.\right. \\
&\left.\left. + \int\frac{[h_1(\mathbf{k},\mathbf{q})]^2}{\varepsilon_\mathbf{k}-\varepsilon_{\mathbf{k+q}}}\Theta^\lessgtr(\mathbf{k})d\mathbf{k}\right\} + \frac{4\pi e^2}{q^2}\left\{[1-G(q)][n_{\text{scr}}(q)]^2 + G(q)[n_{\text{oh}}(q)]^2\right\}\right].
\end{aligned}$$
(4.62)

Similarly, using Eq. (4.51) for $E^q_{\text{vol}}(\Omega)$ in an EDB or FDB metal, Eq. (3.60) for the real-space volume term of a simple metal is generalized to the result

$$\begin{aligned}
E_{\text{vol}}(\Omega) &= E^0_{\text{fe}}(\Omega) + \frac{2\Omega}{(2\pi)^3}\int w_0^{pa}(\mathbf{k})[1+p(\mathbf{k})]\Theta^<(\mathbf{k})d\mathbf{k} \\
&+ \frac{2\Omega}{(2\pi)^3}\int h_1(\mathbf{k})\left[1+p(\mathbf{k}) - \frac{w_0^{pa}(\mathbf{k})+h_1(\mathbf{k})}{\varepsilon_\mathbf{k}-E_d^{\text{vol}}}\right]\Theta^\lessgtr(\mathbf{k})d\mathbf{k} + \frac{9}{10}\frac{(Z^*e)^2}{R_{\text{WS}}} \\
&+ \frac{\pi(Z^*e)^2}{\Omega}\frac{\partial^2 F_N(0,\Omega)}{\partial q^2} - \frac{(Z^*e)^2}{\pi}\int_0^\infty F_N(q,\Omega)dq + \delta E_{\text{oh}}(\Omega) - E^{\text{atom}}_{\text{bind}}(Z).
\end{aligned}$$
(4.63)

For an EDB metal, Eqs. (4.62) and (4.63) can be collapsed and simplified in terms of the total form factor $w(\mathbf{k},\mathbf{q})$ plus the quantities $w_{\text{pa}}(\mathbf{k}) \equiv w_0^{pa}(\mathbf{k}) + h_1(\mathbf{k})$ and $h_2(\mathbf{k})$:

$$\begin{aligned}
F_N(q,\Omega) &= \frac{q^2\Omega^2}{4\pi(Z^*e)^2}\left[-\frac{4}{(2\pi)^3}\int\frac{[w(\mathbf{k},\mathbf{q})]^2}{\varepsilon_\mathbf{k}-\varepsilon_{\mathbf{k+q}}}\Theta^<(\mathbf{k})d\mathbf{k}\right. \\
&\left. + \frac{4\pi e^2}{q^2}\left\{[1-G(q)][n_{\text{scr}}(q)]^2 + G(q)[n_{\text{oh}}(q)]^2\right\}\right]
\end{aligned}$$
(4.64)

4.1.4 Evaluation of the real-space volume term and pair potential

and

$$E_{\text{vol}}(\Omega) = E_{\text{fe}}^0(\Omega) + \frac{2\Omega}{(2\pi)^3} \int w_{\text{pa}}(\mathbf{k})[1 + p(\mathbf{k}) - h_2(\mathbf{k})]\Theta^<(\mathbf{k})d\mathbf{k} + \frac{9}{10}\frac{(Z^*e)^2}{R_{\text{WS}}}$$

$$+ \frac{\pi(Z^*e)^2}{\Omega}\frac{\partial^2 F_N(0,\Omega)}{\partial q^2} - \frac{(Z^*e)^2}{\pi}\int_0^\infty F_N(q,\Omega)dq + \delta E_{\text{oh}}(\Omega) - E_{\text{bind}}^{\text{atom}}(Z).$$

(4.65)

Except for the appearance of $h_2(\mathbf{k})$ in $E_{\text{vol}}(\Omega)$, Eqs. (4.64) and (4.65) are then of the same form as one would have for a simple metal.

As in the treatment of simple metals, the second derivative $\partial^2 F_N(0,\Omega)/\partial q^2$ can be explicitly evaluated in Eqs. (4.63) and (4.65) for $E_{\text{vol}}(\Omega)$. In addition, for both EBD and FDB metals, $n_{\text{scr}}(q)$, $F(q,\Omega)$ and $E_{\text{vol}}(\Omega)$ can be expressed in computationally efficient forms analogous to Eqs. (3.92), (3.93) and (3.95), respectively, for simple metals. To do this, we first write Eq. (4.47) for the screening density in the form

$$n_{\text{scr}}(q) = \left\{ n_{\text{unif}} - \frac{q^2}{4\pi e^2}\left[\bar{w}_0^{\text{core}}(q) + \bar{h}_1(q)\Pi_{10}(q)\right] - [1 - G(q)]n_{\text{oh}}(q) \right\}\chi(q), \quad (4.66)$$

where, using Eq. (3.77), we have noted

$$\bar{w}_0^{\text{core}}(q) = \frac{4\pi e^2}{q^2}n_{\text{unif}} + \bar{w}_0^{\text{ion}}(q). \quad (4.67)$$

Then we can develop Eq. (4.62) for $F_N(q,\Omega)$ directly in terms of $n_{\text{scr}}(q)$:

$$F_N(q,\Omega) = [n_{\text{scr}}(q)/n_{\text{unif}}^*]^2 \chi^{-1}(q) + \frac{q^2}{4\pi e^2}\left\{\overline{\delta[w_0^{\text{core}}]^2}(q) + \overline{\delta w_0^{\text{core}} h_1}(q)\Pi_{10}(q)\right\}$$
$$\times \Pi_0(q)/(n_{\text{unif}}^*)^2 + G(q)\left[n_{\text{oh}}(q)/n_{\text{unif}}^*\right]^2, \quad (4.68)$$

where we recall from Eq. (3.93) that $n_{\text{unif}}^* = (Z^*/Z)n_{\text{unif}}$, and where we have defined the additional smooth functions

$$\overline{\delta[w_0^{\text{core}}]^2}(q) \equiv \overline{[w_0^{\text{core}}]^2}(q) - [\bar{w}_0^{\text{core}}(q)]^2, \quad (4.69)$$

with

$$\overline{[w_0^{\text{core}}]^2}(q) \equiv -\frac{4}{(2\pi)^3}\int \frac{[w_0^{\text{core}}(\mathbf{k},\mathbf{q})]^2}{\varepsilon_{\mathbf{k}} - \varepsilon_{\mathbf{k+q}}}\Theta^<(\mathbf{k})d\mathbf{k}/\Pi_0(q), \quad (4.70)$$

and

$$\overline{\delta w_0^{\text{core}} h_1}(q) \equiv \overline{w_0^{\text{core}} h_1}(q) - 2\bar{w}_0^{\text{core}}(q)\bar{h}_1(q) - [\bar{h}_1(q)]^2\Pi_{10}(q), \quad (4.71)$$

with

$$\overline{w_0^{\text{core}} h_1}(q) \equiv -\frac{4}{(2\pi)^3} \int \frac{2w_0^{\text{core}}(\mathbf{k},\mathbf{q})h_1(\mathbf{k},\mathbf{q}) + [h_1(\mathbf{k},\mathbf{q})]^2}{\varepsilon_\mathbf{k} - \varepsilon_{\mathbf{k}+\mathbf{q}}} \Theta_>^<(\mathbf{k})d\mathbf{k}/\Pi_1(q). \quad (4.72)$$

In Eq. (4.66) for $n_{\text{scr}}(q)$ and Eq. (4.68) for $F_N(q,\Omega)$ we have thereby isolated the main impact of the screening singularity at $q = 2k_F$ in the analytic functions $\chi(q)$, $\Pi_0(q)$ and $\Pi_{10}(q)$. The remaining smooth functions $\bar{w}_0^{\text{core}}(q)$, $\bar{h}_1(q)$, $\overline{[w_0^{\text{core}}]^2}(q)$ and $\overline{w_0^{\text{core}} h_1}(q)$ can then be calculated numerically with good accuracy at all values of q. This in turn allows a very accurate representation of the screening oscillations in the pair potential $v_2(r,\Omega)$ to be calculated from Eq. (4.61). In addition, for an EDB metal, Eqs. (4.66) and (4.68) can be collapsed into simple-metal form. Defining $w_{\text{core}}(\mathbf{k},\mathbf{q}) \equiv w_0^{\text{core}}(\mathbf{k},\mathbf{q}) + h_1(\mathbf{k},\mathbf{q})$ and $\bar{w}_{\text{core}}(q) \equiv \bar{w}_0^{\text{core}}(q) + \bar{h}_1(q)$, Eq. (4.66) for $n_{\text{scr}}(q)$ reduces to the same form as Eq. (3.92), while Eq. (4.68) for $F_N(q,\Omega)$ reduces to the form of Eq. (3.93).

Examples of pair potentials $v_2(r,\Omega_0)$ thereby calculated from Eq. (4.61) for our EDB and FDB prototype metals Ca and Zn are shown in Fig. 4.4, and are compared there with corresponding results obtained in the simple-metal limit without sp-d hybridization and d-state overlap. The impact of the hybridization and overlap on the pair potential is seen to be quite large in both cases, especially in the near-neighbor regime for $r/R_{WS} < 2.5$. In the case of Ca, the shallow first minimum in $v_2(r,\Omega_0)$ is deepened by nearly a factor of four and moved inward to $r/R_{WS} = 1.8$, close to the nearest-neighbor distance of the fcc or ideal hcp structure. In Zn, on the other hand, the similar shallow first minimum in the simple-metal pair potential is moved sharply to positive energy, becoming only a steep shoulder near $r/R_{WS} \simeq 1.8$. As a consequence, sp-d hybridization and d-state overlap can strongly affect fundamental materials properties, especially structural phase stability, as will be discussed further in Sec. 4.5 as well as in Chapter 6. In both metals, the calculated sp-d hybridization and d-state overlap is necessary and sufficient to explain their observed crystal structures: fcc in the case of Ca, and hcp with a high c/a ratio in the case of Zn.

Finally, corresponding computationally efficient representations of the volume term $E_{\text{vol}}(\Omega)$ can be developed for EDB and FDB metals. One proceeds in the same manner as in Chapter 3 for a simple metal, and first uses Eqs. (4.66) and (4.68) to obtain an analytic expansion of $F_N(q,\Omega)$ for small q. Then, evaluating the second derivative of $F_N(q,\Omega)$ at $q = 0$, the general result given by Eq. (4.63) for either an EDB or FDB metal becomes

$$E_{\text{vol}}(\Omega) = E_{\text{fe}}^0(\Omega) - \frac{1}{2}\Omega B_{\text{eg}}(\Omega) + \frac{2\Omega}{(2\pi)^3} \int w_0^{pa}(\mathbf{k})[1+p(\mathbf{k})]\Theta_>^<(\mathbf{k})d\mathbf{k} - Z^\star [\bar{w}_0^{\text{core}}(0)$$

$$+\bar{h}_1(0)\Pi_{10}(0)] + \frac{2\Omega}{(2\pi)^3} \int h_1(\mathbf{k}) \left[1 + p(\mathbf{k}) - \frac{w_0^{pa}(\mathbf{k}) + h_1(\mathbf{k})}{\varepsilon_\mathbf{k} - E_d^{\text{vol}}}\right] \Theta_>^<(\mathbf{k})d\mathbf{k}$$

$$+ \frac{3}{4}\frac{Z}{\varepsilon_F} \left[\delta\overline{[w_0^{\text{core}}]^2}(0) + \delta\overline{w_0^{\text{core}} h_1}(0)\Pi_{10}(0)\right] + \frac{9}{10}\frac{(Z^\star e)^2}{R_{WS}}$$

$$- \frac{(Z^\star e)^2}{\pi} \int_0^\infty F_N^\star(q,\Omega)dq + \delta E_{\text{oh}}^\star(\Omega) - E_{\text{bind}}^{\text{atom}}(Z), \quad (4.73)$$

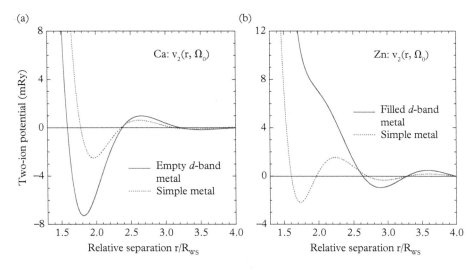

Fig. 4.4 The two-ion GPT pair potential $v_2(r, \Omega_0)$ for prototype empty- and filled-d-band metals at their observed equilibrium volumes, as calculated from Eq. (4.61) using IU electron screening and optimum d basis states (see Sec. 4.2). In each case, the full result with sp-d hybridization and d-state overlap included is compared with that obtained in the simple-metal limit where $\Delta_{dk}^{vol} \to 0$ and $v_{ol}(r, \Omega_0) \to 0$. (a) Ca as an EDB metal at $\Omega_0 = 294.5$ a.u.; (b) Zn as a FDB metal at $\Omega_0 = 102.7$ a.u.

with the definition of $\delta E_{\mathrm{oh}}^{\star}(\Omega)$ in terms of $\delta E_{\mathrm{oh}}(\Omega)$ and γ_{oh} still given by Eq. (3.96). For EDB metals, Eq. (4.65) becomes

$$E_{\mathrm{vol}}(\Omega) = E_{\mathrm{fe}}^0(\Omega) - \frac{1}{2}\Omega B_{\mathrm{eg}}(\Omega) + \frac{2\Omega}{(2\pi)^3} \int w_{\mathrm{pa}}(\mathbf{k})[1 + p(\mathbf{k}) - h_2(\mathbf{k})]\Theta^<(\mathbf{k})d\mathbf{k}$$

$$- Z^\star \overline{w}_{\mathrm{core}}(0) + \frac{3}{4}\frac{Z}{\varepsilon_F}\overline{\delta w_{\mathrm{core}}^2}(0) + \frac{9}{10}\frac{(Z^\star e)^2}{R_{\mathrm{WS}}} - \frac{(Z^\star e)^2}{\pi}\int_0^\infty F_N(q, \Omega)dq$$

$$+ \delta E_{\mathrm{oh}}^\star(\Omega) - E_{\mathrm{bind}}^{\mathrm{atom}}(Z), \qquad (4.74)$$

which differs only in form from the simple-metal result (3.95) by the appearance of the term involving $h_2(\mathbf{k})$.

4.2 Zero-order pseudoatoms and optimized *d* basis states

In the first-principles implementation of the GPT for EDB and FDB metals, a primary consideration is the choice of the localized atomic d-basis states $\phi_d(\mathbf{r})$ that determine the hybridization potential Δ and its matrix elements $\Delta_{\mathbf{k}d} = \langle \mathbf{k}|\Delta|\phi_d\rangle$ and $\Delta_{dd'}^{ij} = \langle \phi_d^i|\Delta^j|\phi_{d'}^j\rangle$. A natural starting point in this process is to consider the corresponding free-atom or free-ion *d* states for the metal in question. For EDB metals, however, the unoccupied *d* states of the free atom are either unbound or too loosely bound to be at all useful in this regard, and the more localized free-ion d states are at best a modest improvement. For FDB metals, on the other hand, the fully occupied d states of the free atom are considerably more localized and potentially useful, although in practice they too are still far from an optimum choice. Because of the nonlinear nature of the xc potential, such free-atom d states lead to an overly complex hybridization potential that doesn't cleanly separate into volume and structure components, $\Delta = \Delta_{\text{vol}} + \Delta_{\text{struc}}$, as one desires in the GPT. Moreover, the tails of free-atom d states only have a rather slow exponential decay in real space, so that the overlap matrix elements $S_{dd'}^{ij}$ and $\Delta_{dd'}^{ij}$ have a much greater spatial extent than is necessary or desirable.

Historically, the solution to the practical problems with free-atom d states was achieved through the construction of a companion zero-order pseudoatom to the GPT, a simple DFT construct that was first introduced by Moriarty (1972c, 1974) and subsequently refined in the three papers that fully developed the DFT representation of GPT (Moriarty, 1977, 1982, 1988a). In the zero-order pseudoatom, one first replaces the valence s and p electrons of the free atom with the corresponding component of the free-electron gas, such that the pseudoatom electron density is a good first approximation to the electron density in the metal:

$$n_{\text{pa}}(r) = n_{\text{unif}} + n_{\text{core}}(r). \quad (4.75)$$

Here the uniform density is taken at it value in the metal, $n_{\text{unif}} = Z/\Omega$, for $r \leq R_{\text{WS}}$, and is set to zero for $r > R_{\text{WS}}$ to preserve charge neutrality. The core density n_{core} is that given by Eq. (2.32) and includes both occupied inner-core states and occupied d states. In the pseudoatom, $n_{\text{core}}(r)$ is self-consistently calculated in the presence of n_{unif}, using atomic boundary conditions on the inner-core states $\phi_c(\mathbf{r})$ in the Schrödinger equation

$$(T + v_{\text{pa}})|\phi_c\rangle = E_c^{\text{pa}}|\phi_c\rangle, \quad (4.76)$$

and applying an additional weak localization potential $v_{\text{loc}}(r)$ to control the radial extent of the d states $\phi_d(\mathbf{r})$, through the corresponding Schrödinger equation

$$(T + v_{\text{pa}} + v_{\text{loc}})|\phi_d\rangle = E_d^{\text{pa}}|\phi_d\rangle. \quad (4.77)$$

The self-consistent pseudoatom potential v_{pa} in Eqs. (4.76) and (4.77) is

$$v_{\text{pa}}(r) = v_{\text{unif}}(r) + v_{\text{ion}}(r), \quad (4.78)$$

where v_{unif} is the Coulomb potential arising from n_{unif} in Eq. (4.75),

$$v_{\text{unif}}(r) = \begin{cases} (Ze^2/R_{\text{WS}})(3 - r^2/R_{\text{WS}}^2)/2 & r < R_{\text{WS}} \\ Ze^2/r & r > R_{\text{WS}} \end{cases} \quad (4.79)$$

and v_{ion} is the remaining ionic potential for the pseudoatom,

$$v_{\text{ion}}(r) = -Z_a e^2/r + v_{\text{core}}(r) + v_{\text{xc}}(r). \quad (4.80)$$

In Eq. (4.80) the xc potential v_{xc} is taken in a form that makes v_{ion} consistent with Eqs. (2.24)–(2.26) for the total ionic potential V_{ion} in the metal:

$$v_{xc}(r) = \mu_{xc}(n_{core}(r) + n_{unif}) - \mu_{xc}(n_{unif}), \qquad (4.81)$$

where $v_{xc}^*(r)$ is the lowest-order approximation to $\mu_{xc}^*(n_{core})$, as given by Eq. (2.26) and evaluated with n_{val} replaced by n_{unif}. The additional localization potential in Eq. (4.77) remains general, and we only assume that it is of such form as not to interfere with the orthonormality between the inner core states and the d states. Also, it is convenient not to add back the constant xc energy $\mu_{xc}(n_{unif})$ for the free-electron gas to v_{unif}. Omitting this constant conveniently places the zero of energy for the pseudoatom eigenvalues, E_c^{pa} and E_d^{pa}, near the bottom of the valence bands in the metal.

In practice, the form of the pseudoatom potential v_{pa} in Eq. (4.78) greatly facilitates the construction of the hybridization potential Δ and its volume and structure components. In terms of the general defining equations for ϕ_d, Δ and δV, Eqs. (2.127)–(2.129), the atomic reference potential establishing ϕ_d in Eq. (2.127) is taken as

$$v_{at} = v_{pa} + v_{loc} - V_0', \qquad (4.82)$$

where $V_0' = V_0 - \mu_{xc}(n_{unif})$, with V_0 a freely chosen constant in the metal potential V. In practice, V_0' is used to adjust the zero of energy to the bottom of the valence energy bands, as was discussed for simple metals in connection with Eqs. (3.34) and (3.35), and for EDB or FDB metals in connection with Eq. (4.52). In most materials, V_0 and $\mu_{xc}(n_{unif})$ nearly cancel, indeed making V_0' a small first-order constant, as is implicitly assumed. Also recall the V_0' appears in the zero-order free-electron energy E_{fe}^0, as defined by Eq. (3.37), and thus contributes to the volume term in the cohesive energy of the metal.

While the constant V_0' shifts important band energies in the GPT formalism such as E_d^{vol}, this constant cancels out in hybridization potential:

$$\Delta = \delta V - \langle \phi_d | \delta V | \phi_d \rangle, \qquad (4.83)$$

where δV is the difference in potential between v_{at} and the full potential in the metal V:

$$\begin{aligned} \delta V(\mathbf{r}) &= v_{at}(r) - V(\mathbf{r}) \\ &= v_{pa}(r) + v_{loc}(r) - [V(\mathbf{r}) + V_0']. \end{aligned} \qquad (4.84)$$

The pseudoatom potential v_{pa} directly cancels against the corresponding component of the full potential V in Eq. (4.84), except for the small structure-independent difference

δV_{unif} between v_{unif} and the total uniform-electron-gas potential V_{unif}. Using Eq. (4.79), it is straightforward to derive an explicit analytic expression for $\delta V_{\text{unif}} = v_{\text{unif}} - V_{\text{unif}}$:

$$\delta V_{\text{unif}}(r) = \begin{cases} 0 & r < R_{\text{WS}} \\ Ze^2 \left[1/r - (3 - r^2/R_{\text{WS}}^2)/(2R_{\text{WS}})\right] & r > R_{\text{WS}} \end{cases}, \quad (4.85)$$

In the principal hybridization matrix element $\langle \mathbf{k} | \Delta | \phi_d \rangle$ one can then cleanly separate Δ and δV into volume and structure components: $\Delta = \Delta_{\text{vol}} + \Delta_{\text{struc}}$ and $\delta V = \delta V_{\text{vol}} + \delta V_{\text{struc}}$. The structure-independent contributions on the RHS of Eq. (4.84) establish the leading volume-dependent potential δV_{vol}:

$$\delta V_{\text{vol}}(r) = v_{\text{loc}}(r) + \delta V_{\text{unif}}(r), \quad (4.86)$$

leaving a small structure-dependent contribution

$$\delta V_{\text{struc}}(\mathbf{r}) = v_{\text{pa}}(r) - [V(\mathbf{r}) + V_0'] - \delta V_{\text{unif}}(r). \quad (4.87)$$

One can readily derive the following explicit form for δV_{struc}:

$$\delta V_{\text{struc}}(\mathbf{r}) = -\sum_{j \neq 0} v_{\text{ion}}(\mathbf{r} - \mathbf{R}_j) - \frac{1}{2} {\sum_{i,j}}' \delta v_{\text{xc}}(\mathbf{r} - \mathbf{R}_i, \mathbf{r} - \mathbf{R}_j)$$
$$- {\sum_{\mathbf{q}}}' S(\mathbf{q}) \frac{4\pi e^2}{q^2} [1 - G(q)][n_{\text{scr}}(q) + n_{\text{oh}}(q)] \exp(i\mathbf{q} \cdot \mathbf{r}), \quad (4.88)$$

where $\delta v_{\text{xc}}(\mathbf{r} - \mathbf{R}_i, \mathbf{r} - \mathbf{R}_j)$ is the lowest-order approximation to $\delta \mu_{\text{xc}}^*(n_i, n_j)$, as discussed in Sec. A2.3 of Appendix A2 and evaluated with n_{val} replaced by n_{unif}, as in Eq. (4.81). Recall from our previous discussion in Sec. 4.1.3 that δV_{struc} only enters the final GPT formalism via Eq. (4.50), through which the matrix elements $E_d^{\text{struc}} = -\langle \phi_d | \delta V_{\text{struc}} | \phi_d \rangle$ and $\Delta_{kd}^{\text{struc}} = \langle \mathbf{k} | \delta V_{\text{struc}} - \langle \phi_d | \delta V_{\text{struc}} | \phi_d \rangle | \phi_d \rangle$ are effectively absorbed into the theory.

The surviving volume component of the hybridization potential,

$$\Delta_{\text{vol}} = \delta V_{\text{vol}} - \langle \phi_d | \delta V_{\text{vol}} | \phi_d \rangle, \quad (4.89)$$

and the central hybridization matrix element $\Delta_{kd}^{\text{vol}} = \langle \mathbf{k} | \Delta_{\text{vol}} | \phi_d \rangle$ can be directly evaluated from Eq. (4.86), once a specific localization potential v_{loc} is chosen. Noting that in terms of Eq. (2.127) for ϕ_d one has $E_d^0 = E_d^{\text{pa}} - V_0'$, and the corresponding volume component of the d-state energy E_d is from Eq. (2.130) just

$$E_d^{\text{vol}} = E_d^0 - \langle \phi_d | \delta V_{\text{vol}} | \phi_d \rangle$$
$$= E_d^{\text{pa}} - \langle \phi_d | \delta V_{\text{vol}} | \phi_d \rangle - V_0'. \quad (4.90)$$

What remains is to evaluate the overlap matrix elements $\Delta_{dd'}^{ij} = \langle \phi_d^i | \Delta^j | \phi_{d'}^j \rangle$, which enter the overlap potential v_{ol} in Eq. (4.55). This requires a special treatment of Δ

and δV in Eqs. (4.83) and (4.84) that emphasizes the local contributions from the sites i and j. If one develops the full potential $V(\mathbf{r})$ in the metal as

$$V(\mathbf{r}) = \sum_i v_{\text{pa}}(\mathbf{r} - \mathbf{R}_i) + \frac{1}{2} \sum_{i,j}{}' \delta v_{\text{xc}}(\mathbf{r} - \mathbf{R}_i, \mathbf{r} - \mathbf{R}_j) + \cdots, \qquad (4.91)$$

then to first order one has for $\Delta_{dd'}(R_{ij}) \equiv \Delta_{dd'}^{ij}$ the central result

$$\Delta_{dd'}(R_{ij}) = \int \phi_d^{\star}(\mathbf{r} - \mathbf{R}_i) \left[v_{\text{loc}}(\mathbf{r} - \mathbf{R}_j) - v_{\text{pa}}(\mathbf{r} - \mathbf{R}_i) \right.$$
$$\left. - \delta v_{\text{xc}}(\mathbf{r} - \mathbf{R}_i, \mathbf{r} - \mathbf{R}_j) \right] \phi_{d'}(\mathbf{r} - \mathbf{R}_j) d\mathbf{r}, \qquad (4.92)$$

as first obtained by Moriarty (1988a). Recall from Eq. (2.133) that $\Delta_{dd'}(R_{ij}) = -h_{ij}^{dd'}(E_d)$, where the matrix elements $h_{ij}^{dd'}$ represent the usual TB bond integrals. Thus, for $d = d'$ the $m = 0$, 1 and 2 components of $-\Delta_{dd}(R_{ij})$ provide the familiar $dd\sigma$, $dd\pi$ and $dd\delta$ bond integrals that determine the d-band structure in the metal.

The external localization potential v_{loc} in Eq. (4.77) is intended to shape and control the spatial extent of the d basis states. Different useful forms for v_{loc} have been proposed and studied (Moriarty, 1972c, 1974, 1977), but in most cases the most effective one is the smooth barrier potential (Moriarty, 1977)

$$v_{\text{loc}}(r) = \begin{cases} 0 & r < R_{\text{WS}} \\ V_{\text{B}}(r/R_{\text{WS}} - 1)^2 & r > R_{\text{WS}} \end{cases}, \qquad (4.93)$$

where the barrier strength V_{B} is a constant. The r^2 dependence of $v_{\text{loc}}(r)$ as $r \to \infty$ gives $\phi_d(\mathbf{r})$ a rapidly decaying Gaussian-like tail, while the overall degree of localization is controlled by the single parameter V_{B}. For FDB metals, where the d states are tightly bound, Eq. (4.93) for $v_{\text{loc}}(r)$ is a robust form with only modest values of V_{B} required. For EDB metals, larger values of V_{B} are needed, and for difficult cases it may be necessary to add a square-well potential to Eq. (4.93) to establish a suitable d basis state.

The arbitrariness in v_{loc} and V_{B} simple reflect the arbitrariness in choosing a d basis state in the GPT, and represents no additional approximation in the theory. In this regard, one can usefully think of V_{B} as imposing an effective boundary condition on the d-basis state $\phi_d(\mathbf{r}) = R_d(r) Y_{2m}(\theta, \phi)$ at $r = R_{\text{WS}}$, where $R_d(r) = P_d(r)/r$. In the limit $V_{\text{B}} \to \infty$, the anti-bonding condition $R_d(R_{\text{WS}}) = 0$ is obtained, or in terms of the logarithmic derivative

$$D_2 \equiv R_{\text{WS}} \frac{d}{dr} \left[\ln R_d(R_{\text{WS}}) \right] = \frac{R_{\text{WS}}}{R_d(R_{\text{WS}})} \frac{dR_d(R_{\text{WS}})}{dr}, \qquad (4.94)$$

one has $D_2 = -\infty$. In the opposite limit $V_{\text{B}} \to 0$, the bonding condition $D_2 = 0$ or $dR_d(R_{\text{WS}})/dr = 0$ is approached although never actually reached in practice.

By construction, the logarithmic derivative D_2 is evaluated with the radial wavefunction $R_d(r)$ of the pseudoatom at energy $E = E_d^{\text{pa}}$. It is also very useful in practice to define a nearby, alternate logarithmic derivative

$$D_2^* \equiv R_{\text{WS}} \frac{d}{dr}\left[\ln R_2(R_{\text{WS}}, \bar{E}_d)\right], \qquad (4.95)$$

where $\bar{E}_d = E_d^{\text{vol}} + V_0' = E_d^{\text{pa}} - \langle \phi_d | \delta V_{\text{vol}} | \phi_d \rangle$ approximates the center of the d bands in the metal, and where $R_2(r, E)$ is the general $\ell = 2$ radial wavefunction one calculates with the pseudoatom potential $v_{\text{pa}}(r)$ for $r \leq R_{\text{WS}}$ at energy E. According to the canonical d-band theory discussed in Chapter 2, the boundary condition $D_2^* = -3$ corresponds to the center of the d bands, $E = \bar{E}_d$, and can be obtained by iterating V_B. This condition establishes the nominal choice of d-basis state for transition metals in the GPT, but EDB and FDB metals usually require a more localized d state corresponding to $|D_2^*| > 3$.

For EDB and FDB metals, the most basic requirements in the selection of the d-basis state and localization potential are the following. The d state must be localized to the degree that the effective strength of the hybridization potential Δ is weak, as measured by the matrix elements $\Delta_{\mathbf{k}d}^{\text{vol}} = \langle \mathbf{k} | \Delta_{\text{vol}} | \phi_d \rangle$ and $\Delta_{dd'}^{ij} = \langle \phi_d^i | \Delta^{ij} | \phi_{d'}^j \rangle$, together with the more general requirement

$$\Delta_{\text{vol}} \gg \Delta_{\text{struc}}. \qquad (4.96)$$

The weakness of $\Delta_{\mathbf{k}d}^{\text{vol}}$ is directly tested by the weak hybridization-gap criterion given by Eq. (2.44),

$$W_{\text{hyb}}/|\varepsilon_F - E_d| \ll 1, \qquad (4.97)$$

which, following the discussion in Sec. 2.7.1 of Chapter 2, we can evaluate as

$$W_{\text{hyb}} = \overline{|\Delta_{\mathbf{k}_F d}^{\text{vol}}|^2}/\varepsilon_F \qquad (4.98)$$

and $E_d = E_d^{\text{vol}}$. Developing the hybridization matrix element $\Delta_{\mathbf{k}d}^{\text{vol}}$ in the form

$$\langle \mathbf{k} | \Delta_{\text{vol}} | \phi_d \rangle = -4\pi \Delta_{\text{vol}}(k) Y_{2m}(\mathbf{k}) \qquad (4.99)$$

defines the useful characteristic function

$$\Delta_{\text{vol}}(k) \equiv \frac{1}{\Omega^{1/2}} \int_0^\infty j_2(kr) \Delta_{\text{vol}}(r) P_d(r) r dr, \qquad (4.100)$$

where j_2 is the $\ell = 2$ spherical Bessel function. This allows an easy evaluation of the average in Eq. (4.98), yielding the simple result $W_{\text{hyb}} = 4\pi \Delta_{\text{vol}}^2(k_F)/\varepsilon_F$.

For EDB metals, one has maximum flexibility in choosing $\phi_d(\mathbf{r})$ while satisfying both Eqs. (4.96) and (4.97), because $\phi_d(\mathbf{r})$ in this case does not affect the zero-order electron density of the bulk metal. Indeed, any d state with $|D_2^*| \geq 3$ easily satisfies the latter requirements, and the simple limiting choice $D_2^* = -\infty$, which maximizes the d-state localization, minimizes $\Delta_{\text{vol}}(k_F)$ and $W_{\text{hyb}}/|\varepsilon_F - E_d|$, while rendering Δ_{struc} negligible. The limit $D_2^* = -\infty$ also provides two further simplifications. First, $\Delta_{dd'}^{ij} \to 0$ in this limit and the overlap potential v_{ol} in the cohesive-energy functional is then entirely negligible. Second, the hybridization function $\Delta_{\text{vol}}(k)$ can be reduced to an exact analytic form

$$\Delta_{\text{vol}}(k) = -\frac{1}{\Omega^{1/2}} R_{\text{WS}} \frac{dP_d(R_{\text{WS}})}{dr} j_2(kR_{\text{WS}}), \quad (4.101)$$

a result first obtained by Moriarty (1974). These simplifications made the $D_2^* = -\infty$ limit the d state of choice for EDB metals in many early applications of the GPT (Moriarty, 1982, 1983; Moriarty and McMahan, 1982; McMahan and Moriarty, 1983). At the same time, this choice produces only lower bounds to $|\Delta_{kd}^{\text{vol}}|$ and $|\Delta_{dd'}^{ij}|$ and not optimum values in terms of calculated materials properties.

For FDB metals, there is generally less flexibility in choosing $\phi_d(\mathbf{r})$ due to the strong zero-order contribution the d states make to the electron density. In this case the choice $D_2^* = -\infty$ can often push the d-bands in the metal way too high in energy, while also maximizing $\Delta_{\text{vol}}(k_F)$ and $W_{\text{hyb}}/|\varepsilon_F - E_d|$. Better guidance is provided by the behavior of the d states in the corresponding free atom. In particular, the choice $D_2 = D_2^{\text{fa}}$, where

$$D_2^{\text{fa}} \equiv R_{\text{WS}} \frac{d}{dr} \left[\ln R_d^{\text{fa}}(R_{\text{WS}}) \right], \quad (4.102)$$

with R_2^{fa} the radial d wavefunction of the free atom, usually gives a good starting point. Calculated DFT values of $|D_2^{\text{fa}}|$ for 3d, 4d and 5d FDB metals are given in Table 4.1, together with pseudoatom values of $|D_2^*|$, if $D_2 = D_2^{\text{fa}}$. The values of D_2^* so obtained reflect the natural d-state localization that the Periodic Table provides, and in particular the strong increasing degree of localization with increasing valence Z. In all cases these D_2^* values also provide the desired weak quantities $\Delta_{\text{vol}}(k_F)$ and $W_{\text{hyb}}/|\varepsilon_F - E_d|$. Moreover, except for the noble metals, the additional requirement (4.96) is also satisfied. In the case of the noble metals, one can alternately choose $D_2^* = D_2^{\text{fa}}$ to achieve the extra needed localization to satisfy Eq. (4.96), as will be done in Sec. 4.3.

The above considerations, of course, do not preclude more accurate optimization of the d-basis states that could be obtained through machine learning. In practice, what one really wants is the d-basis state $\phi_d(\mathbf{r})$ that produces the best overall cohesive-energy functional within the GPT framework we have established for EDB and FDB metals. While a full machine-learning algorithm to accomplish this task is yet to be developed, we can demonstrate its quantitative importance through a relevant example. To do this, we address the easier question of what value of D_2^* produces the best phonon spectrum for a given EDB or FDB metal at its observed atomic volume Ω_0, as measured by the calculated Debye temperature Θ_D via Eq. (7.48) from Chapter 7. As demonstrated in Fig. 4.5 for our Ca and Zn prototypes, this task is readily accomplished by simply calculating Θ_D as a function of $|D_2^*|$ and comparing to experiment, noting in the process

Table 4.1 Free-atom d-state logarithmic derivatives $|D_2^{fa}|$ and, for the nominal choice $D_2 = D_2^{fa}$, pseudoatom values of $|D_2^\star|$ for 3d, 4d and 5d filled d-band metals at their observed equilibrium volumes, as obtained from nonrelativistic DFT calculations for the 3d and 4d elements and semi-relativistic DFT calculations for the 5d elements. Equilibrium atomic volumes are given in a.u.

Metal	Ω_0	$|D_2^{fa}|$	$|D_2^\star|$	Metal	Ω_0	$|D_2^{fa}|$	$|D_2^\star|$
Cu	79.68	2.9622	2.5662	Zn	102.7	3.7534	3.3444
Ag	115.1	3.4578	3.1077	Cd	145.6	4.2206	3.8371
Au	114.1	3.2915	2.9603	Hg	155.1	3.9688	3.6091
Ga	130.5	4.8650	4.5192	Ge	152.8	5.9786	5.3777
In	172.8	5.1434	4.7609	Sn	181.1	5.9176	5.6089
Tl	192.9	4.8332	4.4989	Pb	204.7	5.5263	5.2718

that for both metals there is a significant quantitative sensitivity of Θ_D to $D_2^\star$. For the case of Ca, shown in Fig. 4.5(a), the approximate optimum value of $|D_2^\star|$ so determined is 6.0. In the case of Zn, displayed in Fig. 4.5(b), the approximate optimum value of $|D_2^\star|$ is 3.25, which is close to free-atom guidance of 3.34 from Table 4.1. Figure 4.5

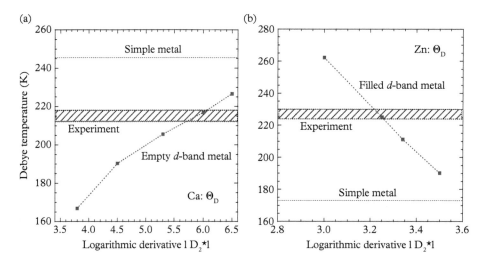

Fig. 4.5 Calculated Debye temperature Θ_D as a function of the logarithmic derivative $|D_2^\star|$ defining the d-basis states for prototype empty- and filled-d-band metals at their observed equilibrium volumes. Shown for comparison in each case is the value of Θ_D obtained in the simple-metal limit with no sp-d hybridization or d-state overlap. (a) fcc Ca as an EDB metal at $\Omega_0 = 294.5$ a.u.; (b) hcp Zn as a FDB metal at $\Omega_0 = 102.7$ a.u.

also shows that both the optimum EDB GPT representation for Ca and optimum FDB GPT representation for Zn provide large improvements over the calculated values of Θ_D obtained in the simple-metal limit without sp-d hybridization or d-state overlap, where Θ_D is overestimated by 14% in Ca and underestimated by 23% in Zn.

In Fig. 4.6 we display the corresponding optimum d-state radial wavefunctions $P_d(r)$ for Ca and Zn, together with the optimum hybridization functions $\Delta_{\text{vol}}(k)$. In the case of Ca, $P_d(r)$ falls sharply to zero beyond about $r = 1.2 R_{\text{WS}}$, leading to large-amplitude oscillations in $\Delta_{\text{vol}}(k)$ for $k > k_{\text{F}}$. These oscillations, of course, have no direct negative consequences for the calculation of either the electron density or the cohesive-energy functional, because all **k**-space integrations in the EDB GPT are for $k \leq k_{\text{F}}$ only. In the case of Zn, $P_d(r)$ decays rapidly but smoothly to zero beyond about $r = 2 R_{\text{WS}}$. The corresponding long-range oscillations in $\Delta_{\text{vol}}(k)$ are now strongly damped, allowing good convergence of the **k**-space integrations in the FDB GPT for $k \geq k_{\text{F}}$.

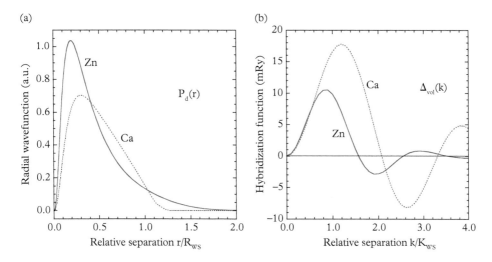

Fig. 4.6 *Optimized d-basis-state wavefunctions and hybridization functions for Ca and Zn at their observed equilibrium volumes of $\Omega_0 = 294.5$ a.u. and 102.7 a.u., respectively. (a) Radial wavefunction $P_d(r)$; (b) hybridization function $\Delta_{\text{vol}}(k)$, as calculated from Eq. (4.100).*

To come full circle, we finally test the ability of the optimum representations for Ca and Zn to describe the underlying H-NFE-TB band structures of these metals, and especially the performance of the Δ_{kd}^{vol} and $\Delta_{dd'}^{ij}$ matrix elements in describing the d bands and their hybridization with the NFE bands. We apply this test here only for the fcc crystal structure to allow direct quantitative comparison with the DFT results of Moruzzi et al. (1978). As discussed in Chapter 2, the general GPT band structure for d-band metals is well described by the H-NFE-TB secular determinant (2.139). Simplified expressions for the fcc band energies at the high symmetry points Γ, X and L in the BZ are given in Sec. A2.5 of Appendix A2 in terms of the quantities w_0, E_d, Δ, $dd\sigma$, $dd\pi$ and $dd\delta$. Evaluated here with E_d^{vol}, Δ_{vol} and $\Delta_{dd'}^{ij}$ from Eqs. (4.89)–(4.92),

Table 4.2 *High-symmetry fcc band-structure energies for Ca and Zn, as obtained from the optimum GPT representations derived from Fig. 4.5, as discussed in the text, and from the DFT calculations of Moruzzi et al. (1978). Volumes are given in a.u. and energies in Ry.*

Metal	Ca	Ca	Ca	Zn	Zn	Zn
Method	GPT	GPT	DFT	GPT	GPT	DFT
Ω	294.5	250.0	250.0	102.7	95.27	95.27
$\|D_2^\star\|$	6.0	5.825	—	3.25	3.18	—
$E_F - \Gamma_1$	0.302	0.326	0.302	0.763	0.810	0.799
$\Gamma_{25'} - \Gamma_1$	0.475	0.517	0.485	0.185	0.221	0.195
$\Gamma_{12} - \Gamma_1$	0.518	0.570	—	0.219	0.260	0.225
$\Gamma_{12} - \Gamma_{25'}$	0.044	0.054	—	0.033	0.039	0.030
$X_1 - \Gamma_1$	0.282	0.300	0.252	0.132	0.157	0.135
$X_5 - \Gamma_1$	0.575	0.640	—	0.238	0.281	0.255
$X_5 - X_1$	0.294	0.340	—	0.105	0.124	0.120
$X_{4'} - \Gamma_1$	0.358	0.403	0.411	0.684	0.715	0.715
$L_1 - \Gamma_1$	0.244	0.269	0.236	0.648	0.704	0.711
$L_{2'} - \Gamma_1$	0.280	0.311	0.305	0.548	0.569	0.554
$L_1 - L_{2'}$	−0.036	−0.042	−0.069	0.100	0.135	0.157

these expressions have been used in Table 4.2 to calculate the GPT band energies for Ca and Zn, both at their observed equilibrium volumes and at the reduced volumes treated by Moruzzi et al. (1978), allowing in each case for the expected volume dependence of $|D_2^\star|$, as discussed in Sec. 4.5. In both metals, there is a qualitatively correct alignment between the GPT and DFT band energies, with the position of the d bands, as measured by $\Gamma_{25'} - \Gamma_1$, located well above the Fermi energy E_F in Ca and near the bottom of the NFE bands at Γ_1 in Zn. The sp-d hybridized band energy X_1 is properly positioned in both metals, and the full width of the occupied d bands in Zn, as measured by $X_5 - X_1$, is well described. Lastly, the sp NFE band energies $X_{4'}$, L_1 and $L_{2'}$ are also well calculated in both metals.

4.3 Modified FDB-GPT treatment for the special case of the noble metals

Conceptually, the noble metals would seem to be ideally suited as filled-d-band metals within the GPT. We know that the atomic configuration of the valence electrons in a noble metal atom is of closed d-shell form, which for the Cu free atom is just $(4s)^1(3d)^{10}$, and similarly $(5s)^1(4d)^{10}$ for Ag and $(6s)^1(5d)^{10}$ for Au. We know further that the observed fcc Fermi surfaces of all three noble metals can be described as a modified free-electron sphere of radius k_F, as calculated for $Z=1$, with small well-defined "necks" of radius $0.14 - 0.19 k_\text{F}$ opening on the sphere surface, each centered along a $<111>$ direction and touching a high-symmetry BZ plane (Coleridge and Templeton, 1972; MacDonald et al., 1982). These Fermi-surface features are clear byproducts of the sp-d hybridization that the FDB GPT attempts to treat. Even before the development of the full GPT, an early FDB treatment of the noble metals using free-atom d-basis states (Moriarty, 1970, 1972b) showed considerable promise from an applications perspective, including a good description of the liquid metal resistivity in Cu (Moriarty, 1970), reasonable phonon spectra for Ag and Au (Moriarty, 1972b), and subsequently a good description of liquid metal structure and thermodynamics in Cu and Ag (Regnaut et al., 1983). At the same time, however, there were clear warning signs of possible problems, especially the large magnitude of the calculated sp-d hybridization contributions, leading to unobserved Kohn anomalies in the phonon spectra, especially for Cu, and the failure to predict the observed fcc structure in any of the three noble metals, with a negative hcp-fcc structural energy difference obtained in each case for at least one value of c/a.

Subsequent in-depth studies of Cu in the context of the DFT-based GPT for FDB metals (Moriarty, 1977, 1982) exposed some of the main underlying technical problems for this material. First, the occupied d bands are relatively close to the Fermi energy E_F and the d-band width itself is a significant fraction of the overall width of the occupied valence bands, as measured by $E_\text{F} - \Gamma_1$. Consequently, $E_\text{F} - \Gamma_1$ is considerably larger than the free-electron Fermi energy ε_F, and using the calculated first-principles GPT energies one finds $E_\text{F} - E_d \gg \varepsilon_\text{F} - E_d^\text{vol}$. Even if E_d^vol is adjusted downward in an ad hoc manner to make $\varepsilon_\text{F} - E_d^\text{vol} \sim E_\text{F} - E_d$, the calculated electron density for Cu remains problematic, with too-large first-order screening and orthogonalization-hole contributions obtained. In particular, the effective valence Z^* is about 1.80 for $D_2^* = -3$, signaling a large $d \rightarrow sp$ transfer of valence electron density. This effect is qualitatively correct, but quantitatively overestimated as a first-order perturbation with $(Z^* - Z)/Z = 0.8$, as has been verified within the GPT by treating Cu more accurately as a transition metal in the construction of the zero-order pseudoatom (Moriarty, 1982). The transition-metal treatment yields a self-consistent valence of $Z = 1.651$ for $D_2^* = -3$. This leads to an effective valence of only $Z^* = 1.579$, with $(Z^* - Z)/Z = -0.0436$ now a properly small first-order quantity, and a correct positioning of the d bands without any need to adjust E_d^vol.

The GPT treatment of the noble metals as transition metals is addressed in Chapter 5, but the conceptual simplicity of an FDB representation remains important because of the significance of the $Z = 1$ Fermi surface. To improve the FDB-GPT representation of the noble metals, the general strategy adopted has been one of using the freedom to choose D_2^* together with a small number of additional parameters needed to adjust the first-principles FDB-GPT formalism. Two important parameters in this regard are a d-energy shift δE_d^shift, which is used to control the position of E_d^vol, and a scaling parameter α_oh, varying between 0 and 1, that multiplies and reduces the magnitude of the too-large orthogonalization-hole correction term δv_oh^{ij}, as given by Eq. (4.60), which enters the overlap potential v_{ol}. Within this framework, different specific schemes have been explored beginning with the work of Moriarty (1982). In the currently favored scheme, one begins with the free-atom guidance on $|D_2^*|$ provided in Table 4.1, and with $\alpha_{\text{oh}} = 0$. Specifically, one here chooses $D_2^* = D_2^{\text{fa}}$, the free-atom logarithmic derivative, to ensure adequately localized d basis states, with $|D_2^*| \cong 2.96$ for Cu, 3.45 for Ag and 3.29 for Au. Next one can derive a simple self-consistent expression for $\delta E_d^{\text{shift}}$ that exactly maintains $\varepsilon_F - E_d^{\text{vol}} = E_F - E_d$, using the fact that $E_F \simeq \varepsilon_F + \langle \mathbf{k}_F | w_0^{\text{pa}} | \mathbf{k}_F \rangle$ implicitly depends on E_d^{vol}. One thereby finds $\delta E_d^{\text{shift}} = -0.3313$ Ry for Cu, -0.1955 Ry for Ag and -0.3610 Ry for Au. The weak sp-d hybridization criteria (4.96) and (4.97) are then adequately satisfied, and a reasonably good description of the fcc band structure is obtained for each metal, as measured by the calculated values of $E_F - E_d$. In particular, one calculates GPT values of $E_F - E_d = 0.20$ Ry for Cu and 0.30 Ry for Ag, in comparison to the DFT estimates of 0.21 Ry and 0.35 Ry, respectively, obtained from the self-consistent band structure results of Moruzzi et al. (1978).

At this point, the pair potentials $v_2(r, \Omega_0)$ derived from this scheme correctly yield positive hcp-fcc and bcc-fcc structural energies, as well as qualitatively reasonable phonon spectra, for all three noble metals. The calculated Debye temperatures Θ_D, however, are still 10–25% too large. There also remain small Kohn anomalies and soft modes in the phonon spectra, features driven by the still too large sp-d hybridization contributions. These unwanted features can be systematically removed by increasing the magnitude of $|\delta E_d^{\text{shift}}|$, and thereby further lowering E_d^{vol}. For example, an acceptable enhancement of the phonon spectra, especially in Cu and Ag, can be accomplished by adding 0.2 Ry to $|\delta E_d^{\text{shift}}|$ for each metal, which reduces Z^* to more desirable levels: 1.3842 in Cu, 1.3667 in Ag and 1.4877 in Au. At the same time, the calculated values of Θ_D are also lowered and, as a last step, can be brought into line with experiment by using small fitted values of the α_{oh} parameter (0.273 for Cu, 0.378 for Ag and 0.113 for Au) without adversely affecting either the quality of the phonon spectra or the calculated structural phase stability.

The final modified-GPT pair potentials for Cu, Ag and Au so obtained are displayed in Fig. 4.7, together with the corresponding form factors for these metals. Both the form factors and the pair potentials of Cu and Ag are quite similar, with a near constant $w(q)$ function for $q/k_F < 1$ followed by a rapid rise to positive energy above $q/k_F = 1.8$, and with shallow first minima in $v_2(r, \Omega_0)$ near $r/R_{\text{WS}} = 1.8$. This behavior is in sharp contrast to that exhibited by Au, with its deep minimum in $w(q)$ near $q/k_F = 1.2$ and its strongly repulsive pair potential below $r/R_{\text{WS}} = 2.8$. For Cu at least, these results represent a clear improvement over the initial modified-GPT scheme developed by Moriarty (1982) for that metal, based on highly localized d basis states. At the same time, the present results are still far from optimized, and important features such as the location, depth and shape of the first minimum in $v_2(r, \Omega_0)$ are sensitive to the parameter details. In the future, machine learning could be used to refine and extend the FDB-GPT treatment of these metals by fully optimizing the variable parameters D_2^*, $\delta E_d^{\text{shift}}$ and α_{oh} as a function of volume.

It should also be mentioned that there have been several independent attempts to develop a parameterized version of the FDB GPT for application to the noble metals.

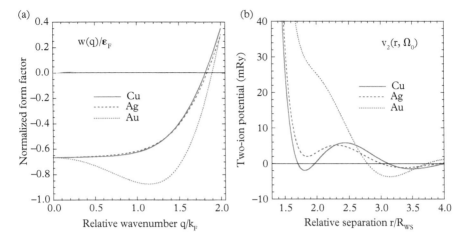

Fig. 4.7 *Modified filled-d-band GPT treatment of the noble metals at their observed equilibrium volumes (see Table 4.1), as discussed in the text. (a) Form factor* $w(q) \equiv \langle \mathbf{k}_F + \mathbf{q}| w |\mathbf{k}_F\rangle$; *(b) Two-ion pair potential* $v_2(r, \Omega_0)$.

The most noteworthy of these parameterizations was developed by Regnaut et al. (1985), using Cu as a prototype. In this approach, the GPT contributions to the nonlocal transition-metal PP w_0 were only very approximately replaced with a simple one-parameter local MP, but the important sp-d hybridization integrals S_{kd} and Δ_{kd}^{vol}, on the other hand, were realistically modeled using an analytic free-atom d state for Cu and a single additional parameter needed to define Δ_{vol}. Including the d-band energy E_d^{vol}, a total of three parameters then constrained the normalized energy-wavenumber characteristic $F_N(q, \Omega)$ as given by Eq. (4.62). The three parameters were fitted to DFT fcc band-structure data and $F_N(q, \Omega)$ numerically evaluated at solid and liquid densities. The remaining overlap potential, however, was not directly parameterized, but instead just replaced in an ad hoc manner by a simple Born-Mayer form $v_{ol}(r) = A\exp(-\lambda r)$, where the additional parameters A and λ were fitted to structure-factor and isothermal compressibility data in liquid Cu. The resulting pair potentials $v_2(r, \Omega)$ calculated in both the solid and the liquid are qualitatively similar to the GPT result for Cu shown in Fig. 4.7(b), with a shallow first minimum in the potential, but one at positive energy. With these potentials, close quantitative agreement with experiment was achieved on the liquid structure factor at two widely separated temperatures.

4.4 Alternate resonant model potential approach

During the same time period that the full DFT version of GPT was being developed and first applied to EDB and FDB metals, an alternate resonant model potential (RMP) approach for such materials was successfully formulated by Dagens (1976, 1977a), with subsequent refinements and additional significant applications to the noble metals (Dagens, 1977b; Upadhyaya and Dagens, 1978, 1979, 1982a; Lam et al., 1983), as well as to Zn and Cd (Upadhyaya and Dagens, 1982b). In part, the RMP approach was inspired by and provides an extension of the earlier DRT MP theory for simple metals discussed in Chapter 3. In the RMP, however, the emphasis is on a strongly energy-dependent MP form for $\ell = 2$ that can be applied to transition-series d states, and, through a perturbation analysis similar to GPT, can yield an energy-wavenumber characteristic and pair potential for EDB and FDB metals.

The RMP theory begins with the assumption that the electron-ion interaction can be represented by a nonlocal RMP for each ion, consisting of a sum of a standard energy-independent simple-metal contribution w_{sm} for $\ell = 0$ and $\ell = 1$ states plus an energy-dependent resonant contribution w_{res} for $\ell = 2$ states. The form factor of the screened RMP is

$$w(\mathbf{k}, \mathbf{q}) \equiv w_{sm}(\mathbf{k}, \mathbf{q}) + w_{res}(\mathbf{k}, \mathbf{q}, \varepsilon_k), \qquad (4.103)$$

where w_{res} has the assumed general form

$$w_{res}(\mathbf{k}, \mathbf{q}, E) = u(\mathbf{k}, \mathbf{q})/[E - \xi(E)]. \qquad (4.104)$$

The quantity ξ determines the position of the d bands E_d. The energy dependence of w_{res} produces a depletion-hole n_{dpl} in the electron density and corresponding negative depletion charge Z_{dpl}, such that the effective valence is $(Z^*)^2 = (Z - Z_{dpl})^2 - Z_{dpl}^2$. If one makes the following nominal identifications with the GPT: $w_0(\mathbf{k}, \mathbf{q}) \to w_{sm}(\mathbf{k}, \mathbf{q})$, $h_1(\mathbf{k}, \mathbf{q}) \to w_{res}(\mathbf{k}, \mathbf{q}, \varepsilon_k)$ and $n_{oh}(q) \to n_{dpl}(q)$, one directly obtains forms for the screening electron density $n_{scr}(q)$ from Eq. (4.32) and the normalized energy-wavenumber characteristics $F_N(q, \Omega)$ from Eq. (4.62).

At the same time, there are some important differences between the GPT and RMP approaches. First, in the GPT the transition-metal PP form factor $w_0(\mathbf{k}, \mathbf{q})$, as given by Eq. (4.33), contains significant d-state contributions that are absent in $w_{sm}(\mathbf{k}, \mathbf{q})$ for the RMP. In addition, the overlap potential $v_{ol}(r, \Omega)$ in the GPT is completely absent in the RMP formalism, based on the assumption that its effect is already contained within w_{res}. One can get insight into the latter assumption by examining the RMP function $u(\mathbf{k}, \mathbf{q})$ in Eq. (4.104), which has the form

$$u(\mathbf{k}, \mathbf{q}) = 4\pi A_d \gamma(k') \gamma(k) \sum_m Y_{2m}^*(\mathbf{k}') Y_{2m}(\mathbf{k}), \qquad (4.105)$$

with $\mathbf{k}' \equiv \mathbf{k} + \mathbf{q}$ and $\gamma(k) = [k_0^2/(k_0^2 - k^2)]j_2(kR_m)$. Here A_d and R_m are parameters and $k_0 R_m = 5.76346$ is the first nonvanishing zero of the $\ell = 2$ spherical Bessel function j_2. Comparing to the function $h_1(\mathbf{k}, \mathbf{q})$ in the GPT, one has the following identification with the GPT hybridization function $\Delta_{vol}(k)$:

$$\Delta_{vol}(k) \to \Delta_{RMP}(k) = \left(\frac{A_d}{4\pi}\right)^{1/2} \gamma(k). \qquad (4.106)$$

4.4 Alternate resonant model potential approach

The quantity $\gamma(k)$ is the Fourier transform of the localized, but otherwise arbitrary, real-space function $j_2(k_0 r)\Theta(1 - r/R_m)$. Dagens maintains a nonoverlapping representation in the RMP by constraining the model radius R_m to be less than half of the nearest-neighbor distance in the metal. In the GPT, this would correspond to an ultra-localized d state $\phi_d(\mathbf{r})$ of approximately the same spatial extent. Localizing $\phi_d(\mathbf{r})$ has the effect of pushing out $\Delta_{\text{vol}}(k)$ in $\mathbf{k}$ space, so the RMP hybridization function $\Delta_{\text{RMP}}(k)$ is even more extended than the $\Delta_{\text{vol}}(k)$ arising from the highly localized $D_2^* = -\infty$ limit of the GPT, as shown in Fig. 4.8(a) for copper. At the same time, it is interesting to note from Fig. 4.8(a) that the magnitudes of $\Delta_{\text{vol}}(k)$ and $\Delta_{\text{RMP}}(k)$ near $k = k_F$ are similar for all three cases compared: RMP and GPT with $D_2^* = -2.96$ and $D_2^* = -\infty$. In this regard, we also point out that in Fig. 8 of Moriarty (1982) there is a factor of $\sqrt{2}$ missing in the plotted $\Delta_{\text{vol}}(k)$ function for the RMP case.

In addition, noting that in the RMP formalism one has

$$Z_{\text{dpl}} = -\frac{20\Omega}{\pi} \int_{k_F}^{\infty} \frac{\Delta_{\text{RMP}}^2(k) k^2 \, dk}{(\varepsilon_k - \xi)^2}, \quad (4.107)$$

which is consistent with the interpretation of Z_{dpl} as $[2\Omega/(2\pi)^3] \int h_2(\mathbf{k})\Theta_>(\mathbf{k}) d\mathbf{k}$ in the GPT with $h_2(\mathbf{k})$ given by Eq. (4.38), the qualitative behavior of $\Delta_{\text{RMP}}(k)$ explains the very large calculated values of Z^* in the RMP: 2.121 for Cu, 2.144 for Ag and 2.360 for Au (Dagens, 1977b). At the same time, this behavior creates the theoretical dilemma of very large first-order screening and depletion-hole densities, a qualitatively similar

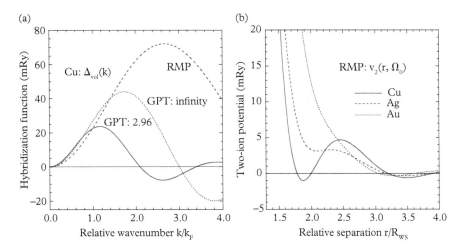

Fig. 4.8 (a) Hybridization function $\Delta_{\text{vol}}(k)$ for copper at $\Omega_0 = 79.68$ a.u., as calculated in the RMP from Eq. (4.106) and in the GPT from Eq. (4.100) for $D_2^* = -2.96$ and from Eq. (4.101) for $D_2^* = -\infty$; (b) optimized RMP pair potentials $v_2(r, \Omega_0)$ for the noble metals at their observed equilibrium volumes (Dagens, 1977b), as in Fig. 4.7 for the GPT.

problem to what is experienced in the GPT for the noble metals with large values of $|D_2^*|$ (Moriarty, 1982). In the RMP approach, Dagens effectively sidesteps this difficulty by not separating zero- and first-order components of the electron density in determining his remaining parameters (e.g., A_d). Instead, he requires only that the total electron density be constrained through the matching of appropriate logarithmic derivatives to an external calculation of the same in the interstitial region $r \geq R_m$ surrounding each ion. To obtain the latter, he introduced a Wigner-Seitz (WS) neutral-atom calculation (Dagens, 1977b), in which the valence s-state wave function, its corresponding energy level $E_s = \Gamma_1$ and the valence sp electron density were established by applying a zero-slope boundary condition at $r = R_{WS}$. He then used band-structure calculations of $E_d - \Gamma_1$ to fix the valence d-state energy E_d needed to calculate the corresponding d-state wave function and electron density for the d electrons. The WS electron-density data so obtained was then input into a companion neutral pseudoatom calculation to establish the RMP parameters.

The final optimized RMP pair potentials $v_2(r, \Omega_0)$ determined for the noble metals (Dagens, 1977b), as calculated from Eq. (4.61) with $v_{ol}(r, \Omega_0) = 0$ and fitted numerically to simple analytic functions by Lam et al. (1983), are plotted in Fig. 4.8(b). The RMP potentials so obtained are seen to be qualitatively similar to the modified-GPT potentials displayed in Fig. 4.7(b), with shallow first minima for Cu and Ag, at negative energy for Cu and positive energy for Ag, and with a strongly repulsive near-neighbor behavior for Au. In his RMP formalism, Dagens made no attempt to treat the corresponding volume term $E_{\text{vol}}(\Omega)$, nor for that matter, did he or his collaborators explore the volume dependence of the noble metal pair potentials. Thus, the full cohesive-energy functional $E_{\text{coh}}(\mathbf{R}, \Omega)$ was not obtained for any of metals studied.

4.5 Trends in first-principles GPT pair potentials with atomic number and volume

In this section, we return to the first-principles GPT for EDB and FDB metals established in Secs. 4.1 and 4.2, and examine trends in the pair potentials $v_2(r, \Omega)$ with atomic number and volume. As with our prototype examples of Ca and Zn, we focus here on the group-IIA and -IIB metals in the Periodic Table. Physically, the corresponding $4d$ elemental metals fcc Sr and hcp Cd behave similarly to fcc Ca and hcp Zn, respectively, such that measured (or in the case of Cd estimated) values of the Debye temperature Θ_D can be used directly to optimize the choice of D_2^*, as was done for Ca and Zn in Sec. 4.2. The approximate optimum values of D_2^* so obtained are -5.0 for Sr and -3.73 for Cd at their respective observed equilibrium atomic volumes. The resulting optimized GPT pair potentials $v_2(r, \Omega_0)$ for Sr and Cd, calculated as a function of r/R_{WS}, are found to be remarkably similar to those for Ca and Zn, respectively, as shown in Fig. 4.9. These results are also consistent with the similar trend displayed by Cu and Ag in Fig. 4.7(b). In the cases of Ca and Sr, the deep minima in $v_2(r, \Omega_0)$ near $r = 1.8 R_{WS}$ are favorable to the predicted and observed fcc structure of these metals. In the cases of Zn and Cd, on the other hand, the contrasting steep shoulder in the potentials near $r = 1.8 R_{WS}$

4.5 Trends in first-principles GPT pair potentials with atomic number and volume

are favorable to a distorted hcp structure with a high c/a ratio, as is both predicted and observed in these metals.

In sharp contrast to their $3d$ and $4d$ counterparts, the $5d$ elements Ba and Hg display unique physical behavior at their observed equilibrium volumes. Unlike fcc Ca and Sr, Ba forms in the bcc structure and with an anomalous phonon spectrum. This behavior is directly driven by the electronic structure of Ba, where the bottom of the $5d$ bands has dropped below the Fermi level, resulting in the onset of transition-metal behavior. In the context of the GPT, a full transition-metal description, with partial filling of the d-bands together with both two-ion and three-ion multi-ion potentials, is both necessary and sufficient to explain the observed structural phase stability and phonon spectrum of this metal (Moriarty, 1986b). The GPT transition-metal treatment of Ba is considered in Chapter 5 and applied to structural phase stability in Chapter 6 and to the bcc phonon spectrum in Chapter 7. Here, for the purpose of direct comparison with Ca and Sr, we consider only an EDB pair potential for Ba calculated with a representative logarithmic derivative of $D_2^* = -5.5$. This potential is displayed in Fig. 4.9(a) and is seen to be qualitatively similar to those for Ca and Sr, but pushed radially outward, with its first minimum at $r = 2.0 R_{\mathrm{WS}}$. The EDB Ba pair potential does still manage to produce a mechanically stable fcc structure with all real phonons, although hcp and not fcc is the structure of lowest total energy, while the observed bcc structure is calculated to be of still higher energy and mechanically unstable.

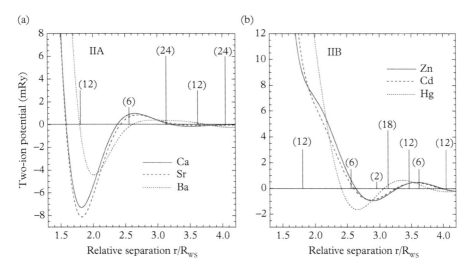

Fig. 4.9 First-principles GPT pair potentials $v_2(r, \Omega_0)$ for the group-IIA and -IIB metals at their observed equilibrium volumes (see Table A1.1). (a) Ca, Sr and Ba in the EDB limit, as obtained with $D_2^* = -6.0$, -5.0 and -5.5, respectively, with the position and number of fcc near neighbors indicated; (b) Zn, Cd and Hg in the FDB limit, as obtained with $D_2^* = -3.25$, -3.73 and -3.51, respectively, with the position and number of ideal hcp near neighbors indicated.

The physical link between the behavior of fcc Ca and Sr on one hand and bcc Ba on the other is to be found at high pressure. As discussed in Chapter 6, pre- and early transition metals experience a continuous $sp \to d$ transfer of electrons under compression as the narrow d bands lower in energy relative to the broad valence sp bands, and this change in electronic structure in turn drives structural phase transitions. Both Ca and Sr are observed to undergo high-pressure fcc $\to$ bcc phase transitions below 20 GPa. The appearance of the bcc structure in this manner for Ca and Sr results from the onset of transition-metal behavior exactly as found for Ba at ambient pressure.

With regard to the remaining $5d$ metal Hg, its physical behavior is separated from that of Zn and Cd by relativistic effects, which lower the valence sp bands relative to the filled d bands. In the context of the GPT, a FDB treatment still applies to Hg with or without a semi-relativistic treatment of the inner core and d states, a treatment that neglects only the spin-orbit interaction. As shown by Moriarty (1988b), in the nonrelativistic limit Hg retains an hcp structure with a high c/a ratio exactly as in Zn and Cd, but with semi-relativistic effects included in the FDB GPT treatment, Hg correctly favors the observed low-temperature bct or β-Hg structure. Moreover, one can use the observed c/a ratio of $1/\sqrt{2} \cong 0.707$ for β-Hg to optimize the choice of D_2^*, which yields $D_2^* = -3.51$ and an appropriately stiff pair potential below $r = 2.5 R_{WS}$, as shown in Fig. 4.9(b). This is an important consideration that produces mechanical stability and all real phonons for the β-Hg structure, quantities that are highly sensitive to the calculated c/a ratio.

The corresponding volume dependence of the pair potentials $v_2(r, \Omega)$ for EDB and FDB metals depends on the implicit volume dependence of the d-basis states and hence on the defining logarithmic derivative $D_2^*(\Omega)$. While in general this requires applying a suitable optimization procedure as a function of volume to determine $D_2^*(\Omega)$, a great deal of useful insight about this function can be gleaned by examining the volume dependence of the free-atom logarithmic derivative D_2^{fa} for our FDB prototype metal Zn. As shown in Fig. 4.10(a), both $D_2^{fa}(\Omega)$ and, with $D_2 = D_2^{fa}$, the corresponding function $D_2^*(\Omega)$ for Zn decrease in magnitude under compression, and are almost exactly linear in the reduced length-scale variable $x = (\Omega/\Omega_0)^{1/3}$. Thus, to an excellent approximation

$$\left|D_2^*(\Omega)\right| = \left|D_2^*(\Omega_0)\right| + \alpha \left[(\Omega/\Omega_0)^{1/3} - 1\right], \qquad (4.108)$$

where α is a positive constant. Note also from Fig. 4.10(a) that this relationship holds over a wide volume range, from $x = 0.76$ or $\Omega/\Omega_0 = 0.44$ to $x = 1.08$ or $\Omega/\Omega_0 = 1.26$, and yields the leading two terms in a general Taylor expansion in powers of $(x - 1)$. In this regard, Eq. (4.108) can be considered to be a good starting point for establishing the optimum form of $\left|D_2^*(\Omega)\right|$ for any metal. For the case of Zn, one has $\left|D_2^*(\Omega_0)\right| = 3.34$ from Table 4.1 and $\alpha = 2.775$ derived from Fig. 4.10(a). Reducing $\left|D_2^*(\Omega_0)\right|$ to its approximate optimum value of 3.25 and keeping α fixed, then yields a good choice of optimum function $\left|D_2^*(\Omega)\right|$ for Zn, which is also displayed in Fig. 4.10(a).

The corresponding optimum volume-dependent potentials $v_2(r, \Omega)$ for Zn are displayed in Fig. 4.10(b). While the near neighbor shoulder at $r = 1.8 R_{WS}$ in the potential deepens under expansion, this shoulder effectively vanishes into a strongly repulsive potential below $r = 2.8 R_{\text{WS}}$ under compression. The magnitude of the characteristic long-range oscillations in $v_2(r, \Omega)$ increases under strong compression, but remain approximately stationary as a function of r/R_{WS}, consistent with the behavior of Friedel oscillations.

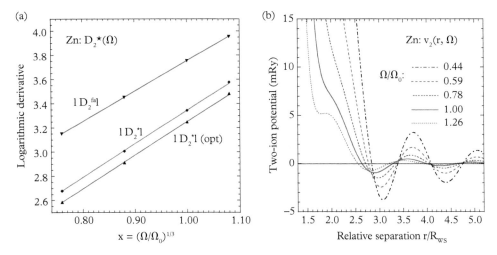

Fig. 4.10 Volume dependence of the logarithmic derivatives and pair potentials for Zn. (a) free-atom function $|D_2^{\text{fa}}(\Omega)|$, corresponding function $|D_2^*(\Omega)|$ in the metal, and inferred approximate optimum function $|D_2^*(\Omega)|$, with the solid points calculated values and the solid lines linear fits to those points in the form of Eq. (4.108); (b) $v_2(r, \Omega)$ calculate from the optimum $|D_2^*(\Omega)|$ at five volumes Ω ranging from 45.08 to 129.37 a.u.

Application of Eq. (4.108) for $|D_2^*(\Omega)|$ to EDB metals requires a separate calculation of the constant α. For application to small volume ranges, such as is needed in Ca and Sr, one useful way to proceed is to use a hidden property of Eq. (4.108), namely, that this equation is consistent with a slowly varying barrier constant V_B as a function of volume in the d-state localization potential v_{loc} establishing $|D_2^*(\Omega)|$. In such situations it is reasonable to establish α by using the simple condition $V_B = $ constant, as has been done implicitly for Ca in Table 4.2.

5
Interatomic Potentials in Transition Metals

In this chapter, we extend our development of QBIPs for d-electron metals from the empty- and filled-d-band limiting cases considered in Chapter 4 to the challenging general case of pure transition metals with partially filled d bands. While the sp-d hybridization that characterizes EDB and FDB metals remains important in transition metals, especially for early and late elements of a given series, the direct multi-ion d-state interactions that arise from the partial d-band filling become dominant as one moves toward the center of each series. In the GPT, we retain a mixed basis set of plane waves and localized d states, which allows us to capture both the sp-d hybridization and the multi-ion d-state interactions from first principles. For elemental transition metals, the cohesive-energy functional is generalized from the volume term plus pair-potential form given by Eq. (3.1) to the full multi-ion cluster expansion anticipated in Chapter 1:

$$E_{\text{coh}}(\mathbf{R}, \Omega) = E_{\text{vol}}(\Omega) + \frac{1}{2N} \sideset{}{'}\sum_{i,j} v_2(ij, \Omega) + \frac{1}{6N} \sideset{}{'}\sum_{i,j,k} v_3(ijk, \Omega)$$
$$+ \frac{1}{24N} \sideset{}{'}\sum_{i,j,k,l} v_4(ijkl, \Omega) + \cdots . \qquad (5.1)$$

As we discussed in Chapter 1, and then formally developed in Chapter 2, this expansion proceeds from a zero-order reference system consisting of N noninteracting, resonant transition-metal ions immersed in a uniform electron gas. Here this reference system is fully implemented through a direct extension of the zero-order pseudoatom construction elaborated in Chapter 4 to transition metals, providing self-consistent values for the sp valence Z and single-site d-state occupation number Z_d as a function of volume. The main underlying small parameters that define the multi-ion potentials in Eq. (5.1) remain the d-state matrix elements of the hybridization potential, Δ_{kd}^{vol} and $\Delta_{dd'}^{ij}$, but now these quantities enter through a unique d-state moments expansion of the d-band structural energy that is central to the full multi-ion cohesive-energy functional. The essential formal elements of the theoretical framework for the transition-metal GPT were elaborated in Sec. 2.7 of Chapter 2, and here, following Moriarty (1988a), we extend and apply that framework to develop multi-ion potentials through four-ion interactions for real materials.

As was pointed out in Chapter 1, the multi-ion potentials in Eq. (5.1) are long-ranged, nonanalytic and multi-dimensional functions that in practice must be recalculated at each usage, greatly limiting the number of atoms that can be efficiently treated with the first-principles GPT in central transition metals. In the early 1990s, this limitation gave rise to the development of a simplified model version of the GPT, the MGPT, for mid-period transition metals with nearly half-filled d bands (Moriarty, 1990a, 1994). The MGPT leverages the relatively weak net contribution of sp-d hybridization to the structural components of E_{coh} in such metals due to the strong destructive interference among the oscillatory tails of the multi-ion potentials. In addition, by introducing canonical d bands, the direct d-state matrix elements $\Delta_{dd'}^{ij}$ and their multi-ion products can be simply modeled, leading to entirely analytic forms for the d-state component of v_2 and the multi-ion potentials v_3 and v_4. These simplifications produced the basic *analytic* MGPT and the first MD simulations with these potentials (Moriarty, 1994). Subsequently, a more general and computationally efficient *matrix* MGPT was developed (Moriarty et al., 2006), allowing noncanonical d bands, on-the-fly potential evaluation and large-scale MD simulation, as well as important extensions to f-electron actinide metals (Chapter 12) and electron-temperature dependent potentials (Chapter 13). The matrix MGPT has also facilitated the optimization and more extensive application of robust multi-ion potentials for prototype bcc transition metals, most notably Ta, Mo and V. Matrix-MGPT applications on these metals have now been performed on a large number of structural, thermodynamic, defect and mechanical properties over wide ranges of pressure and temperature (Yang et al., 2001, 2010; Moriarty et al., 2002a, 2006, 2008, 2012; Haskins et al., 2012; Moriarty and Haskins, 2014; Haskins and Moriarty, 2018). Additional MGPT capabilities under development and discussed here are the re-inclusion of sp-d hybridization, the addition of five- and six-ion potentials to $E_{\text{coh}}(\mathbf{R}, \Omega)$, and the treatment of magnetism in $3d$ metals. These capabilities are viewed as essential to the goal of extending MGPT-MD simulations across the entire $3d$, $4d$ and $5d$ transition-metal series.

Also beginning in the early 1990s was the parallel development of bond-order potentials for elemental transition metals, as reviewed by Aoki et al. (2007) and Drautz et al. (2015). Starting with a free-atom reference state and a basis of localized atomic d states, the focal point of the BOP treatment of transition metals is a real-space TB moment expansion of the attractive d-bonding energy, which is subsequently balanced by an empirical repulsive energy representing the net contributions of the remaining s and p valence electrons plus the hard-core overlap between ions. Also, as in the standard MGPT, there is no explicit treatment of sp-d hybridization in the transition-metal BOP formalism. A number of successful ambient-pressure BOP parameterizations of the d-state bond integrals and repulsive energy have been made for the central bcc transition metals (Mrovec, 2004, 2007; Cák et al., 2014; Lin et al., 2014), as well as for magnetic bcc Fe (Mrovec, 2011), with applications to structural, vibrational and defect properties. Both numerical and simplified analytic BOP representations have now been developed, with the latter approach providing, in principle, more accurate forces and a possible future pathway to large-scale MD simulations.

5.1 GPT multi-ion potentials for metals with partially filled *d* bands

The starting equation for the GPT cohesive energy functional of a pure transition metal with partially filled d bands follows by combining Eqs. (2.36) and (2.37) in their most general form:

$$E_{\text{coh}} = \frac{1}{N} \left\{ \frac{1}{2} \sum_{i,j}{}' \frac{(Ze)^2}{|\mathbf{R}_i - \mathbf{R}_j|} + E_{\text{band}}^{\text{val}}(Z, Z_d) - \frac{1}{2} n_{\text{val}} V_{\text{val}} + n_{\text{val}}[\varepsilon_{\text{xc}}(n_{\text{val}}) - \mu_{\text{xc}}(n_{\text{val}})] \right.$$

$$+ \frac{1}{2} \sum_{i,j}{}' \left[\frac{(Z_a - Z)^2 e^2}{|\mathbf{R}_i - \mathbf{R}_j|} - 2 n_i \frac{(Z_a - Z) e^2}{|\mathbf{r} - \mathbf{R}_j|} + n_i v_j \right] \qquad (5.2)$$

$$+ \sum_{i,j}{}' \left[\frac{(Z_a - Z) Z e^2}{|\mathbf{R}_i - \mathbf{R}_j|} - n_i \frac{Z e^2}{|\mathbf{r} - \mathbf{R}_j|} \right]$$

$$\left. + \frac{1}{2} \sum_{i,j}{}' \left[n_i \mu_{\text{xc}}^*(n_j) + \delta \varepsilon_{\text{xc}}^*(n_i, n_j) - n_{\text{val}} \delta \mu_{\text{xc}}^*(n_i, n_j) \right] \right\} - E_{\text{bind}}^{\text{atom}}(Z, Z_d) + E_{\text{prep}},$$

Equation (5.2) generalizes Eq. (4.1) for EDB and FDB metals by allowing variable, self-consistent values of the valence Z and d-state occupation number Z_d, as well as the corresponding electron densities n_{val} and n_i. As with EDB and FDB metals in Eq. (4.1), the electrostatic and xc terms on the second, third and fourth lines of Eq. (5.2) will contribute to a two-ion d-state overlap potential v_{ol} adding to v_2, and the essential needed input is again the valence band-structure energy $E_{\text{band}}^{\text{val}}$ and the valence electron density n_{val}. The unique d-state moments expansion that arises from $E_{\text{band}}^{\text{val}}$ is essential to the development of multi-ion potentials in both the GPT and MGPT and is discussed in Sec. 5.1.1. The valence electron density is treated in Sec. 5.1.2, including the extension of the zero-order pseudoatom framework to transition metals and the development of modified electron screening. Completion of the cohesive-energy functional is made in Sec. 5.1.3, with application to first-principles potentials considered in Sec. 5.1.4. The important practical issue of multi-ion screening of *sp-d* hybridization contributions is addressed in Sec. 5.1.5.

5.1.1 Essential new elements of the valence band-structure energy

From the transition-metal GPT formalism developed in Sec. 2.7 of Chapter 2, one can combine Eqs. (2.188) and (2.189) to express the valence band-structure energy $E_{\text{band}}^{\text{val}}$ as

$$E_{\text{band}}^{\text{val}}(Z, Z_d) = N\left(\frac{3}{5}Z\varepsilon_{\text{F}} + Z_d(\varepsilon_{\text{F}} - E_d^{\text{vol}}) - \frac{10}{\pi}\int_0^{\varepsilon_{\text{F}}}\delta_2(E)dE\right)$$

$$- \int_0^{\varepsilon_{\text{F}}}\left[\delta N_{sp}(E) + \delta N_d(E)\right]dE - NZ_dE_d^{\text{struc}} + \delta E_{\text{band}}. \quad (5.3)$$

Here $\delta_2(E)$ is the $\ell = 2$ phase shift given by Eq. (2.165) with $E_d = E_d^{\text{vol}}$ and $\Gamma_{dd} = \Gamma_{dd}^{\text{vol}}$, such that $Z_d = (10/\pi)\delta_2(\varepsilon_{\text{F}})$. The important quantities δN_{sp} and δN_d in Eq. (5.3) are, respectively, the main sp and d structural components of the integrated density of states, as defined by Eq. (2.180). The sp component δN_{sp} is given by Eq. (2.181) in terms of the pseudopotential matrix elements $W_{\mathbf{kk'}}$ and leads to the usual first- and second-order PP total-energy contributions to $E_{\text{band}}^{\text{val}}$, which have already been considered in Chapters 3 and 4. The d component δN_d is given in exact form by Eq. (2.187) in terms of the first-order relative dd' coupling strength $T_{dd'}$ between ion sites, as defined by Eq. (2.183).

Here we break $T_{dd'}$ into volume and structure components in the usual way,

$$T_{dd'} = T_{dd'}^{\text{vol}} + T_{dd'}^{\text{struc}}, \quad (5.4)$$

and note that the second-order structural component $T_{dd'}^{\text{struc}}$ will lead to only third-order and higher energy contributions in Eq. (5.3), which we will now neglect. In this manner, Eq. (2.187) is reduced to the form

$$\delta N_d(E) = -\frac{2}{\pi}\text{Im}\left(\sum_d \ln\left[1 - \frac{E_d^{\text{struc}} + \Gamma_{dd}^{\text{struc}} + \Lambda_{dd}^{\text{vol}}}{E - E_d^{\text{vol}} - \Gamma_{dd}^{\text{vol}}}\right] + \ln\left[\det\left(T_{dd'}^{\text{vol}}\right)\right]\right), \quad (5.5)$$

where $\Lambda_{dd}^{\text{vol}}$ is the volume component of Λ_{dd} and

$$T_{dd'}^{\text{vol}}(R_{ij}, E) = \frac{(E - E_d^{\text{vol}})S_{dd'}(R_{ij}) + \Delta_{dd'}(R_{ij}) - \Gamma_{dd'}^{\text{vol}}(R_{ij}, E)}{E - E_d^{\text{vol}} - \Gamma_{dd}^{\text{vol}}(E)}. \quad (5.6)$$

Here det $\left(T_{dd'}^{\text{vol}}\right)$ remains a $5N \times 5N$ determinant spanning all ion sites. In $T_{dd'}^{\text{vol}}(R_{ij}, E)$ we recall that both $S_{dd'}(R_{ij})$ and $\Delta_{dd'}(R_{ij})$ are short-range functions of the distance R_{ij} between ion sites i and j, while $\Gamma_{dd'}^{\text{vol}}(R_{ij}, E)$ is a long-range sp-d hybridization interaction that depends on energy as well as distance. This latter quantity reduces to $\Gamma_{dd}^{\text{vol}}(E)\delta_{dd'}$ in the limit that $R_{ij} \to 0$, so that one has the normalization property $T_{dd}^{\text{vol}}(0, E) = 1$. At finite ion-ion separation distances, $\left|T_{dd'}^{\text{vol}}\right| < 1$ for all R_{ij} of physical interest, and $T_{dd'}^{\text{vol}} \to 0$ as $R_{ij} \to \infty$. We first develop the dominant multi-ion potential series that arises from the second term in Eq. (5.5) involving $T_{dd'}^{\text{vol}}$. The remaining second-order terms in Eq. (5.5) will be treated in Sec. 5.1.3, where the full cohesive-energy functional is assembled.

5.1.1.1 Formal multi-ion d-state potential series

In this section we consider the d-band structural-energy component to $E_{\text{band}}^{\text{val}}$,

$$NE_{\text{struc}}^d \equiv \frac{2}{\pi}\text{Im}\int_0^{\varepsilon_F} \ln\left[\det\left(T_{dd'}^{\text{vol}}\right)\right] dE, \quad (5.7)$$

which results from the main contribution to δN_d in Eq. (5.5). Following Moriarty (1988a), Eq. (5.7) may be formally developed into a multi-ion series. To do this, one first notes that det $\left(T_{dd'}^{\text{vol}}\right)$ in this equation is the determinant of a $5N \times 5N$ matrix made up of symmetric 5×5 blocks. The diagonal blocks are simply 5×5 unit matrices $\mathbf{I}$ resulting from the orthonormality of the d states centered on a given site, while the off-diagonal blocks are 5×5 matrices $\mathbf{T}_{ij}$ linking sites i and j with appropriate components $T_{dd'}^{\text{vol}}$. The determinant det $\left(T_{dd'}^{\text{vol}}\right)$ can then be developed blockwise about its rows and columns and readily folds down into 5×5 form. Specifically, one so obtains the important result

$$\ln\left[\det\left(T_{dd'}^{\text{vol}}\right)\right] = \ln\left[\det\left(\mathbf{I} - \frac{1}{2}{\sum_{i,j}}'\phi_{ij} + \frac{1}{6}{\sum_{i,j,k}}'\phi_{ijk} - \frac{1}{24}{\sum_{i,j,k,l}}'\phi_{ijkl} + \cdots\right)\right], \quad (5.8)$$

where we have defined the matrix products

$$\phi_{ij} \equiv \mathbf{T}_{ij}\mathbf{T}_{ji}, \quad (5.9)$$

$$\phi_{ijk} \equiv \mathbf{T}_{ij}\mathbf{T}_{jk}\mathbf{T}_{ki} + \mathbf{T}_{ik}\mathbf{T}_{kj}\mathbf{T}_{ji} \quad (5.10)$$

and

$$\phi_{ijkl} \equiv \mathbf{T}_{ij}\mathbf{T}_{jk}\mathbf{T}_{kl}\mathbf{T}_{li} + \mathbf{T}_{il}\mathbf{T}_{lk}\mathbf{T}_{kj}\mathbf{T}_{ji} + \mathbf{T}_{ik}\mathbf{T}_{kl}\mathbf{T}_{lj}\mathbf{T}_{ji} + \mathbf{T}_{ij}\mathbf{T}_{jl}\mathbf{T}_{lk}\mathbf{T}_{ki} + \mathbf{T}_{il}\mathbf{T}_{lj}\mathbf{T}_{jk}\mathbf{T}_{ki}$$
$$+ \mathbf{T}_{ik}\mathbf{T}_{kj}\mathbf{T}_{jl}\mathbf{T}_{li} - \phi_{ij}\phi_{kl} - \phi_{ik}\phi_{jl} - \phi_{il}\phi_{jk}. \quad (5.11)$$

5.1.1 Essential new elements of the valence band-structure energy

To proceed further we consider the asymptotic limit of large R_{ij}, where the elements of T_{ij} are arbitrarily small, and use the general expansion for any small, square matrix A:

$$\ln\left[\det\left(\mathbf{I}+\mathbf{A}\right)\right] = \mathrm{Tr}\left(\mathbf{A} - \frac{1}{2}\mathbf{A}\mathbf{A} + \frac{1}{3}\mathbf{A}\mathbf{A}\mathbf{A} - \cdots\right), \quad (5.12)$$

where Tr denotes the trace of the matrix which follows. Expanding the RHS of Eq. (5.8) via Eq. (5.12), collecting terms with a common structural dependence, and then resuming and analytically continuing the result to arbitrary R_{ij} yields our desired multi-ion expansion of the d-band structural energy in the form:

$$NE_{\mathrm{struc}}^d = \frac{1}{2}\sum_{i,j}{}' v_2^d(ij) + \frac{1}{6}\sum_{i,j,k}{}' v_3^d(ijk) + \frac{1}{24}\sum_{i,j,k,l}{}' v_4^d(ijkl) + \cdots. \quad (5.13)$$

The two-ion potential v_2^d in Eq. (5.13) is given by

$$v_2^d(ij) = \frac{2}{\pi}\mathrm{Im}\int_0^{\varepsilon_F} L_{ij}(E)\,dE, \quad (5.14)$$

with

$$L_{ij}(E) \equiv \ln\left[\det\left(\mathbf{I} - \phi_{ij}\right)\right]. \quad (5.15)$$

The corresponding three-ion potential v_3^d is given by

$$v_3^d(ijk) = \frac{2}{\pi}\mathrm{Im}\int_0^{\varepsilon_F}\{L_{ijk}(E) - [L_{ij}(E) + L_{jk}(E) + L_{ki}(E)]\}\,dE, \quad (5.16)$$

with

$$L_{ijk}(E) \equiv \ln\left[\det\left(\mathbf{I} - \{\phi_{ij} + \phi_{jk} + \phi_{ki}\} + \tilde{\phi}_{ijk}\right)\right], \quad (5.17)$$

where $\tilde{\phi}_{ijk}$ is the symmetrized matrix product

$$\tilde{\phi}_{ijk} \equiv \frac{1}{3}\{\phi_{ijk} + \phi_{jki} + \phi_{kij}\}. \quad (5.18)$$

Finally, the additional four-ion potential v_4^d is given by

$$v_4^d(ijkl) = \frac{2}{\pi}\mathrm{Im}\int_0^{\varepsilon_F}\{L_{ijkl}(E) - [L_{ijk}(E) + L_{jkl}(E) + L_{kli}(E) + L_{lij}(E)]$$
$$+ [L_{ij}(E) + L_{jk}(E) + L_{kl}(E) + L_{li}(E) + L_{ki}(E) + L_{lj}(E)]\}\,dE, \quad (5.19)$$

with

$$L_{ijkl}(E) \equiv \ln\left[\det\left(I - \{\phi_{ij} + \phi_{jk} + \phi_{kl} + \phi_{li} + \phi_{ki} + \phi_{lj}\}\right.\right.\\\left.\left. + \{\tilde{\phi}_{ijk} + \tilde{\phi}_{jkl} + \tilde{\phi}_{kli} + \tilde{\phi}_{lij}\} - \tilde{\phi}_{ijkl}\right)\right], \tag{5.20}$$

where $\tilde{\phi}_{ijkl}$ is the symmetrized matrix product

$$\tilde{\phi}_{ijkl} \equiv \frac{1}{4}\{\phi_{ijkl} + \phi_{jkli} + \phi_{klij} + \phi_{lijk}\}. \tag{5.21}$$

While Eq. (5.15) is an exact result for L_{ij}, Eq. (5.17) for L_{ijk} and Eq. (5.20) for L_{ijkl} are exact only for site orientations where all of the two-ion matrices T_{ij}, T_{ki}, etc. involved in these expressions mutually commute. For arbitrary site orientations these matrices will not all commute, and there are correction terms for L_{ijk} and L_{ijkl} that are formally fifth-order and sixth-order, respectively, in the T_{ij}. These corrections are typically quite small and are here neglected.

Useful simplified forms of the general results Eqs. (5.14)–(5.21) can also be derived. For example, one possible approximation is to note from Eq. (5.12) the closely related expansion

$$\det(I + A) = 1 + \text{Tr}(A) + \cdots. \tag{5.22}$$

Using this result in the above expressions for L_{ij}, L_{ijk} and L_{ijkl} removes the determinants, replaces the matrix I with unity and replaces each remaining matrix product with its trace. This approximation is the origin of Eqs. (3) and (4) for v_2^d and v_3^d, respectively, in Moriarty (1985) and was used in all early applications of the transition-metal GPT (Moriarty, 1985, 1986a, 1986b).

More useful approximate results can be obtained by using Eq. (5.12) directly to develop series expansions to any desired order in the T_{ij}. For example, through fourth order one finds the following results:

$$v_2^d(ij) = -\frac{2}{\pi}\text{Im}\int_0^{\varepsilon_F}\text{Tr}\left[\phi_{ij} + \frac{1}{2}\phi_{ij}\phi_{ij} + \cdots\right]dE, \tag{5.23}$$

$$v_3^d(ijk) = -\frac{2}{\pi}\text{Im}\int_0^{\varepsilon_F}\text{Tr}\left[-\phi_{ijk} + \phi_{ij}\phi_{jk} + \phi_{jk}\phi_{ki} + \phi_{ki}\phi_{ij} + \cdots\right]dE \tag{5.24}$$

and

$$v_4^d(ijkl) = -\frac{2}{\pi}\text{Im}\int_0^{\varepsilon_F}\text{Tr}\left[\phi_{ijkl} + \phi_{ij}\phi_{kl} + \phi_{ik}\phi_{jl} + \phi_{il}\phi_{jk} + \cdots\right]dE, \tag{5.25}$$

where it has been noted that $\mathrm{Tr}\left(\tilde{\phi}_{ijk}\right) = \mathrm{Tr}\left(\phi_{ijk}\right)$ and $\mathrm{Tr}\left(\tilde{\phi}_{ijkl}\right) = \mathrm{Tr}\left(\phi_{ijkl}\right)$. Equations (5.23)–(5.25), in fact, represent the effective starting point for developing the simplified MGPT for transition metals, a subject that is discussed in Sec. 5.2. In the next section, we use these results to gain some basic insight into the connection between the multi-ion potentials and TB d-state moments, and in turn, into the convergence of the multi-ion series (5.13).

5.1.1.2 Tight-binding moments and multi-ion series convergence

In this section, we briefly pause in the formal development of the transition-metal GPT to consider the consequences of Eqs. (5.23)–(5.25) for the multi-ion potentials v_n^d in the limit of a simple, but realistic TB model for the $T_{dd'}^{\mathrm{vol}}$ components of the T_{ij} matrices, a model that will also be directly relevant to the MGPT for central transition metals with near half-filled d bands. To establish a simplified model form for $T_{dd'}^{\mathrm{vol}}(R_{ij}, E)$, we temporarily neglect all of the complicating features in Eq. (5.6), namely the d-state overlap and sp-d hybridization contributions, $S_{dd'}(R_{ij})$ and $\Gamma_{dd'}^{\mathrm{vol}}(R_{ij}, E)$, respectively, as well as the energy dependence of the d-state self-energy $\Gamma_{dd}^{\mathrm{vol}}(E)$. This leaves the result

$$T_{dd'}^{\mathrm{vol}}(R_{ij}, E) = \frac{\Delta_{dd'}(R_{ij})}{E - E_0 + i\Gamma_0}, \tag{5.26}$$

where we have defined the energies

$$E_0 \equiv E_d^{\mathrm{vol}} + \mathrm{Re}\left[\Gamma_{dd}^{\mathrm{vol}}(\varepsilon_F)\right] \tag{5.27}$$

and

$$\Gamma_0 \equiv -\mathrm{Im}\left[\Gamma_{dd}^{\mathrm{vol}}(\varepsilon_F)\right]. \tag{5.28}$$

Here E_0 is an energy near the d-band center and Γ_0 is an energy on the order of one-half of the d-band width. The numerator in Eq. (5.26) for $T_{dd'}^{\mathrm{vol}}$ retains the essential TB matrix elements $\Delta_{dd'}(R_{ij})$ that connect sites i and j, while the denominator reflects the modulating effect of d-band filling. With regard to the latter, we can directly relate the d-band occupation number Z_d to E_0 and Γ_0 using Eqs. (2.165) and (2.169):

$$Z_d = (10/\pi)\tan^{-1}\left[\Gamma_0/(E_0 - \varepsilon_F)\right], \tag{5.29}$$

which can also be expressed in the equivalent form

$$(\varepsilon_F - E_0)/\Gamma_0 = -\cot\left(\pi Z_d/10\right). \tag{5.30}$$

Next, the multi-ion-potential contributions to the d-band structural energy E_{struc}^d can be conveniently expressed in terms of moments involving the $\Delta_{dd'}^{ij}$. Specifically, one can define a dimensionless, normalized *partial* moment $M_n^{(s)}$ of order n and linking s

distinct ion sites for each of the terms in Eqs. (5.13) and (5.23)–(5.25). For example, for $n = s = 2$ we define

$$M_2^{(2)} \equiv \frac{1}{2N} {\sum_{i,j}}' \alpha^2 \text{Tr}(\phi_{ij})$$

$$= \frac{1}{2N} {\sum_{i,j}}' \sum_{d,d'} \frac{\Delta_{dd'}(R_{ij}) \Delta_{d'd}(R_{ij})}{\Gamma_0^2}, \qquad (5.31)$$

where $\alpha \equiv (E - E_0 + i\Gamma_0)/\Gamma_0$ is a normalization factor that effectively replaces the denominator of $T_{dd'}^{\text{vol}}$ with Γ_0 in Eq. (5.26). Similarly, we define

$$M_4^{(2)} \equiv \frac{1}{2N} {\sum_{i,j}}' \alpha^4 \text{Tr}\left(\frac{1}{2}\phi_{ij}\phi_{ij}\right), \qquad (5.32)$$

$$M_3^{(3)} \equiv \frac{1}{6N} {\sum_{i,j,k}}' \alpha^3 \text{Tr}(-\phi_{ijk}), \qquad (5.33)$$

$$M_4^{(3)} \equiv \frac{1}{6N} {\sum_{i,j,k}}' \alpha^4 \text{Tr}(\phi_{ij}\phi_{jk} + \phi_{jk}\phi_{ki} + \phi_{ki}\phi_{ij}) \qquad (5.34)$$

and

$$M_4^{(4)} \equiv \frac{1}{24N} {\sum_{i,j,k,l}}' \alpha^4 \text{Tr}(\phi_{ijkl} + \phi_{ij}\phi_{kl} + \phi_{ik}\phi_{jl} + \phi_{il}\phi_{jk}). \qquad (5.35)$$

Each partial moment has a simple graphical representation in which the matrix element $\Delta_{dd'}(R_{ij})$ is denoted by a solid line segment connecting ion sites i and j. Thus $M_n^{(s)}$ is a lattice sum of topologically similar graphs, each involving n line segments liking s sites. The graphs corresponding to the five partial moments defined in Eqs. (5.31)–(5.35) are illustrated in Fig. 5.1. In addition to partial moments, one can also define corresponding *total* moments M_n, which are just the sum of all partial moments of the same order n. Then $M_2 = M_2^{(2)}$, $M_3 = M_3^{(3)}$ and $M_4 = M_4^{(2)} + M_4^{(3)} + M_4^{(4)}$. Apart from the spin factor of 2, which is separately included in Eqs. (5.23)–(5.25) for the multi-ion potentials, and the normalization factor of $[\Gamma_0]^{-n}$, such total moments are the same as the DOS moments defined by Eq. (2.48) and discussed in the context of TB in Sec. 2.6.3 of Chapter 2.

In addition to the partial moments, full evaluation of Eqs. (5.23)–(5.25) for the multi-ion potentials v_n^d requires the calculation of corresponding energy integrals that contain the effects of d-band filling. These integrals depend on the order n and can be evaluated analytically in the following general form:

$$F_n(Z_d) \equiv \text{Im} \int_{-\infty}^{\varepsilon_F} \frac{(n-1)[\Gamma_0]^{n-1}}{[E - E_0 + i\Gamma_0]^n} dE$$

$$= -\text{Im}\left\{[-\cot(\pi Z_d/10) + i]^{-(n-1)}\right\}. \qquad (5.36)$$

5.1.1 Essential new elements of the valence band-structure energy

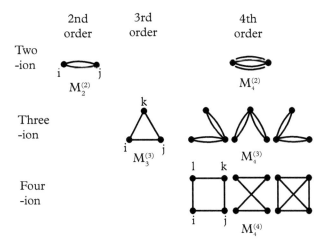

Fig. 5.1 Graphs associated with the partial moments $M_n^{(s)}$ defined in Eqs. (5.31)–(5.35), which contribute directly to the multi-ion potentials v_n^d given by Eqs. (5.23)–(5.25). Each solid line segment represents a matrix element $\Delta_{dd'}^{ij}$ (or, more generally, $T_{dd'}^{\text{vol}}$) linking two ion sites.
From Moriarty (1988a), with publisher (APS) permission.

Here we have extended the lower limit of the energy integrals from 0 to $-\infty$ to compensate for the removal of the energy dependence of the self-energy $\Gamma_{dd}^{\text{vol}}(E)$ in defining E_0 and Γ_0. The dimensionless d-band occupation functions $F_n(Z_d)$ so obtained are oscillatory in their behavior and bounded by the condition $-1 \leq F_n \leq 1$. The specific relevant results for $n = 2$, 3 and 4 are plotted in Fig. 5.2.

Using the functions F_n together with the partial moments $M_n^{(s)}$, it is then straightforward to express the d-band structural energy (5.13) in the form

$$E_{\text{struc}}^d = E_2 + E_3 + E_4 + \cdots, \tag{5.37}$$

where E_2, E_3 and E_4 are, respectively, the two-, three- and four-ion contributions

$$E_2 = -\frac{2}{\pi}\left[M_2^{(2)}F_2(Z_d) + \frac{1}{3}M_4^{(2)}F_4(Z_d) + \cdots\right]\Gamma_0, \tag{5.38}$$

$$E_3 = -\frac{2}{\pi}\left[\frac{1}{2}M_3^{(3)}F_3(Z_d) + \frac{1}{3}M_4^{(3)}F_4(Z_d) + \cdots\right]\Gamma_0 \tag{5.39}$$

and

$$E_4 = -\frac{2}{\pi}\left[\frac{1}{3}M_4^{(4)}F_4(Z_d) + \cdots\right]\Gamma_0. \tag{5.40}$$

Note that at half d-band filling, i.e., $Z_d = 5$, $F_n = 0$ for all odd orders of n, such as $n = 3$. This means that the third-order, three-ion moment $M_3^{(3)}$ contributes nothing to E^d_{struc} for metals with half-filled d bands and, as a result, foretells the special importance of the fourth-order moments $M_4^{(3)}$ and $M_4^{(4)}$ to central transition metals, where $F_4 \simeq -1$.

The bound nature of the d-band filling functions F_n ensures that the convergence of our multi-ion expansion for E^d_{struc} is govern primarily by the behavior of the moments $M_n^{(s)}$. To evaluate the $M_n^{(s)}$ within our simple TB model, we need specific forms for the quantities $\Delta_{dd'}(R_{ij})$ and Γ_0. Here it is convenient to use canonical d bands and appeal to the equivalence between TB and resonance descriptions of the d bands in transition metals, as discussed in Chapter 2. Specifically, we can write the components of $\Delta_{dd'}(R_{ij})$ in the general form of Eqs. (2.106a)–(2.106c) for $dd\sigma$, $dd\pi$ and $dd\delta$, noting the overall minus sign included in the definition of $\Delta_{dd'}^{ij}$. For ions i and j aligned along the z axis

$$\Delta_{dd'}(R_{ij}) = \eta_m \frac{1}{5} \frac{M^2}{R_{ij}^5} \delta_{mm'}, \qquad (5.41)$$

where $\eta_m = 6, -4$ or 1 for $m = 0, \pm 1$ or ± 2, and M is a fixed intra-atomic quantity, as in Eq. (2.101). A corresponding result for Γ_0 can be obtained from Eq. (2.100), noting that Γ_0 is only a half width and replacing κ_{res} with k_F:

$$\Gamma_0 = \left(k_F^5/225\right) M^2. \qquad (5.42)$$

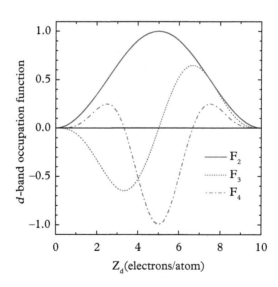

Fig. 5.2 The d-band occupation function $F_n(Z_d)$, as defined by Eq. (5.36), for order $n = 2, 3$ and 4 contributions to the multi-ion potentials.
From Moriarty (1988a), with publisher (APS) permission.

The ratio of Eqs. (5.41) and (5.42), which is independent of M and is the quantity that enters the partial moments $M_n^{(s)}$, can then be expressed in the useful form

$$\frac{\Delta_{dd'}(R_{ij})}{\Gamma_0} = \eta_m C_Z \left[\frac{R_{WS}}{R_{ij}}\right]^5 \delta_{mm'}, \qquad (5.43)$$

where we have used $k_F = (3\pi^2 Z/\Omega)^{1/3}$ to obtain

$$C_Z = \frac{45}{(9\pi Z/4)^{5/3}}. \qquad (5.44)$$

For transition metals, the *sp* valence Z normally has a value between 1 and 2, so that at a given volume Ω, C_Z is a constant in the small range $0.54 \leq C_Z \leq 1.73$.

As a specific relevant example, we have used Eq. (5.43), generalized to a full Slater-Koster matrix, to evaluate the partial moments $M_n^{(s)}$ for bcc, fcc and ideal hcp crystal structures of a model central transition metal with $Z = 1.4$, where $C_Z = 0.987$. To aid in the real-space convergence of each partial moment, we have also multiplied Eq. (5.43) by an appropriate smooth cutoff function $f_{cut}(r)$, as given by Eq. (3.112) with parameters $R_0 = 2.15 R_{WS}$ and $\alpha = 125$. With these parameters, first-neighbor interactions in the fcc and hcp structures and first- and second-neighbor interactions in the bcc structure are fully retained, while more distant interactions become negligible beyond $2.75 R_{WS}$. The partial moments so obtained are given in Table 5.1. As expected, $M_2^{(2)}$ arising from two-ion pair interactions is the largest calculated moment for all three structures considered. The additional two-ion moment $M_4^{(2)}$, however, is almost two orders of magnitude smaller in each case, indicating the rapid convergence of the $M_n^{(2)}$ moments. This directly reflects the effective smallness of $\Delta_{dd'}(R_{ij})/\Gamma_0$ at interatomic separations. Also rapidly converging are the outer moments $M_2^{(2)}$, $M_3^{(3)}$, $M_4^{(4)}$, ... for $s = n$ arising from the so-called ring diagrams in Fig. 5.1. In contrast, the four-ion moment $M_3^{(4)}$ is a substantial fraction of $M_2^{(2)}$, and the sharp falloff from $M_3^{(4)}$ to $M_4^{(4)}$ is driven in part by the phenomenon of destructive interference. That is, the increased multiplicity (i.e., the number of graphs summed) with increasing s for a given n is first balanced in $M_3^{(4)}$ then quickly dominated in $M_4^{(4)}$ by the alternating-sign cancellations of the $m = 0, \pm 1$ and ± 2 couplings. This destructive interference also applies strongly to higher moments. For example, we have found that the sixth-order contributions to $M_6^{(4)}$ arising from graphs analogous to those that define the large three-ion moment $M_4^{(3)}$ are one to two orders of magnitude smaller than $M_4^{(4)}$.

Using the partial moments given in Table 5.1 together with the *d*-band occupation functions $F_n(Z_d)$, one can evaluate relative structural energies of bcc, fcc and hcp phases of our model central transition metal, using Eqs. (5.37)–(5.40). Choosing $Z_d = 4.6$, so that $Z + Z_d = 6$, as for a Group-VIB transition metal like Mo, the bcc-fcc and hcp-fcc structural-energy differences are found to be $-0.029\Gamma_0$ and $-0.00011\Gamma_0$, respectively. These results are consistent with the observed bcc structure of central transition metals

Table 5.1 *Partial TB moments defined by Eqs. (5.31)–(5.35) and evaluated here for the bcc, fcc and ideal hcp structures of a model transition metal, using Eq. (5.43) for $\Delta_{dd'}(R_{ij})/\Gamma_0$ with $Z = 1.4$, as discussed in the text.*

Moment	bcc	fcc	hcp
$M_2^{(2)}$	1.14826	1.10164	1.10164
$M_4^{(2)}$	0.04635	0.03786	0.03786
$M_3^{(3)}$	−0.12134	−0.11832	−0.11832
$M_4^{(3)}$	0.44823	0.41846	0.41689
$M_4^{(4)}$	−0.04484	−0.00636	−0.00538

like Mo, and are also qualitatively consistent with the fourth-moment TB calculations of Aoki (1993) displayed in the left panel of Fig. 1.6.

5.1.2 Valence electron density and self-consistent screening

In this section, we return to the development of the full transition-metal GPT and next consider the evaluation of the valence electron density within the formal framework established in Chapter 2. This requires two main steps. First, the zero-order pseudoatom construction elaborated in Chapter 4 for EDB and FDB metals must be extended to transition metals with partially filled d bands to establish self-consistent values of the d-band occupation number Z_d and the corresponding sp valence Z. Second, the remaining oscillatory screening and orthogonalization-hole components of the total valence electron density n_{val} need to be reduced from the formal results presented in Sec. 2.7.5 of Chapter 2 to useful forms that can be evaluated via first-principles calculation.

5.1.2.1 Zero-order pseudoatoms for transition metals

All of the basic features of the zero-order pseudoatom construction discussed in Sec. 4.2 of Chapter 4 remain intact for transition metals. In particular, there are two main self-consistent loops of calculation, to which a third additional loop for Z_d and Z must now be added. In the first loop, the Schrödinger equations (4.76) and (4.77) for the localized inner-core states and d-basis states, respectively, are solved self-consistently within DFT. In the second loop, the barrier potential V_B contained in the d-state localization potential (4.93) is varied until the desired boundary condition on the d states is achieved. This boundary condition is specified by either by the preferred logarithmic derivative $D_2^\star$ given by Eq. (4.95) or the alternative logarithmic derivative D_2 given by Eq. (4.94). In the third loop, the values of Z_d, Z and ε_F are self-consistently calculated, according to the basic constraints established by Eqs. (2.169)–(2.171) together with Eq. (2.165):

5.1.2 Valence electron density and self-consistent screening

$$Z_d = -\frac{10}{\pi}\tan^{-1}\left(\frac{-\text{Im}\left[\Gamma_{dd}^{\text{vol}}(\varepsilon_F)\right]}{\varepsilon_F - E_d^{\text{vol}} - \text{Re}\left[\Gamma_{dd}^{\text{vol}}(\varepsilon_F)\right]}\right), \qquad (5.45)$$

$$Z = \left(\frac{2m}{\hbar^2}\right)^{3/2}\frac{\Omega}{3\pi^2}\varepsilon_F^{3/2} \qquad (5.46)$$

and $Z + Z_d = $ an integer constant determined by the appropriate column in the Periodic Table. In Eq. (5.45), the full energy dependence of $\Gamma_{dd}^{\text{vol}}(E)$ is retained and this quantity is evaluated in terms of the matrix elements $S_{\mathbf{k}d} = \langle \mathbf{k}|\phi_d\rangle$ and $\Delta_{\mathbf{k}d}^{\text{vol}} = \langle \mathbf{k}|\Delta_{\text{vol}}|\phi_d\rangle$.

Calculated values of the self-consistent sp valence Z, obtained assuming canonical d bands with $D_2^* = -3$, are displayed in Fig. 5.3 for the (nonmagnetic) $3d$ and $4d$ transition series metals at their observed equilibrium volumes (see Table A1.2). Elements at the beginning and end of each series have relatively high values with $Z > 1.5$, while central transition metals have values in the narrow ranges 1.4–1.5 for the $3d$ series and 1.1–1.2 for the $4d$ series. For comparison, we have also plotted in Fig. 5.3, the number of non-d valence electrons contained with the atomic sphere obtained in parallel DFT LMTO-ASA electronic-structure calculations applied to the fcc structure (Moriarty, 1988a). The latter values tend to be rather constant across each series, with most values in the intermediate range 1.3–1.4, except for a sharp dip at Pd in the $4d$ series. Also shown in Fig. 5.3 is the impact of changing the GPT d-state boundary condition to $D_2^* = -2$ in the center of each series. In the $3d$ series elements, Z is thereby *lowered* by less than 0.1 in each case, but in the $4d$ series elements Z is *raised* significantly, especially in Mo and Tc.

The choice of logarithmic derivative D_2^* for transition-metal d states in the GPT must also be consistent with the underlying electronic structure. A good figure of merit in this regard is the unhybridized width of the d-bands W_d, which can be well estimated either from Wigner-Seitz wavefunction boundary conditions applied to the self-consistent pseudoatom potential v_{pa}, or from the corresponding TB matrix elements $\Delta_{dd'}^{ij}$ defined in Eq. (4.92). For WS boundary conditions, one has

$$W_d = E_{\text{top}} - E_{\text{bot}}, \qquad (5.47)$$

where the top E_{top} and bottom E_{bot} of the d bands are determined by antibonding and bonding conditions applied to the radial pseudoatom wavefunction. These conditions are, respectively, $D_2(E_{\text{top}}) = -\infty$ and $D_2(E_{\text{bot}}) = 0$, where the logarithmic derivative $D_2(E)$ is given by Eq. (2.109) with $\ell = 2$. The values of W_d so calculated from zero-order pseudoatom potentials with $D_2^* = -3$ are displayed in Fig. 5.4 for the $3d$ and $4d$ transition metals. These results are confirmed by alternate TB calculations of the unhybridized d-band width in the fcc structure, either from Eq. (2.123) or from the BZ zone-boundary condition (see Sec. A2.5 of Appendix A2)

$$X_5 - X_3 = -6dd\sigma + 4dd\pi + 2dd\delta, \qquad (5.48)$$

190 Interatomic Potentials in Transition Metals

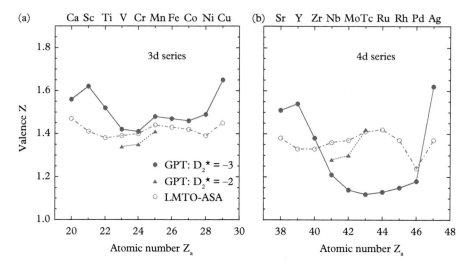

Fig. 5.3 *Self-consistent sp valence Z for 3d and 4d transition metals at their observed equilibrium volumes, as obtained from GPT zero-order pseudoatoms and as inferred from parallel DFT LMTO-ASA electronic-structure calculations. In the latter case, Z represents the number of non-d valence electrons contained with the atomic sphere.*
From Moriarty (1988a), with publisher (APS) permission.

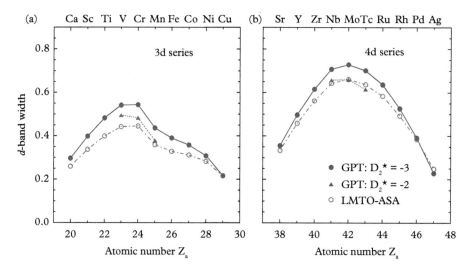

Fig. 5.4 *Width of the d bands for 3d and 4d transition metals at their observed equilibrium volumes, as calculated from self-consistent GPT zero-order pseudoatom potentials and parallel LMTO-ASA potentials, both using Eq. (5.47).*
From Moriarty (1988a), with publisher (APS) permission.

where $dd\sigma$, $dd\pi$ and $dd\delta$ are the $m = 0, \pm 1$ and ± 2 components of $-\Delta_{dd'}^{ij}$ evaluated at the fcc nearest neighbor distance. For all 20 of the 3d and 4d transition metals considered in Fig. 5.4, Eqs. (2.123) and (5.48) agree with Eq. (5.47) for the d-band width to better than 1% for the choice $D_2^\star = -3$ in defining the zero-order pseudoatom.

In addition, the GPT d-band widths yielded by WS boundary conditions applied to the pseudoatom potentials are in generally good agreement with those similarly determined from the self-consistent LMTO-ASA potentials for the fcc structure (Moriarty, 1988a), as is further shown in Fig. 5.4. In particular, the qualitative trends in W_d with atomic number are exactly the same. Quantitatively, however, with $D_2^\star = -3$ the GPT values of W_d are up to 20% larger than the LMTO-ASA values for the central transition metals. Physically, this overestimate results from the position of the d bands, as measured by E_d^{vol}, lying slightly too high in energy. Increasing the d-state boundary condition to $D_2^\star = -2$ for the central elements lowers both E_d^{vol} and W_d, and moves the GPT and LMTO-ASA results into close agreement, as is also shown in Fig. 5.4.

The volume dependence of Z and W_d in transition metals is also revealing. As discussed in Chapter 6, all of the early and central transition metals undergo an sp to d transfer of valence electron density under compression, as the NFE sp bands are pushed up faster in energy than the narrow d bands. This is the so-called $s \to d$ transition. One should expect, therefore, that the zero-order pseudoatom value of Z will decrease with decreasing volume in such metals. This is indeed the case, as illustrated in Fig. 5.5(a) for Mo. Furthermore, in Mo with $D_2^\star = -2$ the correspondence with DFT band-structure predictions is quantitative as well. That is, we find that the extrapolation of the GPT pseudoatom result to $Z = 0$ and the end of the $s \to d$ transition occurs at about the same volume that DFT LMTO-ASA calculations predict that the bottom of the sp bands will move above the Fermi level (Moriarty, 1988a). The reverse transition, in which a d to sp transfer of valence electron density takes place (the corresponding $d \to s$ transition), occurs at the end of each series in the noble metals, as also shown in Fig. 5.5(a) for Cu. Again, the predicted increase in Z with decreasing volume from zero-order pseudoatom calculations is in accord with DFT electronic-structure calculations (Albers et al., 1985).

The volume dependence of the d-band width W_d shows similar characteristic behavior. The canonical d-band theory discussed in Chapter 2 predicts from both Eq. (2.123) and Eq. (2.124), as well as from Eq. (5.48), a volume dependence of the form

$$W_d(\Omega) = W_d(\Omega_0)(\Omega_0/\Omega)^{5/3}. \tag{5.49}$$

This expected behavior is consistent with the choice $D_2^\star = -3$ in the construction of the zero-order pseudoatom. For metals so described, Eq. (5.49) is indeed found to be consistent with the use of Eq. (5.47) for the d-band width, as shown in Fig. 5.5(b) for Cu. For central transition metals that are better described by the choice $D_2^\star = -2$, the volume exponent of 5/3 in Eq. (5.49) must be appropriately reduced to 4/3 to accurately describe $W_d(\Omega)$, as also shown in Fig. 5.5(b) for Mo.

Conceptually, the entire evolution of the transition-metal electronic structure, starting from the free atom, through the promoted atom and the zero-order pseudoatom, to the metal can be characterized as a simple three-step process. This is illustrated in

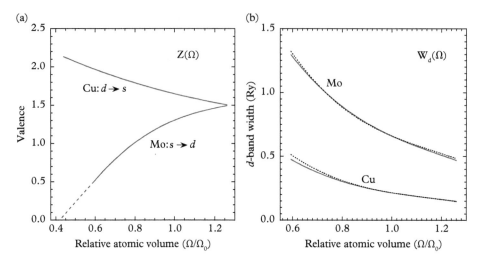

Fig. 5.5 *Volume dependence of the self-consistent valence Z and d-band width W_d for Cu and Mo, as obtained from GPT zero-order pseudoatoms with $D_2^* = -3$ for Cu and $D_2^* = -2$ for Mo. (a) Volume-dependent valence (solid lines). The dashed portion of the Mo curve is an extrapolation to $Z = 0$, where the $s \to d$ transition is completed. (b) Volume-dependent d-band width from Eq. (5.47) (solid lines). The dotted lines represent corresponding results obtained from the general analytic scaling form $W_d(\Omega) = W_d(\Omega_0)(\Omega_0/\Omega)^{p/3}$ with $p = 5$ for Cu and $p = 4$ for Mo.*

Fig. 5.6 in terms of the valence energy levels of Cu, and is qualitatively similar for all transition metals. In the first step, the appropriate transfer of s and d valence electrons is made to create the promoted atom. The atomic s and d orbitals are largely unchanged at this stage, but the corresponding one-electron energy levels are shifted significantly. For a $d \to s$ transfer of electrons, as occurs in Cu, the s and d levels fall in energy because well-localized d electrons are moved to less localized s orbitals, although the preparation energy E_{prep} to create the promoted atom is, of course, positive. In the next step, the zero-atom pseudoatom is created, which permits the s-electron density to be spread out uniformly at a density Z/Ω, while the d-state orbital is self-consistently optimized, but without any large change to the d-electron density. This process transfers electrons from outside the atomic sphere to within, pushing the d level higher in energy to a value E_d^{vol}, and spreading out the s level symmetrically into an occupied free-electron continuum of width ε_{F}. The total electron density as well as the position and occupation of the valence energy levels now simulate the conditions found in the actual metal. The final step is then to allow the pseudoatoms on different sites to interact and relax to complete the full structural details of the energy bands for the given metal crystal structure, which is fcc in the case of Cu. In practice, of course, the values of the valence Z and d-band occupation Z_d are only determined at the zero-order-pseudoatom step, so the promoted atom must actually be constructed after the fact.

5.1.2 Valence electron density and self-consistent screening

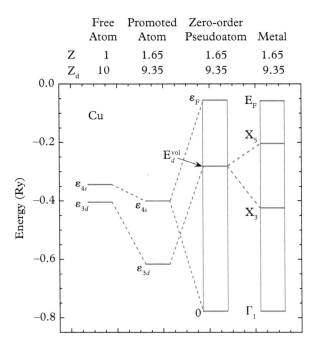

Fig. 5.6 *Evolution of the valence energy levels from the free atom, to the promoted atom, to the zero-order pseudoatom and finally to the metal in Cu. In the latter three stages, $Z = 1.65$ and $Z_d = 9.35$, corresponding to the observed equilibrium volume.*
From Moriarty (1988a), with publisher (APS) permission.

The picture developed in Fig. 5.6 strongly suggests that physical properties of the metal such as the cohesive energy, which depend mostly on volume rather than structure, should already be reasonably well approximated at the zero-order pseudoatom step. This is indeed the case as we now demonstrate. An appropriate expression for the pseudoatom component of the cohesive energy, $E_{\text{coh}}^{\text{pa}}$ includes all of the expected zero- and first-order terms in the full volume term of the metal, E_{vol}, and is defined as

$$E_{\text{coh}}^{\text{pa}}(\Omega) \equiv E_{\text{fe}}^{0}(\Omega) + \frac{2\Omega}{(2\pi)^3} \int \langle \mathbf{k}| w_{\text{pa}} |\mathbf{k}\rangle \Theta^{<}(\mathbf{k}) d\mathbf{k} + Z_d(\varepsilon_F - E_d^{\text{vol}})$$
$$- \frac{10}{\pi} \int_0^{\varepsilon_F} \delta_2(E) dE - E_{\text{bind}}^{\text{atom}}(Z, Z_d) + E_{\text{prep}}. \quad (5.50)$$

The first two terms on the RHS of Eq. (5.50) are contributions from the *sp* valence electrons, as in Eq. (3.60) for simple metals, while the next two terms are the leading contributions of the *d* electrons from Eq. (5.3). The final two atomic terms in Eq. (5.50) derive directly form Eq. (5.2). Calculations of $|E_{\text{coh}}^{\text{pa}}|$ for all 20 of the 3*d* and 4*d* transition metals are shown in Fig. 5.7 and compared with parallel DFT LMTO-ASA calculations of the cohesive energy (Moriarty, 1988a). Qualitatively, all of the observed and DFT-predicted trends, including the double-humped structure of the cohesive energy in each series, are reproduced by Eq. (5.50) for $E_{\text{coh}}^{\text{pa}}$. Quantitatively, there is agreement between Eq. (5.50) evaluated with $D_2^\star = -3$ and the DFT results to within 20% for all 20 elements, with the largest differences resulting from an underestimate of $E_{\text{coh}}^{\text{pa}}$ for the early and mid-period elements of both series. For the mid-period 3d elements, half of difference is removed with the alternate choice of $D_2^\star = -2$. At the same time, there is little sensitivity of $E_{\text{coh}}^{\text{pa}}$ to the choice of $D_2^\star$ for the corresponding mid-period 4d elements.

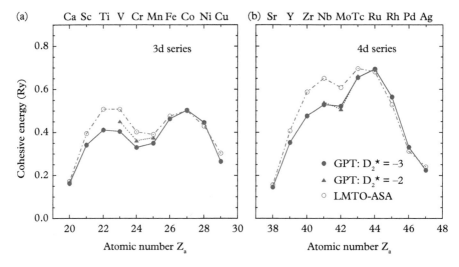

Fig. 5.7 *Magnitude of the cohesive energy for 3d and 4d transition metals at their observed equilibrium volumes, as obtained from self-consistent GPT zero-order pseudoatoms via Eq. (5.50) and from DFT LMTO-ASA electronic-structure calculations. In the DFT calculations, the central elements (V, Cr and Mn in the 3d series and Nb, Mo and Tc in the 4d series) have been treated in the bcc structure and the remaining elements in the fcc structure.*
From Moriarty (1988a), with publisher (APS) permission.

5.1.2.2 Oscillatory screening and orthogonalization-hole components

The most general form of the valence electron density n_{val} for a transition metal in the GPT is given by Eq. (2.193) and consists of a uniform density $n_{\text{unif}} = Z/\Omega$, with Z established from the self-consistent zero-atom pseudoatom, plus charge neutral and oscillatory screening, orthogonalization-hole and Fermi-energy-correction components $\delta n_{\text{scr}}(\mathbf{r})$, $\delta n_{\text{oh}}(\mathbf{r})$ and $\delta n_{\text{band}}(\mathbf{r})$, respectively. As we discussed in Chapter 2, $\delta n_{\text{band}}(\mathbf{r})$ is inherently small and vanishes identically in the desired limit $E_F = \varepsilon_F$, so we shall neglect this component going forward in our treatment of n_{val}. Also, consistent with Eq. (1.27) for the expected general form for the valence electron density in a transition metal, only the leading one-ion components of $\delta n_{\text{scr}}(\mathbf{r})$ and $\delta n_{\text{oh}}(\mathbf{r})$ will be retained.

5.1.2 Valence electron density and self-consistent screening

The GPT form of the one-ion transition-metal screening density is given by Eq. (2.196). In the usual way, one can set $\mathbf{k}' = \mathbf{k} + \mathbf{q}$ on the RHS of that equation to extract a structure factor $S(\mathbf{q})$, so that $\delta n_{\mathrm{scr}}(\mathbf{r})$ can be expressed in terms of its one-ion Fourier transform $n_{\mathrm{scr}}(q)$, as in Eq. (3.24). One thereby obtains

$$n_{\mathrm{scr}}(q) = \frac{4}{(2\pi)^3}\left[\int \frac{w(\mathbf{k},\mathbf{q})}{\varepsilon_{\mathbf{k}} - \varepsilon_{\mathbf{k}+\mathbf{q}}}\Theta^<(\mathbf{k})d\mathbf{k} + \int \frac{h_1^{\mathrm{tm}}(\mathbf{k},\mathbf{q})}{\varepsilon_{\mathbf{k}} - \varepsilon_{\mathbf{k}+\mathbf{q}}}d\mathbf{k}\right], \quad (5.51)$$

where $w(\mathbf{k},\mathbf{q}) \equiv \langle \mathbf{k}+\mathbf{q}|w|\mathbf{k}\rangle$ and we have defined

$$h_1^{\mathrm{tm}}(\mathbf{k},\mathbf{q}) \equiv -\frac{1}{\pi}\mathrm{Im}\int_0^{\varepsilon_F}\sum_d \frac{v'_{\mathbf{k}+\mathbf{q}d}v'_{d\mathbf{k}}}{(E-E_{\mathrm{r}})(E-\varepsilon_{\mathbf{k}})}dE \quad (5.52)$$

with v'_{kd} given by Eq. (2.198) and

$$E_{\mathrm{r}}(E) \equiv E_d^{\mathrm{vol}} + \Gamma_{dd}^{\mathrm{vol}}(E). \quad (5.53)$$

Equations (5.51) and (5.52) generalize Eqs. (4.32) and (4.34), respectively, for EDB and FDB metals. Because of the complex energy dependence of $E_{\mathrm{r}}(E)$, the energy integral in Eq. (5.52) does not have a simple pole structure and must be evaluated numerically to obtain $h_1^{\mathrm{tm}}(\mathbf{k},\mathbf{q})$. As a result, the second integral in Eq. (5.51) is unrestricted over all $\mathbf{k}$ space. Also, it is the simple-metal form of the PP w that appears in the first integral.

The corresponding transition-metal orthogonalization-hole density $\delta n_{\mathrm{oh}}(\mathbf{r})$ has an inner-core component $\delta n_{\mathrm{oh}}^c(\mathbf{r})$ and a d-state component $\delta n_{\mathrm{oh}}^d(\mathbf{r})$ given by Eq. (2.197). It is again convenient to develop $\delta n_{\mathrm{oh}}(\mathbf{r})$ in the general form (3.5), involving an effective valence Z^* and a sum of localized hole densities $n_{\mathrm{oh}}(\mathbf{r}-\mathbf{R}_i)$ from each site i. One thereby obtains for the effective valence

$$Z^* = Z_{\mathrm{c}}^* + \frac{2\Omega}{(2\pi)^3}\int h_2^{\mathrm{tm}}(\mathbf{k})d\mathbf{k}, \quad (5.54)$$

where we have defined

$$Z_{\mathrm{c}}^* \equiv Z + \frac{2\Omega}{(2\pi)^3}\int \langle\mathbf{k}|p_{\mathrm{c}}|\mathbf{k}\rangle\Theta^<(\mathbf{k})d\mathbf{k}, \quad (5.55)$$

and

$$h_2^{\mathrm{tm}}(\mathbf{k}) \equiv -\frac{1}{\pi}\mathrm{Im}\int_0^{\varepsilon_F}\sum_d \frac{v'_{kd}v'_{dk}}{(E-E_{\mathrm{r}})(E-\varepsilon_{\mathbf{k}})^2}dE. \quad (5.56)$$

The companion localized orthogonalization-hole density is

$$n_{\text{oh}}(\mathbf{r}) = n_{\text{oh}}^{\text{c}}(\mathbf{r}) - \frac{2\Omega}{(2\pi)^3} \int \tilde{h}_2^{\text{tm}}(\mathbf{k},\mathbf{r}) d\mathbf{k}, \tag{5.57}$$

where the simple-metal component $n_{\text{oh}}^{\text{c}}(\mathbf{r})$ is given by Eq. (3.39) and we have further defined

$$\tilde{h}_2^{\text{tm}}(\mathbf{k},\mathbf{r}) \equiv -\frac{1}{\pi} \text{Im} \int_0^{\varepsilon_F} \sum_d \left[\frac{\langle \mathbf{r}|\phi_d\rangle v'_{d\mathbf{k}} v'_{\mathbf{k}d} \langle \phi_d|\mathbf{r}\rangle}{(E-E_r)(E-\varepsilon_\mathbf{k})^2} \right.$$
$$\left. - \frac{\langle \mathbf{r}|\phi_d\rangle v'_{d\mathbf{k}} (\langle \mathbf{k}|\mathbf{r}\rangle - S_{\mathbf{k}d}\langle \phi_d|\mathbf{r}\rangle) + \text{c.c.}}{(E-E_r)(E-\varepsilon_\mathbf{k})} \right] dE. \tag{5.58}$$

The same caveats that apply to Eqs. (5.51) and (5.52) also apply to Eqs. (5.54)–(5.58). Both $h_2^{\text{tm}}(\mathbf{k})$ and $\tilde{h}_2^{\text{tm}}(\mathbf{k},\mathbf{r})$ must be evaluated numerically and the integrals in Eqs. (5.54) and (5.57) are unrestricted over all $\mathbf{k}$ space. While invariably one finds $Z_c^* > Z$, as is the case in simple metals and corresponding to a net depletion of valence electrons from the inner-core region, either $Z^* > Z_c^*$ or $Z^* < Z_c^*$ is possible in transition metals. That is, the d-state component of $n_{\text{oh}}(\mathbf{r})$ can correspond to either a net depletion or accumulation of electrons in the outer-core region, depending on d-band filling and other conditions. Typically, we find $Z^* > Z_c^*$ at the beginning and in the middle of each transition series and $Z^* < Z_c^*$ near the end of the series, such that in the noble metals $Z^* \approx Z$.

Finally, starting from Eq. (5.51) one can perform the self-consistent screening calculation for transition metals in an analogous way to that done in Chapter 4 for EDB and FDB metals. This calculation leads to the result

$$n_{\text{scr}}(q) = -\left\{ \bar{w}_{\text{ion}}(q) + \bar{h}_1^{\text{tm}}(q)\Pi_{10}(q) + \frac{4\pi e^2}{q^2}[1-G(q)]n_{\text{oh}}(q) \right\} \Pi_0(q)/\varepsilon(q), \tag{5.59}$$

where we have defined

$$\bar{h}_1^{\text{tm}}(q) \equiv -\frac{4}{(2\pi)^3} \int \frac{h_1^{\text{tm}}(\mathbf{k},\mathbf{q})}{\varepsilon_\mathbf{k} - \varepsilon_{\mathbf{k}+\mathbf{q}}} d\mathbf{k}/\Pi_1^{\text{tm}}(q) \tag{5.60}$$

with

$$\Pi_1^{\text{tm}}(q) \equiv -\frac{4}{(2\pi)^3} \int \frac{1}{\varepsilon_\mathbf{k} - \varepsilon_{\mathbf{k}+\mathbf{q}}} d\mathbf{k}. \tag{5.61}$$

Equation (5.59) generalizes Eq. (4.47) for EDB and FDB metals. In practice, the unrestricted k-space integrals in Eqs. (5.51), (5.54), (5.57), (5.60) and (5.61) are carried out within a large sphere of chosen radius k_{max}, such that $\Pi_{10}(q)$ in Eq. (5.59) is still defined by Eq. (4.46), as for FDB metals.

5.1.3 Completion of the cohesive-energy functional

In this section, we return to the main remaining task of developing the GPT cohesive-energy functional for transition metals. Using the explicit multi-ion d-state potentials v_n^d developed in Sec. 5.1.1.1 for $n = 2$, 3 and 4, and neglecting all remaining three-ion and higher energy contributions to $E_{\text{band}}^{\text{val}}$, we can anticipate a result of the general form

$$E_{\text{coh}}(\mathbf{R}, \Omega) = E_{\text{vol}}(\Omega) + \frac{1}{2N}\sum_{i,j}{}' \left[v_2^d(ij) + v_2^{\text{rest}}(ij) \right] + \frac{1}{6N}\sum_{i,j,k}{}' v_3^d(ijk) + \frac{1}{24N}\sum_{i,j,k,l}{}' v_4^d(ijkl).$$

(5.62)

Thus $v_3 = v_3^d$ and $v_4 = v_4^d$, so that up to and including four-ion interactions the remaining quantities to be evaluated are the second-order contributions beyond $E_{\text{coh}}^{\text{pa}}$ to the volume term E_{vol} and additional second-order pair-potential contributions that comprise v_2^{rest}. The evaluation of these remaining contributions for transition metals requires a detailed treatment parallel to that discussed in Sec. 4.1.3 of Chapter 4 for EDB and FDB metals.

We begin by completing the evaluation of valence band-structure energy $E_{\text{band}}^{\text{val}}$, as given in Eq. (5.3). Consistent with our neglect of $\delta n_{\text{band}}(\mathbf{r})$ in the valence electron density, we now also neglect the corresponding small Fermi-energy correction term δE_{band} in $E_{\text{band}}^{\text{val}}$, which like δn_{band} vanishes identically for $E_F = \varepsilon_F$. Then expanding the first logarithm in Eq. (5.5), retaining only second-order contributions from that term and using the definition of E_{struc}^d in Eq. (5.7), Eq. (5.3) can be readily developed into the form

$$E_{\text{band}}^{\text{val}}(Z, Z_d) = N\left(\frac{3}{5}Z\varepsilon_F + Z_d(\varepsilon_F - E_d^{\text{vol}}) - \frac{10}{\pi}\int_0^{\varepsilon_F}\delta_2(E)dE\right) - \int_0^{\varepsilon_F}\delta N_{sp}(E)dE$$

$$- \frac{2}{\pi}\text{Im}\int_0^{\varepsilon_F}\sum_d\left[\frac{E_d^{\text{struc}}(d\Gamma_{dd}^{\text{vol}}/dE) + \Gamma_{dd}^{\text{struc}}}{E - E_r} + \frac{\Lambda_{dd}^{\text{vol}}}{E - E_r}\right]dE + NE_{\text{struc}}^d.$$

(5.63)

The additional energy integral over δN_{sp} in Eq. (5.63) produces the usual first- and second-order PP terms, and as for EDB and FDB metals, the second-order PP terms can be combined with the second-order terms that arise from $\Lambda_{dd}^{\text{vol}}$. The nominal first-order PP term also produces hidden second-order structure-dependent contributions:

$$-\frac{2}{\pi}\text{Im}\int_0^{\varepsilon_F}\sum_k\frac{W_{kk}}{E-\varepsilon_k}dE = N\left(ZV_0' + \frac{2\Omega}{(2\pi)^3}\int w_{\text{pa}}(\mathbf{k})\Theta^<(\mathbf{k})d\mathbf{k}\right)$$

$$+ n_{\text{unif}}\left(V - \sum_i v_{\text{pa}}^i\right) - \sum_i (n_{\text{oh}}^c)^i\delta V_{\text{struc}}^i. \quad (5.64)$$

The first two leading terms on the RHS of Eq. (5.64) depend only on volume and not structure and combine with first three terms on the RHS of Eq. (5.63) to contribute both to $E_{\text{coh}}^{\text{pa}}$ and E_{vol}. The final structure-dependent term in Eq. (5.64) can be combined with the structure-dependent d-state terms involving E_d^{struc} and $\Gamma_{dd}^{\text{struc}}$ in Eq. (5.63) to produce the real-space orthogonalization-hole interaction

$$-\sum_i n_{\text{oh}}^i \delta V_{\text{struc}}^i = -\sum_i (n_{\text{oh}}^c)^i \delta V_{\text{struc}}^i - \frac{2}{\pi}\text{Im}\int_0^{\varepsilon_F}\sum_d \left[\frac{E_d^{\text{struc}}(d\Gamma_{dd}^{\text{vol}}/dE) + \Gamma_{dd}^{\text{struc}}}{E - E_r}\right]dE. \tag{5.65}$$

This quantity can be processed and reabsorbed into the theory exactly as in Eq. (4.50) for EDB and FDB metals.

Following Moriarty (1988a), and using these results to complete the evaluation of the valence band-structure energy (5.63), Eq. (5.2) for the cohesive-energy functional can then be expressed in a mixed reciprocal-space and real-space form:

$$E_{\text{coh}} = E_{\text{vol}}^q(\Omega) + \sum_{\mathbf{q}}{}' |S(\mathbf{q})|^2 F_0(q,\Omega) + \delta E_{\text{es}}(Z^*,\Omega) + \frac{1}{2N}\sum_{i,j}{}' [v_2^d(ij,\Omega) + v_{\text{ol}}^0(ij,\Omega)]$$

$$+ \frac{1}{6N}\sum_{i,j,k}{}' v_3(ijk,\Omega) + \frac{1}{24N}\sum_{i,j,k,l}{}' v_4(ijkl,\Omega). \tag{5.66}$$

Here the volume term E_{vol}^q is defined as

$$E_{\text{vol}}^q(\Omega) \equiv E_{\text{fe}}^0(\Omega) + \frac{2\Omega}{(2\pi)^3}\int w_{\text{pa}}(\mathbf{k})[1 + p_c(\mathbf{k})]\Theta^<(\mathbf{k})d\mathbf{k} + Z_d(\varepsilon_F - E_d^{\text{vol}}) - \frac{10}{\pi}\int_0^{\varepsilon_F}\delta_2(E)dE$$

$$+ \frac{2\Omega}{(2\pi)^3}\int [h_1^{\text{tm}}(\mathbf{k})p_c(\mathbf{k}) + h_2^{\text{tm}}(\mathbf{k})w_{\text{pa}}(\mathbf{k})]\,d\mathbf{k} + \delta E_{\text{oh}} - E_{\text{bind}}^{\text{atom}}(Z,Z_d) + E_{\text{prep}}, \tag{5.67}$$

the energy-wavenumber characteristic F_0 is given by

$$F_0(q,\Omega) \equiv \frac{2\Omega}{(2\pi)^3}\left[\int \frac{[w(\mathbf{k},\mathbf{q})]^2}{\varepsilon_{\mathbf{k}} - \varepsilon_{\mathbf{k+q}}}\Theta^<(\mathbf{k})d\mathbf{k} + \int \frac{2w(\mathbf{k},\mathbf{q})h_1^{\text{tm}}(\mathbf{k},\mathbf{q})}{\varepsilon_{\mathbf{k}} - \varepsilon_{\mathbf{k+q}}}d\mathbf{k}\right]$$

$$- \frac{2\pi e^2 \Omega}{q^2}\left\{[1 - G(q)][n_{\text{scr}}(q)]^2 + G(q)[n_{\text{oh}}(q)]^2\right\} \tag{5.68}$$

and the overlap potential v_{ol}^0 is of the form

$$v_{\text{ol}}^0(ij,\Omega) \equiv \delta v_{\text{es-xc}}^{ij} + \delta v_{\text{oh}}^{ij} + \delta v_{\text{ke}}^{ij}, \tag{5.69}$$

5.1.3 Completion of the cohesive-energy functional

with $\delta v^{ij}_{\text{es-xc}}$, $\delta v^{ij}_{\text{oh}}$ and $\delta v_{\text{ke}}(R_{ij}) \equiv \delta v^{ij}_{\text{ke}}$ still given by Eqs. (4.59), (4.60) and (A2.50), respectively, exactly as for EDB and FDB metals. In Eq. (5.67), note that $h_1^{\text{tm}}(\mathbf{k}) \equiv h_1^{\text{tm}}(\mathbf{k}, 0)$, as given by Eq. (5.52).

Finally, the second and third terms on the RHS of Eq. (5.66) can be converted to real space in the usual way to produce a cohesive-energy functional in the form of Eq. (5.1). The full transition-metal pair potential v_2 is then

$$v_2(r,\Omega) = v_2^d(r,\Omega) + \frac{(Z^*e)^2}{r}\left[1 - \frac{2}{\pi}\int_0^\infty F_N^0(q,\Omega)\frac{\sin(qr)}{q}dq\right] + v_{\text{ol}}^0(r,\Omega), \tag{5.70}$$

where we have defined a corresponding normalized energy-wavenumber characteristic as $F_N^0(q,\Omega) \equiv -\left[q^2\Omega/2\pi(Z^*e)^2\right]F_0(q,\Omega)$.

A useful alternate representation to Eq. (5.70) for v_2 that more closely parallels the EDB and FDB results discussed in Chapter 4 is also possible. This can be accomplished by first subtracting the long-range hybridization tail v_2^{hyb} from v_2^d to create a short-range overlap potential analogous to Eq. (4.55) for EDB and FDB metals:

$$v_{\text{ol}}(r,\Omega) \equiv v_2^d(r,\Omega) - v_2^{\text{hyb}}(r,\Omega) + v_{\text{ol}}^0(r,\Omega), \tag{5.71}$$

with

$$v_2^{\text{hyb}}(r,\Omega) \equiv -\frac{2}{\pi}\text{Im}\int_0^{\varepsilon_F}\sum_{d,d'}\frac{\Gamma_{dd'}^{\text{vol}}(r,E)\Gamma_{d'd}^{\text{vol}}(r,E)}{(E-E_r)^2}dE. \tag{5.72}$$

One then adds back this tail in reciprocal space to create the total energy-wavenumber characteristic

$$F(q,\Omega) \equiv F_0(q,\Omega) + \frac{2\Omega}{(2\pi)^3}\int\frac{h_{21}^{\text{tm}}(\mathbf{k},\mathbf{q})}{\varepsilon_\mathbf{k} - \varepsilon_{\mathbf{k}+\mathbf{q}}}d\mathbf{k}, \tag{5.73}$$

where

$$h_{21}^{\text{tm}}(\mathbf{k},\mathbf{q}) \equiv -\frac{1}{\pi}\text{Im}\int_0^{\varepsilon_F}\frac{\left(\sum_d v'_{\mathbf{k}+\mathbf{q}d}v'_{d\mathbf{k}}\right)^2}{(E-E_r)^2(E-\varepsilon_\mathbf{k})}dE. \tag{5.74}$$

The pair potential v_2 is then given by the equivalent expression

$$v_2(r,\Omega) = \frac{(Z^*e)^2}{r}\left[1 - \frac{2}{\pi}\int_0^\infty F_N(q,\Omega)\frac{\sin(qr)}{q}dq\right] + v_{\text{ol}}(r,\Omega), \tag{5.75}$$

and, using Eq. (5.73) for $F(q,\Omega)$, the normalized energy-wavenumber characteristic is

$$F_N(q,\Omega) \equiv -\frac{q^2\Omega}{2\pi(Z^*e)^2} F(q,\Omega). \tag{5.76}$$

In subtracting and adding the hybridization tail, the $i = j$ and $\mathbf{q} = 0$ terms are absorbed into the volume term, which is established from the relation

$$E_{\text{vol}}(\Omega) = E_{\text{vol}}^q(\Omega) + (Z^*e)^2 \left[\frac{0.9}{R_{\text{WS}}} + \frac{\pi}{\Omega} \frac{\partial^2 F_N(0,\Omega)}{\partial q^2} - \frac{1}{\pi} \int_0^\infty F_N(q,\Omega) dq \right]$$

$$+ \frac{1}{\pi} \text{Im} \int_0^{\varepsilon_F} \left\{ \sum_d \left(\frac{\Gamma_{dd}^{\text{vol}}(E)}{E - E_{\text{r}}} \right)^2 - \sum_k \frac{\left(\sum_d v'_{kd} v'_{dk} \right)^2}{(E - E_{\text{r}})^2 (E - \varepsilon_{\mathbf{k}})^2} \right\} dE. \tag{5.77}$$

Using Eq. (5.67) for E_{vol}^q and evaluating the second-derivative of F_N in Eq. (5.77), then yields the following result for the real-space volume term E_{vol}:

$$E_{\text{vol}}(\Omega) = E_{\text{fe}}^0(\Omega) - \frac{1}{2}\Omega B_{\text{eg}}(\Omega) + \frac{2\Omega}{(2\pi)^3} \int w_{\text{pa}}(\mathbf{k}) \left[1 + p_c(\mathbf{k})\right] \Theta^<(\mathbf{k}) d\mathbf{k} + Z_d(\varepsilon_F - E_d^{\text{vol}})$$

$$- \frac{10}{\pi} \int_0^{\varepsilon_F} [\delta_2(E) + \delta h_2(E)] \, dE + \frac{2\Omega}{(2\pi)^3} \int [h_1^{\text{tm}}(\mathbf{k}) p_c(\mathbf{k}) + h_2^{\text{tm}}(\mathbf{k}) w_{\text{pa}}(\mathbf{k}) + h_{22}^{\text{tm}}(\mathbf{k})] \, d\mathbf{k}$$

$$- Z^* \left[\bar{w}_{\text{core}}(0) + \bar{h}_1^{\text{tm}}(0) \Pi_{10}(0) \right] + \frac{3}{4} \frac{Z}{\varepsilon_F} \left[\overline{\delta w_{\text{core}}^2}(0) + \overline{\delta w_{\text{core}} h_1^{\text{tm}}}(0) \Pi_{10}(0) \right]$$

$$+ \frac{9}{10} \frac{(Z^*e)^2}{R_{\text{WS}}} - \frac{(Z^*e)^2}{\pi} \int_0^\infty F_N(q,\Omega) dq + \delta E_{\text{oh}}^* - E_{\text{bind}}^{\text{atom}}(Z, Z_d) + E_{\text{prep}}, \tag{5.78}$$

where we have further defined the quantities

$$\delta h_2(E) \equiv -\frac{1}{2} \text{Im} \left(\frac{\Gamma_{dd}^{\text{vol}}(E)}{E - E_{\text{r}}} \right)^2 \tag{5.79}$$

and

$$h_{22}^{\text{tm}}(\mathbf{k}) \equiv -\frac{1}{\pi} \text{Im} \int_0^{\varepsilon_F} \frac{\left(\sum_d v'_{kd} v'_{dk} \right)^2}{(E - E_{\text{r}})^2 (E - \varepsilon_{\mathbf{k}})^2} dE. \tag{5.80}$$

The averages $\bar{w}_{\text{core}}$, $\overline{\delta w_{\text{core}}^2}$ and $\bar{h}_1^{\text{tm}}$ in Eq. (5.78) are defined by Eqs. (3.83), (3.85) and (5.60), respectively. The additional average $\overline{\delta w_{\text{core}} h_1^{\text{tm}}}$ is defined as

$$\overline{\delta w_{\text{core}} h_1^{\text{tm}}}(q) \equiv \overline{w_{\text{core}} h_1^{\text{tm}}}(q) - 2\bar{w}_{\text{core}}(q)\bar{h}_1^{\text{tm}}(q) - [\bar{h}_1^{\text{tm}}(q)]^2 \Pi_{10}(q), \qquad (5.81)$$

with

$$\overline{w_{\text{core}} h_1^{\text{tm}}}(q) \equiv -\frac{4}{(2\pi)^3} \int \frac{2w_{\text{core}}(\mathbf{k},\mathbf{q}) h_1^{\text{tm}}(\mathbf{k},\mathbf{q}) + [h_1^{\text{tm}}(\mathbf{k},\mathbf{q})]^2}{\varepsilon_{\mathbf{k}} - \varepsilon_{\mathbf{k+q}}} d\mathbf{k} / \Pi_1^{\text{tm}}(q). \qquad (5.82)$$

Equations (5.81) and (5.82) are analogous, respectively, to Eqs. (4.71) and (4.72) for EDB and FDB metals.

Figure 5.8 illustrates the behavior of the d-state potential $v_2^d(r,\Omega_0)$ and the full pair potential $v_2(r,\Omega_0)$, together with the impact of their individual components, in the case of nonmagnetic chromium (Cr). At the fcc and bcc near-neighbor distances and below, both v_2^d and v_2 are seen to be attractive, with v_2 considerably more so all the way down to $r/R_{\text{WS}} = 1.4$. As can be inferred from Fig. 5.8(a), together with Eqs. (5.6) and (5.14), the attractive nature of v_2^d below $r/R_{\text{WS}} = 2.1$ is almost entirely due to the direct d-state interactions contained in $T_{dd'}^{\text{vol}}$ through $\Delta_{dd'}$. It is only beyond that point where the impact of sp-d hybridization through $\Gamma_{dd'}^{\text{vol}}$ becomes large in $T_{dd'}^{\text{vol}}$ and produces a substantial long-range tail to v_2^d.

The strongly attractive behavior of the full pair potential v_2 is due in part to the attractive v_{ol}^0 component of v_{ol} in Eq. (5.71), but is mostly the result of the substantial sp-d hybridization contribution to the electron screening, which acts through F_N^0 in Eq. (5.70) and through F_N in Eq. (5.75). At the same time, we can anticipate that the attractive nature of v_2 for central transition metals like Cr will be countered by a corresponding repulsive behavior of the three-ion potential v_3. We will qualitatively illustrate this balancing below in Sec. 5.1.4 and then investigate the issue in more depth quantitatively in Sec. 5.1.5 in connection with the multi-ion screening of sp-d hybridization. In addition, at sufficiently short distances the attractive components of v_2 will be balanced alone by the combined repulsive behavior of the Coulomb potential $(Z^*e)^2/r$ and the increased kinetic energy of overlapping d and inner-core states. This typically results in a deep minimum in the pair potential, which for central transition metals like Cr can lie well inside the near-neighbor distances for metallic structures, as seen in Fig. 5.8(b). But again, the effective repulsive nature of the three-ion potential v_3 makes this minimum physically inaccessible in such cases.

5.1.4 First-principles pair and multi-ion potentials

In this section, we briefly examine first-principles GPT pair and multi-ion potentials across the transition-metal series, with an emphasis on prototypical behavior and significant trends. Additional calculated results for elemental $3d$ and $4d$ transition metals can be found in Moriarty (1988a). In Fig. 5.9 we display the behavior of the two-ion pair potential $v_2(r,\Omega_0)$ for each of the ten members of the $3d$ series of elements, as calculated

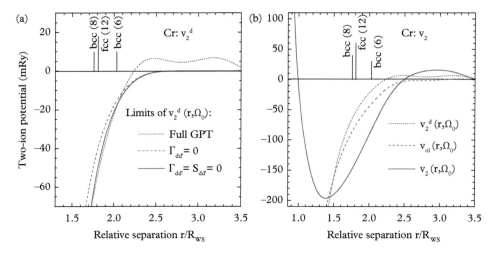

Fig. 5.8 *Two-ion GPT pair potentials $v_2^d(r,\Omega_0)$ and $v_2(r,\Omega_0)$, together with their individual components, for Cr, as calculated at its observed equilibrium volume $\Omega_0 = 80.94$ a.u. with $D_2^\star = -3$. (a) v_2^d as obtained from Eq. (5.14), using Eq. (5.6) for $T_{dd'}^{\text{vol}}$. In the figure, $\Gamma_{dd'} = 0$ is shorthand for $\Gamma_{dd'}^{\text{vol}} = 0$ in Eq. (5.6); (b) v_2 as obtained from Eq. (5.75), with v_{ol} calculated from Eq. (5.71). The location and number of fcc and bcc near neighbors are indicated in each case.*

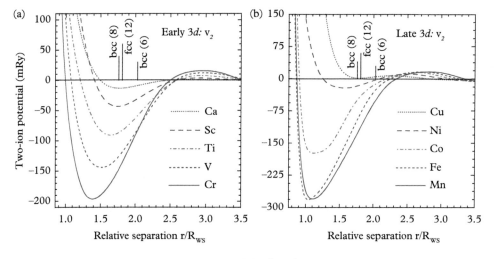

Fig. 5.9 *First-principles two-ion GPT pair potential $v_2(r,\Omega_0)$ across the 3d transition-metal series, as calculated via Eq. (5.75) with $D_2^\star = -3$. (a) Early series metals Ca through Cr; (b) late series metals Mn through Cu. The location and number of fcc and bcc near neighbors are indicated in each case.*
From Moriarty (1988a), with publisher (APS) permission.

with the canonical d-band boundary condition $D_2^\star = -3$. At the beginning of the series in Ca, and at the end of the series in Cu, $v_2(r,\Omega_0)$ has a modest potential well depth with a potential minimum close to the fcc nearest-neighbor position. As expected, this

5.1.4 First-principles pair and multi-ion potentials

behavior is quite similar, respectively, to that of the EDB pair potential for Ca shown in Fig. 4.4(a) and to that of the FDB pair potential for Cu shown in Fig. 4.7(b). In both metals, additional multi-ion contributions to $E_{\mathrm{coh}}(\mathbf{R}, \Omega)$ are expected to be small, although as in the EDB and FDB limits, optimization of the d basis states through $D_2^\star$ is generally important to successful application of the transition-metal (TM) pair potentials. For Ca, the EDB pair-potential treatment developed in Chapter 4 is preferred in the low-pressure region, where the fcc structure and other the ground-state properties of the material are accurately predicted. In Cu, on the other hand, the first-principles TM pair-potential treatment is generally preferred over the adjusted FDB treatment discussed in Chapter 4.

Returning to Fig. 5.9 and moving inward toward the center of the $3d$ series, the well depth of $v_2(r, \Omega_0)$ rapidly and systematically increases in magnitude, as the minimum in the potential moves to smaller r/R_{WS}. In the next elements, Sc and Ni, the magnitude of the well depth remains modest, and neglecting multi-ion potential contributions to $E_{\mathrm{coh}}(\mathbf{R}, \Omega)$ still provides a reasonable description of the metal. In this regard, at the TM pair-potential level of treatment in the GPT both the hcp structure of Sc and the fcc structure of Ni are correctly predicted. Direct calculation of the three-ion potential in Sc confirms that it is small (Moriarty, 1988a), although not necessarily negligible either. Indeed, for the early $5d$ metal Ba, whose three-ion potential v_3 is similar to that of Sc and is displayed in Fig. 5.10(a), the inclusion of v_3 in $E_{\mathrm{coh}}(\mathbf{R}, \Omega)$ is essential to the correct explanation of its bcc stability and phonon spectrum, as briefly mentioned in Chapter 4 and as discussed in more depth in Chapters 6 and 7, respectively.

For the remaining six $3d$ metals shown in Fig. 5.9, the magnitude of the well depth of $v_2(r, \Omega_0)$ becomes large, ranging from nearly 100 mRy in Ti to almost 300 mRy in Mn and Fe. In all of these remaining metals, one thus has the immediate expectation of the major importance of the multi-ion potentials in balancing the strongly attractive nature of the pair potentials. The role of the three-ion potential in achieving that balance is most clear at the center of the series in Cr. Unlike the weakly negative three-ion potentials in Sc and Ba, as shown for Ba in Fig. 5.10(a), v_3 in Cr can be strongly positive, as shown in Fig. 5.10(b), and hence qualitatively a significant repulsive counter to the corresponding attractive pair potential v_2 displayed in Figs. 5.8(b) and 5.9(a). In particular, note that the magnitude of v_3 is generally one to two orders of magnitude larger in Cr than Ba, and that $v_3(d, \theta, \Omega_0)$ for Cr, as defined and illustrated in Fig. 5.10, rapidly increases in magnitude for all angles θ as the interatomic separation distance d decreases.

The behaviors of v_3 and v_4 in the $3d$ mid-period metals V, Cr, Mn and Fe are illustrated and compared in Fig. 5.11 for characteristic planar near-neighbor interactions. As seen in Fig. 5.11(a), the three-ion potential in V has a similarly repulsive shape and magnitude to that of the Cr potential. Likewise, as shown in Fig. 5.11(b), the four-ion potentials in V and Cr are similarly attractive with signature oscillations that are favorable to the observed bcc structure and the large bcc-fcc energy difference found in such group-VB and -VIB metals. Specifically, as was noted in Chapter 1 for the corresponding four-ion MGPT potentials of V, Mo and Ta displayed in Fig. 1.4, v_4 in the full GPT for V and Cr is favorable to the 70.5° and 109.5° near-neighbor bond angles in the bcc structure, but unfavorable to the 90° near-neighbor bond angles in the fcc structure.

The behaviors of v_3 and v_4 for Mn and Fe shown in Fig. 5.11 can be contrasted with those of V and Cr. The three-ion potential v_3 for Mn is generally less repulsive overall and significantly attractive for intermediate angles, while v_3 for Fe is almost entirely attractive.

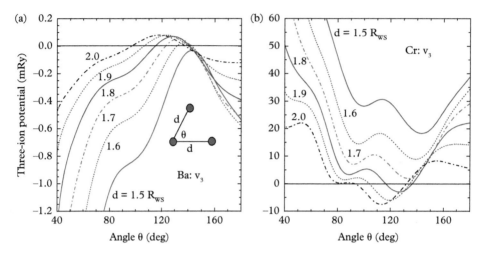

Fig. 5.10 *Contrasting behavior of the three-ion GPT potential $v_3(d, \theta, \Omega_0)$ in the two prototype transition metals Ba and Cr for selected near-neighbor interactions, as calculated using Eq. (5.16) with $D_2^\star = -3$. (a) Early 5d metal Ba; (b) mid-period 3d metal Cr.*
After Moriarty (1988a), with publisher (APS) permission.

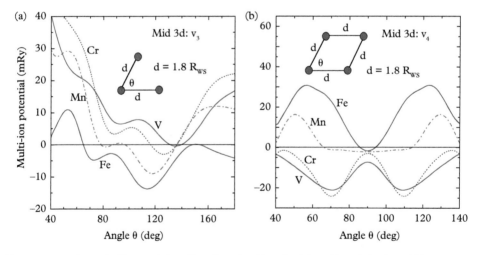

Fig. 5.11 *Multi-ion GPT potentials $v_3(\theta, \Omega_0)$ and $v_4(\theta, \Omega_0)$ of the mid-period 3d transition metals V, Cr, Mn and Fe for near-neighbor planar interactions calculated with $D_2^\star = -3$. (a) $v_3 = v_3^d$, as obtained using Eq. (5.16); (b) $v_4 = v_4^d$, as obtained using Eq. (5.19).*

The four-ion potential v_4 for Mn is flat and near zero for the intermediate angles that distinguish the cubic fcc and bcc structures, while v_4 is almost entirely repulsive for Fe. In addition, v_4 for Fe is out of phase with the four-ion potentials for V and Cr, with a minimum rather than a maximum for 90° near-neighbor bond angles. This behavior is consistent with the predicted hcp, as opposed to bcc, structure of nonmagnetic Fe, and also with the observed hcp structures in the corresponding $4d$ and $5d$ group-VIII elements Ru and Os. A fully adequate GPT treatment of hcp Fe, Ru and Os, however, still likely requires additional five- and six-ion interactions, as was discussed in Chapter 1 in connection with Fig. 1.6.

5.1.5 Multi-ion screening of *sp-d* hybridization

In the central transition metals, the directly calculated contributions of *sp-d* hybridization to the individual pair and multi-ion potentials are substantial, as shown above for the Cr pair potentials v_2^d and v_2 in Fig. 5.8. The quantitative effects of the hybridization on the multi-ion potentials v_3 and v_4 are also quantitatively large, as can be appreciated by comparing the full GPT results for Cr in Fig. 5.11 with the corresponding results obtained in the absence of *sp-d* hybridization shown in Fig. 5.12. For the three-ion potential v_3 the quantitative impact of the hybridization is up to a factor of five, and for the four-ion potential v_4 up to an order of magnitude. Yet, as we saw in the last section, there appears to be a clear tendency in the central transition metals for pair and multi-ion potentials to balance attractive and repulsive contributions, at least qualitatively. In this section, we will examine this trend toward cancellation more quantitatively, again for the prototype case of Cr.

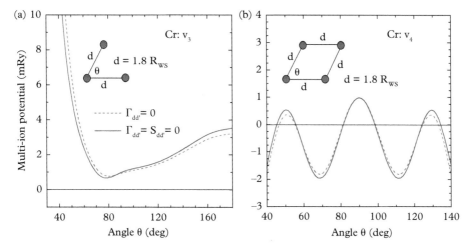

Fig. 5.12 *GPT multi-ion potentials $v_3(\theta, \Omega_0)$ and $v_4(\theta, \Omega_0)$ for Cr, as in Fig. 5.11, but calculated here without sp-d hybridization, i.e., $\Gamma_{dd'} \equiv \Gamma_{dd'}^{vol} = 0$ in Eq. (5.6), and additionally without d-state nonorthogonality, i.e., $S_{dd'} = 0$ in Eq. (5.6). (a) $v_3 = v_3^d$, as obtained using Eq. (5.16); (b) $v_4 = v_4^d$, as obtained using Eq. (5.19).*

One can view such destructive interference phenomenon as multi-ion screening of an effective ion-ion interaction in the metal, and gain some useful insight by examining the effective d-state pair potential

$$v_{\text{eff}}^d(r,\Omega) \equiv v_2^d(r,\Omega) + \langle v_3^d \rangle (r,\Omega) + \langle v_4^d \rangle (r,\Omega), \tag{5.83}$$

where the averages over the three- and four-ion GPT d-state potentials can be defined in the following manner. In general, one expects the statistical average of an n-ion potential $\langle v_n^d \rangle$ to depend directly on a corresponding n-ion correlation function $g_n(\mathbf{R}_1, \mathbf{R}_2 \ldots \mathbf{R}_n)$ representing expected ion positions for the material at a given volume, temperature and phase over which the average is taken. For our limited purposes here, it is convenient to approximate g_n as a simple product of the relevant pair correlation functions, such that $g_n(\mathbf{R}_1, \mathbf{R}_2, \ldots \mathbf{R}_n) \propto g(R_{12})g(R_{23}) \cdots$, where $g(r)$ is the pair correlation function and the product spans the interatomic distances defining the n-ion ensemble. Then following the multi-ion geometry illustrated in Fig. 1.2, with ion positions i,j,k,l denoted as $1,2,3,4$, we define the three- and four-ion d-state averages as

$$\langle v_3^d \rangle (r,\Omega) \equiv \frac{1}{\Omega} \int v_3^d(R_{12}, R_{23}, R_{31}, \Omega) g(R_{23}) g(R_{31}) d\mathbf{R}_3 \tag{5.84}$$

and

$$\langle v_4^d \rangle (r,\Omega) \equiv \frac{1}{\Omega^2} \int \int v_4^d(R_{12}, R_{23}, R_{34}, R_{41}, R_{31}, R_{42}, \Omega) g(R_{23}) g(R_{34}) g(R_{41})$$
$$\times g(R_{31}) g(R_{42}) d\mathbf{R}_3 d\mathbf{R}_4, \tag{5.85}$$

where in both potentials $r = R_{12}$. Note that inside the real-space integrals in Eqs. (5.84) and (5.85), v_3^d and v_4^d are the full first-principles GPT multi-ion functions given by Eqs. (5.16) and (5.19), respectively, with sp-d hybridization included. Also, the potentials v_3^d and v_4^d are expressed in terms of the needed interatomic distances as in Eqs. (1.5) and (1.6), respectively.

In order to evaluate Eqs. (5.83)–(5.85) for Cr, we need an appropriate pair correlation function, which, in principle, can be obtained from an MD simulation for the conditions of interest (see Chapter 8). With sp-d hybridization included, v_3^d and v_4^d are too long-ranged to permit such GPT-MD simulations, but we expect $g(r)$ to be sufficiently generic in character that the inclusion of hybridization is not essential here. Thus, as a good substitute, we have used instead complementary MGPT-MD simulations, which neglect sp-d hybridization, to obtain $g(r)$ at the observed equilibrium volume Ω_0 and selected temperatures in both the bcc solid and in the liquid. These results have then been used in turn to calculate the functions $\langle v_3^d \rangle$, $\langle v_4^d \rangle$ and v_{eff}^d in Cr for the same conditions. Interestingly, these functions are also found to be rather insensitive to both temperature and phase, and are displayed in Fig. 5.13 only for a representative temperature of 2000 K in the bcc solid.

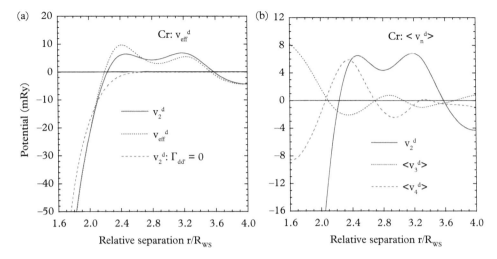

Fig. 5.13 *GPT effective d-state pair potential v_{eff}^d and its components for bcc Cr, calculated at $\Omega_0 = 80.94$ a.u. and 2000 K from Eqs. (5.83)–(5.85), with sp-d hybridization included in the potentials. (a) v_{eff}^d, compared with its major two-ion component v_2^d, and with v_2^d calculated in the absence of hybridization $\left(\Gamma_{dd'} \equiv \Gamma_{dd'}^{\text{vol}} = 0\right)$; (b) components of v_{eff}^d: v_2^d, $\langle v_3^d \rangle$ and $\langle v_4^d \rangle$.*

Figure 5.13(a) shows that $v_{\text{eff}}^d(r, \Omega_0)$ for Cr is dominated by its pair-potential component $v_2^d(r, \Omega_0)$. In addition, by comparing with Fig. 5.8(a), one sees that *sp-d* hybridization itself only dominates v_2^d and hence v_{eff}^d at interatomic separations greater than about $r = 2.1 R_{\text{WS}}$, which is beyond second-nearest neighbors in the bcc structure. In this latter regime, the net hybridization is still strongly oscillatory, so its net effect on the corresponding total-energy contribution can be expected to be significantly smaller than the magnitude of its individual oscillations. At the same time, Fig. 5.13(b) shows in more complete detail that the $\langle v_3^d \rangle$ and $\langle v_4^d \rangle$ components of v_{eff}^d oscillate out of phase in an almost perfect fashion, quantitatively canceling for $r \leq 2.1 R_{\text{WS}}$, while producing only a small net contribution to v_{eff}^d for $r > 2.1 R_{\text{WS}}$.

These results support the picture of strong destructive interference of *sp-d* hybridization contributions to the cohesive-energy functional in central transition metals. Ultimately, this effect is a consequence of the very nonspherical Fermi surfaces in such metals. But any attempt to capture the destructive interference directly in a GPT MD simulation would be not only inherently inefficient but also in practice prohibitively expensive. For such metals, a more viable and so far successful starting point, as adopted in the standard MGPT treatment discussed in Sec. 5.2, is simply to neglect explicit *sp-d* hybridization contributions to the pair and multi-ion potentials entirely. A possible next step beyond that simplification, especially for the neighboring hcp and fcc transition metals, is to re-include the *sp-d* hybridization in the MGPT approximately in terms of an effective pair potential, in a manner similar to Eq. (5.83). Such a capability is currently in the initial stages of development for late-series transition metals, and progress in this direction on a Ni prototype is discussed in Sec. 5.2.4.

5.2 Simplified MGPT potentials for robust atomistic simulations

The simplified MGPT is developed from the full first-principles GPT through a series of systematic approximations applicable to mid-period transition metals with nearly half-filled d bands. First, motivated by the destructive interference exhibited among the long-ranged hybridization tails of the multi-ion GPT potentials in such metals, as discussed in Sec. 5.1.5, all sp-d hybridization contributions beyond those inherent in the zero-order reference system through $\Gamma_{dd'}^{\text{vol}}(E)$ and in the volume term E_{vol} are neglected. Thus, one sets $\Gamma_{dd'}^{\text{vol}} = 0$ in Eq. (5.6) for the T_{ij} matrix components $T_{dd'}^{\text{vol}}$ that define the multi-ion potentials v_n^d. One also removes the d-state nonorthogonality integral $S_{dd'}$ from $T_{dd'}^{\text{vol}}$. In this regard, the relatively large two-ion nonorthogonality contribution to v_2^d, illustrated for Cr in Fig. 5.8, is first separated out from v_2^d and then combined with the overlap potential v_{ol}^0 to create a complementary repulsive "hard-core" potential v_2^{hc}, as discussed in Sec. 5.2.1 below and considered in Sec. A2.7 of Appendix A2. The much smaller corresponding nonorthogonality contributions to v_3^d and v_4^d, as illustrated above for Cr in Fig. 5.12, are neglected by simply setting $S_{dd'} = 0$ in $T_{dd'}^{\text{vol}}$. The net effect of these approximations is to reduce the central d-state interaction matrix T_{ij} that defines the d-state potentials to a short-ranged form with components

$$T_{dd'}^{\text{vol}}(R_{ij}, E) = \frac{\Delta_{dd'}(R_{ij})}{E - E_d^{\text{vol}} - \Gamma_{dd}^{\text{vol}}(E)}, \tag{5.86}$$

which is then analogous to the form assumed in Eq. (5.26) for the simple TB model discussed in Sec. 5.1.1.2.

Also as in the TB model, the d-state potentials v_n^d in the MGPT are represented by the series expansions (5.23)–(5.25), which in the baseline form of the theory are carried out to fourth order in $T_{dd'}^{\text{vol}}$. Written directly in terms of the T_{ij} matrices, one then has

$$v_2^d(ij) = -\frac{2}{\pi}\text{Im} \int_0^{\varepsilon_F} \text{Tr}\left[T_{ij}T_{ji} + \frac{1}{2}(T_{ij}T_{ji})^2\right] dE, \tag{5.87}$$

$$v_3^d(ijk) = -\frac{2}{\pi}\text{Im}\int_0^{\varepsilon_F} \text{Tr}\left[-2\left(T_{ij}T_{jk}T_{ki}\right) + \left(T_{ij}T_{ji}T_{ik}T_{ki} + T_{jk}T_{kj}T_{ji}T_{ij} + T_{ki}T_{ik}T_{kj}T_{jk}\right)\right] dE \tag{5.88}$$

and

$$v_4^d(ijkl) = -\frac{2}{\pi}\text{Im}\int_0^{\varepsilon_F} \text{Tr}\left[2\left(T_{ij}T_{jk}T_{kl}T_{li} + T_{ik}T_{kl}T_{lj}T_{ji} + T_{il}T_{lj}T_{jk}T_{ki}\right)\right] dE. \tag{5.89}$$

The form of $T_{dd'}^{\text{vol}}$ in Eq. (5.86) ensures that the dependence of the d-state potentials upon ion position is independent of energy, and that the energy-dependent denominator in $T_{dd'}^{\text{vol}}$ contributes only, for a given volume, to a small set of well-defined coefficients that depend on d-band filling and width for the five different classes of terms in Eqs. (5.87)–(5.89). This separation of atomic structure from d-band filling and width is again of the same form as assumed in the simple TB model. Finally, as is the case for the multi-ion contributions to the TB moments, the d-band contributions to the potentials v_2^d, v_3^d and v_4^d can be represented graphically by the respective horizontal two-, three- and four-ion diagrams illustrated in Fig. 5.1.

5.2.1 Baseline analytic MGPT for canonical d bands

To establish the baseline analytic MGPT (Moriarty, 1990a), one introduces canonical d bands into the theoretical framework, but in a slightly generalized form that expresses the $\Delta_{dd'}$ matrix elements in Eq. (5.86), for ions i and j aligned along the z axis, as

$$\Delta_{dd'}(R_{ij}) = \alpha_m (R_{\text{WS}}/R_{ij})^p \delta_{mm'}, \tag{5.90}$$

where p is a material-dependent parameter to be optimized. Here the α_m coefficients are independent of atomic structure with fixed ratios between the m components such that $\alpha_0 : \alpha_1 : \alpha_2$ is $6 : (-4) : 1$, exactly as for pure canonical d bands, where $p = 5 = 2\ell + 1$ for $\ell = 2$. In practice, retaining p as a variable parameter allows one to obtain a more accurate description of the radial dependence of $\Delta_{dd'}(R_{ij})$ for real materials, without losing the analytic nature of the canonical d-band description. Using Eqs. (5.86) and (5.90) in Eqs. (5.87)–(5.89), one can then derive entirely analytic representations of the d-state potentials v_2^d, v_3^d and v_4^d.

The two-ion d-state potential v_2^d in the MGPT consists of one second-order and one fourth-order contribution in $\Delta_{dd'}$. These terms are easily evaluated from Eq. (5.87) using the fact that the geometry can be chosen so that T_{ij} is a diagonal matrix, with elements proportional to the RHS of Eq. (5.90). One thereby obtains a result for v_2^d of the general form

$$\begin{aligned} v_2^d(r, \Omega) &= v_a(\Omega)[f(r)]^4 - v_b(\Omega)[f(r)]^2 \\ &= v_a(\Omega)(R_0/r)^{4p} - v_b(\Omega)(R_0/r)^{2p}, \end{aligned} \tag{5.91}$$

where we have defined the dimensionless characteristic radial function

$$f(r) \equiv (R_0/r)^p, \tag{5.92}$$

with $R_0 \equiv 1.8 R_{\text{WS}}$. The two pair-potential coefficients v_a and v_b are material-dependent parameters with the dimensions of energy that depend on d-band filling and width and can also be evaluated within the MGPT framework, as will be discussed later in this section. From such an analysis, one finds for central transition metals that $v_a > 0$ and

$v_b > 0$ with $v_b \gg v_a$, so that v_2^d is an entirely attractive potential at normal interatomic separation distances. This behavior qualitatively conforms to that found in the $\Gamma_{dd'}^{\text{vol}} = S_{dd'} = 0$ limit of the GPT, as illustrated in Fig. 5.8(a) for the case of Cr at $\Omega = \Omega_0$. With the choice of $p = 5$ in Eq. (5.91), one has the pure canonical-d-band potential

$$v_2^d(r, \Omega) = v_a(\Omega)(R_0/r)^{20} - v_b(\Omega)(R_0/r)^{10}. \quad (5.93)$$

This form further allows a good quantitative fit to the first-principles GPT calculation of $v_2^d(r, \Omega_0)$ in Cr with $D_2^* = -3$, when v_a and v_b are treated as free parameters, as shown in Fig. 5.14(a). At the same time, the choice $p = 4$ in Eq. (5.91) provides a similarly good fit to the GPT calculation of $v_2^d(r, \Omega_0)$ in the case of the companion $4d$ metal Mo with $D_2^* = -2.15$ (see Fig. 2 of Moriarty, 1990a).

In addition to v_2^d, one must also construct the full pair potential v_2 in the MGPT. To do this, we return to Eq. (5.70) for $v_2(r, \Omega)$ in the first-principles GPT for transition metals and adapt that result accordingly to the assumptions made in the MGPT. As already indicated, the d-state overlap potential v_{ol}^0, the third term on the RHS of Eq. (5.70), now becomes the hard-core potential v_2^{hc} in the MGPT and is given by Eq. (A2.51). The remaining second potential term on the RHS of Eq. (5.70) involving Z^* and F_N^0 must now be evaluated in the absence of sp-d hybridization. This

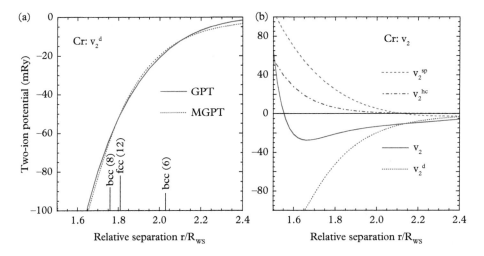

Fig. 5.14 *The MGPT two-ion d-state potential $v_2^d(r, \Omega_0)$ and total pair potential $v_2(r, \Omega_0)$ for Cr. (a) v_2^d, as calculated from Eq. (5.91) with $p = 5$ and compared with the corresponding GPT result obtained with $D_2^* = -3$, but without sp-d hybridization or d-state nonorthogonality, as in Fig. 5.11. (b) v_2, as calculated from Eq. (5.95) with its sp simple-metal v_2^{sp}, hard-core v_2^{hc} and d-state v_2^d components.*

contribution then effectively becomes a simple-metal-like potential for the s and p valence electrons,

$$v_2^{sp}(r,\Omega) = \frac{(Z^*e)^2}{r}\left[1 - \frac{2}{\pi}\int_0^\infty F_N^0(q,\Omega)\frac{\sin(qr)}{q}dq\right], \quad (5.94)$$

which together with $v_2^{hc}(r,\Omega)$ can be evaluated from first principles subject to the transition-metal constraints of specific fixed values for the sp valence Z and the d-state boundary condition D_2^*. The full pair potential in the MGPT is then

$$v_2(r,\Omega) = v_2^{sp}(r,\Omega) + v_2^{hc}(r,\Omega) + v_2^{d}(r,\Omega). \quad (5.95)$$

The calculated two-ion pair potential $v_2(r,\Omega_0)$ and its three components for Cr, obtained with the appropriate physical parameters $p = 5$, $Z = 1.4$ and $D_2^* = -3$, are displayed in Fig. 5.14(b).

The three-ion triplet potential v_3 is physically dominated by d-state electronic-structure contributions, so one takes $v_3 = v_3^d$ in the MGPT, as is done in the first-principles GPT. From Eq. (5.88) one can see that v_3^d consists of one third-order and three fourth-order contributions in $\Delta_{dd'}$. Representing each T_{ij} matrix in Eq. (5.88) with the full Slater-Koster form for d states discussed in Sec. 2.6 of Chapter 2, all of the matrix algebra can be done analytically. This math produces a three-ion triplet potential of the form

$$v_3(r_1,r_2,r_3,\Omega) = v_c(\Omega)f(r_1)f(r_2)f(r_3)L(\theta_1,\theta_2,\theta_3) + v_d(\Omega)\left\{[f(r_1)f(r_2)]^2 P(\theta_3)\right.$$
$$\left. + [f(r_2)f(r_3)]^2 P(\theta_1) + [f(r_3)f(r_1)]^2 P(\theta_2)\right\}, \quad (5.96)$$

where v_c and v_d are additional volume-dependent coefficients. In Eq. (5.96), θ_1, θ_2 and θ_3 are the angles subtended by r_1, r_2 and r_3, respectively, in the triangle connecting the three ions, as illustrated in the inset of Fig. 5.15(a). Here $L(\theta_1,\theta_2,\theta_3) \equiv L(x_1,x_2,x_3)$ and $P(\theta) \equiv P(x)$ are, respectively, the dimensionless angular functions

$$L(x_1,x_2,x_3) = \frac{1}{144}\left[54 + 330 x_1 x_2 x_3 - 105\left(x_1^2 x_2^2 + x_2^2 x_3^2 + x_3^2 x_1^2\right) + 735(x_1 x_2 x_3)^2\right], \quad (5.97)$$

with $x_1 \equiv \cos\theta_1$, $x_2 \equiv \cos\theta_2$ and $x_3 \equiv \cos\theta_3$, and

$$P(x) = \frac{1}{1448}\left(533 + 510x^2 + 405x^4\right), \quad (5.98)$$

with $x \equiv \cos\theta$. Both L and P are universal angular functions that depend only on d symmetry and apply to all transition metals. The behavior of these functions is illustrated in Fig. 5.15(a). The third-order potential term in Eq. (5.96) involving L is the direct analog of the classic triplet potential of Axilrod and Teller (1943), which one obtains in the case of canonical p bands, as discussed in Chapter 12.

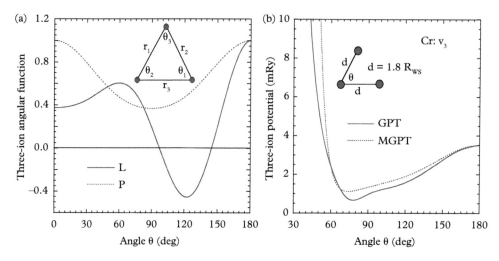

Fig. 5.15 Three-ion MGPT interactions based on canonical d bands. (a) Universal angular functions L and P, as given by Eqs. (5.97) and (5.98), respectively, with $\theta \equiv \theta_3$ and $\theta_1 = \theta_2 = (\pi - \theta)/2$, where the general three-ion interatomic distances r_n and angles θ_n are defined in the inset. (b) MGPT potential $v_3(\theta, \Omega_0)$ for Cr, obtained from Eq. (5.96) with $p = 5$, and compared with the corresponding $\Gamma_{dd'}^{\text{vol}} = S_{dd'} = 0$ GPT result from Fig. 5.12(a).

The three-ion potential coefficients v_c and v_d in Eq. (5.96) are also material-dependent parameters that mostly reflect d-band filling and width. As discussed later in this section, for central transition metals like Cr and Mo with somewhat less than half-filled d bands, one expects v_c to be relatively small in magnitude and negative, while v_d is positive, resulting in an entirely repulsive MGPT potential v_3. This is indeed the calculated GPT behavior obtained in the $\Gamma_{dd'}^{\text{vol}} = S_{dd'} = 0$ limit for Cr with $D_2^\star = -3$, as shown in Fig. 5.12(a), as well as for Mo with $D_2^\star = -2.15$ (Moriarty, 1990a). In both cases, the MGPT quantitatively captures this behavior with v_c and v_d fitted to the GPT results, as shown in Fig. 5.15(b) for Cr with $p = 5$, and in Fig. 5 of Moriarty (1990a) for Mo with $p = 4$.

Finally, for the four-ion quadruplet potential, which is also dominated by d-band contributions, one takes $v_4 = v_4^d$, as done in the GPT. From Eq. (5.89) one sees that there are three fourth-order contributions in $\Delta_{dd'}$ to v_4^d, yielding a four-ion potential form

$$v_4(r_1, r_2, r_3, r_4, r_5, r_6, \Omega) = v_e(\Omega) \, [f(r_1)f(r_2)f(r_4)f(r_5)M(\theta_1, \theta_2, \theta_3, \theta_4, \theta_5, \theta_6)$$
$$+ f(r_3)f(r_2)f(r_6)f(r_5)M(\theta_7, \theta_8, \theta_9, \theta_{10}, \theta_5, \theta_{12})$$
$$+ f(r_1)f(r_6)f(r_4)f(r_3)M(\theta_{11}, \theta_{12}, \theta_5, \theta_6, \theta_3, \theta_4)] \,, \quad (5.99)$$

where M is a third universal angular function that depends only on d symmetry, and the potential coefficient v_e is an additional material-dependent parameter. The function M has also been obtained analytically, but with six independent variables its general form is considerably more complex algebraically than either L or P. The full analytic result is given in Appendix B of Moriarty (1990a), while the specific definitions of the six distances $r_1 \ldots r_6$ and 12 angles $\theta_1 \ldots \theta_{12}$ associated with four-ion geometry are shown in Fig. 5.16. For the special case of coplanar interactions with $r_1 = r_2 = r_4 = r_5$, M can be expressed in terms of a single angle θ as $M(\theta) \equiv M(x)$, with $x = \cos\theta$. For configuration (a) in Fig. 5.16 with $\theta = \theta_1 = \theta_3$ or $\theta = \theta_2 = \theta_4$, one finds

$$M_a(x) = \frac{1}{5792}\left(1757 - 60460x^2 + 327870x^4 - 563500x^6 + 300125x^8\right), \quad (5.100)$$

while for configuration (b) with $\theta = \theta_7 = \theta_{10}$ or $\theta = \theta_8 = \theta_9$ [or for configuration (c) with $\theta = \theta_5 = \theta_6$ or $\theta = \theta_{11} = \theta_{12}$], one obtains

$$M_b(x) = \frac{1}{5792}\left(2172 - 73020x^2 + 373145x^4 - 600250x^6 + 300125x^8\right). \quad (5.101)$$

The behavior of M_a and M_b as a function of angle θ is illustrated in Fig. 5.17(a). Unlike the attractive pair potential v_2 and repulsive triplet potential v_3, the sinusoidal nature of M causes v_4 to oscillate in sign about zero. Again, the MGPT applied to Cr, with $p = 5$ and v_e treated as a free parameter, well reproduces both the magnitude and form of the $\Gamma_{dd'}^{\text{vol}} = S_{dd'} = 0$ GPT v_4 potential, as shown in Fig. 5.17(b). At the same time, there is clearly a small, nearly constant offset between the MGPT and GPT four-ion potentials, an offset that is caused by a negative bias of about -0.53 mRy in the GPT value of v_4. Such a bias is material dependent and essentially vanishes, for example, in the case of Mo, where the corresponding MGPT and GPT four-ion potential are in close quantitative agreement, as shown in Fig. 8 of Moriarty (1990a). In any case, whether or not a small constant bias is present in v_4 is not expected to have any significant effect on either cohesive or structural properties of the material. In particular, the important minimum in v_4 near the prominent $70.5°$ angle contained in the bcc structure and the

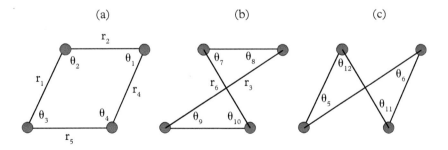

Fig. 5.16 The 6 interatomic distances r_n and 12 angles θ_n defined for configurations (a), (b) and (c) of general four-ion interactions.

maximum at the prominent 90° angle in the fcc structure will produce the same large contribution to the fcc-bcc energy difference in both GPT and MGPT treatments, with or without a bias.

Additional insight into the behavior of the MGPT multi-ion potentials can be gained by independently evaluating the five potential coefficients v_a, v_b, v_c, v_d and v_e directly from Eqs. (5.86)–(5.90). If desired, this can be done without further approximation by evaluating the energy integrals in Eqs. (5.87)–(5.89) numerically, but it is easier and more instructive to follow the TB model of Sec. 5.1.1.2 and replace $T_{dd'}^{\text{vol}}$ defined in Eq. (5.86) with Eq. (5.26). In so doing, the energy dependence of $\Gamma_{dd'}^{\text{vol}}(E)$ is removed and $T_{dd'}^{\text{vol}}(R_{ij}, E)$ can be described in terms of just $\Delta_{dd'}(R_{ij})$ plus defined d-band position and width parameters E_0 and Γ_0. This simplification then allows the energy integrals to be evaluated analytically in terms of the d-band filling functions $F_n(Z_d)$ defined through Eq. (5.36). In addition, by comparing Eq. (5.90) with $\Delta_{dd'}$ evaluated at the canonical d-band condition of $p = 5$ with Eq. (5.43), one establishes the coefficients α_m for $\Delta_{dd'}$ as

$$\alpha_m = \eta_m C_Z \Gamma_0, \tag{5.102}$$

where we recall that $\eta_m = 6$, -4 or 1 for $m = 0$, ± 1 or ± 2, and that C_Z at a given volume is a constant on the order of unity and defined by Eq. (5.44).

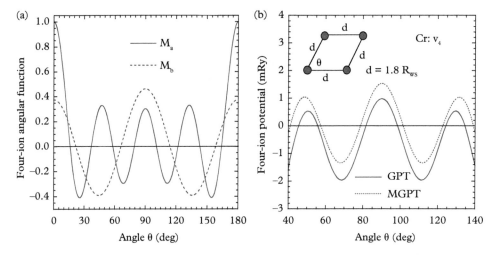

Fig. 5.17 Four-ion MGPT interactions based on canonical d bands. (a) Universal angular functions $M_a(\theta)$ and $M_b(\theta/2)$, as obtained from Eqs. (5.100) and (5.101), respectively. (b) MGPT potential $v_4(\theta, \Omega_0)$ for Cr, as calculated from Eq. (5.99) with $p = 5$, and compared with the corresponding $\Gamma_{dd'}^{\text{vol}} = S_{dd'} = 0$ GPT result from Fig. 5.12(b).

It is then entirely straightforward to evaluate the d-state potential coefficients within the TB model from Eqs. (5.87)–(5.89). One obtains

$$v_b(\Omega) = v_b^0 C_Z^2 F_2(Z_d) \Gamma_0, \tag{5.103}$$

5.2.1 Baseline analytic MGPT for canonical d bands

with $v_b^0 \equiv (140/\pi)(1.8)^{-10} = 0.1248$;

$$v_c(\Omega) = v_c^0 C_Z^3 F_3(Z_d)\Gamma_0, \qquad (5.104)$$

with $v_c^0 \equiv (180/\pi)(1.8)^{-15} = 0.008492$;

$$v_d(\Omega) = -v_d^0 C_Z^4 F_4(Z_d)\Gamma_0, \qquad (5.105)$$

with $v_d^0 \equiv (3620/3\pi)(1.8)^{-20} = 0.003013$;

$$v_a(\Omega) = v_d(\Omega)/2 \qquad (5.106)$$

and

$$v_e(\Omega) = 2v_d(\Omega). \qquad (5.107)$$

The volume dependence of each potential coefficient in the TB model comes indirectly through Z, Z_d and the d-band width parameter Γ_0. At a given volume, each coefficient is proportional to the d-band width and varies with d-band filling according to F_2, F_3 or F_4. From Fig. 5.2 one sees that $F_2(Z_d) \geq 0$, so according to Eq. (5.103), the two-ion potential coefficient v_b is positive for $0 < Z_d < 10$, with a maximum value at $Z_d = 5$ (half-filled d bands). The remaining four coefficients, on the other hand, are predicted to be smaller in magnitude and to oscillate in sign as a function of d-band filling. According to Eq. (5.104), the three-ion potential coefficient v_c is negative for $Z_d < 5$, vanishes for $Z_d = 5$ and is positive for $Z_d > 5$. From Eqs. (5.105)–(5.107) and the behavior of $F_4(Z_d)$, one sees that the remaining coefficients v_a, v_d and v_e are predicted to be positive for $3.33 < Z_d < 6.67$ and maximum at $Z_d = 5$, with the ratios v_a/v_d and v_e/v_d remaining everywhere constant from Eqs. (5.105) and (5.106). Thus, in the TB model there are actually only three independent potential coefficients: v_b, v_c and v_d. This outcome is a direct consequence of the TB requirement that the partial d-state moments s of a given order n, i.e., $M_n^{(s)}$, have a fixed weighting relative to the total moment M_n.

The detailed relationship between the MGPT d-state potential coefficients determined from the simplified TB model applied to a Group-VIB transition metal, and from the unrestricted fitting of the corresponding first-principles GPT potentials for Cr and Mo in the $\Gamma_{dd'}^{\text{vol}} = S_{dd'} = 0$ limit, is elaborated in Table 5.2. Qualitatively, the TB model captures the main features of the coefficients that are revealed in the actual GPT potentials. Specifically, v_a, v_b, v_d and v_e are all positive quantities, while v_c is negative. In addition, v_b is at least an order of magnitude larger than the other four coefficients, while the magnitudes of the ratios v_a/v_d, v_c/v_d and v_e/v_d are no larger than about 2.5. Quantitatively, however, there are clearly significant differences between the model and fitted potential coefficients. These differences mostly reflect the limitations in the TB model, but to some extent they also reflect the nonuniqueness of the potential fitting and the overall sensitivity of the potential coefficients to the manner in which they are established. At the same time, the use of five independent potential coefficients, instead of only three in the TB model, has allowed us to capture accurately the behavior of the first-principles GPT pair and multi-ion d-state potentials in prototype central transition metals. This strongly suggests that the higher-order contributions to the GPT potentials beyond fourth order, which are retained via the use of Eqs. (5.14), (5.16) and (5.19) to calculate the potentials, are well modeled in the MGPT by allowing independent potential coefficients for each of the partial-moment diagrams in Fig. 5.2.

Table 5.2 *MGPT d-state potentials coefficients obtained from the simplified TB model discussed in the text, Eqs. (5.102)–(5.107), with $Z_d = 4.6$, as well as from fitting corresponding first-principles GPT potentials for Cr and Mo at $\Omega = \Omega_0$. The Cr results are those established in Figs. 5.14(a), 5.15(b) and 5.17(b), while the Mo results are from Moriarty (1990a).*

Quantity	TB model	Cr: fit to GPT	Mo: fit to GPT
$D_2^\star$	—	−3.0	−2.15
Z	1.4	1.408	1.402[a]
Z_d	4.6	4.592	4.598[a]
p	5.0	5.0	4.0
v_a/v_d	0.5	1.922	1.122
v_b/v_d	46.15	16.14	13.42
v_c/v_d	−0.770	−1.260	−0.1937
v_e/v_d	2.0	1.261	2.448

[a] constrained value via the zero-of-energy constant V_0''.

What one really desires in the end, of course, is both an optimized and physically reasonable set of MGPT potential coefficients for a given transition metal, with which one can then accurately describe its materials properties. The qualitative guidance enabled by the studies summarized in Table 5.2 provide important insight to, as well as some useful theoretical constraints on, the more general process of selectively fitting MGPT parameters to calculated first-principles physical properties and/or to measured experimental data on specific materials. Historically, this quest has also been greatly aided by the more general and computational efficient formulation of the theory known as the matrix MGPT, which we now discuss in the next section.

5.2.2 Matrix MGPT for large-scale MD, noncanonical d bands and more

While the baseline analytic formulation of the MGPT is conceptually very satisfying and allows maximum transparency in understanding the inner workings of the theory, this formulation does have two important practical limitations. First, the analytic MGPT is not the most computationally inefficient form of the theory, especially for large-scale

5.2.2 Matrix MGPT for large-scale MD, noncanonical d bands and more

atomistic simulations such as MD. This is due in large part to the extreme algebraic complexity of the four-ion angular function M. Second, within the analytic framework established, there are only very limited prospects for extending the theory beyond canonical d bands or beyond four-ion interactions. Both of these limitations are removed in the matrix MGPT.

For transition metals described by canonical d bands, the pair and multi-ion potentials v_2, v_3 and v_4 have the same formal structure in the matrix MGPT as in Eqs. (5.95), (5.96) and (5.99), respectively, for the analytic MGPT. Now, however, the multi-ion angular functions L, P and M in the matrix MGPT are calculated directly by numerical matrix multiplication. To do this, one defines an appropriately normalized 5×5 Slater-Koster matrix $H_{SK}^{dd}(i,j)$ for the d-state coupling between sites i and j. Specifically, to create $H_{SK}^{dd}(i,j)$ for canonical d bands one makes the following parameter substitutions in the d-state SK matrix: $dd\sigma \to -S = -6$, $dd\pi \to -P = 4$ and $dd\delta \to -D = -1$, allowing for the overall minus sign included in the definition of $\Delta_{dd'}^{ij}$. Then, using the three-ion geometry defined in Fig. 5.15(a), proceeding clockwise around the triangle, and denoting the left-side atom as atom 1, one has in terms of matrix products of H_{SK}^{dd},

$$L(x_1, x_2, x_3) = \frac{1}{90} \text{Tr} \left[H_{SK}^{dd}(1,2) H_{SK}^{dd}(2,3) H_{SK}^{dd}(3,1) \right], \quad (5.108)$$

where $90 = S^3 + 2P^3 + 2D^3$, and

$$P(x_2) = \frac{1}{1810} \text{Tr} \left[H_{SK}^{dd}(1,2) H_{SK}^{dd}(2,1) H_{SK}^{dd}(1,3) H_{SK}^{dd}(3,1) \right], \quad (5.109)$$

where $1810 = S^4 + 2P^4 + 2D^4$, with $P(x_1)$ and $P(x_3)$ given by similar expressions. Likewise, using the four-ion geometry defined in Fig. 5.16 and proceeding clockwise starting from the lower left-side atom as atom 1, one has for configuration (a):

$$M(x_1, x_2, x_3, x_4, x_5, x_6) = \frac{1}{1810} \text{Tr} \left[H_{SK}^{dd}(1,2) H_{SK}^{dd}(2,3) H_{SK}^{dd}(3,4) H_{SK}^{dd}(4,1) \right], \quad (5.110)$$

with similar expressions for configurations (b) and (c).

The development and use of efficient algorithms to evaluate energies and forces in MGPT atomistic simulations turns out to be highly commensurate with the evaluation of the angular functions L, P and M via matrix multiplication. For example, one can establish linked multi-ion neighbor lists and corresponding common matrix products for all contributing configurations that can be used to evaluate the three angular functions simultaneously. Such strategies have led to an order-of-magnitude improvement in the performance in MGPT-MD simulations, an advance that is fully leveraged in Chapter 8. In the present context, the improved performance offered by the matrix MGPT in atomistic simulations has been useful in developing optimized MGPT potentials for specific materials, as we discuss in Sec. 5.2.3.

In addition to better efficiency, the matrix MGPT also allows immediate extensions of the theory in three important directions. First, one can introduce general noncanonical d

bands into the theory at no cost by simply adjusting the magnitudes of the fundamental bond-integral ratios $dd\sigma/dd\delta = S/D$ and $dd\pi/dd\delta = P/D$ to any desired values. As is well known, the principal shortcoming of canonical d bands in this regard is that, unlike $dd\sigma$ and $dd\pi$, the magnitude of $dd\delta$ is too large compared to first-principles calculations of this quantity for transition metals. This leads to large calculated first-principles values of $dd\sigma/dd\delta$ and $dd\pi/dd\delta$ across near-neighbor bond lengths compared to the fixed canonical d-band values of 6 and -4, respectively, as is shown in Fig. 5.18 for Cr. At the same time, the corresponding first-principles bond-integral ratio $dd\sigma/dd\pi$ is nearly constant across near-neighbor distances and close in value to the canonical-d-band result of -1.5. In most applications, the bond integrals $dd\sigma$ and $dd\pi$ are much more important than $dd\delta$ to the result obtained, so canonical d bands tend to perform very well. One important exception of note is the treatment of anomalous phonon frequencies that occur in central transition metals. The application of MGPT noncanonical d bands to such transition-metal phonon spectra is considered in Chapter 7.

A second major extension that is enabled by the matrix MGPT is from d bands to p and f bands, including both canonical and noncanonical treatments. Formally, this extension follows by just replacing the 5×5 d-state SK matrix with the corresponding 3×3 p-state or 7×7 f-state matrices. The p-state SK matrix is, of course, well known and sufficiently simple that analytic results for the corresponding multi-ion angular functions can be obtained, if desired. The f-state matrix, on the other hand, is complex and was only in the relatively recent past obtained in general form by McMahan (1998). Applications of these results to f-band actinide metals and first-row p-band metals are considered in Chapter 12. Finally, a third important extension made possible by the matrix MGPT is the development of five- and six-ion multi-ion interatomic potentials. Progress toward that goal for d-band transition metals is discussed in Sec. 5.2.5.

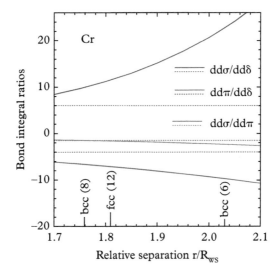

Fig. 5.18 *Ratios of the bond integrals $dd\sigma$, $dd\pi$ and $dd\delta$ for Cr at near-neighbor separations, as obtained from first-principles GPT calculations of $\Delta_{dd'}(R_{ij})$ at $\Omega = \Omega_0$ with $D_2^* = -3$ (solid lines) and compared with canonical d-band values (dotted lines).*

5.2.3 Optimized canonical d-band MGPT potentials for central bcc metals

The development of optimized MGPT potentials for central bcc transition metals has been an ongoing process of continual improvement that began in the early 1990s. The core philosophy adopted has been one of smoothly combining three essential ingredients: (i) theoretical GPT and DFT guidance on the underlying control parameters, including D_2^*, Z, Z_d and the radial exponent p in the bond-integral function $f(r)$; (ii) first-principles GPT calculation of the pair potentials v_2^{sp} and v_2^{hc}; and (iii) selective fitting of the five d-state potential coefficients v_a, v_b, v_c, v_d and v_e to basic physical properties of the material in question. The basic physical properties considered in this regard are those of the observed bcc phase and include the bulk modulus, vacancy formation energy, shear elastic constants, zone-boundary phonon frequencies and the Debye temperature. The input properties are taken from both accurately measured data (the preferred input near ambient conditions) and first-principles DFT calculations (preferred at high pressure). The working hypothesis here is that if the fitted properties of the observed phase are indeed well accommodated by the theory, then the fundamental transferability of the interatomic potentials guaranteed by the first-principles GPT will extend to the MGPT as well. Since only input data on a single phase is used to establish the potentials, their assumed transferability can be immediately tested by the predicted structural phase stability of the metal and comparison with first-principles DFT results. More generally, the expectation is that optimized MGPT potentials so established will be applicable to a wide range of structural, thermodynamic, defect and mechanical properties of not only the bcc phase, but any metallic solid phases that might exist, and of the liquid.

In the first generation of MGPT potential development (Moriarty, 1990a, 1990b), initial results were obtained for Nb and Mo prototypes (nominally designated as Nb1 and Mo1), but the main focus was on Mo, for which there already existed a large set of complementary first-principles DFT LMTO-ASA data (Moriarty, 1988a), including a full cohesion curve and the results displayed in Figs. 5.3, 5.4 and 5.7. Guided by the TB and GPT studies summarized above in Table 5.2, several simplifying assumptions were made, including fixed values of $p = 4$ in the radial function $f(r)$ and $v_c/v_d = -0.1937$ together with constant values of v_a/v_d and v_e/v_d as a function of volume. In addition, the volume term was constrained to vary with volume as the d-band width, $E_{\text{vol}}(\Omega) = E_{\text{vol}}(\Omega_0)(\Omega_0/\Omega)^{p/3}$, and the long-ranged tails of $f(r)$, as well as the pair potentials $v_2^{sp}(r,\Omega)$ and $v_2^{hc}(r,\Omega)$, were screened with a smooth cutoff function. Within this framework, five separate schemes to evaluate the four remaining potential coefficients v_a, v_b, v_d and v_e were tested and evaluated. In each case, the four coefficients were first fitted to measured bcc physical properties at normal density, one of which was always the observed bulk modulus $B_{\text{tot}} = 264$ GPa, which in this case exactly matched the calculated DFT value. The volume dependencies of v_b and v_d were then calculated directly from the DFT values of $E_{\text{coh}}(\Omega) - E_{\text{coh}}(\Omega_0)$ and $B_{\text{tot}}(\Omega)$ such that the rigorous compressibility sum rule (see Chapter 7),

$$B_{\text{tot}} = \Omega \frac{d^2 E_{\text{coh}}}{d\Omega^2} = (C_{11} + 2C_{12})/3, \quad (5.111)$$

was satisfied at each volume, where C_{11} and C_{12} in Eq. (5.111) were obtained from radial and tangential force-constant derivatives of the potentials, v_2, v_3 and v_4. All five schemes produced generally good quality results on cohesion, vacancy formation, structural phase stability, elastic moduli and their pressure derivatives, phonon frequencies, the Debye temperature and the Grüneisen parameter.

Subsequently, the most promising of these first-generation Mo1 schemes was selected for further development and more extensive application in a second generation of MGPT potentials (Moriarty, 1994), designated as Mo2. In this scheme the constraining physical data were chosen to be the observed vacancy formation energy E_{vac} and elastic constants C_{11} and C_{44}, in addition to the bulk modulus B_{tot}. Several numerical improvements were also added at this stage, the most important of which were a more accurate treatment of the four-ion radial force-constant function needed in the evaluation of elastic moduli and phonon frequencies, and stronger screening of the bond-orbital radial function $f(r)$. The former improvement resulted in larger values of the v_e potential coefficient, which in turn significantly improved the calculated fcc-bcc structural energy differences as a function of volume, moving them close to the DFT values. The stronger screening reduced the overall range of the multi-ion potentials, making MGPT-MD simulations feasible for the first time. Combining quasiharmonic lattice dynamics for the bcc solid with MD simulation of the high-temperature solid and the liquid allowed calculation of the thermodynamic properties and free energies of both phases. This led to the first attempted MGPT calculation of a high-pressure melt curve. To reconcile theory with the observed melting point in Mo at ambient pressure, however, required the addition of large ad hoc electron-thermal free-energy components for both the solid and the liquid, which were obtained from a simplified TB model. Necessary improvements in the melting calculation to accommodate the actual strong electron-phonon coupling at high temperature in metals like Mo were only later developed in the form of temperature-dependent MGPT potentials (Moriarty et al., 2012), which are discussed in Chapter 13. At low temperature, on the other hand, the Mo2 potentials were also applied to a wide range of defect and mechanical properties, including ideal shear strength, point defect energies, and screw dislocation core structure and mobility (Xu and Moriarty, 1996, 1998), as well as grain-boundary atomic structure (Campbell et al., 1999).

In the late 1990s, attention turned to Ta as a second important transition-metal prototype for MGPT potential development and application. The second-generation Mo2 scheme was generalized and significantly improved, with the previous simplifying, but still uncertain, assumptions concerning E_{vol}, v_a/v_d, v_c/v_d and v_e/v_d removed entirely and all quantities determined as a function of volume as at normal density. This improvement was made possible by the calculation of a large DFT FP-LMTO database on the zero-temperature equation of state, shear elastic moduli, zone-boundary phonon frequencies and vacancy formation energy of bcc Ta extending to 1000 GPa (Söderlind et al., 1998, 2000). This theoretical data was then smoothly blended with corresponding measured data across a uniform volume mesh. In this process, the observed Debye temperature at normal density was scaled to lower and higher pressure by means of the measured Grüneisen parameter and the DFT zone-boundary phonons. Also introduced

5.2.3 Optimized canonical d-band MGPT potentials for central bcc metals

at this point was a smoother cutoff function in the form of Eq. (3.112) with parameters $R_0 = 2.15 R_{WS}$ and $\alpha = 75$ for the pair potentials $v_2^{sp}(r,\Omega)$ and $v_2^{hc}(r,\Omega)$ and with the same R_0 and $\alpha = 125$ for the bond-integral function $f(r)$. In this third-generation MGPT potential scheme applied to Ta (Moriarty et al., 1999), designated as Ta3, the compressibility sum rule (5.111) remained in force and the six quantities E_{vol}, v_a, v_b, v_c, v_d and v_e were fitted to the cohesive energy E_{coh}, the bulk modulus B_{tot}, the *unrelaxed* vacancy formation energy $E_{\text{vac}}^{\text{u}}$, the shear elastic moduli $C' = (C_{11} - C_{12})/2$ and C_{44}, plus the Debye temperature Θ_D at each of 18 volumes spanning a pressure range of -30 to 1000 GPa.

The Ta3 MGPT potentials were successfully applied to generalized stacking-fault energy surfaces (Moriarty et al., 1999) and grain-boundary atomic structure (Campbell et al., 2000), as well as to bcc and liquid thermodynamic properties (Moriarty et al., 2002a), where unlike the case of Mo, weak electron-phonon coupling at high temperature was displayed in Ta, allowing for a normal calculation of melting. At the same time, it was found that many important properties, including the screw dislocation core structure, point defect formation and migration energies, and the high-pressure melt curve itself, were sensitive to the treatment of the input vacancy formation data. Heretofore, no important distinction had been made between unrelaxed and relaxed formation energies, but this was found to be no longer adequate in the case of Ta. This issue was resolved in the development of a closely related fourth-generation scheme applied to the Ta potentials (Yang et al., 2001; Moriarty et al., 2002a), designated as Ta4. The sole change in the Ta4 scheme compared to the Ta3 scheme was the rigid upward shift of the input $E_{\text{vac}}^{\text{u}}(\Omega)$ curve by 0.15eV, the amount required to match the fully relaxed vacancy formation energy to experiment at normal density. This upgrade resulted in an improved description of all of the identified sensitive properties and increased the reliability in applying the Ta4 MGPT potentials more generally to defects and multiscale materials modeling of mechanical properties (Moriarty et al., 2006; Yang et al., 2010).

The Ta4 MGPT potential scheme set the current paradigm for determining E_{vol} and the five d-state potential coefficients from basic physical data in central bcc transition metals. Subsequent fifth- and sixth-generation MGPT potential schemes have added only numerical refinements to the Ta4 scheme. Beginning with the fifth-generation of MGPT potentials, we used a denser volume mesh that is linear in the reduced volume variable $x = (\Omega/\Omega_0)^{1/3}$ with a default increment of $\Delta x = 0.01$. At this point, we revisited the case of Mo and developed improved Mo5.2 MGPT potentials extending to 410 GPa in pressure (Moriarty et al., 2002b, 2006), using an upgraded DFT FP-LMTO database (Söderlind et al., 1994a, 2000). As with Ta4, the Mo5.2 potentials have also now been widely applied to defects and multiscale materials modeling (Moriarty et al., 2002b, 2006; Yang et al., 2010).

Finally, in the sixth generation of MGPT potentials, we began representing the input physical property data with smooth analytic forms to provide the most reliable variation in the data from one volume to another. We also added a third V prototype to Mo and Ta, developing V6.1 MGPT potentials that extend to 230 GPa in pressure (Moriarty et al., 2008), with the aid of available DFT FP-LMTO data (Söderlind et al., 2000;

Landa et al., 2006a). One final sixth-generation refinement was a more accurate numerical fit to the hard-core potential $v_2^{\mathrm{hc}}(r, \Omega)$ at each volume considered. Refined Ta6.8x potentials that extend to 420 GPa in pressure (Moriarty and Haskins, 2014) include this improvement. These latter potentials have been used to study polymorphism in Ta (Haskins et al., 2012; Haskins and Moriarty, 2018), and are transferable to at least 10 different possible phases of the metal. Table 5.3 summarizes the MGPT parameters and potential coefficients for V6.1, Mo5.2, Ta4 and Ta6.8x, which are the four most widely applied potentials to date. All four of these potentials are available on LAMMPS as part of the USER MGPT package. Regarding basic control parameters, note that we have fixed Z and Z_d at normal-density values for each metal, while $D_2^*(\Omega)$ and $p(\Omega)$ have been allowed to increase slowly with volume, with $p(\Omega_0) = 5$ for Mo instead of the values near 4 used earlier. Note also that qualitatively the theoretical expectations on the behavior of the equilibrium potential coefficients from Table 5.2 continue to hold, with v_a, v_b, v_d and v_e all positive quantities, while v_c is negative and v_b is much larger in magnitude

Table 5.3 *Representative canonical d-band MGPT parameters and potential coefficients for the central transition metals V, Mo and Ta obtained from a combination of theoretical considerations, first-principles GPT calculation and the selective fitting of the coefficients to basic physical properties, as discussed in the text. Pressure ranges are in given GPa and the potential coefficient v_d is given in mRy.*

Quantity	V	Mo	Ta	Ta
Version	6.1	5.2	4	6.8x
Pressure range	−23 to 230	−37 to 410	−30 to 1000	−30 to 420
$D_2^*(\Omega_0)$	−3.0	−3.0	−3.0	−3.0
D_2^* range	−3.1 to −2.6	−3.1 to −2.6	−3.1 to −2.0	−3.1 to −2.4
Z	1.5[a]	1.4[a]	2.0[a]	2.0[a]
Z_d	3.5[a]	4.6[a]	3.0[a]	3.0[a]
$p(\Omega_0)$	5.0	5.0	4.0	4.0
p range	4.8 to 5.4	4.8 to 5.4	3.8 to 4.8	3.8 to 4.6
at Ω_0 :				
v_a/v_d	4.678	1.014	0.880	0.896
v_b/v_d	54.63	17.36	31.73	31.81
v_c/v_d	−0.055	−2.877	−1.168	−1.159
v_e/v_d	7.671	6.756	5.226	5.228
v_d	0.636	2.919	1.532	1.532

[a] constrained value via the zero-of-energy constant V_0''.

5.2.3 Optimized canonical d-band MGPT potentials for central bcc metals 223

than the other four coefficients. For the case of Ta6.8x, the volume dependence of $p(\Omega)$ and the five potential coefficients is illustrated in Fig. 5.19(a). The volume dependence of the pair potential $v_2(r, \Omega)$ and the multi-ion potentials $v_3(\theta, \Omega)$ and $v_4(\theta, \Omega)$ for Ta6.8x is displayed in Figs. 5.19(b), 5.20(a) and 5.20(b), respectively. Recall also that the pair and multi-ion MGPT potentials for V6.1, Mo5.2 and Ta4 at their respective equilibrium volumes were shown previously in Fig. 1.4 of Chapter 1. In addition, it should be pointed that out sixth-generation potentials V6.8 and Mo6.8 have now been developed over the same pressure ranges as V6.1 and Mo 5.2, respectively, and at the same level of treatment as Ta6.8x. To date, however, neither the V6.8 or Mo6.8 potentials have been used in published applications.

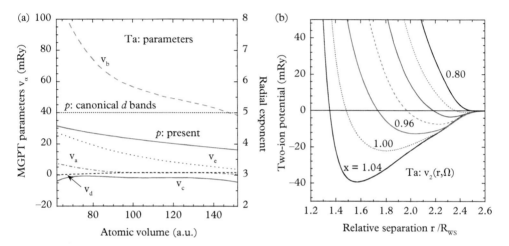

Fig. 5.19 *Volume dependence of the MGPT potential parameters and pair potential for Ta6.8x. (a) Radial function exponent p in the bond-orbital function f(r) (right vertical scale) and the d-band potential coefficients v_a, v_b, v_c, v_d and v_e (left vertical scale). (b) Two-ion pair potential v_2 for selected volumes, in x increments of 0.04 with $x = (\Omega/\Omega_0)^{1/3}$.*
From Moriarty and Haskins (2014), with publisher (APS) permission.

Finally, it is important to emphasize that the basic underlying control parameters of the MGPT potentials developed so far, namely, $D_2^\star$, Z, Z_d and p, really have not as yet been fully optimized, but only assigned reasonable values. Moreover, the use of more general noncanonical d bands has only been considered to date in difficult special cases such as Mo phonons (see Chapter 7) and related electron-temperature-dependent melting (see Chapter 13). Full optimization with respect to both the basic MGPT control parameters and with respect to noncanonical d-band character remains a future challenge for which machine learning is required. Machine learning will also be needed to implement the proposed re-inclusion of explicit sp-d hybridization contributions to the MGPT potentials as one moves outward from the central transition metals, and to implement five- and six-ion potentials for the mid-period metals themselves. We discuss current progress on the latter two challenges in Secs. 5.2.4 and 5.2.5.

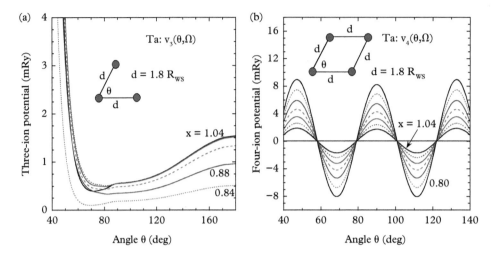

Fig. 5.20 *Angular and volume dependence of the Ta6.8x multi-ion MGPT potentials v_3 [panel (a)] and v_4 [panel (b)] for near-neighbor configurations at selected volumes, in x increments of 0.04 with $x = (\Omega/\Omega_0)^{1/3}$.*
From Moriarty and Haskins (2014), with publisher (APS) permission.

5.2.4 Re-inclusion of *sp-d* hybridization and extension to late-series metals

As we have discussed, the importance of *sp-d* hybridization in transition metals increases as one moves outward from the mid-period elements toward the beginning and end of each series. Thus, to extend MGPT across all transition metals requires the re-inclusion of the *sp-d* hybridization at some acceptable level of approximation. In the full GPT, one can attempt to screen the individual components of the *d*-state interaction matrix T_{ij}, as given by Eq. (5.6), where the *sp-d* hybridization is contained in the $\Gamma^{\text{vol}}_{dd'}(R_{ij}, E)$ term. This type of screening is easily implemented, but the resulting multi-ion potentials generally remain too long-ranged to allow efficient atomistic simulations. A more promising, but still tentative, strategy for the MGPT is rather to fold-down the multi-ion hybridization interactions into an effective pair potential. One way this can be done is as follows. First, we assume that the inclusion of *sp-d* hybridization contributions into the MGPT potentials begins in the same general form as in the GPT, such that schematically

$$v_2 = v_2^{sp} + v_2^{hc} + v_2^{sp-d} + v_2^d$$
$$v_3 = v_3^{sp-d} + v_3^d$$
$$v_4 = v_4^{sp-d} + v_4^d$$
$$\vdots$$
$$v_n = v_n^{sp-d} + v_n^d, \tag{5.112}$$

5.2.4 Re-inclusion of sp-d hybridization and extension to late-series metals

where we have departed from the full GPT notation and broken out explicit multi-ion sp-d hybridization potentials v_n^{sp-d} such that v_2^{sp}, v_2^{hc} and v_n^d have exactly the same meaning as in Eq. (5.95) of the standard MGPT. Next, we fold down v_3^{sp-d}, v_4^{sp-d} and any higher potentials by averaging these quantities in the manner of Sec. 5.1.5 and combine the results with v_2^{sp-d} to form the effective pair potential of the form

$$v_{\text{eff}}^{sp-d} \equiv \left[v_2^{sp-d} + \left\langle v_3^{sp-d} \right\rangle + \left\langle v_4^{sp-d} \right\rangle + \cdots \right] f_{\text{scr}}. \quad (5.113)$$

Here the averaging of the three- and four-ion potentials has the meaning

$$\left\langle v_n^{sp-d} \right\rangle \equiv \left\langle v_n \right\rangle - \left\langle v_n^d \right\rangle, \quad (5.114)$$

and the term f_{scr} is an additional chosen screening function that must be developed and optimized through appropriate machine learning. The averaging in Eq. (5.114) is to be performed with first-principles GPT functions, such that v_n^d is the multi-ion potential arising from only the direct d-state interactions, as given in Eq. (5.86).

The implementation and testing of this proposed MGPT scheme are still at an early stage. We have initially focused on the simplifying case of late-series transition metals and a Ni prototype, for which the central-force sp-d hybridization is relatively strong but the multi-ion interactions are effectively weak and negligible. The remaining pair potential in this case can then be expressed in the form

$$v_2(r, \Omega) = v_2^{sp}(r, \Omega) + v_2^{hc}(r, \Omega) + v_2^{sp-d}(r, \Omega) + v_2^d(r, \Omega), \quad (5.115)$$

where for notational convenience we have now absorbed the screening function $f_{\text{scr}}(r, \Omega)$ into the definition of $v_2^{sp-d}(r, \Omega)$. The corresponding cohesive energy functional then reverts to the standard form of volume term plus pair potential as in Eq. (3.1). One can process Eq. (5.115) for the pair potential v_2 in a manner that is similar to the standard MGPT, with v_2^{sp} and v_2^{hc} calculated from the first-principles GPT and v_2^d taken in the form of Eq. (5.91), with volume-dependent parameters v_a, v_b and p. For the additional hybridization term in Eq. (5.115), the bare unscreened form of v_2^{sp-d} is calculated from the first-principles GPT, while the screening function f_{scr} is taken in the form of the smooth cutoff function f_{cut} given by Eq. (3.112), with parameters R_0 and $\alpha_{\text{scr}} \equiv \alpha$.

In our first wide-volume MGPT scheme to implement Eq. (5.115) for nickel (denoted as Ni2), we have chosen a screening parameter $R_0 = 2.15 R_{\text{WS}}$, and have set appropriate control parameters Z and $D_2^*(\Omega)$ to define the zero-order pseudoatom needed in the calculation of v_2^{sp}, v_2^{hc} and v_2^{sp-d}. We then fitted as a function of volume the five remaining quantities E_{vol}, v_a, v_b, p and α_{scr}. Guided by our work on the central bcc transition metals, these quantities were fitted to a blend of experimental and DFT data on the cohesive energy, bulk modulus, unrelaxed vacancy formation energy, shear modulus C' and Debye temperature at each of 18 volumes spanning a pressure range of -26 to 100 GPa. This fit maintained the compressibility sum rule (5.111) on a volume mesh linear in $x = (\Omega/\Omega_0)^{1/3}$, with the observed equilibrium volume $\Omega_0 = 73.82$ a.u. The constraining Ni DFT data used here were calculated by L.H. Yang with his efficient first-principles plane-wave PP method (Yang, 2000), and are reported in Vashishta et al. (2008).

The Ni2 MGPT parameters so established are displayed in Fig. 5.21 as a function of volume. Regarding the control parameters, we have fixed Z and Z_d at physically reasonable and near-optimum normal-density values of 1.225 and 8.775, respectively, while $|D_2^*(\Omega)|$ has been allowed to increase with increasing volume from 2.65 to 3.50. The d-state pair potential coefficients $v_a(\Omega)$ and $v_b(\Omega)$ have the qualitatively expected form of positive increasing functions with decreasing volume, where $v_b \gg v_a$. The bond-integral radial function exponent $p(\Omega)$ varies modestly between 4.65 and 5.45 for volumes above about 65 a.u., but then decreases rapidly for lower volumes. Finally, the screening function $\alpha_{\text{scr}}(\Omega)$ is nearly constant over the whole volume range considered, varying only between 0.37 and 0.41.

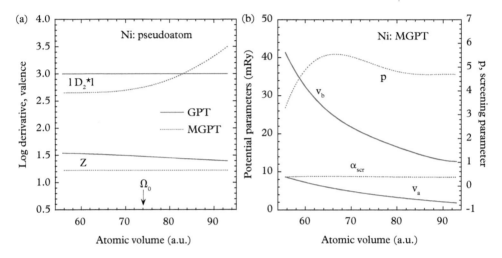

Fig. 5.21 *MGPT parameters for Ni2 potentials as a function of volume. (a) Chosen sp valence $Z = 1.225$ and corresponding d-state logarithmic derivative $|D_2^*|$ defining the Ni zero-order pseudoatom, as compared with nominal first-principles GPT values for $D_2^* = -3$. (b) Potential parameters v_a and v_b (read left vertical axis), together with the exponent p of the bond-integral radial function $f(r)$ and the screening parameter α_{scr} (read right vertical axis).*

The MGPT Ni2 normal-density pair potential $v_2(r, \Omega_0)$ calculated from Eq. (5.115) with the above parameters is shown in Fig. 5.22(a) and compared there with the corresponding first-principles GPT potential for $D_2^* = -3$. The MGPT and GPT potentials are clearly very similar in both magnitude and shape, and both potentials correctly stabilize the observed fcc structure of Ni. Because of the introduction of the screening function, however, the effective range of the MGPT potential is about half that of the GPT potential, making possible much more efficient atomistic simulations. Figure 5.22(b) illustrates the four separate components of the Ni2 MGPT potential and confirms that beyond near-neighbor interactions the full potential is indeed dominated by the screened sp-d hybridization interactions contained in $v_2^{sp-d}(r, \Omega_0)$.

Overall, detailed applications of the Ni2 MGPT potentials have produced a reasonable description of most basic physical properties, including an excellent result for the

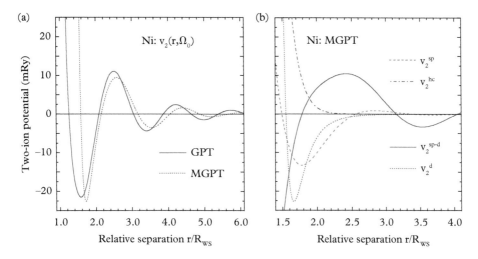

Fig. 5.22 *MGPT pair potential and its components for Ni2 at normal density. (a) Full MGPT potential compared with the corresponding first-principles GPT potential. (b) Potential components v_2^{sp}, v_2^{hc}, v_2^{sp-d} and v_2^d.*

generalized stacking fault energy surface, which is discussed in Chapter 9. A refined next-generation Ni3 scheme has also been developed that adds an additional screening parameter and allows the remaining shear modulus C_{44} to be fit, as is done in the standard MGPT for central transition metals. This refinement improves the calculated phonon spectrum, which is discussed in Chapter 7, and also lowers the magnitudes of bcc-fcc and hcp-fcc structural energies closer to DFT values. Going forward, however, the major challenge is to evaluate the adequacy of the multi-ion *sp-d* hybridization averaging scheme envisioned in Eqs. (5.113) and (5.114) for the central hcp transition metals to the left and right of the bcc metals, such as group-IVB metals like Ti, where v_3 and v_4, and possibly v_5 and v_6, will be important.

5.2.5 Development of five- and six-ion potentials for mid-period metals

In Fig. 1.6 of Chapter 1, we showed that while TB calculations carried out to the fourth *d*-band moment were adequate to explain the observed bcc stability of the central group-V and group-VI metals, TB calculations must be taken to the sixth *d*-band moment to explain structural phase stability across the transition-metal series, and in particular the hcp stability observed to the left and right of the mid-period bcc metals. Thus, one must expect the same level of importance of five- and six-ion potentials in the GPT and MGPT. In this section, we briefly discuss our progress in this direction for the MGPT as applied to mid-period transition metals.

Formally, the GPT multi-ion expansion of the d-band structural energy developed in Sec. 5.1.1.1 through four-ion interactions can be systematically extended to five- and six-ion interactions by generalizing Eq. (5.8) to the form

$$\ln\left[\det\left(T_{dd'}^{\text{vol}}\right)\right] = \ln\left[\det\left(\mathbf{I} - \frac{1}{2}\sum_{i,j}{}'\phi_{ij} + \frac{1}{6}\sum_{i,j,k}{}'\phi_{ijk} - \frac{1}{24}\sum_{i,j,k,l}{}'\phi_{ijkl} \right.\right.$$
$$\left.\left. + \frac{1}{120}\sum_{i,j,k,l,m}{}'\phi_{ijklm} - \frac{1}{720}\sum_{i,j,k,l,m,n}{}'\phi_{ijklmn} + \cdots\right)\right], \quad (5.116)$$

where ϕ_{ijklm} and ϕ_{ijklmn} are, respectively, five- and six-ion functionals of the 5×5 d-state matrices T_{ij} coupling different sites, as in Eqs. (5.10)–(5.12) for two-, three- and four-ion interactions. The mathematical details of this analysis are algebraically quite complex, but fortunately, we will not require all of these details for our discussion here. Rather we will focus on the new qualitative features that emerge and, in particular for the MGPT, on the leading contributions to the potentials v_5 and v_6, which can be obtained much more simply from a graphical analysis of the most important five- and six-ion interactions.

Implicit in Eq. (5.116) is the nominal requirement that all contributions to the MGPT potentials should, in principle, be taken to sixth order in the T_{ij} matrices for internal consistency. The fifth- and sixth-order contributions to v_2^d, v_3^d and v_4^d require only the extension of the existing expansions in Eqs. (5.23)–(5.25) to sixth order in the T_{ij}. Qualitatively, however, these contributions add no obvious important missing physics, and quantitatively, these contributions are expected to be small in magnitude, so we will neglect them here. On the other hand, the importance the fourth-order, four-ion partial moment $M_4^{(4)}$ and corresponding potential v_4^d to bcc phase stability suggests a possible similar importance of $M_5^{(5)}$ and $M_6^{(6)}$, and of the corresponding potentials v_5^d and v_6^d, to hcp phase stability. Graphically, all of these latter contributions arise from qualitatively similar order-n ring diagrams. For multi-ion interactions, an order-n ring diagram is a continuous path of n line segments linking n ions, as illustrated in Fig. 5.1 for $M_3^{(3)}$ and $M_4^{(4)}$. The basic geometrical properties of order-n ring diagrams through order six are summarized in Table 5.4. The total number of line segments needed to connect all pairs of an n-ion ensemble is just $n(n-1)/2$, yielding six lines for $n = 4$, 10 lines for $n = 5$ and 15 lines for $n = 6$. This leads to three ring diagrams for $n = 4$, while much larger totals of 12 and 30 ring diagrams emerge for $n = 5$ and $n = 6$, respectively. Representative ring diagrams for the latter two cases are illustrated in Fig. 5.23.

The expected importance of the leading ring-diagram contributions to the multi-ion potentials v_5 and v_6 can be demonstrated by extending the simple TB model introduced in Sec. 5.1.1.2 to include the five- and six-ion ring-term dimensionless moments $M_5^{(5)}$ and $M_6^{(6)}$. In analogy to Eq. (5.35) for $M_4^{(4)}$, these higher moments have definitions of the general forms

5.2.5 Development of five- and six-ion potentials for mid-period metals

Table 5.4 *Interatomic coordinates, lines segments and order-n ring diagrams contributing to the MGPT n-ion multi-ion potentials through six-ion interactions.*

n	Relative coordinates	Line segments	Order-n ring diagrams	Qualitative description
2	3	1	1	1 line
3	6	3	1	1 triangle
4	9	6	3	1 square, 2 hour glasses
5	12	10	12	1 pentagon, 5 fish, 5 birds, 1 star
6	15	15	30	1 hexagon, 6 fish, 12 barbecues, 6 lawn chairs, 3 gem pairs, 2 pie slices

$$M_5^{(5)} \equiv \frac{1}{120N} {\sum_{i,j,k,l,m}}' \alpha^5 \text{Tr}\left(T_{ij}T_{jk}T_{kl}T_{lm}T_{mi} + \cdots\right) \tag{5.117}$$

and

$$M_6^{(6)} \equiv \frac{1}{720N} {\sum_{i,j,k,l,m,n}}' \alpha^6 \text{Tr}\left(T_{ij}T_{jk}T_{kl}T_{lm}T_{mn}T_{ni} + \cdots\right), \tag{5.118}$$

where in each moment one is summing, in a linear fashion, over all of the associated ring diagrams, with the components of the T_{ij} matrices in the simplified form of Eq. (5.26). The corresponding five- and six-ion contributions to the d-band structural energy E_{struc}^d in Eq. (5.37) are then, respectively,

$$E_5 = -\frac{2}{\pi}\left[\frac{1}{4}M_5^{(5)}F_5(Z_d) + \cdots\right]\Gamma_0 \tag{5.119}$$

and

$$E_6 = -\frac{2}{\pi}\left[\frac{1}{5}M_6^{(6)}F_6(Z_d) + \cdots\right]\Gamma_0, \tag{5.120}$$

where the d-band occupation functions $F_5(Z_d)$ and $F_6(Z_d)$ are defined by Eq. (5.36) for $n = 5$ and $n = 6$, respectively. The behavior of these latter two functions is illustrated in Fig. 5.24.

Using Eqs. (5.117) and (5.118), we have evaluated the ring-term partial moments $M_5^{(5)}$ and $M_6^{(6)}$ for the bcc, fcc and ideal hcp structures of the model transition metal defined in Sec. 5.1.1.2, with assumed values of $Z = 1.4$ and $Z_d = 4.6$, and with

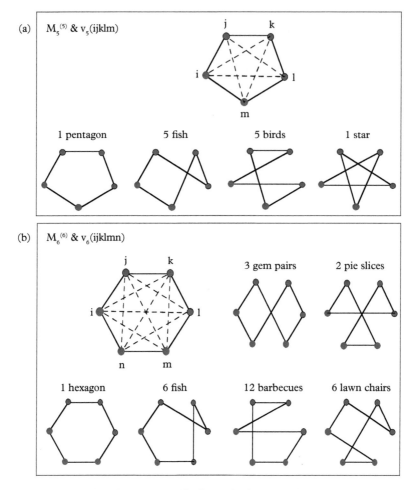

Fig. 5.23 *Representative ring diagrams contributing to the five- and six-ion TB moments and MGPT multi-ion potentials. (a) Diagrams for $M_5^{(5)}$ and v_5; (b) diagrams for $M_6^{(6)}$ and v_6.*

$\Delta_{dd'}(R_{ij})/\Gamma_0$ given by Eq. (5.43). In addition, to simplify the calculation of $M_5^{(5)}$ and $M_6^{(6)}$, we have replaced the smooth truncation of $\Delta_{dd'}(R_{ij})$ used in Table 5.1 with a sharp truncation of this function beyond first neighbors in the fcc and hcp structures and past second neighbors in the bcc structure. The results for $M_5^{(5)}$ and $M_6^{(6)}$ so obtained are given in Table 5.5. For comparison, we have also included in Table 5.5 an evaluation of $M_4^{(4)}$ under the same sharp truncation conditions. In the bcc structure, the magnitude of $M_5^{(5)}$ is found to be about one-quarter that of $M_4^{(4)}$, while the magnitude of $M_6^{(6)}$ is about one-fifth that of $M_5^{(5)}$. In the fcc and hcp structures, the magnitudes of $M_5^{(5)}$ and $M_6^{(6)}$ range from about one-third to two-thirds that of $M_4^{(4)}$.

5.2.5 Development of five- and six-ion potentials for mid-period metals

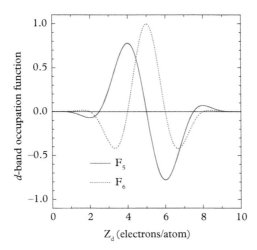

Fig. 5.24 *The d-band occupation functions $F_5(Z_d)$ and $F_6(Z_d)$ for five- and six-ion interactions, as defined by Eq. (5.36).*

Table 5.5 *Multi-ion ring-term partial TB moments $M_n^{(n)}$ evaluated for the bcc, fcc and ideal hcp structures of a model transition metal with $Z = 1.4$ and $Z_d = 4.6$, using Eq. (5.43) for $\Delta_{dd'}(R_{ij})/\Gamma_0$. To simplify the calculations, $\Delta_{dd'}(R_{ij})$ has been truncated past first neighbors in the fcc and hcp structures and past second neighbors in the bcc structure.*

Moment	bcc	fcc	hcp
$M_4^{(4)}$	−0.04306	−0.00655	−0.00619
$M_5^{(5)}$	0.01046	0.00200	0.00409
$M_6^{(6)}$	−0.00218	−0.00250	−0.00295

We have also calculated corresponding bcc-fcc and hcp-fcc structural energies with the five- and six-ion contributions obtained from Eqs. (5.119) and (5.120), respectively. For consistency in these calculations, we have applied sharp truncation to $\Delta_{dd'}(R_{ij})$ for all partial moments retained. For the two-, three- and four-ion moments, this procedure keeps the bcc-fcc energy at a value $-0.029\Gamma_0$, while the much smaller hcp-fcc energy is increased in magnitude to $-0.00022\Gamma_0$. Adding the five- and six-ion contributions then results in bcc-fcc and hcp-fcc structural energies of $-0.030\Gamma_0$ and $-0.00033\Gamma_0$, respectively. The small change in the bcc-fcc energy is not qualitatively significant, but the 50% increase in the magnitude of the hcp-fcc energy is consistent with the expected impact of the five- and six-ion partial moment contributions to hcp stability.

232   Interatomic Potentials in Transition Metals

Within the ring-diagram framework we have established here, it is then straightforward to extend the MGPT to include five- and six-ion d-state potentials. In analogy with Eq. (5.89) for $v_4^d(ijkl)$, the five-ion d-state potential is given by

$$v_5^d(ijklm) = -\frac{2}{\pi}\mathrm{Im}\int_0^{\varepsilon_F} \mathrm{Tr}\left[-2\left(T_{ij}T_{jk}T_{kl}T_{lm}T_{mi} + T_{ij}T_{jl}T_{lk}T_{km}T_{mi} + T_{ij}T_{jk}T_{km}T_{ml}T_{li}\right.\right.$$

$$+ T_{il}T_{lk}T_{kj}T_{jm}T_{mi} + T_{ij}T_{jm}T_{ml}T_{lk}T_{ki} + T_{ik}T_{kj}T_{jl}T_{lm}T_{mi} + T_{ik}T_{kj}T_{jm}T_{ml}T_{li}$$

$$+ T_{ik}T_{kl}T_{lj}T_{jm}T_{mi} + T_{ij}T_{lm}T_{mk}T_{ki}T_{ij} + T_{il}T_{lj}T_{jk}T_{km}T_{mi} + T_{ij}T_{jm}T_{mk}T_{kl}T_{li}$$

$$\left.\left. + T_{ik}T_{km}T_{mj}T_{jl}T_{li}\right)\right] dE \tag{5.121}$$

and the six-ion d-state potential is given by

$$v_6^d(ijklmn) = -\frac{2}{\pi}\mathrm{Im}\int_0^{\varepsilon_F} \mathrm{Tr}\left[2\left(T_{ij}T_{jk}T_{kl}T_{lm}T_{mn}T_{ni} + T_{ij}T_{jl}T_{lk}T_{km}T_{mn}T_{ni}\right.\right.$$

$$-+ T_{ij}T_{jk}T_{km}T_{ml}T_{ln}T_{ni} + T_{ij}T_{jk}T_{kl}T_{ln}T_{nm}T_{mi} + T_{im}T_{ml}T_{lk}T_{kj}T_{jn}T_{ni}$$

$$+ T_{ij}T_{jn}T_{nm}T_{ml}T_{lk}T_{ki} + T_{ik}T_{kj}T_{jl}T_{lm}T_{mn}T_{ni} + T_{ik}T_{kj}T_{jn}T_{nm}T_{ml}T_{li}$$

$$+ T_{il}T_{lk}T_{kj}T_{jn}T_{nm}T_{mi} + T_{ik}T_{kl}T_{lj}T_{jm}T_{mn}T_{ni} + T_{ik}T_{kl}T_{lm}T_{mj}T_{jn}T_{ni}$$

$$+ T_{ij}T_{jl}T_{lm}T_{mk}T_{kn}T_{ni} + T_{ij}T_{jl}T_{lm}T_{mn}T_{nk}T_{ki} + T_{ij}T_{jk}T_{km}T_{mn}T_{nl}T_{li}$$

$$+ T_{il}T_{lj}T_{jk}T_{km}T_{mn}T_{ni} + T_{im}T_{mj}T_{jk}T_{kl}T_{ln}T_{ni} + T_{ij}T_{jm}T_{mk}T_{kl}T_{ln}T_{ni}$$

$$+ T_{ij}T_{jn}T_{nk}T_{kl}T_{lm}T_{mi} + T_{ij}T_{jk}T_{kn}T_{nl}T_{lm}T_{mi} + T_{ij}T_{jm}T_{mn}T_{nl}T_{lk}T_{ki}$$

$$+ T_{ij}T_{jl}T_{lk}T_{kn}T_{nm}T_{mi} + T_{im}T_{ml}T_{lj}T_{jk}T_{kn}T_{ni} + T_{ik}T_{kj}T_{jm}T_{ml}T_{ln}T_{ni}$$

$$+ T_{il}T_{lm}T_{mk}T_{kj}T_{jn}T_{ni} + T_{ij}T_{jn}T_{nm}T_{mk}T_{kl}T_{li} + T_{ij}T_{jm}T_{ml}T_{lk}T_{kn}T_{ni}$$

$$+ T_{il}T_{lk}T_{kj}T_{jm}T_{mn}T_{ni} + T_{ij}T_{jk}T_{kn}T_{nm}T_{ml}T_{li} + T_{ij}T_{jm}T_{mn}T_{nk}T_{kl}T_{li}$$

$$\left.\left. + T_{il}T_{lm}T_{mj}T_{jk}T_{kn}T_{ni}\right)\right] dE \tag{5.122}$$

The 12 matrix products defining v_5^d in Eq. (5.121) and the 30 matrix products defining v_6^d in Eq. (5.122) correspond to the respective five- and six-ion ring diagrams illustrated in Fig. 5.23. As in Eq. (5.89) for v_4^d, the $T_{dd'}^{\mathrm{vol}}(R_{ij}, E)$ components of the T_{ij} matrices in Eqs. (5.121) and (5.122) are restored to the form given by Eq. (5.86), with the full energy dependence retained and only the nonorthogonality and sp-d hybridization terms neglected. Also, as in Eq. (5.89) for v_4^d, the overall sign of v_5^d in Eq. (5.121) and of v_6^d in Eq. (5.122) follows directly from Eq. (5.116). In addition, the spin factor of two multiplying each matrix product in the potential definitions of v_4^d, v_5^d and v_6^d reflects the fact that each ring diagram can be traversed in either a clockwise or counterclockwise fashion.

Finally, the five- and six-ion MGPT potentials can be further developed and cast in terms of radial and angular function in the manner of Secs. 5.2.1 and 5.2.2. Analogous

5.2.5 Development of five- and six-ion potentials for mid-period metals

to the treatment of the four-ion MGPT potential, we extract the common energy dependence and other structure-independent factors from the T_{ij} matrices in Eqs. (5.121) and (5.122) and absorb these factors into volume-dependent potential coefficients, $v_f(\Omega)$ for the five-ion potential $v_5 = v_5^d$ and $v_g(\Omega)$ for the six-ion potential $v_6 = v_6^d$. The remaining bond-integral $\Delta_{dd'}(R_{ij})$ in each T_{ij} matrix is then represented by the product of a radial function $f(R_{ij})$ and a normalized 5×5 SK matrix $H_{SK}^{dd}(i,j)$, as in the matrix MGPT discussed in Sec. 5.2.2. Lastly, we adopt the numerical notation of Secs. 5.2.1 and denote atoms i, j, k, l, m, n as atoms $1, 2, 3, 4, 5, 6$. We also retain the corresponding notation for four-ion interatomic separation distances from Fig. 5.16, namely $r_1 \equiv R_{ij}$, $r_2 \equiv R_{jk}$, $r_3 \equiv R_{ki}$, $r_4 \equiv R_{kl}$, $r_5 \equiv R_{li}$ and $r_6 \equiv R_{jl}$. For five-ion interactions, we add the four remaining interatomic distances $r_7 \equiv R_{lm}$, $r_8 \equiv R_{mi}$, $r_9 \equiv R_{km}$ and $r_{10} \equiv R_{jm}$. The five-ion MGPT potential may then be expressed in the general form

$$v_5(r_1, r_2, \ldots r_{10}, \Omega) = v_f(\Omega) \left[f(r_1) f(r_2) f(r_4) f(r_7) f(r_8) N(\theta_1, \theta_2 \ldots) + \cdots \right], \quad (5.123)$$

where the matrix representation of the angular function N is, analogous to Eq. (5.129) for four-ion angular function M,

$$N(x_1, x_2, \ldots) = \frac{1}{5730} \text{Tr} \left[H_{SK}^{dd}(1,2) H_{SK}^{dd}(2,3) H_{SK}^{dd}(3,4) H_{SK}^{dd}(4,5) H_{SK}^{dd}(5,1) \right], \quad (5.124)$$

with $5730 = S^5 + 2P^5 + 2D^5$ for canonical d bands. The lone representative five-ion configuration treated in Eqs. (5.123) and (5.124) corresponds to the first term on the RHS of Eq. (5.121), which is the pentagon ring term in Fig. 5.23(a). Elaborating the same details for the remaining 11 five-ion configurations contributing to the RHS of Eq. (5.123), denoted above only as $+\cdots$, is lengthy but entirely straightforward.

To consider six-ion interactions, we must define the five additional interatomic distances $r_{11} \equiv R_{mn}$, $r_{12} \equiv R_{ni}$, $r_{13} \equiv R_{kn}$, $r_{14} \equiv R_{jn}$ and $r_{15} \equiv R_{ln}$. The corresponding six-ion MGPT potential is then of the general form

$$v_6(r_1, r_2, \ldots r_{15}, \Omega) = v_g(\Omega) \left[f(r_1) f(r_2) f(r_4) f(r_7) f(r_{11}) f(r_{12}) Q(\theta_1, \theta_2 \ldots) + \cdots \right], \quad (5.125)$$

where the angular function Q has the matrix representation

$$Q(x_1, x_2, \ldots) = \frac{1}{54850} \text{Tr} \left[H_{SK}^{dd}(1,2) H_{SK}^{dd}(2,3) H_{SK}^{dd}(3,4) H_{SK}^{dd}(4,5) \right.$$
$$\left. H_{SK}^{dd}(5,6) H_{SK}^{dd}(6,1) \right], \quad (5.126)$$

with $54850 = S^6 + 2P^6 + 2D^6$ for canonical d bands. As above, the single six-ion configuration treated in Eqs. (5.125) and (5.126) corresponds to the first term on the RHS of Eq. (5.122), which is the hexagon ring term in Fig. 5.23(b). Again, elaborating the same details for the remaining 29 six-ion configurations contributing to the RHS of Eq. (5.125) is lengthy but straightforward.

It should be noted that our continuing use of interatomic separation distances, as opposed to Cartesian coordinates, as the independent variables of v_5 and v_6 is a matter of both convenience and efficiency. As indicated in Table 5.4, the number of line segments needed to specify the relative ion positions on an n-ion ensemble is less than the number of relative coordinates needed up to $n = 6$, where the two representations then both require 15. In particular, all needed direction cosines in the angular functions N and Q continue to be provided by the respective 10 and 15 interatomic distances defined for five- and six-ion interactions. One should further note that although four-, five- and six-ion MGPT potentials are formally functions of 6, 10 and 15 interatomic distances, respectively, each of these potentials is made up of a linear sum of individual pieces that each depend on only 4, 5 and 6 distances, respectively. The present focus on ring-term contributions thus avoids the eventual mathematical conundrum of trying to represent uniquely a multi-centered potential function of more than six atomic centers by interatomic distances alone. For example, with an $n = 7$ ion ensemble, 18 Cartesian coordinates are needed to define the relative atomic positions, and while there are then 21 interatomic distances, only 7 at a time would be needed for each ring-term contribution.

The above MGPT formalism for v_5 and v_6 still awaits implementation for real transition metals. This requires at a minimum the establishment of robust, material-dependent potential coefficients $v_f(\Omega)$ and $v_g(\Omega)$. For the study of anomalous phonons in the mid-period transition metals, the introduction of noncanonical d bands is also likely to be important. As in the standard matrix MGPT, noncanonical d bands can be introduced at no additional computational cost by simply altering the constants S and P in the SK d-state matrix H_{SK}^{dd} as well as in the normalizing coefficients $(S^5 + 2P^5 + 2D^5)^{-1}$ and $(S^6 + 2P^6 + 2D^6)^{-1}$ for the angular functions N and Q, respectively. For applications beyond the central transition metals, re-inclusion of sp-d hybridization may also be needed, as discussed in Sec. 5.2.4.

5.3 Bond-order potentials for transition metals

5.3.1 Localized d-state moments expansion for the bond energy

As introduced in Chapter 1, the BOP cohesive energy for an elemental transition metal consists of an attractive bonding contribution E_{bond}, arising directly from the partially filled d bands of the metal, and a balancing repulsive energy contribution E_{rep} arising from the remaining valence s and p electrons plus the hard-core overlap of the ions:

$$E_{\text{coh}}(\mathbf{R}) = E_{\text{bond}}(\mathbf{R}) + E_{\text{rep}}(\mathbf{R}) . \qquad (5.127)$$

The central d-bond energy E_{bond} is developed quantum mechanically in an orthogonal TB representation, starting from a free-atom reference state and using a basis of localized atomic d-states ϕ_d^i, which are analogous to the basis d states used in the GPT and MGPT. The additional repulsive energy E_{rep} is treated only empirically with a pair-functional form, as described in Sec. 5.3.3.

5.3.1 Localized d-state moments expansion for the bond energy

One begins by defining the bond energy (per atom) in an *on-site* TB representation involving the local density of states n_d^i for site i and orbital ϕ_d^i as

$$E_{\text{bond}} \equiv \frac{2}{N} \sum_i \sum_d \int_{-\infty}^{E_F} (E - \varepsilon_d^i) n_d^i(E) dE. \tag{5.128}$$

In terms of the closely related definitions made in Chapter 2, the mean d-band energy E_d given by Eq. (2.93) is the average of the individual site energies ε_d^i,

$$E_d = \frac{1}{N} \sum_i \varepsilon_d^i, \tag{5.129}$$

while the local density of states ρ_d^i given by Eq. (2.113) is an orbital sum of the n_d^i,

$$\rho_d^i(E) = \sum_d n_d^i(E). \tag{5.130}$$

Next, analogous to Eq. (2.146), one can write n_d^i in terms of the appropriate on-site Green's function G_{dd}^{ii} as

$$n_d^i(E) = -\frac{1}{\pi} \operatorname{Im} G_{dd}^{ii}(E + i0^+). \tag{5.131}$$

One can then course-grain the local electronic structure by using the TB recursion method (Haydock et al., 1972, 1980; Finnis, 2003) to express $G_{dd}^{ii}(E)$, with the energy E referenced to the site energy ε_d^i, as a continued fraction of the form

$$G_{dd}^{ii}(E) = \cfrac{1}{E - \varepsilon_d^i - a_0 - \cfrac{b_1^2}{E - \varepsilon_d^i - a_1 - \cfrac{b_2^2}{E - \varepsilon_d^i - \cdots}}}, \tag{5.132}$$

where the recursion coefficients a_m and b_m can be algebraically related to the local d-state moments μ_n^{id} of the LDOS n_d^i through the equations $a_0 = 0$ and

$$\mu_2^{id} = b_1^2$$
$$\mu_3^{id} = a_1 b_1^2$$
$$\mu_4^{id} = a_1^2 b_1^2 + b_1^4 + b_1^2 b_2^2.$$
$$\vdots \tag{5.133}$$

Here b_1 is a measure of the root-mean-square width of the LDOS $n_d^i(E)$, a_1 is the amount of skewing of the LDOS center away from ε_d^i, and b_2 measures the bimodality of $n_d^i(E)$.

The moments μ_n^{id} themselves can be calculated directly in terms of the bond integrals $h_{ij}^{dd'}$. Noting from the definition of μ_n^i in Eq. (2.16) that

$$\mu_n^i = \sum_d \mu_n^{id}, \qquad (5.134)$$

one can infer from Eqs. (2.118)–(2.120) the explicit results

$$\begin{aligned}
\mu_2^{id} &= {\sum_j}' \sum_{d'} h_{ij}^{dd'} h_{ji}^{d'd} \\
\mu_3^{id} &= {\sum_{j,k}}' \sum_{d',d''} h_{ij}^{dd'} h_{jk}^{d'd''} h_{ki}^{d''d} \\
\mu_4^{id} &= {\sum_{j,k,l}}' \sum_{d',d'',d'''} h_{ij}^{dd'} h_{jk}^{d'd''} h_{kl}^{d''d'''} h_{li}^{d'''d} \\
&\vdots
\end{aligned} \qquad (5.135)$$

Each moment μ_n^{id} has a graphical representation involving the possible self-returning paths connecting the ion positions that are of exactly n steps starting and ending on the site i in orbital ϕ_d^i. The specific paths corresponding to μ_2^{id}, μ_3^{id} and μ_4^{id} are, respectively, those in the 2nd-, 3rd- and 4th-order graphs displayed in Fig. 5.1.

The continued-fraction expansion for G_{dd}^{ii} and n_d^i can be smoothly terminated at any level $m = L$ by approximating the remaining recursion coefficients as constants $a_m = a_\infty$ and $b_m = b_\infty$ for $m \geq L$, as discussed by Aoki et al. (2007). In practice, this is usually done so as to include some desired number of moments into the calculation. Then to obtain the d-bond energy from Eq. (5.128) requires first establishing the Fermi energy E_F from the corresponding conservation of electrons condition

$$Z_d = \frac{2}{N} \sum_i \sum_d \int_{-\infty}^{E_F} n_d^i(E) dE. \qquad (5.136)$$

The number of d electrons per atom Z_d is not an internally determined quantity here, but is rather a chosen input parameter from which the Fermi energy is determined. In both Eq. (5.128) and Eq. (5.136), the site energy ε_d^i is also self-consistently adjusted to ensure local charge neutrality, which is both an excellent approximation for metals and consistent with Eq. (1.27) for the electron density. This bond-energy framework was that used by Aoki (1993) in Fig. 1.6 to calculate transition-metal bcc-fcc and hcp-fcc structural energies as functions of both Z_d and M, the number of moments μ_n^{id} retained in Eq. (5.132) for E_{bond}, where the individual moments were evaluated with a canonical d-band model of the bond integrals $h_{ij}^{dd'}(R_{ij}) \propto \eta_m (R_{\text{WS}}/R_{ij})^5$. As discussed in Chapter 1, retaining six moments in these calculations provides good physical convergence for the structural energies, but up to 18 moments are needed to obtain correspondingly good mathematical convergence.

While the on-site TB representation can provide reliable values of E_{bond} for total-energy calculations, this representation does not provide a useful way of calculating the

5.3.1 Localized d-state moments expansion for the bond energy

corresponding forces needed to move the ions. The latter task requires the equivalent *inter-site* representation of the bond energy,

$$E_{\text{bond}}(\mathbf{R}) = \frac{1}{N} \sum_{i,j}{}' \sum_{d,d'} h_{ij}^{dd'} \Theta_{ji}^{d'd}, \qquad (5.137)$$

where the bond order $\Theta_{ij}^{dd'}$ of the dd' bond between ions i and j is calculated from the corresponding inter-site Green's function $G_{dd'}^{ij}$ as

$$\Theta_{ij}^{dd'} = -\frac{2}{\pi} \text{Im} \int_{-\infty}^{E_F} G_{dd'}^{ij}(E) dE. \qquad (5.138)$$

The bond-order $\Theta_{ij}^{dd'}$ is formally the off-diagonal ($i \ne j$) matrix element of the more general density matrix $\rho_{ij}^{dd'}$ (Finnis, 2003), and its physical meaning can be understood qualitatively as follows. If one defines bonding $|+\rangle$ and antibonding $|-\rangle$ states by the relationships

$$|\pm\rangle \equiv \left(|\phi_d^i\rangle \pm |\phi_{d'}^j\rangle\right)/\sqrt{2}, \qquad (5.139)$$

then $G_{dd'}^{ij}$ in Eq. (5.138) can be expressed in terms of the difference between on-site Green's functions $G_{dd'}^{++}$ and $G_{dd'}^{--}$ as

$$G_{dd'}^{ij} = \frac{1}{2}(G_{dd'}^{++} - G_{dd'}^{--}). \qquad (5.140)$$

Using Eq. (5.140) in Eq. (5.138), the bond order is just

$$\Theta_{ij}^{dd'} = \frac{1}{2}(N_+ - N_-), \qquad (5.141)$$

where N_+ is the number of electrons in the bonding state $|+\rangle$ and N_- is the number of electrons in the antibonding state $|-\rangle$.

Using the inter-site representation of E_{bond}, it is straightforward to calculate Hellman-Feynman forces on the ions, provided, of course, that $G_{dd'}^{ij}$ and $\Theta_{ij}^{dd'}$ are first well converged. The d bonding component of force $\mathbf{F}_k$ on the kth ion is then just

$$\mathbf{F}_k(\mathbf{R}) = -N \nabla_k E_{\text{bond}}(\mathbf{R})$$

$$= -\sum_{i,j}{}' \sum_{d,d'} \left(\nabla_k h_{ij}^{dd'}\right) \Theta_{ji}^{d'd}, \qquad (5.142)$$

noting that by first-order perturbation theory $\nabla_k \Theta_{ji}^{d'd} = 0$ (Aoki et al., 2007). Thus, calculating Hellman-Feynman forces in the inter-site representation requires only the additional step of differentiating the SK matrix for the bond integrals $h_{ij}^{dd'}$.

However, it was discovered in the 1980s that the direct use of Eq. (5.140) to calculate the inter-site Green's function $G_{dd'}^{ij}$ by the recursion method leads to very poor convergence and therefore cannot be used to obtain either reliable energies or forces. Pettifor (1989) and Aoki (1993) solved this problem with the development of a stable and rapidly convergent many-atom expansion method to calculate $G_{dd'}^{ij}$ and $\Theta_{ij}^{dd'}$, a method that has since become a cornerstone of the BOP formalism. In contrast to the covalent orbitals $|\pm\rangle$ that lead to Eq. (5.140) for $G_{dd'}^{ij}$, in the BOP approach applied to transition metals (Aoki et al., 1993, 2007) one begins with a starting basis orbital $|0\rangle$ that is a general admixture of d orbitals on sites i and j,

$$|0\rangle = c_i |\phi_d^i\rangle + c_j e^{i\theta} |\phi_{d'}^j\rangle, \tag{5.143}$$

where $\lambda = \cos\theta$ is a phase factor and $c_i^2 + c_j^2 = 1$. The corresponding Green's function $G_{dd'}^{00}$ can be written in terms on on-site and inter-site components as

$$G_{dd'}^{00}(\lambda) = c_i^2 G_{dd'}^{ii} + c_j^2 G_{dd'}^{jj} + 2c_i c_j \lambda G_{dd'}^{ij}. \tag{5.144}$$

Thus, the inter-site Green's function $G_{dd'}^{ij}$ can be formally expressed as the first derivative of $G_{dd'}^{00}$ with respect to λ,

$$G_{dd'}^{ij} = \frac{1}{2c_i c_j} \frac{dG_{dd'}^{00}(\lambda)}{d\lambda}. \tag{5.145}$$

Next, one considers a continued-fraction expansion of $G_{dd'}^{00}$ such that its dependence on λ is folded into the expansion coefficients $\{a_n^\lambda, b_n^\lambda\}$. Then by chain-rule differentiation

$$\frac{dG_{dd'}^{00}(\{a_n^\lambda, b_n^\lambda\})}{d\lambda} = \sum_{n=0}^{\infty} \frac{\partial G_{dd'}^{00}}{\partial a_n^\lambda} \frac{\partial a_n^\lambda}{\partial \lambda} + \sum_{n=1}^{\infty} \frac{\partial G_{dd'}^{00}}{\partial b_n^\lambda} \frac{\partial b_n^\lambda}{\partial \lambda}. \tag{5.146}$$

In turn, the derivatives of $G_{dd'}^{00}$ with respect to a_n^λ and b_n^λ can be expressed in terms of Green's functions $G_{dd'}^{0n}$ defined along the recursion chain by using Dyson's equation:

$$\frac{\partial G_{dd'}^{00}}{\partial a_n^\lambda} = G_{dd'}^{0n} G_{dd'}^{n0} \tag{5.147}$$

and

$$\frac{\partial G_{dd'}^{00}}{\partial b_n^\lambda} = 2 G_{dd'}^{0(n-1)} G_{dd'}^{n0}. \tag{5.148}$$

The derivatives of a_n^λ and b_n^λ with respect to λ in Eq. (5.146) can be expressed in terms of moments μ_r^0 defined by the basis orbital $|0\rangle$,

$$\frac{\partial a_n^\lambda}{\partial \lambda} = \sum_{r=1}^{2n+1} \frac{\partial a_n^\lambda}{\partial \mu_r^0} \frac{\partial \mu_r^0}{\partial \lambda} = 2c_i c_j \sum_{r=1}^{2n+1} \frac{\partial a_n^\lambda}{\partial \mu_r^0} \zeta_{r+1}^{id, jd'} \tag{5.149}$$

5.3.1 Localized d-state moments expansion for the bond energy

and

$$\frac{\partial b_n^\lambda}{\partial \lambda} = \sum_{r=1}^{2n} \frac{\partial b_n^\lambda}{\partial \mu_r^0} \frac{\partial \mu_r^0}{\partial \lambda} = 2c_i c_j \sum_{r=1}^{2n} \frac{\partial b_n^\lambda}{\partial \mu_r^0} \zeta_{r+1}^{id,jd'}, \qquad (5.150)$$

where

$$\mu_r^0(\lambda) = c_i^2 \mu_r^{id} + c_j^2 \mu_r^{jd'} + 2c_i c_j \lambda \zeta_{r+1}^{id,jd'}. \qquad (5.151)$$

The new ingredient here is the so-called interference term $\zeta_{r+1}^{id,jd'}$, which represents a path of r steps starting on ion i in the orbital ϕ_d^i and ending on ion j in the orbital $\phi_{d'}^j$.

Using Eqs. (5.145)–(5.151) in Eq. (5.138) and taking the limit $\lambda \to 0$, yields an exact many-ion expansion for the bond order of the form

$$\Theta_{ij}^{dd'} = 2 \left[\sum_{n=0}^{\infty} \chi_{0n,n0} \delta a_n^{id,jd'} + 2 \sum_{n=1}^{\infty} \chi_{0n,(n-1)0} \delta b_n^{id,jd'} \right], \qquad (5.152)$$

where the response function $\chi_{0m,n0}$ is defined by

$$\chi_{0m,n0} \equiv \frac{1}{\pi} \text{Im} \int_{-\infty}^{E_F} G_{0m}(E) G_{n0}(E) dE, \qquad (5.153)$$

and the additional coefficients $\delta a_n^{id,jd'}$ and $\delta b_n^{id,jd'}$ are linear functions of the $\zeta_{r+1}^{id,jd'}$. For transition metals, the fastest convergence of Eq. (5.152) for the bond order is in the limit $c_j \to 0$ and $c_i \to 1$, in which case $\mu_r^0 \to \mu_r^{id}$ in Eq. (5.151) and

$$\delta a_0^{id,jd'} = \zeta_2^{id,jd'}$$
$$\delta b_1^{id,jd'} = [1/2(\mu_2^{id})^{1/2}] \zeta_3^{id,jd'}$$
$$\delta a_1^{id,jd'} = (1/\mu_2^{id}) \zeta_4^{id,jd'} - [\mu_3^{id}/(\mu_2^{id})^2] \zeta_3^{id,jd'} - 2\zeta_2^{id,jd'}.$$
$$\vdots \qquad (5.154)$$

The rapid convergence of Eq. (5.152) for the transition-metal σ, π and δ bond orders as function of Z_d was demonstrated by Aoki (1993), as illustrated in Fig. 5.25. These calculations were carried out for the same bond integrals assumed in Fig. 1.6.

Several special points of interest should be mentioned in connection with the above BOP formalism for a convergent bond order and bond energy with corresponding Hellman-Feynman forces, and its relationship to GPT and MGPT. First, the response function $\chi_{0m,n0}$ appearing in Eq. (5.152) for the bond order plays a qualitatively similar role in the BOP theory to that of the d-band filling function $F_n(Z_d)$ in the simplified

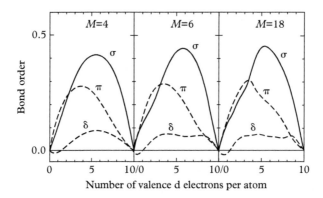

Fig. 5.25 *Calculated σ, π and δ bond orders from Eq. (5.152) for a d-bonded fcc transition metal as a function of d-band filling and the number M of d-state moments μ_n^{id} retained in a BOP treatment based on canonical d bands, as in Fig. 1.6.*
From Aoki (1993), with publisher (APS) permission.

TB model defined in Sec. 5.1.1.2, a model that has provided the path forward to the development of the MGPT. In the original paper establishing the BOP, Pettifor (1989) defined a normalized, dimensionless response function $\hat{\chi}_{0m,n0}$, whose specific shapes for $\hat{\chi}_{00,00}$, $\hat{\chi}_{00,10}$ and $\hat{\chi}_{01,10}$ as a function of band filling (albeit for s states rather than d states) are remarkably similar to the shapes of F_2, F_3 and F_4, respectively, as displayed in Fig. 5.2 as a function of d-band filling.

A second point of interest concerns the specific bonding conditions needed for the fastest convergence of the bond order in Eq. (5.152). It would appear to be significant that this is found to occur for atom-based bonding conditions, $c_i = 1$ and $c_j = 0$, for the case of transition metals, as opposed to the covalent bonding condition, $c_i = c_j = 1/\sqrt{2}$, which is found to produce the fastest convergence in case of semiconductors (Aoki et al., 2007). These findings are consistent with the viewpoint expressed in Chapter 1 that in transition metals the d bonding is directional but only weakly covalent, with no measurable bond charges, while in semiconductors, the sp^3 bonding is both directional and strongly covalent, with experimentally measurable bond charges. Also in this regard, the covalent bonding and antibonding conditions given by Eq. (5.139), which are imposed in the on-site representation of $G_{dd'}^{ij}$ in Eq. (5.140), would also seem to help explain the resulting poor convergence with that approach in the case of transition metals.

A final point of interest is the more general comparison of BOP and MGPT simulation methods for transition metals. Some of the major features of that comparison are summarized in Table 5.6. In the GPT and MGPT, one begins with a plane-wave, localized d-state mixed basis set and a self-consistent resonant-ion reference system. This allows one to separate the cohesive energy into a large volume term and a smaller structural-energy component that takes the form of a cluster series of volume-dependent multi-ion potentials. In the MGPT, the d-band components of the multi-ion potentials

5.3.1 Localized d-state moments expansion for the bond energy

Table 5.6 *Comparison of some of the major features of the MGPT and BOP simulation methods as applied to bulk transition metals. Values in parenthesis are the number of moments retained.*

Method	Reference system	d-state moments expansion	sp-d hybridization	sp-electron/ hard-core repulsion	Forces	Large scale MD
MGPT	Resonant ion	Linear (4–6)	Partial	First-principles GPT	Term by term	Yes
BOP	Free atom	Nonlinear (9–18)	No	Empirical	Hellman-Feynman	No

are provided by a normalized but linear d-state moments expansion, while the additional repulsive sp valence-electron and hard-core-overlap contributions to the pair potential are retained from the first-principles GPT. The remaining sp-d hybridization energy contributions in the MGPT are retained only in the volume term and in the pair potential for late series metals, whereas for central transition metals hybridization contributions to the multi-ion potentials are currently neglected. In the BOP approach, on the other hand, one begins with a localized d-state basis set and a free-atom reference system. This allows a direct treatment of the major d-band component of the cohesive energy in a pure real-space TB d-bonding framework, which specifically takes the form of a convergent, nonlinear d-state moments expansion of the bond order. Additional sp electron and hard-core overlap contributions to the cohesive energy are only treated with empirical potentials, and sp-d hybridization contributions are neglected in the central transition metals.

State-of-the-art MGPT applications to structural, thermodynamic, defect and mechanical properties of bulk central transition metals all include two-, three- and four-ion potentials, corresponding to the retention of four d-state moments. The future extension of MGPT to five- and six-ion potentials, with the corresponding retention of five and six d-state moments, is possible and currently in progress, as discussed in Sec. 5.2.5. The baseline analytic MGPT representation provides maximum transparency and insight in applications but is restricted to no more than four-ion potentials and requires the use of canonical d bands. The more general matrix MGPT representation allows the treatment of full noncanonical d bands while at the same time providing optimum computational efficiency, with a term-by-term calculation of forces that can be used in both static and dynamic simulations, including large-scale MD. State-of-the-art BOP applications to structural phase stability and defects in central transition metals with noncanonical d bands, converged bond energies and accurate Hellman-Feynman forces can be performed with as few as nine d-state moments. This achievement has been made possible by the incorporation of a small degree of electron temperature into the calculation of the bond order to speed numerical convergence (Horsfield et al., 1996a, 1996b), as is done in DFT calculations. This has allowed static simulations of point and extended defects; it also, in principle, allows the direct treatment of surfaces.

However, corresponding dynamic simulations, such as MD, have not proven possible with the standard BOP approach, although some progress in that direction has been made recently with the development of so-called analytic BOPs, as described in the next section.

5.3.2 Simplified analytic bond-order potentials

The standard BOP formalism for transition metals presented in the last section requires a full numerical implementation for applications to real materials. Drautz and Pettifor (2006, 2011) more recently developed an alternative analytic BOP representation that offers, in principle, the possibilities of simplification, improved computational efficiency and more accurate forces for a given number of moments retained. The analytic BOP theory begins by defining a reference system whose LDOS n_d^{i0} is given by a continued-fraction expansion with constant recursion coefficients $a_n = a_\infty$ and $b_n = b_\infty$, where a_∞ and b_∞ are intended as chosen input parameters to be optimized for the transition-metal system of interest. In the reduced energy variable

$$\varepsilon = \frac{E - a_\infty}{2b_\infty} \tag{5.155}$$

one has $n_d^{i0}(E) = n_d^{i0}(\varepsilon)/(2b_\infty)$ between $E = a_\infty \pm 2b_\infty$, where $n_d^{i0}(\varepsilon)$ is the semi-elliptic density of states:

$$n_d^{i0}(\varepsilon) = \frac{2}{\pi}\sqrt{1-\varepsilon^2} \tag{5.156}$$

representing a single band of d states residing between $\varepsilon = \pm 1$.

Next, using Chebyshev polynomials of the second kind, $P(\varepsilon)$, which represent a complete set in same $-1 \le \varepsilon \le 1$ interval and are orthonormal functions with respect to the reference LDOS $n_d^{i0}(\varepsilon)$, one seeks an expansion of the LDOS for the system of interest in the form

$$n_d^i(\varepsilon) = \frac{2}{\pi}\sqrt{1-\varepsilon^2}\left[\sum_{m=0}^{n_{\max}} \sigma_m^{id} P_m(\varepsilon)\right], \tag{5.157}$$

where $n_{\max}$ is a chosen integer maximum, such that in the limit $n_{\max} \to \infty$ the expansion is expected to be an exact representation of the LDOS for physically reasonable values of a_∞ and b_∞. The expansion coefficients σ_m^{id} in Eq. (5.157) can be evaluated directly from the d-state moments μ_n^{id}. Noting that the Chebyshev polynomials of the second kind have the simple form

$$P_m(\varepsilon) = \sum_{n=0}^{m} p_{mn}\varepsilon^n, \tag{5.158}$$

one can show that

$$\sigma_m^{id} = \sum_{n=0}^{m} p_{mn} \hat{\mu}_n^{id}, \qquad (5.159)$$

where $\hat{\mu}_n^{id}$ is the dimensionless d-state moment

$$\hat{\mu}_n^{id} \equiv \frac{1}{(2b_\infty)^n} \sum_{k=0}^{n} \frac{(-a_\infty)^k n!}{k!(n-k)!} \mu_{n-k}^{id}. \qquad (5.160)$$

Thus, including terms in the analytic DOS expansion (5.157) up to and including n_{max} exactly includes d-state moments up to and including $\mu_{n_{max}}^{id}$.

Using Eq. (5.157) for n_d^i, one can analytically integrate Eq. (5.136) for the number of d electron per atom to obtain the result

$$Z_d = \frac{2}{N} \sum_i \sum_d \sum_{m=0}^{n_{max}} \sigma_m^{id} \hat{\chi}_{m+1}(\phi_F), \qquad (5.161)$$

where ϕ_F is the Fermi-energy phase

$$\phi_F \equiv \cos^{-1}\left[E_F - a_\infty/(2b_\infty)\right] \qquad (5.162)$$

and $\hat{\chi}_n(\phi_F)$ is a dimensionless response function, which for $n = 1$ is given by

$$\hat{\chi}_1(\phi_F) = 1 - \frac{\phi_F}{\pi} + \frac{1}{2\pi}\sin(2\phi_F), \qquad (5.163)$$

and for $n \geq 2$ is given by

$$\hat{\chi}_n(\phi_F) = \frac{1}{\pi}\left[\frac{\sin(n+1)\phi_F}{n+1} - \frac{\sin(n-1)\phi_F}{n-1}\right]. \qquad (5.164)$$

These response functions are quite similar to those originally defined by Pettifor (1989) in the standard BOP. Additionally, one can analytically integrate Eq. (5.128) for the on-site bond energy to obtain

$$E_{bond} = \frac{2b_\infty}{N} \sum_i \sum_d \sum_{m=0}^{n_{max}} \sigma_m^{id} \left[\hat{\chi}_{m+2}(\phi_F) - \gamma_0^{id} \hat{\chi}_{m+1}(\phi_F) + \hat{\chi}_m(\phi_F)\right], \qquad (5.165)$$

where $\gamma_0^{id} = (\varepsilon_d^i - a_\infty)/b_\infty$ and $\hat{\chi}_0 = 0$.

Derivation of the bond order, inter-site bond energy and Hellman-Feynman forces in the analytic BOP requires a somewhat more complex analysis, as detailed by Drautz and Pettifor (2006). The derivatives of the Green's function $G_{dd'}^{00}$ with respect to the recursion coefficients a_n^λ and b_n^λ in Eq. (5.146) of the standard BOP must be replaced with derivatives with respect to the moments $\mu_n^0(\lambda)$ in the analytic BOP. Equation (5.151) in the standard BOP then becomes

$$\mu_n^0(\lambda) = c_i^2 \mu_n^{id} + c_j^2 \mu_n^{jd'} + 2c_i c_j \lambda \xi_n^{id,jd'}, \tag{5.166}$$

where $\xi_n^{id,jd'}$ is now the interference path that determines the inter-site Green's function $G_{dd'}^{ij}$ in the analytic BOP, according to

$$G_{dd'}^{ij}(E) = \sum_n \frac{1}{2c_i c_j} \frac{dG_{dd'}^{00}(E)}{d\mu_n^0} \frac{d\mu_n^0(\lambda)}{d\lambda} = \sum_n \xi_n^{id,jd'} \frac{dG_{dd'}^{00}(E)}{d\mu_n^0}. \tag{5.167}$$

Translated into reduced energy units, Eq. (5.167) becomes

$$G_{dd'}^{ij}(\varepsilon) = \sum_n \hat{\xi}_n^{id,jd'} \frac{dG_{dd'}^{00}(\varepsilon)}{d\hat{\mu}_n^0}, \tag{5.168}$$

where $\hat{\xi}_k^{id,jd'}$ is the dimensionless interference path

$$\hat{\xi}_k^{id,jd'} = \frac{1}{(2b_\infty)^k} \sum_{n=0}^k \frac{(-a_\infty)^{n-k} n!}{k!(n-k)!} \xi_n^{id,jd'}, \tag{5.169}$$

with the constraint $\xi_n^{id,jd'} = 0$ for $\phi_d^i \neq \phi_{d'}^j$. To guarantee the Hermiticity of $G_{dd'}^{ij}$ for finite $n_{\max}$, one then introduces an average inter-site Green's function

$$\bar{G}_{dd'}^{ij}(E) = \frac{1}{2}\left[G_{dd'}^{ij}(E) + G_{d'd}^{ji}(E)\right]. \tag{5.170}$$

This leads in turn to an average bond order

$$\bar{\Theta}_{dd'}^{ij} = \frac{1}{2}\left[\Theta_{dd'}^{ij} + \Theta_{d'd}^{ji}\right], \tag{5.171}$$

where

$$\Theta_{dd'}^{ij} = 2 \sum_{m=1}^{n_{\max}} \sum_{n=0}^m p_{mn} \hat{\xi}_n^{id,jd'} \hat{\chi}_{m+1}(\phi_F) + \cdots. \tag{5.172}$$

Here the $+\cdots$ in Eq. (5.172) represent additional small constraint terms that have been added to ensure that for a given $n_{\max}$ the inter-site bond energy

$$E_{\text{bond}}(\mathbf{R}) = \frac{1}{N} \sum_{i,j}{}' \sum_{d,d'} \bar{\Theta}_{ij}^{dd'} h_{ji}^{d'd}, \tag{5.173}$$

is equivalent to the on-site bond energy given by Eq. (5.165).

5.3.2 Simplified analytic bond-order potentials

Finally, to obtain analytic BOP forces for a given n_{max} in the useful form

$$\mathbf{F}_k(\mathbf{R}) = -\sum_{i,j}{}' \sum_{d,d'} \tilde{\Theta}_{ij}^{dd'} \left(\nabla_k h_{ji}^{d'd} \right) \tag{5.174}$$

requires the modified bond order

$$\tilde{\Theta}_{dd'}^{ij} = 2 \sum_{m=0}^{n_{\text{max}}} \sum_{n=0}^{m} p_{mn} n \hat{\xi}_{n-1}^{id,jd'} \left[\hat{\chi}_{m+2}(\phi_F) + \hat{\chi}_m(\phi_F) \right]. \tag{5.175}$$

In the limit $n_{\text{max}} \to \infty$, however, $\tilde{\Theta}_{ij}^{dd'} \to \bar{\Theta}_{ij}^{dd'} = \Theta_{ij}^{dd'}$ and Eq. (5.174) gives exact Hellman-Feynman forces.

For both the standard numerical BOP and the analytic BOP, evaluation of the d-state moments is the most computationally expensive part of calculating energies and forces. With the analytic BOP, however, there is in addition a possible path forward to further simplification. Specifically, Drautz and Pettifor (2006) showed that the dimensionless moments $\hat{\mu}_n^{id}$, as well as the corresponding LDOS expansion coefficients σ_n^{id}, can be expressed analytically in terms of the small parameters

$$\gamma_n = (a_n - a_\infty)/b_\infty \tag{5.176}$$

and

$$\delta_n = (b_n^2 - b_\infty^2)/b_\infty^2, \tag{5.177}$$

where a_n and b_n are the recursion coefficients for the system under consideration. Using the simple representative choice of analytic-BOP input parameters $a_\infty = a_1$ and $b_\infty = b_1$, where $\gamma_1 = \delta_1 = 0$, Drautz and Pettifor derived exact analytic results for $\hat{\mu}_n^{id}$ and σ_n^{id} through $n_{\text{max}} = 6$. Applied to the bcc, fcc and hcp canonical d bands of a model transition metal, this sixth-moment treatment produced in turn good descriptions of the LDOS, bond energy and bcc-fcc and hcp-fcc structural energies as a function of d-band filling. Moreover, if one makes a further *linear approximation*, by retaining only contributing terms to $\hat{\mu}_n^{id}$ and σ_n^{id} that are first-order in γ_n and δ_n, then the bcc, fcc and hcp LDOSs in the important energy range $-0.8 < \varepsilon < 0.8$ are still adequately described, with only four parameters, $\gamma_0, \gamma_2, \delta_2$ and δ_3, needed for each phase. At the same time, the recursion coefficients a_n and b_n themselves still require evaluation of the full moments μ_n^{id}, which might be approximated by only ring-term contributions for $n > 4$, as proposed for the MGPT. This suggests combining a linear treatment $\hat{\mu}_n^{id}$ with a ring-term treatment of μ_n^{id} as a possible path forward to large-scale BOP-MD simulations in central transition metals. In this regard, it has already been established that small-scale MD simulations (with < 1000 atoms) on topological-close-packed transition-metal phases such as A15 are feasible with the analytic BOP at the sixth-moment level (Hammerschmidt et al., 2011).

5.3.3 Parameterization of bond integrals and the repulsive energy

Both the numerical BOP and analytic BOP representations in transition metals are parameterizations of an orthogonal TB treatment of the d bands, without any actual specification of either the electron potential in the Hamiltonian or the localized d basis states ϕ_d^i. The impact of the latter quantities on the bond energy, bond order and forces in the BOP is consequently folded entirely into the fitted bond-integral functions $dd\sigma(R_{ij})$, $dd\pi(R_{ij})$ and $dd\delta(R_{ij})$ that establish the full SK matrix $h_{ij}^{dd'}$ at the observed lattice constant or equilibrium volume of the metal. In practice, these functions have been modeled in the five-parameter form first proposed by Goodwin et al. (1989):

$$dd\beta_m(R_{ij}) = dd\beta_m(R_{nn}) \left(\frac{R_{nn}}{R_{ij}}\right)^{n_a} \exp\left\{ n_b \left[\left(\frac{R_{nn}}{R_c}\right)^{n_c} - \left(\frac{R_{ij}}{R_c}\right)^{n_c} \right] \right\}, \quad (5.178)$$

where $\beta_m = \sigma, \pi$ and δ for $m = 0, 1$ and 2. Here R_{nn} is the nearest-neighbor distance for the equilibrium crystal structure of the metal, and for each m, the five adjustable parameters are $dd\beta_m(R_{nn})$, n_a n_b, n_c and R_c.

In the first transition-metal application of the numerical BOP (Girshick et al., 1998), canonical d bands were used to establish the bond-integral functions for hcp Ti, corresponding to a choice of $n_a = 5$ and $n_b = 0$ in Eq. (5.178) with $dd\sigma : dd\pi : dd\delta$ taken in the fixed ratios $-6 : 4 : -1$ and normalized to give the correct d-band width. Subsequently, beginning with the work of Mrovec et al. (2004, 2007) on applications of the numerical BOP to bcc Mo and W, the preferred strategy has been to establish the parameters in Eq. (5.178) by fitting to the actual DFT electronic structure of the material in question, as established by self-consistent TB-LMTO calculations. This procedure, however, is rather complex and includes the effects of nonorthogonality and sd electron screening on the bond integrals in the manner on Nguyen-Manh et al. (2000). Variants of the approach have also been used, including the scheme of Cawkwell et al. (2006) for fcc Ir and the refined scheme of Lin et al. (2014) for bcc V, Nb, Ta, Cr, Mo and W, both with the numerical BOP, as well as the alternate scheme of Cák et al. (2014) for bcc Nb, Ta, Mo and W, with the analytic BOP. In the latter three schemes, it is assumed that $n_b = n_a$ in Eq. (5.178), so the number of adjustable parameters is reduced from five to four.

The DFT-fitted BOP bond integrals (5.178) for bcc V, Mo and Ta, as determined by Lin et al. (2014) at the observed equilibrium volumes of these metals, are plotted in panels (a), (b) and (c) of Fig. 5.26. It is interesting to compare these results with corresponding first-principles GPT bond integrals determined from the standard relationship

$$dd\beta_m(R_{ij}) = -\Delta_{dd'}(R_{ij})\delta_{mm'}, \quad (5.179)$$

where the ions i and j are aligned along the z axis. Here as usual, $\Delta_{dd'}(R_{ij})$ is evaluated from Eq. (4.92) with the needed potential components and d basis states calculated

5.3.3 Parameterization of bond integrals and the repulsive energy

directly from the self-consistent zero-order pseudoatom for each metal with a boundary condition of $D_2^\star = -3$. The GPT bond integrals so obtained for V, Mo and Ta, which unlike the BOP results do not depend on crystal structure, are plotted in panels (d), (e) and (f) of Fig. 5.26. The BOP and GPT bond-integral results are clearly similar in shape, magnitude and ordering among the three metals, although for Mo and Ta the GPT bond integrals are nearly the same, while in the BOP results they are well separated. Also note that for actual BOP applications in atomistic calculations and simulations, the raw bond integrals given by Eq. (5.178) are always multiplied by a cutoff function that sharply terminates the $dd\beta_m(R_{ij})$ past a chosen radial distance $R_{ij} = R_{\text{cut}}$. For the central bcc metals, R_{cut} is usually taken at the closest convenient point past the second-neighbor distance. Thus, in practice, the longer-range behavior of $dd\sigma(R_{ij})$ and $dd\pi(R_{ij})$ for BOP Ta, as displayed in Figs. 5.26(a) and 5.26(b), is actually not retained in applications.

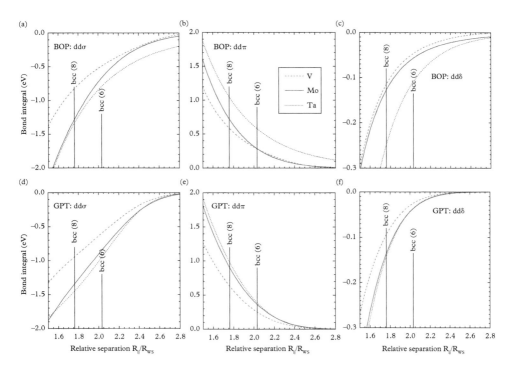

Fig. 5.26 *DFT-fitted BOP (Lin et al., 2014) and first-principles GPT bond-integral functions, $dd\sigma(R_{ij})$, $dd\pi(R_{ij})$ and $dd\delta(R_{ij})$ for V, Mo and Ta at their observed equilibrium volumes. Vertical lines indicate location and number of first and second neighbors in the bcc structure. (a) BOP $dd\sigma$; (b) BOP $dd\pi$; (c) BOP $dd\delta$; (d) GPT $dd\sigma$; (e) GPT $dd\pi$; (f) GPT $dd\delta$.*

The second part of the BOP parameterization concerns the empirical treatment of the repulsive energy E_{rep} in Eq. (5–131) for the BOP cohesive-energy functional of a transition metal. In both the numerical and analytic BOP, E_{rep} is parameterized as sum of an environment-dependent pair-functional contribution E_{env} and an additional pair-potential contribution E_{pair}:

$$E_{rep}(\mathbf{R}) = E_{env}(\mathbf{R}) + E_{pair}(\mathbf{R}). \qquad (5.180)$$

The practical goal in all cases is to model $E_{env}(\mathbf{R})$ and $E_{pair}(\mathbf{R})$ such that when combined with $E_{bond}(\mathbf{R})$, a good description of the equilibrium cohesive and elastic properties of the metal is achieved. The environment-dependent pair functional has most often been taken in the form a screened Yukawa-like potential

$$E_{env}(\mathbf{R}) = \frac{1}{2N} {\sum_{i,j}}' \frac{B}{R_{ij}} \exp\left[-0.5(\lambda_i + \lambda_j)(R_{ij} - 2R_s)\right], \qquad (5.181)$$

where the screening parameter λ_i depends on the local environment through a nonlinear embedding function given by

$$\lambda_i = \lambda_0 + \left[\sum_{k \neq i} C \exp(-\nu R_{ik})\right]^{1/m}. \qquad (5.182)$$

Here there are six adjustable parameters: B, R_s λ_0, C, ν and m. Lin et al. (2014), on the other hand, have proposed a modified form of Eq. (5.181), with the Coulomb potential $1/R_{ij}$ replaced by $\exp(-\mu R_{ij})$ and with the argument in the exponential term replaced by $-(\lambda_i + \lambda_j)(R_{ij} - R_s)$, while setting $\lambda_0 = 0$ and $m = 1$ in Eq. (5.182). These changes leave five adjustable parameters: B, μ, R_s, C and ν. In treating cubic transition metals with either the unmodified or modified forms of Eqs. (5.181) and (5.182), the normal strategy is to use the available adjustable parameters to fit the observed Cauchy pressure $P_C = C_{12} - C_{44}$, lattice constant and cohesive energy. As in the case of the bond-integral functions, the range of interatomic interactions in Eqs. (5.181) and (5.182) is controlled with a sharp cutoff function, where R_{cut} is applied past second-neighbor interactions in the central bcc metals.

The remaining pair-potential energy E_{pair} in Eq. (5.180) is usually modeled as a sum of cubic splines:

$$E_{pair}(\mathbf{R}) = \frac{1}{2N} {\sum_{i,j}}' \sum_k A_k (R_k - R_{ij})^3 \Theta(R_k - R_{ij}), \qquad (5.183)$$

where the sum over k typically extends to 4–6 terms and Θ is the usual step function. The node points R_k and coefficients A_k are treated here as adjustable parameters, which in cubic transition metals are used to fit the remaining observed shear elastic constants C_{44} and $C' = (C_{11} - C_{12})/2$. The mathematical character of the cubic splines ensures that the pair potential and its first two radial derivatives are everywhere continuous and can be constructed to vanish identically at the cutoff radius R_{cut}. In some treatments, e.g., Cák et al. (2014), an additional short-range repulsive pair potential is added to the cubic-spline sum in Eq. (5.183). This additional potential has the form $A(R_0 - R_{ij})^m/R_{ij}^n$, where A, R_0, m and n are adjustable parameters.

5.4 Inclusion of magnetism in bond-order and MGPT potentials

The $3d$ transition metals Cr, Mn, Fe, Co and Ni are all observed to be magnetic, so the extension of BOP and GPT/MGPT methodology to treat magnetism is both natural and important. As the fundamental underpinning of such QBIPs for ordinary nonmagnetic metals is DFT, the fundamental underpinning of QBIPs for magnetic metals is the corresponding spin density functional theory (SDFT; von Barth and Hedin, 1972; Jones and Gunnarsson, 1989), in which the total energy becomes a functional of both the spin-up and spin-down electron densities, $n^\uparrow(\mathbf{r})$ and $n^\downarrow(\mathbf{r})$, or equivalently, the total electron density $n(\mathbf{r}) = n^\uparrow(\mathbf{r}) + n^\downarrow(\mathbf{r})$ and the total electron spin density $m(\mathbf{r}) = n^\uparrow(\mathbf{r}) - n^\downarrow(\mathbf{r})$. The development of a first-principles SDFT version of GPT is entirely feasible, but has not yet been attempted and remains a future challenge. At the same time, the main physics of ferromagnetism in prototype transition metals like bcc Fe is already found in the classic Stoner model of itinerant magnetism (Stoner, 1938, 1939), in which the exchange interaction splits the d bands into spin-up and spin-down components, as is indicated schematically in Fig. 5.27. To a first approximation, and relative to the nonmagnetic d-band energies $E_d(\mathbf{k})$, the spin-up and spin-down d bands have respective energies

$$E_d^\uparrow(\mathbf{k}) = E_d(\mathbf{k}) - \frac{1}{2}m_d I \tag{5.184}$$

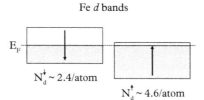

Fig. 5.27 *Schematic representation of the occupation of the spin-down and spin-up d bands in ferromagnetic bcc Fe.*

and

$$E_d^\downarrow(\mathbf{k}) = E_d(\mathbf{k}) + \frac{1}{2}m_d I, \tag{5.185}$$

where I is the effective exchange integral and m_d is the average magnetic moment,

$$m_d = N_d^\uparrow - N_d^\downarrow, \tag{5.186}$$

with $N_d^\uparrow$ the number of d electrons per atom in the spin-up bands and $N_d^\downarrow$ the number of d electrons per atom in the spin-down bands. For the observed ferromagnetic bcc α phase of Fe, $m_d = 2.2$ electrons/atom, corresponding to $N_d^\uparrow \sim 4.6$ electrons/atom and $N_d^\downarrow \sim 2.4$ electrons/atom. The theoretical Stoner criterion for a stable ferromagnetic state is $I\rho(E_F) > 1$, where $\rho(E_F)$ is the total density of states at the Fermi level.

Building on the TB Stoner model of Liu et al. (2005) and the numerical BOP formalism of Sec. 5.3.1, first Mrovec et al. (2011) and later in a refined treatment Lin et al. (2016) developed magnetic bond-order potentials for Fe. The cohesive-energy functional in these approaches is taken as a sum of bond, magnetic and repulsive energy contributions:

$$E_{\text{coh}}(\mathbf{R}) = E_{\text{bond}}(\mathbf{R}) + E_{\text{mag}}(\mathbf{R}) + E_{\text{rep}}(\mathbf{R}). \tag{5.187}$$

In the on-site representation, the bond energy given by Eq. (5.128) now becomes

$$E_{\text{bond}} = \frac{1}{N} \sum_i \sum_d \int_{-\infty}^{E_F} \left[(E - \varepsilon_d^{i\uparrow}) n_d^{i\uparrow}(E) + (E - \varepsilon_d^{i\downarrow}) n_d^{i\downarrow}(E) \right] dE, \tag{5.188}$$

where the spin-polarized band energies of Eqs. (5.184) and (5.185) become local spin-polarized site energies

$$\varepsilon_d^{i\uparrow} = \varepsilon_d^i - \frac{1}{2} m_d^i I \tag{5.189}$$

and

$$\varepsilon_d^{i\downarrow} = \varepsilon_d^i + \frac{1}{2} m_d^i I, \tag{5.190}$$

with local magnetic moments m_d^i. The corresponding LDOSs $n_d^{i\uparrow}(E)$ and $n_d^{i\downarrow}(E)$ in Eq. (5.188) are assumed to be rigidly shifted down and up, respectively, by $m_d^i I/2$, so the qualitative picture of Fig. 5.27 is preserved. The explicit presence of a local magnetic moment on each atom makes it particularly easy to configure and address any desired co-linear magnetic configuration including both ferromagnetic and antiferromagnetic ones.

In practice, both E_{bond} and the corresponding Hellman-Feynman forces for magnetic Fe can still be evaluated using the inter-site BOP representation discussed in Sec. 5.3.1. In this regard, separate bond orders need to be constructed for the spin-up and spin-down components of Eq. (5.188). This task can be accomplished in the same manner as above in Eq. (5.152), but only one set of bond integrals, $dd\sigma(R_{ij})$, $dd\pi(R_{ij})$ and $dd\delta(R_{ij})$, is required in the process because the spin-up and spin-down LDOSs are identical in shape. Mrovec et al. (2011) initially established the needed bond integrals for Fe without inclusion of the effects of nonorthogonality or sd screening, but these effects were later added in the refined Lin et al. (2016) treatment, as had been done previously for Mo and W (Mrovec et al., 2004, 2007). Also in the latter treatment, the bond integrals were fitted with a simplified version of Eq. (5.178), where $n_a = 0$ and $n_b = n_c = 1$.

The magnetic energy E_mag in Eq. (5.187) was taken in the standard second-order form

$$E_\text{mag} = -\frac{1}{4} \sum_i \left[\left(m_d^i\right)^2 - m_\text{at}^2 \right], \qquad (5.191)$$

where m_at is the corresponding magnetic moment of the free atom. The effective exchange integral was determined to be $I = 0.8$ eV by Mrovec et al. (2011), based on DFT calculations, but also allowing a small adjustment to obtain the correct magnetic behavior, and this value of I was maintained in the Lin et al. (2016) treatment as well. In both treatments, the total exchange splitting on each atom, $\Delta_d^i = m_d^i I$, was determined self consistently.

The final repulsive energy E_rep in Eq. (5.187) was evaluated empirically, as in the BOP treatment of nonmagnetic transition metals. However, in Mrovec et al. (2011) the usual pair-functional contribution to E_rep was omitted and only a pair-potential contribution was used, which consisted of Eq. (5.183) plus an ad hoc repulsive core term. This treatment was upgraded in Lin et al. (2016) to the pair-functional scheme of Lin et al. (2014) discussed in Sec. 5.3.3, allowing the complete and normal fitting of the lattice constant, cohesive energy and elastic moduli of ferromagnetic Fe. Both treatments of the full cohesive-energy functional $E_\text{coh}(\mathbf{R})$ were successfully applied to describe the structural phase stability of ferromagnetic bcc Fe, as well as important aspects of point-defect energetics and dislocation behavior.

Returning to Fig. 5.27 and the general treatment of ferromagnetism in transition metals, it has long been recognized that the simple d-band filling arguments that explain structural phase stability across the nonmagnetic transition metals, as discussed in Chapter 1, can be extended to the late magnetic transition metals Fe, Co and Ni as well. Specifically, Söderlind et al. (1994b) used FP-LMTO DFT calculations to show that, while the majority spin-up d bands in these latter metals are nearly filled and have only a small impact on structural phase stability, the partial filling of the minority spin-down d bands alone establish their respective bcc, hcp and fcc structures. This argument is especially compelling in the case of Fe, where the spin-down d bands are essentially half filled, as shown in Fig. 5.27 and is the case, of course, for mid-period bcc transition metals. This picture is particularly relevant to the MGPT and immediately suggests a simplified treatment of ferromagnetic Fe. Since the spin-up d bands are nearly filled, they are qualitatively Ni like and thus in the MGPT should be well treated at the two-ion pair-potential level and contribute only a small positive bcc-fcc energy difference to structural phase stability. The spin-down bands, on the other hand, are nearly half filled, so they are qualitatively Mo like and thus in the MGPT should be well treated at the four-ion potential level and contribute to the large negative bcc-fcc energy difference needed for bcc phase stability. Consequently, one expects an MGPT cohesive-energy functional of the useful form

$$E_{\text{coh}}(\mathbf{R},\Omega) = E_{\text{vol}}^{\uparrow}(\Omega) + E_{\text{vol}}^{\downarrow}(\Omega) + \frac{1}{2N}{\sum_{i,j}}' \left[v_2^{\uparrow}(ij,\Omega) + v_2^{\downarrow}(ij,\Omega)\right]$$

$$+ \frac{1}{6N}{\sum_{i,j,k}}' v_3^{\downarrow}(ijk,\Omega) + \frac{1}{24N}{\sum_{i,j,k,l}}' v_4^{\downarrow}(ijkl,\Omega). \qquad (5.192)$$

The development of such a treatment for ferromagnetic Fe looks entirely feasible. If successful, one would expect this approach to allow large-scale MGPT-MD simulations, as has been the case for central bcc transition metals.

Another noteworthy development has been the extension of the analytic BOP formalism to the general treatment of magnetism in transition metals by Drautz and Pettifor (2011). This includes a DFT-based theoretical roadmap that extends beyond the simple Stoner model and, in principle, has the capability to address complex antiferromagnetic and noncollinear spin configurations in addition to ferromagnetic states. The analytic BOP formalism so developed has been successfully applied to the challenging case of the six observed complex magnetic phases of Mn by Drain et al. (2014).

6

Structural Phase Stability and High-Pressure Phase Transitions

In this and subsequent chapters of this book, we consider important applications of the quantum-based interatomic potentials developed in Chapters 3, 4 and 5 for elemental nontransition metals, in the SM, EDB and FDB limits, and for transition metals in the PFDB limit. In this chapter, we address the basic subject of $T = 0$ structural phase stability, including as a fundamental test of prototype QBIPs, calculation of the ground-state crystal structure amongst various competing alternate structures. In this process, we also investigate total-energy variations along important deformation paths connecting different higher-energy structures, and the prediction of pressure-induced solid-solid phase transitions. Extension of these considerations to finite temperature, including temperature-induced solid-solid phase transitions, melting and the prediction of pressure-temperature phase diagrams is addressed in Chapters 7 and 8.

6.1 Useful basic concepts and computational tools

6.1.1 Separation of cohesion and structure

In the case of volume-dependent QBIPs, such as those provided by the GPT, MGPT, DRT and RMP methods discussed in Chapters 3–5, one can always represent the cohesive-energy functional $E_{\text{coh}}(\mathbf{R}, \Omega)$ of a metal in the universal form of a large volume term $E_{\text{vol}}(\Omega)$, which provides the majority of the metallic cohesion at normal density, plus a generally smaller structural energy contribution $E_{\text{struc}}(\mathbf{R}, \Omega)$ involving pair and multi-ion potentials, which alone determines the crystal structure at constant volume:

$$E_{\text{coh}}(\mathbf{R}, \Omega) = E_{\text{vol}}(\Omega) + E_{\text{struc}}(\mathbf{R}, \Omega), \tag{6.1}$$

where in general for any elemental metal

$$E_{\text{struc}}(\mathbf{R}, \Omega) = \frac{1}{2N} {\sum_{i,j}}' v_2(ij, \Omega) + \frac{1}{6N} {\sum_{i,j,k}}' v_3(ijk, \Omega) + \frac{1}{24N} {\sum_{i,j,k,l}}' v_4(ijkl, \Omega) + \cdots. \tag{6.2}$$

This separation of cohesion and structure is a useful basic property of both nontransition and transition metals alike. Calculation of the cohesion curve, i.e., E_{coh} vs. Ω for a given structural phase of the material, requires, of course, a robust volume term E_{vol}, and this is provided in the GPT and MGPT methods. Calculation of the $T = 0$ crystal structure for a material at a given volume, however, requires only an adequate knowledge of the volume-dependent interatomic potentials v_2, v_3, $v_4 \cdots$, something that quality QBIPs can provide, including those from the DRT and RMP methods, as well as from the GPT and MGPT methods. The direct isolation of a small structural energy E_{struc} is especially advantageous in metals, where structural energy differences between competing metallic structures can be two to three orders of magnitude smaller than the cohesive energy itself. This small energy scale is demonstrated in Table 6.1 for ten prototype simple and transition metals, which have been treated at normal density by GPT and MGPT calculations in the appropriate SM, EDB, FDB and TM limits. For each metal, the calculated cohesive energy, volume term and the structural energy E_{struc} of the observed phase are given in the table together with the predicted fcc-bcc energy difference

$$\Delta E_{struc} \equiv E_{struc}^{fcc} - E_{struc}^{bcc}. \tag{6.3}$$

As shown in the companion result, Fig. 6.1, the equilibrium cohesive energy itself for each element in Table 6.1 is well calculated and captures the near order of magnitude increase in E_{coh} between simple metals like K and central transition metals like Ta. But the systematics of structural phase stability among these ten elements is not directly revealed by the variation of E_{coh} from element to element. Nor is such systematics even revealed in

Table 6.1 GPT/MGPT calculated cohesive energy E_{coh} and its volume E_{vol} and structure E_{struc} components for ten prototype metals at equilibrium conditions, with $\Omega = \Omega_0$. The quantity ΔE_{struc} is the calculated fcc-bcc structural energy difference, as given by Eq. (6.3). Here E_{coh}, E_{vol} and E_{struc} are expressed in Ry, while ΔE_{struc} is expressed in mRy.

Metal	Phase	Treatment	E_{coh}	E_{vol}	E_{struc}	$|\Delta E_{struc}|$	$|\Delta E_{struc}/E_{coh}|$
Na	bcc	GPT: SM	−0.0985	−0.0883	−0.0101	0.057	0.0006 0.06%
K	bcc	GPT: SM	−0.0752	−0.0687	−0.0065	0.034	0.0004 0.04%
Mg	hcp	GPT: SM	−0.1171	−0.1118	−0.0053	1.278	0.0109 1.1%
Ca	fcc	GPT: EDB	−0.1323	−0.0920	−0.0404	1.271	0.0096 1.0%
Al	fcc	GPT: EDB	−0.2842	−0.3116	+0.0275	7.322	0.0258 2.6%
Cu	fcc	GPT: FDB	−0.2663	−0.2893	+0.0230	2.520	0.0095 1.0%
Zn	hcp	GPT: FDB	−0.0976	−0.1485	+0.0509	5.639	0.0578 5.8%
V	bcc	MGPT: TM	−0.3900	−0.3893	−0.0007	12.54	0.0322 3.2%
Mo	bcc	MGPT: TM	−0.5010	−0.4570	−0.0440	30.64	0.0611 6.1%
Ta	bcc	MGPT: TM	−0.5953	−0.4853	−0.1100	13.23	0.0222 2.2%

6.1.1 Separation of cohesion and structure

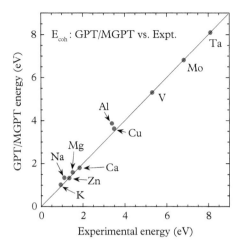

Fig. 6.1 *GPT/MGPT calculations of the equilibrium cohesive energy E_{coh} for ten prototype metals in their observed crystal structures at $\Omega = \Omega_0$, as also treated in Table 6.1, and here compared with experiment. The GPT results for Na, K, Mg, Ca, Al, Cu and Zn represent first-principles calculations, while in the MGPT results for V, Mo and Ta, the value of E_{coh} is constrained to experiment.*

the behavior of E_{struc} itself, which varies in magnitude and sign somewhat randomly from element to element, with the group-VB transition metals V and Ta possessing both the smallest and largest values. The magnitude of the structural energy difference $|\Delta E_{\text{struc}}|$, on the other hand, is neatly ordered into three well-defined categories: the alkali metals Na and K, which have exceptionally small values $|\Delta E_{\text{struc}}| \sim 0.03 - 0.06$ mRy; the central transition metals V, Mo and Ta with relatively large values $|\Delta E_{\text{struc}}| \sim 10 - 35$ mRy; and the remaining metals Mg, Ca, Al, Cu and Zn with smaller values $|\Delta E_{\text{struc}}| \sim 1 - 8$ mRy. But as shown in the final column of Table 6.1, if ΔE_{struc} is normalized by the cohesive energy, the latter two categories then completely overlap and narrow in range, with $|\Delta E_{\text{struc}}/E_{\text{coh}}| \sim 1 - 6\%$ for all eight metals. Thus, apart from the unique situation in the alkali metals, we expect the magnitude of structural energy differences relative to the cohesive energy to be similar in all remaining metals.

More generally, QBIP calculation of specific structural-energy differences can be used as a fundamental test of theory in cases where comparison against experiment and/or accurate DFT calculation is possible. Useful experimental information in this regard may be available in the form of observed pressure- or temperature-induced phase transitions. DFT calculation of structural-energy differences is readily possible, but usually proceeds by direct total-energy subtraction between two large numbers, without any separation of cohesion and structure, and is consequently limited by issues of numerical precision. Near equilibrium conditions, typical DFT structural-energy uncertainties in metals are in the range $0.1 - 1.0$ mRy/atom. This level of accuracy is adequate for most transition metals and many nontransition metals as well, but extremely challenging for cases like Na and K.

Simplification of DFT structural-energy calculations is possible, however, by use of the so-called Andersen force theorem (Mackintosh and Andersen, 1980). In this approach, one freezes the self-consistent electron potential at its value for one of the two structures to be compared, so that all but the valence band-structure and electrostatic contributions to the DFT total energies drop out of the energy difference. Thus, for example, ΔE_{struc} for the fcc-bcc structural energy difference would be given by

$$\Delta E_{\text{struc}} = \int_0^{E_F} E\rho_{\text{fcc}}(E)dE - \int_0^{E_F} E\rho_{\text{bcc}}(E)dE + \Delta(\delta E_{\text{es}}), \qquad (6.4)$$

where the fcc DOS ρ_{fcc} and the bcc DOS ρ_{bcc} are to be calculated from the same self-consistent electron potential. In Eq. (6.4), the electrostatic energy correction δE_{es} for each structure is defined in analogy with Eq. (3.21) as

$$\delta E_{\text{es}} \equiv \frac{1}{2}[n(R_{\text{WS}})\Omega e]^2 \frac{1.8 - \alpha_{\text{es}}}{R_{\text{WS}}}, \qquad (6.5)$$

where $n(R_{\text{WS}})$ is the calculated electron density at the atomic-sphere radius. The Andersen force theorem does not separate cohesion and structure, but it does reduce structural energy difference to subtraction between two smaller quantities, and hence improves precision, but not necessarily overall quantitative accuracy. Skriver (1985) showed that this approach could qualitatively explain structural phase stability in metals across most of the Periodic Table, but he did comment that the method does tend to overestimate the magnitude of structural energy differences, so the quantitative challenges with the alkali metals and other systems remain.

On the other hand, robust volume-dependent QBIPs such as GPT and DRT potentials that do separate cohesion and structure are not precision limited and can produce meaningful results even for Na and K that can be compared with experiment, as we discuss further in Sec. 6.2. In addition, for prototype metals such as those listed in Table 6.1, calculated GPT and MGPT structural-energy differences are generally found to be reliable over wide ranges of volume, as we have already demonstrated for the case of Mg fcc-hcp and bcc-hcp energies in Fig. 3.9(b), where favorable comparison with both experiment and DFT is achieved. Moreover, because of the expected transferability of GPT and MGPT QBIPs, this reliability extends to complex structures as well, and we give a number of additional examples for both simple and complex structures in Secs. 6.2 and 6.3.

Other important QBIPs, but ones that do not separate cohesion and structure, such as transition-metal BOPs, require full total-energy subtraction, or use of the Andersen force theorem, to obtain structural energy differences. Consequently, both numerical and analytic BOP calculations of structural-energy differences have similar numerical precision issues as in the case of DFT calculations. One thus expects nominal structural-energy uncertainties on the order of 0.1 − 1.0 mRy/atom, which normally present no significant problem for mid-period transition metals.

6.1.2 Transformation paths connecting multiple structures

Calculation of a particular structural-energy difference under equilibrium environmental conditions provides only a single number and hence only limited insight about structural phase stability in general. Considerably more insight can be gained by performing such calculations as a continuous function of an environmental variable such as atomic volume, or at fixed volume, as a continuous function of an appropriate strain parameter that transforms one structure into another. The former exercise explores structural behavior under pressure, an important subject we shall return to in Sec. 6.3. The latter exercise can be used to establish a series of possible phase transformation paths connecting a family of two or more structures, while at the same time, giving one useful information about the mechanical and thermodynamic stability of each phase on the path.

The simplest and often the most useful of the possible constant-volume transformation paths is one that connects the bcc and fcc structures along the so-called Bain deformation path (Bain, 1924). The Bain path is established by applying a homogeneous tetragonal strain to the bcc structure, which in turn generates the more general body-centered tetragonal (bct) family of structures, with lattice parameters a and c, where the c/a axial ratio is the effective strain parameter s. If a is the lattice parameter in the [100] and [110] directions, and c is the lattice parameter in the [001] direction, then the Lagrange strain tensor for finite deformation along the Bain path (Paidar et al., 1999) has components

$$\varepsilon_{11} = \varepsilon_{22} = (s^{-2/3} - 1)/2$$
$$\varepsilon_{33} = (s^{4/3} - 1)/2 \qquad (6.6)$$
$$\varepsilon_{12} = \varepsilon_{13} = \varepsilon_{23} = 0.$$

For $s = c/a = 1$ one has the starting bcc structure, while for $s = c/a = \sqrt{2}$ one has transformed to the fcc structure.

In the regime of most physical interest, $0.5 \leq c/a \leq 2.0$, there are normally at least three energy minima and maxima along the Bain path for metals, and in the most common situations two of the three extrema correspond to the bcc and fcc structures, while the third is a more general bct structure. In addition, the energy variation around each minimum or maximum is that which establishes the tetragonal shear elastic constant

$$C' = (C_{11} - C_{12})/2 \qquad (6.7)$$

for the corresponding structure (see Chapter 7). Thus, the positive curvature of the Bain path at an energy minimum yields a positive C' and quite often a mechanically stable structure. Conversely, the negative curvature of the Bain path at an energy maximum yields a negative C' and hence a mechanically unstable structure.

Examples of common behaviors along the Bain path in nontransition and transition metals are illustrated in Figs. 6.2 and 6.3. In Fig. 6.2 we compare and contrast the Bain paths of the group-IIA metals Mg and Ca, as calculated from first-principles GPT potentials. In both metals the overall shape of the Bain path is similar, with a mechanically stable fcc structure of lowest energy and a bcc structure about 1 mRy higher in energy. In the case of Ca, the fcc structure is calculated to be the lowest-energy phase overall in agreement with experiment, a result driven by the strong sp-d hybridization present in that EDB metal, as discussed in Chapter 4. In contrast, Mg is well described in the GPT as a prototype simple metal with both a predicted and observed ground-state hcp structure. Also, for Ca note from the inset in Fig. 6.2 that the bcc structure as well is predicted to be mechanically stable, but that the bcc minimum is on a sub-0.01 mRy energy scale and thus corresponds to a relatively small positive value of the C' shear elastic modulus. This behavior is qualitatively consistent with soft bcc phonons in Ca and the observed temperature-induced fcc $\rightarrow$ bcc phase transition in that metal, a transition that is discussed in Chapter 7. Again, in contrast to the behavior found in Ca, the normal-density bcc structure in Mg is mechanically unstable, which leads to only a predicted bct minimum at $c/a \cong 0.95$. Under modest pressure, however, bcc Mg is mechanically stabilized, and at a high pressure of ~ 50 GPa there is both a predicted and observed hcp $\rightarrow$ bcc phase transition, as previously mentioned in connection with Fig. 3.9(a) in Chapter 3, and as discussed further in Sec. 6.3.2.

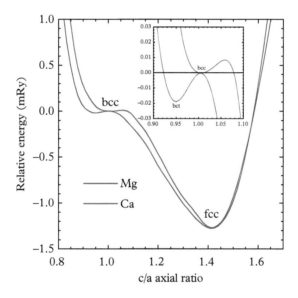

Fig. 6.2 *Predicted Bain path for prototype group-IIA metals Mg and Ca at normal density, as obtained from first-principles GPT calculations in the SM and EDB limits, respectively.*

Additional typical Bain paths for both transition and nontransition metals are illustrated in Fig. 6.3. In Fig. 6.3(a), we display prototype MGPT results for the central bcc transition metals Ta and Mo. The Bain path in these metals is characterized by a deep bcc energy minimum and corresponding high mechanical and thermodynamic stability, with large positive values of C', together with a high-energy, mechanically unstable fcc structure and a nearby metastable bct structure with $c/a \simeq 1.5 - 1.7$. This behavior is generic in group-VB and -VIB transition metals, and is obtained in MGPT calculations on V as well, in the DFT and numerical BOP calculations on V, Nb, Ta, Cr, Mo and W

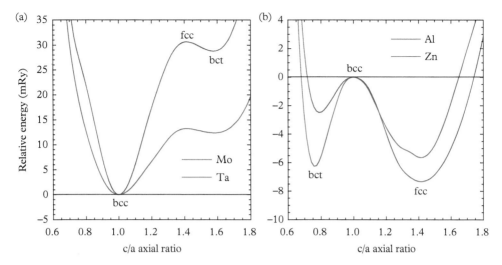

Fig. 6.3 *Calculated Bain path for prototype transition and nontransition metals at normal density. (a) Group-VB and -VIB transition metals Ta and Mo, as obtained from optimized MGPT calculations; (b) group-IIB and IIIA metals Zn and Al, as obtained from first-principles GPT calculations in the FDB and EDB limits, respectively.*

by Lin et al. (2014), and in the DFT and analytic BOP calculations on Nb, Ta, Mo and W by Cák et al. (2014). In addition, for all the bcc metals there is a strong correlation between the magnitude of the C' shear modulus and the magnitude of the fcc-bcc energy difference $|\Delta E_{\text{struc}}|$. The physical origin of this correlation is d-band filling, which dominantly impacts both quantities, with the values of C' and $|\Delta E_{\text{struc}}|$ significantly larger in the VIB metals than the VB metals, as can be appreciated from Fig. 6.3(a). More generally, Wills et al. (1992) showed via DFT calculations that in the $5d$ transition metals this correlation extends not only to the bcc metals (Ta and W), but to the mechanically stable fcc structure of all remaining metals (La, Hf, Re, Os, Ir, Pt and Au) as well.

Figure 6.3(b) illustrates the Bain path for the group-IIB metal Zn and the group-IIIA metal Al, as obtained from first-principles GPT calculations. The Al path is qualitatively similar to that of Mg, with a stable fcc structure of lowest energy, an unstable bcc structure and a metastable bct structure with $c/a \cong 0.79$. Unlike Mg, however, fcc is calculated to be the lowest-energy structure of Al overall, in agreement with experiment. The Bain path in Zn is similar to that of Al, but with two interesting differences. First, the bct minimum in Zn at $c/a \cong 0.76$ is now the lowest-energy minimum along the entire Bain bath, although hcp with a high c/a ratio is both the predicted and observed equilibrium structure of Zn. Second, one finds a small "wiggle" in the Bain path at $c/a = 1.3 - 1.4$ as one approaches the fcc minimum, suggesting that one is near the onset of mechanical instability in the fcc structure and the emergence of a second bct structure. These effects can be regarded as precursors to the behavior that controls structural phase stability in the $5d$ group-IIB metal Hg, whose observed low-temperature phase is bct with $c/a = 1/\sqrt{2} = 0.7071$ and is known as the β-Hg structure. A full GPT treatment of structural phase stability in Hg at both ambient and high pressure was published by Moriarty (1988b). Highlights and refinements of that work are discussed in Sec. 6.3.2.

With regard to structural phase stability in Hg and other metals, a second interesting constant-volume transformation path is the trigonal path that connects the bcc, simple cubic (sc) and fcc structures by a homogeneous deformation corresponding to the uniform extension of the [111] crystal axis. This deformation path is formally described by the Lagrange strain tensor given by Eq. (6.6), except with the x, y and z axes parallel to $[1\bar{1}0]$, $[11\bar{2}]$ and [111] directions, respectively, and where the strain parameter s is 1 for bcc, 2 for sc and 4 for fcc (Paidar et al., 1999). More generally, the trigonal deformation path generates the simple rhombohedral (rhom) family of structures, which are more easily described by the standard rhombohedral angle θ, with $\theta = 60°$ for the fcc structure, $\theta = 90°$ for the sc structure and $\theta = 109.5°$ for the bcc structure. Also included in the rhom family of structures is the observed high-temperature phase of Hg at $\theta = 70.5°$, which is known as the α-Hg structure.

Additional constant-volume transformation paths and/or families of structures of interest can also be defined. A second tetragonal path, starting with the diamond structure rather than bcc, yields the so-called beta-tin family of structures as a function of the c/a axial ratio. This family includes both the ideal and observed form of the β-Sn structure for $c/a = \sqrt{2/15} = 0.3651$ and 0.384, respectively, the diamond structure for $c/a = 1.0$ and the ideal and observed high-pressure Cs-IV structure for $c/a = \sqrt{14} = 3.6417$ and 3.724, respectively. Also of interest are transformation paths between cubic and hexagonal structures, but these paths are always more complex and require "shuffles" of atomic positions in addition to homogeneous deformation. The classic Burgers path between bcc and hcp (Burgers, 1934) attempts to model the actual bcc → hcp phase transition observed in Zr at ambient pressure. This model contains three main elements: a shear of the bcc structure to create a lattice of hexagonal symmetry; a shift in position of the atoms in every second basal plane of the hexagonal lattice to create the hcp structure; and a homogeneous contraction of the hcp structure to match the observed change in volume for Zr. Paidar et al. (1999) have defined a strain tensor and constant-volume transformation path connecting bcc with hcp that captures the first two elements of the Burgers path.

6.1.3 Simplified calculation of the total enthalpy difference between two structures at finite pressure

The convenience of treating phase stability and phase transitions at constant volume in elemental metals can be readily extended to finite pressure by capitalizing on the fact that the volume difference $\Delta\Omega$ between any two metallic phases that are held at constant temperature and pressure is small. Typically, at the average volume $\bar{\Omega}$ of a solid-solid phase transition, $\Delta\Omega/\bar{\Omega}$ is 1% or less. For solid-solid phase transitions at $T = 0$, we can establish the specific simplifications that a small $\Delta\Omega/\bar{\Omega}$ allows in the following manner.

6.1.3 Simplified calculation of the total enthalpy difference between two structures at finite pressure

At $T = 0$ and constant pressure P in an elemental metal, the thermodynamically most stable phase is the one that has the lowest total enthalpy per ion H, as given by

$$H(P) = E(\Omega) + P\Omega, \tag{6.8}$$

where E is the corresponding total energy per ion, or in our present QBIP nomenclature the cohesive energy E_{coh}. At the phase-transition boundary separating an initial phase i from a final phase f, one must thus have $H_f = H_i$, so that

$$\Delta H = \Delta E + P\Delta\Omega = 0, \tag{6.9}$$

where $\Delta H = H_f(P) - H_i(P)$, $\Delta E = E_f(\Omega_f) - E_i(\Omega_i)$ and $\Delta\Omega = \Omega_f - \Omega_i$. For $\Delta H > 0$ the initial phase i is stable, while for $\Delta H < 0$ the final phase f is stable.

Next, one can usefully simplify Eq. (6.9) for small $\Delta\Omega$. Assuming only that $E_i(\Omega)$ and $E_f(\Omega)$ are continuous and slowly varying functions of volume Ω, one can then Taylor expand $E_i(\Omega)$ and $E_f(\Omega)$ about the average volume of the phase transition $\bar{\Omega}$:

$$E_i(\Omega_i) = E_i(\bar{\Omega}) + \frac{dE_i(\bar{\Omega})}{d\Omega}(\bar{\Omega} - \Omega_i) + \frac{1}{2}\frac{d^2E_i(\bar{\Omega})}{d\Omega^2}(\bar{\Omega} - \Omega_i)^2 + \cdots$$
$$= E_i(\bar{\Omega}) - P(\bar{\Omega} - \Omega_i) + O\left((\Delta\Omega)^2\right) \tag{6.10}$$

and

$$E_f(\Omega_f) = E_f(\bar{\Omega}) + \frac{dE_f(\bar{\Omega})}{d\Omega}(\bar{\Omega} - \Omega_f) + \frac{1}{2}\frac{d^2E_f(\bar{\Omega})}{d\Omega^2}(\bar{\Omega} - \Omega_f)^2 + \cdots$$
$$= E_f(\bar{\Omega}) - P(\bar{\Omega} - \Omega_f) + O\left((\Delta\Omega)^2\right), \tag{6.11}$$

where we have set $P \cong P_i \cong P_f$ and further noted that $(\Delta\Omega)^2 \cong 4(\bar{\Omega} - \Omega_i)^2 \cong 4(\bar{\Omega} - \Omega_f)^2$. Subtracting Eq. (6.10) from Eq. (6.11) then gives a result for ΔE of the form

$$\Delta E = E_f(\bar{\Omega}) - E_i(\bar{\Omega}) - P\Delta\Omega + O\left((\Delta\Omega)^2\right). \tag{6.12}$$

Neglecting the second-order correction terms in Eq. (6.12) of order $(\Delta\Omega)^2$ and substituting the remaining terms into Eq. (6.9) cancels the $P\Delta\Omega$ contribution and gives the desired simplified result for locating a finite-pressure phase boundary with transition volume $\Omega_T = \bar{\Omega}$:

$$\Delta H = E_f(\Omega_T) - E_i(\Omega_T) = 0. \tag{6.13}$$

This result is valid to first order in $\Delta\Omega$. In Eq. (6.13) the pressure P has dropped out completely, and the location of the phase boundary is established only by the intersection of the total-energy functions $E_i(\Omega)$ and $E_f(\Omega)$, which determines Ω_T. As a final step, the average transition pressure $P_T = \bar{P}$ can be calculated as

$$P_T(\Omega_T) = -\frac{1}{2}\left(\frac{dE_i(\Omega_T)}{d\Omega} + \frac{dE_f(\Omega_T)}{d\Omega}\right). \qquad (6.14)$$

In our present QBIP nomenclature, E_i is replaced by E_{coh}^i and E_f by E_{coh}^f in Eqs. (6.13) and (6.14). Also, for volume-dependent QBIPs obeying Eqs. (6.1) and (6.2), the volume term $E_{vol}(\Omega)$ drops out of Eq. (6.13), and one is left with

$$\Delta H = E_{struc}^f(\Omega_T) - E_{struc}^i(\Omega_T) = 0. \qquad (6.15)$$

In this case, the volume term still makes the major contribution to the transition pressure, however, and Eq. (6.14) becomes

$$P_T(\Omega_T) = -\frac{dE_{vol}(\Omega_T)}{d\Omega} - \frac{1}{2}\left(\frac{dE_{struc}^i(\Omega_T)}{d\Omega} + \frac{dE_{struc}^f(\Omega_T)}{d\Omega}\right). \qquad (6.16)$$

In practice, Eqs. (6.13)–(6.16) represent a convenient and quantitatively reliable approach to address $T = 0$ phase stability and phase transitions in metals at high pressure. These results can also be readily generalized to finite temperature in terms of appropriate free energies, as we discuss in Chapter 7.

6.2 QBIP-predicted structures and structural energies of the elements

With the aid of the available good-quality QBIPs discussed in Chapters 3–5 and the simplifying computational tools discussed in Sec. 6.1, we are now in a position to address the question of structural phase stability in the Periodic Table more systematically. In this section, we consider and test $T = 0$ QBIP predictions of equilibrium crystal structures and structural energies in elemental nontransition and transition metals near ambient pressure conditions.

6.2.1 Nontransition metals

In Table 6.2 we summarize a selected blend of the most significant historical and current QBIP calculations of crystal structure in the traditional nontransition metals, including the group-IA alkalis, the group-IIA alkaline earths, the group-IB noble metals, and the group-IIB divalent and -IIIA trivalent metals. As we discussed in Chapter 3, with the possible exception of Li, the alkali metals are well treated as NFE simple metals. In the prototype elements Na and K, the spread in energy amongst fcc, hcp and bcc structures is found to be less than 0.07 mRy and all three structures are calculated to be mechanically stable with both GPT and DRT potentials. The same is true for DRT Rb and

Cs. In Li, on the other hand, the spread in structural energies increases to 0.5 mRy and the fcc structure is calculated to be mechanically unstable with both GPT and DRT potentials. In the simple-metal limit of GPT, Li is calculated to be stable in both the hcp and bcc structures, with the hcp structure of lowest overall energy. However, the calculated bcc Debye temperature is some 35% too high, leading to an over-predicted temperature-induced hcp → bcc transition at ∼ 400 K. This result is to be compared with the observed martensitic bcc → cp transition in Li at ∼ 75 K (Young, 1991). In DRT Li, hcp is also calculated to be mechanically stable and the structure of lowest energy, but the bcc structure is mechanically unstable with a small negative value of the C' shear elastic constant. In contrast, the bcc elastic moduli and phonon spectrum in both GPT and DRT Na are well calculated, and in the case of GPT Na the estimated hcp → bcc transition temperature is only ∼ 100 K. This latter value is reasonably consistent with the observed martensitic bcc → cp transition in Na at ∼ 35 K (Young, 1991), especially considering the large hysteresis in temperature observed for the reverse cp → bcc transition, which begins at ∼ 45 K and ends at ∼ 70 K (Martin, 1960). A similar low-temperature hcp → bcc transition is estimated to occur in GPT potassium at ∼ 70 K, and corresponding cp → bcc transitions are expected in GPT Rb and Cs, and in DRT K, Rb and Cs at even lower temperatures. However, at ambient pressure no structure other than bcc has even been observed in the heavy alkali metals above 5 K.

The observed low-temperature close-packed structures of Li and Na were originally identified to be hcp, but subsequent neutron-diffraction studies in the 1980s suggested that the actual cp structure was instead the same as that found in the rare-earth metal Sm (Young, 1991). However, in each case the observed phase had a significant concentration of stacking faults, so this may not be a completely settled question. The Sm structure, which is often denoted as 9R, together with the double hcp (dhcp) structure found in the rare-earth metals and the additional triple hcp (thcp) structure, are differentiated from hcp and fcc by their stacking sequences in close-packed hexagonal planes. Hcp and fcc have respective two-layer AB... and three-layer ABC... stacking, while dhcp, thcp and 9R have four-layer ABAC..., six-layer ABCACB... and nine-layer ABCBCACAB... stacking sequences, respectively. The "R" denotes the rhombohedral primitive cell of the 9R structure. With the same first-principles GPT approach used in Table 6.2 for Li, Na and K, we have also calculated structural energies for the ideal dhcp, thcp and 9R structures in each metal. In all three metals the ordering of the calculated structural energies is from highest to lowest: fcc, thcp, 9R, dhcp and hcp, with a very small energy spread of 0.025 mRy for Li, 0.008 mRy for Na and 0.007 mRy for K amongst the five structures. The calculated 9R-hcp structural energy differences are 0.011 mRy for Li, 0.004 mRy for Na and 0.003 mRy for K. These latter energy differences, however, are probably too small to be considered significant.

As also shown in Table 6.2, for applicable alkaline-earth, noble and group-IIB metals the equilibrium crystal structures are well calculated with the GPT, DRT and RMP potentials. Even in the difficult case of Be, with its strong nonlocal PP discussed in Chapter 3, the simple-metal GPT treatment produces a mechanically stable hcp structure of lowest energy with a c/a axial ratio less than ideal, in agreement with observation. At the same time, calculated elastic constants and phonon frequencies in Be are generally

Table 6.2 *QBIP-calculated equilibrium crystal structures in nontransition metals from GPT, DRT and RMP potentials, as compared with experimental observation. Temperatures T_{QBIP} and T_{expt} are given in K. Values in parentheses are c/a axial ratios. Experimental data are from Pearson (1967), except for Li and Na, which are from Young (1991), and for Zn, which are from Almqvist and Stedman (1971).*

Metal	T_{QBIP}	GPT treatment	GPT	DRT[a]	RMP	Experiment	T_{expt}
Li	0	SM	hcp(1.63)[a]	hcp(1.63)	–	cp/bcc	75</300
Na	0/300	SM	hcp(1.63)/bcc[a]	hcp(1.63)/bcc	–	cp/bcc	35</300
K	0/300	SM	hcp(1.63)/bcc[a]	fcc/bcc	–	bcc	5
Rb	0/300	SM	hcp(1.63)/bcc[a]	fcc/bcc	–	bcc	5
Cs	0/300	SM	hcp(1.63)/bcc[a]	hcp(1.63)/bcc	–	bcc	5
Be	0	SM	hcp(1.60)[a]	–	–	hcp(1.57)	~300
Mg	0	SM	hcp(1.62)[a,b]	hcp(1.63)	–	hcp(1.62)	298
Ca	0	EDB	fcc[a,c]	–	fcc[e]	fcc	299
Sr	0	EDB	fcc[a,c]	–	–	fcc	298
Ba	0	TM	bcc[d]	–	–	bcc	5
Cu	0	FDB	fcc[a]	–	fcc[e]	fcc	293
Ag	0	FDB	fcc[a]	–	fcc[e]	fcc	291
Au	0	FDB	fcc[a]	–	fcc[e]	fcc	291
Zn	0	FDB	hcp(1.76)[a]	–	hcp(1.70)[f]	hcp(1.83)	80
Cd	0	FDB	hcp(1.71)[a]	–	hcp(1.72)[f]	hcp(1.89)	294
Hg	0	FDB	bct(0.69)[g]	–	–	bct(0.707)	77
Al	0	EDB	fcc[a]	fcc	–	fcc	298
Ga	0	FDB	β-Ga[a]	–	–	α-Ga	4.2
In	0	FDB	fcc[a]	–	–	fct(1.08)	4.2
Tl	0	FDB	bcc[a]	–	–	hcp(1.60)	291

[a] Present calculation, with VWN exchange-correlation and IU electron screening.
[b] Althoff et al. (1993).
[c] Moriarty (1973).
[d] Moriarty (1986b).
[e] Dagens (1977a, 1977b).
[f] Upadhyaya and Dagens (1982b).
[g] Moriarty (1988b).

overestimated in the GPT SM limit, with the calculated Debye temperature some 25% too high. In contrast, Mg is the prototype NFE hcp simple metal with its comparatively weak nonlocal PP, allowing for an accurate quantitative treatment of a wide range of properties including structural phase stability in both the GPT and DRT formalisms. In the remaining d-band metals, it is sp-d hybridization, d-state overlap and partial d-band occupation interactions that control structural phase stability, as we have discussed in Chapter 4. In the EDB metals Ca and Sr, sp-d hybridization is primarily responsible for their observed fcc structure instead of the hcp structure assumed by Be and Mg. In the FDB metals Zn and Cd, sp-d hybridization plus d-state overlap favor an hcp structure with a high c/a axial ratio over both ideal hcp and the metastable bct structure encountered along the Bain path, as shown for zinc in Fig. 6.3(b). In the heavy $5d$ FDB metal Hg, however, relativistic energy shifts reverse that balance to favor bct over hcp (Moriarty, 1988b). In the corresponding alkaline-earth metal Ba, additional partial d-band occupation, arising from the bottom of the $5d$ band dropping below the Fermi level, results in PFDB transition-metal behavior and additional angular forces, with necessary two- and three-ion GPT potentials, as discussed in Chapter 5. The TM representation of Ba correctly favors the observed bcc structure over the hcp and fcc structures in the lighter alkaline-earth metals (Moriarty, 1986b). In the alternate RMP approach, the fcc structure of Ca and the hcp structures of Zn and Cd are also correctly calculated. Finally, the modified FDB GPT treatment of the noble metals, as discussed in Chapter 4, stabilizes the observed fcc structure of these materials, as does the alternate RMP treatment.

For the additional group-IIIA elements covered in Table 6.2, the observed fcc phase stability of the prototype trivalent metal Al is accurately described by both GPT and DRT potentials. The structural phase stabilities of the remaining IIIA elements, Ga, In and Tl however, have always been challenging to explain in terms of conventional metals physics. The $3d$ element Ga assumes a unique orthorhombic structure, usually denoted as α-Ga or Ga I, which has only one nearest neighbor at a very short distance of $1.46 R_{\mathrm{WS}}$. Physically, this single nearest neighbor corresponds to a Ga_2 dimer, so that the α-Ga structure can be characterized as a weakly metallic, low-density molecular-like crystal, a point of view confirmed by DFT calculations (Gong et al., 1991). In early PP research, however, Heine (1968) and Inglesfield (1968) tried to explain the α-Ga phase in terms of local second-order PP perturbation theory, using an empirically adjusted PP. Not surprisingly, this treatment required an effective pair potential with a deep minimum in the vicinity of the nearest-neighbor distance. But first-principles DFT-based theory suggests that such a potential is not very realistic for an NFE metal. Indeed, as illustrated in Fig. 6.4(a), the GPT pair potentials of all the group-IIIA metals are repulsive below $r = 2.0 R_{\mathrm{WS}}$, and only Al has even a local minimum below that point. Thus the α-Ga structure cannot be explained in either the SM or FDB limits of the GPT. Possibly this outcome could be altered by adding appropriate localized p states to the GPT basis states to accommodate the Ga_2 covalent bonding, but that has yet to be attempted.

Some additional insight into structural phase stability in Ga can be gained by noting that the low-density α-Ga structure results in a melting curve with a negative slope as a function of pressure. Thermodynamics tells us that liquid Ga thus has a higher density than that of solid α-Ga. Inglesfield (1968) noted that experimentally there is a second, metastable Ga phase, denoted as β-Ga, which is obtained by rapidly quenching the liquid to low temperature. The originally determined structure of β-Ga was also orthorhombic, but with two nearest neighbors at a more metallic-like distance of $1.63 R_{WS}$. Although β-Ga does not appear in the observed equilibrium phase diagram, we find with FDB GPT calculations that this structure has the lowest energy of nine candidate structures for Ga, including α-Ga and fcc, as indicated in Table 6.2. The structure of β-Ga was later refined with a more accurate XRD analysis to a monoclinic form (Bosio and Defrain, 1969), but we did not repeat the GPT calculation with this refinement.

The observed face-centered tetragonal (fct) structure of the $4d$ trivalent metal In is also a long-standing puzzle. This is a distorted fcc structure that can be formed by a small 8% elongation of one of the cubic axes of fcc, such that fct has a c/a axial ratio of 1.08 relative to fcc, which is equivalent to bct with $c/a = 1.53$. In this case, GPT in the FDB limit goes a long way towards explaining this behavior, as can be seen by examining the Bain path. Except in the vicinity of fcc, the Bain path in In is well converged in real space using a standard smooth potential cutoff scheme discussed in Chapter 3, as shown in Fig. 6.4(b). Near the bct value of $c/a = \sqrt{2} \cong 1.41$ for the fcc structure, however, the effective range of the potential extends to infinity, requiring an alternate reciprocal-space treatment of the energy to obtain good convergence. The latter treatment yields a nearly flat energy minimum, centered at fcc, but extending from about $c/a = 1.27$ to about $c/a = 1.55$, as can also be seen in Fig. 6.4(b). The flatness of this minimum results in an fcc C' shear elastic constant that is near zero, which points to the onset of fcc instability. This behavior appears to be driven by the significant spatial extent of the $4d$ basis states in In, as is captured in the FDB GPT treatment. In contrast, the corresponding SM GPT treatment of In produces a more normal rounded minimum in the Bain path at

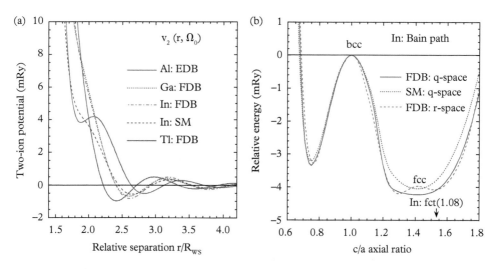

Fig. 6.4 *(a) First-principles GPT pair potentials for the group-IIIA metals calculated in EDB, FDB and SM limits, as indicated. (b) GPT Bain path for In calculated to convergence in reciprocal space (q-space) in the FDB and SM limits and in real space (r-space) using the smooth potential cutoff scheme with $R_{cut}/R_{WS} = 8.5$ defined in Table 3.2 of Chapter 3.*

fcc, as further shown in Fig. 6.4(b). Finally, it is interesting to note that the real-space FDB treatment of the Bain path displayed in Fig. 6.4(b), although not fully converged, resolves the fcc instability by producing a local maximum at fcc and local minima on either side. The minimum at higher c/a is actually quite close to that of the observed fct structure for In.

Lastly, the $5d$ trivalent metal Tl is observed to have yet a fourth different equilibrium structure for the group-IIIA elements, which is hcp. With a semi-relativistic FDB GPT treatment, however, a bcc ground state is calculated instead, although hcp is only about 0.6 mRy higher in energy. There is an observed hcp → bcc phase transition in Tl at high temperature, so our calculation of a mechanically stable bcc structure is presumably correct. It is possible in this case that a fully relativistic treatment is needed to get the $T = 0$ structural phase-stability ordering correct, as is known to be the case from DFT calculations in the neighboring $5d$ metal Pb (Christensen et al., 1986).

6.2.2 Transition metals

With the exception of the magnetic $3d$ element Mn, the observed crystal structures in transition metals are simple hcp, bcc or fcc phases that follow from the partial filling of the d bands in these materials and, allowing only for the ferromagnetic ordering in bcc Fe and hcp Co, are well explained by DFT. The additional magnetic elements Cr and Ni require only a nonmagnetic treatment to explain their respective bcc and fcc observed structures. This simplicity of structure provides useful general guidance for QBIP development. In the GPT, as in DFT, structure is not a necessary input into the theory, but is predictable from the first-principles potentials. In transition metals, however, such a capability is only really straightforward to apply at the beginning and end of each series, where at most two- and three-ion potentials are needed. Success here has been specifically demonstrated in the cases of Ba, Sc and Ni, where the respective observed bcc, hcp and fcc structures are correctly calculated, as indicated in Table 6.3. In the remaining central transition metals, however, one must deal with the substantial convergence issues that arise in attempting to use the full long-ranged GPT multi-ion potentials, where sp-d hybridization is included, to obtain accurate structural energies with up to six-ion interactions. This more difficult task remains a future challenge.

In the MGPT and BOP approaches, on the other hand, only the direct short-ranged d-state interactions are retained, which avoids the convergence issues of the GPT for central transition metals. At the same time, in these latter methods input of the observed ground- state structure and the corresponding elastic constants into the theory is an essential part of the parameterization used in developing the potentials, as we have discussed at length in Chapter 5. This procedure, of course, effectively insures both the thermodynamic and mechanical stability of the observed structure. Nonetheless, we also list the calculated MGPT and BOP structures for specifically treated transition metals in Table 6.3 as a convenient means of pointing to and summarizing all of the work relevant to phase stability that has been performed to date. For both the MGPT and BOP methods, the important tests on structural phase stability come in calculating energy

Table 6.3 *QBIP-calculated $T = 0$ equilibrium crystal structures in transition metals from GPT, MGPT and BOP potentials, together with the experimental observation of these structures at temperature T_{expt} in K from Pearson (1967). Here Cr and Ni are treated in the nonmagnetic limit.*

Metal	GPT	MGPT	Numerical BOP	Analytic BOP	Experiment	T_{expt}
Ba	bcc[a]	–	–	–	bcc	5
Sc	hcp[b]	–	–	–	hcp	~ 300
Ti	–	–	hcp[c]	–	hcp	298
V	–	bcc[d]	bcc[e]	–	bcc	~ 300
Nb	–	bcc[f]	bcc[e]	bcc[g]	bcc	293
Ta	–	bcc[h]	bcc[e]	bcc[g]	bcc	~ 300
Cr	–	–	bcc[e]	–	bcc	293
Mo	–	bcc[i]	bcc[e,j]	bcc[g]	bcc	293
W	–	–	bcc[e,k]	bcc[g]	bcc	298
Mn	–	–	–	α-Mn[l]	α-Mn	~ 300
Fe	–	–	bcc[m]	–	bcc	293
Ir	–	–	fcc[n]	–	fcc	293
Ni	fcc[o]	fcc[p]	–	–	fcc	~ 300

[a]Moriarty (1986b).
[b]Present calculation with $D_2^* = -3$ and $Z = 1.615$.
[c]Girshick et al. (1998).
[d]Moriarty et al. (2008).
[e]Lin et al. (2014).
[f]Moriarty (1990b).
[g]Cák et al. (2014).
[h]Yang et al. (2001); Moriarty et al. (2002a); Moriarty and Haskins (2014).
[i]Moriarty et al. (2002b, 2006).
[j]Mrovec et al. (2004).
[k]Mrovec et al. (2007).
[l]Drain et al. (2014).
[m]Lin et al. (2016).
[n]Cawkwell et al. (2006).
[o]Present calculation with $D_2^* = -3$ and $Z = 1.482$.
[p]Present calculation with $D_2^* = -2.75$ and $Z = 1.225$.

differences between different possible higher-energy structures, including the transformation paths defined in Sec. 6.1.2 connecting bcc, fcc and hcp phases, as illustrated in Fig. 6.3(a) for the Bain paths of MGPT Mo and Ta. Also of interest are any additional known metastable phases such as A15 in the group-VB and -VIB metals.

Perhaps the ideal prototype transition-metal element for testing structural phase stability in the MGPT and BOP methods is the group-VIB element Mo, with its nearly half-filled d bands in equilibrium, its large bcc shear elastic moduli, ensuring substantial fcc-bcc and hcp-bcc energy differences, and the apparent minimal consequences from neglected sp-d hybridization for this metal. Moreover, Mo is free of both the antiferromagnetic magnetism of the corresponding $3d$ metal Cr as well as possible relativistic effects in the corresponding $5d$ metal W. In Fig. 6.5 we compare calculated fcc-bcc, hcp-bcc and A15-bcc energy differences for Mo as a function of atomic volume, as obtained from the Mo5.2 MGPT multi-ion potentials and from corresponding FP-LMTO DFT calculations. The MGPT potentials capture both the quantitative magnitude of the DFT energy differences near the observed equilibrium volume Ω_0, as well as the downward trends in the fcc-bcc and hcp-bcc energies under strong compression, which point to an impending bcc → cp high-pressure phase transition. The latter trends are driven in part by the continuous $s \to d$ valence electron transfer in transition-series metals at high pressure, a phenomenon that was introduced in Chapter 5 and is further elaborated in Sec. 6.3. Also, as expected, the much smaller hcp-fcc energy difference in Mo is underestimated by the MGPT multi-ion potentials, which do not include higher d-state moment contributions beyond the fourth moment. Similar calculations and tests of MGPT multi-ion potentials have also been done on the group-VB prototype Ta, although in that case with a greater emphasis on structures of potential importance to the high-P, T phase diagram (Haskins et al., 2012; Haskins and Moriarty, 2018), as is discussed in Chapter 8. Quantitative MGPT and DFT structural energy differences for both Mo and Ta at normal density are given in Table 6.4.

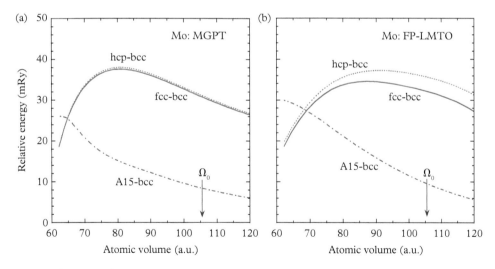

Fig. 6.5 *Calculated fcc-bcc, hcp-bcc and A15-bcc structural energy differences as a function of atomic volume in Mo. (a) MGPT energy differences as obtained from the Mo5.2 multi-ion potentials (Moriarty et al., 2002b, 2006); (b) DFT energy differences as obtained from the FP-LMTO method (Söderlind et al., 1994a; Söderlind, 2001).*

In contrast to MGPT, numerical and analytic BOP potentials calculated to date for Mo, Ta and other nonmagnetic bcc transition metals (Mrovec et al., 2004, 2007; Cák et al., 2014; Lin et al., 2014) have only been constructed for use near normal density. Tests of structural phase stability, via comparison to corresponding DFT calculations, have been performed on tetragonal (Bain), trigonal or rhombohedral, hexagonal and orthorhombic transformation paths covering the fcc, hcp, sc and bct structures. Additional BOP and DFT calculations of A15-bcc energy differences were also performed. Representative BOP and PAW DFT results for Mo and Ta are given in Table 6.4 for

Table 6.4 *MGPT, BOP and DFT calculations of fcc, ideal hcp, A15, Pnma and hex–ω structural-energy differences relative to the equilibrium bcc ground state of the prototype central transition metals Ta and Mo (in mRy). Here the MGPT calculations have been performed at their standard four-ion or four-moment level and the BOP calculations at their standard nine-moment level.*

Metal	Method	fcc	hcp	A15	Pnma	hex--ω
Ta:						
	MGPT: Ta4[a]	13.3	13.4	2.2	–	16.8
	MGPT: Ta6.8x[b]	13.2	13.4	2.2	8.2	16.8
	Numerical BOP[c]	20.6	32.7	8.0	–	–
	Analytic BOP[d]	19.8	25.4	–	–	–
	DFT: FP-LMTO[e]	18.9	23.7	3.9	–	–
	DFT: PAW[c,f]	18.2[c]	24.3[c]	1.8[c]	8.2[f]	17.6[f]
Mo:						
	MGPT: Mo5.2[g]	30.9	31.3	8.5	–	40.0
	MGPT: Mo6.8[h]	30.6	31.0	9.4	–	39.9
	Numerical BOP[c]	31.1	32.9	17.0	–	–
	Analytic BOP[d]	30.1	32.0	–	–	–
	DFT: FP-LMTO[i]	32.2	35.7	9.7	–	–
	DFT: PAW[c]	31.2	34.7	6.6	–	–

[a]Yang et al. (2001); Moriarty et al. (2002a).
[b]Moriarty and Haskins (2014); Haskins et al. (2012); Haskins and Moriarty (2018).
[c]Lin et al. (2014).
[d]Cák et al. (2014).
[e]Söderlind and Moriarty (1998).
[f]Yao and Klug (2013).
[g]Moriarty et al. (2002b, 2006).
[h]Moriarty (unpublished) and present calculations.
[i]Söderlind et al. (1994a) and Söderlind (2001).

comparison with each other and with corresponding FP-LMTO DFT and MGPT calculations.

With regard to Table 6.4 itself, one should first note that the large fcc-bcc and hcp-bcc DFT energies calculated by separate FP-LMTO and PAW methods are in excellent agreement and within 1 mRy of each other for both Mo and Ta. The much smaller A15-bcc DFT energies, on the other hand, are 2–3 mRy lower in the PAW calculation than in the FP-LMTO calculation for both metals, which is a significant uncertainty. In the case of Mo, the MGPT and BOP fcc-bcc and hcp-bcc energies are in good agreement with each other and with the DFT results. However, it is interesting to note that the small hcp-fcc energy in Mo is 3.5 mRy for both DFT calculations, 1.8–1.9 mRy for the numerical and analytic BOP potentials, and 0.4 mRy for the two MGPT potentials. This is the expected ordering of the hcp-fcc energies, with possibly the remaining factor of two difference between the DFT and BOP results due to the neglect of sp-d hybridization. A more important BOP issue, however, is the factor of 2–4 overestimate of the A15-bcc energy in both Mo and Ta for the numerical BOP potentials. This problem is quite likely related to the small A15 nearest-neighbor distance of only $1.61 R_{\mathrm{WS}}$ and the repulsive part of the BOP potentials. The MGPT potentials fare much better in this regard, with A15-bcc energies that are quite reasonable and intermediate in value between the two DFT results in both Mo and Ta. Regarding the fcc-bcc and hcp-bcc energies in Ta, there is generally poorer agreement with DFT than in the case of Mo. In particular, the numerical BOP hcp-bcc energy is some 8–9 mRy too large, while the MGPT fcc-bcc and hcp-bcc energies are 5–11 mRy too small. The specific reasons for these shortcomings are unclear, but in the case of the MGPT Ta potentials, they do perform much better on a range of structures of interest to the high-P, T diagram, including A15, *Pnma* and hex-ω.

6.3 High-pressure phase stability and pressure-induced phase transitions

Of the specific QBIP methods that we have treated or mentioned in this book, only GPT and MGPT have ventured into the interesting world of high-pressure physics and the calculation of materials properties far from normal density, including structural phase stability. High-pressure phase stability is important for two basic reasons. First, it provides a great deal of additional insight about the nature of structural phase stability for both given materials and for the Periodic Table as a whole, as structural behavior at normal density or ambient pressure is often only an isolated snapshot of what is happening more globally and of what is actually possible. Some of the important specific implications for the universal $s \to d$ transition are discussed in Sec. 6.3.1. Second, investigating high-pressure behavior also brings with it the possibility of discovering new, previously unknown phases of a material. The GPT and MGPT methods have so impacted high-pressure physics in a number of specific cases, which are summarized and discussed in Sec. 6.3.2.

6.3.1 *sp-d* electron transfer across the Periodic Table and systematic trends

In both transition metals and lanthanide series rare-earth metals, structural phase stability is controlled to a large extent by the number of d-electrons per atom, Z_d. For transition metals, of course, this behavior is a natural consequence of the dominant d-bonding present, as we have discussed. For the trivalent rare-earth metals beyond La, on the other hand, this simplification is an indirect consequence of strong electron correlation among the corresponding f valence electrons that effectively makes them nonbonding with no direct contribution to crystal structure. In either the transition or lanthanide series, simple d-resonance or TB rigid-band models of the electronic structure (e.g., Pettifor, 1970b; Duthie and Pettifor, 1977) that assign a fixed d-band DOS to each candidate crystal structure and treat Z_d as the single variable parameter can explain most of the observed trends with atomic number and pressure, and these models are supported by first-principles DFT calculations via the Andersen force theorem (Skriver, 1985). Due to their more compressible nature, the rare-earth metals have been systematically studied in high-pressure experiments. In these metals, Z_d is found to increase both with *decreasing* atomic number through the series (right to left in the Periodic Table) and with increasing pressure for a given element, such that for either variation, the same hcp $\rightarrow$ 9R $\rightarrow$ dhcp $\rightarrow$ fcc sequence of close-packed structures is predicted, in agreement with experiment (Benedict et al., 1986; Young, 1991). In transition metals, however, Z_d is increased by *increasing* atomic number, and, except for the late members of each series, also by the application of high pressure. As we have already noted in connection with Fig. 1.6, the variation of Z_d with atomic number largely explains the observed hcp $\rightarrow$ bcc $\rightarrow$ hcp $\rightarrow$ fcc sequence of structures across the nonmagnetic $4d$ and $5d$ transition series. The general increase in Z_d with increasing pressure that is shared in common between the rare earth and transition metals results from an sp to d transfer of valence electrons under compression, and is usually referred to as an electronic s-d or $s \rightarrow d$ transition (McMahan, 1986). This transition arises from the fact that the spatially extended s and p valence states feel the effects of high pressure more strongly than do the more localized d states. Thus, under compression, the corresponding s and p bands rise faster in energy than do the d bands, transferring electrons from filled s and p states to empty d states in the process.

In general, the systematics of high-pressure phase stability in transition metals has been less well studied experimentally than in rare-earth metals, especially in the relatively incompressible mid-period elements, where at the very least hundreds of GPa in pressure are required to induce any phase changes in these materials. At the same time, as a rough guide, it is possible to make elementary structural predictions of high-pressure TM behavior based on the observed structures at ambient pressure, albeit with a number of important caveats. Assuming that as a practical matter, *sp-d* electron transfer can only induce changes in Z_d of 1 or less under compression, the consequences of high pressure for elements in a given TM column of the Periodic Table can be inferred by examining the column immediately to the right of the one under consideration. Starting with group-IIIB metals (Sc, Y and Lu in place of La), both they and the IVB metals

(Ti, Zr and Hf) to their right are hcp, so no high-pressure s-d transition to bcc, the next structure in the TM sequence, is expected, and none is observed. The metals Sc, Y and Lu are also trivalent rare-earth metals, however, and, therefore, are expected to undergo instead the canonical sequence of rare-earth high-pressure transitions. This behavior is observed in Y and Lu, although not in the case of Sc (Young, 1991).

Turning to the group-IVB metals, their VB neighbors (V, Nb and Ta) to the right are bcc, so a high-pressure s-d transition from hcp to bcc is expected. In this case, however, there is a well-known intermediate hex-ω phase. In high-pressure DAC experiments, the full sequence hcp $\rightarrow$ hex$-\omega$ $\rightarrow$ bcc is observed in Zr and Hf (Xia et al., 1990a, 1990b), as well as in Ti, where, however, the high-pressure bcc phase is found to closely compete with three additional metastable structures (Ahuja et al., 2004). Next, for the group-VB metals, their VIB neighbors (Cr, Mo and W) to the right are also bcc, so no normal s-d driven phase transitions are expected, and none has been observed in either Nb or Ta. In the case of V, however, a related high-pressure transition that is characterized as a rhombohedral lattice distortion of the bcc structure and has been observed in DAC experiments at 69 GPa (Ding et al., 2007). This latter phase transition results from a unique elastic instability in the bcc structure, and is of special interest here because it was correctly calculated and explained from existing MGPT multi-ion potentials for V soon after it was reported, as is discussed in Sec. 6.3.2.

Moving on to the group-VIB bcc metals Cr, Mo and W, the nonmagnetic $4d$ and $5d$ members of their VIIB neighbors (Mn, Tc and Re) to the right are hcp, so an s-d driven bcc $\rightarrow$ hcp transition is expected. No such transition has yet been observed in Cr, Mo or W, but in the prototype case of Mo, which DFT calculations predict will have the lowest bcc-hcp transition pressure of the three metals (Moriarty, 1992; Söderlind, 1994a), there has been significant effort in this direction. First, in shock measurements on Mo, Hixson et al. (1989) found an apparent break in the sound velocity near 210 GPa and 4100 K, prior to melting at 390 GPa and 10,000 K, indicating a possible solid-solid phase transition. Supporting *nonrelativistic* DFT calculations with the LMTO-ASA method and LDA exchange-correlation predicted a bcc $\rightarrow$ hcp $T = 0$ phase transition at 320 GPa in Mo, which at the time seemed to be a reasonable match to the sound velocity measurement. However, it was soon realized that because of the very low compressibility of Mo, the magnitude of the bcc $\rightarrow$ hcp transition pressure P_T is highly sensitive to the predicted transition volume Ω_T and hence to the details of the calculation. A follow-up *semi-relativistic* LMTO-ASA treatment of Mo (Moriarty, 1992) produced a reduced size of the ion core, lowering Ω_T and raising P_T to 420 GPa, with an additional hcp $\rightarrow$ fcc transition predicted at 620 GPa. Subsequent *fully relativistic* DFT calculations by Söderlind et al. (1994a), as displayed in Fig. 6.5(b) and using the more accurate FP-LMTO method of Wills et al. (2010), found the same sequence of structures, but further raised P_T to 520 GPa and the hcp-fcc transition pressure to 740 GPa. Additional DFT calculations by Christensen et al. (1995), using the alternate FP-LMTO method of Methfessel et al. (2000), however, have predicted that the hcp structure is actually bypassed at high pressure in favor of fcc in Mo, and in favor of dhcp in W (Ruoff et al., 1998), with respective bcc-fcc and bcc-dhcp transition pressures of 570 and 650 GPa.

Meanwhile, ultrahigh-pressure DAC experiments on Mo, first to 416 GPa (Ruoff et al., 1990) and then to 560 GPa (Ruoff et al., 1992), have found no evidence of a phase transition. Moreover, recent more accurate sound velocity measurements under shock compression in Mo by Nguyen et al. (2014) have found no evidence of a break in the data prior to the onset of melting near 380 GPa, thus calling into question any solid-solid phase transition at high temperature below that pressure. Consequently, the fundamental challenges of finding and correctly explaining s-d driven bcc $\rightarrow$ cp high-pressure phase transitions in Cr, Mo and W remain.

Finally, in the group-VIII elements, there is the possibility of an s-d driven hcp $\rightarrow$ fcc transition in the iron-group elements, where Ru and Os are hcp, as is Fe above 13 GPa in the ε phase (Young, 1991), while the $4d$ and $5d$ members of the neighboring cobalt group (Co, Rh and Ir) to the right are fcc, as is Co above 210 GPa (Yoo et al., 2000). No hcp $\rightarrow$ fcc transition has been observed in ε-Fe, Ru or Os, however, nor to our knowledge has one been predicted by DFT calculations. The remaining cobalt-group and nickel-group (Ni, Pd and Pt) members are all fcc, except for ferromagnetic hcp Co, as are the noble metals, so no s-d driven phase transition is expected in these elements. Indeed, as we noted in Chapter 5, the sp to d electron transfer under pressure ends and reverses to an d to sp electron transfer at the end of the transition metal series, impacting nickel-group metals like Pt (Holmes et al., 1989) and noble metals like Cu (Moriarty, 1988a; Nellis et al., 1988), but without any additional phase transitions reported in these two metals. There have been both predicted and observed high-pressure phase transitions in Au, however, and most recently the observation of a high-pressure bcc phase in shock XRD experiments (Briggs et al., 2019; Sharma et al., 2019).

In addition to transition and rare-earth metals, the s-d transition also exerts its influence back though the heavy alkaline-earth metals (Ca, Sr and Ba) and the heavy alkali metals (K, Rb and Cs), and can extend even to the third-period simple metals (Na, Mg, Al and Si). At ambient pressure, we have already seen the impact of empty d bands from above though sp-d hybridization on the GPT potentials and physical properties of Ca, Sr and Al, as well as the additional impact of partial d-band filling in Ba. Under pressure, sp-d electron transfer enhances this behavior and extends it to the heavy alkali and other third-period metals, driving a host of phase transitions in the process. In the alkali and heavy alkaline-earth metals at room temperature, Degtyareva (2010) has cataloged some 37 observed high-pressure phase transitions below 200 GPa. At the lowest pressures, these transitions are among the usual high-symmetry metallic structures: bcc $\rightarrow$ fcc in the alkali metals, at pressures ranging from 2.4 GPa in Cs to 65 GPa in Na; fcc $\rightarrow$ bcc in Ca at 20 GPa and Sr at 3.5 GPa, signaling the onset of transition-metal behavior, as in Ba; and bcc $\rightarrow$ hcp in Ba at 5.5 GPa. At higher pressures, however, the remaining observed phase transitions are to more complex, lower-symmetry structures, which in some cases are even insulating phases. In the cases of Li and Na, Neaton and Ashcroft (1999, 2001) predicted via DFT PAW calculations that ion pairing into effective dimers would occur at sufficiently large compression and produce such complex structural behavior, reminiscent of the behavior of α-Ga. Overall, a number of electronic-structure mechanisms are potentially in play for the alkali and heavy alkaline-earth metals under pressure, including sp-d electron transfer, ion pairing and band-gap formation. Consequently, there has been a great deal of DFT-based theoretical effort to explain the actual observed complex phases in these metals, as discussed by Young (1991) and Degtyareva (2010). At least some of these transitions should also fall within the expected domain of GPT calculation, although none yet has been attempted in that manner. In this regard, for the case of the alkalis a finite temperature treatment would be desirable, while in the heavy alkaline earths, and perhaps in the heavy alkalis as well, a full multi-ion transition-metal treatment is needed, as was done previously for bcc Ba at ambient pressure (Moriarty, 1986b).

6.3.2 Successful GPT and MGPT predictions of new high-pressure phases

In addition to the above phase transitions for the alkali and heavy alkaline-earth metals, a separate series of s-d driven high-pressure phase transitions in the third-period simple metals Na, Mg, Al and Si up to 800 GPa earlier had been predicted by Moriarty and McMahan (1982) and McMahan and Moriarty (1983), on the basis of first-principles GPT and LMTO-ASA calculations. Five of these latter transitions occurring below 400 GPa have subsequently been confirmed by experiment, as well as by numerous additional DFT calculations. These results are summarized in Table 6.5.

The GPT and LMTO high-pressure phase-transition calculations on the third period metals were carried out using the constant-volume formalism described in Sec. 6.1.3, with transition volumes Ω_T determined by use of Eqs. (6.15) and (6.13), respectively. Corresponding transition pressures P_T at these volumes were established by LMTO-ASA calculations. Both the GPT and LMTO treatments used LDA

Table 6.5 *Experimental and theoretical confirmation of five previously unknown high-pressure phase transitions in the third-period simple metals first predicted by Moriarty and McMahan (1982) and McMahan and Moriarty (1983) via first-principles GPT and LMTO-ASA calculations.*

Metal	Transition	Method	Ω_0 (a.u.)	Ω_T/Ω_0	P_T (GPa)	T_T (K)
Na	hcp → bcc	GPT: EDB[a]	255.2	0.86	1	0
		DFT: PP[b]	–	–	1	0
		Expt.[c]	–	–	~ 0.1–0.2	0
Mg	hcp → bcc	GPT: EDB[a]	156.8	0.58	50	0
		GPT: SM[d]	156.8	0.58	54	0
		DFT: LMTO[a]	156.8	0.56	57	0
		DFT: PP[e]	156.8	0.57	60	0
		DFT: PAW[f]	–	–	57, 56	0
		DFT: PP[g]	–	–	65	0
		Expt: DAC[h]	156.8	0.59±0.02	50±6	300
		Expt: DAC[i]			51±7	300

Continued

Table 6.5 *Continued*

Al	fcc → hcp	GPT: EDB[a]	112.0	0.43	360	0
		DFT: LMTO[a]	112.0	0.58	120	0
		DFT: PP[j]	112.0	0.48	240	0
		DFT: LCGTO[k]	112.0	0.51	205±20	0
		DFT: FP-LMTO[l]	–	–	170	0
		DFT: FPLAPW[m]	–	–	192	0
		Expt: DAC[n]	112.0	0.51	217±10	297
		Expt: DRC[o]	–	–	216±9	~ 830
Al	hcp → bcc	GPT: EDB[a]	112.0	0.37	560	0
		DFT: LMTO[a]	112.0	0.51	200	0
		DFT: PP[h]	112.0	0.41	420	0
		DFT: LCGTO[k]	112.0	0.36	565±60	0
		DFT: FP-LMTO[l]	–	–	360	0
		Expt: DRC[o]	–	–	321±12	~ 1000
		Expt: DAC[p]	–	–	> 320	300
Si	hcp → fcc	GPT: EDB[a]	135.1	0.48	80	0
		DFT: LMTO[a]	135.1	0.50	76	0
		DFT: PP[q]	135.1	0.46	116	0
		Expt: DAC[r]	135.1	0.48	79±2	300

[a] McMahan and Moriarty (1983). The LMTO pressure scale was used to obtain GPT pressures.
[b] Dacorogna and Cohen (1986).
[c] Extrapolation to zero K of measured cp-bcc phase boundary in Na: Chernyshov et al. (1983) and Vaks et al. (1989).
[d] Althoff et al. (1993). Real-space GPT calculations with a self-consistent pressure scale.
[e] Wentzcovitch and Cohen (1988).
[f] Mehta et al. (2006). Both LDA and GGA calculations were performed.
[g] Liu et al. (2009). GGA calculations were performed.
[h] Olijnyk and Holzapfel (1985).
[i] Stinton et al. (2014).
[j] Lam and Cohen (1983).
[k] Boettger and Trickey (1995, 1996).
[l] Sin'ko and Smirnov (2002).
[m] Jona and Marcus (2006).
[n] Akahama et al. (2006).
[o] Polsin et al. (2017).
[p] Fiquet et al. (2019).
[q] Chang and Cohen (1985).
[r] Duclos et al. (1987, 1990).

exchange-correlation, as did all of the subsequent DFT calculations cited in Table 6.5, except where noted. In Moriarty and McMahan (1982), the GPT calculations were performed on Na, Mg and Al in the simple-metal limit, using the reciprocal-space formalism described in Chapter 3, while the LMTO calculations were performed on Mg and Al only in the ASA limit, but both with and without occupied d states. High-pressure transition sequences of hcp → bcc → hcp in Na, hcp → bcc → fcc in Mg and fcc → hcp → bcc in Al were predicted. In Mg and Al, the same sequence and qualitative features of the transitions were obtained by the GPT and LMTO-ASA treatments, so long as occupied d states were included in the LMTO calculations. This convincingly demonstrated that indeed an $sp \to d$ transfer of electrons with increasing pressure is driving these phase transitions. At the same time, the ability of GPT in the SM limit to calculate the same transitions demonstrated that the d bands themselves are quite free-electron like.

McMahan and Moriarty (1983) extended and refined these calculations, including the addition of Si as a third-period simple metal in both the GPT and LMTO treatments. To capture the effects of sp-d hybridization, the GPT treatment was enhanced to the EDB level of description, as elaborated in Chapter 4. This change had minimal impact in Na and Mg below 100 GPa, but was a useful improvement for these metals above 100 GPa, as well as in Al at all pressures, as we previously noted at normal density in Fig. 3.8(c). Qualitatively, however, the same sequences of phase transitions were predicted in Na, Mg and Al. In Si above the semiconductor to metal transition and the attainment of an hcp simple-metal structure near 42 GPa, both GPT-EDB and LMTO-ASA calculations predicted the high-pressure phase-transition sequence hcp → fcc → bcc, with again occupied d states included in the LMTO treatment. As indicated in Table 6.5, the experimentally observed transition volumes and pressures for the hcp → bcc transitions in Na and Mg and the hcp → fcc transition in Si are in good accord with the GPT and LMTO predictions. For the fcc → hcp and hcp → bcc transitions in Al, on the other hand, there is a significantly wider spread between the predicted GPT and LMTO pressures, although in both cases the theoretical results bracket the observed transition pressures. Subsequent more accurate DFT calculations have generally improved the quantitative calculation of these latter pressures, although for the hcp → bcc transition there is still a 200 GPa spread among PP, LCGTO and FP-LMTO pressures. Also in this regard, it seems likely that an optimum choice of d states in the GPT-EDB treatment of Al, as already accomplished at $\Omega = \Omega_0$, will also improve the quantitative calculation of high-pressure transition volumes and pressures.

In addition to the five predicted high-pressure phase transitions for the third-period simple metals listed in Table 6.5 that have been confirmed experimentally, two more of the predicted transitions at higher pressures for these metals by McMahan and Moriarty (1983) may be within future experimental reach. These predictions are for a bcc → fcc transition in Mg and for the reverse fcc → bcc transition in Si. In the case of Mg, however, there is a wide gap between the calculated GPT-EDB and LMTO-ASA transition pressures at 180 and 790 GPa, respectively. This transition has also been studied recently by Li et al. (2010) and by Tsuppayakorn-aek et al. (2020) using DFT calculations with the PAW method and a PBE xc potential. Their predicted transition pressures of 456 GPa and 489 GPa, respectively, split the difference between the GPT and LMTO values. For the fcc-bcc transition in Si the GPT and LMTO predicted pressures are in closer agreement at 360 and 250 GPa, respectively.

An additional successful GPT prediction of a new high-pressure phase was made in the interesting case of hcp Hg. In this instance, the GPT theoretical prediction (Moriarty, 1988b) and the DAC experimental discovery (Schulte and Holzapfel, 1988) of this phase were simultaneous and completely independent, with the cited papers published, by author agreement, back-to-back in *Physics Letters*. In the GPT research, several unique features of the structural phase stability in Hg were explained for the first time, and in particular, how and why these features differ from the other group-IIB metals Zn and Cd. Five families of structures (bct, rhom, hcp, hex and beta-tin) were studied in Hg at $T = 0$ over a wide volume range: $0.35 \leq \Omega/\Omega_0 \leq 1.12$. These structures were treated in both the SM and FDB limits of the GPT, and both with and without relativistic effects included. In the nonrelativistic SM limit, the predicted ground-state structure is ideal hcp. In the nonrelativistic FDB limit, with sp-d hybridization and d-state overlap included, the predicted ground state becomes a high c/a hcp structure, as is also calculated and observed for Zn and Cd. However, only in the semi-relativistic FDB limit, with relativistic core and d-state energy shifts included, is the observed β-Hg ground-state bct structure correctly obtained, with a calculated c/a ratio of 0.69 at $\Omega = \Omega_0$. With the same semi-relativistic GPT-FDB treatment, a zero-temperature $E_{\text{struc}} - E_{\text{struc}}^{\text{fcc}}$ vs Ω/Ω_0 phase diagram over the whole volume range, involving eight candidate structures [including α-Hg (rhom)], was also calculated (see Fig. 4 of Moriarty, 1988b). At expanded volumes the lowest-energy structure was found to be simple cubic (sc), but as one approaches $\Omega/\Omega_0 = 1$, β-Hg (bct) was stabilized and remained the lowest-energy structure under compression until a bct ($c/a = 0.745$) to hcp ($c/a = 1.86$) phase transition was indicated at $\Omega_T/\Omega_0 = 0.61$ and $P_T = 99$ GPa. The hcp structure then remained the stable phase under further compression to a pressure of 1 TPa (1000 GPa), with a limiting c/a value of 1.83.

The corresponding Schulte and Holzapfel (1988) discovery of the high-pressure hcp phase in Hg was obtained with DAC-XRD measurements at room temperature as the final structure in an observed α (rhom) → β (bct) → α (reentrant rhom) → δ (hcp) sequence of phase transitions in the solid up to 39 GPa in pressure. The δ-Hg hcp structure became the dominant observed phase at pressures above 37 GPa, with a measured c/a axial ratio of 1.76. In subsequent more complete DAC-XRD measurements of the Hg phase diagram (Schulte and Holzapfel, 1993), extending up to 67 GPa in pressure and for temperatures in the range 150–500 K, it was discovered that the previously identified reentrant α (rhom) phase was actually a new orthorhombic structure, designated as γ-Hg. Between 150 and 400 K and for pressures up to 50 GPa, the correct high-pressure sequence of phase transitions in the solid is then α (rhom) → β (bct) → γ (ortho) → δ (hcp). With the α-β phase line beginning at 100 K at ambient pressure and ending in a α-β-γ triple point near 400 K and 8 GPa, this sequence extrapolates to α (rhom) → γ (ortho) → δ (hcp) at temperatures above 400 K and to β (bct) → γ (ortho) → δ (hcp) at temperatures below 100 K. At $T = 0$, the estimated ortho-hcp transition pressure is 42 GPa. Recent room temperature DAC-XRD studies on Hg to 200 GPa by Takemura et al. (2015) have confirmed the γ (ortho) → δ (hcp) phase transition near 37 GPa and found the hcp structure to remain the stable phase at all higher pressures, with its measured c/a ratio declining from about 1.74 at 37 GPa to a near ideal value of 1.63 at 196 GPa.

Recent refinement of the semi-relativistic GPT-FDB treatment of Hg through the use of optimized d-basis states obtained in the manner described in Sec. 4.5 of Chapter 4 now allows a more accurate calculation of transition volumes and pressures. In this regard, the logarithmic derivative $D_2^\star(\Omega)$ defining the optimized d-basis states for Hg can be obtained from Eq. (4.108) with parameters $|D_2^\star(\Omega_0)| = 3.51$ and $\alpha = 3.5225$, and the calculated function $|D_2^\star(\Omega)|$ is qualitatively similar in behavior to that displayed for Zn in Fig. 4.10(a). As shown in Fig. 6.6, application of the refined GPT-FDB treatment of Hg to the $T = 0$ high-pressure bct → hcp phase transition raises the transition volume Ω_T from $0.61\,\Omega_0$ to $0.77\,\Omega_0$ and lowers the transition pressure P_T from 99 to 49 GPa. This latter improved pressure value is then more quantitatively commensurate with the DAC measurements of transition pressures in Hg and bodes well for a general semi-relativistic GPT-FDB treatment of the entire P, T phase diagram, including the α, β, γ and δ solid phases and the liquid.

A third noteworthy example of a successful GPT/MGPT prediction of a high-pressure phase transition concerns the group-VB transition metal V. The bcc to rhombohedral transition in V was briefly discussed in Sec. 6.3.1, and its discovery and

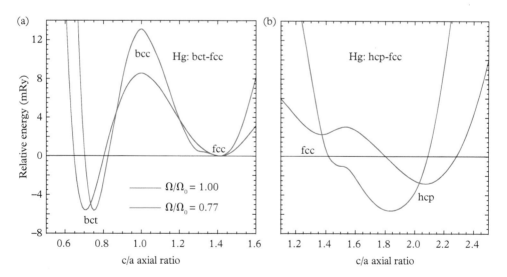

Fig. 6.6 *Optimized semi-relativistic GPT-EDB energetics for the calculated bct → hcp high-pressure phase transition in Hg at $T = 0$, where $\Omega_T/\Omega_0 = 0.77$ and $P_T = 49$ GPa. Plotted are (a) bct-fcc and (b) hcp-fcc structural energy differences as a function of c/a axial ratio at normal density ($\Omega = \Omega_0$), where bct is the stable structure, and at the transition volume $\Omega = \Omega_T$.*

explanation represent another case where there was close coupling between theory and experiment. Interested in the high-pressure behavior of the superconducting transition temperature in vanadium, Suzuki and Otani (2002) performed DFT calculations of the V phonon spectrum up to 150 GPa. They found a large Kohn anomaly develop under increasing pressure in the transverse branch of the [100] phonon spectrum, which beyond 130 GPa resulted in imaginary phonon frequencies and hence an unstable bcc structure. This behavior was confirmed by later FP-LMTO calculations of the elastic constants in V under pressure by Landa et al. (2006a), who found that the bcc C_{44} shear modulus becomes negative beyond 130 GPa, as shown in Fig. 6.7(a). The specific physical origin of this bcc lattice instability in V was soon thereafter shown to be the result of Fermi-surface nesting (Landa et al., 2006b). In addition, the unique C' and C_{44} FP-LMTO elastic moduli for V displayed in Fig. 6.7(a) were also soon after used to develop the V6.1 MGPT potentials (Moriarty et al., 2008), as discussed in Chapter 5.

Motivated by the theoretical developments, DAC experimentalists were interested to re-examine the high-pressure phase stability of bcc V. While earlier measurements had not detected a phase transition in this metal, in new DAC-XRD measurements, Ding et al. (2007) discovered a unique bcc to rhombohedral transition in V beginning at 69 GPa, with the rhom structure remaining stable up to the highest pressure investigated, which was 155 GPa. The bcc → rhom transition was then quickly confirmed by independent DFT-PAW and MGPT calculations. In the DFT-PAW study (Lee et al., 2007), the onset of the transition was calculated at 84 GPa and a reentrant transition back to bcc was predicted at 280 GPa. The MGPT results (Moriarty et al., 2008) were obtained directly from the existing V6.1 multi-ion potentials. Representative MGPT rhom-bcc energy differences in V obtained at three relevant pressures are displayed in Fig. 6.7(b).

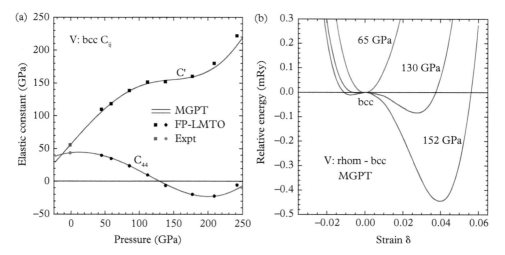

Fig. 6.7 *MGPT bcc shear elastic moduli (a) and rhom-bcc energy difference as a function of bcc strain δ along the rhombohedral direction under pressure (b), in vanadium.*

6.3.2 Successful GPT and MGPT predictions of new high-pressure phases

At 65 GPa, the observed bcc structure prior to the phase transition is clearly calculated to be the stable phase, while at 130 and 152 GPa, the pressure-induced rhombohedral structure is correctly calculated to be the lowest-energy phase. This success demonstrated the advanced ability of the MGPT method to predict complex structural phase stability from complex elastic behavior.

It should also be mentioned that a very recent DAC-XRD high-pressure study of the bcc → rhom transition in vanadium by Akahama et al. (2021), which combines room-temperature measurements with high-temperature laser annealing, suggests a modified picture for this transition. While with increasing pressure these authors detect the rhom structure beginning in the 30–69 GPa range, they find that the bcc structure is restored at each pressure investigated by laser annealing, until one reaches 189 GPa. They speculate that 189 GPa is the true equilibrium transition pressure and that below this point the rhom structure is only a metastable phase, initially stabilized by nonhydrostatic pressure effects in the DAC. Further investigation of this conjecture would seem warranted.

7
Elastic Moduli and Phonons

In this chapter we consider the QBIP calculation of two additional fundamental properties of crystalline metals: second-order elastic constants and quasiharmonic normal-mode phonon frequencies, both of which we address here for nontransition metals as well as transition metals. As we noted in Chapter 6, elastic moduli are closely coupled to structural phase stability, but more generally, they underpin the treatment of mechanical properties in solids, including such things as dislocation mobility, plasticity and strength. Similarly, lattice vibrations can alter structural phase stability at finite temperature leading to temperature-induced phase transitions, but more generally, they underpin the treatment of the thermodynamic properties of solids, including specific heat and thermal expansion, all properties that depend on the ion-thermal energy, pressure and free energy. Here we will be concerned with the evaluation of elastic constants and phonon frequencies at zero temperature, and their application to metal properties below the Debye temperature in the quasiharmonic limit. Corrections for high-temperature effects such as thermoelasticity and anharmonic lattice vibrations are considered in Chapter 8, and the full treatment of defects and mechanical properties is given in Chapters 9 and 11.

7.1 Quasiharmonic lattice dynamics for QBIP applications

The standard treatment of harmonic lattice dynamics is discussed in many references, including, for example, Born and Huang (1954), Wallace (1972), Ashcroft and Mermin (1976) and Finnis (2003). Here we generalize the standard approach to treat an explicit volume-dependent total-energy functional suitable for QBIP application at either normal density or high pressure. Such a treatment is commonly referred to as *quasiharmonic lattice dynamics* (QHLD) and encompasses both the calculation of volume-dependent phonon frequencies and, in the long wavelength limit, the calculation of second-order elastic constants, which we specifically treat in Sec. 7.3. One may anticipate from the outset that QHLD for a given material will depend on the symmetry of the crystal structure under consideration in the same manner as it does for the electronic band structure. For simplicity and clarity, we first consider the case of an elemental metal whose crystal structure can be represented by a primitive or Bravais lattice with one atom per primitive cell (e.g., sc, hex, fcc, bcc, bct, rhom, ...). The total potential-energy functional of such a metal at zero temperature and constant volume Ω is just

$$U_{\text{tot}}(\mathbf{R}) = E_{\text{tot}}(\mathbf{R}, \Omega) = NE_{\text{coh}}(\mathbf{R}, \Omega) = NE_{\text{coh}}(\mathbf{R}_1, \mathbf{R}_2, \ldots \mathbf{R}_N, \Omega). \tag{7.1}$$

To proceed, one then introduces small displacements $\mathbf{u}_1, \mathbf{u}_2, \ldots \mathbf{u}_N$ of the N ions away from their equilibrium positions, such that for the ith ion

$$\mathbf{u}_i = \mathbf{R}_i - \mathbf{R}_i^0, \tag{7.2}$$

where $\mathbf{R}_i^0$ is the equilibrium position of the ion at volume Ω. One next develops a three-dimensional Taylor expansion of $U_{\text{tot}}(\mathbf{R})$ in the small displacements $\mathbf{u}_i$, which takes the general form

$$U_{\text{tot}}(\mathbf{R}) = U_{\text{tot}}(\mathbf{R}_0) + \sum_i \mathbf{u}_i \cdot \nabla_i U_{\text{tot}}(\mathbf{R}_0) + \frac{1}{2} \sum_{i,j} (\mathbf{u}_i \cdot \nabla_i)(\mathbf{u}_j \cdot \nabla_j) U_{\text{tot}}(\mathbf{R}_0) + \cdots. \tag{7.3}$$

In Eq. (7.3) the first term on the RHS is the total energy in equilibrium, $E_{\text{tot}}(\mathbf{R}_0, \Omega)$, while the second term on the RHS vanishes identically because the force on each ion in equilibrium is zero: $\mathbf{F}_i(\mathbf{R}_0) = -\nabla_i U_{\text{tot}}(\mathbf{R}_0) = 0$. Neglecting all third- and higher-order anharmonic terms in Eq. (7.3), we can then recast that equation into the directly useful second-order form for QHLD:

$$E_{\text{tot}}(\mathbf{R}, \Omega) = E_{\text{tot}}(\mathbf{R}_0, \Omega) - \frac{1}{2} \sum_{\substack{i,j \\ \mu,\nu=x,y,z}} u_{i\mu} K_{\mu\nu}(\mathbf{R}_i - \mathbf{R}_j, \Omega) u_{j\nu}, \tag{7.4}$$

where $u_{i\mu}$ and $u_{j\nu}$ are Cartesian components of $\mathbf{u}_i$ and $\mathbf{u}_j$, respectively, and where we have defined the 3×3 *force-constant* matrix $\mathbf{K}(\mathbf{R}_i - \mathbf{R}_j, \Omega)$ with components

$$K_{\mu\nu}(\mathbf{R}_i - \mathbf{R}_j, \Omega) \equiv -\left. \frac{\partial^2 E_{\text{tot}}(\mathbf{R}, \Omega)}{\partial r_{i\mu} \partial r_{j\nu}} \right|_{\mathbf{R}=\mathbf{R}_0}, \tag{7.5}$$

with $r_{i\mu}$ and $r_{j\nu}$ representing the Cartesian components of $\mathbf{R}_i$ and $\mathbf{R}_j$, respectively. The matrix $\mathbf{K}$ is both real and symmetric, with $K_{\mu\nu} = K_{\nu\mu}$. This matrix also has two additional important symmetry properties:

$$K_{\mu\nu}(\mathbf{R}_i - \mathbf{R}_j, \Omega) = K_{\mu\nu}(\mathbf{R}_j - \mathbf{R}_i, \Omega), \tag{7.6}$$

which follows from the inversion symmetry of any Bravais lattice, and

$$\sum_{i,j} \mathbf{K}(\mathbf{R}_i - \mathbf{R}_j, \Omega) = 0, \tag{7.7}$$

which follows from the translational invariance of the lattice (Wallace, 1972).

For a nonprimitive crystal structure that is represented by a unit cell with $n_b \geq 2$ basis atoms (e.g., hcp, with $n_b = 2$; hex-ω, with $n_b = 3$; A15, with $n_b = 8$; ...), we must generalize our notation and description accordingly. In this regard, we let $\mathbf{R}_i \to \mathbf{R}_{i\alpha}$ and $\mathbf{R}_j \to \mathbf{R}_{j\beta}$, where $\mathbf{R}_{i\alpha}$ denotes the position of basis atom α in unit cell i, and $\mathbf{R}_{j\beta}$ denotes the position of basis atom β in unit cell j. The force-constant matrix $\mathbf{K}(\mathbf{R}_{i\alpha} - \mathbf{R}_{j\beta}, \Omega)$ is now $3n_b \times 3n_b$ in size with components given by

$$K_{\alpha\mu,\beta\nu}(\mathbf{R}_{i\alpha} - \mathbf{R}_{j\beta}, \Omega) \equiv -\left.\frac{\partial^2 E_{\text{tot}}(\mathbf{R}, \Omega)}{\partial r_{i\alpha\mu} \partial r_{j\beta\nu}}\right|_{\mathbf{R}=\mathbf{R}_0}. \tag{7.8}$$

Depending upon the crystal structure, there may or may not be inversion symmetry in the full matrix $\mathbf{K}$, but translational invariance is maintained, with Eq. (7.7) generalized to the form

$$\sum_{i,j} \sum_{\alpha,\beta} \mathbf{K}(\mathbf{R}_{i\alpha} - \mathbf{R}_{j\beta}, \Omega) = 0. \tag{7.9}$$

It is, of course, possible to represent any crystal structure, primitive or nonprimitive, with a unit cell that is larger than its primitive cell, a so-called supercell, together with an appropriate larger number of basis atoms. This is often convenient in computer simulations such as MD. For example, one may use a simple cube to represent the unit cell of the cubic fcc and bcc structures, with $n_b = 4$ and $n_b = 2$ basis atoms, respectively.

7.1.1 Dynamical matrix and the calculation of normal-mode phonon frequencies

To establish the normal modes of vibration, we need to set up and solve $3N$ equations of motion for the harmonic oscillation of the ions about their equilibrium positions. Again, for simplicity and clarity, we do this first for the case of a Bravais lattice. Making use of

7.1.1 Dynamical matrix and the calculation of normal-mode phonon frequencies

Eq. (7.4), the equation of motion for the displacement component $u_{i\mu}$ of ion i with mass M_a is just

$$M_a \ddot{u}_{i\mu} = -\frac{\partial E_{\text{tot}}(\mathbf{R}, \Omega)}{\partial u_{i\mu}} = \sum_{\substack{j \\ \nu=x,y,z}} K_{\mu\nu}(\mathbf{R}_i - \mathbf{R}_j, \Omega) u_{j\nu} . \tag{7.10}$$

In vector notation Eq. (7.10) becomes

$$M_a \ddot{\mathbf{u}}_i = \sum_j \mathbf{K}(\mathbf{R}_i - \mathbf{R}_j, \Omega) \mathbf{u}_j . \tag{7.11}$$

With the usual periodic boundary conditions applied to the crystal, Eq. (7.11) has a solution of the general form

$$\mathbf{u}_i(\mathbf{R}_i, t) = \mathbf{p}_\lambda(\mathbf{q}) e^{-i[\mathbf{q} \cdot \mathbf{R}_i - \omega_\lambda(\mathbf{q}) t]} , \tag{7.12}$$

where $\mathbf{q}$ is a vector in the reciprocal lattice that is contained within the BZ of the crystal structure, $\mathbf{p}_\lambda(\mathbf{q})$ is a polarization vector, to be determined, that describes the direction in which the ions move, and $\omega_\lambda(\mathbf{q})$ is a corresponding angular vibrational frequency, also to be determined. Inserting Eq. (7.12) into Eq. (7.11), summing both sides over the site index i and multiplying both sides by N^{-1}, produces a symmetrical three-dimensional eigenvalue problem to determine the normal-mode values of $\mathbf{p}_\lambda(\mathbf{q})$ and $\omega_\lambda(\mathbf{q})$:

$$\left\{ M_a [\omega_\lambda(\mathbf{q})]^2 - \mathbf{D}(\mathbf{q}) \right\} \mathbf{p}_\lambda(\mathbf{q}) = 0 , \tag{7.13}$$

where $\mathbf{D}(\mathbf{q})$ is the so-called *dynamical matrix* defined by

$$\mathbf{D}(\mathbf{q}) \equiv -\frac{1}{N} \sum_{i,j} \mathbf{K}(\mathbf{R}_i - \mathbf{R}_j, \Omega) e^{i \mathbf{q} \cdot (\mathbf{R}_i - \mathbf{R}_j)} . \tag{7.14}$$

Note that $\mathbf{D}(\mathbf{q})$ is a 3×3 Hermitian matrix with all real eigenvalues $M_a[\omega_\lambda(\mathbf{q})]^2$, although as defined in Eq. (7.14), each ij component of $\mathbf{D}(\mathbf{q})$ is a complex number. If desired, however, one can translate the crystal symmetries of $\mathbf{K}$ into corresponding symmetries of $\mathbf{D}$, creating all real components of the latter matrix in the process. To do this here, we use the property of inversion symmetry, as given by Eq. (7.6), together with the property of translational invariance, as given by Eq. (7.7), to write

$$\mathbf{D}(\mathbf{q}) = -\frac{1}{2N} \sum_{i,j} \mathbf{K}(\mathbf{R}_i - \mathbf{R}_j, \Omega) \left[e^{i \mathbf{q} \cdot (\mathbf{R}_i - \mathbf{R}_j)} + e^{-i \mathbf{q} \cdot (\mathbf{R}_i - \mathbf{R}_j)} - 2 \right]$$

$$= \frac{1}{N} \sum_{i,j} \mathbf{K}(\mathbf{R}_i - \mathbf{R}_j, \Omega) \left[1 - \cos\{\mathbf{q} \cdot (\mathbf{R}_i - \mathbf{R}_j)\} \right] . \tag{7.15}$$

For each value of **q**, the solution of Eq. (7.13) is accomplished by diagonalizing the dynamical matrix $\mathbf{D}(\mathbf{q})$, yielding three normal-mode vibrational or phonon frequencies $\omega_\lambda(\mathbf{q})$ and three corresponding polarization vectors $\mathbf{p}_\lambda(\mathbf{q})$ for $\lambda = 1, 2, 3$. The phonon frequencies $\omega_\lambda(\mathbf{q})$ are *acoustic* modes and, writing $\mathbf{q} = q\hat{\mathbf{q}}$, have the simple properties that first $[\omega_\lambda(\mathbf{q})]^2$ is an even function of q, and second $\omega_\lambda(\mathbf{q}) \propto q$ as $q \to 0$, properties that are evident from the form of $\mathbf{D}(\mathbf{q})$ in Eq. (7.15). The magnitudes of the polarization vectors $\mathbf{p}_\lambda(\mathbf{q})$ can be normalized to unity and then satisfy the mutual orthogonality condition

$$\mathbf{p}_\lambda(\mathbf{q}) \cdot \mathbf{p}_{\lambda'}(\mathbf{q}) = \delta_{\lambda\lambda'} \quad \lambda, \lambda' = 1, 2, 3. \tag{7.16}$$

Along certain high-symmetry directions in the BZ, the three acoustic branches of the phonon spectrum will be divided into one *longitudinal* branch, with the ion motion and $\mathbf{p}_\lambda(\mathbf{q})$ parallel to the direction of wave propagation, and two *transverse* branches, with the ion motion and $\mathbf{p}_\lambda(\mathbf{q})$ perpendicular to the wave-propagation direction.

For the more general case of a crystal structure with a nonprimitive lattice, represented by N_c unit cells and n_b basis atoms in each cell, Eq. (7.13) continues to hold for the normal-mode frequencies and polarizations, but with the index λ now running from 1 to $3n_b$ and the dynamical matrix increasing to $3n_b \times 3n_b$ in size with components

$$D_{\alpha\mu,\beta\nu}(\mathbf{q}) \equiv -\frac{1}{N_c} \sum_{i,j}{}' K_{\alpha\mu,\beta\nu}(\mathbf{R}_{i\alpha} - \mathbf{R}_{j\beta}, \Omega) e^{i\mathbf{q}\cdot(\mathbf{R}_{i\alpha} - \mathbf{R}_{j\beta})}, \tag{7.17}$$

where we have used the same notation as in Eq. (7.8). In this case for a given wavevector **q**, the $3n_b$ phonon frequencies and polarization vectors obtained from Eq. (7.13) will consist of 3 acoustic modes and $3n_b - 3$ *optical* modes. The optical branches of the phonon spectrum generally involve higher lying vibrational frequencies than do the acoustic branches, with nonzero values of $\omega_\lambda(\mathbf{q})$ at $q = 0$.

To calculate the QHLD phonon spectrum for a real material, one requires QBIP input on the force-constant matrix $\mathbf{K}(\mathbf{R}_i - \mathbf{R}_j, \Omega)$, as given by Eqs. (7.5) and (7.8). Consistent with our treatment of volume-dependent QBIPs in Chapters 3–5, we consider a total-energy functional in the general multi-ion potential form

$$E_{\text{tot}}(\mathbf{R}, \Omega) = NE_{\text{vol}}(\Omega) + \frac{1}{2}\sum_{i,j}{}' v_2(ij, \Omega) + \frac{1}{6}\sum_{i,j,k}{}' v_3(ijk, \Omega) + \frac{1}{24}\sum_{i,j,k,l}{}' v_4(ijkl, \Omega) + \cdots \tag{7.18}$$

to obtain the needed second derivatives of E_{tot} in important prototype cases. Again, for simplicity and clarity, we assume a primitive lattice, where **K** has simple Cartesian-component contributions from $v_2, v_3, v_4, \ldots$, which we respectively denote as

$$K_{\mu\nu} = K_{\mu\nu}^{(2)} + K_{\mu\nu}^{(3)} + K_{\mu\nu}^{(4)} + \cdots. \tag{7.19}$$

We first treat the two-ion pair-potential contribution $K_{\mu\nu}^{(2)}$ for central-force nontransition metals in Sec. 7.1.2. This treatment directly applies to first-principles GPT pair potentials in the SM, EDB and FDB limits, as well as to DRT and RMP pair potentials. We then consider the contributions $K^{(2)}_{\mu\nu}$, $K^{(3)}_{\mu\nu}$ and $K^{(4)}_{\mu\nu}$ derived from MGPT multi-ion potentials for transition metals in Sec. 7.1.3.

7.1.2 Tangential and radial force-constant functions for pair potentials

Recall that at constant volume, the two-ion pair potential in Eq. (7.18) depends only on the radial distance between atoms: $v_2(ij, \Omega) = v_2(R_{ij}, \Omega)$. This greatly simplifies the calculation of the force-constant components $K_{\mu\nu} = K^{(2)}_{\mu\nu}$, which can then be separated into first and second radial derivatives of the pair potential and geometrical lattice factors, as follows. From Eq. (7.5), and Eq. (7.18) terminated at the pair-potential level, one has

$$-\frac{\partial^2 E_{\text{tot}}(\mathbf{R}, \Omega)}{\partial r_{i\mu} \partial r_{j\nu}} = -\frac{\partial^2}{\partial r_{i\mu} \partial r_{j\nu}} \left[\frac{1}{2} \sum_{i,j}{}' v_2(R_{ij}, \Omega) \right] = -\frac{\partial^2 v_2(R_{ij}, \Omega)}{\partial r_{i\mu} \partial r_{j\nu}}$$

$$= -\frac{\partial v_2(R_{ij}, \Omega)}{\partial R_{ij}} \frac{\partial^2 R_{ij}}{\partial r_{i\mu} \partial r_{j\nu}} - \frac{\partial^2 v_2(R_{ij}, \Omega)}{\partial R_{ij}^2} \frac{\partial R_{ij}}{\partial r_{i\mu}} \frac{\partial R_{ij}}{\partial r_{j\nu}}. \quad (7.20)$$

To evaluate the partial derivatives of R_{ij} with respect to its components on the second line of Eq. (7.20), we note that

$$R_{ij} = \left[(x_i - x_j)^2 + (y_i - y_j)^2 + (z_i - z_j)^2 \right]^{1/2} \equiv \left[\sum_{\mu=x,y,z} (r_{i\mu} - r_{j\mu})^2 \right]^{1/2}. \quad (7.21)$$

Then in Eq. (7.20) one has for the first-derivative terms

$$\frac{\partial R_{ij}}{\partial r_{i\mu}} = \frac{r_{i\mu} - r_{j\mu}}{R_{ij}} \equiv \frac{r_{ij\mu}}{R_{ij}} \quad \text{and} \quad \frac{\partial R_{ij}}{\partial r_{j\nu}} = -\frac{r_{i\nu} - r_{j\nu}}{R_{ij}} \equiv -\frac{r_{ij\nu}}{R_{ij}}, \quad (7.22)$$

while in the second-derivative term for $\nu = \mu$

$$\frac{\partial^2 R_{ij}}{\partial r_{i\mu} \partial r_{j\mu}} = -\left[1 - \frac{(r_{i\mu} - r_{j\mu})^2}{R_{ij}^2} \right] \frac{1}{R_{ij}} \equiv -\left[1 - \frac{r_{ij\mu}^2}{R_{ij}^2} \right] \frac{1}{R_{ij}} \quad (7.23)$$

and for $\nu \neq \mu$

$$\frac{\partial^2 R_{ij}}{\partial r_{i\mu} \partial r_{j\nu}} = \frac{(r_{i\mu} - r_{j\mu})(r_{i\nu} - r_{j\nu})}{R_{ij}^2} \frac{1}{R_{ij}} \equiv \frac{r_{ij\mu} r_{ij\nu}}{R_{ij}^2} \frac{1}{R_{ij}}. \quad (7.24)$$

Defining general two-ion tangential and radial force-constant functions for the pair potential as

$$K_{\tan}^{(2)}(r,\Omega) \equiv \frac{1}{r}\frac{\partial v_2(r,\Omega)}{\partial r} \tag{7.25}$$

and

$$K_{\rad}^{(2)}(r,\Omega) \equiv \frac{\partial^2 v_2(r,\Omega)}{\partial r^2}, \tag{7.26}$$

respectively, one can then express the components of the full force-constant matrix $\mathbf{K}$ in forms that can be readily evaluated for real materials. One has for the $\nu = \mu$ diagonal elements

$$K_{\mu\mu}(\mathbf{R}_i - \mathbf{R}_j, \Omega) = K_{\mu\mu}^{(2)} = K_{\tan}^{(2)}(R_{ij}, \Omega) + \left[K_{\rad}^{(2)}(R_{ij}, \Omega) - K_{\tan}^{(2)}(R_{ij}, \Omega)\right]\frac{r_{ij\mu}^2}{R_{ij}^2}\bigg|_{R=R_0} \tag{7.27}$$

and for the $\nu \neq \mu$ off-diagonal elements

$$K_{\mu\nu}(\mathbf{R}_i - \mathbf{R}_j, \Omega) = K_{\mu\nu}^{(2)} = \left[K_{\rad}^{(2)}(R_{ij}, \Omega) - K_{\tan}^{(2)}(R_{ij}, \Omega)\right]\frac{r_{ij\mu}r_{ij\nu}}{R_{ij}^2}\bigg|_{R=R_0}. \tag{7.28}$$

It should also be noted that one can quite easily extend Eqs. (7.27) and (7.28) to the case of a nonprimitive crystal structure by increasing the size of the force-constant matrix to $3n_b \times 3n_b$ and simply generalizing the notation in the manner of Eq. (7.8). Thus, one has on the RHS of Eqs. (7.27) and (7.28) the substitution $R_{ij} \to R_{i\alpha,j\beta}$ with the corresponding Cartesian components $r_{ij\mu} \to r_{(i\alpha,j\beta)\mu}$ and $r_{ij\nu} \to r_{(i\alpha,j\beta)\nu}$.

In practice, to calculate the long-ranged tangential and radial force-constant functions with sufficient accuracy for use in Eqs. (7.27) and (7.28), one needs to evaluate the radial derivatives of the pair potential v_2 analytically in Eqs. (7.25) and (7.26). To do this, we make use of the general form of $v_2(r,\Omega)$ given by Eq. (4.61) for EDB and FDB metals, but now with that result conveniently written in terms of the $\ell = 0$ spherical Bessel function $j_0(qr)$ to simplify the calculation of the needed derivatives under the integral sign:

$$v_2(r,\Omega) = \frac{(Z^\star e)^2}{r}\left[1 - \frac{2r}{\pi}\int_0^\infty F_N(q,\Omega)j_0(qr)dq\right] + v_{\ol}(r,\Omega). \tag{7.29}$$

Of course, in the limit of vanishing sp-d hybridization contributions to $F_N(q,\Omega)$ and a vanishing d-state overlap potential, with $v_{\ol}(r,\Omega) = 0$ in Eq. (7.29), this equation is also equivalent to the GPT result given by Eq. (3.59) in the SM limit. And with the additional assumption $Z^\star = Z$, Eq. (7.29) is furthermore equivalent to the DRT simple-metal result given by Eq. (3.57).

Differentiating the first two terms in Eq. (7.29) analytically with respect to r then yields

$$K_{\tan}^{(2)}(r,\Omega) = -\frac{(Z^*e)^2}{r^3}\left[1 - \frac{2r^2}{\pi}\int_0^\infty qF_N(q,\Omega)j_1(qr)dq\right] + \frac{1}{r}\frac{\partial v_{\text{ol}}(r,\Omega)}{\partial r} \quad (7.30)$$

and

$$K_{\text{rad}}^{(2)}(r,\Omega) = \frac{2(Z^*e)^2}{r^3}\left[1 - \frac{r^2}{\pi}\int_0^\infty qF_N(q,\Omega)\{2j_1(qr) - (qr)j_0(qr)\}dq\right] + \frac{\partial^2 v_{\text{ol}}(r,\Omega)}{\partial r^2}. \quad (7.31)$$

Regarding the additional short-ranged d-state overlap potential $v_{\text{ol}}(r,\Omega)$, which is needed for GPT-EDB and -FDB metals, recall that this potential is expressed quite generally by Eq. (4.55). While not an analytic result with respect the ion-ion separation distance r, this equation for $v_{\text{ol}}(r,\Omega)$ at constant volume can be fit very accurately with a polynomial expansion of the form

$$v_{\text{ol}}(r,\Omega) = v_0 P(x)e^{\beta x} = v_0\left[1 + \gamma_1 x + \gamma_2 x^2 + \cdots + \gamma_{10}x^{10}\right]e^{\beta x}, \quad (7.32)$$

with $x = 1.8 - r/R_{\text{WS}}$. The desired analytic first and second radial derivatives of $v_{\text{ol}}(r,\Omega)$ appearing in Eqs. (7.30) and (7.31), respectively, are then readily obtained.

To illustrate the general behavior of $K_{\tan}^{(2)}(r,\Omega)$ and $K_{\text{rad}}^{(2)}(r,\Omega)$ for nontransition metals we have plotted these functions at normal density for the prototype cases of Ca and Zn in Fig. 7.1. Two qualitatively distinctive features are seen for both metals. First, at short ion-ion separation distances r, near and below the nearest-neighbor interatomic distance, $K_{\tan}^{(2)}(r,\Omega_0)$ is negative and monotonically decreasing at a rapid rate, while $K_{\text{rad}}^{(2)}(r,\Omega_0)$ is positive and monotonically increasing at a similar rapid rate. Second, at essentially all interatomic separation distances $r = R_{ij}$, one finds $|K_{\text{rad}}(R_{ij},\Omega_0)| \gg |K_{\tan}(R_{ij},\Omega_0)|$, so that more generally, one expects $K_{\text{rad}}^{(2)}(R_{ij},\Omega)$ contributions to dominate quantitatively in the evaluation of the force-constant matrix components, Eqs. (7.27) and (7.28), and hence in the dynamical matrix $\mathbf{D}(\mathbf{k})$. This finding is consistent with the heuristic notion that $K_{\text{rad}}^{(2)}$ arises primarily from "bond-stretching" forces, while $K_{\tan}^{(2)}$ arises from "bond-bending" forces.

7.1.3 Tangential and radial force-constant functions for multi-ion potentials

The above evaluation of the force-constant matrix in terms of tangential and radial force-constant functions for a nontransition-metal pair potential can be extended to the more general case of the transition-metal multi-ion potentials entering Eq. (7.18). Regarding practical applications in transition metals, most of the current focus has been on the

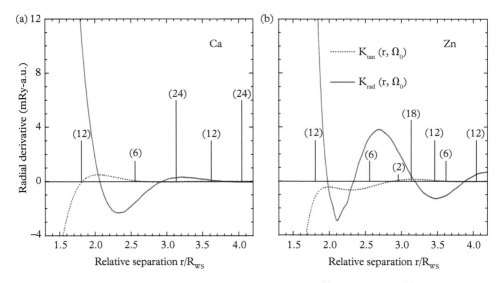

Fig. 7.1 *Tangential and radial force-constant functions, $K_\text{tan} = K_\text{tan}^{(2)}$ and $K_\text{rad} = K_\text{rad}^{(2)}$, for two prototype nontransition metals at their equilibrium volumes, as calculated from first-principles GPT potentials. (a) Ca treated as an EDB metal, with the positions and number of its fcc near neighbors indicated. (b) Zn treated as an FDB metal, with the position and number of its hcp near neighbors indicated.*

simplified MGPT pair and multi-ion potentials, but it is important to note that the formal analysis can also be applied to the underlying first-principles GPT potentials, as we now discuss. Retaining two-, three- and four-ion contributions to either the GPT or MGPT transition-metal total energy given by Eq. (7.18), expressed in the shorthand form

$$E_\text{tot} = NE_\text{vol} + E_\text{tot}^{(2)} + E_\text{tot}^{(3)} + E_\text{tot}^{(4)}, \tag{7.33}$$

one can then proceed to evaluate the corresponding components of the force-constant matrix $K_{\mu\nu}^{(2)}$, $K_{\mu\nu}^{(3)}$ and $K_{\mu\nu}^{(4)}$ defined in Eq. (7.19). Again, for simplicity, we consider the case of a primitive lattice, with the extension to a nonprimitive crystal structure entirely straightforward.

In the first-principles GPT for transition metals, Eqs. (7.27)–(7.32) above still apply to the calculation of $K_{\mu\nu}^{(2)}$, with the additional effects of partial d-band filling contained in the generalized forms of $F_\text{N}(q, \Omega)$ and $v_\text{ol}(r, \Omega)$ defined in Chapter 5 via Eqs. (5.66)–(5.76). To calculate $K_{\mu\nu}^{(2)}$ in the MGPT, on the other hand, recall that the two-ion pair potential has the simplified form given by Eq. (5.99):

$$v_2(r, \Omega) = v_2^{sp}(r, \Omega) + v_2^{hc}(r, \Omega) + v_2^{d}(r, \Omega), \tag{7.34}$$

where the sp simple-metal potential v_2^{sp} and the hard-core potential v_2^{hc} are analogous, respectively, to the first two and third terms in Eq. (7.29), with F_N evaluated in the SM

7.1.3 Tangential and radial force-constant functions for multi-ion potentials

limit and v_{ol} replaced by v_2^{hc}. Also recall from Eq. (5.95) that the additional short-ranged two-ion d-state potential v_2^d in Eq. (7.33) has the characteristic analytic form

$$v_2^d(r, \Omega) = v_a(\Omega)(R_0/r)^{4p} - v_b(\Omega)(R_0/r)^{2p}, \qquad (7.35)$$

with $R_0 \equiv 1.8 R_{\text{WS}}$ and p a chosen constant parameter. In place of Eqs. (7.30) and (7.31), the MGPT two-ion tangential and radial force-constant functions are then given by

$$K_{\text{tan}}^{(2)}(r, \Omega) = -\frac{(Z^*e)^2}{r^3}\left[1 - \frac{2r^2}{\pi}\int_0^\infty qF_{\text{N}}(q, \Omega)j_1(qr)dq\right] + \frac{1}{r}\frac{\partial v_2^{\text{hc}}(r, \Omega)}{\partial r}$$
$$- \frac{2p}{r^2}\left[2v_a(\Omega)(R_0/r)^{4p} - v_b(\Omega)(R_0/r)^{2p}\right] \qquad (7.36)$$

and

$$K_{\text{rad}}^{(2)}(r, \Omega) = \frac{2(Z^*e)^2}{r^3}\left[1 - \frac{r^2}{\pi}\int_0^\infty qF_{\text{N}}(q, \Omega)\{2j_1(qr) - (qr)j_0(qr)\}dq\right] + \frac{\partial^2 v_2^{\text{hc}}(r, \Omega)}{\partial r^2}$$
$$+ \frac{2p}{r^2}\left[2v_a(\Omega)(4p+1)(R_0/r)^{4p} - v_b(\Omega)(2p+1)(R_0/r)^{2p}\right]. \qquad (7.37)$$

These results can then be used directly in Eqs. (7.27) and (7.28) to evaluate $K_{\mu\nu}^{(2)}$, with the hard-core potential v_2^{hc} in practice fitted to an analytic function in the same manner as v_{ol} in Eq. (7.32).

To calculate the multi-ion components $K_{\mu\nu}^{(3)}$ and $K_{\mu\nu}^{(4)}$ to the full force-constant matrix $K_{\mu\nu}$ in either the GPT or MGPT, one needs to develop the spatial derivatives of $E_{\text{tot}}^{(3)}$ and $E_{\text{tot}}^{(4)}$ in a similar manner to that done in Eq. (7.20) for $E_{\text{tot}}^{(2)}$:

$$-\frac{\partial^2 E_{\text{tot}}^{(3)}(\mathbf{R}, \Omega)}{\partial r_{i\mu}\partial r_{j\nu}} = -\frac{\partial^2}{\partial r_{i\mu}\partial r_{j\nu}}\left[\frac{1}{6}\sum_{i,j,k}{}' v_3(R_{ij}, R_{jk}, R_{ki}, \Omega)\right] = -\sum_k{}' \frac{\partial^2 v_3(R_{ij}, R_{jk}, R_{ki}, \Omega)}{\partial r_{i\mu}\partial r_{j\nu}}$$

$$= -\sum_k{}'\left\{\frac{\partial v_3(ijk, \Omega)}{\partial R_{ij}}\frac{\partial^2 R_{ij}}{\partial r_{i\mu}\partial r_{j\nu}} + \frac{\partial^2 v_3(ijk, \Omega)}{\partial R_{ij}^2}\frac{\partial R_{ij}}{\partial r_{i\mu}}\frac{\partial R_{ij}}{\partial r_{j\nu}}\right.$$
$$+ \frac{\partial^2 v_3(ijk, \Omega)}{\partial R_{ij}\partial R_{jk}}\frac{\partial R_{ij}}{\partial r_{i\mu}}\frac{\partial R_{jk}}{\partial r_{j\nu}} + \frac{\partial^2 v_3(ijk, \Omega)}{\partial R_{jk}\partial R_{ki}}\frac{\partial R_{ki}}{\partial r_{i\mu}}\frac{\partial R_{jk}}{\partial r_{j\nu}}$$
$$\left.+ \frac{\partial^2 v_3(ijk, \Omega)}{\partial R_{ki}\partial R_{ij}}\frac{\partial R_{ki}}{\partial r_{i\mu}}\frac{\partial R_{ij}}{\partial r_{j\nu}}\right\} \qquad (7.38)$$

and

$$-\frac{\partial^2 E_{\text{tot}}^{(4)}(\mathbf{R},\Omega)}{\partial r_{i\mu}\partial r_{j\nu}} = -\frac{\partial^2}{\partial r_{i\mu}\partial r_{j\nu}}\left[\frac{1}{24}\sum_{i,j,k,l}' v_4(R_{ij},R_{jk},R_{kl}\ldots,\Omega)\right]$$

$$= -\frac{1}{2}\sum_{k,l}' \frac{\partial^2 v_4(R_{ij},R_{jk},R_{kl}\ldots,\Omega)}{\partial r_{i\mu}\partial r_{j\nu}}$$

$$= -\frac{1}{2}\sum_{k,l}' \left\{ \frac{\partial v_4(ijkl,\Omega)}{\partial R_{ij}} \frac{\partial^2 R_{ij}}{\partial r_{i\mu}\partial r_{j\nu}} + \frac{\partial^2 v_4(ijkl,\Omega)}{\partial R_{ij}^2} \frac{\partial R_{ij}}{\partial r_{i\mu}} \frac{\partial R_{ij}}{\partial r_{j\nu}} \right.$$

$$+ \frac{\partial^2 v_4(ijkl,\Omega)}{\partial R_{ij}\partial R_{jk}} \frac{\partial R_{ij}}{\partial r_{i\mu}} \frac{\partial R_{jk}}{\partial r_{j\nu}} + \frac{\partial^2 v_4(ijkl,\Omega)}{\partial R_{jk}\partial R_{ki}} \frac{\partial R_{ki}}{\partial r_{i\mu}} \frac{\partial R_{jk}}{\partial r_{j\nu}}$$

$$+ \frac{\partial^2 v_4(ijkl,\Omega)}{\partial R_{ki}\partial R_{ij}} \frac{\partial R_{ki}}{\partial r_{i\mu}} \frac{\partial R_{ij}}{\partial r_{j\nu}} + \frac{\partial^2 v_4(ijkl,\Omega)}{\partial R_{li}\partial R_{ij}} \frac{\partial R_{li}}{\partial r_{i\mu}} \frac{\partial R_{ij}}{\partial r_{j\nu}}$$

$$+ \frac{\partial^2 v_4(ijkl,\Omega)}{\partial R_{li}\partial R_{jk}} \frac{\partial R_{li}}{\partial r_{i\mu}} \frac{\partial R_{jk}}{\partial r_{j\nu}} + \frac{\partial^2 v_4(ijkl,\Omega)}{\partial R_{ij}\partial R_{lj}} \frac{\partial R_{ij}}{\partial r_{i\mu}} \frac{\partial R_{lj}}{\partial r_{j\nu}}$$

$$\left. + \frac{\partial^2 v_4(ijkl,\Omega)}{\partial R_{ki}\partial R_{lj}} \frac{\partial R_{ki}}{\partial r_{i\mu}} \frac{\partial R_{lj}}{\partial r_{j\nu}} + \frac{\partial^2 v_4(ijkl,\Omega)}{\partial R_{li}\partial R_{lj}} \frac{\partial R_{li}}{\partial r_{i\mu}} \frac{\partial R_{lj}}{\partial r_{j\nu}} \right\}. \quad (7.39)$$

As was done in Eq. (7.20), we have separated the RHSs of Eqs. (7.38) and (7.39) into radial first and second radial derivatives of v_3 and v_4, respectively, and into geometrical lattice factors. The lattice factors can be evaluated exactly as in Eq. (7.21)–(7.24). The radial derivatives of the multi-ion potentials can be re-expressed in terms of generalized tangential and radial force-constant functions defined for the n-ion potential v_n as

$$K_{\tan}^{(n)}(r_1,\Omega) \equiv \frac{1}{r_1}\frac{\partial v_n(r_1,r_2,\ldots,r_m,\Omega)}{\partial r_1} \quad (7.40)$$

and

$$K_{\text{rad}}^{(n)}(r_1,r_k,\Omega) \equiv \frac{\partial^2 v_n(r_1,r_2,\ldots r_k\ldots r_m,\Omega)}{\partial r_1\partial r_k}. \quad (7.41)$$

For $n = 3$, corresponding to the three-ion potential v_3, one has the conditions $m = 3$ and $k \leq 3$ in Eqs. (7.40) and (7.41). For $n = 4$, corresponding to the four-ion potential v_4, one has $m = 6$ and $k \leq 6$. Also note that for $n = 2$ and $k = m = 1$, corresponding to the two-ion pair potential v_2, Eqs. (7.40) and (7.41) are equivalent to Eqs. (7.25) and (7.26), respectively.

7.1.3 Tangential and radial force-constant functions for multi-ion potentials

For the diagonal $\nu = \mu$ elements of the three-ion force-constant matrix component $K^{(3)}_{\mu\nu}$ one finds

$$K^{(3)}_{\mu\mu} = \sum_{k}{}' \left\{ K^{(3)}_{\text{tan}}(R_{ij},\Omega) + \left[K^{(3)}_{\text{rad}}(R_{ij},R_{ij},\Omega) - K^{(3)}_{\text{tan}}(R_{ij},\Omega)\right]\frac{r^2_{ij\mu}}{R^2_{ij}} - K^{(3)}_{\text{rad}}(R_{ij},R_{jk},\Omega)\frac{r_{ij\mu}r_{jk\mu}}{R_{ij}R_{jk}} \right.$$

$$\left. + K^{(3)}_{\text{rad}}(R_{jk},R_{ki},\Omega)\frac{r_{jk\mu}r_{ki\mu}}{R_{jk}R_{ki}} - K^{(3)}_{\text{rad}}(R_{ki},R_{ij},\Omega)\frac{r_{ki\mu}r_{ij\mu}}{R_{ki}R_{ij}} \right\}\Bigg|_{\mathbf{R}=\mathbf{R}_0} \quad (7.42)$$

and for the $\nu \neq \mu$ off-diagonal elements

$$K^{(3)}_{\mu\nu} = \sum_{k}{}' \left\{ \left[K^{(3)}_{\text{rad}}(R_{ij},R_{ij},\Omega) - K^{(3)}_{\text{tan}}(R_{ij},\Omega)\right]\frac{r_{ij\mu}r_{ij\nu}}{R^2_{ij}} - K^{(3)}_{\text{rad}}(R_{ij},R_{jk},\Omega)\frac{r_{ij\mu}r_{jk\nu}}{R_{ij}R_{jk}} \right.$$

$$\left. + K^{(3)}_{\text{rad}}(R_{jk},R_{ki},\Omega)\frac{r_{jk\mu}r_{ki\nu}}{R_{jk}R_{ki}} - K^{(3)}_{\text{rad}}(R_{ki},R_{ij},\Omega)\frac{r_{ki\mu}r_{ij\nu}}{R_{ki}R_{ij}} \right\}\Bigg|_{\mathbf{R}=\mathbf{R}_0}. \quad (7.43)$$

Similarly, for the diagonal $\nu = \mu$ elements of the four-ion force-constant matrix component $K^{(4)}_{\mu\nu}$ one obtains

$$K^{(4)}_{\mu\mu} = \frac{1}{2}\sum_{k,l}{}' \left\{ K^{(4)}_{\text{tan}}(R_{ij},\Omega) + \left[K^{(4)}_{\text{rad}}(R_{ij},R_{ij},\Omega) - K^{(4)}_{\text{tan}}(R_{ij},\Omega)\right]\frac{r^2_{ij\mu}}{R^2_{ij}} \right.$$

$$- K^{(4)}_{\text{rad}}(R_{ij},R_{jk},\Omega)\frac{r_{ij\mu}r_{jk\mu}}{R_{ij}R_{jk}} + K^{(4)}_{\text{rad}}(R_{jk},R_{ki},\Omega)\frac{r_{jk\mu}r_{ki\mu}}{R_{jk}R_{ki}}$$

$$- K^{(4)}_{\text{rad}}(R_{ki},R_{ij},\Omega)\frac{r_{ki\mu}r_{ij\mu}}{R_{ki}R_{ij}} - K^{(4)}_{\text{rad}}(R_{li},R_{ij},\Omega)\frac{r_{li\mu}r_{ij\mu}}{R_{li}R_{ij}}$$

$$+ K^{(4)}_{\text{rad}}(R_{li},R_{jk},\Omega)\frac{r_{li\mu}r_{jk\mu}}{R_{li}R_{jk}} + K^{(4)}_{\text{rad}}(R_{ij},R_{lj},\Omega)\frac{r_{ij\mu}r_{lj\mu}}{R_{ij}R_{lj}}$$

$$\left. - K^{(4)}_{\text{rad}}(R_{ki},R_{lj},\Omega)\frac{r_{ki\mu}r_{lj\mu}}{R_{ki}R_{lj}} - K^{(4)}_{\text{rad}}(R_{li},R_{lj},\Omega)\frac{r_{li\mu}r_{lj\mu}}{R_{li}R_{lj}} \right\}\Bigg|_{\mathbf{R}=\mathbf{R}_0} \quad (7.44)$$

and for the $\nu \neq \mu$ off-diagonal elements

$$K^{(4)}_{\mu\nu} = \frac{1}{2}\sum_{k,l}{}' \left\{ \left[K^{(4)}_{\text{rad}}(R_{ij},R_{ij},\Omega) - K^{(4)}_{\text{tan}}(R_{ij},\Omega)\right]\frac{r_{ij\mu}r_{ij\nu}}{R^2_{ij}} - K^{(4)}_{\text{rad}}(R_{ij},R_{jk},\Omega)\frac{r_{ij\mu}r_{jk\nu}}{R_{ij}R_{jk}} \right.$$

$$+ K^{(4)}_{\text{rad}}(R_{jk},R_{ki},\Omega)\frac{r_{jk\mu}r_{ki\nu}}{R_{jk}R_{ki}} - K^{(4)}_{\text{rad}}(R_{ki},R_{ij},\Omega)\frac{r_{ki\mu}r_{ij\nu}}{R_{ki}R_{ij}} - K^{(4)}_{\text{rad}}(R_{li},R_{ij},\Omega)\frac{r_{li\mu}r_{ij\nu}}{R_{li}R_{ij}}$$

$$+ K^{(4)}_{\text{rad}}(R_{li},R_{jk},\Omega)\frac{r_{li\mu}r_{jk\nu}}{R_{li}R_{jk}} + K^{(4)}_{\text{rad}}(R_{ij},R_{lj},\Omega)\frac{r_{ij\mu}r_{lj\nu}}{R_{ij}R_{lj}} - K^{(4)}_{\text{rad}}(R_{ki},R_{lj},\Omega)\frac{r_{ki\mu}r_{lj\nu}}{R_{ki}R_{lj}}$$

$$\left. - K^{(4)}_{\text{rad}}(R_{li},R_{lj},\Omega)\frac{r_{li\mu}r_{lj\nu}}{R_{li}R_{lj}} \right\}\Bigg|_{\mathbf{R}=\mathbf{R}_0}. \quad (7.45)$$

In the full first-principles GPT, the multi-ion potentials v_3 and v_4 are both long ranged, due to *sp-d* hybridization, and nonanalytic, so the calculation of the corresponding tangential and radial force-constant functions must be done numerically. Thus, the full GPT calculation and implementation of $K_{\mu\nu}^{(3)}$ and $K_{\mu\nu}^{(4)}$ is very challenging, and in practice this has limited application to the simplest cases such as Ba, where only two- and three-ion forces need to be taken into account (Moriarty, 1986b). In the standard MGPT, however, *sp-d* hybridization is neglected, and the remaining short-ranged multi-ion potentials can be expressed in analytic form, as given by Eq. (5.100) for v_3 and Eq. (5.103) for v_4. In practice, it is still convenient and most efficient to calculate the needed values of $K_{\tan}^{(n)}$ and $K_{\rad}^{(n)}$ numerically, and this has proven to be a very successful strategy in applications to the central bcc transition metals.

7.1.4 Important caveats and alternate approaches

While the above QHLD formalism is a straightforward and computationally efficient approach for QBIP phonon calculations in metals, it does introduce a hidden, but well-known, approximation inherent in its application at constant volume. Specifically, except for purely transverse phonon modes, the displacement of atoms in a general lattice vibration is not a volume-conserving process. The local-volume effects that occur as a result are generally small in metals, however, except at long wavelengths, where they can significantly impact the calculated elastic constants, a shortcoming that we will address in Sec. 7.3. For shorter-wavelength phonons, on the other hand, the local volume effects are mostly negligible, such that in practice both good quality phonon spectra and accurate thermodynamic properties can be obtained in most cases, as we will demonstrate below in Secs. 7.2, 7.4 and 7.5.

At the same time, it should be noted that it is possible to go beyond the constant-volume assumption, as well as the total-energy functional (7.18), in calculating the needed force constants for QHLD, using robust numerical methods developed over the past 30 years that allow both DFT and QBIP calculation of phonon band structures. These latter methods fall into two broad categories: methods based on the linear response of the crystal to small atomic displacements and direct force-constant methods using numerical calculations of energies and forces associated with the displacement field. In the linear response approach, the force constants can be directly related to the inverse dielectric function, or equivalent susceptibility functions, describing the response of the material to small atomic displacements (Varma and Weber, 1979; Finnis et al., 1984; Beroni et al., 2001; Finnis, 2003). But in practice such methods can be complicated and depend on the detailed electronic structure of the material. In the present context, they have mostly been implemented in TB or DFT treatments of transition metals.

In the direct approach, on the other hand, one introduces supercells and uses accurate forces to build upon the simple concept of a frozen phonon. In an isolated frozen-phonon calculation, the atomic displacements for a short-wavelength, zone-boundary phonon mode can be produced in a small-sized supercell, and the change in total energy, calculated as a function of the magnitude of the displacement, can then be fit to a quadratic displacement form to obtain the corresponding phonon frequency. This isolated atomic process becomes more computationally difficult for longer-wavelength, lower-symmetry individual phonons, however, which can require large supercells. To obtain the full phonon spectrum, a more attractive alternative to the use of frozen-phonon energies is to use accurate atomic forces combined with chosen small ion displacements. To do this, the usual QHLD equation of motion given by Eq. (7.10) for a primitive lattice can be replaced by the static force equation

$$F_{i\mu} = -\frac{\partial E_{\text{tot}}(\mathbf{R}, \Omega)}{\partial u_{i\mu}} = \sum_{\substack{j \\ \nu=x,y,z}} K_{\mu\nu}(\mathbf{R}_i - \mathbf{R}_j, \Omega) u_{j\nu}, \qquad (7.46)$$

where $F_{i\mu}$ is the μth component of force on ion i for small ion displacements $u_{j\nu}$ from equilibrium. For a small chosen displacement of a single ion j within the supercell, the corresponding component of the force-constant matrix is just

$$K_{\mu\nu}(\mathbf{R}_i - \mathbf{R}_j, \Omega) = \frac{F_{i\mu}}{u_{j\nu}}, \qquad (7.47)$$

where now the force $F_{i\mu}$ can be calculated directly by an external DFT or QBIP method. The process can be repeated for as many displacements as needed, when combined with crystal symmetry, to derive the entire force-constant matrix $\mathbf{K}$. However, to remain tractable, such methods must discard distant interatomic interactions beyond a certain range and seek convergence of the whole process within that range. In practice, this has proven to be possible within DFT using only moderate-sized supercells, as first demonstrated by Frank et al. (1995) for the alkali metals. The method can be readily generalized to nonprimitive lattices and to nonmetals, as shown by Kresse et al. (1995) for carbon.

A number of such direct force-constant phonon methods have been developed and made available in the form of open-source computer codes. One popular code is PHON written by Alfè (2009), which allows the use of external forces derived from any source, including both DFT and QBIPs. This code has been used, for example, to calculate BOP phonon spectra for bcc transition metals by Cák et al. (2014), results that are discussed in Sec. 7.2.2. Another popular direct method due to Tadano et al. (2014) is implemented in the multi-purpose code ALAMODE, which can be interfaced with both DFT electronic-structure and QMD codes as well as with the LAMMPS MD code. This code also allows both DFT and QBIP direct force-constant evaluation and phonon band-structure calculation. Using ALAMODE, Skinner et al. (2019) have calculated GPT, aGPT and DFT phonon-spectra for both hcp and bcc structures in Mg. The ALAMODE aGPT results, which explicitly include local volume effects on the phonon frequencies, were found to be in close agreement with the corresponding GPT results, confirming the expected smallness of the local volume effects on the phonon frequencies. Likewise, the ALAMODE GPT and DFT phonon spectra were found to be in good agreement with each other as well as with independent GPT phonon spectra obtained from the QHLD formalism of Sec. 7.1, the results of which for hcp Mg are discussed in Sec. 7.2.1.

7.2 Calculated quasiharmonic phonon spectra for elemental metals

In this section, using available GPT, DRT, RMP and MGPT potentials, we return to the QHLD formalism that we developed in Sec. 7.1 and apply it to calculate phonon spectra for prototype nontransition and transition metals, and then test the results by direct comparison to experiment.

7.2.1 Nontransition metals

In the case of simple, EDB and FDB metals, the QBIP phonon frequencies at a given volume Ω are determined entirely from force-constant functions established by the two-ion pair potential $v_2(r, \Omega)$, as elaborated above in Sec. 7.1.2. Here we consider calculated phonon spectra for eight such prototype bcc (Na, K, Rb), fcc (Al, Ca, Cu) and hcp (Mg, Zn) metals at or near Ω_0, as obtained with current GPT and DRT potentials and with historical RMP potentials. In the case of the GPT results, the metals Na, K and Mg have been treated in the SM limit, Al and Ca in the EDB limit, and Cu and Zn in the FDB limit. In the latter four metals, optimized d basis states have been used, as described in Chapter 4. The DRT potentials for Na, K, Rb and Al have been obtained with the optimized MP parameters listed in Table 3.1. The GPT and DRT potentials all use IU screening with VWN exchange-correlation, as described in Chapter 3.

In Fig. 7.2 we have plotted calculated GPT phonon spectra along major symmetry directions for the $3d$ metals bcc K, fcc Ca and hcp Zn, and compared the results with experimental neutron diffraction data. Good to excellent agreement between theory and experiment is seen for all three metals. In the case of K, this agreement is representative of that achieved for all of the alkali metals treated with either the GPT or DRT potentials, as is demonstrated in the companion Table 7.1, where GPT, DRT and experimental zone-boundary frequencies and Debye temperatures are favorably compared for bcc Na, K and Rb, as well as for fcc Al, and in Table 7.2 for hcp Mg.

In this regard, we define the Debye temperature $\Theta_D(\Omega)$ here as a precise measure of the average phonon frequency in the ground state of the metal at volume Ω by the relation

$$k_B \Theta_D(\Omega) = \frac{8}{9} E_{ph}^0(\Omega), \quad (7.48)$$

where E_{ph}^0 is the zero-point vibrational energy

$$E_{ph}^0(\Omega) = \frac{1}{2N} \sum_{q,\lambda} h\nu_\lambda(\mathbf{q}, \Omega), \quad (7.49)$$

7.2.1 Nontransition metals

with $\hbar\omega_\lambda$ replaced by $h\nu_\lambda$, and where $\nu_\lambda(\mathbf{q}, \Omega)$ is a calculated or measured frequency of the λ phonon branch at wavevector $\mathbf{q}$ in the first BZ. The evaluation of Eq. (7.49) with experimental phonon frequencies is most easily accomplished by using an empirical force-constant fit to the measured phonon spectrum. Historically, such force-constant fits have often accompanied the reported measured phonon data, especially in cubic metals, as is the case for all six of the bcc or fcc metals considered in Table 7.1.

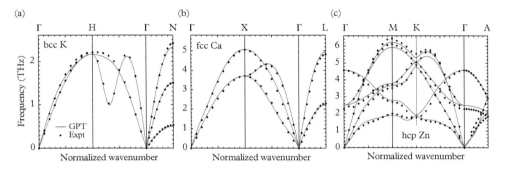

Fig. 7.2 *Phonon spectra for prototype 3d-series nontransition metals at or near normal density, as calculated from first-principles GPT potentials. (a) Bcc K treated in the SM limit at $\Omega_0 = 481.3$ a.u. (volume at 5 K), and compared against the experimental data of Cowley et al. (1966) at 9 K; (b) fcc Ca treated in the EDB limit at $\Omega_0 = 294.5$ a.u., and compared against the experimental data of Heiroth et al. (1986) at room temperature; (c) hcp Zn treated in the FDB limit at $\Omega = 100.8$ a.u. (volume at 80 K), and compared against the experimental data of Almqvist and Stedman (1971) at 80 K.*

The precise quantitative values of Θ_D established by Eqs. (7.48) and (7.49) should also be compared and contrasted with the historical values of the Debye temperature derived from experimental specific heat data. As discussed by Ashcroft and Mermin (1976), such derivation has been done in terms of the highly simplified Debye phonon model itself and results in an artificial, and sometimes rather large, temperature variation in the reported values of the Debye temperature. Most often cited are either a low-temperature value, Θ_D^{LT}, or a mean-temperature value, Θ_D^{MT}, which is obtained at the temperature where the measured specific heat is one-half the limiting high-temperature value of $3k_B$. Available experimental data for both nontransition and transition metals on Θ_D^{LT}, Θ_D^{MT} and Θ_D are listed in Tables A1.3 and A1.4, respectively. For most metals, but not all, one finds $\Theta_D^{LT} \geq \Theta_D^{MT}$. Also, for many of the prototype cubic metals, including K, Al, Cu, Ag, Ta and Mo, one finds $\Theta_D^{MT} \cong \Theta_D$. Experimental values of Θ_D for hcp metals have yet to be calculated, but for good prototypes like Mg and Zn, we expect the latter equality will also hold, and this has been assumed in Fig. 3.9(a) and in Table 7.2.

Returning to the phonon spectra displayed in Fig. 7.2, the excellent agreement seen in fcc Ca between theory and experiment represents a significant quantitative improvement over the early GPT-EDB phonon-spectrum calculations on Ca and Sr by Moriarty (1983). This improvement reflects the importance of optimizing the choice of d basis states used in establishing the cohesive-energy functional and pair potential. At the

298 *Elastic Moduli and Phonons*

Table 7.1 *Zone-boundary phonon frequencies (in THz) and the Debye temperature (in K) for six prototype bcc and fcc nontransition metals, as calculated from GPT, DRT and RMP potentials.*

Metal	BZ Pt	Phonon	GPT[a]	DRT[b]	RMP[c]	Experiment	T_{expt}(K)
Bcc:							
Na	H:	L,T[100]	3.57	3.54	–	3.58±0.04[d]	90
	N:	L[110]	3.63	3.60	–	3.82±0.07	
		T_1[110]	0.85	0.82	–	0.93±0.02	
		T_2[110]	2.56	2.54	–	2.56±0.05	
		Θ_D	158	157	–	163	
K	H:	L,T [100]	2.16	2.17	–	2.21±0.02[e]	9
	N:	L[110]	2.31	2.29	–	2.40±0.04	
		T_1[110]	0.50	0.54	–	0.55±0.01	
		T_2[110]	1.51	1.50	–	1.50±0.02	
		Θ_D	97	97	–	100	
Rb	H:	L,T [100]	1.34	1.37	–	1.385±0.015[f]	12
	N:	L[110]	1.48	1.48	–	1.50±0.02	
		T_1[110]	0.29	0.33	–	0.34±0.02	
		T_2[110]	0.92	0.91	–	0.96±0.03	
		Θ_D	61	61	–	62	
Fcc:							
Al	X:	L[100]	9.09	8.63	–	9.46[g]	300
		T[100]	6.01	5.39	–	5.75	
	L:	L[111]	9.15	8.66	–	9.71	
		T[111]	4.45	4.26	–	4.15	
		Θ_D	391	371	–	391	
Ca	X:	L[100]	5.07	–	–	5.04[h], 4.52±0.08[i]	293, 295
		T[100]	3.69	–	–	3.71, 3.63±0.06	
	L:	L[111]	4.93	–	–	4.82, 4.61±0.08	
		T[111]	2.23	–	–	2.30, 2.36±0.06	
		Θ_D	217	–	–	215, 211	

Cu	X:	L[100]	6.95	–	7.33	7.20±0.20[j]		298
		T[100]	5.57	–	5.54	5.09±0.15		
	L:	L[111]	6.81	–	7.30	7.29±0.20		
		T[111]	3.20	–	3.35	3.41±0.10		
		Θ_D	313	–	–	313		

[a]Present calculated GPT results at the observed volume Ω_0 (Table A1.1), as obtained for Na, K and Rb in the SM limit, for Al and Ca in the EDB limit, and for Cu in the modified FDB limit.
[b]Present calculated DRT results at the observed volume Ω_0, based on the DRT MP parameters for Na, K, Rb and Al listed in Table 3–1.
[c]Upadhyaya and Dagens (1978, 1979), as obtained for Cu at the observed volume $\Omega_0 = 79.68$ a.u. with their RMP2 potential.
[d]Woods et al. (1962).
[e]Cowley et al. (1966).
[f]Copley and Brockhouse (1973).
[g]Gilat and Nicklow (1966), as obtained from a composite eighth-neighbor force-constant fit to experimental data from several sources.
[h]Heiroth et al. (1986).
[i]Stassis et al. (1983).
[j]Nicklow et al. (1967).

same time, on the experimental side in fcc Ca, there remains a small unresolved discrepancy near the X point in the BZ between the polycrystalline data of Heiroth et al. (1986), which are shown in Fig. 7.2(b) and listed in Table 7.1, and the single-crystal data of Stassis et al. (1983), also listed in Table 7.1. The former data is preferred here only because there is some significant fluctuation and relatively large error bars in the Stassis et al. data near the X point. Otherwise, the two sets of phonon data are in good agreement, with only a small 4 K difference in the respective experimental values of Θ_D.

Also significant is the good agreement between the fully optimized GPT-FDB theory and experiment displayed in Fig. 7.2(c) for the hcp Zn phonon spectrum, which is considerably more complex than for fcc or bcc nontransition metals. In this regard, the present GPT treatment of the Zn phonon frequencies includes the optional refinement of a self-consistent d-state energy shift $\delta E_d^{\text{shift}}$ that exactly maintains $\varepsilon_F - E_d^{\text{vol}} = E_F - E_d$, as discussed in Chapter 4. Historically, the first calculations of the Zn hcp phonon spectrum with only simple-metal treatments were mostly unsuccessful. The early reciprocal-space GPT treatments of Panitz et al. (1974) and Cutler et al. (1975), however, using free-atom d states, both with and without sp-d hybridization, but neglecting d-state overlap, showed definite promise. The subsequent real-space RMP calculations of Upadhyaya and Dagens (1982) provided further improvement in this regard, but as is demonstrated in Table 7.2, their success is still well short of that provided by the present optimized GPT-FDB treatment.

Table 7.2 *High-symmetry phonon frequencies (in THz) and the Debye temperature (in K) for the prototype hcp nontransition metals Mg and Zn, as calculated from GPT, DRT and RMP potentials.*

Metal	BZ Pt	Phonon	GPT[a]	DRT[b]	RMP[c]	Experiment	T_{expt}(K)
Mg:		c/a	1.62	1.63	–	1.62	298
	Γ:	Γ_3^+	7.69	7.03	–	7.23[d]	290
		Γ_5^+	3.80	3.88	–	3.73	
	K:	K_1	6.65	6.22	–	6.28	
		K_5	6.16	5.91	–	5.89	
		K_3	5.62	5.58	–	5.44	
		K_6	5.24	4.90	–	5.00	
	M:	M_1^+	7.21	6.80	–	6.89	
		M_2^-	6.91	6.69	–	6.72	
		M_4^-	6.24	5.84	–	5.89	
		M_3^-	5.54	5.53	–	5.44	
		M_3^+	4.11	4.01	–	3.89	
		M_4^+	3.73	3.83	–	3.67	
	A:	A_1	5.55	5.17	–	5.22	
		A_3	2.82	2.99	–	2.78	
		Θ_D	329	320	–	318[e]	
Zn:		c/a	1.76	–	1.705	1.83	80
	Γ:	Γ_3^+	4.58	–	4.71	4.57±0.02[f]	80
		Γ_5^+	2.50	–	2.38	2.29±0.02	
	K:	K_1	5.30	–	5.90	5.55±0.03	
		K_5	5.05	–	5.76	5.04±0.03	
		K_3	4.78	–	5.62	4.89±0.03	
		K_6	1.90	–	2.19	1.79±0.02	
	M:	M_1^+	6.20	–	6.96	6.44±0.03	
		M_2^-	5.91	–	6.82	6.11±0.02	
		M_3^-	3.79	–	4.31	3.72±0.02	
		M_4^+	3.64	–	3.80	3.52±0.02	
		M_4^-	2.79	–	3.19	2.70±0.02	
		M_3^+	1.91	–	1.63	2.02±0.02	

A:	A_1	2.96	–	3.30	2.92±0.02
	A_3	2.10	–	1.89	1.86±0.01
	Θ_D	232	–	–	234[e]

[a] Present calculated GPT results, as obtained for Mg in the SM limit at $\Omega_0 = 156.8$ a.u., and for Zn in the FDB limit at $\Omega = 100.8$ a.u.
[b] Present calculated results, based on the DRT MP parameters for Mg listed in Table 3–1, applied at $\Omega_0 = 156.8$ a.u.
[c] Upadhyaya and Dagens (1982b), at $\Omega = 102.0$ a.u.
[d] Pynn and Squires (1972), as obtained from an eighth-neighbor force-constant fit to the measured experimental data.
[e] Experimental Θ_D^{MT} value only, from Table A1.3.
[f] Almqvist and Stedman (1971).

In contrast to hcp Zn, the phonon spectrum of hcp Mg is well described as a simple metal in pseudopotential perturbation theory, as first shown historically in the reciprocal-space OMP and PP treatments of Shaw and Pynn (1969) and of King and Cutler (1971), respectively. In real-space QBIP phonon calculations on Mg, the robustness of the SM limit of the GPT in this regard was first demonstrated by Althoff et al. (1993). In particular, when theory and experiment are compared at ambient pressure, as in Fig. 4 of the Althoff et al. (1993) paper, remarkable agreement is found between the GPT and experimental phonon spectra, including close agreement between the calculated GPT Debye temperature of $\Theta_D = 317$ K and the available experimental value of $\Theta_D^{MT} = 318$ K. For the GPT treatment, this represents a calculation at a slightly expanded volume of $\Omega = 160.4$ a.u., where zero pressure is predicted, as opposed to the observed normal-density volume of $\Omega_0 = 156.8$ a.u., where Θ_D is calculated to be 3.7% higher at 329 K. At the latter volume, there is still good agreement (within 2–4%) among GPT, DRT and experimental phonon frequencies, as shown in Table 7.2. Moreover, there is close agreement between the calculated GPT and DRT Debye temperatures in Mg over a large volume range extending to high pressure, as was shown in Fig. 3.9(a).

Finally, as a small point of reference connecting the Mg phonon results in Table 7.2 with those previously given in Table 3.2 of Chapter 3, we note that the $\mathbf{q} = 0$ hcp phonons Γ_3^+ and Γ_5^+ are frequently referred to as the longitudinal optic and transverse optic phonons, respectively. These phonon frequencies are denoted as ν_{LO} and ν_{TO}, respectively, in Table 3.2. Also, note that slightly different c/a axial ratios for Mg are used in Table 3.2 and Table 7.2.

7.2.2 Transition metals

Due to the frequent presence of strong Kohn-like anomalies, accurate calculation of phonon spectra in transition metals, especially in the central bcc metals, is considerably more challenging than for nontransition metals. But even in the pre-transition alkaline-earth metals, the occurrence of such anomalies can also serve as a clear fingerprint of transition-metal behavior, as is the case in the $5d$ metal Ba. As we have discussed in

Chapter 4, the influence of d electrons on the properties of the alkaline-earth metals increases with atomic number, and one indicator of transition-metal behavior in the heavy alkaline earths is the appearance of the bcc crystal structure at low temperature. In the normally fcc metals Ca and Sr, this situation only occurs at high pressure, but in Ba at ambient pressure, bcc is already the observed ground-state structure. The corresponding anomalous behavior in the Ba phonon spectrum appears in the longitudinal L[100] branch, which is observed to lie entirely below the T[100] transverse branch, as first discovered in neutron scattering measurements by Mizuki et al. (1985).

Independent first-principles DFT and GPT investigations of the anomalous Ba phonon spectra followed shortly thereafter by Chen et al. (1986) and Moriarty (1986b), respectively. Their results both pointed to transition-metal behavior in Ba driven by the lowering, sp-d hybridization with and partial occupation of the $5d$ energy bands. The calculated $5d$-band influence on the L[100] and T[100] phonon branches, and its comparison with experiment are summarized in Fig. 7.3. When Ba is treated as an sp-bonded simple metal in the GPT, with the nearby $5d$ bands ignored and a nominal valence of $Z = 2$ assigned, the metal is predicted to have an hcp crystal structure, as is observed for the light alkaline earths Be and Mg, and as is predicted in the same SM limit for Ca and Sr as well. Nonetheless, the Ba bcc structure is mechanically stable in this limit and normal phonon branches are found, with the L[100] branch above and well separated from the T[100] branch, as shown in the left panel of Fig. 7.3. As also shown in the same panel, this result is supported both qualitatively and quantitatively by the corresponding DFT frozen-phonon calculations of Chen et al. (1986), performed at the BZ zone midpoint, with the $5d$ bands removed to higher energy. When Ba is treated as an empty-d-band metal in the GPT to capture the effects of sp-d hybridization with the nearby $5d$ bands, the lowest-energy crystal structure is found to be fcc, which is the observed and, in the same EDB limit, the calculated structure of both Ca and Sr. While the Ba bcc structure actually becomes mechanically unstable in this limit, the L[100] and T[100] phonon branches remain qualitatively normal and are modestly lowered in frequency, as shown in the second panel of Fig. 7.3. Finally, when the $5d$ bands are allowed to self-consistently lower in energy and partially fill in the transition-metal limit of the GPT, with calculated values of $Z = 1.25$ and $Z_d = 0.75$, and with both two-ion pair and three-ion triplet potentials included in the treatment, the bcc structure is again mechanically stabilized and becomes the lowest energy structure. The calculated GPT phonon frequencies are then dramatically lowered to the experimental range, with the L[100] branch now properly located below the T[100] branch, and again in agreement with the corresponding midpoint DFT frozen-phonon calculations of Chen et al. (1986) performed with the $5d$ bands fully included. These comparisons are displayed in the final two panels of Fig. 7.3. The remainder of the GPT-TM Ba phonon spectrum calculated by Moriarty (1986b) was found to be qualitatively normal, but with lowered vibrational frequencies, and also in good agreement with the measurements of Mizuki et al. (1985).

With regard to the success of the first-principles GPT-TM treatment of the anomalous Ba L[100] phonons, the relative high importance of the directional three-ion potential v_3, which is derived solely from sp-d hybridization and d-band filing contributions, as opposed to the central-force two-ion potential v_2, should be emphasized. This distinction is explicitly demonstrated in Fig. 6 of Moriarty (1986b). In this regard, one can note from Fig. 5.10(a) of Chapter 5 that $v_3(d, \theta, \Omega_0)$ in Ba is dominantly attractive,

7.2.2 Transition metals 303

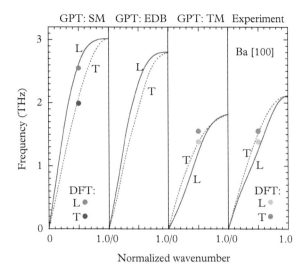

Fig. 7.3 *Origin of the anomalous L[100] phonon branch in bcc Ba, indicating the importance of partial filling of the 5d energy bands and the need for a transition-metal treatment in the first-principles DFT and GPT calculations of Chen et al. (1986) and Moriarty (1986b), respectively. The corresponding experimental measurements of Mizuki et al. (1985) are displayed in the last panel as part of a seventh-neighbor force-constant fit to the observed phonon spectrum.*

and although small in magnitude, strongly distance and angle dependent. This behavior is in sharp contrast to that in the central bcc transition metals, where v_3 is dominantly repulsive and much larger in magnitude, as shown in Fig. 5.10(b) for Cr.

Moving on to the important group-VB and -VIB bcc transition metals, the tendency for phonon anomalies generally increases in these materials, but at the same time, the anomalies that arise are largely generic in character, with two noteworthy specific varieties in the metals of each group. In the group-VB metals, there is a strong dip in the L[100] phonon branch beyond the midpoint in the BZ, with that branch becoming either degenerate with the T[100] branch or moving slightly below it as one approaches the zone boundary at the H point. This behavior is seen, for example, in the experimental data of Woods (1964) for Ta, which is plotted in Fig. 7.4(a). The second major anomaly in the group-VB metals is the equally strong stiffening of the T_2[110] transverse branch above the T_1[110] transverse branch beyond the midpoint in the BZ, such that at normal density the T_2[110] and longitudinal L[110] phonon frequencies become accidentally degenerate, or nearly so, at the zone boundary N point. This behavior is also illustrated in Fig. 7.4(a) for Ta. The same anomalies are seen in the measured phonon spectra of V (Colella and Batterman, 1970; Bosak et al., 2008) and of Nb (Powell et al., 1977).

While the canonical d-band MGPT multi-ion potentials we have developed for V and Ta do not capture either of these anomalies, the calculated phonon spectra are otherwise in good agreement with experiment, especially for tantalum, as is shown in Fig. 7.4(a) for the Ta6.8x potentials. The same level of agreement is also found between calculated GPT and DFT phonon frequencies in Ta over wide ranges of volume and pressure. This is demonstrated in Fig. 7.5, where excellent agreement is displayed as a function of volume for the zone-boundary L, T[100] frequency at the H point, as well as the

$T_1[110]$ and $L[110]$ frequencies at the N point. In addition, the MGPT-DFT offset for the remaining $T_2[110]$ N-point frequency is rather constant as a function of volume and thus the error diminishes percentagewise under compression. At the same time, the large and accurate $L[110]$ N-point and $L, T[100]$ frequencies have dominant statistical weight in most phonon averages, such as the Debye temperature $\Theta_D(\Omega)$, which is constrained to experiment at normal density in the MGPT and remains accurate at high pressure. As a result, the Ta6.8x potentials generally produce very good thermodynamic properties, as discussed in Sec. 7.4.

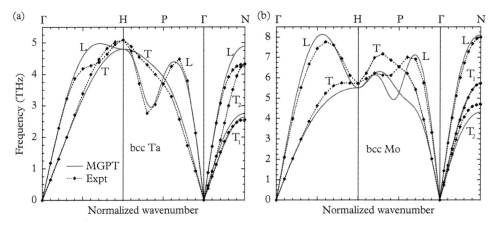

Fig. 7.4 *Phonon spectra for prototype group-VB and -VIB bcc transition metals at or near normal density. (a) Ta, as calculated from Ta6.8x MGPT potentials at $\Omega_0 = 121.6$ a.u., based on canonical d bands; (b) Mo, as calculated from Mo12.t0 MGPT potentials at $\Omega = 102.0$ a.u., based on noncanonical d bands. Experimental data shown for comparison are from seventh-neighbor force-constant fits to the neutron measurements of Woods (1964) in Ta and Powell et al. (1977) in Mo, both made at 296 K.*

In the group-VIB bcc transition metals, on the other hand, the two main observed phonon anomalies are first, the high midzone peak in the $L[100]$ branch combined with a low H-point frequency in the same branch, and second, the nearly flat $L[111]$ branch between the H and P points in the BZ. These features, which are in strong contrast to the behavior in the group-V transition metals, are especially prominent in Cr and Mo, as can be seen in the experimental neutron scattering data of Powell et al. (1977) for Mo, which is plotted in Fig. 7.4(b). Again, a canonical d-band MGPT treatment is not adequate to explain either of these anomalies, which are necessarily coupled at the H point. While a high midzone maximum frequency in the $L[100]$ branch is obtained with both the Mo5.2 and Mo6.8 MGPT potentials, there is little drop-off at the H point, where the L, $T[100]$ frequency is calculated to be about 60% too large, as shown in Table 7.3. Unlike the anomalies in the group-VB metals, however, this shortcoming does adversely affect phonon averages and thermodynamics properties in a significant way. A very useful

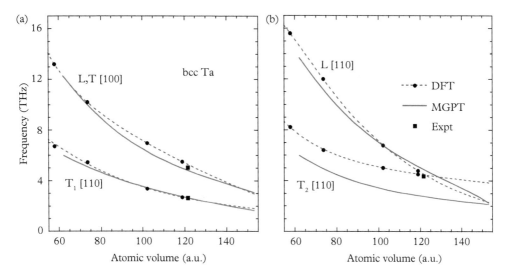

Fig. 7.5 *Volume dependence of the longitudinal (L) and transverse (T, T_1, T_2) zone-boundary phonon frequencies in bcc Ta, as calculated from MGPT Ta6.8x potentials and compared with first-principles DFT calculations (Söderlind and Moriarty, 1998) and experiment (Woods, 1964). (a) Degenerate H-point L[100] and T[100] phonons, and N-point T_1[110] phonon; (b) N-point L[110] and T_2[110] phonons.*
From Haskins et al. (2012), with publisher (APS) permission.

alternative in this case is the more general *noncanonical d*-band MGPT treatment for molybdenum due to Moriarty et al. (2012), denoted here as Mo12.t0, with additional TB parameters

$$c_0 \equiv \alpha_0/\alpha_2 = dd\sigma/dd\delta \qquad (7.50)$$

and

$$c_1 \equiv \alpha_1/\alpha_2 = dd\pi/dd\delta, \qquad (7.51)$$

which are to be used in Eq. (5.94) for $\Delta_{dd'}(R_{ij})$. These additional parameters allow one to capture both the peak height of the L[100] phonon branch as well as the low value of the H point frequency, as is demonstrated in Fig. 7.4(b). At the same time, the accuracies of the three N-point zone-boundary phonon frequencies are also improved, as shown in Table 7.3. The Mo12.t0 representation, which permits a greatly improved description of thermodynamic properties, was developed at $T = 0$ in connection with special electron-temperature-dependent MGPT potentials and the calculation of the Mo melt curve. These potentials and their application to melting are discussed in Chapter 13.

Table 7.3 *MGPT and BOP calculations of high-symmetry phonon frequencies (in THz) and Debye temperatures (in K) for bcc Ta and Mo. The MGPT calculations have been performed at their standard four-ion or four-moment level and the BOP calculations at their standard nine-moment level.*

Metal	Method	H: L,T	P: L,T	N: L	N: T1	N: T2	Θ_D
Ta:							
	MGPT: Ta4[a]	4.80	3.91	4.88	2.64	2.75	226
	MGPT: Ta6.8x[b]	4.80	3.91	4.89	2.64	2.75	226
	Numerical BOP[c]	5.50	4.31	4.25	3.69	3.81	–
	Analytic BOP[d]	6.41	4.35	3.59	3.82	3.24	–
	Experiment[e]	5.03±0.07	3.78±0.06	4.35±0.08	2.63±0.08	4.35±0.06	226
Mo:							
	MGPT: Mo5.2[f]	8.71	6.15	5.99	5.69	3.10	367
	MGPT: Mo6.8[g]	8.91	6.32	6.22	6.00	3.23	376
	MGPT: Mo12.t0[h]	5.50	5.28	8.10	5.73	4.32	354
	Numerical BOP[c]	7.65	6.53	6.00	6.59	4.00	–
	Analytic BOP[d]	8.00	7.07	4.20	6.67	2.00	–
	Experiment[i]	5.52±0.04	6.49±0.05	8.14±0.10	5.73±0.06	4.46±0.06	376

[a] Yang et al. (2001); Moriarty et al. (2002a).
[b] Moriarty and Haskins (2014); Haskins and Moriarty (2012, 2018).
[c] Lin et al. (2014).
[d] Cák et al. (2014).
[e] Woods (1964).
[f] Moriarty et al. (2002b, 2006).
[g] Present calculations.
[h] Moriarty et al. (2012).
[i] Powell et al. (1977).

Several other unique features of the phonon spectra in group-VIB transition metals should also be mentioned. First, note that unlike the phonon spectra of nontransition bcc metals such as K, displayed in Fig. 7.2(a), and of group-VB metals such as Ta, displayed in Fig. 7.4(a), the $T_1[110]$ phonon branch lies entirely above the $T_2[110]$ branch in both Cr and Mo, as shown for Mo in Fig. 7.4(b). This is sometimes referred to as an additional anomaly in group-VIB metals, but the behavior is simply a reflection of the corresponding shear elastic constants C' and C_{44}, which determine the long-wavelength slopes of the $T_1[110]$ and $T_2[110]$ phonon branches, respectively, with $C' > C_{44}$ in these metals. Second, the flat plateau observed in the L[111] branch of Cr and Mo is also found in the T[111] branch of Cr, as revealed in the neutron scattering data of Shaw and Muhlestein (1971). Finally, the phonon spectra of the third member of the group-VIB metals W is observed to be generally less anomalous than those of either Cr or Mo, with a only a modest-sized midzone peak in the L[100] branch and a smooth shallow minimum in the L[111] branch between the H and P points, which is halfway in depth between the flat plateaus in Cr and Mo and the normal deep minimum in V, Nb and Ta, as well as in Ba and all bcc alkali metals. Also, unlike Cr, Mo and the other bcc transition metals, W is elastically isotropic with $C' \simeq C_{44}$, resulting in nearly degenerate $T_1[110]$ and $T_2[110]$ phonon branches. These unique features of W are all found in the neutron scattering data of Larose and Brockhouse (1976) and of Chen and Brockhouse (1964).

Thus, with the possible exception of W, the MGPT results discussed above for Ta and Mo suggest that a full noncanonical d-band description of the group-VB and -VIB transition metals is a necessary, and quite likely sensitive, ingredient that is required to explain their observed anomalous phonon spectra. This conclusion is consistent with the earlier real-space TB investigation of these anomalies by Finnis et al. (1984). Using a linear response lattice-dynamics method and retaining two-, three- and four-ion interactions, as in the MGPT, these authors were able to explain the phonon spectra of the $3d$ bcc metals V and Cr, as well as ferromagnetic bcc Fe, by fitting the available TB parameters to the observed spectra.

The MGPT and TB success stories, however, are in rather sharp contrast to the more marginal-quality results obtained in recent BOP calculations of the group-VB and -VIB transition-metal phonon spectra by Lin et al. (2014) and Cák et al. (2014), which do not accurately describe either the observed anomalies or the spectra as a whole. In this regard, it should be recalled from Chapter 5 that although the BOP potentials are also based on noncanonical d bands, for a given transition metal, they are only directly informed by the elastic constants and general aspects of the DFT electronic structure, and not by the detailed Fermi surface or any short-wavelength properties of the phonon spectrum itself.

To be more specific about their shortcomings, the BOP bcc phonon results, which were obtained using direct phonon methods, as discussed in Sec. 7.14, are compared in Table 7.3 at high-symmetry BZ points with both MGPT calculations and experimental data for the prototype cases of Ta and Mo. In Ta, the numerical BOP calculations do not capture either of the observed anomalies: neither the midzone dip in the L[100] branch nor the stiffening of the $T_2[110]$ branch and degeneracy at the N point with the L[110] branch. The analytic BOP calculations in Ta do find the L[100] dip, but at the same time, they overestimate the H-point L, T[100] frequency by almost 30%. Moreover, there is no stiffening of the $T_2[110]$ branch and an incorrect ordering of the N-point frequencies. In the case of Mo, the numerical and analytic BOP calculations produce only modest L[100] peak structures, and with large overestimates of the H-point frequency by 38% and 45%, respectively. The numerical BOP calculations do capture the flat L[111] plateau between the H and P points, but the ordering of the N-point frequencies is incorrect in both BOP treatments, with too low L[110] and $T_2[110]$ frequencies, especially in the analytic BOP results.

As ones move to the right in the Periodic Table away from the midperiod bcc metals, the occurrence and importance of phonon anomalies generally declines. In ferromagnetic bcc Fe, for example, the observed phonon spectrum is considerably more regular (Brockhouse et al., 1967; Minkiewicz et al., 1967), and qualitatively rather similar to that of nontransition bcc metals, without any obvious strong anomalies. Accordingly, the numerical BOP methodology of Lin et al. (2016) is able to provide both a qualitatively and quantitatively accurate description of the room temperature phonon spectrum. At the same time, due to their ferromagnetic nature, the Fe phonon frequencies do have a significant temperature dependence above 300 K, as observed up to 1423 K by Vallera (1981) and as first explained by Hasegawa et al. (1987).

Likewise, in the case fcc Ni, the observed phonon spectrum is entirely regular without anomalies (Birgeneau et al., 1964), and qualitatively similar to that of nontransition fcc metals, such as Ca in Fig. 7.2(b). Using refined next-generation Ni3 MGPT potentials, which incorporate improved screening of sp-d hybridization contributions, as described in Sec. 5.2.4 of Chapter 5, a good account of the normal modes of vibration in this metal both at normal density and under high compression can be obtained. This is demonstrated in Fig. 7.6.

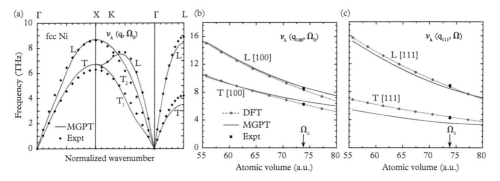

Fig. 7.6 *Phonon spectra for fcc Ni and its volume dependence, as obtained from refined Ni3 MGPT potentials and the DFT frozen-phonon calculations of Yang (2000, 2008). (a) Full MGPT spectrum calculated at $\Omega_0 = 73.82$ a.u., compared with experiment; (b) calculated MGPT and DFT volume dependence of the L[100] and T[100] zone-boundary frequencies at the X point; (c) calculated MGPT and DFT volume dependence of the L[111] and T[111] zone-boundary frequencies at the L point. Experimental data shown for comparison are from a fourth-neighbor force-constant fit to the neutron measurements of Birgeneau et al. (1964), made at 296 K.*

7.3 Elastic moduli for QBIP applications

Next, we turn from the full phonon spectrum to the long-wavelength limit and the closely related subjects of linear elasticity and the calculation of elastic moduli for crystalline metals. We begin in Sec. 7.3.1 with the fundamental definition of second-order elastic constants as the coefficients relating stress to strain in linear elasticity, and discuss a basic method of calculation of cubic elastic moduli directly from the total-energy functional. Then, in Sec. 7.3.2 we make connection with the long-wavelength limit of the QHLD formalism of Sec. 7.1 for volume-dependent QBIPs to calculate the elastic moduli in terms the same tangential and radial force-constant functions, and more generally the full force-constant matrix, that are used to obtain the phonon spectrum. These results provide exact elastic moduli for the bcc transition metals, and exact shear moduli for cubic nontransition metals. Finally, in Sec. 7.3.3 we discuss an alternative accurate method to obtain the bulk modulus and the calculation of local-volume corrections for nonshear elastic moduli in fcc and bcc nontransition metals, as well as the more complex corrections for hcp metals.

7.3.1 Elastic constants from stress-strain in linear elasticity

As in the case of lattice dynamics, elasticity in solids is discussed in many references. In his book, Wallace (1972) gives a rather complete formal discussion of elasticity theory as it applies to crystalline materials, including the treatment of second-, third- and fourth-order elastic constants. Regarding the linear elasticity and second-order elastic constants of specific interest here, Finnis (2003) also gives a useful complementary introduction, as applied to cubic systems.

We begin by considering a crystalline metal initially in equilibrium at $T = 0$, to which a small homogeneous strain is introduced by a uniform applied stress. Different measures of strain are possible, but the simplest one required for our discussion can be expressed by a symmetric 3×3 strain matrix that takes a given ion located at a position $\mathbf{R}$ to a position $\mathbf{R} + \mathbf{u}$, such that in the limit $|\mathbf{u}| \to 0$ the dimensionless Cartesian components of the strain matrix are formally given by

$$\eta_{\mu\nu} = \frac{1}{2}\left(\frac{\partial u_\mu}{\partial r_\nu} + \frac{\partial u_\nu}{\partial r_\mu}\right). \tag{7.52}$$

In linear elasticity, the corresponding applied stress components needed to create such a strain field are then

$$\sigma_{\alpha\beta} = \sum_{\substack{\mu=x,y,z \\ \nu=x,y,z}} C_{\alpha\beta\mu\nu} \eta_{\mu\nu}. \tag{7.53}$$

where the 81 coefficients $C_{\alpha\beta\mu\nu}$ are the elastic constants to be determined. Of course, crystal symmetry greatly reduces the number of independent constants in real materials to just 3 in cubic crystal systems, to 5 in hexagonal systems, to 6 or 7 in tetragonal systems as well as in rhombohedral (or trigonal) systems, to 9 in orthorhombic systems, to 13 in monoclinic systems and to 21 in triclinic systems. Useful corresponding point-group symmetry information defining these crystal systems is given in Table 1 of Wallace (1972, p. 29).

Symmetry also allows one to usefully simplify the notation in Eq. (7.53). Because the strain matrix is symmetric, pairs of indices can only appear in one of six unique ways.

Using compact Voigt notation, one makes the following substitutions in the indices of all quantities:

$$xx \to 1 \quad yy \to 2 \quad zz \to 3 \quad zy = yz \to 4 \quad zx = xz \to 5 \quad yx = xy \to 6. \tag{7.54}$$

Then, Eq. (7.53) reduces to the form

$$\sigma_i = \sum_{j=1}^{6} C_{ij} \eta_j, \tag{7.55}$$

where C_{ij} is now a symmetric 6×6 matrix.

To actually evaluate the elastic constants for a real metal, we assume the system is govern by a known total-energy functional $E_{\text{tot}}(\mathbf{R}) = NE_{\text{coh}}(\mathbf{R})$, which we then expand to second order in strain to obtain:

$$E_{\text{tot}}(\mathbf{R}) = E_{\text{tot}}(\mathbf{R}_0) + \frac{1}{2} N\Omega \sum_{i,j=1}^{6} C_{ij} \eta_i \eta_j, \tag{7.56}$$

Here the first-order term vanishes because of the zero-stress equilibrium condition, and the elastic constants now appear as the second strain derivatives of E_{tot},

$$C_{ij} = \frac{1}{N\Omega} \frac{\partial^2 E_{\text{tot}}(\mathbf{R})}{\partial \eta_i \partial \eta_j} \bigg|_{\mathbf{R}=\mathbf{R}_0}. \tag{7.57}$$

To proceed further, one can use the fact that the strain field is homogeneous and thus the same in each primitive or other chosen unit cell of the crystal. If there are N_c ions per unit cell, then the change in energy per cell is just

$$\Delta E = \frac{N_c}{N} [E_{\text{tot}}(\mathbf{R}) - E_{\text{tot}}(\mathbf{R}_0)] \equiv E_{\text{cell}}(\eta) - E_{\text{cell}}(0) = \frac{1}{2} \Omega_c \sum_{i,j=1}^{6} C_{ij} \eta_i \eta_j, \tag{7.58}$$

with $\Omega_c = N_c \Omega$. For a cubic crystal, the elastic constant matrix has the simple form

$$[C_{ij}] = \begin{pmatrix} C_{11} & C_{12} & C_{12} & 0 & 0 & 0 \\ C_{12} & C_{11} & C_{12} & 0 & 0 & 0 \\ C_{12} & C_{12} & C_{11} & 0 & 0 & 0 \\ 0 & 0 & 0 & C_{44} & 0 & 0 \\ 0 & 0 & 0 & 0 & C_{44} & 0 \\ 0 & 0 & 0 & 0 & 0 & C_{44} \end{pmatrix}. \tag{7.59}$$

It then follows from Eq. (7.58) that

$$\frac{\Delta E}{\Omega_c} = \frac{1}{2} C_{11} \left(\eta_1^2 + \eta_2^2 + \eta_3^2\right) + C_{12} \left(\eta_1 \eta_2 + \eta_2 \eta_3 + \eta_3 \eta_1\right) + \frac{1}{2} C_{44} \left(\eta_4^2 + \eta_5^2 + \eta_6^2\right). \tag{7.60}$$

7.3.1 Elastic constants from stress-strain in linear elasticity

Next, to determine C_{11}, C_{12} and C_{44} one needs to choose a suitable set of strains for which the corresponding values of ΔE can be readily calculated from the total-energy functional. There is more than one way to do this, but the following approach is perhaps the most straightforward, and also provides good transparency into the process.

One can represent the crystal by either its normal primitive cell or, for maximum clarity here, by a small cubic unit cell, where, $N_c = 1$ for sc, $N_c = 2$ for bcc, $N_c = 4$ for fcc and $N_c = 8$ for A15. In either case, we first choose $\eta_1 = \eta_2 = \eta_3 = \varepsilon$ and $\eta_4 = \eta_5 = \eta_6 = 0$, corresponding to a small uniform expansion or contraction of the unit cell from its equilibrium volume $N_c\Omega_0$ to a volume $\Omega_c(\varepsilon)$, where

$$\Omega_c(\varepsilon) = (1+\varepsilon)^3 N_c \Omega_0. \tag{7.61}$$

Equation (7.60) is thus reduced to the form

$$\frac{\Delta E}{\Omega_c} = \frac{3}{2}(C_{11} + 2C_{12})\varepsilon^2. \tag{7.62}$$

Note that the positions of the basis atoms in the unit cell move accordingly under the strain field with new components

$$r_{\alpha\mu}(\varepsilon) = (1+\varepsilon)r^0_{\alpha\mu}, \tag{7.63}$$

for $\alpha = 1,\ldots N_c$ and $\mu = x, y, z$.

Next, to evaluate the LHS of Eq. (7.62) in a compatible manner, one can develop $E_{\text{cell}}(\varepsilon)$ in a Taylor series in the strain variable ε:

$$E_{\text{cell}}(\varepsilon) = E_{\text{cell}}(0) + \left.\frac{\partial E_{\text{cell}}(\varepsilon)}{\partial \varepsilon}\right|_{\varepsilon=0}\varepsilon + \frac{1}{2}\left.\frac{\partial^2 E_{\text{cell}}(\varepsilon)}{\partial \varepsilon^2}\right|_{\varepsilon=0}\varepsilon^2 + \cdots$$

$$= E_{\text{cell}}(0) + \frac{9\Omega_c}{2}B_0\varepsilon^2 + \cdots, \tag{7.64}$$

where again the first-order term on the RHS vanishes, since the equilibrium pressure is

$$P_0 \equiv P^0_{\text{tot}} = -\frac{1}{3\Omega_c}\left.\frac{\partial E_{\text{cell}}(\varepsilon)}{\partial \varepsilon}\right|_{\varepsilon=0} = -\left.\frac{dE_{\text{coh}}(\mathbf{R})}{d\Omega}\right|_{\mathbf{R}=\mathbf{R}_0} = 0, \tag{7.65}$$

and where we identify B_0 as the equilibrium bulk modulus of the crystal:

$$B_0 \equiv B^0_{\text{tot}} = \frac{1}{9\Omega_c}\left.\frac{\partial^2 E_{\text{cell}}(\varepsilon)}{\partial \varepsilon^2}\right|_{\varepsilon=0} = \Omega\left.\frac{d^2 E_{\text{coh}}(\mathbf{R})}{d\Omega^2}\right|_{\mathbf{R}=\mathbf{R}_0}. \tag{7.66}$$

Substituting the second line of Eq. (7.64) back into Eq. (7.62), dividing both sides by $9\varepsilon^2/2$ and taking the limit $\varepsilon \to 0$, yields the well-known result

$$B_0 = \frac{1}{3}(C_{11} + 2C_{12}). \tag{7.67}$$

To actually calculate B_0 here for a real material, one can first numerically generate a smooth $E_{\text{cell}}(\varepsilon)$ vs. ε curve in a series of small steps within a symmetric interval, say $-\varepsilon_{\max} \leq \varepsilon \leq \varepsilon_{\max}$, taking care that the basis atoms in the unit cell are fully relaxed at each step according to Eq. (7.63). The resulting $E_{\text{cell}}(\varepsilon)$ curve can then be fit with a low-order polynomial in ε, allowing one to extract the second-order coefficient and use Eq. (7.66) to obtain B_0. Even so, the accuracy of the bulk modulus so calculated can be expected to be sensitive to the numerical precision with which this procedure is carried out for a given total-energy functional. In practice, often a better and more reliable way to obtain B_0 is to extract it from the corresponding equation of state (EOS) obtained over a larger volume range, as we discuss below in Sec. 7.3.3. The latter is the preferred procedure in GPT calculations and also for many DFT calculations of elastic moduli and EOS as well.

Returning to Eq. (7.60), we then need to choose strains to obtain the remaining shear elastic moduli $C' = (C_{11} - C_{12})/2$ and C_{44}, which both arise from volume-conserving strains. The quantity $C_{11} - C_{12}$ can be isolated on the RHS of Eq. (7.60) by choosing $\eta_1 = \varepsilon$, and to conserve volume,

$$\eta_2 = \frac{1}{(1+\varepsilon)} - 1 = -\varepsilon + \varepsilon^2 - \cdots, \qquad (7.68)$$

with $\eta_3 = \eta_4 = \eta_5 = \eta_6 = 0$. These strains correspond to a *tetragonal* shear of the crystal in a (110) plane at constant volume. Retaining terms to order ε^2, Eq. (7.60) then becomes

$$\frac{\Delta E}{\Omega_c} = (C_{11} - C_{12})\,\varepsilon^2 = 2C'\varepsilon^2. \qquad (7.69)$$

Now the basis atoms in the unit cell have moved to new positions with components

$$r_{\alpha x}(\varepsilon) = (1+\varepsilon) r^0_{\alpha x} \quad r_{\alpha y}(\varepsilon) = (1+\varepsilon)^{-1} r^0_{\alpha y} \quad r_{\alpha z}(\varepsilon) = r^0_{\alpha z} \qquad (7.70)$$

for $\alpha = 1, \ldots N_c$. One can again evaluate ΔE by making a Taylor expansion of $E_{\text{cell}}(\varepsilon)$ in the strain ε, exactly as in the first line of Eq. (7.64). Noting that here the first-order term in ε vanishes by symmetry, one arrives at the result

$$C' = \frac{1}{4\Omega_c} \left. \frac{\partial^2 E_{\text{cell}}(\varepsilon)}{\partial \varepsilon^2} \right|_{\varepsilon=0}. \qquad (7.71)$$

Finally, to obtain the shear constant C_{44}, choose $\eta_4 = \varepsilon$ and $\eta_1 = \eta_2 = \eta_3 = \eta_5 = \eta_6 = 0$, which corresponds to a shear of the unit cell in a (100) plane while maintaining constant

7.3.1 Elastic constants from stress-strain in linear elasticity

volume. Equation (7.60) now becomes

$$\frac{\Delta E}{\Omega_c} = \frac{1}{2} C_{44} \varepsilon^2, \tag{7.72}$$

with basis-atom position components

$$r_{\alpha x}(\varepsilon) = r_{\alpha x}^0 + \varepsilon r_{\alpha y}^0 \quad r_{\alpha y}(\varepsilon) = r_{\alpha y}^0 \quad r_{\alpha z}(\varepsilon) = r_{\alpha z}^0 \tag{7.73}$$

for $\alpha = 1, \ldots N_c$. In the same manner as for C', one is led to the result

$$C_{44} = \frac{1}{\Omega_c} \left. \frac{\partial^2 E_{\text{cell}}(\varepsilon)}{\partial \varepsilon^2} \right|_{\varepsilon=0}, \tag{7.74}$$

As in the case of the bulk modulus B_0, the shear moduli C' and C_{44} can be calculated for real materials by numerically generating the appropriate $E_{\text{cell}}(\varepsilon)$ vs. ε curves, fitting the results with low-order polynomials in ε, and extracting from each fit a second-order coefficient for use in Eq. (7.71) to obtain C' and for use in Eq. (7.74) to obtain C_{44}. In the case of volume-dependent QBIPs, however, the preferred treatment of such shear moduli is to convert the strain derivatives in Eqs. (7.71) and (7.74) to the volume-conserving tangential and radial force constant functions defined in Sec. 7.1, which can be directly summed over lattice positions. This latter approach is considered below in Sec. 7.3.2.

While we have focused above on the calculation of elastic moduli in equilibrium at zero pressure, the treatment can be readily extended to arbitrary pressure, and is so considered by Wallace (1972). This extension is particularly simple if one follows the convention used in the field of high-pressure physics, namely, that of measuring strain as the *relative* strain from the uniformly compressed (or expanded) state at volume Ω, and not the total strain from the equilibrium volume Ω_0. With that convention in place, the mathematical treatment of elastic moduli under pressure is essentially the same as that given above, and the moduli simply become volume-dependent quantities: $B_0 \equiv B_{\text{tot}}(\Omega_0) \to B_{\text{tot}}(\Omega)$ and $C_{ij}(\Omega_0) \to C_{ij}(\Omega)$. This is the definition of elastic moduli adopted in this book. Wallace (1972), on the other hand, introduces the additional symbols $B_{ij}(\Omega)$, which he refers to as stress-strain coefficients, for elastic moduli so defined.

Finally, there is the additional issue of temperature dependence of the elastic moduli that should be mentioned. Wallace (1972) carefully distinguishes between isothermal elastic moduli obtained at constant temperature, denoted as C_{ij}^T and the quantities normally calculated by theory, and adiabatic elastic moduli obtained at constant entropy S, denoted as C_{ij}^S and the quantities measured in ultrasonic experiments. These are the same quantities only at $T = 0$, where, in fact, our formal theoretical treatment has been made. However, the entire low-temperature regime of 300 K and below is of direct interest, since most measurements of elastic constants are made at room temperature. In this regime, the difference between C_{ij}^T and C_{ij}^S is small and usually neglected, but both C_{ij}^T and C_{ij}^S will decrease in magnitude with increasing temperature, amounting to a 5–10%

difference between $T = 0$ and $T = 300$ K values for most metals, but with a larger difference for soft metals like Na, K and Pb and a smaller difference for stiff metals like V, Ta and Mo. In all cases, however, the main impact of temperature on the elastic moduli is indirect and comes though thermal expansion and the non-negligible volume dependence of the $C_{ij}(\Omega)$. In this regard, when one is using or making a comparison with elastic constants measured at room temperature, one should use a reference equilibrium volume Ω_0 corresponding to that temperature in calculating the $C_{ij}(\Omega_0)$, and not a value of Ω_0 corresponding to $T = 0$.

7.3.2 Elastic constants and the long-wavelength limit of QHLD

We now return to the QHLD formalism of Sec. 7.1 developed for volume-dependent QBIPs, and make contact between the long-wavelength limit of that formalism and the stress-strain definition of elastic moduli made in the last section. We begin by considering the basic functions $P_{\text{tot}}(\Omega)$, $B_{\text{tot}}(\Omega)$ and $C_{ij}(\Omega)$ for a cubic nontransition metal in a primitive lattice (e.g., bcc or fcc), where the total-energy functional $E_{\text{tot}}(\mathbf{R}, \Omega)$ is given by Eq. (7.18) carried to the pair-potential level. To calculate the pressure and bulk modulus from $E_{\text{tot}}(\mathbf{R}, \Omega)$, one needs the following first- and second-derivative relations inferred from the discussion in Sec. 7.3.1:

$$\frac{d}{d\Omega} = \frac{1}{3\Omega}\frac{\partial}{\partial \varepsilon} \rightarrow \frac{1}{3\Omega}R\frac{\partial}{\partial R} + \frac{\partial}{\partial \Omega} \quad (7.75)$$

and

$$\Omega\frac{d^2}{d\Omega^2} = \frac{1}{9\Omega}\left(-2\frac{\partial}{\partial \varepsilon} + \frac{\partial^2}{\partial \varepsilon^2}\right) \rightarrow \frac{1}{9\Omega}\left(-2R\frac{\partial}{\partial R} + R^2\frac{\partial^2}{\partial R^2}\right) + \Omega\frac{\partial^2}{\partial \Omega^2}, \quad (7.76)$$

where R is a general interatomic separation distance. These relations follow directly from Eqs. (7.63)–(7.66). Using Eq. (7.75), one can immediately derive the large virial component of the pressure arising from the first radial derivative of the pair potential summed over ion positions at volume Ω:

$$P_{\text{vir}}(\Omega) = -\frac{1}{6\Omega}\frac{1}{N}\sum_{i,j}{}' R_{ij}^2 K_{\text{tan}}^{(2)}(R_{ij}, \Omega), \quad (7.77)$$

where the tangential force-constant function $K_{\text{tan}}^{(2)}(r, \Omega)$ is given by Eq. (7.25). Similarly, using Eq. (7.76), one can derive the corresponding large virial component of the bulk modulus as

$$B_{\text{vir}}(\Omega) = \frac{1}{18\Omega}\frac{1}{N}\sum_{i,j}{}' R_{ij}^2 \left[-2K_{\text{tan}}^{(2)}(R_{ij}, \Omega) + K_{\text{rad}}^{(2)}(R_{ij}, \Omega)\right], \quad (7.78)$$

7.3.2 Elastic constants and the long-wavelength limit of QHLD 315

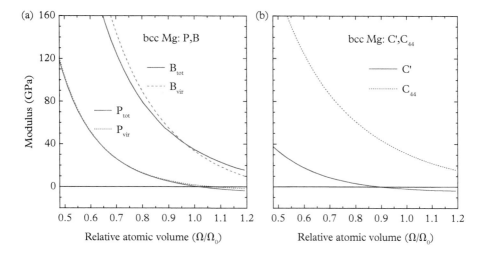

Fig. 7.7 *Pressure, bulk modulus, and shear elastic moduli for bcc Mg as a function of volume calculated from first-principles GPT potentials. (a) Virial pressure P_{vir} from Eq. (7.77) compared with the total pressure P_{tot}, and the virial bulk modulus B_{vir} from Eq. (7.78) compared with the total bulk modulus B_{tot}; (b) shear moduli C' from Eq. (7.81) and C_{44} from Eq. (7.82). In both panels, $\Omega_0 = 156.8$ a.u. is the observed equilibrium volume for hcp Mg.*

where the additional radial force-constant function $K_{\text{rad}}^{(2)}(r, \Omega)$ is given by Eq. (7.26). Thus P_{vir} and B_{vir} here are calculated at the same level of approximation as used to obtain the force-constant matrix **K** in Eqs. (7.27) and (7.28) of the QHLD treatment of nontransition metals.

In addition to the large virial components P_{vir} and B_{vir}, there are, of course, additional contributions to the total pressure and bulk modulus arising from the partial volume derivatives on the RHS of Eqs. (7.75) and (7.76). These latter contributions are generally much smaller in magnitude, however. This is demonstrated in Fig. 7.7 for the illustrative case of high-pressure bcc Mg, considered here over a large volume range. In panel (a), the total calculated pressure $P_{\text{tot}} = -dE_{coh}/d\Omega$ and bulk modulus $B_{\text{tot}} = \Omega d^2 E_{coh}/d\Omega^2$ obtained from direct numerical evaluation, are compared with P_{vir} and B_{vir}, respectively. The virial pressure P_{vir} is generally a good approximation to P_{tot}, although the small difference at large volumes above $0.9\Omega_0$ will affect the location of the zero-pressure volume. The quantitative difference between B_{vir} and B_{tot} is significantly larger, but nonetheless, except at volumes above $1.1\Omega_0$, this volume correction is never more than about 15%. To keep track of the latter correction in the discussion below we denote it as ΔB and write the total bulk modulus as

$$B_{\text{tot}}(\Omega) = B_{\text{vir}}(\Omega) + \Delta B(\Omega), \qquad (7.79)$$

where ΔB is formally given by

$$\Delta B(\Omega) = \Omega \left\{ \frac{d^2 E_{\text{vol}}(\Omega)}{d\Omega^2} + \frac{1}{2N} {\sum_{i,j}}' \frac{\partial^2 v_2(R_{ij}, \Omega)}{\partial \Omega^2} \right\}. \quad (7.80)$$

Next, we consider the corresponding treatment of the volume-conserving shear moduli C' and C_{44}. For these moduli, we need the general second-derivative relationship

$$\frac{\partial^2}{\partial \varepsilon^2} \to \frac{\partial^2 R^2}{\partial \varepsilon^2} \frac{\partial}{\partial R^2} + \left(\frac{\partial R^2}{\partial \varepsilon} \right)^2 \frac{\partial^2}{\partial (R^2)^2}, \quad (7.81)$$

which follows directly from chain-rule differentiation. The specifics of the first and second strain derivatives of R^2 in Eq. (7.81) are readily worked out from Eq. (7.70) for C' and from Eq. (7.73) for C_{44}. Using these results together with the symmetry equivalence of the x, y and z directions in cubic crystals, Eq. (7.71) for C' transforms to

$$C'(\Omega) = C'_{\text{vir}}(\Omega) = \frac{1}{2\Omega} \frac{1}{N} {\sum_{i,j}}' \left[\frac{1}{3} R_{ij}^2 K_{\text{tan}}^{(2)}(R_{ij}, \Omega) + \frac{1}{2}(r_{ijx}^4 - r_{ijx}^2 r_{ijy}^2) M_{\text{rad}}^{(2)}(R_{ij}, \Omega) \right] \quad (7.82)$$

and Eq. (7.74) for C_{44} transforms to

$$C_{44}(\Omega) = C_{44}^{\text{vir}}(\Omega) = \frac{1}{2\Omega} \frac{1}{N} {\sum_{i,j}}' \left[\frac{1}{3} R_{ij}^2 K_{\text{tan}}^{(2)}(R_{ij}, \Omega) + r_{ijx}^2 r_{ijy}^2 M_{\text{rad}}^{(2)}(R_{ij}, \Omega) \right], \quad (7.83)$$

where we have defined

$$M_{\text{rad}}^{(2)}(r, \Omega) \equiv \left[K_{\text{rad}}^{(2)}(r, \Omega) - K_{\text{tan}}^{(2)}(r, \Omega) \right] / r^2, \quad (7.84)$$

and $R_{ij}^2 = r_{ijx}^2 + r_{ijy}^2 + r_{ijz}^2$, as denoted in the QHLD formalism of Sec. 7.1. Note also that because of volume conservation, Eq. (7.82) for C' and Eq. (7.83) for C_{44} are exact results for our assumed total-energy functional, so no correction term is needed as in the case of the bulk modulus. The predicted behavior of the shear moduli as a function of volume for our bcc Mg example is illustrated in Fig. 7.7(b). In this regard, and as shown by Wallace (1972), the elastic-moduli conditions for mechanical stability of a cubic crystal at a given volume Ω are that $B_{\text{tot}}(\Omega)$, $C'(\Omega)$ and $C_{44}(\Omega)$ are all positive quantities. Thus, as demonstrated in Fig. 7.7, bcc Mg is predicted to be mechanically stable for volumes below about $0.9\Omega_0$. Recall from Chapter 3 and Fig. 3.9, as well as Chapter 6 and Table 6.5, that bcc is both the GPT-predicted and observed high-pressure phase of Mg above about 50 GPa, with an hcp $\to$ bcc transition volume near $0.6\Omega_0$.

The exact results (7.82) and (7.83) for the C' and C_{44} shear elastic moduli of cubic nontransition metals also provide an opportunity for a useful quantitative test of volume-dependent QBIPs at normal density, where comparison with experiment is possible.

7.3.2 Elastic constants and the long-wavelength limit of QHLD

In Table 7.4 we compare calculated values of $C' = C'(\Omega_0)$, $C_{44} = C_{44}(\Omega_0)$ and the elastic anisotropy ratio

$$A = \frac{C_{44}}{C'}, \qquad (7.85)$$

obtained from GPT, DRT and RMP potentials for six prototype bcc and fcc metals, with corresponding values obtained from ultrasonic experiments. As we previously pointed out in Sec. 3.5 of Chapter 3, because of the presence of long-range Friedel oscillations in the QBIPs of nontransition metals, the elastic moduli of such materials are among the most difficult properties to converge numerically. Nonetheless, in the present GPT and DRT calculations reported in Table 7.4, adequate real-space convergence was obtained in all but two cases: the GPT calculation of the small value of C' in Ca and the GPT calculation of the relatively large value of C_{44} in Cu. In those two cases, we have substituted in Table 7.4 formally equivalent values obtained from reciprocal-space calculations. In this regard, the RMP elastic moduli for Ca and Cu calculated previously by Dagens (1977a) and reported in Table 7.4 were also obtained in reciprocal space. With those caveats acknowledged, the overall agreement between theory and experiment is seen to be good for all six metals considered. In particular, the near isotropic elasticity observed in Al and the contrasting high elastic anisotropy in Na, K, Rb, Ca and Cu is well accounted for in the theoretical results. Also, we note that complete experimental C_{ij} data at $\Omega = \Omega_0$ on bcc and fcc nontransition metals are summarized in Table A1.5.

Returning to our theoretical formalism, and using Eqs. (7.78), (7.79) and (7.82), together with the requirements

$$B_{\text{tot}}(\Omega) = \frac{1}{3}\left[C_{11}(\Omega) + 2C_{12}(\Omega)\right] \qquad (7.86)$$

and

$$C'(\Omega) = \frac{1}{2}\left[C_{11}(\Omega) - C_{12}(\Omega)\right], \qquad (7.87)$$

allows one to determine specific forms for the final two remaining elastic moduli $C_{11}(\Omega)$ and $C_{12}(\Omega)$. These results are given by

$$C_{11}(\Omega) = C_{11}^{\text{vir}}(\Omega) + \Delta B(\Omega), \qquad (7.88)$$

with

$$C_{11}^{\text{vir}}(\Omega) = \frac{1}{2\Omega}\frac{1}{N}\sum_{i,j}{}'\left[\frac{1}{3}R_{ij}^2 K_{\text{tan}}^{(2)}(R_{ij},\Omega) + r_{ijx}^4 M_{\text{rad}}^{(2)}(R_{ij},\Omega)\right], \qquad (7.89)$$

and

$$C_{12}(\Omega) = C_{12}^{\text{vir}}(\Omega) + \Delta B(\Omega), \qquad (7.90)$$

Table 7.4 Shear elastic moduli C' and C_{44} (in GPa), and the elastic anisotropy $A = C_{44}/C'$, for six bcc and fcc nontransition metals, as calculated from GPT, DRT and RMP potentials.

Metal	Modulus	GPT[a]	DRT[b]	RMP[c]	Experiment	T_{expt}(K)
Bcc:						
Na	C'	1.07	0.61	–	0.72[d]	4
	C_{44}	6.29	5.77	–	6.27	
	A	5.9	9.5	–	8.7	
K	C'	0.47	0.42	–	0.38[e]	4
	C_{44}	2.97	2.73	–	2.86	
	A	6.3	6.5	–	7.6	
Rb	C'	0.31	0.30	–	0.27[f]	4
	C_{44}	2.28	2.03	–	2.21	
	A	7.4	6.8	–	8.1	
Fcc:						
Al	C'	28.0	16.0	–	23.1[g]	298
	C_{44}	35.5	18.6	–	28.2	
	A	1.3	1.2	–	1.2	
Ca	C'	4.8	–	5.2	4.8[h]	295
	C_{44}	17.0	–	17.0	16.3	
	A	3.5	–	3.3	3.4	
Cu	C'	19.1	–	21.0	23.5[i]	300
	C_{44}	75.6	–	74.5	75.4	
	A	4.0	–	3.5	3.2	

[a]Present calculated GPT results at the observed volume Ω_0 (Table A1.1), as obtained for Na, K and Rb in the SM limit, for Al and Ca in the EDB limit, and for Cu in the modified FDB limit.
[b]Present calculated DRT results at the observed volume Ω_0, based on the DRT MP parameters for Na, K, Rb and Al listed in Table 3.1.
[c]Dagens (1977a).
[d]Diederich and Trivisonno (1966), as extrapolated to 4 K from data in range 78–195 K.
[e]Marquardt and Trivisonno (1965).
[f]Gutman and Trivisonno (1967), as extrapolated to 4 K from data in range 78–170 K.
[g]Kamm and Alers (1964).
[h]Stassis et al. (1983).
[i]Overton and Gaffney (1955).

7.3.2 Elastic constants and the long-wavelength limit of QHLD

with

$$C_{12}^{\text{vir}}(\Omega) = \frac{1}{2\Omega}\frac{1}{N}\sum_{i,j}{}' \left[-\frac{1}{3}R_{ij}^2 K_{\tan}^{(2)}(R_{ij},\Omega) + r_{ijx}^2 r_{ijy}^2 M_{\text{rad}}^{(2)}(R_{ij},\Omega) \right]. \tag{7.91}$$

The virial components $C_{11}^{\text{vir}}(\Omega)$ and $C_{12}^{\text{vir}}(\Omega)$, together with $B_{\text{vir}}(\Omega)$, $C'(\Omega)$ and $C_{44}(\Omega)$, collectively represent the QHLD values of the cubic elastic moduli. In addition, note that only the bulk modulus correction term $\Delta B(\Omega)$ is necessary to correct both $C_{11}(\Omega)$ and $C_{12}(\Omega)$, and hence provide exact results for not only $B_{\text{tot}}(\Omega)$ but all the cubic $C_{ij}(\Omega)$. In addition, using Eqs. (7.83), (7.90) and (7.91), the equilibrium QHLD value of the Cauchy pressure given by Eq. (1.18), is generalized to the form

$$P_C(\Omega) = C_{12}(\Omega) - C_{44}(\Omega) = 2P_{\text{vir}}(\Omega) + \Delta B(\Omega). \tag{7.92}$$

To come full circle in our discussion here, we also need to connect the elastic moduli determined above to the long-wavelength acoustic branches of the QHLD phonon spectrum for fcc and bcc metals. At small wavenumber q in the λ branch, for a given volume Ω, one has quite generally

$$\omega_\lambda(\mathbf{q}) = c_\lambda(\mathbf{q})q, \tag{7.93}$$

where $c_\lambda(\mathbf{q})$ is the corresponding phonon wave speed or sound velocity in the direction $\mathbf{q}$. Along high-symmetry directions in the crystal, the sound velocities have simple relationships to the elastic moduli. In general, one has

$$c_\lambda(\mathbf{q}) = [C_\lambda(\mathbf{q})/\rho]^{1/2}, \tag{7.94}$$

where $C_\lambda(\mathbf{q})$ is an effective elastic modulus for wave propagation in the λ phonon branch in the $\mathbf{q}$ direction, and ρ is the material density. For wave propagation in the [100] direction of a cubic crystal, one has for the longitudinal (L) and degenerate transverse (T) branches the values $C_L(\mathbf{q}) = C_{11}$ and $C_T(\mathbf{q}) = C_{44}$, respectively. For wave propagation in the [110] direction, one has $C_L(\mathbf{q}) = (C_{11} + C_{12} + 2C_{44})/2$, $C_{T_1}(\mathbf{q}) = C'$ and $C_{T_2}(\mathbf{q}) = C_{44}$. For wave propagation in the [111] direction: $C_L(\mathbf{q}) = (C_{11}+2C_{12}+4C_{44})/3$ and $C_T(\mathbf{q}) = (2C'+C_{44})/3$. Of course, in the baseline QHLD equations, C_{11} and C_{12} are replaced by C_{11}^{vir} and C_{12}^{vir}, respectively, in each of the longitudinal moduli, so that the correction beyond QHLD is just $C_L(\mathbf{q}) \to C_L(\mathbf{q}) + \Delta B$ in all three directions. Due to the smallness of ΔB and the square root in Eq. (7.94), this correction represents only a modest adjustment in the QHLD longitudinal sound velocities and long-wavelength phonon frequencies.

The specific results obtained above, of course, only apply to cubic crystals, but the treatment itself can be readily generalized beyond nontransition metals at the pair-potential level. In this regard, we first note that the smallness of the bulk modulus correction term ΔB demonstrated in Fig. 7.7 for bcc Mg is not simply a fortuitous result, but is expected from the rigorous compressibility sum rule in condensed matter physics (Pines and Nozieres, 1966; Finnis, 2003). The compressibility sum rule is usually discussed in the context of the long-wavelength behavior of electron-gas response functions, such as those discussed in Chapter 3, but the sum rule itself applies more generally to all solids and liquids. In the present context, it requires that for a full, valid total-energy functional of a given material, such as $E_{\text{tot}}(\mathbf{R}, \Omega)$ given by Eq. (7.18), the compressibility, or its inverse, the bulk modulus, is the same, whether calculated from the method of homogeneous deformation (i.e., second volume derivatives of E_{tot}) or from the method of long waves (i.e., the virial elastic moduli defining the sound velocity coefficients c_λ). That is, as one moves beyond pair potentials and adds higher-order, multi-ion interactions to E_{tot}, the bulk modulus correction ΔB should ultimately vanish:

$$\Delta B(\Omega) = \Omega \frac{d^2 E_{\text{vol}}(\Omega)}{d\Omega^2} + \Omega \frac{\partial^2}{\partial \Omega^2} \left\{ \frac{1}{2N} {\sum_{i,j}}' v_2(ij, \Omega) + \frac{1}{6N} {\sum_{i,j,k}}' v_3(ijk, \Omega) \right.$$

$$\left. + \frac{1}{24} {\sum_{i,j,k,l}}' v_4(ijkl, \Omega) + \cdots \right\} \to 0. \qquad (7.95)$$

In the context of PP perturbation theory applied to a simple metal, this has been explicitly demonstrated. Specifically, using a local MP treatment of the total energy taken to include both third- and fourth-order terms in the PP, Brovman and Kagan (1969, 1974) showed that $B_{\text{tot}} = B_{\text{vir}}$. More importantly and more generally, the compressibility sum rule is verified experimentally for all metals, where it is observed that the bulk modulus determined from static EOS experiments, the so-called static bulk modulus $B_s = B_{\text{tot}}$, is the same as the bulk modulus derived from dynamic ultrasonic measurements, the so-called dynamic bulk modulus, $B_d = B_{\text{vir}}$.

In the MGPT applied to the central transition metals at the level of four-ion interactions, the compressibility sum rule is assumed to hold with $\Delta B = 0$, as in Eqs. (5.115) and (7.95), so that $P_{\text{tot}} = P_{\text{vir}}$, $B_{\text{tot}} = B_{\text{vir}}$ and $C_{ij} = C_{ij}^{\text{vir}}$. The virial pressure treated at the four-ion level is generalized to the form

$$P_{\text{vir}}(\Omega) = -\frac{1}{6\Omega} \frac{1}{N} {\sum_{i,j}}' R_{ij}^2 K_{\tan}(R_{ij}, \Omega), \qquad (7.96)$$

where

$$K_{\tan}(R_{ij}, \Omega) = K_{\tan}^{(2)}(R_{ij}, \Omega) + {\sum_k}' K_{\tan}^{(3)}(R_{ij}, \Omega) + \frac{1}{2} {\sum_{k,l}}' K_{\tan}^{(4)}(R_{ij}, \Omega), \qquad (7.97)$$

with the components $K_{\tan}^{(n)}$ defined by Eq. (7.40). The corresponding viral elastic moduli C_{ij}^{vir} for any primitive lattice can be expressed quite generally in terms of the four-ion force-constant matrix $K_{\mu\nu} = K_{\mu\nu}^{(2)} + K_{\mu\nu}^{(3)} + K_{\mu\nu}^{(4)}$, with the $K_{\mu\nu}^{(n)}$ defined by Eqs. (7.27), (7.28) and (7.42)–(7.45). Writing $K_{\mu\nu}$ as $K_{\mu\nu}(R_{ij}, \Omega)$ for additional clarity, one obtains the following compact and symmetric general forms:

7.3.2 Elastic constants and the long-wavelength limit of QHLD

$$C_{11}^{\text{vir}}(\Omega) = \frac{1}{2\Omega} \frac{1}{N} {\sum_{i,j}}' \left[r_{ijx}^2 K_{xx}(R_{ij}, \Omega) \right] \tag{7.98}$$

$$C_{22}^{\text{vir}}(\Omega) = \frac{1}{2\Omega} \frac{1}{N} {\sum_{i,j}}' \left[r_{ijy}^2 K_{yy}(R_{ij}, \Omega) \right] \tag{7.99}$$

$$C_{33}^{\text{vir}}(\Omega) = \frac{1}{2\Omega} \frac{1}{N} {\sum_{i,j}}' \left[r_{ijz}^2 K_{zz}(R_{ij}, \Omega) \right] \tag{7.100}$$

$$C_{12}^{\text{vir}}(\Omega) = \frac{1}{2\Omega} \frac{1}{N} {\sum_{i,j}}' \left[2 r_{ijx} r_{ijy} K_{xy}(R_{ij}, \Omega) - r_{ijx}^2 K_{yy}(R_{ij}, \Omega) \right] \tag{7.101}$$

$$C_{13}^{\text{vir}}(\Omega) = \frac{1}{2\Omega} \frac{1}{N} {\sum_{i,j}}' \left[2 r_{ijx} r_{ijz} K_{xz}(R_{ij}, \Omega) - r_{ijx}^2 K_{zz}(R_{ij}, \Omega) \right] \tag{7.102}$$

$$C_{23}^{\text{vir}}(\Omega) = \frac{1}{2\Omega} \frac{1}{N} {\sum_{i,j}}' \left[2 r_{ijy} r_{ijz} K_{yz}(R_{ij}, \Omega) - r_{ijy}^2 K_{zz}(R_{ij}, \Omega) \right] \tag{7.103}$$

$$C_{44}^{\text{vir}}(\Omega) = \frac{1}{4\Omega} \frac{1}{N} {\sum_{i,j}}' \left[r_{ijy}^2 K_{zz}(R_{ij}, \Omega) + r_{ijz}^2 K_{yy}(R_{ij}, \Omega) \right] \tag{7.104}$$

$$C_{55}^{\text{vir}}(\Omega) = \frac{1}{4\Omega} \frac{1}{N} {\sum_{i,j}}' \left[r_{ijz}^2 K_{xx}(R_{ij}, \Omega) + r_{ijx}^2 K_{zz}(R_{ij}, \Omega) \right] \tag{7.105}$$

and

$$C_{66}^{\text{vir}}(\Omega) = \frac{1}{4\Omega} \frac{1}{N} {\sum_{i,j}}' \left[r_{ijx}^2 K_{yy}(R_{ij}, \Omega) + r_{ijy}^2 K_{xx}(R_{ij}, \Omega) \right], \tag{7.106}$$

with

$$B_{\text{vir}}(\Omega) = \frac{1}{9} \left[C_{11}^{\text{vir}}(\Omega) + C_{22}^{\text{vir}}(\Omega) + C_{33}^{\text{vir}}(\Omega) + 2 \left\{ C_{12}^{\text{vir}}(\Omega) + C_{13}^{\text{vir}}(\Omega) + C_{23}^{\text{vir}}(\Omega) \right\} \right]. \tag{7.107}$$

For any primitive lattice, Eqs. (7.98)–(7.107) represent virial elastic moduli that are fully consistent with the QHLD treatment of the phonon spectrum considered in Sec. 7.1. In the case of central transition metals treated at the four-ion level in the MGPT with $\Delta B = 0$, these are also exact results for the bcc elastic moduli, where, of course, $C_{11} = C_{22} = C_{33}$, $C_{12} = C_{13} = C_{23}$ and $C_{44} = C_{55} = C_{66}$. For individual MGPT transition metals, this means that the experimental $C_{ij}(\Omega_0)$ data used in establishing potential parameters is fully reproduced by Eqs. (7.98)–(7.107). Finally, in the case of cubic non-transition metals treated at the pair-potential level with GPT, DRT or RMP potentials, Eqs. (7.98), (7.101) and (7.104) produce the exact equivalents of Eqs. (7.89), (7.91) and (7.83) for C_{11}^{vir}, C_{12}^{vir} and C_{44}^{vir}, respectively.

It is also straightforward to generalize Eqs. (7.98)–(7.107) to a nonprimitive lattice with N_c unit cells and n_b basis atoms in each cell. The force-constant matrix becomes $3n_b \times 3n_b$ in size, and in each equation, one has the usual substitutions $K_{\mu\nu} \to K_{\alpha\mu,\beta\nu}$, $R_{ij} \to R_{i\alpha,j\beta}$ and $r_{ij\mu} \to r_{(i\alpha,j\beta)\mu}$. For both primitive and nonprimitive lattices that involve variable structural parameters (e.g., c/a, b/a, etc.) and are thus not determined by symmetry alone, the virial elastic moduli so obtained do not include the additional possible effects of sublattice relaxation of ion positions under strain. Such relaxation effects must be treated separately, as discussed in Sec. 7.3.3 below for the hcp structure. Lastly, in both the MGPT QHLD formalism of Sec. 7.1.3 and in the treatment of the corresponding virial elastic moduli here, it is clearly possible to extend the total-energy functional and force-constant matrix to include the additional five- and six-ion interactions envisioned in Chapter 5.

7.3.3 EOS calculation of the bulk modulus and local environment corrections

In this section, we turn to the consideration of two additional elastic-moduli issues raised by the discussion in the preceding sections: an alternate EOS calculation of the total bulk modulus B_{tot} plus the correction term $\Delta B = B_{\text{tot}} - B_{\text{vir}}$, and the related topic of the full calculation of C_{ij}, including relaxation effects, for nonprimitive hcp metals. The use of simplified analytic forms to represent the basic EOS pressure-volume relationship,

$$P_{\text{tot}}(\Omega) = -\frac{dE_{coh}(\mathbf{R}, \Omega)}{d\Omega}, \qquad (7.108)$$

for real materials has a long history in high-pressure physics, with many specific forms having been developed. In the present context, one particularly useful representation is the so-called universal equation of state (UEOS) of Vinet et al. (1987, 1989), aka the Vinet EOS, which has the analytic form

$$P_{\text{ueos}}(x) = 3B_0 \frac{(1-x)}{x^2} \exp\left[\eta(1-x)\right], \qquad (7.109)$$

with $x = (\Omega/\Omega_0)^{1/3}$. This form has been shown to describe accurately the observed room-temperature or 300-K isotherm of a wide range of materials, including both metals and nonmetals alike, up to pressures of several hundred GPa. Here Ω_0 is the chosen input equilibrium volume, while B_0 and η are parameters determined by fitting to input pressure-volume data. Physically, $B_0 = B_{\text{tot}}(\Omega_0)$ is the equilibrium bulk modulus, and the parameter η is

$$\eta = \frac{3}{2}(B_0' - 1), \qquad (7.110)$$

where B_0' is the corresponding first pressure derivative of B_0. Thus in practice accurate values of B_0 and B_0' can be deduced from measured high-pressure DAC data on $P_{\text{tot}}(\Omega)$.

7.3.3 EOS calculation of the bulk modulus and local environment corrections

Table 7.5 Comparison of several different possible GPT-based calculations of the bulk modulus $B_0 \equiv B_{\text{tot}}(\Omega_0)$, its first pressure derivative B_0', and the virial correction term $\Delta B = B_0 - B_{\text{vir}}^0$ in hcp Mg, all performed at its observed room-temperature equilibrium volume $\Omega_0 = 156.8$ a.u. and axial ratio $c/a = 1.62$. The quantities ΔP, B_0 and ΔB are given in GPa.

Method	$T(K)$	ΔP	B_0	ΔB	B_0'
Numerical: $B_{\text{tot}} = \Omega d^2 E_{\text{coh}}/d\Omega^2$	0	–	35.6	−4.8	–
Analytic: UEOS fit of P_{tot}	0	0.0	35.6	−4.8	4.38
Analytic: UEOS fit of $P_{\text{tot}} + \Delta P$	0	−0.714	35.6	−4.8	4.38
Analytic: UEOS fit of 300-K P_{tot}	300	0.0	35.3	–	4.43
Experiment	300	–	35.2[a]	–	4.05[b]

[a]Slutsky and Garland (1957).
[b]Schmunk and Smith (1959).

The UEOS analytic form (7.109) can also be used effectively in representing first-principles DFT or GPT EOS data, as shown by Moriarty (1995) for the 300-K isotherms of Al, Cu, Mo and Pb, and as we demonstrate more generally for GPT Mg in Table 7.5 and Fig. 7.8. In this regard, several important and useful features of Eq. (7.109) should be noted. First, the UEOS fit can be successfully applied to any low-temperature isotherm, and in particular at both $T = 0$ and $T = 300$ K, as done in Table 7.5. Second, the choice of reference volume Ω_0 in Eq. (7.109) is a matter of convenience only and does *not* have to correspond to the zero-pressure equilibrium condition for the input $P_{\text{tot}}(\Omega)$ function. Thus, if desired, the same value of Ω_0 can be used at all temperatures, as has also been done in Table 7.5. The physical reason for this flexibility is that the bulk modulus and all higher volume or pressure derivatives depend only on the shape of the $P_{\text{tot}}(\Omega)$ curve and not its absolute value. This requirement can be readily verified in the actual UEOS fit by self-consistently adding a constant pressure ΔP to $P_{\text{tot}}(\Omega)$ such that $P_{\text{tot}}(\Omega_0) + \Delta P = 0$, a change achieved by letting $E_{\text{coh}}(\mathbf{R}, \Omega) \to E_{\text{coh}}(\mathbf{R}, \Omega) - \Delta P(\Omega - \Omega_0)$ in Eq. (7.108), with $\Delta P = -0.714$ GPa for Mg. As also shown in Table 7.5, this change has no effect on the calculated UEOS parameters B_0 and B_0'. Finally, we note that although only $P_{\text{tot}}(\Omega)$ is fitted in the UEOS process, both it and $B_{\text{tot}}(\Omega)$ are accurately represented over the entire volume range of the fit, as demonstrated in Fig. 7.8. In this regard, the range of the UEOS fit of $P_{\text{tot}}(\Omega)$ used in both Table 7.5 and Fig. 7.8 extends from -3.6 GPa, corresponding to $\Omega/\Omega_0 = 1.19$, to 124 GPa, corresponding to $\Omega/\Omega_0 = 0.47$.

To move beyond B_{tot} and calculate the full values of the C_{ij} for an hcp crystal, including both local volume effects and relaxation of ion positions under strain, one must return to the stress-strain formalism of Sec. 7.3.1. Equation (7.60) for the change in energy of the crystal in the presence of the possible set of strains is now generalized to the form

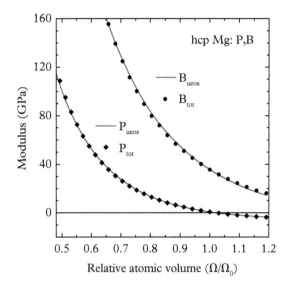

Fig. 7.8 *Analytic UEOS representations of first-principles GPT calculations of the total pressure $P_{tot}(\Omega)$ and bulk modulus $B_{tot}(\Omega)$ for hcp Mg at $T=0$. The UEOS curves, $P_{ueos}(\Omega)$ and $B_{ueos}(\Omega)$, have been obtained from Eq. (7.109) with a two-parameter (B_0 and B'_0) least-squares fit of $P_{tot}(\Omega)$ only.*

$$\frac{\Delta E}{\Omega_c} = \frac{1}{2}C_{11}\left(\eta_1^2 + \eta_2^2\right) + \frac{1}{2}C_{33}\eta_3^2 + C_{12}\eta_1\eta_2 + C_{13}\left(\eta_2\eta_3 + \eta_3\eta_1\right)$$
$$+ \frac{1}{2}C_{44}\left(\eta_4^2 + \eta_5^2\right) + \frac{1}{2}C_{66}\eta_6^2, \tag{7.111}$$

together with the symmetry requirement

$$C_{66} = C' = \frac{1}{2}\left(C_{11} - C_{12}\right). \tag{7.112}$$

Thus, there are five independent elastic moduli for the hcp structure. There are, of course, multiple possible ways to choose the six specific strains in Eq. (7.111) to isolate and calculate five independent combinations of the elastic moduli. Skinner et al. (2019) have developed a specific GPT-based stress-strain treatment for hcp Mg at normal density, which allows full relaxation of ion positions in the basal plane while maintaining a constant c/a axial ratio. The elastic moduli calculated from this treatment are given in Table 7.6, and compared there both with the corresponding virial elastic moduli calculated from Eqs. (7.98)–(7.107), and with experiment.

In general, there are three independent shear elastic moduli for the hcp structure that conserve volume under strain:

(i) C_{66} and C'. The strains that produce these shear moduli change the angles between the equilateral basal-plane lattice vectors, while keeping the basal-plane area unchanged. The elastic constant C_{66} corresponds to a shear on the (100)

7.3.3 EOS calculation of the bulk modulus and local environment corrections

Table 7.6 *Calculated first-principles GPT elastic constants (in GPa) for hcp Mg at its observed equilibrium volume $\Omega_0 = 156.8$ a.u. and axial ratio $c/a = 1.62$.*

Treatment	C_{11}	C_{33}	C_{12}	C_{13}	C_{44}	C_{66}	C_{hex}	B_0
Virial C_{ij}^{vir}	73.3	63.6	27.8	24.6	19.5	22.8	21.65	40.5
Stress-strain C_{ij}[a]	63.9	62.6	25.2	21.1	19.5	19.4	21.65	36.1
Experiment[b]	59.4	61.6	25.6	21.4	16.4	16.9	20.4	35.2

[a] Skinner et al. (2019); see Chapter 11.
[b] Slutsky and Garland (1957).

plane in a [110] direction, while elastic constant C' corresponds to a shear on the (110) plane in a $[1\bar{1}0]$ direction. The equality of C_{66} and C' produces transverse isotropy, making the basal plane elastically isotropic. At the same time, internal relaxation effects can lower the calculated values of C_{66} and C' significantly, as first analyzed for Mg in reciprocal space by Cousins (1970) using Shaw's OMP. The 15% reduction of C_{66} below C_{66}^{vir} indicated in Table 7.6 for GPT Mg is in good accord with the 12% reduction predicted by Cousins.

(ii) C_{44}. The corresponding strain shears the hcp crystal on the basal plane, leaving the plane unaffected but tilting the c axis with respect to the basal plane, without internal relaxation effects, so $C_{44} = C_{44}^{\text{vir}}$, as indicated for GPT Mg in Table 7.6.

(iii) The third additional hexagonal shear modulus is given by

$$C_{\text{hex}} = \frac{1}{6}\left(C_{11} + 2C_{33} + C_{12} - 4C_{13}\right). \tag{7.113}$$

This modulus is produced from a strain that stretches (or compresses) the c axis while at the same time contracting (or expanding) the basal plane uniformly, so as to maintain constant volume. When calculated at a constant c/a ratio, however, as done by Skinner et al. (2019) for Mg, one finds $C_{\text{hex}} = C_{\text{hex}}^{\text{vir}}$, as is verified in Table 7.6. As can also be inferred from Table 7.6, Mg is observed to be elastically isotropic more generally, with values of $C_{33}/C_{11} = 1.04$, $C_{44}/C_{66} = 0.97$ and $C_{\text{hex}}/C_{66} = 1.21$. This behavior is well described by the stress-strain calculations of Skinner et al. (2019), with corresponding values of $C_{33}/C_{11} = 0.98$, $C_{44}/C_{66} = 1.01$ and $C_{\text{hex}}/C_{66} = 1.12$.

An additional quantity of interest is, of course, the bulk modulus. Under the constraint of a constant c/a axial ratio, the general relationship between B_{tot} for an hcp crystal and the corresponding C_{ij} is the same as in Eq. (7.107) between B_{vir} and C_{ij}^{vir} for the same structure. In this limit, one can derive the following two ideal forms for B_{tot} of an hcp crystal:

$$\begin{aligned} B_{\text{tot}} &= \frac{1}{9}\left(2C_{11} + C_{33} + 2C_{12} + 4C_{13}\right) \\ &= \frac{1}{3}\left(2C_{11} + C_{33} - 2C' - 2C_{\text{hex}}\right). \end{aligned} \tag{7.114}$$

Evaluating Eq. (7.114) with the Skinner et al. (2019) stress-strain values of C_{ij} from Table 7.6, gives values of $B_0 = B_{\text{tot}}(\Omega_0) = 36.1$ GPa and $\Delta B = -4.4$ GPa, as compared with values of $B_0 = 35.6$ GPa and $\Delta B = -4.8$ GPa from Table 7.5. This verifies that the assumption of a constant c/a ratio implicit in Eq. (7.114) is a good approximation for Mg. This behavior, however, is in sharp contrast to that observed in group-IIB metals, such as Zn, which have high and significantly strain-dependent c/a ratios (Ledbetter, 1977). In both Zn and Cd, the experimental bulk modulus derived from Eq. (7.114), and given in Table A1.6, is about 25% higher than the more accurate EOS-based result given in Table A1.3.

7.4 Thermodynamic properties in the QHLD limit

In this section, we return to the quasiharmonic phonons of Secs. 7.1 and 7.2, and apply our QHLD results to the calculation of basic thermodynamic properties of metals. In general, the finite-temperature thermodynamic properties of a given solid phase of a metal can be determined from a corresponding Helmholtz free-energy function $A_{\text{tot}}(\Omega, T)$ for that phase. In the conventional weak-coupling limit of condensed-matter theory, the free energy $A_{\text{tot}}(\Omega, T)$ expressed in units of energy per ion, can be calculated as a sum of well-defined cold ($T = 0$), ion-thermal and electron-thermal contributions:

$$A_{\text{tot}}(\Omega, T) = E_0(\Omega) + A_{\text{ion}}(\Omega, T) + A_{\text{el}}(\Omega, T), \qquad (7.115)$$

The cold energy E_0 is just the cohesive-energy contribution from the ground state (or other phase in question) of the solid, so for volume-dependent QBIPs one normally has

$$E_0(\Omega) = E_{\text{coh}}(\mathbf{R}, \Omega). \qquad (7.116)$$

The ion-thermal free energy A_{ion} in Eq. (7.115) for the same phase will have a leading second-order quasiharmonic-phonon contribution plus a smaller collective higher-order contribution due to anharmonic vibrations at high temperature:

$$A_{\text{ion}}(\Omega, T) = A_{\text{ion}}^{\text{qh}}(\Omega, T) + A_{\text{ion}}^{\text{ah}}(\Omega, T). \qquad (7.117)$$

The quasiharmonic contribution can be expressed directly in terms of the volume-dependent QHLD phonon frequencies $\nu_\lambda(\mathbf{q}, \Omega)$ for the crystal structure in question by the following well-known result based on Bose-Einstein statistics:

$$A_{\text{ion}}^{\text{qh}}(\Omega, T) = \frac{k_B T}{N} \sum_{\mathbf{q},\lambda} \ln\left\{2 \sinh\left[h\nu_\lambda(\mathbf{q}, \Omega)/(2k_B T)\right]\right\}, \qquad (7.118)$$

where the sum is over all wavevectors $\mathbf{q}$ and phonon branches λ in the first BZ of the reciprocal lattice. The smaller anharmonic free energy $A_{\text{ion}}^{\text{ah}}$ in Eq. (7.117) requires a separate, specialized treatment, which we consider in Chapter 8 together with the treatment of the additional small electron-thermal contribution A_{el} in Eq. (7.115), arising from the excitation of electrons above the Fermi level of the metal. Setting $A_{ion}^{ah} = A_{el} = 0$ for now, then yields the total free energy of the solid in the QHLD limit:

$$A_{tot}(\Omega, T) = E_0(\Omega) + A_{ion}^{qh}(\Omega, T). \tag{7.119}$$

To proceed further, one needs corresponding expressions for the total internal energy $E_{tot}(\Omega, T)$ per ion and the total internal pressure $P_{tot}(\Omega, T)$. These quantities can be obtained from the general thermodynamic relationships

$$E_{tot}(\Omega, T) = -T^2 \frac{\partial \{A_{tot}(\Omega, T)/T\}}{\partial T} \tag{7.120}$$

and

$$P_{tot}(\Omega, T) = -\frac{\partial A_{tot}(\Omega, T)}{\partial \Omega}, \tag{7.121}$$

respectively. Using Eqs. (7.118) and (7.119) in Eq. (7.120) for E_{tot}, one obtains

$$E_{tot}(\Omega, T) = E_0(\Omega) + E_{ion}^{qh}(\Omega, T), \tag{7.122}$$

where the quasiharmonic ion-thermal energy is

$$E_{ion}^{qh}(\Omega, T) = \frac{1}{N} \sum_{q,\lambda} \left[n_{q\lambda} + 1/2\right] h\nu_\lambda(\mathbf{q}, \Omega), \tag{7.123}$$

with $n_{q\lambda}$ the familiar Bose-Einstein occupation number $1/\{\exp[h\nu_\lambda(\mathbf{q}, \Omega)/k_B T - 1]\}$. In Eq. (7.123) one can see that the second term on the RHS is just the zero-point vibrational energy $E_{ph}^0(\Omega)$ defined by Eq. (7.49). In the zero-temperature limit $E_{ion}^{qh}(\Omega, 0) = E_{ph}^0(\Omega)$, although this is only a small 1–2% or less correction to $E_0(\Omega)$, and a correction we have not included in $T = 0$ calculations up to this point. Conversely, in the high-temperature limit where $n_{q\lambda} \to k_B T/h\nu_\lambda(q, \Omega)$, one recovers from Eq. (7.123) the well-known classical result $E_{ion}^{qh}(\Omega, T) = 3k_B T$.

Similarly, using Eqs. (7.118) and (7.119) in Eq. (7.121) for P_{tot}, one obtains

$$P_{tot}(\Omega, T) = P_0(\Omega) + P_{ion}^{qh}(\Omega, T), \tag{7.124}$$

where the cold pressure is just

$$P_0(\Omega) = -\frac{dE_0(\Omega)}{d\Omega} \tag{7.125}$$

and the quasiharmonic ion-thermal pressure is given by

$$P_{ion}^{qh}(\Omega, T) = \frac{1}{N\Omega} \sum_{q,\lambda} \gamma_\lambda(\mathbf{q}, \Omega) \left[n_\lambda(\mathbf{q}, \Omega) + 1/2\right] h\nu_\lambda(\mathbf{q}, \Omega). \tag{7.126}$$

In Eq. (7.126) the dimensionless quantity $\gamma_\lambda(\mathbf{q}, \Omega)$ is the so-called mode or phonon Grüneisen parameter defined here by the relation

$$\gamma_\lambda(\mathbf{q}, \Omega) \equiv -\frac{\Omega}{\nu_\lambda(\mathbf{q}, \Omega)} \frac{\partial \nu_\lambda(\mathbf{q}, \Omega)}{\partial \Omega}. \tag{7.127}$$

In the high-temperature limit, the thermal pressure given by Eq. (7.126) reduces to the result $P_{\text{ion}}^{\text{qh}}(\Omega, T) = 3k_B T \gamma_{\text{ion}}^{\text{qh}}(\Omega)/\Omega$, where $\gamma_{\text{ion}}^{\text{qh}}(\Omega)$ is the quasiharmonic value of the thermodynamic Grüneisen parameter, given in this limit by the phonon average

$$\gamma_{\text{ion}}^{\text{qh}}(\Omega) = \frac{1}{3N} \sum_{\mathbf{q},\lambda} \left[-\frac{\Omega}{\nu_\lambda(\mathbf{q}, \Omega)} \frac{\partial \nu_\lambda(\mathbf{q}, \Omega)}{\partial \Omega} \right]. \tag{7.128}$$

Both Eq. (7.127) for $\gamma_\lambda(\mathbf{q}, \Omega)$ and Eq. (7.128) for $\gamma_{\text{ion}}^{\text{qh}}(\Omega)$ can be evaluated for a given material from knowledge of the full phonon spectrum as a function of volume, and then used in turn to obtain $P_{\text{ion}}^{\text{qh}}(\Omega, T)$. While this procedure can provide useful physical insight about the involved quantities, it requires more calculation than really necessary, however. In practice, a much more efficient way to obtain the ion-thermal pressure is first to evaluate Eq. (7.118) for $A_{\text{ion}}^{\text{qh}}(\Omega, T)$ as a function of volume and temperature, and then numerically differentiate the result with respect to volume to obtain

$$P_{\text{ion}}^{\text{qh}}(\Omega, T) = -\frac{\partial A_{\text{ion}}^{\text{qh}}(\Omega, T)}{\partial \Omega}. \tag{7.129}$$

This approach bypasses the lengthy step of calculating volume derivatives of the individual phonon frequencies.

The thermodynamic Grüneisen parameter $\gamma(\Omega, T)$ can then be readily calculated as follows. In its full general form, this quantity can be defined as

$$\gamma(\Omega, T) \equiv \Omega \left(\frac{\partial P_{\text{tot}}}{\partial E_{\text{tot}}} \right)_\Omega = \Omega \frac{\partial P_{\text{tot}}(\Omega, T)}{\partial T} \Big/ \frac{\partial E_{\text{tot}}(\Omega, T)}{\partial T}, \tag{7.130}$$

which in the quasiharmonic limit reduces to the form

$$\gamma(\Omega, T) = \Omega \frac{\partial P_{\text{ion}}^{\text{qh}}(\Omega, T)}{\partial T} \Big/ \frac{\partial E_{\text{ion}}^{\text{qh}}(\Omega, T)}{\partial T}. \tag{7.131}$$

At high temperature, of course, both $E_{\text{ion}}^{\text{qh}}$ and $P_{\text{ion}}^{\text{qh}}$ become linear in temperature, leading to the useful, simplified result

$$\gamma_{\text{ion}}^{\text{qh}}(\Omega) = \frac{\Omega P_{\text{ion}}^{\text{qh}}(\Omega, T)}{E_{\text{ion}}^{\text{qh}}(\Omega, T)}. \tag{7.132}$$

In practice, the temperature dependence of the RHS of Eq. (7.132) becomes negligible above room temperature, where this result is equivalent to Eq. (7.128), but much easier to evaluate.

In Fig. 7.9 we display the calculated volume dependence of the basic thermodynamic quantities $\Theta_D(\Omega)$ and $\gamma_{\text{ion}}^{\text{qh}}(\Omega)$ for representative simple and transition metals, as obtained from GPT and MGPT potentials, respectively. In the highly compressible simple metal potassium, the Debye temperature $\Theta_D(\Omega)$ rises in a near linear fashion under compression, but in the less compressible metals Ta and Mg, the rise becomes steeper with decreasing volume, where at two-fold compression the Debye temperature has increased by about a factor of 2.5 over its normal-density value in both metals. The Grüneisen parameter, on the other hand, decreases under compression, and in simple thermodynamic models this is often assumed to be linear, with $\gamma_{\text{ion}}^{\text{qh}} \propto \Omega/\Omega_0$. As shown in Fig. 7.10(b) for Mg and K, however, the decrease of $\gamma_{\text{ion}}^{\text{qh}}$ under compression is smooth and monotonic but much more rapid, varying approximately as $(\Omega/\Omega_0)^\alpha$, with $\alpha \cong 0.83$ for Mg and $\alpha \cong 0.60$ for K. For the case of tantalum, the behavior of $\gamma_{\text{ion}}^{\text{qh}}(\Omega)$ is more complex and impacted by transition-metal d-band physics. Finally, also note that in all three metals, the calculated normal-density values of $\Theta_D(\Omega_0)$ and $\gamma_{\text{ion}}^{\text{qh}}(\Omega_0)$ agree well with experiment.

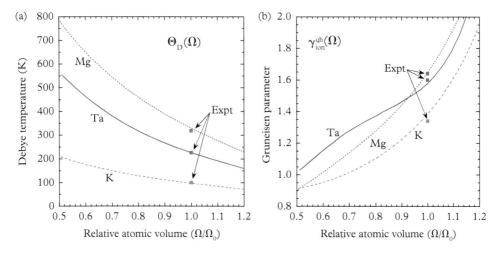

Fig. 7.9 *Volume dependence of basic thermodynamic quantities for the simple metals K and Mg, as calculated with first-principles GPT potentials, and for the transition metal Ta, as calculated with Ta6.8x multi-ion MGPT potentials. (a) Debye temperature $\Theta_D(\Omega)$ from Eqs. (7.48) and (7.49), with the experimental data from Tables A1.3 and A1.4. (b) Ion-thermal component of the Grüneisen parameter at room temperature, as obtained in the quasiharmonic limit from Eq. (7.132), with the experimental data from Katahara et al. (1979) for Ta and Gschneidner (1964) for K and Mg.*

More generally, important thermodynamic derivatives such as the specific heat and the thermal expansion coefficient or expansivity can be calculated in the QHLD limit as a function of temperature and compared with experiment. The specific heat at constant volume Ω is defined as

$$C_\Omega \equiv \left(\frac{\partial E_{\text{tot}}}{\partial T}\right)_\Omega = \frac{\partial E_{\text{tot}}(\Omega, T)}{\partial T} \rightarrow \frac{\partial E_{\text{ion}}^{\text{qh}}(\Omega, T)}{\partial T}, \qquad (7.133)$$

where the latter simplified form of the RHS of Eq. (7.133) is for the quasiharmonic limit. The corresponding specific heat at constant pressure, which is the quantity usually measured experimentally, is given by the well-known thermodynamic relationship

$$C_P = C_\Omega + T\Omega\beta^2 B_T, \qquad (7.134)$$

where β is the thermal expansion coefficient at constant pressure,

$$\beta \equiv \frac{1}{\Omega}\left(\frac{\partial \Omega}{\partial T}\right)_{P_{\text{tot}}}, \qquad (7.135)$$

and B_T is the isothermal bulk modulus

$$B_T \equiv B_{\text{tot}}(\Omega, T) = -\Omega\frac{\partial P_{\text{tot}}(\Omega, T)}{\partial \Omega}. \qquad (7.136)$$

Using the thermodynamic identity

$$-\left(\frac{\partial \Omega}{\partial T}\right)_{P_{\text{tot}}}\left(\frac{\partial P_{\text{tot}}}{\partial \Omega}\right)_T = \left(\frac{\partial P_{\text{tot}}}{\partial T}\right)_\Omega, \qquad (7.137)$$

the product of β and B_T has the simple form

$$\beta B_T = \frac{\partial P_{\text{tot}}(\Omega, T)}{\partial T} \rightarrow \frac{\partial P_{\text{ion}}^{\text{qh}}(\Omega, T)}{\partial T}, \qquad (7.138)$$

where the latter expression on the RHS of Eq. (7.138) is for the quasiharmonic limit. Comparing Eqs. (7.133) and (7.138) with Eq. (7.130), one arrives at the well-known equivalent form of the Grüneisen parameter

$$\gamma = \frac{\Omega\beta B_T}{C_\Omega}, \qquad (7.139)$$

a form that is often considered to be the experimental definition of γ.

In the context of QBIPs for simple and transition metals, prototype QHLD calculations of the zero-pressure specific heat C_P and thermal expansion coefficient β have been performed by Althoff et al. (1993) on hcp Mg, using first-principles GPT potentials, and by Moriarty and Haskins (2014) on bcc Ta, using the Ta6.8x multi-ion MGPT potentials. The Ta calculations were part of a larger study carried out to high temperature that also included anharmonic and electron-thermal effects, as discussed in Chapter 8.

The Ta QHLD results below 500 K are displayed in Fig. 7.10 and compared with experiment. As expected, the temperature variations of C_P and β are qualitatively similar, rising rapidly from zero values at $T = 0$ to nearly constant values above the Ta Debye temperature, $\Theta_D = 226$ K. Quantitatively, there is excellent agreement with experiment for both quantities up to Θ_D, and beyond that point the experimental data begins to rise only slowly above the QHLD predictions. In the case of Mg, the GPT quasi-harmonic calculation of C_P is in close agreement with experiment all the way to about 800 K, which is ~2.5 Θ_D. A similarly successful calculation of β, however, also requires an accurate zero-pressure volume, which is overestimated in the first-principles GPT treatment of hcp Mg by a small but significant amount (2.7%). The problem is easily corrected, however, by adding a small constant pressure of 1.8 GPa to the equation of state at all volumes. This correction, plus the *ad hoc* addition of a small anharmonic contribution above the Debye temperature, then yields an accurate calculation of β also all the way up to about 800 K, as shown in Fig. 8 of Althoff et al. (1993).

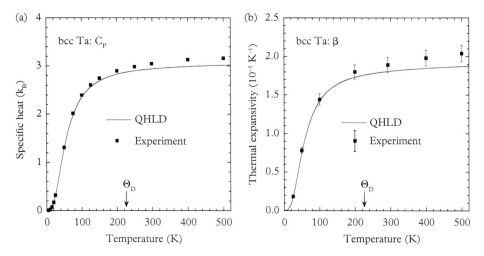

Fig. 7.10 *Thermodynamic derivatives in bcc Ta at zero pressure, as calculated from QHLD with Ta6.8x multi-ion MGPT potentials and compared with experimental data. (a) Constant-pressure specific heat C_P, with the measured data from Hultgren et al. (1973). (b) Thermal expansion coefficient β, with the measured data from Touloukian et al. (1975). The Debye temperature $\Theta_D = 226$ K is indicated.*
From Moriarty and Haskins (2014), with publisher (APS) permission.

7.5 Temperature-induced solid-solid phase transitions

In this section, we discuss the additional application of QHLD to temperature-induced phase transitions in metals. This topic is especially important for metals on the LHS of the Periodic Table in the first four columns that have an fcc, hcp or other cp structure at $T = 0$. All such metals are known to have a temperature-induced transition to bcc prior to melt, either at ambient pressure, or in the case of Mg at high pressure. In treating such phase transitions, one can again capitalize on the fact that the volume change $\Delta\Omega$ between any two transforming metallic phases that are held at constant temperature and pressure is small. This allows the simplified constant-volume treatment we established in Chapter 6 at $T = 0$ to be readily extended to finite temperature.

At constant pressure P and temperature T, the thermodynamically most stable phase of a metal is the one that has the lowest Gibbs free energy per ion G_{tot}, as given by

$$G_{\text{tot}}(P, T) = A_{\text{tot}}(\Omega, T) + P\Omega, \qquad (7.140)$$

where $P = P_{\text{tot}}$ and A_{tot} is the corresponding Helmholtz free energy per ion, expressed quite generally by Eq. (7.115). At the pressure-temperature phase boundary separating an initial (e.g., low-temperature) phase i from a final (e.g., high-temperature) phase f, one must thus have $G_{\text{tot}}^f = G_{\text{tot}}^i$, so that

$$\Delta G_{\text{tot}} = \Delta A_{\text{tot}} + P\Delta\Omega = 0, \qquad (7.141)$$

where $\Delta G_{\text{tot}} = G_{\text{tot}}^f(P, T) - G_{\text{tot}}^i(P, T)$ and $\Delta A_{\text{tot}} = A_{\text{tot}}^f(\Omega_f, T) - A_{\text{tot}}^i(\Omega_i, T)$. For $\Delta G_{\text{tot}} > 0$ the initial phase i is stable, while for $\Delta G_{\text{tot}} < 0$ the final phase f is stable.

As done previously with the analogous zero-temperature change in enthalpy in Eq. (6.9), one can usefully simplify Eq. (7.141) for small $\Delta\Omega$. The derivation exactly parallels that given in Sec. 6.1.3 at $T = 0$. In this case one obtains the following generalized result, valid to first order in $\Delta\Omega$:

$$\Delta G_{\text{tot}} = A_{\text{tot}}^f(\Omega_T, T) - A_{\text{tot}}^i(\Omega_T, T) = 0. \qquad (7.142)$$

Again the pressure P itself has dropped out completely from Eq. (7.142), and the location of the phase boundary is established only by the intersection of the free-energy functions $A_{\text{tot}}^f(\Omega_T, T)$ and $A_{\text{tot}}^i(\Omega_T, T)$, which determines the average transition volume Ω_T at temperature T. The average transition pressure can be calculated as

$$P_T(\Omega_T, T) = -\frac{1}{2}\left(\frac{dA_{\text{tot}}^i(\Omega_T, T)}{d\Omega} + \frac{dA_{\text{tot}}^f(\Omega_T, T)}{d\Omega}\right). \qquad (7.143)$$

Equations (7.142) and (7.143) apply quite generally to the mechanically stable solid phases of any metal for which a valid free-energy function $A_{\text{tot}}(\Omega, T)$ can be constructed, including the general weak-coupling form given by Eq. (7.115), and for sufficiently low temperatures the corresponding quasiharmonic form given by Eq. (7.119). The latter starting point offers some additional simplification as well in the context of volume-dependent QBIPs. In particular, the large volume term in the cold-energy component

E_0 of A_{tot} again drops out of Eq. (7.142). One can then rewrite Eq. (7.142) as a balance between the cold structural-energy contributions and ion-thermal phonon contributions:

$$\Delta G_{\text{tot}} = \Delta E_{\text{struc}} + \Delta A_{\text{ion}}^{\text{qh}} = 0, \quad (7.144)$$

where

$$\Delta E_{\text{struc}} = E_{\text{struc}}^{f}(\Omega_{\text{T}}) - E_{\text{struc}}^{i}(\Omega_{\text{T}}) \quad (7.145)$$

and

$$\Delta A_{\text{ion}}^{\text{qh}} = A_{\text{ion}}^{\text{qh}-f}(\Omega_{\text{T}}, T) - A_{\text{ion}}^{\text{qh}-i}(\Omega_{\text{T}}, T). \quad (7.146)$$

For pressure-induced phase transitions at finite temperature, the driving force is still ΔE_{struc}, but now mitigated by $\Delta A_{\text{ion}}^{\text{qh}}$. For temperature-induced phase transitions, on the other hand, the driving force is now a substantial negative value of $\Delta A_{\text{ion}}^{\text{qh}}$ arising from the final-phase quasiharmonic phonons, which must overcome the inherently positive value of ΔE_{struc}. For temperature-induced transitions from closed-packed structures to bcc, this outcome is a consequence of the relative high entropy $S_{\text{ion}}^{\text{qh}}$ of the bcc structure arising from the soft $T_1[110]$ phonon branch, and where one has

$$A_{\text{ion}}^{\text{qh}} = E_{\text{ion}}^{\text{qh}} - TS_{\text{ion}}^{\text{qh}}. \quad (7.147)$$

Such soft bcc phonons occur for all the metals in columns IA, IIA, IIIB and IVB of the Periodic Table, as illustrated for the alkali metals in Fig. 7.2(a) and Table 7.1.

In the context of QBIPs, the first GPT QHLD calculations of temperature-induced phase transitions making use of Eqs. (7.140)–(7.147) helped to explain, in the EDB limit, the observed fcc → bcc transitions for both Ca and Sr (Moriarty, 1973). Although these initial GPT-EDB calculations were carried out in reciprocal space with early-generation d-basis states and only used approximate BZ summations in the evaluation of the fcc and bcc quasiharmonic free energies $A_{\text{ion}}^{\text{qh}}(\Omega, T)$ via Eq. (7.118), the results clearly showed the importance of sp-d hybridization to the phase transitions. Not only is the hybridization necessary to explain the low-temperature fcc phase of both Ca and Sr, as we discussed in Chapter 4, but it is also necessary to explain qualitatively the large observed magnitudes of the transition temperature, 721 K in Ca and 830 K in Sr at ambient pressure. In the GPT-EDB calculations, over half of the corresponding predicted transition temperatures, 555 K in Ca and 625 K in Sr, resulted from the sp-d hybridization. In the case of Ca, a qualitatively correct pressure-temperature phase diagram to 2 GPa was obtained, with a positive slope of the fcc-bcc phase line. This outcome is consistent with the increase in the sp-d hybridization under pressure. Unexplained in the EDB limit, however, is the observed negative slope of the fcc-bcc phase line in the case of Sr. This latter behavior probably signals the beginning of transition-metal physics and the filling of the $4d$ bands, which results in a room-temperature bcc structure at a pressure of only about 3.5 GPa.

Two decades later, the interesting case of the hcp-bcc phase line in Mg was similarly investigated with GPT QHLD calculations in the SM limit (Althoff et al., 1993; Moriarty and Althoff, 1995), although in this case with a real-space analysis, using first-principles pair potentials. Magnesium is unique among the metals of the LHS of the Periodic Table in that the bcc structure is mechanically unstable near ambient pressure; as a consequence, low-pressure melting takes place instead out of its normal-density hcp structure. What has happened in Mg is that the bcc $T_1[110]$ phonon branch becomes so soft near normal density that imaginary phonon frequencies and a corresponding negative C' shear modulus result. Mechanical stability of the bcc structure is achieved under modest pressure of 5–10 GPa, however, and this stability increases with increasing pressure on through the observed hcp $\rightarrow$ bcc phase transition near 50 GPa at room temperature. As we have discussed in Chapter 6 and summarized in Table 6.5, the latter transition is well accounted for in both first-principles GPT and DFT calculations. In a more complete study of the pressure-temperature phase diagram of Mg, including both the large hcp and bcc solid domains and the corresponding hcp-liquid and bcc-liquid melt curves, Moriarty and Althoff (1995) first predicted the full temperature dependence of the hcp-bcc phase line. This result is displayed in Fig. 7.11 and compared there with a recent improved version of this calculation, as well as with two independent DFT QHLD calculations of the same phase line, obtained using the PAW electronic-structure method together with PHON calculations of the hcp and bcc phonon spectra (Mehta et al., 2006).

As shown in Fig. 7.11, at relatively high pressure above 35 GPa all four calculated hcp-bcc phase lines are similar and consistent with room-temperature DAC data, with each line beginning at $T = 0$ between 52 and 58 GPa in pressure and rising sharply with a negative PT slope. Below 35 GPa, however, where all four phase lines are predicted to be above 1000 K, the extreme softness of the bcc phonons lead to more sensitivity to the details of the QHLD calculations, including the treatment of exchange and correlation. Both GPT QHLD calculations have used the same basic Mg pair potentials with IU screening, which one will recall from Fig. 2.2 incorporate LDA xc at long wavelengths together with a many-body electron-gas correction at short wavelengths. The present GPT calculation, however, corrects two numerical shortcomings in the original Moriarty and Althoff (1995) treatment by using the smooth potential cutoff scheme discussed in Sec. 3.5.2 of Chapter 3 together with a much finer volume mesh in defining the GPT potentials. These improvements raise the calculated hcp-bcc phase line by some 200 K near 20 GPa. The present GPT phase line is then close in value to the baseline DFT QHLD result obtained with LDA xc. The second DFT QHLD calculation, which used GGA xc, pushes the hcp-bcc phase even much higher in temperature, with a peak value of about 1700 K at 17 GPa. Also in this regard, it should be mentioned that both DFT calculations have incorporated effective electron-thermal contributions to the calculated phase lines through the use a temperature-dependent electronic structure. At the same time, of course, possible corrections due to additional anharmonic temperature effects at these high temperatures, which are at 2–3 Θ_D, are not included in any of the four calculations. Such anharmonic effects will be treated for the GPT bcc phase

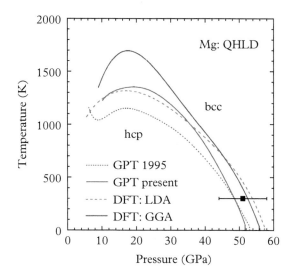

Fig. 7.11 *Comparison of four calculated GPT and DFT hcp-bcc phase lines for Mg in the QHLD limit. Shown are the original GPT result of Moriarty and Althoff (1995), its improved present version discussed in the text, both calculated with IU xc, and the DFT results of Mehta et al. (2006) calculated with LDA and GGA xc. All four results are consistent with the DAC-observed hcp $\rightarrow$ bcc transition at room temperature near 50 GPa (Olijnyk and Holzapfel, 1985; Stinton et al., 2014), displayed as a solid square with error bars.*

in Chapter 8, where we consider the entire Mg pressure-temperature phase diagram in light of recent static and dynamic measurements.

Finally, we conclude this chapter by adding several additional comments about the numerical accuracy and further extension of the central result derived here for calculating finite-temperature phase lines in metals, Eq. (7.142). Using this result as a starting point, we have developed a very efficient numerical algorithm to iterate back to a full Helmholtz free-energy construction that is equivalent to the exact Gibbs criterion describing the phase transition, Eq. (7.141). This capability can be used as a convenient means of checking the numerical accuracy of Eq. (7.142) for any particular solid-solid phase transition. In the case of the GPT hcp-bcc phase lines for Mg shown in Fig. 7.11, we find that there is a negligible difference between the results obtained from the use of Eq. (7.142) and the results obtained from the full Helmholtz free-energy construction. Moreover, in applications to more general phase transitions where there is a significant and non-negligible volume change, such as the solid-liquid melting transition, one can still use Eq. (7.142) as the starting point in the same algorithm to obtain the Helmholtz free-energy construction needed to calculate the phase transition accurately. This is, in fact, the manner in which the GPT and MGPT free-energy-based melting calculations described in Chapters 8 and 13 have been performed.

8
High-Temperature Properties, Melting and Phase Diagrams

In this chapter, we extend our treatment of thermodynamic properties, phase stability and phase transitions in metals to high temperature, including anharmonic vibrational effects in the solid, liquid-state structure and energetics, and melting. In the weak-coupling limit, our $T = 0$ QBIPs still apply for the cold and ion-thermal components of all quantities, supplemented by independent electron-thermal contributions derived from electronic-structure considerations. Additional statistical-mechanics tools are needed, however, to treat the large ion-thermal contributions at high temperature efficiently, including the novel technique of reversible-scaling MD (RSMD) simulation for both the solid and liquid, and variational perturbation theory (VPT) in the liquid. With these and other MD tools, accurate total energies, pressures and free energies can be obtained at high temperature over wide volume ranges to complete the calculation of thermodynamic properties, to predict pressure-temperature phase diagrams, including the melting curve, and to make contact with laser-heated DAC and dynamic shock-wave experiments. It is also possible to treat related topics such as thermoelasticity at high temperature for a given solid phase, and to search for new solid phases at high-P, T conditions.

8.1 Important QBIP computational tools at high temperature

We first consider four complementary QBIP computational tools that can be used to advantage for high-temperature applications. The first is large-scale MD, applied in bulk metals under either equilibrium or nonequilibrium conditions, and which, together with fast algorithms to evaluate multi-ion forces, can be used in both GPT-MD and MGPT-MD simulations up to at least 50–100 million atoms. The second tool is the specialized technique of RSMD, which can be used to obtain accurate solid and liquid ion-thermal free energies very efficiently over wide ranges of volume and temperature. In the solid, RSMD simulation replaces the traditional approximation methods of anharmonic perturbation theory and self-consistent phonons. For a given set of QBIP forces, RSMD simulation provides both an exact treatment of anharmonic effects in the solid all the way up to melt, and liquid ion-thermal free energies from melt to very high temperature, where proper normalization can be provided by an independently calculated reference free energy. The third tool we discuss is simplified VPT applied at high temperatures, which can be used to obtain quick estimates of the melting curve and, more importantly, to calculate the reference high-T free energy of the liquid needed in RSMD simulation for accurate free-energy melt-curve calculations. Finally, the fourth tool we discuss is the dynamic two-phase MD melting simulation, which provides a fast alternate approach to obtain the equilibrium melt curve, and is particularly useful for comparing the relative thermodynamic stability of different competing high-temperature solid phases.

8.1.1 Molecular dynamics simulation with fast algorithms

It is of interest to consider both single-phase and two-phase MD simulations with QBIP forces. Single-phase simulations of the bulk solid or liquid in equilibrium can be performed in a large computational cell of chosen size and shape containing N ions of atomic mass M_a held at constant volume $V = N\Omega$, with PBCs applied to all faces of the cell, and maintained at a constant temperature T by a variable-friction thermostat. The equation of motion for the ith ion in the cell is then of the form

$$\ddot{\mathbf{R}}_i = \mathbf{F}_i/M_a - \xi \dot{\mathbf{R}}_i, \tag{8.1}$$

where $\dot{\mathbf{R}}_i \equiv d\mathbf{R}_i/dt$ is the instantaneous velocity of the ion and $\ddot{\mathbf{R}}_i \equiv d^2\mathbf{R}_i/dt^2$ is its instantaneous acceleration. Quite generally, the force on the ion $\mathbf{F}_i$ is from Eqs. (1.1) and (1.2) just

$$\mathbf{F}_i(\mathbf{R}) = -\nabla_i E_{\text{tot}}(\mathbf{R}_1, \mathbf{R}_2 \ldots \mathbf{R}_N), \tag{8.2}$$

while ξ is a time-dependent friction coefficient, to be specified, that connects the physical system with an external heat bath with which it can exchange thermal energy. An idealized constant NVT or *canonical* ensemble with a Boltzmann distribution of ion velocities in equilibrium can be achieved with the well-known Nosé-Hoover thermostat (Nosé, 1984; Hoover, 1985, 1991; Frenkel and Smit, 2002). For this thermostat, the friction coefficient ξ satisfies the additional coupled differential equation

$$\dot{\xi} = \frac{1}{M_s} \left[\sum_{i=1}^{N} M_a \dot{\mathbf{R}}_i \cdot \dot{\mathbf{R}}_i - 3Nk_B T \right], \tag{8.3}$$

where $\dot{\xi} \equiv d\xi/dt$ and M_s is a thermal "mass" parameter, to be specified, which controls the rate at which heat is added to or removed from the physical system. The optimum value of the parameter M_s is system dependent, and needs to be chosen such that the effective total energy of the physical system plus the heat bath is conserved during the simulation. The latter condition will normally be satisfied for a range of values of M_s, and within that range further optimization criteria can be applied, if desired, to minimize the time to achieve thermal equilibrium.

In practice, a simpler and more computationally efficient thermostat, especially for high-temperature solid and liquid *NVT* MD simulations, is the parameter-free Gaussian thermostat, based on Gauss' principle of least constraint (Hoover et al., 1982; Evans, 1983; Allen and Tildesley, 1987; Hoover, 1991) and first validated against the Nosé-Hoover thermostat by Evans and Holian (1985). With the Gaussian thermostat, one has in place of Eq. (8.3) the following explicit form for the friction coefficient ξ:

$$\xi = \sum_{i=1}^{N} \dot{\mathbf{R}}_i \cdot (\mathbf{F}_i/M_\mathrm{a}) \Big/ \sum_{i=1}^{N} \dot{\mathbf{R}}_i \cdot \dot{\mathbf{R}}_i. \qquad (8.4)$$

Other forms of the thermostat in Eq. (8.1) are also possible. In the LLNL parallel molecular-dynamics code ddcMD (Glosli and Streitz, 2002), for example, which facilitates both two-phase melting simulations and rapid resolidification simulations (see Secs. 8.1.4 and 8.4, respectively), a stochastic Langevin thermostat (Allen and Tildesley, 1987) is used.

The *NVT* canonical ensemble for MD simulations is, of course, especially convenient for volume-dependent QBIPs, where in the force equation (8.2) $E_\mathrm{tot}(\mathbf{R}) \to E_\mathrm{tot}(\mathbf{R}, \Omega)$ with the total-energy functional given in its most general form by Eq. (7.18). The force $\mathbf{F}_i$ then naturally breaks up into pair and multi-ion components:

$$\mathbf{F}_i = \mathbf{F}_i^{(2)} + \mathbf{F}_i^{(3)} + \mathbf{F}_i^{(4)} + \cdots. \qquad (8.5)$$

The practical challenge is to reduce each force component $\mathbf{F}_i^{(n)}$ to its simplest possible computational form. Using the symmetry of the corresponding potential $v_n(ijk\ldots, \Omega)$ with respect to the interchange of any two indices, Moriarty (1994) showed that one could reduce $\mathbf{F}_i^{(n)}$ to the general form

$$\mathbf{F}_i^{(n)} = -\frac{1}{(n-2)!} \sum_{j,k,\ldots}' \frac{\partial v_n(ijk\ldots, \Omega)}{\partial R_{ij}} \hat{\mathbf{R}}_{ij}, \qquad (8.6)$$

where $\hat{\mathbf{R}}_{ij}$ is the unit vector $\mathbf{R}_{ij}/R_{ij}$. In practice, it is convenient to re-express the force components as

$$\mathbf{F}_i^{(n)} = \sum_j{}' \mathbf{F}_{ij}^{(n)}, \qquad (8.7)$$

and exploit the additional symmetry property that $\mathbf{F}_{ji}^{(n)} = -\mathbf{F}_{ij}^{(n)}$. This provides a factor of two computational-time savings when accumulating the forces during a given time step in an MD simulation. For $n = 2, 3$ and 4, the force components $\mathbf{F}_{ij}^{(n)}$ are given by

8.1.1 Molecular dynamics simulation with fast algorithms

$$\mathbf{F}_{ij}^{(2)} = -\frac{\partial v_2(ij, \Omega)}{\partial R_{ij}} \hat{\mathbf{R}}_{ij}, \tag{8.8}$$

$$\mathbf{F}_{ij}^{(3)} = -\sum_{k}{}' \frac{\partial v_3(ijk, \Omega)}{\partial R_{ij}} \hat{\mathbf{R}}_{ij} \tag{8.9}$$

and

$$\mathbf{F}_{ij}^{(4)} = -\frac{1}{2}\sum_{k,l}{}' \frac{\partial v_4(ijkl, \Omega)}{\partial R_{ij}} \hat{\mathbf{R}}_{ij} = -\sum_{k<l}{}' \frac{\partial v_4(ijkl, \Omega)}{\partial R_{ij}} \hat{\mathbf{R}}_{ij}, \tag{8.10}$$

where in Eq. (8.10) the second form of that equation provides an additional factor of two time savings in an MGPT-MD simulation.

One can further note that the pair and multi-ion virial components of the instantaneous pressure, $P_{\text{vir}}^{(n)}$, can be expressed directly in terms of the force components $\mathbf{F}_{ij}^{(n)}$:

$$P_{\text{vir}}^{(n)} = -\frac{1}{6\Omega} \frac{1}{N} \sum_{i,j}{}' \mathbf{R}_{ij} \cdot \mathbf{F}_{ij}^{(n)} = -\frac{1}{3\Omega} \frac{1}{N} \sum_{i<j}{}' \mathbf{R}_{ij} \cdot \mathbf{F}_{ij}^{(n)}. \tag{8.11}$$

The virial components of the pressure can then be combined with the remaining direct volume components,

$$P_{\text{vol}} = -dE_{\text{vol}}/d\Omega \tag{8.12}$$

and

$$P_{\text{vol}}^{(n)} = -\frac{1}{n!} \frac{1}{N} \sum_{i,j,k,\dots}{}' \frac{\partial v_n(ijk\dots, \Omega)}{\partial \Omega}, \tag{8.13}$$

to calculate the total instantaneous pressure

$$P_{\text{tot}} = P_{\text{vol}} + \sum_{n} \left(P_{\text{vir}}^{(n)} + P_{\text{vol}}^{(n)} \right). \tag{8.14}$$

In a given MD simulation, one wants to calculate the instantaneous energy, pressure and forces simultaneously during a time step. This necessitates evaluating the potentials v_n and their derivatives $\partial v_n/\partial R_{ij}$ and $\partial v_n/\partial \Omega$ on the fly for each encountered ion configuration in the simulation.

In the case of nontransition metals treated with, for example, GPT, DRT or RMP pair potentials, and requiring only a two-ion force component $\mathbf{F}_{ij}^{(2)}$, the derivative $\partial v_2/\partial R_{ij}$

in Eq. (8.8) can be evaluated analytically. In the case of GPT pair potentials, one obtains from Eq. (7.30) the general result

$$-\frac{\partial v_2(r,\Omega)}{\partial r} = \frac{(Z^\star e)^2}{r^2}\left[1 - \frac{2r^2}{\pi}\int_0^\infty qF_N(q,\Omega)j_1(qr)dq\right] - \frac{\partial v_{\text{ol}}(r,\Omega)}{\partial r}. \quad (8.15)$$

Computationally for a GPT-MD simulation, it is most efficient to pre-tabulate the RHS of Eq. (8.15) as a function of r for the desired volume Ω, and then obtain needed values in the simulation by interpolation on the table. Similarly, the derivative $\partial v_2/\partial\Omega$ entering $P_{\text{vol}}^{(2)}$ in Eq. (8.13) for $n=2$ can be calculated numerically, pre-tabulated and needed values obtained by interpolation. Similarly, for the pair-potential component of the MGPT total-energy functional for a central transition metal, one has from Eq. (7.36)

$$-\frac{\partial v_2(r,\Omega)}{\partial r} = \frac{(Z^\star e)^2}{r^2}\left[1 - \frac{2r^2}{\pi}\int_0^\infty qF_N(q,\Omega)j_1(qr)dq\right] - \frac{\partial v_2^{\text{hc}}(r,\Omega)}{\partial r}$$
$$+ \frac{2p}{R_0}\left[2v_a(\Omega)(R_0/r)^{4p+1} - v_b(\Omega)(R_0/r)^{2p+1}\right], \quad (8.16)$$

with an analogous tabulation-interpolation strategy used to obtain $\partial v_2/\partial R_{ij}$, and a parallel strategy used to obtain $\partial v_2/\partial\Omega$, in MGPT-MD simulations.

Efficient evaluation of the multi-ion potential derivatives $\partial v_3/\partial R_{ij}$ in Eq. (8.9) for $\mathbf{F}_{ij}^{(3)}$ and $\partial v_4/\partial R_{ij}$ in Eq. (8.10) for $\mathbf{F}_{ij}^{(4)}$ is more challenging, however. In early versions of MGPT-MD codes for central transition metals (Moriarty, 1994; Moriarty et al., 2002a), these derivatives were evaluated numerically on the fly. Due to the precise nature of v_3 and v_4 in the baseline analytic MGPT, this evaluation is indeed accurate, and these codes produced a number of useful high-temperature thermodynamic results for Mo and Ta. However, for large-scale MGPT-MD a more computationally efficient code was needed. Taking full advantage of the matrix multiplication used to define v_3 and v_4 in the matrix MGPT, Glosli (2001) developed a fast algorithm to calculate the multi-ion forces $\mathbf{F}_i^{(3)}$ and $\mathbf{F}_i^{(4)}$ analytically. This algorithm produced an immediate order of magnitude improvement in performance, and soon became the main engine powering the parallel LLNL ddcMD code for large-scale MGPT-MD simulations (Moriarty et al., 2006; Streitz et al., 2006a, 2006b). A decade later, Opplestrup (2015) developed an alternate, but equally efficient, fast algorithm to implement multi-ion MGPT potentials, which led to the USER-MGPT package (Opplestrup and Moriarty, 2018) in the open-source parallel LAMMPS molecular-dynamics code (Plimpton, 1995).

It should also be pointed out that volume-dependent QBIPs such as GPT and MGPT may be used for MD simulations in other ensembles, as well as in alternate representations, depending on the application. The basic *microcanonical* or *NVE* ensemble, which replaces constant temperature T by constant energy E, is achieved by simply turning off the thermostat altogether and setting $\xi = 0$ in Eq. (8.1). The more complex *isothermal-isobaric NPT* ensemble, which replaces constant volume V with constant pressure P, is achieved by adding a barostat to work in tandem with the thermostat. The barostat allows the volume of the computational cell to fluctuate so as to maintain constant pressure, while, as the same time, either maintaining the shape of the cell or allowing the shape of the cell to fluctuate as well. Both GPT-MD and MGPT-MD *NPT* simulations in a fixed-shape cell are relatively easy to perform. One simply needs to update the potentials, as well as the energy, pressure and forces, at each time step. It is also possible to perform aGPT-MD simulations, with or without defects present, in effective *NVE*, *NVT* or *NPT* ensembles.

In addition to providing thermodynamic quantities such as energies and pressures at finite temperature, MD simulations can also provide valuable structural information about both the solid and the liquid. The familiar and useful pair correlation function, $g(r)$, aka the pair distribution function, can be formally defined as the following thermal average to be evaluated at constant volume:

$$g(r) = \frac{\Omega}{N} \left\langle \sum_{i,j}{}' \delta(\mathbf{r} - \mathbf{R}_{ij}) \right\rangle. \tag{8.17}$$

This function gives the probability of finding a pair of ions in the metal a distance r apart, relative to the probability expected for a random distribution of noninteracting free ions at the same density. It can be evaluated in a standard way for use in an MD simulation code, as discussed by Allen and Tildesley (1987).

Similarly, one can define an angular correlation function $b(\theta)$. This quantity is taken to be the thermal average of the number of three-ion bond angles a given ion makes with its near neighbors. Specifically, a bond angle θ for ion i with neighboring ions j and k is defined as

$$\theta = \cos^{-1}\left[\mathbf{R}_{ij} \cdot \mathbf{R}_{ik}/(R_{ij}R_{ik})\right], \tag{8.18}$$

where both R_{ij} and R_{ik} are required to be less than some given distance R_{max}. We normally choose $R_{\text{max}} = 2.3 R_{\text{WS}}$, so that both first- and second-neighbor bond angles are included in $b(\theta)$ for the bcc structure. The angular correlation function so defined is an especially useful tool for investigating the local atomic structure about an ion in a transition metal with strong angular forces.

The representative behaviors of $g(r)$ and $b(\theta)$ for the bcc and liquid structures of a central transition metal at high-temperature melting conditions are illustrated in Fig. 8.1 for the case of Ta. In the pair correlation function depicted in panel (a), note that the first peak is almost the same in the bcc solid and in the liquid, indicating that on average first and second neighbor distances are also nearly the same. However, for more distant neighbors, the bcc solid retains some visible shell structure even at melting, while this is essentially lost in the liquid. Similarly, in the angular correlation function depicted in panel (b), some near-neighbor bond-angle structure is retained in the bcc solid at melt but is lost in the liquid. The detailed evolution of the bcc solid $b(\theta)$ function from low to high temperature is shown in Fig. 2 of Moriarty (1994) for the similar case of Mo.

Finally, one should note that in the liquid the reciprocal-space structure factor $S(q)$ can be calculated through a Fourier transform of the pair correlation function:

$$S(q) = 1 + \frac{4\pi}{\Omega} \int_0^\infty g(r) j_0(qr) r^2 dr, \qquad (8.19)$$

where j_0 is again the $\ell = 0$ spherical Bessel function. The liquid structure factor can and has been measured experimentally at ambient pressure for a number of metals where the melting temperature is below 2000 K. Unfortunately, the latter condition is not met for the refractory bcc transition metals.

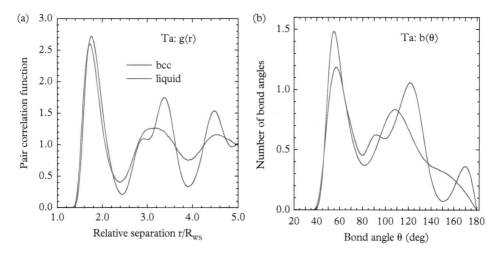

Fig. 8.1 *Pair correlation function (a) and angular correlation function (b) in bcc and liquid Ta at representative high-pressure melting conditions: $T_m = 4000\,K$ and $P_m = 13.8\,GPa$, as evaluated at an average atomic volume $\Omega = 121.6\,a.u.$ with Ta6.8x multi-ion potentials.*

8.1.2 Reversible-scaling MD for ion-thermal free energies

The efficient calculation of accurate free energies in solids and liquids from MD simulations is a challenging problem of long-standing interest and importance in condensed matter physics. Numerous specific computational schemes have been developed that depend to varying degrees on the principles of thermodynamic integration (TI) (Frenkel and Smit, 2002), adiabatic switching (AS) (Watanabe and Reinhardt, 1990), and/or thermodynamic perturbation theory (TPT) (Ashcroft and Stroud, 1978; Alfè et al., 2002). Particularly challenging is the problem of efficiently calculating free energies in real materials over wide ranges of volume or pressure and temperature in multiple phases. This capability is required to obtain multiphase equations of state and equilibrium phase boundaries to high (1–2%) accuracy from a given set of QBIP forces. The simplifying and highly efficient method of RSMD proposed by de Koning et al. (1999), which marries the principles of TI and AS, has now been developed in detail by Moriarty and Haskins (2014) for GPT and MGPT potentials applied to metals in the weak-coupling limit, allowing one to obtain free-energy differences as a function of temperature at constant volume in a given phase from a single MD simulation.

To begin, one first notes that the total Helmholtz free energy of the metal per ion $A_{\text{tot}}(\Omega, T)$, in any given phase of interest, can be expressed as a functional of the corresponding total internal energy per ion $E_{\text{tot}}(\Omega, T)$. In particular, formal thermodynamic integration of Eq. (7.120) for E_{tot} at constant volume, from a reference temperature T_{ref} to the desired temperature T, gives the general result

$$A_{\text{tot}}(\Omega, T) = (T/T_{\text{ref}})A_{\text{tot}}(\Omega, T_{\text{ref}}) - T \int_{T_{\text{ref}}}^{T} [E_{\text{tot}}(\Omega, T')/(T')^2]dT'. \tag{8.20}$$

In the weak-coupling limit of interest here, A_{tot} is, at the same time, a sum of cold, ion-thermal and electron-thermal components, as expressed in Eq. (7.115). It follows that all three components separately satisfy a relationship of the same form as Eq. (8.20). In particular, for the unknown temperature-dependent ion- and electron-thermal free-energy components, one has the useful results

$$A_{\text{ion}}(\Omega, T) = (T/T_{\text{ref}})A_{\text{ion}}(\Omega, T_{\text{ref}}) - T \int_{T_{\text{ref}}}^{T} [E_{\text{ion}}(\Omega, T')/(T')^2]dT' \tag{8.21}$$

and

$$A_{\text{el}}(\Omega, T) = (T/T_{\text{ref}})A_{\text{el}}(\Omega, T_{\text{ref}}) - T \int_{T_{\text{ref}}}^{T} [E_{\text{el}}(\Omega, T')/(T')^2]dT', \tag{8.22}$$

where

$$E_{\text{tot}}(\Omega, T) = E_0(\Omega) + E_{\text{ion}}(\Omega, T) + E_{\text{el}}(\Omega, T), \tag{8.23}$$

with $A_0(\Omega) = E_0(\Omega)$.

The focus of the RSMD simulation method is on the calculation of $A_{\text{ion}}(\Omega, T)$ over an extended temperature range at constant volume, using Eq. (8.21). This is done in the $A_{\text{el}} = 0$ limit, with $A_{\text{el}}(\Omega, T)$ determined and added independently for specific cases, as discussed in Sec. 8.2.1. The RSMD approach takes direct advantage of the fact that in the classical statistical mechanics of an MD simulation for an N-ion system, the ion-thermal free energy $A_{\text{ion}}(\Omega, T)$ depends on the total potential energy functional $U(\mathbf{R}) \equiv E_{\text{tot}}(\mathbf{R}_1, \mathbf{R}_2, \cdots \mathbf{R}_N)$ establishing the interatomic forces on the ions only through the Boltzmann factor $\exp(-U/k_B T)$. If one runs an MD simulation at a fixed simulation temperature T_{ref} for a scaled potential-energy function λU, then the corresponding free energy of the unscaled system is thereby determined for a temperature $T = T_{\text{ref}}/\lambda$. In an RSMD simulation, the scaling factor λ is allowed to vary slowly and linearly with time,

$$\lambda(t) = 1 + \frac{\lambda_f - 1}{t_S} t, \tag{8.24}$$

starting from an initial value of $\lambda = 1$, corresponding to the reference temperature T_{ref}, to a final value of $\lambda_f = \lambda(t_S) = T_{\text{ref}}/T_f$, where T_f is the final temperature for which the free energy is to be obtained. If the switching time t_S between those limits is long, then the RSMD simulation process is adiabatic and the temperature integral in Eq. (8.21) can be evaluated in terms of an equivalent time integral W_{ion} as follows:

$$-T \int_{T_{\text{ref}}}^{T} [E_{\text{ion}}(\Omega, T')/(T')^2] dT' = \frac{W_{\text{ion}}}{\lambda} + \frac{3}{2} k_B T \ln \lambda - \left(1 - \frac{1}{\lambda}\right) E_0, \tag{8.25}$$

where

$$W_{\text{ion}}(\Omega, T(t)) = \frac{1}{N} \int_0^t \frac{d\lambda(t')}{dt'} U(\mathbf{R}(t')) \, dt'$$

$$= \frac{1}{N} \frac{\lambda_f - 1}{N_S} \sum_{n=1}^{n_S(t)} U(\mathbf{R}(t_n)). \tag{8.26}$$

In the second line of Eq. (8.26), the time integral has been evaluated using the histogram method, assuming a small simulation time step h and N_S time steps, such that $t_S = N_S h$. In practice, one can take h on the order of 0.2 fs and N_S in the range 10^5–10^7, depending on the problem addressed and the numerical accuracy required, so t_S is in the range of 20 – 2000 ps. Finally, inserting Eq. (8.25) back into Eq. (8.21) and using $\lambda^{-1} = T/T_{\text{ref}}$ one has the general classical result

$$A_{\text{ion}}(\Omega, T) = \frac{T}{T_{\text{ref}}} [E_0(\Omega) + A_{\text{ion}}(\Omega, T_{\text{ref}}) + W_{\text{ion}}(\Omega, T)] - \frac{3}{2} k_B T \ln \frac{T}{T_{\text{ref}}} - E_0(\Omega). \tag{8.27}$$

The use of this result in the special cases of a stable solid phase, the liquid and a metastable solid phase is discussed below.

8.1.2.1 RSMD applied to a stable solid phase

The ion-thermal free energy A_{ion} for the stable solid-phase ground state of a metal can be conveniently separated into a sum of quasiharmonic and anharmonic components, as in Eq. (7.117), with the quasiharmonic component $A_{\text{ion}}^{\text{qh}}$ expressed directly

8.1.2 Reversible-scaling MD for ion-thermal free energies

in terms of volume-dependent phonon frequencies via Eq. (7.118). The remaining anharmonic free-energy component $A_{\text{ion}}^{\text{ah}}$ can be obtained from the general RSMD result for A_{ion}, given by Eq. (8.27), in the following manner. Because Eq. (8.27) is based on classical Boltzmann statistics, to correctly isolate $A_{\text{ion}}^{\text{ah}}(\Omega, T)$ one must subtract from that equation the classical harmonic component of the ion-thermal free energy, which is $3k_B T(1 - \ln T)$. Assuming only that $A_{\text{ion}}^{\text{ah}}(\Omega, T_{\text{ref}}) = 0$, so that $A_{\text{ion}}(\Omega, T_{\text{ref}}) = 3k_B T_{\text{ref}}(1 - \ln T_{\text{ref}})$ in Eq. (8.27), one is then left with an explicit equation for the anharmonic free energy:

$$A_{\text{ion}}^{\text{ah}}(\Omega, T) = \frac{T}{T_{\text{ref}}}[E_0(\Omega) + W_{\text{ion}}(\Omega, T)] + \frac{3}{2}k_B T \ln \frac{T}{T_{\text{ref}}} - E_0(\Omega). \quad (8.28)$$

We note that physically the anharmonic free energy will vanish at zero temperature, so that $A_{\text{ion}}^{\text{ah}}(\Omega, 0) = 0$, and further that $A_{\text{ion}}^{\text{ah}}$ remains negligibly small for temperatures below the Debye temperature, $T < \Theta_D$. On the other hand, as a practical matter RSMD simulation cannot be carried down to very low temperatures and is most efficient and accurate numerically when $T \geq T_{\text{ref}} \geq \Theta_D$. As a compromise, we assume a volume-dependent reference temperature in Eq. (8.28) of the form $T_{\text{ref}}(\Omega) = x_D \Theta_D(\Omega)$, where x_D is a chosen fixed parameter and the Debye temperature is defined by Eq. (7.48). In practice, we find that an optimum choice for x_D is in the range $0.5 \leq x_D \leq 0.8$. For such a choice, $A_{\text{ion}}^{\text{ah}}(\Omega, T)$ can be evaluated as a function of temperature at a given volume using Eq. (8.28) with a single RSMD simulation ranging from $T = x_D \Theta_D$ to $T = T_{\text{max}}^{\text{sol}}$, the maximum temperature of interest in the solid at that volume. To maintain good physical accuracy, $T_{\text{max}}^{\text{sol}}$ should not be taken too far above the solidus melt point, which can be estimated for this purpose by the volume-dependent Lindemann melt temperature

$$T_L(\Omega) = T_m^0 [\Theta_D(\Omega)/\Theta_D(\Omega_m^0)]^2 (\Omega/\Omega_m^0)^{2/3}, \quad (8.29)$$

with T_m^0 the observed or calculated stable-solid melt temperature at solidus volume Ω_m^0. One can then take $T_{\text{max}}^{\text{sol}}(\Omega) = x_L^{\text{max}} T_L(\Omega)$, with x_L^{max} a chosen parameter in the range $1 \leq x_L^{\text{max}} \leq 2$, depending on the magnitude of T_L. At expanded volumes, where T_L is small, a value in the vicinity of $x_L^{\text{max}} = 2$ can be safely used, while at high pressure, where T_L is large, a value near $x_L^{\text{max}} = 1$ is usually required.

To extend the RSMD anharmonic free-energy calculation of $A_{\text{ion}}^{\text{ah}}(\Omega, T)$ over a wide volume range, an effective strategy is to first calculate a small set (typically, 4 to 8) of constant-volume $A_{\text{ion}}^{\text{ah}}$ isochores via Eq. (8.28) that span the volume range of interest. These isochores should be uniformly spaced in the reduced volume variable $x = (\Omega/\Omega_0)^{1/3}$, and be separated by a reasonable volume spacing of at least $\Delta x = 0.04$. Each simulated isochore can then be smoothly fitted with a polynomial temperature expansion of the form

$$A_{\text{ion}}^{\text{ah}}(\Omega, T) = A_2 T^2 + A_3 T^3 + A_4 T^4 + A_5 T^5 + A_6 T^6 + \cdots, \quad (8.30)$$

where the quantities $A_n \equiv A_n(\Omega)$ are volume-dependent fitting coefficients. The leading T^2 term in this expansion ensures the proper behavior of $A_{\text{ion}}^{\text{ah}}$ at low temperature and corrects the small error incurred by starting each RSMD simulation at finite temperature. The number of terms that need to be retained in the expansion is system-dependent and can be optimized to obtain the best smooth fit. Retaining the same number of terms and taking into account the general requirement of thermodynamic consistency,

$$E_{\text{ion}}^{\text{ah}}(\Omega, T) = -T^2 \frac{\partial \{A_{\text{ion}}^{\text{ah}}(\Omega, T)/T\}}{\partial T}, \tag{8.31}$$

the corresponding anharmonic component of the ion-thermal internal energy is then given by the expression

$$E_{\text{ion}}^{\text{ah}}(\Omega, T) = -A_2 T^2 - 2A_3 T^3 - 3A_4 T^4 - 4A_5 T^5 - 5A_6 T^6 - \cdots. \tag{8.32}$$

The smooth values of $A_{\text{ion}}^{\text{ah}}$ and $E_{\text{ion}}^{\text{ah}}$ so calculated on the selected isochores can finally be extended to any volume within the range of interest through numerical interpolation.

Examples of RSMD simulations of $A_{\text{ion}}^{\text{ah}}(\Omega, T)$ along representative isochores, and the fitting of the results with Eq. (8.30), are displayed in Fig. 8.2 for bcc Mg and Ta. The isochores and temperature ranges considered are relevant to thermodynamic properties and high-pressure melting up to 100 GPa in Mg and up to 400 GPa in Ta, applications that are discussed in Secs. 8.2.3 and 8.3. Note that the sign, magnitude and general behavior of $A_{\text{ion}}^{\text{ah}}(\Omega, T)$ are very different for these two metals. In the NFE simple metal Mg,

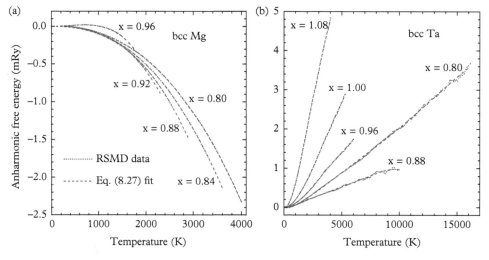

Fig. 8.2 *Representative anharmonic free-energy isochores $A_{\text{ion}}^{\text{ah}}(\Omega, T)$ in the prototype metals Mg and Ta, as calculated by RSMD simulation at constant values of $x = (\Omega/\Omega_0)^{1/3}$, and then analytically fit with Eq. (8.30). (a) Simple metal Mg treated with first-principles GPT pair potentials in its high-pressure bcc phase, with $\Omega_0 = 156.8$ a.u.; (b) bcc transition metal Ta treated with MGPT Ta6.8x multi-ion potentials, with $\Omega_0 = 121.6$ a.u.*
Panel (b) from Moriarty and Haskins (2014), with publisher (APS) permission.

the GPT-calculated $A_{\text{ion}}^{\text{ah}}$ energies are negative, small and smoothly varying in magnitude, and confined to a narrow band of values as a function of temperature. An excellent fit of the results only requires terms through T^4 in Eq. (8.30). In contrast, for the d-bonded transition metal Ta, the MGPT-calculated $A_{\text{ion}}^{\text{ah}}$ energies are positive, as well as significantly larger in magnitude and widely varying in value as a function of temperature. To accommodate both the large temperature curvatures of the isochores below about 3000 K and the lower curvatures at higher temperature, fitting terms through T^6 are needed in Eq. (8.30) for Ta.

8.1.2.2 RSMD applied to the liquid

In the liquid, the central RSMD result, Eq. (8.27), can be used directly to obtain the temperature dependence of $A_{\text{ion}}^{\text{liq}}(\Omega, T)$ for a given volume, but the added challenge is to independently calculate the required reference free energy $A_{\text{ion}}^{\text{liq}}(\Omega, T_{\text{ref}})$ at that volume as well. To accomplish the latter task, we take T_{ref} to be a high temperature at the top of the range of interest in the liquid, $T_{\text{ref}} = T_{\text{max}}^{\text{liq}}$, and above the maximum expected melt temperature in the volume range of interest. We next introduce an appropriate reference system, with potential energy function $U_{\text{ref}}(\mathbf{R})$, whose free energy $A_{\text{ion}}^{\text{ref}}(\Omega, T_{\text{ref}})$ is accurately known at the required conditions. One can then calculate $A_{\text{ion}}^{\text{liq}}(\Omega, T_{\text{ref}})$ with equal accuracy by a smooth thermodynamic integration from the reference system to the true system:

$$A_{\text{ion}}^{\text{liq}}(\Omega, T_{\text{ref}}) = A_{\text{ion}}^{\text{ref}}(\Omega, T_{\text{ref}}) + \frac{1}{N} \int_0^1 \langle U(\mathbf{R}) - U_{\text{ref}}(\mathbf{R}) \rangle_\lambda \, d\lambda, \qquad (8.33)$$

where λ is a scaling parameter varying between 0 and 1, and the quantity $\langle U(\mathbf{R}) - U_{\text{ref}}(\mathbf{R}) \rangle_\lambda$ in the integrand is a thermal average at volume Ω and temperature T_{ref} in the canonical ensemble of the mixed potential energy function

$$U_\lambda(\mathbf{R}) = \lambda U(\mathbf{R}) + (1 - \lambda) U_{\text{ref}}(\mathbf{R}). \qquad (8.34)$$

To make the integration in Eq. (8.33) computationally efficient, however, the reference system used must be a good match to the physical system under consideration, so that $\langle U(\mathbf{R}) - U_{\text{ref}}(\mathbf{R}) \rangle_\lambda = \langle dU_\lambda/d\lambda \rangle_\lambda$ varies slowly and smoothly as a function of λ, and only a few points are needed to evaluate the integral accurately.

In most metals, the thermodynamics of the high-temperature liquid is dominated by short-range repulsive forces, and a useful reference system to consider in the context of Eq. (8.33) is the inverse 12th power potential, r^{-12}, a pair potential that has been extensively studied via computer simulations and for which an accurate free energy and pair correlation function are known (Young and Rogers, 1984). This choice also allows one to use the complementary VPT technique to optimize the coefficient of the r^{-12} potential and to determine the corresponding values of $A_{\text{ion}}^{\text{ref}}(\Omega, T_{\text{ref}})$ in Eq. (8.33), as is discussed in Sec. 8.1.3.

Once the reference free energy $A_{\text{ion}}^{\text{liq}}(\Omega, T_{\text{ref}})$ has been established for use in Eq. (8.33), that equation can be used directly to calculate the liquid ion-thermal free energy

$A_{\text{ion}}^{\text{liq}}(\Omega, T)$ as a function of temperature at a given volume from the chosen reference temperature $T = T_{\text{ref}} = T_{\text{max}}^{\text{liq}}$ down to some desired minimum temperature $T = T_{\text{min}}^{\text{liq}}$ below the melt curve. It is again very often convenient to express the latter temperature as a specific fraction $x_{\text{L}}^{\text{min}}$ of the volume-dependent Lindemann melt temperature, such that $T_{\text{min}}^{\text{liq}}(\Omega) = x_{\text{L}}^{\text{min}} T_{\text{L}}(\Omega)$. The optimum value of $x_{\text{L}}^{\text{min}}$ depends to some extent on the application at hand. For melting from the stable solid phase, a value in the range $0.5 \leq x_{\text{L}}^{\text{min}} \leq 0.8$ usually works well.

The RSMD liquid free energy calculation of $A_{\text{ion}}^{\text{liq}}(\Omega, T)$ can be efficiently extended to a wide volume range in a similar manner to that for the anharmonic free energy. One again first calculates a selected set of equally spaced $A_{\text{ion}}^{\text{liq}}$ isochores within the volume range of interest, and then fits each isochore with a temperature expansion that is now of the form

$$A_{\text{ion}}^{\text{liq}}(\Omega, T) = C_0 \tau - C_1 \tau \ln \tau + C_2 \tau(\tau - 1) + C_3 \tau(\tau^2 - 1) + C_4 \tau(\tau^3 - 1)$$
$$+ C_5 \tau(\tau^4 - 1) + C_6 \tau(\tau^5 - 1) + \cdots, \quad (8.35)$$

where $\tau = T/T_{\text{ref}}$ and where the quantities $C_n \equiv C_n(\Omega)$ are volume-dependent fitting coefficients that include the constraint $C_0(\Omega) = A_{\text{ion}}^{\text{liq}}(\Omega, T_{\text{ref}})$. The corresponding thermodynamically consistent ion-thermal component of the total internal energy is then given by

$$E_{\text{ion}}^{\text{liq}}(\Omega, T) = C_1 \tau - C_2 \tau^2 - 2C_3 \tau^3 - 3C_4 \tau^4 - 4C_5 \tau^5 - 5C_6 \tau^6 - \cdots, \quad (8.36)$$

which is then analogous to Eq. (8.32) for $A_{\text{ion}}^{\text{ah}}$, except for the necessary addition of a term linear in the temperature. While Eqs. (8.35) and (8.36) can be used directly in this form, if six or more terms are retained in these expansions, it is desirable to introduce the following additional specific constraint on $E_{\text{ion}}^{\text{liq}}$ at the reference temperature:

$$E_{\text{ion}}^{\text{liq}}(\Omega, T_{\text{ref}}) = C_1 - C_2 - 2C_3 - 3C_4 - 4C_5 - 5C_6 - \cdots. \quad (8.37)$$

To impose Eq. (8.37), one needs to calculate an accurate value of $E_{\text{ion}}^{\text{liq}}(\Omega, T_{\text{ref}})$ for each of the selected isochore volumes with an independent GPT-MD or MGPT-MD simulation. Then combining Eq. (8.37) with Eqs. (8.35) and (8.36) not only ensures that both $A_{\text{ion}}^{\text{liq}}$ and $E_{\text{ion}}^{\text{liq}}$ are accurately maintained along the $T = T_{\text{ref}}$ isotherm, but also produces a robust fit of $A_{\text{ion}}^{\text{liq}}$ along all of the RSMD isochores. The values of $A_{\text{ion}}^{\text{liq}}$ and $E_{\text{ion}}^{\text{liq}}$ so calculated can then be extended to any volume within the chosen range by numerical interpolation.

8.1.2.3 RSMD applied to a metastable solid phase

Metastable solid phases that are fully mechanically stable at low temperature with all real $T = 0$ phonon frequencies present no special problem, and such cases can be treated in the same manner as the stable solid ground state, using Eqs. (8.28)–(8.32) to obtain

$A_{\text{ion}}^{\text{ah}}$ and $E_{\text{ion}}^{\text{ah}}$. The more challenging case is the one of a metastable solid phase that is mechanically unstable at low temperature and is only mechanically stabilized at high temperature by anharmonic phonon-phonon interactions, with or without the aid of additional electron-thermal effects. Such a phase has at least some imaginary phonon frequencies at $T = 0$, and hence its quasiharmonic ion-thermal free energy, $A_{\text{ion}}^{\text{qh}}(\Omega, T)$ as given by Eq. (7.118), does not exist and can't be used to establish the reference energy $A_{\text{ion}}^{\text{sol}}(\Omega, T_{\text{ref}})$ that is required in Eq. (8.27). For this case, a special procedure can be used in which one determines $A_{\text{ion}}^{\text{sol}}(\Omega, T_{\text{ref}})$ along the solidus melt line of the metastable phase in the $A_{\text{el}} = 0$ limit. In that limit, the melt curve for the metastable phase can be determined directly by MD simulation using the two-phase coexistence method discussed in Sec. 8.1.4. From the two-phase melt calculation, one can extract the volume-dependent reference solidus melt temperature $T_{\text{m}}^{\text{sol}}(\Omega)$, melt pressure $P_{\text{m}}^{\text{sol}}(\Omega)$, and melt energy $E_{\text{m}}^{\text{sol}}(\Omega)$ over the entire volume range of interest. Along the melt line, the melt temperature T_{m}, melt pressure P_{m}, and Gibbs free energy, $G = A + P\Omega$, of the liquid and of the metastable solid must be equal, so that $P_{\text{m}} = P_{\text{m}}^{\text{sol}} = P_{\text{m}}^{\text{liq}}$, $T_{\text{m}} = T_{\text{m}}^{\text{sol}} = T_{\text{m}}^{\text{liq}}$, and the Helmholtz free energy of the metastable solid along the solidus melt line can be calculated as

$$A_{\text{sol}}(\Omega_{\text{sol}}, T_{\text{m}}) = A_{\text{liq}}(\Omega_{\text{liq}}, T_{\text{m}}) + P_{\text{m}}(\Omega_{\text{liq}} - \Omega_{\text{sol}}). \tag{8.38}$$

For given values of T_{m}, P_{m} and the solidus volume Ω_{sol} on the melt curve, the corresponding liquidus volume Ω_{liq} and free energy $A_{\text{liq}}(\Omega_{\text{liq}}, T_{\text{m}})$ can be obtained via interpolation on the liquid equation of state (calculated in the $A_{\text{el}} = 0$ limit). With $T_{\text{ref}} = T_{\text{m}}$ and $\Omega = \Omega_{\text{sol}}$, Eq. (8.38) can be used directly to establish the reference energy

$$A_{\text{ion}}^{\text{sol}}(\Omega, T_{\text{ref}}) = A_{\text{sol}}(\Omega_{\text{sol}}, T_{\text{m}}) - E_0(\Omega_{\text{sol}}) \tag{8.39}$$

needed in Eq. (8.27). One can then apply Eq. (8.27) both upward in temperature from $T = T_{\text{m}}$ to the desired maximum $T = T_{\text{max}}^{\text{sol}}$ and downward in temperature from $T = T_{\text{m}}$ to an allowable minimum $T = T_{\text{min}}^{\text{sol}}$, which must be established independently for the metastable phase in question. Once $A_{\text{ion}}^{\text{sol}}(\Omega, T)$ is thereby calculated for the volumes and temperatures of interest, the electron-thermal component $A_{\text{el}}(\Omega, T)$ can finally be added to it to establish a total free energy and equation of state for the metastable phase.

8.1.3 Variational perturbation theory for liquid metals

Variational perturbation theory or VPT, which is part of a more general set of TPT methods, is an approximation technique for liquid metals and other fluids that can be used to establish close upper and lower bounds on the free energy (Ashcroft and Stroud, 1978; Young and Rogers, 1984; Boercker and Young, 1989). In the standard formulation of VPT, it is assumed that the dominant repulsive forces present in the high-temperature liquid metal are established by an effective pair potential v_2^{eff}. For our purposes, and without any loss of generality, we can take the corresponding VPT liquid potential energy $U_{\text{liq}}^{\text{VPT}}(\mathbf{R})$ in the volume-dependent QBIP form

$$U_{\text{liq}}^{\text{VPT}}(\mathbf{R}) = NE_{vol}(\Omega) + \frac{1}{2}{\sum_{i,j}}' v_2^{\text{eff}}(R_{ij}, \Omega). \tag{8.40}$$

For nontransition metals, of course, one may simply take $v_2^{\text{eff}} = v_2$, the established QBIP potential of choice, so no further approximation is needed. For transition metals treated with multi-ion MGPT potentials, on the other hand, it is necessary to construct a thermally averaged effective pair potential at each volume and temperature of interest in the form

$$v_2^{\text{eff}}(R_{ij}, \Omega) = v_2(R_{ij}, \Omega) + \frac{1}{3}\left\langle {\sum_k}' v_3(ijk, \Omega) \right\rangle + \frac{1}{12}\left\langle {\sum_{k,l}}' v_4(ijkl, \Omega) \right\rangle. \tag{8.41}$$

For the liquid RSMD application discussed above to obtain $A_{\text{ion}}^{\text{liq}}(\Omega, T_{\text{ref}})$ in Eq. (8.33), it is important to note that the thermal averages over v_3 and v_4 are needed only at the reference temperature T_{ref}. Illustrative comparisons of v_2^{eff} and the optimized r^{-12} reference potential v_{ref} determined at $T = T_{\text{ref}}$ and $\Omega = \Omega_0$ for the prototype metals Mg and Ta are displayed in Fig. 8.3.

In VPT applied to the r^{-12} reference system, the use of the well-known Gibbs-Bogolyubov inequality (Ashcroft and Stroud, 1978) provides a rigorous upper bound (ub) on the liquid free energy A_{liq} (in the limit $A_{\text{el}} = 0$):

$$A_{\text{liq}}(\Omega, T) \equiv E_0(\Omega) + A_{\text{ion}}^{\text{liq}}(\Omega, T) \leq A_{\text{liq}}^{\text{ub}}(\Omega, T), \tag{8.42}$$

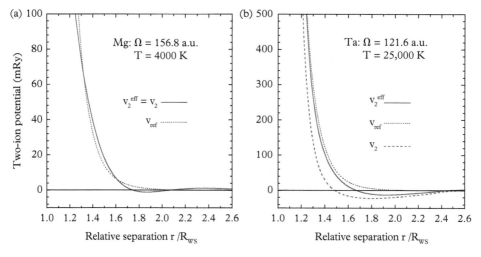

Fig. 8.3 *Effective pair potential v_2^{eff} in prototype high-temperature liquid metals compared with the corresponding optimum VPT reference potential v_{ref} [Eq. (8.53)], as determined under the same physical conditions: $T = T_{\text{ref}}$ and $\Omega = \Omega_0$. (a) Simple metal Mg with GPT $v_2^{\text{eff}} = v_2$; (b) transition metal Ta with MGPT-based v_2^{eff} given by Eq. (8.41).*
Panel (b) from Moriarty and Haskins (2014), with publisher (APS) permission.

where the upper bound on the free energy is given by

$$A_{\text{liq}}^{\text{ub}}(\Omega, T) = E_{\text{vol}}(\Omega) + A_{\text{ref}}(z, T) + (2\pi/\Omega) \int_0^\infty g_{\text{ref}}(r, z)[v_2^{\text{eff}}(r, \Omega) - v_{\text{ref}}(r, z)]r^2 dr. \tag{8.43}$$

In Eq. (8.43), the quantities A_{ref}, g_{ref} and v_{ref}, are, respectively, the known reference free energy, the pair-correlation function, and the pair potential of the r^{-12} reference system (Young and Rogers, 1984), with

$$A_{\text{ref}}(z, T) = \sum_{n=1}^{5} \frac{1}{n} B_{n+1} z^n k_B T + A_{\text{gas}}(\Omega, T), \tag{8.44}$$

$$v_{\text{ref}}(r, z) = \varepsilon(\sigma/r)^{12} \tag{8.45}$$

and the dimensionless variational parameter z given by

$$z = (\sigma^3/\sqrt{2}\Omega)(\varepsilon/k_B T)^{1/4}. \tag{8.46}$$

In Eq. (8.44), the quantities B_{n+1} are the calculated virial coefficients—$B_2 = 3.6296$, $B_3 = 7.5816$, $B_4 = 9.9792$, $B_5 = 8.4520$ and $B_6 = 4.4$—while A_{gas} is the ideal-gas free energy: $A_{\text{gas}}(\Omega, T) = -[\ln(\Omega/\lambda_{\text{ion}}^3) + 1]k_B T$, with $\lambda_{\text{ion}} = \sqrt{(2\pi\hbar^2)/(M_{\text{ion}} k_B T)}$ and M_{ion} the ion mass.

At each of the volumes of interest and temperature $T = T_{\text{ref}}$, the variational parameter z can be chosen to minimize the RHS of Eq. (8.42), yielding a function $z^{\text{ub}}(\Omega)$, for which $A_{\text{liq}}^{\text{ub}}(\Omega, T_{\text{ref}})$ becomes a close upper bound to the true liquid free energy $A_{\text{liq}}(\Omega, T_{\text{ref}})$ over the entire volume range. A corresponding lower bound (lb) on the free energy can also be established within VPT by reversing the roles of the physical system and the reference system (Boercker and Young, 1989). One so obtains

$$A_{\text{liq}}(\Omega, T) \geq A_{\text{liq}}^{\text{lb}}(\Omega, T) \tag{8.47}$$

where the lower bound on the free energy is given by

$$A_{\text{liq}}^{\text{lb}}(\Omega, T) = E_{\text{vol}}(\Omega) + A_{\text{ref}}(z) + (2\pi/\Omega) \int_0^\infty g_{\text{eff}}(r, \Omega)[v_2^{\text{eff}}(r, \Omega) - v_{\text{ref}}(r, z)]r^2 dr. \tag{8.48}$$

In Eq. (8.48), the quantity $g_{\text{eff}}(r, \Omega)$ is the physical-system pair correlation function for $v_2^{\text{eff}}(r, \Omega)$ and must be determined from independent GPT-MD or MGPT-MD simulations at the volumes of interest and $T = T_{\text{ref}}$. The variational parameter z can now be chosen to maximize the RHS of Eq. (8.47) at each volume, yielding $z^{\text{lb}}(\Omega)$ and making $A_{\text{liq}}^{\text{lb}}(\Omega, T_{\text{ref}})$ a close lower bound to $A_{\text{liq}}(\Omega, T_{\text{ref}})$ across the entire volume range or interest.

An approximate value for $A_{\text{ion}}^{\text{liq}}(\Omega, T_{\text{ref}})$ can be calculated by simply averaging the upper and lower bounds of $A_{\text{liq}}(\Omega, T_{\text{ref}})$, and then subtracting the cold energy:

$$A_{\text{ion}}^{\text{liq}}(\Omega, T_{\text{ref}}) \cong \frac{1}{2}[A_{\text{liq}}^{\text{ub}}(\Omega, T_{\text{ref}}) + A_{\text{liq}}^{\text{lb}}(\Omega, T_{\text{ref}})] - E_0(\Omega). \quad (8.49)$$

While Eq. (8.49) often provides a good first estimate, the more reliable and accurate procedure to obtain $A_{\text{ion}}^{\text{liq}}(\Omega, T_{\text{ref}})$ is to calculate first the average of z^{ub} and z^{lb},

$$\bar{z}(\Omega) = \frac{1}{2}[z^{\text{ub}}(\Omega) + z^{\text{lb}}(\Omega)], \quad (8.50)$$

and then use this result to define the final details of the optimum r^{-12} reference system to be used in Eq. (8.33) for $A_{\text{ion}}^{\text{liq}}(\Omega, T_{\text{ref}})$. One thereby has as final input for Eq. (8.33)

$$A_{\text{ion}}^{\text{ref}}(\Omega, T_{\text{ref}}) = A_{\text{ref}}(\bar{z}, T_{\text{ref}}), \quad (8.51)$$

with A_{ref} given by Eq. (8.44), and

$$U_{\text{ref}}(\mathbf{R}) = \frac{1}{2}\sum_{i,j}{}' v_{\text{ref}}(R_{ij}, \bar{z}) = \frac{1}{2}\sum_{i,j}{}' \frac{4(\bar{z}\Omega)^4 k_B T_{\text{ref}}}{R_{ij}^{12}}. \quad (8.52)$$

One should note that the optimum reference potential can also be written as

$$v_{\text{ref}}(r, \bar{z}) = \frac{4(4\pi\bar{z}/3)^4 k_B T_{\text{ref}}}{(r/R_{\text{WS}})^{12}}. \quad (8.53)$$

It should be further noted in the case of transition metals that the approximate potential energy function $U(\mathbf{R}) \cong U_{\text{liq}}^{\text{VPT}}(\mathbf{R})$ for the physical system assumed in Eq. (8.40) is used here only to obtain $\bar{z}$, and that once $\bar{z}$ is calculated, the full potential-energy function $U(\mathbf{R}) \equiv E_{\text{tot}}(\mathbf{R}_1, \mathbf{R}_2, \cdots \mathbf{R}_N)$ is to be used in Eq. (8.33).

8.1.4 Two-phase melting simulations and other dynamic methods

Our primary dynamic two-phase melt method (Haskins et al., 2012; Hood, 2012; Moriarty et al., 2012) is a refined and robust version of the solid-liquid coexistence method of Morris et al. (1994), where melting is determined directly from the interatomic potentials via MD simulation. This method has been implemented for QBIP potentials in the ddcMD code. For full temperature-dependent MGPT potentials, which are discussed in Chapter 13, both ion- and electron-thermal contributions to melting are included, while for the more standard $T = 0$ GPT and MGPT potentials considered in this chapter, the melt curves so obtained correspond to solid and liquid free energies in the $A_{\text{el}} = 0$ limit. In either case, our two-phase melt approach replaces the constant-volume coexistence in the Morris et al. method with the desired thermodynamic condition of constant-pressure coexistence. This improvement allows one to capture the expected change in volume that occurs when a metal melts under constant-pressure conditions. In our two-phase approach, equilibrated subcells of equal size and shape, and maintained at equal constant pressure, are placed in contact and movement of the solid-liquid interface is monitored with a sensitive order parameter as a function of pressure for a trial melt temperature T_m. The usual PBCs are applied to all remaining faces of the subcells. The equilibrium melting pressure $P_m(T_m)$ is achieved when the solid-liquid interface remains stationary and neither the solid nor the liquid phase is growing at the expense of the other.

For melting out of a solid crystalline phase, our two-phase melt procedure is carried out in three main steps for each melt temperature T_m considered. The first step is to run separate MD subcell simulations to determine accurate pressure-volume relations at temperature T_m for both the solid and the liquid. These relations are needed to equalize solid and liquid pressures in the full simulation cell. The next step is to initialize and run large two-phase simulations for a range of atomic volumes, each corresponding to a solid trial pressure $P_{tr}(\Omega, T_m)$. Prior to each of these latter simulations, one first randomly removes from (or adds to, if necessary) the liquid subcell a small number of atoms necessary to match the solid pressure P_{tr}, and then equilibrates the solid and liquid subcells separately. In the two-phase simulations one then allows all of the atoms in the full cell to move freely in an extended MD simulation at pressure P_{tr} to accumulate statistics needed to indicate any movement of the solid-liquid interface.

In the final step of two-phase melting, one uses the accumulated statistics to trace the movement of the solid-liquid interface as a function of $P_{tr}(\Omega, T_m)$. This task is readily accomplished with the aid of the order parameter $O_\ell(t)$ created by Steinhardt et al. (1981), which is defined as

$$O_\ell(t) = 4\pi \sum_{m=-\ell}^{\ell} |\langle Y_{\ell m} \rangle|^2 / (2\ell + 1), \qquad (8.54)$$

where

$$\langle Y_{\ell m} \rangle = {\sum_{i,j}}' \Theta(r_{cut} - R_{ij}) Y_{\ell m}(\mathbf{R}_i - \mathbf{R}_j), \qquad (8.55)$$

with Θ the usual step function and r_{cut} a chosen cutoff radius. Specifically, it has been found in the present application that one needs only to accumulate statistics for the $\ell = 6$ order parameter $O_6(t)$ as a function of time t. This parameter is near zero for the liquid phase and is a nonzero constant for the solid phase, so $O_6(t)$ will tend towards one or the other as a given two-phase simulation proceeds. In practice, one obtains nearly linear plots for $O_6(t)$ versus t until the metal either completely melts or is completely solidified. For the range of atomic volumes considered, the slopes of the $O_6(t)$ curves can be plotted versus volume Ω and then fitted to a quadratic equation. The atomic volume that produces a zero slope is taken as the solid melting volume Ω_m at temperature T_m, such that $P_m = P_{tr}(\Omega_m, T_m)$. The whole three-step two-phase melting process is then repeated for as many melt temperatures as desired to obtain the $T_m(P_m)$ melt curve.

The respective two-phase and RSMD free-energy melt curves obtained in the $A_{el} = 0$ limit for bcc tantalum with Ta6.8x MGPT potentials to 400 GPa (Haskins et al., 2012) are compared in Fig. 8.4(a). The two-phase and free-energy melt curves so obtained are essentially identical below 100 GPa and differ only slightly by less than 2% above 100 GPa, a difference that is within the numerical error in either calculation. In this regard, the two-phase melt result shown in Fig. 8.4(a) is based on nine calculated melt pressures with an initial total of 78,732 atoms used in each simulation. This total corresponds to a $54 \times 27 \times 27$ full two-phase simulation cell, based on a standard bcc cubic unit cell with two atoms. While for two-phase melting out of the bcc structure, net size effects beyond 256 atoms ($8 \times 4 \times 4$ full cell) are small and essentially negligible, the use of a large number of atoms in the simulation cell greatly reduces the statistical error bars on each individual melt point, so that a smooth melt curve is obtained directly from the calculated points. The free-energy melt result in Fig. 8.4(a), on the other hand, was obtained from 250-atom MGPT-RSMD simulations on the bcc solid and the liquid, using a standard Helmholtz free-energy construction to obtain the melt curve (see Sec. 8.3).

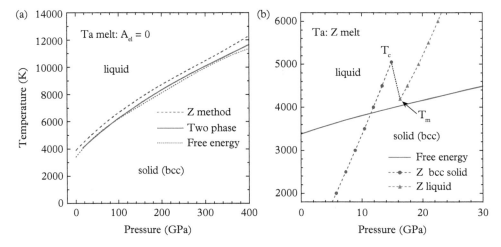

Fig. 8.4 *Melt curve of tantalum from the observed bcc solid phase, as calculated by different methods in the $A_{el} = 0$ limit with Ta6.8x MGPT potentials. (a) High-pressure curve obtained from the Z, two-phase and free-energy methods discussed in the text; (b) detailed low-pressure Z-melt calculation at constant atomic volume of $\Omega_0 = 121.6$ a.u.*
From Haskins et al. (2012), with publisher (APS) permission.

A complementary melting approach that can be used to advantage in studying melt size effects, especially for more complex noncubic solids, is the Z method of Belonoshko et al. (2006). The Z method attempts to connect the homogeneous melting of a solid at a critical temperature T_c, as obtained from superheating the material in an MD simulation at constant volume with PBCs, to the true equilibrium melt temperature T_m. The underlying hypothesis of the Z method is that for such conditions the total energy of the solid $E_{\text{tot}}^{\text{sol}}$ at $T = T_c$ is equal to the total energy of the liquid $E_{\text{tot}}^{\text{liq}}$ at $T = T_m$:

$$E_{\text{tot}}^{\text{sol}}(\Omega, T_c) = E_{\text{tot}}^{\text{liq}}(\Omega, T_m). \tag{8.56}$$

Belonoshko et al. (2006) proposed a simple dynamic approach to implement the content of Eq. (8.56): perform a small sequence of MD simulations in an *NVE* ensemble, adding kinetic energy to increase the temperature in the solid phase until T_c is reached, and then watch the system spontaneously drop in temperature during melt to T_m in the liquid. Adding additional kinetic energy increases the temperature of the system in the liquid, such that on a pressure-temperature phase diagram the entire thermodynamic path followed resembles the letter Z.

To cleanly separate possible dynamic effects from actual size effects in the melting, however, Haskins et al. (2012) developed an alternate *static* version of the Z method. In this approach to Z melting, one uses a series of MGPT-MD simulations in an *NVT* ensemble to determine the functions $E_{\text{tot}}^{\text{sol}}(\Omega, T)$ and $E_{\text{tot}}^{\text{liq}}(\Omega, T)$, together with the critical temperature T_c, as accurately as possible in the $A_{\text{el}} = 0$ limit for a given volume Ω of the material. One then solves Eq. (8.56) for T_m. An example of this approach is illustrated in Fig. 8.4(b) for bcc Ta, where 250 atoms were used in each MGPT-MD simulation at a volume of $\Omega_0 = 121.6$ a.u. The melting temperature so determined is about 5% higher than that determined from free energies. A similar overestimate of T_m has been found all along the high-pressure bcc melt curve, as shown in Fig. 8.4(a), where the Z-method results in that case were obtained with 2000-atom MGPT-MD simulations. One thereby concludes that the Z method as applied to bcc Ta provides only a close upper bound to the equilibrium melt curve. At the same time, the Z method confirms that size-independent melt in bcc Ta is achieved with only a 250-atom simulation. Haskins et al. (2012) also found that the Z method produced a close upper bound to two-phase melting results for the hex-ω structure, but not for the A15, fcc or hcp structures, where Z-method melt curves were found to be 25–35% higher than corresponding two-phase results. Since the Z method is equally applicable to both small and large computational cells, it has proven to be a convenient approach to study melt size effects in hex-ω and in the orthorhombic structure *Pnma*, as part of an in-depth investigation of possible high-*P*, *T* polymorphism in Ta (Haskins et al., 2012; Haskins and Moriarty, 2018), as is discussed in Sec. 8.4.

8.2 Equation of state and high-temperature thermodynamic properties

The equation of state of a material is commonly identified with the total pressure as a function of volume (or density) and temperature, $P_{\text{tot}}(\Omega, T)$, as given by Eq. (7.121). In the weak-coupling limit, P_{tot} is most usefully calculated in component form, starting from Eq. (7.115) for the total free energy $A_{\text{tot}}(\Omega, T)$. One thus has corresponding pressure components

$$P_{\text{tot}}(\Omega, T) = P_0(\Omega) + P_{\text{ion}}(\Omega, T) + P_{\text{el}}(\Omega, T), \tag{8.57}$$

where the cold pressure $P_0(\Omega)$, aka the cold curve, is calculated from the volume derivative of $E_0(\Omega)$ using Eq. (7.125), the ion-thermal pressure is calculated as

$$P_{\text{ion}}(\Omega, T) = -\frac{\partial A_{\text{ion}}(\Omega, T)}{\partial \Omega} \tag{8.58}$$

and the electron-thermal pressure is calculated as

$$P_{\text{el}}(\Omega, T) = -\frac{\partial A_{\text{el}}(\Omega, T)}{\partial \Omega}. \tag{8.59}$$

Generally speaking, these volume derivatives must be computed numerically for real materials. For any solid phase of a metal, of course, one has both quasiharmonic and anharmonic components to the ion-thermal free energy A_{ion}, as expressed by Eq. (7.117). The quasiharmonic pressure $P_{\text{ion}}^{\text{qh}}(\Omega, T)$ is calculated from the volume derivative of $A_{\text{ion}}^{\text{qh}}(\Omega, T)$ using Eq. (7.129), with $A_{\text{ion}}^{\text{qh}}$ given by Eq. (7.118). If the anharmonic free energy is obtained with the RSMD method of Sec. 8.1.2, the anharmonic pressure

$$P_{\text{ion}}^{\text{ah}}(\Omega, T) = -\frac{\partial A_{\text{ion}}^{\text{ah}}(\Omega, T)}{\partial \Omega} \tag{8.60}$$

can be calculated by differentiating the individual volume coefficients $A_n(\Omega)$ on the RHS of Eq. (8.30) for $A_{\text{ion}}^{\text{ah}}$. Likewise, in the liquid if the ion-thermal free energy is obtained with the RSMD method, the liquid ion-thermal pressure;

$$P_{\text{ion}}^{\text{liq}}(\Omega, T) = -\frac{\partial A_{\text{ion}}^{\text{liq}}(\Omega, T)}{\partial \Omega} \tag{8.61}$$

can be calculated by differentiating the individual volume coefficients $C_n(\Omega)$ the RHS of Eq. (8.35). Finally, the remaining electron-thermal free-energy component $A_{\text{el}}(\Omega, T)$ appearing in Eq. (7.115) for A_{tot} and in Eq. (8.59) for $P_{\text{el}}(\Omega, T)$ requires a separate independent treatment, as is now discussed in Sec. 8.2.1.

8.2.1 Electron-thermal free energy in metals

For metals in the weak-coupling regime, the simplest treatment of the additional electron-thermal contributions to the thermodynamic functions is derived from the standard low-temperature expansion of the total internal energy $E_{\text{tot}}(\Omega, T)$, in the $A_{\text{ion}} = 0$ limit, to obtain the leading T^2 contribution to $E_{\text{el}}(\Omega, T)$. Inserting the latter contribution back into Eq. (8.19) for $A_{\text{el}}(\Omega, T)$ with a $T_{\text{ref}} = 0$ reference temperature, one then recovers the well-known results

$$A_{\text{el}}(\Omega, T) = -E_{\text{el}}(\Omega, T) = -\frac{\pi^2}{6}\rho(E_{\text{F}})(k_{\text{B}} T)^2, \tag{8.62}$$

where $\rho(E_{\text{F}})$ is the $T = 0$ density of electronic states per ion at the Fermi energy E_{F}. Based on the magnitude of the expansion parameter $k_{\text{B}} T/E_{\text{F}}$ alone, Eq. (8.62) is potentially valid to high temperatures, at least in the range $T < 0.1 T_{\text{F}}$, where $k_{\text{B}} T_{\text{F}} = E_{\text{F}}$. In real materials, this can mean temperatures up to and even above 10,000 K. To the extent that $\rho(E_{\text{F}})$ depends weakly on atomic structure, as is the case for good NFE simple metals, Eq. (8.62) can thus be used in solid and liquid phases alike. In the case of NFE simple metals, one can further approximate $\rho(E_{\text{F}})$ by the free-electron density of states per ion $\rho_0(\varepsilon_F)$:

$$\rho(E_{\text{F}}) \cong \rho_0(\varepsilon_F) = \left(\frac{2m}{\hbar^2}\right)^{3/2} \frac{\Omega}{2\pi^2} \varepsilon_{\text{F}}^{1/2} = \frac{3}{2}\frac{Z}{\varepsilon_F}, \tag{8.63}$$

where in the latter two equalities we have used Eqs. (2.41) and (2.170), respectively. Using Eq. (8.59), the remaining electron-thermal pressure $P_{\text{el}}(\Omega, T)$ that is consistent with Eqs. (8.62) and (8.63) is just

$$P_{\text{el}}(\Omega, T) = \frac{\pi^2}{6}\frac{\partial \rho(E_{\text{F}})}{\partial \Omega}(k_{\text{B}} T)^2. \tag{8.64}$$

In general, the volume derivative in Eq. (8.64) must be evaluated numerically, but in the case of NFE simple metals where Eq. (8.63) applies, this derivative can be taken analytically:

$$\frac{\partial \rho(E_{\text{F}})}{\partial \Omega} \cong \frac{\partial \rho_0(\varepsilon_F)}{\partial \Omega} = \frac{1}{\Omega}\frac{Z}{\varepsilon_F}. \tag{8.65}$$

In most transition metals, on the other hand, $\rho(E_{\text{F}})$ is strongly material and structure dependent, so Eqs. (8.62) and (8.64) are only adequate in the solid ground-state structure up to modest temperatures. At high temperatures in the solid and in the liquid, alternate methods need to be used.

A more accurate treatment of A_{el} and E_{el} for high-temperature transition-metal phases has been developed by Moriarty et al. (2002a) in terms of configuration-averaged DFT calculations of the electronic entropy $S_{\text{el}} = -\partial A_{\text{el}}/\partial T$ in the hot solid and liquid, using atomic configurations obtained with MGPT-MD simulations. In the prototype case of Ta, the DFT values of electronic entropy thereby obtained in both the hot solid and the liquid can be well fitted in the form

$$S_{\text{el}}(\Omega, T) = \alpha(\Omega)T + \gamma(\Omega)T^3 + \cdots, \qquad (8.66)$$

where α and γ are volume-dependent fitting coefficients. The thermodynamically consistent expressions for the electron-thermal free energy and internal energy are then

$$A_{\text{el}}(\Omega, T) = -\frac{1}{2}\alpha(\Omega)T^2 - \frac{1}{4}\gamma(\Omega)T^4 + \cdots \qquad (8.67)$$

and

$$E_{\text{el}}(\Omega, T) = \frac{1}{2}\alpha(\Omega)T^2 + \frac{3}{4}\gamma(\Omega)T^4 + \cdots. \qquad (8.68)$$

In the vicinity of the high-pressure melt curve for tantalum, the resulting $A_{\text{el}}(\Omega, T)$ and $E_{\text{el}}(\Omega, T)$ functions differ little in the hot solid and in the liquid, and these functions can be adequately approximated by only the first T^2 expansion term in Eqs. (8.67) and (8.68), respectively. In that limit, one recovers the form of Eq. (8.59), but with $\rho(E_{\text{F}})$ replaced by an effective volume-dependent density of states $\rho_{\text{eff}}(\Omega) = 3\alpha(\Omega)/(\pi k_{\text{B}})^2$. Partly because the electron-thermal contribution to the hot solid and to the liquid are nearly the same, together they have been found to impact the calculated melt curve in Ta by only a small amount ($< 5\%$) (Moriarty et al., 2002a; Haskins et al., 2012). At the same time, the electron-thermal contribution to sensitive thermodynamic derivatives such as the specific heat and the thermal expansion coefficient can be much larger in both the solid and the liquid, as we further discuss in Sec. 8.2.3.

A third useful option for calculating electron-thermal contributions in metals, especially at very high temperatures $T > 0.1 T_{\text{F}}$, is the atom-in-jellium self-consistent-field model due to Liberman (1979), which was first implemented in the LANL code INFERNO. This model attempts to approximate finite-temperature band theory and is conceptually similar to the LMTO-ASA method, except that free-electron boundary conditions are applied at the atomic sphere in place of rigorous electronic-structure boundary conditions based on crystal symmetry. Also, in INFERNO one solves a fully relativisitic Dirac equation up to extreme temperatures and densities, where the model is smoothly joined to the well-known Thomas-Fermi-Dirac model. For the metals under consideration here, INFERNO effectively treats broad sp bands as NFE-like and narrow d bands as resonances in the electron gas. The electron-thermal free-energy and internal energy are calculated as total-energy differences between temperature T and $T = 0$:

$$A_{\text{el}}(\Omega, T) = A_{\text{tot}}(\Omega, T) - A_{tot}(\Omega, 0) \qquad (8.69)$$

and

$$E_{\text{el}}(\Omega, T) = E_{\text{tot}}(\Omega, T) - E_{\text{tot}}(\Omega, 0), \qquad (8.70)$$

respectively. In recent years, a somewhat more computationally robust version of the INFERNO model, with improved thermodynamic consistency, has been developed by Wilson et al. (2006) and is implemented in the LLNL code PURGATORIO.

8.2.2 Shock physics and the Hugoniot

The EOS of metals at high pressure can be readily probed in both static, low-temperature DAC experiments and in dynamic, high-temperature shock-wave (SW) experiments. In typical laboratory SW experiments, a high-velocity impactor, driven by a gas gun, impacts a sample target material initially at rest under ambient conditions. Absolute EOS data for the target can be derived from the measured shock velocity u_s created in the sample and the measured mass or particle velocity u_p behind the shock front. Specifically, the total pressure P_{tot}, the density ρ or atomic volume Ω and the total internal energy per ion E_{tot} can be calculated directly from the well-known Rankine-Hugoniot conservation laws of mass, momentum and energy, here applied in the form

$$P_{tot} = \rho_0 u_s u_p, \tag{8.71}$$

$$\rho_0/\rho = \Omega/\Omega_0 = 1 - u_p/u_s \tag{8.72}$$

and

$$E_{tot} = E_{tot}^i + \frac{1}{2} P_{tot}(\Omega_0 - \Omega), \tag{8.73}$$

where E_{tot}^i is the initial total energy per ion at ambient conditions, with $P_{tot}^i = 0$. In such a SW experiment, only the temperature of the shocked target sample is unmeasured and must be supplied by theory. The pressure-density or pressure-volume relationship obtained from Eqs. (8.71) and (8.72) is known as the principal Hugoniot $P_{Hug} = P_{tot}$.

In Fig. 8.5 we compare full-theory MGPT calculations of the room-temperature (300-K) isotherm P_{300} and the corresponding Hugoniot P_{Hug} for a tantalum prototype, as obtained by Moriarty and Haskins (2014), with measured DAC and gas-gun SW data, respectively. Note that under compression, the Hugoniot moves increasingly above the isotherm in value because the Hugoniot temperature is rising with increasing pressure. That is, for $\rho > \rho_0$ or $\Omega < \Omega_0$, one necessarily has $P_{Hug} > P_{300}$. Melting on the Ta Hugoniot out of the bcc phase occurs near 300 GPa, but is only visible as a clear break in the curve in pressure-temperature coordinates, as is discussed in Sec. 8.3.2 in connection with the P, T phase diagram. As can be seen in Fig. 8.5(a), the calculated MGPT 300-K isotherm for bcc Ta is in good agreement with both the DAC data of Cynn and Yoo (1999), measured to 174 GPa, and the DAC data of Dewaele et al. (2004), measured to 89 GPa. Similarly, the calculated bcc Hugoniot, whose temperature ranges from 300 K to about 10,000 K, is in good agreement with the SW data of Mitchell and Nellis (1981). Figure 8.5(b) displays the calculated Hugoniot in both the bcc solid and the liquid, extending to 600 GPa. Good agreement with the higher-pressure shock data of Mitchell and Nellis and with the additional data of Holmes et al. (1989) is maintained, although above 400 GPa, the calculated Hugoniot pressure is slightly underestimated by theory.

Regarding the MGPT treatment of the Ta bcc and liquid EOSs used to obtain Fig. 8.5, these quantities have been constructed within the above weak-coupling framework for $P_{\text{tot}}(\Omega, T)$, beginning with Eq. (8.57). Specifically, the cold and ion-thermal components of the bcc and liquid EOSs have been calculated entirely with Ta6.8x multi-ion potentials, using at finite temperature quasiharmonic phonons together with RSMD anharmonic simulations in the bcc solid and corresponding RSMD simulations in the liquid. The additional electron-thermal EOS components have been calculated using the hybrid DFT-MGPT scheme discussed in Sec. 8.2.1.

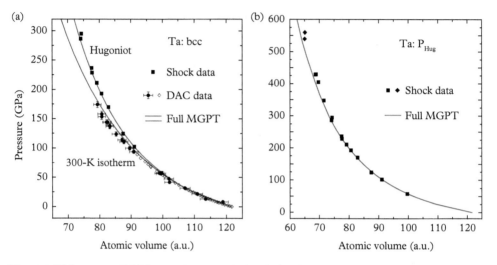

Fig. 8.5 *High-pressure EOS in tantalum, as calculated with the full MGPT theory discussed in the text, and as compared to experiment. (a) Room-temperature (300-K) isotherm and principal Hugoniot in the bcc phase, with the static DAC data from Cynn and Yoo (1999) (solid circles) and Dewaele et al. (2004) (open diamonds), and the dynamic shock data from Mitchell and Nellis (1981) (solid squares); (b) principal Hugoniot P_{Hug} across both the bcc solid and the liquid, with additional shock data in the liquid from Holmes et al. (1989) (solid diamonds).*
From Moriarty and Haskins (2014), with publisher (APS) permission.

Still higher pressures in metals can be reached with double-shock gas-gun and nuclear-impedance-match (NIM) experiments (Nellis et al., 1988; Mitchell et al., 1991). In a laboratory double-shock experiment, the materials configuration consists of an impactor, a target metal and in addition a higher-density fixed anvil. A second shock is created in the moving target when it strikes the anvil, yielding a higher pressure-density point, but one at a lower temperature that is intermediate between the 300-K isotherm and the Hugoniot. As in a single-shock gas-gun experiment, both u_s and u_p data in the target are measured, so *absolute* values of P_{tot}, ρ, Ω and E_{tot} are still obtained. In a NIM experiment, on the other hand, as is the case in a DAC measurement, only *relative* EOS data are obtained. In a DAC experiment, volume is the measured quantity and pressure is obtained relative to a chosen EOS standard, which is typically a second calibrated material included in the cell together with the sample under investigation. Similarly, in an explosively driven NIM experiment, only shock velocities in the sample and an adjacent reference material are measured. The mass velocity u_p and pressure P_{tot} in the shocked sample are determined in this case by matching the shock impedance, $\rho_0 u_s$, with that of a theoretical EOS for the reference material, as discussed by Moriarty (1995). In LLNL NIM experiments (Mitchell et al., 1991), Al was chosen to be the reference material to investigate the Hugoniots of Cu, Mo and Pb in the ultrahigh pressure range 0.5 to 3.0 TPa (5–30 Mbar). Using first-principles equations of state constructed for all four metals, Moriarty (1995) successfully calibrated the reference metal Al and analyzed the experimental NIM data on Cu, Mo and Pb. Each of these EOSs consisted of a DFT cold curve, GPT/MGPT quasiharmonic-phonon and VPT liquid treatments of the ion-thermal EOS component, and an INFERNO treatment of the electron-thermal EOS component.

8.2.3 Thermodynamic derivatives

The quasiharmonic-phonon treatment of thermodynamic derivatives at low and modest temperatures up to and somewhat beyond the Debye temperature Θ_D was considered in Chapter 7. Here, by including both high-temperature anharmonic and electron-thermal contributions in the solid, those treatments can be extended all the way up to melt. In particular, the Moriarty and Haskins (2014) MGPT-based calculations of the zero-pressure specific heat C_P and the thermal expansion coefficient β for bcc Ta, presented in Fig. 7.10 to 500 K, are now extended to 3500 K and the results analyzed in Fig. 8.6.

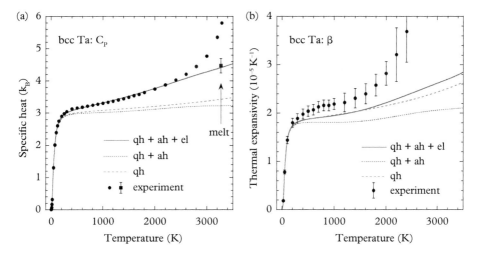

Fig. 8.6 *High-temperature thermodynamic derivatives in bcc tantalum at zero pressure, as calculated via Ta6.8x multi-ion MGPT potentials, showing the impact of quasiharmonic (qh), anharmonic (ah) and electron-thermal (el) contributions, and compared with experimental data. (a) Constant-pressure specific heat C_P, with measured data from Hultgren et al. (1973) (solid circles) and the additional point at melt derived from isobaric expansion data (solid square); (b) thermal expansion coefficient β, with the measured data from Touloukian et al. (1975) (solid circles).*

From Moriarty and Haskins (2014), with publisher (APS) permission.

Regarding the specific heat C_P of bcc Ta considered in Fig. 8.6(a), the full-theory MGPT treatment, which includes quasiharmonic, RSMD anharmonic and DFT-MGPT electron-thermal contributions, is in close agreement with the experimental data of Hultgren et al. (1973) up to about 2600 K. Interestingly, the anharmonic vibrational contribution to this calculated result is very small all the way to melt, while the much larger electron-thermal contribution is clearly essential to the quantitative agreement above room temperature. The observed steep rise in C_P above 2600 K recorded in the Hultgren et al. (1973) data is not predicted by the present theory, but this feature is well known to be sensitive to impurities and other experimental factors, so this shortcoming is not necessarily significant. In particular, note that the present full-theory result for C_P at melt is in good agreement with the alternate experimental value derived from isobaric expansion measurements (Shaner et al., 1977; Gathers, 1983; Berthault et al., 1986).

Regarding the thermal expansion coefficient β of bcc Ta considered in Fig. 8.6(b), the present full-theory MGPT result is in close agreement with the measured data of Touloukian et al. (1975) up to about 400 K and still in reasonable agreement to 1500 K. Above 1000 K, however, the data have large error bars, and in fact, only provisional values of the thermal expansion coefficient were obtained above 2000 K. As to the theory, it is interesting to note that the anharmonic and electron-thermal contributions to β become large beyond about 400 K, but they are almost exactly cancelling all the way to about 1600 K, with only a small net positive contribution beyond the quasiharmonic result up to melt.

The closely related Grüneisen parameter, $\gamma = \Omega \beta B_T / C_\Omega$ from Eq. (7.139), can as a result also display some significant high-temperature behavior. As pointed out in Chapter 7, γ remains nearly constant at high T for all metals when treated in the quasiharmonic approximation. However, as shown in Fig. 8.7 for bcc Mg and bcc Ta, anharmonic-vibrational contributions, in addition to electron-thermal contributions, can lower the value of γ substantially at high temperature as one proceeds toward melt. In the case of bcc Ta, note that this behavior is indeed quite consistent with that displayed by C_P and β in Fig. 8.6. In particular, because $\gamma \propto \beta$, the large and increasing anharmonic lowering of β below the quasiharmonic limit at high temperature, as seen in Fig. 8.6(b), is also necessarily reflected in the behavior of γ in Fig. 8.7(b).

In the case of tantalum, the MGPT-based treatment of C_P and β at zero pressure has also been extended into the high-temperature liquid by Moriarty and Haskins (2014), including RSMD ion-thermal and DFT-MGPT electron-thermal contributions. These results are displayed in Fig. 8.8 and compared there with experimental isobaric expansion data. For both C_P and β the electron-thermal contributions are very large, amplifying the trends already established for the bcc solid in Fig. 8.6. In this regard, the calculated value of C_P at melt rises from 4.4 k_B in the bcc solid to 5.6 k_B in the liquid, which is in good agreement with the temperature-corrected isobaric data of Shaner et al. (1977) data reported by Gathers (1983). On the other hand, Berthault et al. (1986) find a much smaller rises of C_P to only 4.6 k_B in the liquid in their isobaric measurements. In a similar manner, the calculated value of β at melt rises from $2.7 \times 10^{-5} \text{K}^{-1}$ in the bcc solid to $4.0 \times 10^{-5} \text{K}^{-1}$ in the liquid. In this case, the liquid result at melt is in close agreement with Berthault et al. (1986) data, but some 20% less than inferred from Gathers (1983).

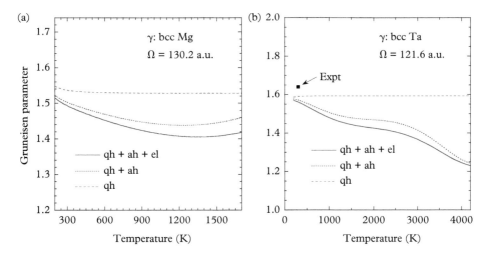

Fig. 8.7 *Temperature dependence of the Grüneisen parameter γ in bcc Mg and Ta at constant volume, showing the impact of quasiharmonic (qh), anharmonic (ah) and electron-thermal (el) contributions. (a) Moderately compressed magnesium ($\Omega = 0.94\Omega_0$), where the bcc structure is mechanically stable, as calculated from first-principles GPT pair potentials; (b) ambient density tantalum ($\Omega = \Omega_0$) as calculated from Ta6.8x MGPT multi-ion potentials. The measured room-temperature value of γ by Katahara et al. (1979) is shown for comparison (solid square).*
Panel (b) from Moriarty and Haskins (2014), with publisher (APS) permission.

8.2.4 Thermoelasticity and sound velocity

Next, we consider the related issue of thermoelasticity in metals and the evaluation of the temperature dependence of elastic moduli all the way up to melting conditions. Unlike the low-temperature regime discussed in Chapter 7, it is now necessary to distinguish between adiabatic and isothermal elastic moduli, C_{ij}^S versus C_{ij}^T in the simplified Voigt notation, with $i,j = 1 - 6$. In analogy with Eq. (7.57) at $T = 0$, the second-order isothermal elastic moduli at temperature T are given by

$$C_{ij}^T = \frac{1}{N\Omega} \left.\frac{\partial^2 A_{\text{tot}}(\mathbf{R}, T)}{\partial \eta_i \partial \eta_j}\right|_{\mathbf{R}=\mathbf{R}_0}, \quad (8.74)$$

where the total-energy functional $E_{\text{tot}}(\mathbf{R})$ is now replaced by the corresponding total free-energy functional $A_{\text{tot}}(\mathbf{R}, T)$. In the weak-coupling limit, $A_{\text{tot}}(\mathbf{R}, T)$ can be divided up into the familiar cold, ion-thermal and electron-thermal components, so the isothermal elastic constants can be similarly calculated:

$$C_{ij}^T = C_{ij}^0 + C_{ij}^{\text{ion}} + C_{ij}^{\text{el}}. \quad (8.75)$$

364 *High-Temperature Properties, Melting and Phase Diagrams*

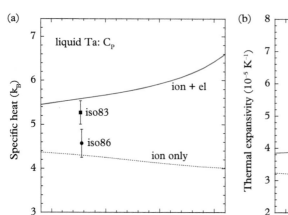
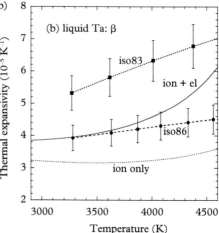

Fig. 8.8 *High-temperature thermodynamic derivatives in liquid tantalum at zero pressure, as calculated via Ta6.8x multi-ion MGPT potentials, both with (ion + el) and without (ion only) electron-thermal contributions, and compared to the isobaric expansion data of Gathers (1983) (iso83) and Berthault et al. (1986) (iso86). (a) Constant-pressure specific heat C_P, with the measured data at melt; (b) thermal expansion coefficient or expansivity β.*
From Moriarty and Haskins (2014), with publisher (APS) permission.

For bcc transition metals, with Ta as a prototype, Orlikowski et al. (2006) developed a first-principles formalism to evaluate the three components on the RHS of Eq. (8.75) as a function of temperature and pressure from closely coupled DFT and MGPT calculations, with C_{ij}^{ion} evaluated in the quasiharmonic approximation. The cold elastic constants C_{ij}^{0} as a function of pressure to 1000 GPa (10 Mbar) were taken directly from the FP-LMTO calculations of Söderlind and Moriarty (1998). Recall from Chapter 5 that these same data were used to constrain the tantalum MGPT potentials beginning with versions Ta3 and Ta4. The FP-LMTO calculations were then repeated as a function of temperature over the same pressure range, up to the MGPT melting curve, as obtained by Moriarty et al. (2002a), to produce the electron-thermal contribution C_{ij}^{el}.

The remaining ion-thermal components C_{ij}^{ion} were calculated from strain derivatives of the quasiharmonic phonon frequencies. In the four-index elasticity notation for these latter moduli, with $i,j = 1 - 3$, one has, starting from Eq. (7.118) for $A_{\text{ion}}^{\text{qh}}$, the result

$$C_{ijkl}^{\text{ion}} = \frac{1}{\Omega} \frac{\partial^2 A_{\text{ion}}^{\text{qh}}}{\partial \eta_{ij} \partial \eta_{kl}}$$

$$= \frac{1}{\Omega N} \sum_{\mathbf{q},\lambda} h\nu_{\mathbf{q},\lambda} \left\{ \xi_{\mathbf{q},\lambda,ijkl} \left[n_{\mathbf{q},\lambda} + 1/2 \right] + \gamma_{\mathbf{q},\lambda,ij} \gamma_{\mathbf{q},\lambda,kl} T \frac{\partial n_{\mathbf{q},\lambda}}{\partial T} \right\}, \quad (8.76)$$

where we have introduced the shorthand notation $\nu_{\mathbf{q},\lambda} \equiv \nu_\lambda(\mathbf{q},\Omega)$ and $n_{\mathbf{q},\lambda} \equiv n_\lambda(\mathbf{q},\Omega)$, and further defined the strain derivatives

$$\gamma_{\mathbf{q},\lambda,ij} \equiv -\frac{1}{\nu_{\mathbf{q},\lambda}}\frac{\partial \nu_{\mathbf{q},\lambda}}{\partial \eta_{ij}} \tag{8.77}$$

and

$$\xi_{\mathbf{q},\lambda,ijkl} \equiv \frac{1}{\nu_{\mathbf{q},\lambda}}\frac{\partial^2 \nu_{\mathbf{q},\lambda}}{\partial \eta_{ij} \partial \eta_{kl}}. \tag{8.78}$$

Note that the quantity $\gamma_{\mathbf{q},\lambda,ij}$ is a generalized Grüneisen parameter. Orlikowski et al. (2006) evaluated Eqs. (8.76)–(8.78) using the Ta4 MGPT multi-ion potentials together with the general Lagrangian definition of finite strains (Wallace, 1972). In the latter definition, the displacement field of the ions is given by $\mathbf{u} = \mathbf{X} - \mathbf{x}$, between the reference equilibrium configuration of ions $\mathbf{X}$ and the strained configuration $\mathbf{x}$. The 3×3 displacement-field gradient tensor then has components $u_{ij} = \partial u_i / \partial X_j$ and the corresponding Lagrangian strains are defined as

$$\eta_{ij} \equiv \frac{1}{2}\left(u_{ij} + u_{ji} + \sum_{k=1}^{3} u_{ik}u_{kj}\right). \tag{8.79}$$

To make contact with experimental ultrasonic measurements, one needs the adiabatic elastic moduli C^S_{ijkl}, which are related to the isothermal elastic moduli C^T_{ijkl} by the following general relationship (Wallace, 1972, but using the present convention adopted in Chapter 7, where $C_{ijkl} \equiv B_{ijkl}$, the relative stress-strain coefficients of Wallace):

$$C^S_{ijkl} - C^T_{ijkl} = \frac{T\Omega}{C_\Omega}\left(\frac{\partial \sigma_{ij}}{\partial T}\right)_\varepsilon \left(\frac{\partial \sigma_{kl}}{\partial T}\right)_\varepsilon, \tag{8.80}$$

with σ_{ij} the stress needed to produce the strain η_{ij}, and where the subscript ε denotes evaluation of the derivative at a constant strain rate. In isotropic materials, such as cubic crystals, the latter derivatives have the simplified form

$$\left(\frac{\partial \sigma_{ij}}{\partial T}\right)_\varepsilon = -\left(\frac{\partial P_{\text{tot}}}{\partial T}\right)_\Omega \delta_{ij}. \tag{8.81}$$

Using Eq. (8.81) in Eq. (8.80) together with Eqs. (7.135)–(7.137), and reverting back to Voigt notation, one can derive the following useful results for cubic metals:

$$C^S_{11} - C^T_{11} = C^S_{12} - C^T_{12} = \frac{T\Omega}{C_\Omega}\beta^2 B_T^2, \tag{8.82}$$

$$C^S_{44} = C^T_{44}, \tag{8.83}$$

$$C'^S = C'^T = (C^T_{11} - C^T_{12})/2, \tag{8.84}$$

and noting that $B_T = (C_{11}^T + 2C_{12}^T)/3$, the adiabatic bulk modulus is

$$B_S = B_T + \frac{T\Omega}{C_\Omega}\beta^2 B_T^2. \tag{8.85}$$

In addition, when displaying experimental elastic constants or when comparing theory with experiment, as we do in Fig. 8.9, it is customary to drop the superscript "S" and write $C_{ij} = C_{ij}^S$ and $C' = C'^S$.

As shown in Fig. 8.9(a), the DFT-based FP-LMTO/MGPT treatment of Orlikowski et al. (2006) rather well explains the observed temperature dependence of the adiabatic elastic constants in bcc Ta at ambient pressure all the way up to the vicinity of melt. In particular, the observed values of C_{11}, C_{12}, C_{44} and C' drop in magnitude by 25%, 10%, 29% and 46%, respectively, between $T = 0$ and $T = 3000$ K, compared to corresponding calculated decreases of 22%, 10%, 27% and 41%. At the same time, the Ta4 MGPT potentials themselves provide a good description of the pressure dependence of C_{11}, C_{12} and C_{44} at $T = 300$ K, as shown in Fig. 8.9(b). In this case, the experimental DAC data used for comparison with theory comes from two specialized techniques: the stress and angle-resolved X-ray diffraction (SAX) method of Cynn and Yoo (2000), and the impulsive stimulated light scattering (ISLS) method of Zaug et al. (2000) and Crowhurst et al. (2001).

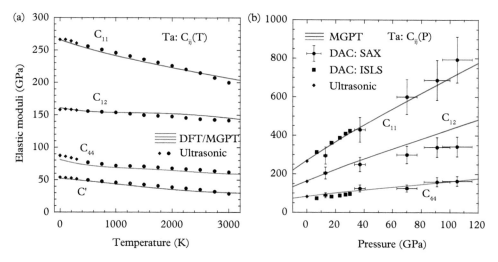

Fig. 8.9 *Temperature and pressure dependence of the adiabatic elastic constants of bcc Ta, as obtained from DFT and MGPT calculations, and as compared to experiment. (a) $C_{ij}(T)$ at zero pressure from the combined FP-LMTO/MGPT treatment of Orlikowski et al. (2006), denoted as DFT/MGPT, together with the ultrasonic data from Featherston and Neighbours (1963, solid diamonds), and from Walker and Bujard (1980, solid circles); (b) $C_{ij}(P)$ at 300 K from the Ta4 MGPT multi-ion potentials, with the DAC-SAX data from Cynn and Yoo (2001), the DAC-ISLS data from Zaug and Crowhurst (2001) and the ultrasonic data from Katahara et al. (1979).*

8.2.4 Thermoelasticity and sound velocity

At high pressures and temperatures, contact with shock experiments can be made through the measured sound velocity, which can be obtained in both the bulk solid and the liquid, and is frequently used as a diagnostic to detect shock melting (e.g., Brown and Shaner, 1984 in Ta; Hixson et al., 1989 in Mo). In such an experiment, the sound velocity of a macroscopic, polycrystalline sample is measured along the Hugoniot of the material. On the theoretical side, the single-crystal adiabatic elastic moduli can be calculated along the same path and used to obtain the polycrystalline sound velocity though the adiabatic bulk modulus B_S and an appropriate average shear modulus G. For the latter one can use an average of the rigorous and tight Hashin-Strickman bounds, g_1 and g_2, as calculated for an isotropic polycrystal (Simmons and Wang, 1971):

$$G = (g_1 + g_2)/2, \tag{8.86}$$

where

$$g_1 = C' + \frac{3}{5/(C_{44} - C') - 4\beta_1}, \tag{8.87}$$

with

$$\beta_1 = -3 \frac{B_S + 2C'}{5C'(3B_S + 4C')}, \tag{8.88}$$

and

$$g_2 = C_{44} + \frac{2}{5/(C' - C_{44}) - 4\beta_2}, \tag{8.89}$$

with

$$\beta_2 = -3 \frac{B_S + 2C_{44}}{5C_{44}(3B_S + 4C_{44})}. \tag{8.90}$$

The longitudinal sound velocity in the solid isotropic polycrystal is then

$$v_s^L = \sqrt{(B_S + 4G/3)/\rho}, \tag{8.91}$$

where ρ is the density. In the high-temperature liquid, where all shear moduli vanish and $G = 0$, v_s^L is reduced to the bulk sound velocity

$$v_s^B = \sqrt{B_S/\rho}. \tag{8.92}$$

Thus, in a shock-wave experiment, the pressure at which $v_s^L \to v_s^B$ along the Hugoniot gives the shock melting point. In Fig. 8.10 we compare the measured shock melting so obtained in Ta by Brown and Shaner (1984) with the corresponding calculations of v_s^L

and v_s^B by Orlikowski et al. (2006). The agreement between theory and experiment is clearly very good with the onset of shock melting observed near 300 GPa.

The Orlikowski et al. (2006) treatment of high-temperature elastic moduli in bcc Ta does not address additional anharmonic effects, however. While it is possible, in principle, to extend our RSMD simulation method to calculate these effects for elastic moduli, this is a significant challenge beyond the current state of the art and has not yet been attempted. It is currently possible, however, to treat cold and ion-thermal contributions to the elastic moduli together, including anharmonic effects, in alternate and efficient Monte Carlo (MC) simulations. Greeff and Moriarty (1999) developed such a simulation method that is specifically adapted to the GPT total-energy functional at the pair-potential level, and successfully applied the method to both hcp and bcc Mg. In this approach, the elastic moduli are simulated in a canonical MC ensemble at a given temperature and volume, using exact statistical formulas that include anharmonic effects as well as internal relaxation for the hcp structure. For simulation temperatures ranging from 200 to 900 K at zero pressure, all five independent hcp Mg elastic constants were studied and found to decrease linearly with temperature, with the calculated slopes in good agreement with experiment at 300 K, as shown in Table 8.1 Orlikowski (2007) later extended the GPT MC simulation method of Greeff and Moriarty (1999) to MGPT multi-ion potentials in an attempt to study anharmonic effects on the elastic moduli of bcc tantalum using the Ta4 potentials. While the MGPT-MC simulations themselves were successfully implemented and carried out for C_{11}, C_{12} and C_{44} over wide ranges

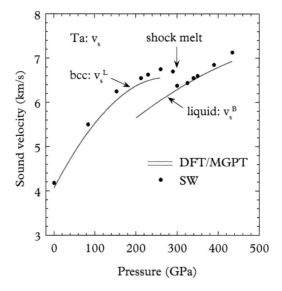

Fig. 8.10 *Sound velocities in bcc and liquid Ta along the Hugoniot, as obtained from DFT/MGPT calculations by Orlikowski et al. (2006), and as measured in SW experiments by Brown and Shaner (1984). The onset of shock melting is observed to occur near 300 GPa.*

Table 8.1 *Temperature derivatives* $(\partial C_{ij}/\partial T)_P$ *at* $P = 0$ *and* $T = 300K$ *for hcp Mg (in units of* 10^{-3} *GPa/K), obtained from first-principles GPT MC simulations and compared with experiment.*

Treatment	C_{11}	C_{33}	C_{12}	C_{13}	C_{44}
GPT MC[a]	−24±4	−26±2	−5±2	0±2	−10.2±0.6
Experiment[b]	−18	−20	−2	−2	−9.6

[a]Greeff and Moriarty (1999).
[b]Simmons and Wang (1971).

of temperature and volume in bcc Ta, the extraction of definitive quantitative results on anharmonic effects proved to be difficult and sensitive to the method of analysis. In particular, the quasiharmonic and anharmonic ion-thermal contributions to the elastic moduli are not separated in the MC simulations, and can only be distinguished with the introduction of further assumptions and approximations into the analysis of the simulation results. In the analysis that was performed on bcc Ta by Orlikowski (2007), Eq. (8.75) for C_{ij} was expanded out in the form

$$C_{ij}^T = C_{ij}^0 + C_{ij}^{\text{ion}} + C_{ij}^{\text{el}}$$
$$= C_{ij}^0(1 + \alpha_{ij}^{qh} + \alpha_{ij}^{ah}) + C_{ij}^{\text{el}}, \quad (8.93)$$

where the volume-dependent quasiharmonic and anharmonic coefficients, α_{ij}^{qh} and α_{ij}^{ah}, of the ion-thermal contribution were assumed to vary with temperature as T and T^2, respectively. With $C_{ij}^0(\Omega)$ and $C_{ij}^{\text{el}}(\Omega, T)$ independently established, as in Orlikowski et al. (2006), $\alpha_{ij}^{qh}(\Omega, T)$ and $\alpha_{ij}^{ah}(\Omega.T)$ were then determined by best fits to the MGPT-MC simulation data. At an ambient pressure volume of $\Omega = 121.6$ a.u. and a near-melt temperature of 3500 K, the anharmonic ion-thermal coefficients α_{11}^{ah} and α_{44}^{ah} were found to be negligible, while $\alpha_{12}^{ah} = 0.07$. The corresponding quasiharmonic coefficients have values $\alpha_{11}^{qh} = -0.21$, $\alpha_{12}^{qh} = -0.10$ and $\alpha_{44}^{qh} = -0.02$. In contrast, at a typical high-pressure volume of $\Omega = 57.7$ a.u. and a near-melt temperature of 17,000 K, it was found that $\alpha_{11}^{ah} = 0.21$, $\alpha_{12}^{ah} = -0.05$ and $\alpha_{44}^{ah} = -0.20$, with $\alpha_{11}^{qh} = -0.40$, $\alpha_{12}^{qh} = 0.18$ and $\alpha_{44}^{qh} = 0.45$. Thus, except where α_{ij}^{ah} is negligible, α_{ij}^{qh} and α_{ij}^{ah} tend to have opposite signs, leading to large cancellations in $\alpha_{ij}^{qh} + \alpha_{ij}^{ah}$ and quantitative sensitivity in the analysis. Nonetheless, at ambient conditions one can at least calculate the corresponding adiabatic elastic constants and compare with experiment, as is done in Table 8.2. Both the MGPT-MC treatment with anharmonic effects included and the DFT/MGPT treatment at the quasiharmonic level (Orlikowski et al., 2006) produce reasonable agreement with experiment.

Table 8.2 *Adiabatic elastic moduli $C_{ij} \equiv C_{ij}^S$ and B_S at $P = 0$ and $T = 300K$ for bcc Ta (in GPa) as obtained from MC simulations with quasiharmonic (QH) plus anharmonic (AH) contributions.*

Treatment	C_{11}	C_{12}	C_{44}	C'	B_S
MC: QH+AH	256	163	74	47	197
Previous[a]: QH	258	155	74	52	189
Experiment[b]	261	157	82	52	192

[a]Orlikowski et al. (2006).
[b]Featherston and Neighbours (1963).

8.3 Melting and the pressure-temperature phase diagram

The study of pressure-temperature phase diagrams in elemental metals is a subject of high current research interest, with forefront experimental and theoretical studies, as well as remaining uncertainties and controversy. Historically, the experimental investigation of such phase diagrams was quite limited. Early studies of melt curves in metals consisted of piston-cylinder or Bridgman-cell (BC) static measurements up to 4–10 GPa in simple metals or, alternately in transition metals, dynamic isobaric expansion measurements yielding the zero-pressure melting slope. In selected metals, these data could be combined with high-pressure shock melting measurements, but these required empirical modeling of the melt temperature to place the data on the P, T phase diagram. In addition, the study of pressure-driven solid-solid transitions had been mostly confined to room temperature, and the study of temperature-induced phase transitions to ambient pressure. The situation has changed dramatically in the past two decades, however, with the advanced development of the laser-heated DAC for static high-P, T measurements to 100 GPa or more in pressure and 5000 K or more in temperature, together with other recent advances, such as the use of in situ XRD in dynamic SW experiments (e.g., Beason et al., 2020) as well as in dynamic ramp-compression (DRC) experiments (e.g., Polsin et al., 2017). Our two prototype metals in this chapter, Mg and Ta, have been among the most studied metals in recent years with regard to their P, T phase diagrams, and we shall again focus most of our discussion in this section on these elements.

On the theoretical side, it has been possible to calculate realistic quantum-based melting curves in simple metals and series-end transition metals since the 1980s, using QBIPs together with QHLD free energies in the solid and VPT free energies in the liquid. This was done successfully with first-principles GPT potentials for Al (Moriarty et al., 1984), Cu (Moriarty, 1986a) and Mg (Moriarty and Althoff, 1995). Now with accurate RSMD calculations of solid anharmonic and liquid ion-thermal free-energy contributions, plus electronic-structure-based electron-thermal contributions, full free energies can be obtained in both the solid and the liquid, for nontransition and transition metals alike. Full free-energy high-pressure bcc melt curves for GPT Mg and MGPT Ta are illustrated in Fig. 8.11, together with the impact of anharmonic and electron-thermal contributions. Electron thermal only affects the melt curves of these two metals by a small amount, an amount that is negligible in the case of Mg, as shown in Fig. 8.11(a). In the case of Ta, displayed in Fig. 8.11(b), the electron-thermal contribution is negative and increases in magnitude with increasing pressure, but only amounts to about a 200 K decrease of the 11,200 K melt point at 400 GPa. The solid anharmonic contribution to melting is much larger in both metals, but opposite in sign, raising the high-pressure melt curve by about 10% in Mg, but lowering the melt curve by about the same percentage in Ta above 100 GPa. The difference in sign is a direct reflection of the respective sign differences in the anharmonic free energy, as shown in Fig. 8.2. Finally, note from Fig. 8.11(a) that, because of compensating errors in the solid and liquid free energies, the baseline QHLD solid plus upper-bound VPT liquid melting approach in Mg is actually quite accurate up to about 60 GPa, but at higher pressures increasingly overestimates the melt temperature.

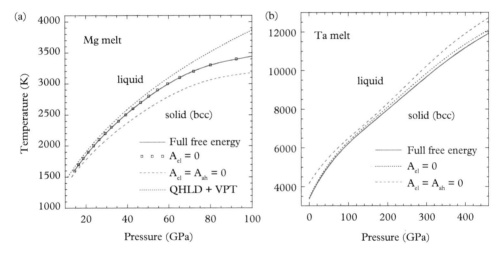

Fig. 8.11 *Full free-energy derived high-pressure melt curves out of the bcc phase for Mg and Ta, showing the impact of anharmonic (A_{ah}) and electron-thermal (A_{el}) components. (a) Mg, as calculated from first-principles GPT potentials. Shown for reference is the baseline QHLD solid plus (ub) VPT liquid treatment, as previously used by Moriarty and Althoff (1995). (b) Ta, as calculated from Ta6.8x multi-ion MGPT potentials.*
Panel (b) from Moriarty and Haskins (2014), with publisher (APS) permission.

A full GPT-based P, T phase diagram for Mg is illustrated in Fig. 8.12, and compared there with recent static and dynamic experimental data. The theoretical phase diagram consists of hcp, bcc and liquid phases, with the hcp phase and the hcp-bcc phase line treated in the QHLD approximation, as in Fig. 7.11, and the hcp-liquid melt transition to 1600 K treated in the QHLD solid plus upper-bound VPT liquid approximation, as in Moriarty and Althoff (1995). The liquid phase above 1600 K, the bcc solid phase above the hcp-bcc phase line, and the bcc-liquid melt curve are treated with full free energies, as in Fig. 8.11(a). The hcp-liquid melt curve is in excellent agreement with the Bridgman-cell measurements of Errandonea (2010). The bcc-liquid phase line is also in generally good agreement with the earlier laser-heated DAC data of Errandonea et al. (2001b), although the latter data trend moderately higher in temperature than the present theory predicts above 30 GPa. Further note that the calculated hcp-liquid and bcc-liquid melt curves join smoothly without a change in slope, and this is confirmed by experiment.

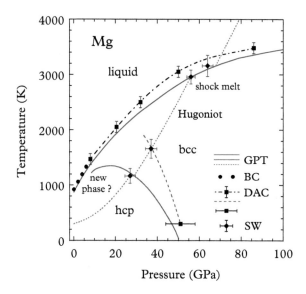

Fig. 8.12 *Magnesium pressure-temperature phase diagram and Hugoniot to 100 GPa, as predicted from the present first-principles GPT interatomic potentials, and compared with static and dynamic high-pressure measurements: BC (Errandonea, 2010), DAC (Olijnyk and Holzapfel, 1985; Errandonea et al., 2001b; Stinton et al., 2014) and SW (Beason et al., 2020).*

Also shown in Fig. 8.12 is the calculated multi-phase principal Hugoniot, with clear and expected breaks in the curve at the hcp-bcc and bcc-liquid boundaries. In this regard, a small pressure shift of -1.66 GPa has been added in all calculated phases to ensure that both the calculated and measured Hugoniot starting points correspond exactly to the observed room-temperature, ambient-pressure density of 1.738 g/cc. In particular, the very recent SW data of Beason et al. (2020) on Mg, which measure phase (through XRD) as well as Hugoniot pressure as a function of density, then align very closely with theory. Specifically, using the present theory to assign temperature to the measured SW pressure-density points, all four points agree as to measured and predicted phase, with the lowest three points lying exactly on the calculated Hugoniot, in the hcp, bcc and mixed bcc-liquid phase on the melt curve, respectively. The latter point corresponds to the onset of shock melting at 56 GPa and 2960 K.

8.3 Melting and the pressure-temperature phase diagram

There remain three major unsettled issues on the Mg phase diagram that require further investigation. First, although there is general agreement that the hcp-bcc phase line has a negative temperature slope, the magnitude of this slope and the extent of the phase line at low pressure are still unresolved. The present GPT QHLD hcp-bcc line, which does not extend fully to the melt curve, would seem to represent an effective lower bound, as the measured hcp SW point is located just under this curve. It is possible that yet to be calculated anharmonic vibrational contributions could raise this line and allow it to extend to the melt curve. At the other extreme, the proposed DAC hcp-bcc phase line by Stinton et al. (2014) would seem to represent an effective upper bound, since this phase line intersects the measured bcc SW point.

The second, possibly related, but still unresolved issue is whether or not there is a third high-temperature solid phase in Mg beneath the melt curve at low pressure, as we have indicated with a question mark in Fig. 8.12. Errandonea et al. (2003a) reported synchrotron XRD evidence for such a phase, data obtained with a resistively heated multi-anvil apparatus and measurements above 1000 K in the pressure range between 4 and 20 GPa. They identified this phase to be a dhcp structure. However, there has been no subsequence experimental or theoretical confirmation of an equilibrium dhcp phase in Mg. In similar synchrotron XRD, laser-heated DAC experiments on Mg to 25 GPa and 1900 K, Cynn et al. (2004) found no evidence of a dhcp structure. Yao and Klug (2012) showed, with DFT calculations on Mg that at $T = 0$ and up to ~ 50 GPa, that the dhcp structure is both higher in total energy than hcp and mechanically unstable with imaginary phonons, so that this structure cannot be part of the equilibrium phase diagram at room temperature. With complementary metadynamic simulations on Mg performed at high temperature and 15 GPa, they did find that the dhcp structure is mechanically stabilized through anharmonic lattice vibrations and becomes energetically competitive with hcp and bcc, but they found no evidence that dhcp is part of the equilibrium phase diagram. Finally, in recent laser heated DAC experiments with angle-dispersive XRD, Stinton et al. (2014) did find evidence for a third solid structure near 6 GPa and 1200 K. They could not determine the specific crystal structure found, however, and only established that it was not dhcp.

The third major unsettled issue in the Mg phase diagram concerns the behavior of the experimental melt curve above 45 GPa. In their DAC studies, Stinton et al. (2014) proposed an alternative to the high-pressure end of the Errandonea et al. (2001b) melt curve illustrated in Fig. 8.12. This alternative melt curve extends nearly linearly upward in temperature past 4000 K to end at a single additional measured melt point of 4400 K and 105 GPa, beyond the P, T boundaries of Fig. 8.12. This higher melt curve is also supported by two additional measured bcc phase points at 89 GPa and 3980 K and 97 GPa and 4320 K, respectively, both well above the Errandonea et al. (2001b) melt curve.

An important consideration in all three of the above issues, as well as in phase-diagram studies of metals in general, is the extent to which reported DAC measurements represent the true equilibrium behavior of the material. Hysteresis effects involving both pressure and temperature are well known to occur, adding uncertainty as to location of equilibrium phase lines. Also, the method of melt measurement (e.g., optical vs XRD) can impact the result. In the solid, nonequilibrium metastable structures can coexist with equilibrium structures in a given sample, masking the true phase diagram. Contamination leading to chemical reactions within the DAC can also seriously distort what is being measured. The latter is particularly important in the case of carbide-forming bcc transition metals such as Ta, whose present phase diagram and Hugoniot are displayed in Fig. 8.13.

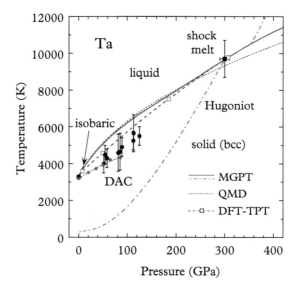

Fig. 8.13 *Tantalum pressure-temperature phase diagram and Hugoniot to 420 GPa, as predicted from Ta6.8x MGPT multi-ion interatomic potentials (Moriarty and Haskins, 2014), and compared to experimental isobaric (Shaner et al., 1977), DAC (Dewaele et al., 2010; Karandikar and Boehler, 2016) and SW (Brown and Shaner, 1984; Dai et al., 2009) melt data, as well as to QMD (Burakovsky et al., 2010) and DFT-TPT (Taioli et al., 2007) melt calculations.*
From Haskins and Moriarty (2018), with publisher (APS) permission.

In early laser-heated DAC measurements of high-pressure melting in bcc transition metals up to 100 GPa, obtained by the optical laser-speckle method, unexpected flat melt curves, with $T_m(P_m)$ nearly constant, were found for V, Ta, Cr, Mo and W (Errandonea et al., 2001a). These results were in sharp contrast to the normal steep melting curves inferred in Ta and Mo from dynamic isobaric data providing the initial $P_m = 0$ melting slope (Shaner et al., 1977) and from measurements of shock melting at high pressure (Brown and Shaner, 1984; Hixson et al., 1989). This latter perspective was supported by the initial MGPT calculation of the Ta melt curve, with the early Ta3 multi-ion potentials (Moriarty et al., 2002a). Thus began a long-running and still continuing controversy about the nature of melting and the phase diagram in bcc transition metals.

Next, the DAC flat-melting-curve picture was confirmed in Ta by Errandonea et al. (2003b) and then later in Mo by Santamaría-Perez et al. (2009), using XRD to detect melting in place of the optical laser-speckle method, which had been questioned as a result of the controversy. At about the same time, a theoretical study by Wu et al. (2009)

on Ta suggested the possible importance of shear stresses in the DAC, which might cloud the interpretation of the experimental results. Using MD simulations with the Ta4 MGPT potentials, these workers found a shear-induced transformation from bcc to a partially ordered, partially disordered solid phase at similar P, T conditions to the DAC melt measurements. They interpreted this transformation as a Bingham-like plastic flow, and possibly linked to high P, T polymorphism in the solid. At the same time, strong theoretical support for the steep-bcc-melting-curve picture was provided in the case of Ta by the DFT-based melting calculations of Taioli et al. (2007), Burakovsky et al. (2010) and Moriarty and Haskins (2014), results that are all displayed in Fig. 8.13.

Then on the experimental side, there finally came two improved laser-heated DAC measurements of bcc melting in Ta by Dewaele et al. (2011) and Karandikar and Boehler (2016) that yielded significantly higher melt temperatures as a function of pressure than the earlier DAC measurements. In these breakthrough experiments, both groups identified and addressed the important issue of chemical contamination in the DAC and resulting chemical reactions with the Ta sample. These workers differed, however, in their respective approaches to the melt-temperature measurement itself, with Dewaele et al. (2011) finding higher melt temperatures, but with large error bars, than Karandikar and Boehler (2016). Their respective results are also displayed on the Ta phase diagram in Fig. 8.13. While the improved DAC melt measurements on Ta have moved closer to the theoretical bcc melt curves, there is still an unexplained gap between the two.

Regarding the three theoretical Ta bcc melt curves illustrated in Fig. 8.13, they derive respectively from the accurate Ta6.8x MGPT bcc and liquid free energies of Moriarty and Haskins (2014), from the QMD Z-melt simulations of Burakovsky et al. (2010), and from the DFT-TPT calculations of Taioli et al. (2007). The MGPT and QMD melt curves closely agree with each other up to a pressure of ~250 GPa, as well as with the initial low-pressure melting slope obtained from isobaric expansion data (Shaner et al., 1977). The MGPT and DFT-TPT melt curves agree well with each other above ~180 GPa and with the experimental shock melting point at 300 GPa and 9700 K (Brown and Shaner, 1984, with additional pyrometry temperature measurements from Dai et al., 2009), which is also right on the calculated MGPT bcc Hugoniot. In the DAC pressure regime below about 135 GPa, the DFT-TPT melt curve is closest to the new DAC data and within the large error bars of the Dewaele et al. (2011) data. In all of the DAC measurements to date on Ta, no other solid phase than bcc has ever been reported. It remains an open question as to whether or not other solid phases could exist in the Ta equilibrium phase diagram. The question of possible polymorphism in the central transition metals is discussed in the next section.

8.4 Rapid solidification and polymorphism in transition metals

Historically, a limited amount of polymorphism has been known to exist in the elemental central transition metals, including several well-established allotropes in bcc metals. In the case of tantalum, there is a second tetragonal phase known as β-Ta, which occurs on free surfaces and also in thin films at ambient conditions. This phase is technologically significant because it actually interferes with the epitaxial growth of bcc single crystals. However, β-Ta is not part of the bulk equilibrium phase diagram. In tungsten, there is an additional cubic A15 phase known as β-W, which is metastable in the bulk metal and can coexist with the bcc structure, but again is not part of the equilibrium phase diagram. Thirdly, in the case of vanadium there is the more recent discovery of a high-pressure rhombohedral phase in the bulk metal that appears at 69 GPa, where the bcc structure becomes mechanically unstable, as was discussed in Chapter 6.

At high temperature in bulk Ta, it has also been observed that the A15 structure can be solidified from the supercooled liquid (Cortella et al., 1993). The competitive nature of bcc and A15 in Ta has been investigated extensively with DFT and MGPT calculations, including static $T = 0$ structural-energy differences (see Table 6.4) and dynamic high-T MGPT-MD rapid resolidification (RR) simulations (Streitz and Moriarty, 2002; Streitz et al., 2006a), as well as recent full MGPT free-energy bcc and A15 melt curves (Haskins et al., 2012; Haskins and Moriarty, 2018). Mechanical stability of the A15 structure has been confirmed at low temperature by MGPT QH phonon calculations to 420 GPa, and at high temperature and high-P, T conditions by MGPT-MD simulations. The calculated bcc and A15 MGPT melt curves are displayed in Fig. 8.14(a) below 80 GPa. The bcc melt curve lies above that of A15, indicating that bcc is the more thermodynamically stable phase and hence that A15 is still only metastable at high temperature. But also note from Fig. 8.14(a) that this outcome depends crucially on anharmonic and

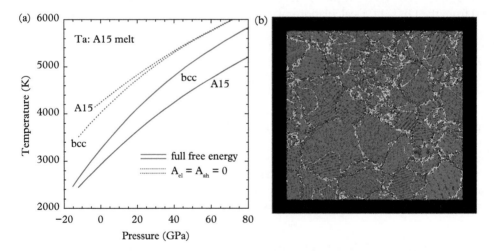

Fig. 8.14 *Melting of metastable A15 phase and bcc grain-boundary (GB) formation in tantalum. (a) A15 and bcc melt curves from full MGPT free energies and in the quasiharmonic limit; (b) rich size-independent bcc GB structure obtained at 175 GPa and 5000 K from a 16,384,000-atom rapid-resolidification MGPT-MD simulation.*

Panel (a) from Haskins and Moriarty (2018) and panel (b) from Streitz et al. (2006a), with publisher (APS) permissions.

electron-thermal free-energy contributions favoring bcc. If these contributions are set to zero, and one repeats the melt calculations in the QH limit, the A15 melt curve overtakes that of bcc below about 40 GPa, so A15 becomes the stable phase in that pressure regime.

The close competition between bcc and A15 at high temperature in Ta was played out in real time through the MGPT-MD RR simulations of Streitz and Moriarty (2002). These simulations were performed using the Ta4 potentials in an NVT ensemble on a small sample of $N = 1000$ atoms compressed slowly at $T = 5000$ K from the liquid, across the melt curve, into the solid to high pressure over a long simulation time of 840 ps or 0.84 ns. The initial sample volume $V = N\Omega$ corresponded to a liquid pressure of $P = 10$ GPa and the final compressed volume to solid pressure of $P = 400$ GPa (see the Ta phase diagram in Fig. 8.13). Once the melt curve was crossed in this simulation, small fluctuating pockets of both bcc and A15 crystals formed and competed until about 800 ps when the final coalescence of the bcc structure occurred. Thus, the mere presence of a competitive metastable A15 structure in the simulation dramatically impacted the kinetics of bcc coalescence.

In subsequent MGPT-MD RR simulations on tantalum by Streitz et al. (2006a), the details of the bcc coalescence were studied in much larger samples sizes ranging from 65,536 to 32,768,000 (32M) atoms. To shorten the time until the onset of coalescence, the initial volume was decreased to produce a starting pressure of about 40 GPa at 5000 K, very close to the melt curve, and the Ta4 potentials were slightly tweaked to increase the A15-bcc energy difference. The simulation time to the onset of coalescence was thereby decreased to an average of 250 ps, and the final bcc pressure reduced to a maximum of 175 GPa. These simulations produced detailed bcc grain boundary (GB) structures, where the average maximum cluster size of an individual grain was found to become independent of the simulation size at 8M atoms. A snapshot of the rich GB structure formed in a 16M-atom MGPT-MD RR simulation is shown in Fig. 8.14(b).

Interest in possible additional polymorphism in the unexplored high-P, T regions of the phase diagrams of central transition metals has increased in recent years, motivated in large part by the melting curve controversy of the past two decades that was discussed in Sec. 8.3. In this regard, Burakovsky et al. (2010) used small-cell (< 150 atoms) QMD Z-melt simulations to predict a hexagonal omega (hex-ω) phase in Ta at high temperatures and pressures above 70 GPa. Shortly thereafter, Haskins et al. (2012) used the refined Ta6.8x MGPT potentials, which allow accurate treatment of much larger system sizes, to examine both the mechanical and the thermodynamic stability of hex-ω, as well as the bcc, A15, fcc and hcp structures also considered by Burakovsky et al. While at low temperature only the bcc and A15 structures are calculated to be mechanically stable, at high temperature strong, but size- and structure-dependent, anharmonic effects produce mechanical stability in fcc, hcp and hex-ω as well, allowing relative thermodynamic stability to be determined by MGPT-MD melting simulations in all five structures. First, using the Z-melt method to perform this task, Haskins et al. discovered important melt size effects for the non-cubic hcp and hex-ω structures, requiring in each case a minimum of ~500 atoms in the solid to obtain a reliable melt curve. In the small-cell limit of less than 150 atoms, and with the same cell sizes and choice of hex-ω c/a ratio used in the QMD simulations, the MGPT-MD bcc and hex-ω Z-melt curves were found to be in good agreement with the QMD results, thus confirming the prediction of Burakovsky et al. in that limit. However, with a much larger cell size of 3000 atoms for the hex-ω structure, its melt curve moved entirely below that of bcc for all pressures up to 420 GPa, so bcc is actually the more stable phase. Finally, using the accurate two-phase melt method in the large-cell limit, with $\sim 40,000$ atoms per phase, MGPT-MD simulations showed conclusively that, among the five phases considered, bcc is the most thermodynamically stable phase in Ta over the entire 0–420 GPa pressure range.

An interesting and important caveat about the high-temperature mechanical stability of the hex-ω structure is its unexpected complexity, as elaborated in detail by Haskins et al. (2012). For low values of $c/a < 0.54$, normal metastability is indeed realized by anharmonic effects in MGPT-MD simulations. For larger, and more normal, values of $c/a > 0.54$, however, the high-T mechanical stability, as measured in MD simulations by the calculated stress tensor, is poor and a partial transformation to bcc occurs. The transformation to bcc is incomplete because of the hex-ω boundary conditions imposed by the MD simulations, so the physical inference is that the hex-ω structure actually remains mechanically unstable in the high-T, $c/a > 0.54$ regime. This conclusion is supported by the DFT self-consistent phonon calculations of Yao and Klug (2013) and of Liu et al. (2013). Such calculations impart a temperature dependence to the phonon frequencies and thereby capture at least part of the anharmonic vibrational effects at high T. These authors find that imaginary phonon frequencies, and hence mechanical instability, persist in the hex-ω structure at high-P, T conditions up to 400 GPa and 6000 K and up to 300 GPa and 7000 K, respectively.

Current theoretical research on polymorphism and the equilibrium phase diagram in Ta has moved on from candidate cubic and hexagonal structures, and towards the possibility of lower-energy orthorhombic phases occurring at high pressures and temperatures. This interest follows additional DFT calculations by Yao and Klug (2013) and by Liu et al. (2013), as well as small-cell QMD simulations by Burakovsky et al. (2014), indicating a possible high-P, T *Pnma* phase in Ta. In sharp contrast to hex-ω, DFT-QH phonon calculations by Yao and Klug and by Liu et al. on *Pnma* have produced all real phonon frequencies to 400 GPa and 250 GPa, respectively, and hence low-temperature mechanical stability at the pressures of interest. Using small-cell (64 atom) DFT-based metadynamics simulations, which include anharmonic effects, Yao and Klug further studied the possibility of a high-P, T bcc to *Pnma* phase transition, and elaborated a plausible Burgers-like transition path between the two structures. Subsequently, Burakovsky et al. used ~500-atom QMD, Z-melt simulations on *Pnma* Ta to predict a bcc to *Pnma* transition at high temperature above 200 GPa.

Haskins and Moriarty (2018) have recently extended the MGPT Ta polymorphism studies discussed above to *Pnma* and three other promising orthorhombic structures: *Fddd*, *Pmma* and *Cmcm* or α-U. All of these structures either are observed phases in heavy metals or have been predicted to be competitive phases in bcc transition metals. The more competitive nature of these four orthorhombic structures compared to hex-ω can be immediately appreciated by calculating optimized MGPT total energies for all five structures at $T = 0$, using Eq. (7.18) for $E_{tot}(\mathbf{R}, \Omega)$ through four-ion interactions.

To do this for each phase, E_{tot} must be minimized as a function of atomic volume Ω with respect to the appropriate b/a and c/a axial ratios and all internal structural parameters. The results so obtained by Haskins and Moriarty are plotted in Fig. 8.15(a) as total-energy differences per atom relative to bcc, $E_{\text{tot}} - E_{\text{tot}}^{\text{bcc}}$. Note that the four orthorhombic total energies are closely spaced and indeed well below that of hex-ω, with the *Fddd* and *Pnma* energies the lowest for atomic volumes greater than about 70 a.u., corresponding to pressures less than 300 GPa.

Regarding the results shown in Fig. 8.15(a), one can also convert total energy per atom as a function of atomic volume to total enthalpy per atom as a function of total pressure, $H_{\text{tot}} = E_{\text{tot}} + P_{\text{tot}}\Omega$. This conversion has been made in Fig. 8.15(b) for the hex-ω and *Pnma* structures to allow a direct comparison with the DFT calculations of Yao and Klug (2013), in units of enthalpy relative to bcc per atom. At low pressure, the MGPT and DFT results agree very closely. Above about 250 GPa, the MGPT relative hex-ω and *Pnma* enthalpies trend somewhat below the DFT values, but the agreement remains quite reasonable up to the maximum $T = 0$ pressure considered of 420 GPa.

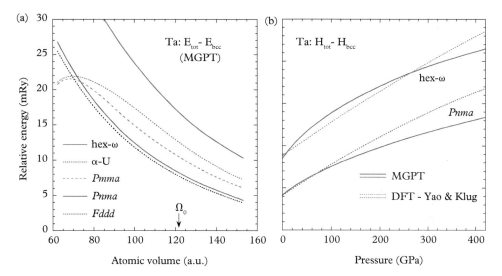

Fig. 8.15 *Optimized MGPT hex-ω and orthorhombic total energies and enthalpies per atom at $T = 0$ relative to bcc in tantalum, as obtained using Ta6.8x potentials. (a) hex-ω, α-U, Pmma, Pnma and Fddd relative total energies, with $\Omega_0 = 121.6$ a.u., the observed bcc equilibrium volume, and $E_{bcc} \equiv E_{tot}^{bcc}$; (b) hex-$\omega$ and Pnma relative total enthalpies, where $H_{bcc} \equiv H_{tot}^{bcc}$, compared with the DFT results of Yao and Klug (2013).*
From Haskins and Moriarty (2018), with publisher (APS) permission.

Haskins and Moriarty (2018) further showed through MGPT QH phonon calculations that the *Fddd* and *Pnma* structures in Ta are mechanically stable over the full volume and pressure ranges considered in Fig. 8.15, while *Pmma* and α-U, like hex-ω, are unstable over the same ranges. It was further shown that the four calculated mechanically stable structures at $T = 0$, i.e., bcc, A15, *Fddd* and *Pnma*, have very similar

Debye temperatures over the same volume range. The mechanical stability of the *Fddd* and *Pnma* structures at modest temperatures is readily confirmed by MGPT-MD simulations, but interestingly Haskins and Moriarty observed spontaneous *Fddd* → bcc and *Pnma* → bcc temperature-induced phase transitions above a pressure-dependent critical temperature in each case. The maximum critical temperatures found were 1450 K at 80 GPa in *Fddd* and 1250 K at 160 GPa in *Pnma*. The physical origin of this phenomenon is readily explained in terms of the total-energy differences between the orthorhombic and bcc structures displayed in Fig. 8.15(a), which imply a large thermodynamic potential for these transitions to occur. Specifically, at modest temperatures *Fddd* and *Pnma* are vibrating in relatively shallow potential energy wells that are at much higher energy than bcc. As the amplitude of vibration increases with increasing temperature, a maximum possible amplitude will be reached before the material spontaneously "spills over" into the lower-energy bcc vibrational well. This simple picture is confirmed by the actual *Fddd* → bcc and *Pnma* → bcc transition paths determined by Haskins and Moriarty (2018).

At high temperature near melt, strong anharmonic effects again play a leading role in the mechanical and relative thermodynamic stability of the four candidate orthorhombic structures in Ta considered by Haskins and Moriarty (2018). The anharmonic component of the ion-thermal energy was found to be in the range of 10–25% for all four structures, enhancing the mechanical stability of *Fddd* and *Pnma* and fully establishing mechanical stability for *Pmma* and α-U. The melting curves of all four structures are competitive with that of bcc, but large size effects were found as well. These melt size effect now range up to ~1000–4000 atoms, depending on the melt method and the simulation cell shape, and again can significantly alter phase prediction at high pressure. Using the Z-melt method, Haskins and Moriarty showed that with an appropriate shape-corrected cell of ~500 atoms the MGPT-MD *Pnma* melt curve moved above that of bcc at high pressure near 280 GPa, qualitatively matching the QMD, Z-melt prediction of a *bcc* → *Pnma* phase transition near 200 GPa by Burakovsky et al. (2014). However, not until the cell size was increased to 900 atoms did the MGPT-MD *Pnma* melt curve fully converge. At that point, the *Pnma* melt curve dropped completely below the bcc melt curve, eliminating the phase transition.

With the more accurate two-phase melt method, however, the converged MGPT-MD orthorhombic Ta melt curves move closer to that of bcc. In the large-cell limit with $\sim 40{,}000$ atoms per phase, only the *Pmma* melt curve lies entirely below that of bcc over the 0–420 GPa pressure range, while for *Pnma*, *Fddd* and α-U, their melt curves touch the bcc melt curve at one or more pressures in the vicinity of 100 GPa, as shown in Fig. 8.16 for *Pnma* and *Fddd*. Haskins and Moriarty (2018) thus concluded that for Ta at high-P, T conditions, in the approximate pressure range 90–140 GPa, *Pnma*, *Fddd* and/or α-U could coexist with bcc and possibly be observed experimentally. This conjecture awaits experimental investigation. Interestingly, however, in the neighboring Group-VB metal niobium, a recent combined experimental and theoretical investigation by Errandonea et al. (2020) now claims to have found a high-P, T *Pnma* phase below 120 GPa.

Finally, two other recent investigations relevant to the question of polymorphism and the phase diagram in bcc transition metals also should be mentioned. Kraus et al. (2021) have reported in situ XRD experiments of shock compressed Ta in the pressure range from 201 to 343 GPa along the principal Hugoniot. Over this range only bcc, mixed bcc-liquid and liquid phases were found, which is completely consistent with DFT and MGPT calculations, where any possible appearance of the competitive A15, *Pnma*, *Fddd* or α-U structures is predicted to be below 200 GPa and well above the Hugoniot in temperature at those pressures. In the extended analysis of their data, Kraus et al. further went on to assume that the first appearance of liquid fluctuations in the mixed two-phase region corresponds to the full onset of macroscopic shock melting, leading to a P, T shock melting point of only 254 GPa and 8070 K. Although this alternative shock melting point lines up well with the DAC melting data of Dewaele et al. (2011) at lower pressures, it is at odds both with the currently accepted shock melting point centered at 295 GPa and 9700 K and with the three DFT and MGPT equilibrium Ta melt curves, as displayed in Fig. 8.13.

The second recent paper of interest is by Gal (2021) and concerns the importance of the pressure-transmitting medium used in the DAC transition-metal melting experiments, which typically has been argon. Gal notes that under pressure and above 3000 K that Ar is fully in the liquid state, conditions that well match the DAC melting data on the metals Mo, Ta and W. His main conjecture is that a liquid-liquid phase transition in Ar under those conditions would reduce the thermal pressure developed in the DAC, leading to the observed nearly flat melting curves. He claims to be able to model the DAC melting behavior with existing argon and metal experimental data.

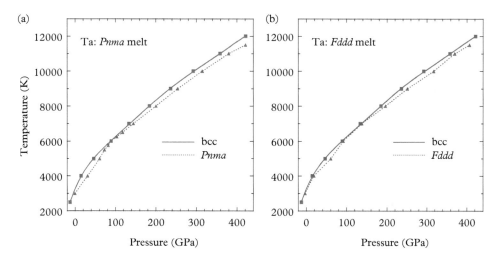

Fig. 8.16 *High-pressure melt curves for the Pnma and Fddd tantalum orthorhombic phases compared with that of bcc, as obtained from MGPT two-phase MD melt simulations, using Ta6.8x multi-ion potentials and large simulation cells. (a) Pnma and (b) Fddd.*
From Haskins and Moriarty (2018), with publisher (APS) permission.

9
Defects and Mechanical Properties

In this chapter we turn to the important subject of point and line defects in bulk metals, and their impact on mechanical properties. Historically, the desire for an atomistic description of the structure, motion and interaction of bulk defects such as vacancies, interstitials, dislocations and grain boundaries has been the main motivating factor for the development of empirical interatomic potentials. But all point and line defects, in addition to the bulk elasticity in metals that underpins them, are significantly affected by quantum mechanics, and hence the fundamental importance of QBIPs. This is especially so for advanced and increasingly complex applications, such as the multiscale materials modeling of crystal plasticity and strength. And while the principles discussed in this chapter can be applied to all metals, we will concentrate here mostly on the central bcc transition metals, where significant QBIP applications have been made with both MGPT and BOP potentials. In addition, for bulk central transition metals there is minimal concern about any additional local volume corrections to defect energies, and we assume that such corrections can be safely neglected. This situation is in sharp contrast to that in simple and other nontransition metals, where, for example, the corresponding bulk GPT treatment of the vacancy formation energy will underestimate its value due to the absence of local volume corrections. The needed local-density aGPT treatment to correct this problem for vacancies, and more generally to obtain a wide range of bulk and surface defect energies for nontransition metals, is discussed in Chapter 11.

9.1 Point defect formation and migration energies

We first consider the basic problems of the formation and migration energies of monovacancies and self-interstitials in a metal. Relative to the perfect crystalline metal with N identical atoms, the equilibrium defect formation energy E_D^f is formally defined as

$$E_\mathrm{D}^\mathrm{f} = E_\mathrm{tot}[N_\mathrm{D}, \Omega_\mathrm{D}] - N_\mathrm{D} E_\mathrm{coh}[N, \Omega_0], \tag{9.1}$$

where $N_\mathrm{D} = N - 1$ for a single vacancy and $N_\mathrm{D} = N + 1$ for a single self-interstitial atom or SIA. Here $E_\mathrm{tot}[N_\mathrm{D}, \Omega_\mathrm{D}]$ is the total energy of the metal with the defect present, containing N_D atoms with a total volume of Ω_D (to be determined), while $E_{coh}[N, \Omega_0]$ is the bulk cohesive energy at the reference equilibrium atomic volume Ω_0. The corresponding defect formation volume, i.e., the additional volume created by the presence of the defect, is just

$$\Omega_D^f = \Omega_D - N_D \Omega_0. \tag{9.2}$$

Note that for volume-dependent QBIPs such as MGPT, $E_{tot}[N_D, \Omega_D] \equiv E_{tot}^D(\mathbf{R}, \Omega)$, where $\Omega = \Omega_D/N_D$, while $E_{coh}[N, \Omega_0] \equiv E_{coh}(\mathbf{R}_0, \Omega_0) = (1/N)E_{tot}(\mathbf{R}_0, \Omega_0)$. Also, in this limit it follows from Eq. (9.2) that $\Omega_D^f/\Omega_0 = N_D(\Omega/\Omega_0 - 1)$.

For bcc transition metals, the QBIP calculation of E_D^f usually proceeds by first creating a large cubic supercell to represent the bulk material, based on the conventional two-atom bcc unit cell with lattice constant a, and which contains a chosen number of N atoms, with PBCs applied to the supercell. To perform a bcc vacancy calculation via Eq. (9.1) with the MGPT multi-ion potentials, a $5 \times 5 \times 5$ supercell corresponding to $N = 250$, and containing 249 atoms, is normally adequate to obtain good convergence. To perform a corresponding bcc self-interstitial calculation with similar good convergence, on the other hand, can require a supercell as large as $8 \times 8 \times 8$ corresponding to $N = 1024$, and containing 1025 atoms (Xu and Moriarty, 1996).

In calculating E_D^f directly from Eq. (9.1), the RHS of that equation is to be evaluated at a constant pressure P_0 corresponding to the volume Ω_0 in the bulk metal. If Ω_0 represents the calculated $T = 0$ equilibrium atomic volume, then one has $P_0 = 0$. If instead one takes Ω_0 to be the observed room temperature atomic volume, as we have normally done in this book, then P_0 is the calculated bulk pressure at that volume, which is typically on the order of -1 GPa. In either case, once an atom is removed from or added to the supercell to create the defect, one must then vary the supercell volume Ω_D in the $E_{tot}[N_D, \Omega_D]$ term of Eq. (9.1), while simultaneously relaxing all atomic positions, until the target pressure P_0 is reached. The fully relaxed defect formation energy, either $E_D^f = E_{vac}^f$ for a vacancy or $E_D^f = E_{int}^f$ for a self-interstitial, is thereby obtained.

In the case of a vacancy, it is also useful and instructive to consider the computationally easier calculation of the *unrelaxed* formation energy from Eq. (9.1). In this limit, one knows from the outset all atomic positions in the supercell and that $\Omega_D = N\Omega_0$, so the lattice constant to be used in calculating $E_{tot}[N_D, \Omega_D]$ is the same as for the bulk metal. In addition, E_{vac}^f then represents a rigorous upper bound to the true relaxed vacancy formation energy. Typically, this upper bound is only 10% or so higher in magnitude than the fully relaxed value of E_{vac}^f, as demonstrated for the case of Ta in Table 9.1.

For volume-dependent QBIPs such as MGPT, an equivalent but even more useful calculation of the unrelaxed vacancy formation energy proceeds by evaluating Eq. (9.1) at a constant atomic volume Ω_0 in the limit $N \to \infty$. To do this, one compresses the vacancy supercell, so that $\Omega_D/(N-1) = \Omega_0$ at large N. Noting that the volume term $E_{vol}(\Omega_0)$ in the total energy then exactly cancels between the two terms on the RHS of Eq. (9.1), one has

Table 9.1 *Calculated MGPT vacancy formation energy E_{vac}^f (in eV) and formation volume Ω_{vac}^f/Ω_0 for Ta near equilibrium conditions (atomic volume Ω_0 in a.u. and pressure P_0 in GPa), as obtained with the different possible treatments discussed in the text, using Ta6.8x potentials.*

Treatment	Ω_0	P_0	E_{vac}^f	Ω_{vac}^f/Ω_0
Unrelaxed: Eq. (9.8)	121.6	−1.16	3.20	0.0
Unrelaxed: Eq. (9.1)	121.6	−1.16	3.22	0.0
Relaxed: Eq. (9.1)	121.6	−1.16	2.95	0.54
Relaxed: Eq. (9.1)	120.95	0.0	2.95	0.54

$$E_{vac}^f = E_{tot}^{vac}(\mathbf{R}, \Omega_0) - \frac{(N-1)}{N} E_{tot}(\mathbf{R}_0, \Omega_0)$$

$$= E_{struc}^{vac}(\mathbf{R}, \Omega_0) - \frac{(N-1)}{N} E_{struc}(\mathbf{R}_0, \Omega_0), \quad (9.3)$$

where for a bcc transition metal treated through four-ion interactions one has

$$E_{struc}^{vac}(\mathbf{R}, \Omega_0) = \frac{1}{2} \sum_{i,j=1}^{N-1}{}' v_2(R_{ij}, \Omega_0) + \frac{1}{6} \sum_{i,j,k=1}^{N-1}{}' v_3(R_{ij}, R_{jk}, \cdots, \Omega_0)$$

$$+ \frac{1}{24} \sum_{i,j,k,l=1}^{N-1}{}' v_4(R_{ij}, R_{jk}, \cdots, \Omega_0) \quad (9.4)$$

and

$$E_{struc}(\mathbf{R}_0, \Omega_0) = \frac{1}{2} \sum_{i,j=1}^{N}{}' v_2(R_{ij}^0, \Omega_0) + \frac{1}{6} \sum_{i,j,k=1}^{N}{}' v_3(R_{ij}^0, R_{jk}^0, \cdots, \Omega_0)$$

$$+ \frac{1}{24} \sum_{i,j,k,l=1}^{N}{}' v_4(R_{ij}^0, R_{jk}^0, \cdots, \Omega_0). \quad (9.5)$$

In Eq. (9.4) for $E_{struc}^{vac}(\mathbf{R}, \Omega_0)$ at large N, each interatomic distance R_{ij} can be scaled to the equilibrium distance R_{ij}^0 as follows:

$$R_{ij} = R_{ij}^0 + \delta R_{ij} = \left(\frac{N-1}{N}\right)^{1/3} R_{ij}^0$$

$$= R_{ij}^0 - \frac{1}{3N} R_{ij}^0 + \cdots, \quad (9.6)$$

so in the large N limit $\delta R_{ij} = -R_{ij}^0/(3N)$. Using this result, one can Taylor expand the interatomic potentials v_n appearing in Eq. (9.4) in the general form

$$v_n(R_{ij}^0 + \delta R_{ij}, \cdots, \Omega_0) = v_n(R_{ij}^0, \cdots, \Omega_0) + \frac{\partial v_n(R_{ij}^0, \cdots, \Omega_0)}{\partial R_{ij}} \delta R_{ij} + \cdots$$

$$= v_n(R_{ij}^0, \cdots, \Omega_0) - \frac{1}{3N} R_{ij}^0 \frac{\partial v_n(R_{ij}^0, \cdots, \Omega_0)}{\partial R_{ij}} + \cdots . \quad (9.7)$$

Finally, using Eq. (9.7) in Eq. (9.4) then allows one to combine common terms in the second line of Eq. (9.3) for the unrelaxed value of $E_{\text{vac}}^{\text{f}}$, yielding the final result

$$E_{\text{vac}}^{\text{f}} = -E_{\text{struc}}(\mathbf{R}_0, \Omega_0) + \Omega_0 P_{\text{vir}}(\Omega_0), \quad (9.8)$$

where P_{vir} is the bulk virial pressure that we defined in Eq. (7.96) of Chapter 7. Also, in Eq. (9.8) we have now conveniently absorbed a factor of $1/N$ on the RHS of Eq. (9.5) for the bulk definition of E_{struc}, making that quantity fully consistent with our general definition in Eq. (6.2) of Chapter 6. Equations (9.1) and (9.8) evaluated for the unrelaxed vacancy formation energy are quantitatively equal within numerical error, as shown in Table 9.1 for Ta. Also, by including only pair-potential interactions in evaluating E_{struc} and P_{vir}, Eq. (9.8) then directly confirms Eq. (1.19) of Chapter 1.

In addition to the important fact that Eq. (9.8) only requires *bulk* energies and pressures to make a good estimate of a fundamental defect property, the equilibrium value of $E_{\text{vac}}^{\text{f}}$, is the equally important realization that this result can be immediately generalized to any volume in the same material. That is, the equilibrium volume Ω_0 itself plays no essential role in the derivation of Eq. (9.8), so the unrelaxed vacancy formation energy at any expanded or compressed volume Ω is just

$$E_{\text{vac}}^{\text{f}}(\Omega) = -E_{\text{struc}}(\mathbf{R}, \Omega) + \Omega P_{\text{vir}}(\Omega). \quad (9.9)$$

This result has been used to advantage in developing MGPT potentials for bcc transition metals over extended volume ranges, as discussed in Chapter 5.

The accurate treatment of self-interstitials via Eq. (9.1) brings on additional challenges, since a priori the atomistic configuration producing the lowest SIA formation energy for a given bcc transition metal is not known. The six highest symmetry configurations most often studied theoretically are the following: the three two-atom dumbbells formed around the center of the bcc unit cell and aligned along the <100>, <110> and <111> directions, respectively, and the three configurations formed by placing the SIA at one of the remaining highest-symmetry points in bcc unit cell. The latter three positions are the <111> crowdion site at $(0.25, 0.25, 0.25)a$ in the unit cell, halfway between nearest neighbors; the octahedral site at $(0.50, 0.50, 0.00)a$, halfway between second nearest neighbors; and the tetrahedral site at $(0.50, 0.25, 0.00)a$. Experimentally, x-ray diffuse scattering experiments on Mo in the late 1970s (Ehrhart, 1978) initially pointed to the $<110>$ dumbbell as the configuration with the lowest SIA formation energy, and this observation was supported by most subsequent calculations with empirical potentials, as well as by the first MGPT calculations with the early Mo2 multi-ion potentials (Xu and Moriarty, 1996). This was, in fact, the prevailing point of view on SIAs in the group-VB and -VIB transition metals for three decades, until the surprising DFT calculations of Nguyen-Manh et al. (2006) and Derlet et al. (2007). These authors found that for all six VB and VIB metals that the SIA configuration of lowest energy always had <111> symmetry, with very closely spaced formation energies for the <111> dumbbell and crowdion configurations. Moreover, in the group-VB metals (V, Nb, Ta) they found a significantly larger energy separation between the <111> configurations and the <110> dumbbell configuration than in the group-VIB metals (Cr, Mo, W), where the <111> and <110> dumbbell configurations were found to be linked by a soft bending mode. In more recent DFT calculations on the group-VIB metals, Ma and Dudarev (2019b) find that the lowest energy SIA configuration in all three metals is actually a symmetry-broken $<11\xi>$ dumbbell configuration, where ξ is a material-dependent irrational number varying between 0.36 and 0.48 in their $5 \times 5 \times 5$ supercell calculations.

It is also of interest to calculate the corresponding migration energy barriers for the monovacancy and for the low-formation-energy SIAs in the bcc lattice. To do this in the case of the vacancy, one moves a nearest-neighbor atom from its equilibrium site toward the vacancy site in a <111> direction until a maximum in total energy is reached, which corresponds to a saddle point on the total-energy surface, such that $E_{\text{vac}}^{\text{saddle}} = E_{\text{tot}}^{\text{max}}$. The vacancy migration energy is then

$$E_{\text{vac}}^{\text{m}} = E_{\text{vac}}^{\text{saddle}} - E_{\text{vac}}^{\text{f}}. \qquad (9.10)$$

Calculation of the self-interstitial migration energy $E_{\text{int}}^{\text{m}}$ is more complex, however. This quantity depends not only on the symmetry character of the SIA, but also on the fact that multiple migration paths may be possible. To our knowledge, the only case that has been studied in detail for bcc transition metals with either QBIP or DFT calculations is the migration energy of the <110> dumbbell self-interstitial. Using the early Mo2 MGPT potentials, Xu and Moriarty (1996) considered three possible SIA migration paths, depicted in Fig. 4 of their paper. Two of the paths involved single and double jumps, respectively, of the SIA to neighboring (001) atomic planes, and the third involved a jump plus rotation to a neighboring (010) plane. The latter path provided the lowest migration energy for Mo; this path was also considered by Yang et al. (2001) in MGPT and DFT calculations on Ta.

In Table 9.2 we compare selected monovacancy and single self-interstitial properties for bcc V, Ta and Mo, as calculated from MGPT, BOP and GAP potentials, as well as from several different DFT methods, in each metal. Calculated values of the vacancy formation energy $E_{\text{vac}}^{\text{f}}$ and migration energy $E_{\text{vac}}^{\text{m}}$ are also compared with the historical experimental data for these quantities, as reviewed by Schultz (1991). The agreement between theory and experiment for these latter quantities is generally quite good, with the observed trends of higher values of $E_{\text{vac}}^{\text{f}}$ in Ta and Mo than in V and increasing values of $E_{\text{vac}}^{\text{m}}$ in going from V to Ta to Mo well calculated. Good agreement for these quantities is also obtained among the different theoretical methods, as is also the case for the relative vacancy formation volume $\Omega_{\text{vac}}^{\text{f}}/\Omega_0$, which interestingly, is predicted to be in the narrow range 0.50–0.67 for all three metals. Also note in this regard, that the basic four-ion canonical d-band MGPT treatment of bcc transition metals, as here best represented by the V6.1, Ta4, Ta6.8x and Mo5.2 potentials, appears to be fully adequate to describe monovacancy properties.

With regard to the self-interstitial properties considered in Table 9.2, we have included the <111> and <110> dumbbell formation energies, $E_{\text{int}}^{\text{f-111}}$ and $E_{\text{int}}^{\text{f-110}}$, respectively, as well as the few available calculations of the <110> migration energy $E_{\text{int}}^{\text{m-110}}$ and relative formation volume $\Omega_{\text{int}}^{\text{f-110}}/\Omega_0$. Note that the BOP and GAP SIA results qualitatively capture the DFT prediction of $E_{\text{int}}^{\text{f-111}} < E_{\text{int}}^{\text{f-110}}$ in group-VB and -VIB transition metals, while the currently available MGPT potentials do not. Nonetheless, in the case of group-VB metals, the canonical d-band MGPT potentials V6.1, Ta4 and Ta 6.8x all produce good quantitative values of $E_{\text{int}}^{\text{f-110}}$. But at the same time, $E_{\text{int}}^{\text{f-110}}$ for the group-VIB metal Mo is overestimated by some 3 eV with both the Mo5.2 and the early Mo2 canonical d-band potentials. This latter problem is readily corrected, however, with the more general *noncanonical* d-band Mo12.t0 potentials (see Chapter 13), which were also invoked in Chapter 7 to address strong anomalies in the Mo phonon spectrum (see Fig. 7.4). Moreover, note that the Mo12.t0 potentials bring $E_{\text{int}}^{\text{f-111}}$ within about 0.2 eV of $E_{\text{int}}^{\text{f-110}}$.

Table 9.2 *MGPT, BOP, GAP and DFT calculations of selected bcc transition-metal vacancy and self-interstitial formation energies (in eV) and volumes (in Ω_0), and migration energies (in eV). Here the superscript "111" denotes the $<111>$ dumbbell SIA and the superscript "110" denotes the $<110>$ dumbbell SIA.*

Metal	Method	$E_{\text{vac}}^{\text{f}}$	$E_{\text{vac}}^{\text{m}}$	$\Omega_{\text{vac}}^{\text{f}}$	$E_{\text{int}}^{\text{f-111}}$	$E_{\text{int}}^{\text{f-110}}$	$E_{\text{int}}^{\text{m-110}}$	$\Omega_{\text{int}}^{\text{f-110}}$
V:								
	MGPT: V6.1[a,b]	2.15	0.36	0.53	4.96	4.21	—	0.34
	Numerical BOP[c]	2.19	—	—	—	—	—	—
	GAP[d]	2.56	0.39	0.55	2.80	3.06	—	—
	DFT: PP[e]	2.48	—	0.52	—	—	—	—
	DFT: FP-LMTO[e]	2.55	—	—	—	—	—	—
	DFT: PLATO[f]	2.51	0.62	—	3.37	3.65	—	—
	DFT: PAW[g]	2.55	0.65	0.53	—	—	—	—
	Experiment[h]	2.1–2.2	0.5	—	—	—	—	—
Ta:								
	MGPT: Ta4[i]	3.08	0.78	0.53	—	6.37	0.50	0.40
	MGPT: Ta6.8x[j,b]	2.95	0.79	0.54	7.31	6.17	—	0.51
	Numerical BOP[c]	3.23	—	—	—	—	—	—
	Analytic BOP[k]	3.08	—	—	5.48	5.98	—	—
	GAP[d]	3.09	0.62	0.53	4.84	5.47	—	—
	DFT: PP[e,i]	3.10	0.90	0.60	—	6.65	0.60	0.20
	DFT: FP-LMTO[e]	3.10	0.74	—	—	—	—	—
	DFT: PLATO[f]	3.14	1.48	—	5.83	6.38	—	—
	DFT: PAW[g]	2.95	0.76	0.56	—	—	—	—
	Experiment[h]	2.8–3.1	0.7	—	—	—	—	—

Continued

Table 9.2 *Continued*

Metal	Method	E^{f}_{vac}	E^{m}_{vac}	Ω^{f}_{vac}	E^{f-111}_{int}	E^{f-110}_{int}	E^{m-110}_{int}	Ω^{f-110}_{int}
Mo:								
	MGPT: Mo2[l]	2.9	1.6	—	—	10.9	0.76	—
	MGPT: Mo5.2[m,b]	3.19	1.15	0.57	—	11.3	—	0.90
	MGPT: Mo12.t0[n,b]	2.90	—	0.56	8.41	8.18	—	0.92
	Numerical BOP[c]	2.88	—	—	—	—	—	—
	Analytic BOP[k]	2.78	—	—	8.77	9.62	—	—
	GAP[d]	2.84	1.28	0.67	7.56	7.61	—	—
	DFT: PP[e]	2.85	—	0.50	—	—	—	—
	DFT: FP-LMTO[e]	2.90	—	—	—	—	—	—
	DFT: PLATO[f]	2.96	1.28	—	7.42	7.58	—	—
	DFT: PAW[g]	2.83	1.16	0.62	7.52	7.63	—	—
	Experiment[h]	3.0–3.2	1.3–1.6	—	—	—	—	—

[a]Moriarty et al. (2008).
[b]Present calculations.
[c]Lin et al. (2014).
[d]Byggmästar et al. (2020).
[e]Söderlind et al. (2000).
[f]Nguyen-Manh et al. (2006); Derlet et al. (2007).
[g]Ma and Dudarev (2019a); Ma and Dudarev (2019b).
[h]Schultz (1991).
[i]Yang et al. (2001); Moriarty et al. (2002a).
[j]Moriarty and Haskins (2014); Haskins et al. (2012); Haskins and Moriarty (2018).
[k]Cák et al. (2014).
[l]Moriarty (1994); Xu and Moriarty (1996).
[m]Moriarty et al. (2002b, 2006).
[n]Moriarty et al. (2012).

It is entirely possible that a noncanonical d-band treatment coupled with the inclusion of additional higher-order five- and six-ion potential contributions (see Chapter 5) could bring MGPT SIA formation energies fully in line with the DFT results.

9.2 Salient elastic and deformation properties of bcc transition metals

We next discuss some important basic ambient- and high-pressure mechanical properties of bcc transition metals that underpin our QBIP-based treatment of dislocations and single-crystal plasticity in Secs. 9.3 and 9.4, and can also be benchmarked from experiment and/or first-principles electronic-structure calculations. These properties include the shear elastic moduli, the ideal shear strength, and generalized stacking-fault energy

(GSFE) or γ surfaces. The pressure-dependent shear elastic moduli of a material establish the detailed character of the elasticity field in which individual dislocations move and interact, as well as serving as fundamental constraints on MGPT and BOP multi-ion interatomic potentials, as we have discussed in Chapter 5. The complementary ideal shear strength provides a fundamental upper-bound limit on material strength in the absence of dislocations. The relevant low-energy γ surfaces for the {110} and {211} slip planes of interest in bcc plastic flow impose general constraints on dislocation character and are also very useful validation tests for QBIPs in treating dislocations and plasticity. In this section and in Sec. 9.3 that follows, unless otherwise indicated, all transition-metal MGPT calculations discussed have been obtained with the Ta4, Mo5.2 and/or V6.1 canonical-d-band multi-ion potentials (see Chapter 5), as reported by Yang et al. (2001, 2010) and Moriarty et al. (2002a, 2002b).

9.2.1 Shear elastic moduli and their pressure dependence

In our treatment of high-pressure mechanical properties, we assume that the bcc solid is subject to a stress tensor of the general form

$$S_{ij} = -P\delta_{ij} + \tau_{ij}, \qquad (9.11)$$

where P is the uniform isotropic pressure in the material and τ_{ij} is a small additional applied deviatoric shear stress. In this regard, we assume that the loading path is such that the material is first uniformly compressed to pressure P through either static (e.g., DAC) or dynamic (e.g., shock) means and then τ_{ij} is applied in some unspecified manner. In bcc transition metals at the modest temperature conditions we will consider, the pressure $P = P(\Omega, T)$ has a strong dependence on the atomic volume Ω of the metal, but a relatively weak dependence on temperature T. Thus, for the purposes of our discussion, it is adequate to replace $P(\Omega, T)$ with the zero-temperature EOS $P_0(\Omega)$. Similarly, for the shear elastic moduli, we will consider only calculated zero-temperature values, as we treated in Chapter 7, together with any additional validating room-temperature experimental data.

Calculated $T = 0$ MGPT high-pressure shear elastic moduli $C'(P)$ and $C_{44}(P)$ are displayed in Fig. 9.1 for Ta to 1000 GPa and Mo to 400 GPa, together with constraining FP-LMTO and experimental data. The latter data was among that used in the development of the Ta4, Ta6.8x and Mo5.2 MGPT potentials, as discussed in Chapter 5. The experimental data includes ultrasonic measurements of the moduli at ambient conditions from Katahara et al. (1979), and in the case of Mo, the high-resolution inelastic x-ray scattering (HRIXS) measurements at 37 GPa of Farber et al. (2006). Corresponding elastic-moduli calculations and data for V to 250 GPa were illustrated in Fig. 6.6(a) of Chapter 6. Observed equilibrium values of the shear moduli for all three metals are listed in Table 9.3. The shear moduli of these metals generally increase with increasing pressure, except for C' in Mo above 300 GPa and C_{44} in V above 20 GPa. The latter behaviors in Mo and V reflect the nearby high-pressure phase transitions discussed in Chapter 6: the predicted bcc to cp phases above 500 GPa in Mo (Söderlind et al., 1994a; Christiansen et al., 1995) and the observed bcc to rhom phase transition at 69 GPa in

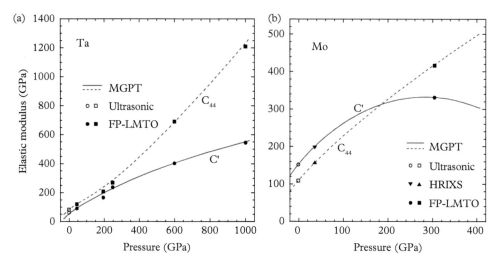

Fig. 9.1 *MGPT calculated high-pressure shear elastic moduli in Ta [left panel (a)] and Mo [right panel (b)], together with constraining DFT FP-LMTO and experimental ultrasonic and HRIXS data discussed in the text.*

V (Ding et al., 2007). In the case of V, note from Fig. 6.6(a) that C_{44} becomes negative and the bcc structure mechanically unstable above 120 GPa.

Also, of special interest here for bcc screw dislocation motion is the effective shear modulus along the <111> slip direction,

$$G_{111} = \frac{2C' + C_{44}}{3}, \qquad (9.12)$$

and the corresponding anisotropy ratio, $A = C_{44}/C'$, where a value of $A = 1.0$ would denote an elastically isotropic solid. Observed equilibrium values of G_{111} and A for Ta, Mo and V are given in Table 9.3, while MGPT calculated results on the pressure dependence of these quantities to 400 GPa are displayed in Fig. 9.2. Note that over the latter pressure range, G_{111} is nearly linear in pressure for the case of Ta, but has a more complex behavior for Mo and V due to the proximity of the noted phase transitions in these metals. The pressure dependence of A over the same range also shows somewhat complex behavior, with a minimum of about 1.2 in the case of Ta near 140 GPa, a substantial increase with pressure from a value of 0.72 at ambient to about 1.6 at 400 GPa in Mo and a very rapid decrease with pressure from a value of 0.78 at ambient to zero near 120 GPa in V.

9.2.2 Ideal shear strength

A related fundamental mechanical property we wish to consider is the ideal shear strength of a bcc transition metal, as defined and first calculated by Paxton et al. (1991) for V, Nb, Cr, Mo and W with the DFT FP-LMTO method. Specifically, these authors identified the ideal strength of the uniformly compressed perfect crystal with the maximum shear stress τ_c required for a continuous homogeneous deformation of

9.2.2 Ideal shear strength 391

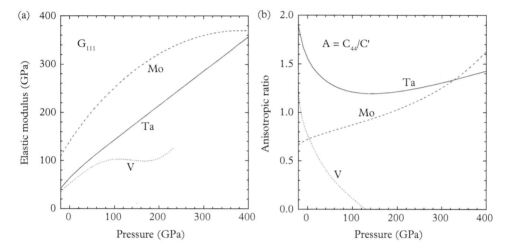

Table 9.3 *Observed elasticity properties of Ta, Mo and V at their respective room-temperature equilibrium volumes, with stress values in GPa, as obtained from the experimental data of Katahara et al. (1979).*

Quantity	Ta	Mo	V
C_{44}	82.5	108.9	43.4
C'	52.5	151.6	55.4
G_{111}	62.5	137.4	51.4
A	1.57	0.72	0.78

Fig. 9.2 *MGPT calculated high-pressure behavior of the <111> shear modulus G_{111} [panel (a)] and the anisotropic ratio A [panel (b)] for Ta, Mo and V.*

the crystal into itself via the observed twinning mode. For bcc metals, this mode can be specified by a shear direction $\eta = [\bar{1}\bar{1}1]$ and a normal plane K = (112). In the absence of tensile relaxation normal to K, which has been shown to be small for bcc transition metals (Paxton et al., 1991; Morris et al., 2000) the atomic positions during the deformation can be directly related to the relative amount of shear x/s along the twinning path, where $s = 1/\sqrt{2}$ is the maximum shear displacement per unit length. In particular, the unrelaxed ideal strength calculation can be carried out entirely using a single atom per unit cell and PBCs, allowing for easy application of full DFT electronic-structure methods, as well as QBIPs. In the case of Ta, self-consistent FP-LMTO calculations of the unrelaxed ideal shear strength so defined were performed at a few selected volumes in the 0–1000 GPa pressure range by Söderlind and Moriarty (1998) and also repeated by Yang et al. (2001) at the observed equilibrium volume Ω_0 for comparison with corresponding MGPT calculations. Yang et al. (2010) later supplemented these results with extensive MGPT and DFT PP calculations as a function of pressure over the range of 0–400 GPa in Ta and Mo and 0–100 GPa in V.

In all cases, one calculates a symmetric energy barrier along the twinning path at constant volume, as given by

$$W(x, \Omega) = \frac{E_{\text{tot}}[x, \Omega] - E_{\text{tot}}[0, \Omega]}{N}, \qquad (9.13)$$

where the barrier height is W_c at $x = s/2$. The corresponding stress along this path is

$$\tau(x, \Omega) = \frac{1}{\Omega} \frac{\partial W(x, \Omega)}{\partial x}. \qquad (9.14)$$

The ideal shear strength is then defined as the maximum calculated stress on the twinning path, $\tau_c \equiv \tau(x_c, \Omega)$, where x_c is the critical shear separating regimes of elastic and plastic deformation of the crystal. In the PP calculations, $W(x, \Omega)$ has been calculated at intervals of $x/s = 0.025$ in the range $0 \leq x \leq s/2$ and the curve extended to $x = s$ by symmetry. The result has then been fitted and differentiated analytically to obtain $\tau(x, \Omega)$ via Eq. (9.14). In the MGPT calculations, $W(x, \Omega)$ has been calculated at smaller intervals of 0.01 over the full range $0 \leq x \leq s$, and then a smooth $\tau(x, \Omega)$ curve has been obtained directly from numerical differentiation. The MGPT and PP results obtained for $W(x, \Omega_0)$ and $\tau(x, \Omega_0)$ in Ta are compared in Fig. 9.3. Calculated MGPT, FP-LMTO and PP values of the barrier height W_c, critical stress τ_c and relative critical shear x_c/s are compared in Table 9.4 for Ta, Mo and V at their observed equilibrium volumes. The overall agreement for these quantities among the three methods is quite reasonable and especially good between the MGPT and FP-LMTO methods.

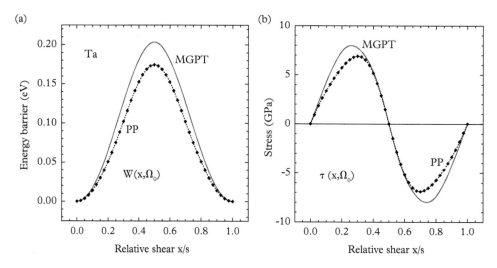

Fig. 9.3 *Ideal shear strength of Ta at its observed equilibrium volume $\Omega_0 = 121.6$ a.u., as calculated by the MGPT and DFT PP methods (Yang et al., 2001, 2010). (a) Symmetric energy barrier $W(x, \Omega_0)$; (b) corresponding shear stress $\tau(x, \Omega_0)$.*

Table 9.4 *Calculated ideal shear strength properties of Ta, Mo and V at their observed equilibrium volumes (121.6 105.1 and 93.23 a.u., respectively), as obtained with the MGPT, FP-LMTO and PP methods.*

Metal	Method	W_c (eV)	x_c/s	τ_c (GPa)
Ta:				
	MGPT: Ta4[a]	0.20	0.26	8.0
	DFT: FP-LMTO[a]	0.18	0.26	6.5
	DFT: PP[b]	0.17	0.30	6.9
Mo:				
	MGPT: Mo5.2[b]	0.46	0.27	21.6
	DFT: FP-LMTO[c]	0.42	0.26	19.2
	DFT: PP[b]	0.34	0.25	16.0
V:				
	MGPT: V6.1[b]	0.15	0.28	7.9
	DFT: FP-LMTO[c]	0.15	0.26	7.3
	DFT: PP[b]	0.09	0.30	5.0

[a]Yang et al. (2001).
[b]Yang et al. (2010).
[c]Paxton et al. (1991).

Under high pressure, the MGPT potentials for Ta, Mo and V also fully capture the qualitative behavior of the ideal strength in these metals, and with about the same level of quantitative agreement with the DFT electronic-structure calculations as at normal density. In Fig. 9.4(a), we compare MGPT, FP-LMTO and PP results for the critical stress τ_c in Ta as a function of pressure to 400 GPa. All three results show an approximate linear dependence of τ_c on pressure, as one expects from the linear variation of G_{111} with pressure for Ta displayed in Fig. 9.2(a). The MGPT-calculated scaling behavior of τ_c/G_{111} with pressure for Ta and Mo to 400 GPa and V to 100 GPa is plotted Fig. 9.4(b). In contrast to the case of Ta, where τ_c/G_{111} remains nearly constant at a value of about 0.12, τ_c/G_{111} for Mo and V displays a noticeable decrease with increasing pressure. In part this reflects the nonlinear variation of G_{111} with pressure for these metals [Fig. 9.2(a)], although in both Mo and V the variation with pressure is clearly somewhat different for τ_c and G_{111}. At the same time, the variations in τ_c/G_{111} from one material to another, as well as under high pressure for a given material, are confined to the small range from 0.12 to 0.16 for all three metals.

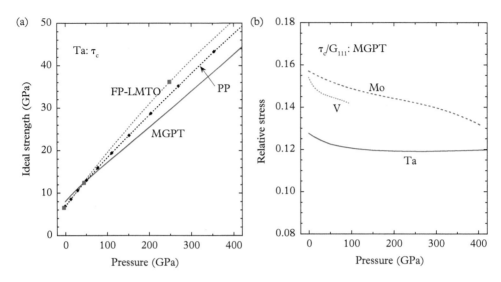

Fig. 9.4 *Calculated ideal shear strength in bcc metals at high pressure. (a) Critical stress τ_c in Ta, as obtained with the MGPT, FP-LMTO and PP methods; (b) relative stress τ_c/G_{111} for Ta, Mo and V, as obtained with the MGPT method.*

From Yang et al. (2010), with publisher (Elsevier) permission.

9.2.3 Generalized stacking-fault energy surfaces

To model bcc screw dislocation behavior accurately, an even more important validation test concerns the GSFE or γ surfaces for the {110} and {211} slip planes. As first defined by Vitek (1974), the γ surface is an energy profile of two semi-infinite blocks of bulk crystal rigidly displaced relative to each other by a vector u in a chosen fault plane, with atomic relaxation allowed only perpendicular to the plane. One can calculate ambient- and high-pressure γ-surface energies at constant atomic volume using an appropriate supercell with PBCs. If desired, this can be done using two fault surfaces per supercell, so that the full translational symmetry of the bulk crystal is preserved. Alternately, one can use one fault surface per triclinic supercell with two constant lattice translation vectors and a variable vector inclined along the displacement direction u. In this way, the number of atoms needed to define the supercell is reduced by half, making first-principles DFT electronic-structure calculations of high-symmetry features of the γ surface much more tractable. As previously done in the case of Ta at ambient pressure (Yang et al., 2001), at high pressure Yang et al. (2010) followed the latter approach and used a supercell consisting of at least 12 atomic planes perpendicular to the fault surface and with one half of the cell shifted by the displacement vector u. In the <111> direction, $u = \alpha b$, where $0 \leq \alpha \leq 1.0$ and b is the Burgers vector. The same approach can be applied to MGPT, FP-LMTO or PP calculations at any pressure.

Complete {110} and {211} γ surfaces calculated for bcc Ta at its equilibrium volume with the Ta4 MGPT potentials by Yang et al. (2001, 2010) are displayed in Fig. 9.5. Supercell size is not a limitation in the MGPT calculations, and larger cells consisting

9.2.3 Generalized stacking-fault energy surfaces

of 32 planes (96 atoms) for the {110} surface and 96 atomic planes (96 atoms) for the {211} surface were used to ensure full convergence. Qualitatively, the calculated γ surfaces display the well-known general features expected for bcc metals (Vitek, 1974). In particular, the {110} surface is fully symmetric, while the {211} surface reveals the well-known twinning–anti-twinning asymmetry along the <111> direction characteristic of bcc materials.

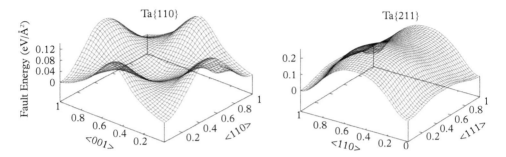

Fig. 9.5 *Calculated γ surfaces for Ta at its equilibrium volume $\Omega = \Omega_0$, as obtained by MGPT Ta4 multi-ion potentials. Left panel: the {110} surface; right panel: the {211} surface. No stable stacking faults in the form of local minima are found on either surface, and all extrema are either maxima or saddle points. These same qualitative features are also maintained in Ta at high pressure to at least 400 GPa, as well as in Mo to 400 GPa and V to 100 GPa.*
From Yang et al. (2001), with publisher (Taylor & Francis) permission and courtesy of L.H. Yang.

In order to validate the MGPT potentials quantitatively for both ambient- and high-pressure dislocation studies, high-symmetry slices of the {110} and {211} γ surfaces along the <111> direction have been calculated in Ta, Mo and V using both the MGPT and PP methods. In particular, such calculations along <111> provide a very sensitive test of the quality of the MGPT potentials because the stacking-fault energies involved are small and similar in magnitude to those encountered in the formation and motion of $a/2$ <111> screw dislocations. Calculated MGPT and PP results at selected pressures for Ta, Mo and V are shown in Figs. 9.6, 9.7 and 9.8, respectively. Here the displacement parameter α can be conveniently written as x/b, where $b = \sqrt{3}a/2$ is the magnitude of the Burgers vector. In the cases of Ta and Mo, the maxima in these curves at $x = b/2$, which is commonly defined as the unstable stacking-fault energy γ_{usf}, increase monotonically with pressure. This is shown more directly in Fig. 9.9(a), where complete MGPT results for $\gamma_{\text{usf}}^{110}$ and $\gamma_{\text{usf}}^{211}$ are plotted as a function of pressure up to 400 GPa. At each of the selected pressures in Fig. 9.6 for Ta and Fig. 9.7 for Mo, the MGPT curves conform closely to the PP points, and the quantitative agreement is within 7%. In this regard, on the {211} γ surface of Mo, the expected twinning–anti-twinning asymmetry is clearly evident in both the MGPT and PP results, and is very accurately predicted by the MGPT potentials. We conclude, therefore, that the Ta4 and Mo5.2 MGPT multi-ion potentials should be reliable for the calculation of the $a/2$ <111> screw dislocation properties in Ta and Mo up to at least 400 GPa.

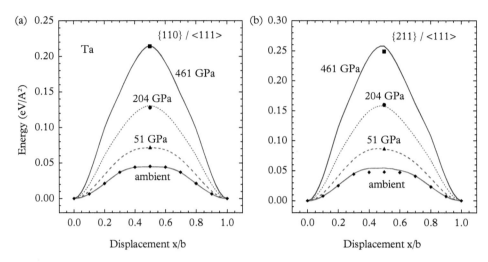

Fig. 9.6 *Calculated slices of the {110} and {211} γ surfaces in Ta at four selected pressures obtained using the MGPT (solid, dashed and dotted lines) and DFT PP (solid points) methods. (a) The {110} surface along the <111> direction; (b) the {211} surface along the <111> direction.*
From Yang et al. (2010), with publisher (Elsevier) permission.

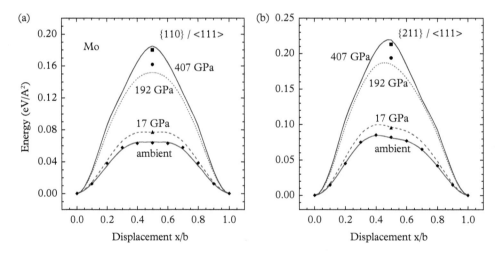

Fig. 9.7 *Calculated slices of the {110} and {211} γ surfaces in Mo at four selected pressures obtained using the MGPT (solid, dashed and dotted lines) and DFT PP (solid points) methods. (a) The {110} surface along the <111> direction; (b) the {211} surface along the <111> direction.*
From Yang et al. (2010), with publisher (Elsevier) permission.

9.2.3 Generalized stacking-fault energy surfaces

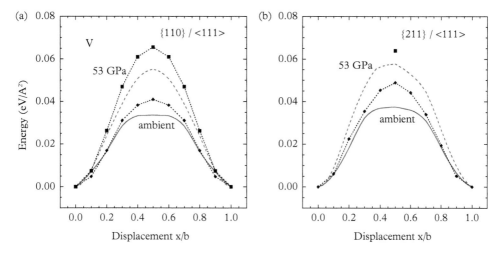

Fig. 9.8 *Calculated slices of the {110} and {211} γ surfaces in V at two selected pressures obtained using the MGPT (solid and dashed lines) and DFT PP (dotted lines and solid points) methods. (a) The {110} surface along the <111> direction; (b) the {211} surface along the <111> direction.*
From Yang et al. (2010), with publisher (Elsevier) permission.

The case of vanadium shows somewhat more complicated behavior. In this metal, the calculated MGPT and PP stacking-fault energies increase monotonically with pressure only up to about 80 GPa such that above that point the γ surface actually lies lower in energy than that at 53 GPa. This is shown more clearly in Fig. 9.9(a), where the complete set of MGPT unstable stacking-fault energies $\gamma_{\text{usf}}^{110}$ and $\gamma_{\text{usf}}^{211}$ for V up to 230 GPa are plotted. In addition, up to at least 53 GPa, the calculated PP stacking-fault energies are systematically higher than those obtained with the MGPT V6.1 potentials by 10–25%, as shown in Fig. 9.8. This makes the MGPT calculation of $a/2 <111>$ dislocation properties in V less certain quantitatively than in either Ta or Mo.

Also note that Fig. 9.9(a) further shows that the {110} unstable stacking-fault energy $\gamma_{\text{usf}}^{110}$ is systematically smaller than the {211} fault energy $\gamma_{\text{usf}}^{211}$ for all pressures up to 400 GPa in Ta and Mo and up to 230 GPa in V. This has important implications for the motion of the $a/2 <111>$ screw dislocations on {110} and {211} slip planes. At all pressures considered here, the screw dislocations in Ta, Mo and V prefer to move on {110} planes for Ta, Mo and V, although, as discussed by Yang et al. (2001), this can happen in more than one way, such that at larger length scales slip may effectively appear to occur on either {110} or {211} planes. Also, the scaled fault energy $\gamma_{\text{usf}}^{110}/(G_{111}b)$ displays scaling properties similar to those found for the scaled ideal strength τ_c/G_{111}, except that the roles of Ta and Mo are reversed, as shown in Fig. 9.9(b). In particular, note that the scaled fault energy is nearly constant as a function of pressure in the case of Mo, but clearly decreases with pressure for Ta and V.

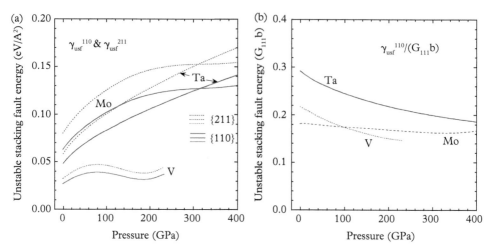

Fig. 9.9 Unstable stacking-fault energies for Ta, Mo and V, as calculated with the MGPT method. (a) Fault energies γ_{usf}^{110} and γ_{usf}^{211} for the {110} and (211) γ surfaces, respectively; (b) scaled stacking-fault energy $\gamma_{usf}^{110}/(G_{111}b)$ for the {110} surface.
From Yang et al. (2010), with publisher (Elsevier) permission.

9.3 Screw dislocation atomic structure and mobility in bcc transition metals

In this section, we consider atomistic simulations of important strength-related properties controlled by the dominant $a/2$ <111> screw dislocations in bcc transition metals. Such simulations performed with QBIPs provide detailed information on the equilibrium core structure, kink-pair formation and activation energies, and the Peierls stress for these dislocations at low temperature, as well phonon-drag-limited dislocation mobility under applied shear stress at higher temperatures. These calculations are carried out at both ambient-pressure conditions and at high pressures up to 1000 GPa. This atomistic information can then be distilled into specific dislocation velocity laws that provide direct multiscale-modeling input into microscale DD simulations of the single-crystal yield stress, simulations that will be considered in Sec. 9.4.

9.3.1 Green's function simulation method for dislocation calculations

Atomistic simulation of individual dislocation structure and motion requires special treatment due to the long-ranged ($\sim 1/r$) elastic field associated with them. Traditionally, fixed boundary conditions (FBCs) have been most often used in such dislocation simulations, where exterior atoms are frozen at their bulk lattice positions and distant atomic positions in the computational cell are established by the conditions of linear anisotropic

9.3.1 Green's function simulation method for dislocation calculations

elasticity. This requires very large simulation cells in practice, but this method is always problematic with respect to force build-up between fixed and relaxed atomic regions. An elegant and practical solution to the latter problem is to use so-called flexible boundary conditions. In particular, Rao et al. (1998) have developed an advanced Green's function version of such boundary conditions for both two-dimensional (2D) and three-dimensional (3D) dislocation simulations, denoted as Green's function boundary conditions (GFBCs). In this method, a buffer layer is introduced between the fixed outer and inner relaxed atomistic regions of the simulation cell, allowing one to update dynamically the boundary conditions of the simulation, while dramatically reducing the size of the atomistic region. Using the GFBC approach, Yang et al. (2001, 2010) developed a specialized Green's function atomistic simulation method to implement multi-ion MGPT potentials and calculate the pressure-dependent properties of $a/2$ <111> screw dislocations in Ta, Mo and V through MS and MD simulations.

In the flexible GFBC method, the simulation cell is divided into three regions, denoted as atomistic, Green's function (GF) and continuum. In the outer continuum region, the atomic positions are initially determined according to the anisotropic elastic displacement field (Stroh, 1958, 1962) for a dislocation line defect at the center of the atomistic region, and then are relaxed by GF methods according to the forces in GF region. Complete atomistic relaxation is performed in the atomistic region according to the interatomic forces generated from the MGPT total-energy functional, Eq. (7.18). Forces developed in the GF region, as relaxation is achieved in the atomistic region, are then used to relax those atoms in *all three* regions by the 2D or 3D elastic and lattice GF solutions for line or point forces. The atomistic and GF relaxations are iterated until all force components on each atom are sufficiently small (10^{-4} eV/Å or less), and the final few steps must also be performed by direct atomistic relaxation for the atomistic and GF regions to ensure there is no force build-up in these two regions.

In the actual MGPT-GFBC simulation code developed by Yang et al. (2001, 2010), a spatial domain decomposition scheme is implemented for all three regions of calculation, as illustrated schematically in Fig. 9.10. The small domain cells defined in this scheme are connected via a cell-linked-list method such that each cell has a fixed number of neighboring cells. This reduces the number of unnecessary interatomic separations considered in evaluating the MPGT potentials, which is crucial to their efficient application. In general, there are three major computational issues that need to be addressed: (1) the geometry of the simulation cell, which is purely cylindrical for a straight dislocation and in the form of a series of displaced cylindrical disks for a kink; (2) the fact that there are three regions in the full simulation cell, so that a connectivity algorithm for information passing between different regions is therefore necessary; and (3) the large effective cutoff radius $R_{cut} = 4.25 R_{WS}$ for the four-ion MGPT potentials, which means there is a large overhead associated with the number of atoms per cell if the conventional domain-cell partition is considered (i.e., if each cell covers a volume R_{cut}^3, then it contains ~16 atoms). To solve these problems, a so-called layered-cake decomposition is used to split the three regions in the full simulation cell, so that each region has its own domain cell-linked list. To reduce the overhead associated with the number of atoms per cell, the cell sizes are reduced by a factor of eight, therefore the average number of atoms per cell is ~2. In addition, this approach allows a better description of the cylindrical geometry involved in the simulation when a cubic domain decomposition is used. As shown by Yang et al. (2001), the performance of the MGPT-GFBC simulation code is thereby increased by an order of magnitude as compared to conventional domain-decomposition methods.

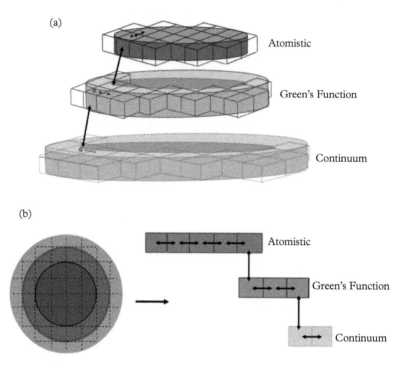

Fig. 9.10 *Schematic representation of the domain decomposition scheme used to implement flexible Green's function boundary conditions in the MGPT-GFBC atomistic simulation code for dislocation calculations. (a) The three main computational regions separated into a layered-cake structure for a cylindrical coordinate system such that each region has its own domain decomposition. (b) To ensure the connectivity between regions and compatibility with parallel computing platforms, the domain cells are mapped into three one-dimensional arrays with cell-linked pointers between the cells and overlap regions.*
From Yang et al. (2001), with publisher (Taylor & Francis) permission and courtesy of L.H. Yang.

To take advantage of the scalable architectures of modern HPC computer platforms, a mapping algorithm was also developed for massively parallel computers in the MGPT-GFBC code. Specifically, a 3D to 1D mapping list is built at the beginning of the simulation. This mapping list ensures the connectivity between different regions, so that no information is lost during the simulation. The logic behind this algorithm is that the MGPT-GFBC simulation is always performed in a 1D computational domain regardless of the physical geometry involved in the calculation. This is particular useful when dealing with complicated geometries such as kinked dislocation structures or dislocation-dislocation interactions.

It should also be mentioned that the GFBC method, as applied to bcc dislocation-core properties, has also been adapted to DFT calculations by Woodward and Rao (2001, 2002). The main operational difference with the above MGPT-GFBC approach is the need for a much smaller atomistic region in Fig. 9.10 to realize an efficient DFT computational scheme. In their applications to Ta and Mo, these authors claim that adequate convergence was obtained with just 63 atoms in the atomistic region, where DFT Hellman-Feynman forces were used to optimize atomic positions, and with the use of only 270 atoms in all three GF regions combined. In principle, the GFBC method could be adapted to BOP dislocation-core calculations as well, but in the initial applications to Mo, W and Fe by Mrovec et al. (2004, 2007, 2011) only FBCs have been used. A third approach to dislocation-core calculations should also be mentioned, namely, the treatment of a chosen 2D periodic array of dislocation dipoles, i.e., two oppositely directed dislocations, instead of the single isolated dislocation treated in the FBC and GFBC methods. This method allows efficient cancellation of the long-ranged elastic field of an isolated dislocation and the use of PBCs. As discussed by Rodney et al. (2017), a quadrupolar array of dipoles is usually the simplest preferred arrangement. Such a treatment was first used in DFT calculations on Ta and Mo by Ismail-Beigi and Arias (2000), with a small periodic cell containing only 90 atoms. A more complex hexagonal array of dipoles was used in the DFT calculations of Frederiksen and Jacobsen (2003) on Mo and Fe, an array that is also discussed by Woodward (2005).

9.3.2 Equilibrium dislocation core structures

It has long been recognized that the mobility of an $a/2 <111>$ screw dislocation in a bcc lattice is severely restricted by the atomic structure of its core. Around a given <111> direction, the bcc structure has three-fold symmetry. Each such <111> zone contains three {110} and three {112} planes that are potential slip planes in the bcc structure, as well as admitting the possibility of a 3D spreading of the core structure along <112> directions on the {110} planes when the screw dislocation is formed. Detailed descriptions of extended core structures in bcc transition metals were obtained from three decades of pioneering atomistic studies, initially with only empirical potentials (Duesbery et al., 1973; Vitek, 1974; Duesbery, 1984a, 1984b; Duesbery and Vitek, 1998), but then later with QBIPs (Xu[1] and Moriarty, 1996, 1998; Rao and Woodard, 2001; Yang et al., 2001; Mrovec et al., 2004) and DFT calculations (Ismail-Beigi and Arias, 2000; Woodward and Rao, 2001, 2002; Frederiksen and Jacobsen, 2003). These and most subsequent QBIP and DFT studies have indicated a relatively compact core with an extension of up to a few Burgers vectors in length and a corresponding high Peierls barrier to the movement of the dislocation under an applied stress. In this section, we discuss comprehensive atomistic simulations of basic $a/2 <111>$ screw dislocation core properties at both ambient and high pressure in bcc Ta, Mo and V, as obtained by Yang et al. (2001, 2010), using refined MGPT potentials and the Green's-function methodology discussed above. Our focus will be on the zero-temperature calculation of core properties that are fundamental to an understanding of dislocation structure and mobility. The properties considered include: the atomic structure of the equilibrium dislocation core in the absence of any additional shear stress; the nature and energetics of isolated kinks and mobile kink pairs that can be formed from this core in the low shear-stress limit; and the magnitude and orientation dependence of the Peierls stress required to move the rigid dislocation in the high shear-stress limit.

The $a/2$ <111> screw dislocation in a bcc lattice has one or more stable core configurations located at the center of gravity of three <111> atomic rows forming a triangular prism. Around these three rows the near-neighbor atoms are located on a helix that winds up in a clockwise or counter-clockwise manner, depending on the location of the elastic center and the sign of the Burgers vector, so that two different types of core configurations can be obtained (Xu and Moriarty, 1996). One configuration is isotropic and of high energy, and may or may not be stable. This is usually referred to as the "hard" core. The other configuration is of low energy and is normally the stable ground-state structure. This latter configuration is the so-called "easy" core. In general, the "easy" core can exhibit three-fold <112>/{110} directional spreading in two geometrically distinct, but energetically equivalent ways, resulting in a doubly degenerate ground-state core structure with D3 symmetry and two possible orientations. Under certain circumstances, however, this directional spreading may vanish and an isotropic nondegenerate core with a higher, but still three-fold C3 symmetry results. It is now widely believed, based on DFT and QBIP studies in the past two decades, that the latter nondegenerate easy core is the actual ground state in most, if not all, of the bcc transition metals at ambient pressure (Rodney et al., 2017). It should also be noted in this regard that one can connect the degenerate and nondegenerate easy cores physically through a continuous core-polarization variable assigned to the degenerate easy core structure. When the core polarization vanishes, the degenerate and nondegenerate core structures are identical, as further discussed below.

Yang et al. (2001, 2010) have studied the pressure dependence of the easy-core ground state of the $a/2$ <111> screw dislocations in bcc Ta, Mo and V, using MGPT-GFBC simulations carried out from somewhat below ambient pressure up to very high pressures. For these simulations, a 2D GFBC approach in cylindrical geometry has been used, with PBCs applied along the z axis in the <111> direction, with a period $b = \sqrt{3}a/2$ at constant atomic volume $\Omega = a^3/2$. In this procedure, an infinite $a/2$ <111> screw dislocation is first introduced by displacing all atoms in the simulation according to anisotropic elasticity, using the sextic formalism of Stroh (1958). The atomic positions of the core atoms are then allowed to relax within the GFBC simulation cell. Radially outward from the cylinder axis, the atomistic, GF and continuum regions of the simulation cell, as depicted in Fig. 9.10, each require a minimum shell thickness of $4.25R_{WS} \cong 2.42b$, the effective cutoff radius for the MGPT potentials. In practice, however, a radius for the atomistic region of about $20b$ is needed to characterize accurately the fully relaxed core structure.

The qualitative aspects of the calculated core structures are most easily displayed and discussed by constructing the standard differential displacement map originally due to

Vitek (1974). In this approach, the <111> screw components of the relative displacement of neighboring atoms due to the dislocation (i.e., the total relative displacement in the z direction less that in the perfect lattice) is represented by an arrow between the two atoms. The calculated screw-component differential displacement maps for Ta at two widely different pressures are shown in Fig. 9.11. The left panel of that figure displays the isotropic (nondegenerate) core structure of tantalum that is calculated near ambient pressure, while the right panel shows the strong directional spreading in the degenerate core structure obtained in the same metal at a pressure of 1000 GPa. Corresponding differential displacement maps can be constructed for the edge components of the dislocation as well, but the magnitude of the edge displacements for Ta were found to be 10–100 times smaller than that of the screw components (Yang et al., 2001), and we do not consider these further here.

Qualitatively, the core structures displayed in Fig. 9.11 are representative of the isotropic and the directionally spread cores that Yang et al. (2001, 2010) have obtained for Ta, Mo and V. The degree of three-fold directional spreading of the degenerate core can be quantified by the core polarization **p**, which measures the simultaneous translation of the three central atoms nearest to the core center (Duesbery et al., 1973; Seeger and Würthrich, 1976). This translation is parallel to the dislocation line but in the opposite sense for the two different degenerate core orientations. By symmetry, the core-polarization magnitude $p = |\mathbf{p}|$ can only vary from zero to $b/6$. At $p = 0$, the two degenerate core configurations coincide and the fully symmetric, isotropic core structure with a higher three-fold symmetry is obtained. At $p = b/6$, on the other hand, a fully polarized degenerate core is obtained with a maximum three-fold spreading along <211> directions. Core polarizations calculated by Yang et al. (2001, 2010) as functions of volume and pressure are plotted in Fig. 9.12 for Ta and Mo to 400 GPa and for V to 75 GPa. In the cases of Ta and Mo near the observed equilibrium density ($\Omega = \Omega_0$) and

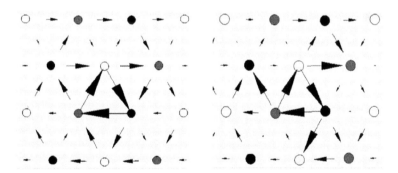

Fig. 9.11 *Differential displacement maps of the relaxed dislocation core structures of Ta at low and high pressure, as calculated from MGPT-GFBC atomistic simulations. Left panel: isotropic (nondegenerate) core structure near ambient pressure; right panel: directionally spread (degenerate) core structure at 1000 GPa.*

Left panel from Yang et al. (2001), with publisher (Taylor & Francis) permission. Courtesy of L.H. Yang.

ambient pressure, as well as at expanded volumes corresponding to negative pressures, the calculated core polarization is less than about 0.01 ($b/6$), which is the expected level of accuracy of these results, and thus consistent with a nondegenerate isotropic core structure. Under compression, however, the core polarization rises rapidly in both cases and attains a value of 0.5 ($b/6$) near pressures of 400 GPa. In contrast to Ta and Mo, the dislocation core of V is already partially polarized at ambient pressure and p rises only to a maximum value of about 0.275 ($b/6$) near 32 GPa and then descends rapidly toward zero. This descent occurs close to the bcc → rhom phase transition observed at 69 GPa (Ding et al., 2007), and also calculated by the same V6.1 MGPT potentials (Moriarty et al., 2008; see Chapter 6). As in Ta and Mo, at sufficiently expanded volumes and negative pressure in V, the polarization p does tend toward zero. In all three metals, the transition between a nondegenerate isotropic core with p = 0 and a degenerate directionally spread core with finite polarization appears to be continuous and not a first-order phase transition.

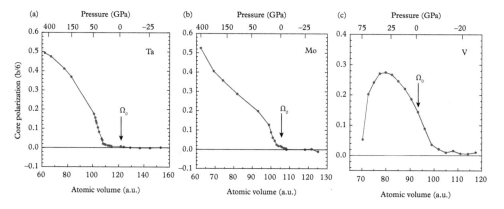

Fig. 9.12 *Volume and pressure dependence of the core polarization for Ta, Mo and V, as obtained from MGPT-GFBC atomistic simulations. (a) Ta calculated with Ta4 potentials; (b) Mo calculated with Mo5.2 potentials; (c) V calculated with V6.1 potentials.*
Panels (a) and (b) from Moriarty et al. (2006), with publisher (Springer) permission.

The above results show clearly that the screw dislocation core structure in bcc transition metals can be a materials-specific property that depends both on the chemical element and on environmental factors such as pressure. The MGPT-GFBC predictions of an isotropic core structure in Ta and Mo near ambient pressure are consistent with published first-principles DFT calculations of the core structure for these metals, obtained using small computational cells and GFBCs or PBCs (Ismail-Beigi and Arias, 2000; Woodward and Rao, 2002; Frederiksen and Jacobsen, 2003; Weinberger et al., 2013; Dezerald et al., 2014), as well as with BOP calculations on Mo using FBCs (Mrovec et al., 2004). The present Yang et al. (2001, 2010) MGPT-GFBC results for the Ta and Mo core structures at ambient pressure also supersede the previously reported results of Moriarty et al. (1999) on Ta and those of Xu and Moriarty (1996, 1998) on Mo. These authors both found, in contrast, a substantial three-fold spreading of the core for each metal, but these findings were based on the preliminary, early-generation MGPT Ta3 and Mo2 potentials, respectively, and FBCs applied at normal density.

It is also of interest to contrast the QBIP and DFT results of the past two decades on core structure with the earlier systematic study of the Group-VB and -VIB metals at normal density by Duesbery and Vitek (1998), obtained using modified radial-force FS potentials (Ackland and Thetford, 1987; Ackland and Vitek, 1990). These authors found isotropic cores for the VB metals, including V and Ta, and three-fold spread cores for the VIB metals, including Mo. They correlated these trends directly with the corresponding FS-calculated behavior of the {110}/<111> γ surfaces, proposing the empirical rule that if

$$\gamma(b/3) > 2\gamma(b/6), \tag{9.15}$$

then an isotropic core structure will be calculated, and conversely, if $\gamma(b/3) < 2\gamma(b/6)$ then a three-fold spread core will be calculated. Subsequently published DFT and BOP results on bcc transition metals at ambient pressure conform to the first half of the rule in that the {110}/<111> γ surface always obeys Eq. (9.15) and only isotropic core structures are calculated. The MGPT and DFT PP {110}/<111> γ surfaces displayed in Figs. 9.6, 9.7 and 9.8 for Ta, Mo and V, respectively, also obey Eq. (9.15) at both ambient pressure and high pressure. However, as we have seen from Fig. 9.12, only near ambient pressure in Ta and Mo are corresponding MGPT isotropic core structures calculated. This makes the remaining unique case of V, with its partially polarized core at ambient pressure, a potentially interesting exception to the trend. In particular, we note that in the DFT dislocation studies of Weinberger et al. (2013) and Dezerald et al. (2014), while ambient-pressure isotropic core structures are indeed obtained for V, as well as for Ta and Mo, the underlying elasticity in V is poorly calculated, with the C_{44} shear constant underestimated by 50% in both studies. In sharp contrast, ambient-pressure elasticity is accurately maintained with the MGPT V6.1 potentials, as previously shown in Fig. 6.7(a) and discussed in Sec. 9.2.1. Further investigation of the impact of elasticity on the core structure of V would seem to be warranted.

9.3.3 Movement under shear stress: kink-pair formation and the Peierls stress

At finite temperature and low applied shear stress in Eq. (9.11), the motion of the screw dislocation in the bcc lattice normally occurs by the thermally assisted formation and migration of kink pairs. For low shear-stress conditions, the individual kinks in a kink pair are well separated and weakly interacting, so kink-pair formation can be modeled by just looking at isolated left and right kink formation. In this limit, the nature and atomic structure of the possible kinks is closely related to the unstressed dislocation core, which for full generality we assume here is of doubly degenerate form, with a fixed core polarization. As we have discussed above, the doubly degenerate core structure of the rigid $a/2$ <111> screw dislocation can have two energetically equivalent configurations with opposite polarizations, here denoted for clarity as positive p and negative n, but each with a core polarization magnitude p. As a result, there are different possible kinks and kink-pair configurations involving p and n segments that can be formed. In addition, p and n segments can co-exist on the same dislocation line in the form of a so-called antiphase defect (APD). This further increases the multiplicity of possible kinks and kink pairs, as has been fully elaborated by Yang et al. (2001) in the case of Ta.

Here we assume that an isolated left (l) or right (r) kink of a kink pair consists of two semi-infinite segments of p or n orientation separated by a kink height h. The symmetry of the bcc lattice allows six distinct and nondegenerate kinks (Duesbery, 1983a, 1983b). These kinks are of character nln (degenerate with plp), nrn (degenerate with prp), nlp, nrp, pln and prn. As previously demonstrated in the case of Ta (Yang et al., 2001), the lowest energy kink pair in the absence of a pre-existing APD has the character pln-nrp. In Ta at ambient pressure and zero applied shear stress, this kink pair was found to have a calculated formation energy E_{kp}^f of 0.96 eV, which is in close agreement with the empirically derived zero-stress activation enthalpy of 1.02 eV used in microscale DD simulations to account for the observed yield stress (Tang et al., 1998). For this reason, the pln-nrp kink pair was adopted as the appropriate model for kink-pair formation in bcc transition metals, and it has been so used for Ta, Mo and V at both ambient and high pressure (Yang et al., 2001, 2010).

To model an isolated pln or nrp kink accurately, one can work at constant volume and set up the GFBC simulation cell in the form of a long compliant cylinder made up of unit disks of width b and radius $20b$ (for the atomistic region) and a total length 60–$80b$ centered on the dislocation line. A transition region of 10–$15b$ is allowed across the kink height h, where the kink is fully relaxed in the MGPT-GFBC simulation. To form a closed 3D cage, the two ends of the cylinder are capped with GF and continuum regions. The z axis of the compliant cylinder is taken parallel to a [111] dislocation line direction, while the y axis can be taken parallel to [1$\bar{1}$0] and the x axis parallel to [$\bar{1}\bar{1}$2]. The smallest repeat translation vector for the rigid screw dislocation core in the bcc lattice is $(a/3)[\bar{1}\bar{1}2]$ on a {110} plane, and this defines the elementary kink height h with magnitude $\sqrt{6}a/3$. Yang et al. (2001, 2010) have considered only kinks formed within this geometry. Kinks formed on other planes such as {211} have significantly larger kink heights and therefore are either unstable or have much larger kink formation energies (Xu and Moriarty, 1996, 1998; Duesbery, 1999).

Yang et al. (2001, 2010) further assumed that the process of kink-pair formation is limited by the isolated kink formation energies E_{pln}^f and E_{nrp}^f, and not the competing process of kink migration, which is controlled by the secondary Peierls stresses needed to move the left- and right-hand kinks. In this regard, Yang et al. (2001) showed in the case of Ta at ambient pressure that the secondary Peierls stresses are one to two orders of magnitude smaller than the corresponding Peierls stress for the rigid screw dislocation itself, so both kinks are expected to be mobile with the left kink moving faster than the right kink. Consequently, one expects the dislocation velocity at low shear stress τ to be controlled by the kink-pair formation energy E_{kp}^f rather than any small kink migration barriers. In the $\tau = 0$ limit, E_{kp}^f is then just obtained as a sum of left- and right-kink formations energies:

$$E_{kp}^f = E_{pln}^f + E_{nrp}^f. \tag{9.16}$$

9.3.3 Movement under shear stress: kink-pair formation and the Peierls stress

The individual left and right kink formation energies are most efficiently calculated by summing the unit disk contributions in the GFBC compliant cylinder across the transition region in the kink. That is, in each unit disk one subtracts from the atomistic total energy with the kink present the corresponding total energy for the perfect straight dislocation. This procedure provides a cancellation of total-energy errors and leads to kink-formation energy values that are accurate to 0.05–0.10 eV or 5–10%. Calculated values of E_{kp}^f at selected pressures in Ta, Mo and V are plotted in Fig. 9.13(a). As expected, E_{kp}^f is monotonically increasing with pressure, but in a significantly nonlinear way in each case. The scaled kink-pair formation energy $E_{kp}^f/(G_{111}b^3)$ displays a somewhat soother and slowly varying pressure dependence in the cases of Ta and Mo, as shown in Fig. 9.13(b). In any case, the calculated values of E_{kp}^f constrain the low shear-stress limit of the full stress dependent activation enthalpy for dislocation motion, which is discussed in Sec. 9.3.4.

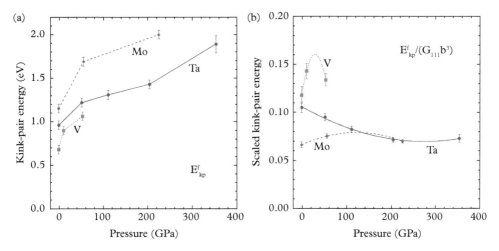

Fig. 9.13 *Pressure dependence of the pln-nrp kink pair formation energy for Ta, Mo and V, as calculated from MGPT-GFBC atomistic simulations. (a) Full kink-pair energy E_{kp}^f; (b) scaled kink-pair energy $E_{kp}^f/(G_{111}b^3)$.*
From Yang et al. (2010), with publisher (Elsevier) permission.

We next turn to the high shear-stress limit and the calculation of the Peierls stress τ_P for the rigid $a/2<111>$ screw dislocation. Bcc metals are known to slip predominantly on {110} and/or {112} planes at low temperatures, but, as discussed by Ito and Vitek (2001), this slip does not follow the simplified Schmid law. In this regard, a rather complex orientation dependence of the slip geometry and the yield stress has been experimentally observed in Ta (Takeuchi et al., 1972). Consequently, one expects that there is a strong dependence of the critical resolved shear stress (CRSS) needed to move the rigid screw dislocation on the orientation of the applied stress. In the context of MGPT-GFBC simulations, this orientation dependence was first investigated by Yang et al. (2001) in bcc Ta at ambient pressure, applying both pure glide shear stresses and selected shears with a uniaxial stress component. For applied stresses on either a {110} plane or in the twinning direction of a {112} plane, it was found that the addition of a uniaxial stress component in either compression or tension always raises the CRSS. For this reason, we confine our attention here to applied shear stresses and the CRSS in our Ta and Mo prototypes at ambient pressure. For the present purposes, we identify the Peierls stress τ_P with the minimum CRSS as a function of shear stress orientation, and then discuss the pressure dependence of τ_P for that orientation.

In order to determine the Peierls stress in a self-consistent and accurate manner, the MGPT-GFBC simulations of the CRSS have been performed at conditions of constant stress, rather than at constant volume, and start from the relaxed equilibrium core structure as determined in Sec. 9.3.2. The simulations again utilize PBCs along the screw axis, so they are strictly 2D zero-temperature calculations. For a given applied stress orientation, the CRSS is assumed to be reached when the dislocation moves at least one lattice spacing on the maximum resolved shear stress plane. In a bcc crystal along a given <111> direction, there are three {110} planes and three {112} planes, mutually intersecting every 30°. Because of the twinning–anti-twinning asymmetry in the bcc lattice, unique values of the CRSS can exist on different planes ranging in orientation from $\chi = -30°$ (twinning orientation on {211}) to $\chi = 30°$ (anti-twinning orientation on {211}), with χ being the angle measured from a given {110} slip plane. The MGPT predictions of the CRSS over this orientation range for Ta and Mo, as calculated at ambient pressure are displayed in Fig. 9.14. Both Ta and Mo reveal a significant twinning–anti-twinning asymmetry, with the minimum CRSS occurring at $\chi = 0°$ for both metals. These latter results represent the defined MGPT values of τ_P at ambient pressure: 0.577 GPa for Ta and 0.860 GPa for Mo. The corresponding result for V is 0.360 GPa.

Also plotted in Fig. 9.14 are experimental estimates of the CRSS based on the observed yield stress at the indicated stress orientations. In the case of Ta at $\chi = -10°$ and Mo at $\chi = -30°$, the calculated MGPT CRSS values are nearly a factor of two greater than the experimental estimates, while in Mo at $\chi = 0°$ the MGPT result is only about 20% higher. However, experiment here does not represent a direct measurement of the CRSS, and consequently, the precise relationship between theory and experiment here remains an open question. At the same time, other quantum-based calculations of the CRSS in Ta and Mo produce generally higher values than MGPT, as is also shown in Fig. 9.14. These results include the small-cell DFT PP-GFBC calculations of Woodward and Rao (2002) and the larger-cell BOP-FBC calculations of Mrovec et al. (2004). The extent to which these differences represent a sensitivity of the calculations to cell size and boundary conditions, as well as to the interatomic forces used, remains a further open question to be investigated.

Yang et al. (2010) have also calculated the pressure dependence of the Peierls stress via MGPT-GFBC simulations of the $\chi = 0°$ CRSS for selected pressures up to 1000 GPa in Ta, up to 400 GPa in Mo and up to 53 GPa in V. The resulting scaled Peierls stress τ_P/G_{111} is plotted for Ta, Mo and V in Fig. 9.15 as a function of pressure, together with the corresponding average values of τ_P/G_{111}, which are 0.0102 in Ta, 0.0059 in Mo and 0.0068 in V. Although there are significant fluctuations from these latter averages

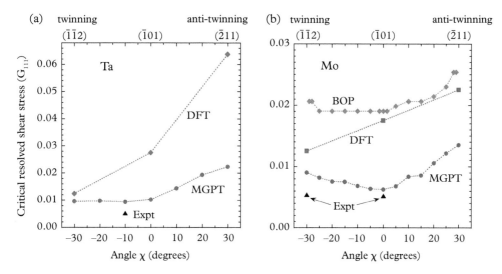

Fig. 9.14 Orientation dependence of the critical resolved shear stress (CRSS) in Ta and Mo at ambient pressure, as obtained from MGPT (Yang et al., 2001; Moriarty et al., 2002b), DFT (Woodward and Rao, 2002) and BOP (Mrovec et al., 2004) calculations, and compared with experimental estimates based on the observed yield stress (Hollang et al., 1997; Suzuki et al., 1999). (a) Ta, with the MGPT results calculated using Ta4 potentials; (b) Mo, with the MGPT results calculated using Mo5.2 potentials.

in the case of Ta and Mo at pressures below 150 GPa, they are representative values over the entire pressure range for all three metals. Also shown in Fig. 9.15 for comparison are the corresponding values of the scaled ideal strength τ_c/G_{111} for Ta, Mo and V, which in each case is more than an order of magnitude larger in value.

9.3.4 Kink-pair activation enthalpy and high-temperature mobility

To enable corresponding DD simulations of single-crystal plasticity, we next consider the calculation of the full activation enthalpy associated with activated screw dislocation motion as a function of applied shear stress, as well as dislocation motion above the Peierls stress in the phonon drag regime.

The pressure and shear-stress dependent activation enthalpy $\Delta H(P, \tau)$ for dislocation motion provides the necessary connection between the $\tau = 0$ kink-pair formation energy E_{kp}^f and the high shear-stress $\tau = \tau_P$ limit where the rigid screw dislocation moves without kink formation. The calculation of $\Delta H(P, \tau)$ requires an atomistic simulation of kink formation under both pressure and shear stress, which is convenient to perform at constant atomic volume rather than constant total stress. To do so, we first consider the thermodynamic enthalpy of N simulation atoms at zero temperature, $H = N(E + P\Omega)$, and manipulate the required enthalpy change into a useful form for constant-volume calculations. In this regard, Hirth (2005) has argued that at high pressure there should be an explicit contribution to the activation enthalpy at constant pressure arising from the "$P\Omega$" term in H, but as we now show, such a contribution drops out in a constant-volume formulation.

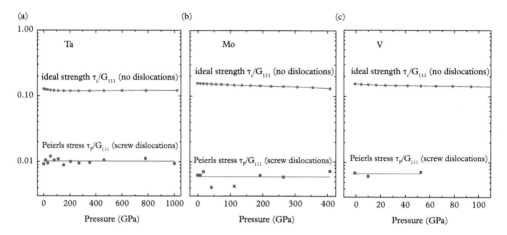

Fig. 9.15 *Pressure dependence of the scaled Peierls stress τ_P/G_{111} in Ta, Mo and V, as calculated from MGPT-GFBC atomistic simulations (solid square points) and also displayed as an average of the calculated points (solid lines). Shown for comparison is the corresponding scaled ideal strength τ_c/G_{111} (solid circles and solid lines).*

We begin in the $\tau = 0$ limit, where the change in enthalpy at $T = 0$ to form either a kinked or unkinked screw dislocation at constant pressure P can be written

$$\Delta H(P) = N[E_2(\Omega_2) - E_1(\Omega_1) + P\Delta\Omega], \quad (9.17)$$

with E the average energy per atom and $\Delta\Omega = \Omega_2 - \Omega_1$. Here the subscript "1" refers to the initial state and the subscript "2" to the final state. While the total volume change $N\Delta\Omega$ may be significant if N is large, the average change in atomic volume $\Delta\Omega$ is small, so that one may perform a Taylor series expansion of the term $E_2(\Omega_2)$ about the volume Ω_1:

$$E_2(\Omega_2) = E_2(\Omega_1) + \frac{dE_2(\Omega_1)}{d\Omega}\Delta\Omega + \cdots$$
$$= E_2(\Omega_1) - P\Delta\Omega + \cdots. \quad (9.18)$$

To obtain the second line of Eq. (9.18), we have re-expanded the derivative term in the first line about Ω_2 and noted that at constant pressure

$$P = -\frac{dE_1(\Omega_1)}{d\Omega} = -\frac{dE_2(\Omega_2)}{d\Omega}. \quad (9.19)$$

9.3.4 Kink-pair activation enthalpy and high-temperature mobility 411

Using Eq. (9.18) in Eq. (9.17), one finds that the pressure terms cancel and one is left with the result, correct to first order in $\Delta\Omega$,

$$\Delta H(P) = N\Delta E(\Omega_1) = N[E_2(\Omega_1) - E_1(\Omega_1)]. \tag{9.20}$$

The leading correction to this result is of the order $(\Delta\Omega)^2$ and negligible.

We can immediately identify $N\Delta E$ in Eq. (9.20) with the kink-pair formation energy at constant volume E_{kp}^f in Eq. (9.16). Following Yang et al. (2001), one can then generalize Eq. (9.20) to the case of two attractively interacting kinks separated by a distance λ and held in (unstable) equilibrium under an applied shear stress τ to obtain the desired total activation enthalpy:

$$\Delta H(P, \tau) = E_{kp}^f(\Omega) + E_{int}(\lambda) - \tau(\lambda)\lambda h b, \tag{9.21}$$

where E_{kp}^f remains the constant-volume kink-pair formation energy at infinite separation and E_{int} is the additional interaction energy at separation λ. In the small shear stress limit $\tau < 0.2\tau_P$, the kink-kink separation λ is larger than the kink width ($\sim 7b$ for Ta) while $E_{int}(\lambda)$ varies as λ^{-1} and $\tau(\lambda)$ varies as $\lambda^{-1.5}$ (Yang et al., 2001), making it possible to evaluate the final two terms in Eq. (9.21).

For larger shear stresses $\tau > 0.2\tau_P$, Yang et al. (2010) additionally developed a special complementary atomistic simulation procedure to evaluate $\Delta H(P, \tau)$. In this procedure, a self-consistent 3D atomistic model of kink-pair formation and migration is constructed involving three new steps in the MGPT-GFBC simulations: First, a straight $a/2$ <111> screw dislocation is constructed and is then fully relaxed under a trial applied shear stress. The straight screw dislocation line is lifted in energy above the valley of the Peierls potential, and the degree of lifting depends on the magnitude of the applied shear stress. Next, a 3D kink-pair model is constructed from this reference configuration. In this construction, the kink separation distance λ is treated as a fixed parameter, which is chosen to approximate the separation distance at which the kink pair is just balanced by the applied stress. Under the constraint of fixed λ, the kink pair configuration is then fully relaxed. The total energy is calculated by summing over the atom-to-atom energy difference between the relaxed 3D configuration and the straight screw dislocation under the same applied shear stress. This produces the sum of the first two terms on the RHS of Eq. (9.21). Finally, the shear stress τ for kink-pair formation at the separation λ and the work done by that stress is calculated using the trapezoid model of Koizumi et al. (1994). This gives an appropriate generalization of the final term in Eq. (9.21). This approach has been successfully applied here to calculate $\Delta H(P, \tau)$ for shear stresses up to $0.9\tau_P$.

Using the above procedure, Yang et al. (2010) have calculated a full kink-pair activation enthalpy curve $\Delta H(P, \tau)$ at a total of four selected pressures in Ta and three pressures each in Mo and V. Representative results for Ta and Mo are displayed in Fig. 9.16, where it is noted that the result for Ta at ambient pressure was first reported in Moriarty et al. (2006). In this figure, we have plotted both individual points obtained from MGPT-GFBC atomistic simulations and smooth analytic fits to the results.

412 *Defects and Mechanical Properties*

The latter analytic fits provide a convenient means to directly input atomistic activation enthalpy data into DD plasticity and yield strength simulations, as is further discussed in Sec. 9.4. The individual simulation points at shear stresses above $0.2\tau_P$ have significant error bars of up to 0.1 eV, but the high stress part of the curve is well constrained by the requirement that $\Delta H(P, \tau)$ vanish at $\tau = \tau_P$, so quite regular fits can be obtained.

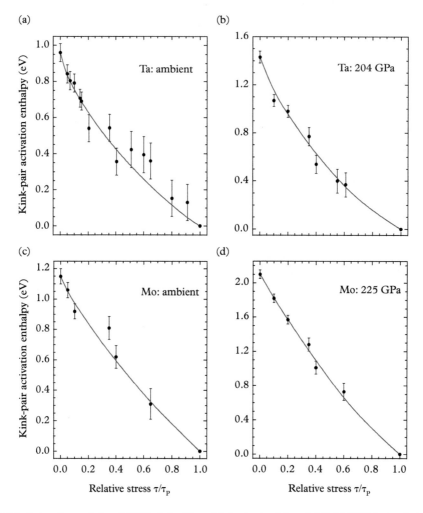

Fig. 9.16 *Activation enthalpy $\Delta H(P, \tau)$ for Ta and Mo obtained from MGPT-GFBC simulations (solid points with error bars) and an analytic fit to these data (solid lines). (a) Ta at ambient pressure; (b) Ta at 204 GPa; (c) Mo at ambient pressure; (d) Mo at 225 GPa.*
Panel (a) from Moriarty et al. (2006), with publisher (Springer) permission.

At finite temperature, a dislocation moves under the influence of thermal fluctuations, and as the temperature rises, the possibility of forming double kinks is increased (Hirth and Lothe, 1982). When the applied shear stress is high and approaches τ_P, multiple kinks begin to be formed and it actually becomes possible to see them appear in dynamic atomistic simulations. To demonstrate this phenomenon in bcc transition metals, Yang et al. (2010) performed a large-scale MGPT-GFBC-MD simulation of $a/2 <111>$ screw dislocation motion at 300 K in Ta, under pure shear loading at a stress level about 10% below the Peierls stress. For larger shear stresses $\tau > \tau_P$, the resistance to dislocation motion comes entirely from thermal vibrations and the screw dislocation velocity $v_s(P, T, \tau)$ becomes linear in the applied stress with a phonon-drag mobility that depends on pressure and temperature. To study dislocation motion above the Peierls stress at a given pressure and temperature, Yang et al. (2010) performed similar large-scale MGPT-GFBC-MD simulations in Ta as a function of applied shear stress. The usual cylindrical simulation cell geometry was used, except that the atomistic region of the cylinder was taken to be appropriately larger with a radius of $50b$ and length of $400b$, so that about 4 million atoms could be simulated in each case. In these simulations, the screw dislocation was initially placed at the center of the simulation box. The simulation cell was then pre-strained at the plastic strain corresponding to the applied shear stress τ, so that the simulation could be run at constant volume rather than constant total stress. The MD was carried out by integrating the equations of motion for the atoms in the atomistic region using a time step of 1 fs at constant temperature, which was maintained using the Nosé-Hoover thermostat discussed in Chapter 8. The displacements of the atoms in the Green's function region of the simulation cell were updated every 10 MD time steps while the atoms in the continuum region were kept fixed.

Using this computational scheme, Yang et al. (2010) focused the MGPT-GFBC-MD simulations on bcc Ta within the applied stress range from 1.05 τ_P to 1.25 τ_P and the pressure range from ambient to 400 GPa. Within these ranges it was found that the phonon-drag mobility was approximately linear in the scaled temperature, $T/T_m(P)$, where $T_m(P)$ is the pressure-dependent melt temperature as determined from prior MGPT calculations on Ta (Moriarty et al., 2002a). As a result of this simplification, at each pressure treated only temperatures of approximately 0.3 $T_m(P)$ and 0.6 $T_m(P)$ were considered in the MD simulations. The actual temperature values used were 900 and 1800 K at ambient pressure, 1545 and 3090 K at 50 GPa, and 2580 and 5200 K at 230 GPa. The results of these simulations for $v_s(P, T, \tau)$ are plotted in Fig. 9.17 together with least-squares linear analytic fits to the simulation data. In these simulations, the screw dislocation was found to glide on a {110} plane at all pressures, temperatures and applied stress levels considered. The analytic fits displayed in Fig. 9.17 show that the velocity data is well represented by a form that is linear in the pressure P as well as in both $T/T_m(P)$ and τ/τ_P. These fits are used in Sec. 9.4 to provide an analytic form of the phonon-drag mobility suitable for DD simulations.

9.4 Multiscale modeling of single-crystal plasticity in bcc transition metals

At the microscale, DD simulations implement the equations of continuum elasticity theory to track the motion and interaction of individual dislocations under an applied stress, leading to the development of a dislocation microstructure and single-crystal plastic deformation. In the present multiscale modeling strategy for bcc transition metals, the primary atomistic input supplied to the DD simulations is the dislocation velocity of individual $a/2$ <111> screw dislocations segments as a function of pressure, temperature and applied shear stress. As a practical matter, all computational DD methods require an analytic representation of dislocation mobility, and in this section, we first consider the specific forms chosen to represent the atomistic results on activation enthalpy and dislocation velocity for $a/2$ <111> screw dislocations that were discussed in Sec. 9.3.4. We then elaborate the atomistically informed DD simulations of yield stress and plasticity for Ta and Mo as a function of pressure, temperature and strain rate carried out by Yang et al. (2010). In this regard, these DD simulations have been partially performed with the baseline lattice-based code developed for bcc metals by Tang et al. (1998, 2005), but more extensively performed with the general node-based ParaDiS (Parallel Dislocation Simulator) code developed at LLNL by Bulatov and co-workers (Bulatov and Cai, 2006; Arsenlis et al., 2007).

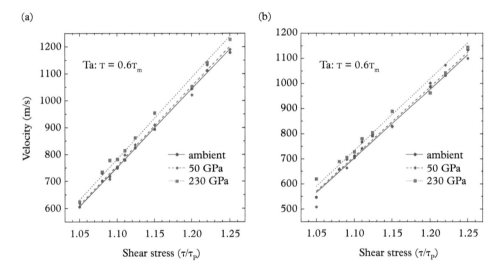

Fig. 9.17 *Screw dislocation velocity $v_s(P, T, \tau)$ above the Peierls stress in the phonon-drag regime for Ta, as calculated by MGPT-GFBC-MD simulations (solid points), with linear analytic fits to the simulation data (solid and dashed lines). (a) $T = 0.3 T_m(P)$; (b) $T = 0.6 T_m(P)$.*
From Yang et al. (2010), with publisher (Elsevier) permission.

As illustrated in Fig. 9.16 for Ta and Mo, the calculated atomistic activation enthalpy data for $a/2$ <111> screw dislocation motion in the thermally activated regime below the Peierls stress τ_P are well represented by the analytic form

$$\Delta H(P, \tau) = \Delta H_0(P) \left[1 - \left(\frac{\tau}{\tau_P} \right)^p \right]^q, \quad (9.22)$$

where the physical parameters

$$\Delta H_0(P) = E_{kp}^f(\Omega)\big|_{P=P_0(\Omega)} \qquad (9.23)$$

and τ_P are both directly calculated quantities, and the additional numerical parameters p and q have been determined by a least-squares fit to the atomistic simulation data for $\Delta H(P, \tau)$. Values of the quantities $\Delta H_0(P)$, τ_P, p and q so determined are listed in Table 9.5 for the ten cases in which full activation enthalpy data for Ta, Mo and V have been calculated by Yang et al. (2001, 2010).

Equation (9.22) is used in both the lattice-based and ParaDiS DD codes, albeit with some modified treatment of the internal parameters $\Delta H_0(P)$, τ_P, p and q in the latter case. In the lattice-based DD code, the parameters in Table 9.5 have been used directly, except for the values of τ_P in Ta, which have been scaled by a factor of 0.5 to account for the apparent overestimate of the Peierls stress relative to experiment noted in Fig. 9.14. In ParaDiS, however, some additional modeling has also been introduced to smooth the pressure dependence of the internal parameters, as discussed by Yang et al. (2010). In both DD codes, the screw dislocation velocity ν_s is calculated as a function of pressure, temperature and applied stress τ in the general form

Table 9.5 *Directly calculated* $(\Delta H_0, \tau_P)$ *and fitted* (p, q) *parameters entering the analytic activation enthalpy function* $\Delta H(P, \tau)$ *for Ta, Mo and V, as given by Eq. (9.22).*

Metal	P (GPa)	ΔH_0 (eV)	p	q	τ_P (GPa)
Ta:					
	ambient	0.96	0.71	1.10	0.577
	51	1.22	0.85	1.34	1.283
	204	1.43	0.81	1.27	2.158
	354	1.89	0.84	1.31	3.139
Mo:					
	ambient	1.15	0.84	1.06	0.860
	55	1.69	0.93	1.14	1.349
	225	2.10	0.97	1.23	2.035
V:					
	ambient	0.68	0.74	1.12	0.360
	9.7	0.90	0.78	1.10	0.369
	53	1.06	0.82	1.14	0.633

$$\nu_s(P, T, \tau) = \nu_0(P, T, \tau) \exp\left[-\frac{\Delta H(P, \tau)}{k_B T}\right], \qquad (9.24)$$

where it is assumed that $\Delta H = 0$ above the thermally activated regime with $\tau > \tau_P$, and where the pre-factor $\nu_0(P, T, \tau)$ is a tailored and method-dependent function, which is discussed in detail by Yang et al. (2010). In the baseline lattice-based DD code, $\nu_0(P, T, \tau) \rightarrow \nu_0(P)$, a pressure-dependent constant for $\tau < \tau_P$, and a single constant ν_0 for $\tau > \tau_P$. In ParaDiS, on the other hand, for $\tau > \tau_P$, the function $\nu_0(P, T, \tau)$ captures the full high-temperature dislocation velocity behavior obtained from the atomistic phonon-drag simulations described in Sec. 9.3.4 and illustrated in Fig. 9.17.

Yang et al. (2010) have discussed a series of DD simulations in Ta and Mo for a range of pressure, temperature and strain-rate conditions, results obtained using both the lattice-based and ParaDiS simulation codes. The simulations all focus on the initial yield behavior rather than later stage strain hardening. The simulated ParaDiS stress-strain curves for Ta are shown in Fig. 9.18 at ambient pressure and at 30 GPa, for temperatures of 300, 600 and 1000 K, at strain rates of 1/s and 1000/s. All of the simulated responses show an initial elastic behavior, followed by the onset of plastic deformation when the flow stress is high enough to move the dislocations. The dislocation densities also go through orders of magnitude increase in these simulations. The corresponding resolved yield stress values obtained are shown in Fig. 19.19(a). The resolved yield stresses clearly show a strong dependence on strain rate, pressure and temperature. Higher strain rates,

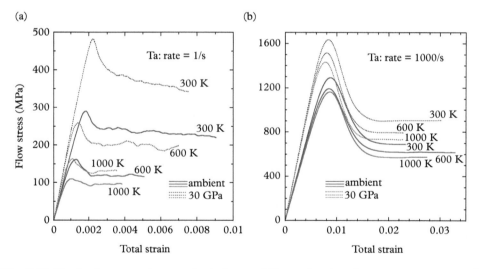

Fig. 9.18 *Simulated stress-strain curves in single-crystal Ta as a function of pressure, temperature and strain rate, as obtained with the ParaDiS DD code. The nearly steady-state flow stress values at the end of the simulations are used to obtain the resolved yield stress values in Fig. 9.19(a). (a) Strain rate of 1/s; (b) strain rate of 1000/s.*
From Yang et al. (2010), with publisher (Elsevier) permission.

higher pressures and lower temperatures all produce higher yield stress values. This is expected for plastic deformation processes dominated by thermally activated dislocation motion. The pressure dependence comes from the fact that both the activation enthalpy and the Peierls stress increase in magnitude at higher pressures. The temperature dependence is direct through the dislocation velocity function, Eq. (9.24). As for the strain-rate dependence, higher stress is required to move the dislocations with a higher deformation rate. The dislocations multiply faster and also move faster at a higher strain rate. The simulations show that the dislocation density at yield for the higher strain rate is about 25 times the density at the lower strain rate. This means the main effect responding to the high strain rate is through the velocity speed up rather than density multiplication alone.

In the case of Ta, calculations of the temperature dependence of the yield stress have also been performed over a wider pressure range using the lattice-based DD code, as earlier reported in Moriarty et al. (2006). These results are plotted in Fig. 9.19(b) and cover pressures as high as 204 GPa. These simulations were performed at a quasistatic strain rate of 10^{-3}/s, which for ambient pressure allows a close comparison with accurate experimental data (Wasserbach, 1986). While the factor of two reduction of the Peierls stress in the activation enthalpy $\Delta H(P, \tau)$ for Ta was motivated in part by a desire to normalize to experiment at one point on the ambient-pressure curve, the full temperature dependence of the experimental data is nonetheless well captured by the simulation.

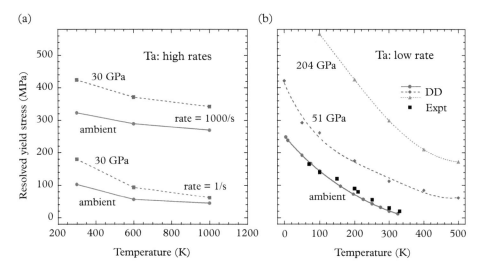

Fig. 9.19 *The resolved yield stress for single-crystal Ta under various pressure, temperature and strain-rate conditions, as calculated from DD simulations. (a) High strain rates of 1/s and 1000/s corresponding to the stress-strain curves in Fig. 9.18 and obtained from ParaDiS simulations; (b) low quasistatic strain rate of 10^{-3}/s, as obtained from lattice-based DD simulations, with the experimental data from Wasserbach (1986).*
Panel (b) from Moriarty et al. (2006), with publisher (Springer) permission.

Using ParaDiS, verification DD simulations were performed at ambient pressure with the same parameters and loading conditions and showed reasonable consistency with the lattice-based DD results.

ParaDiS DD simulations have also been carried out for single-crystal Mo at ambient pressure for temperatures of 300, 600 and 1000 K and at 225 GPa for temperatures of 600 and 1000 K (Yang et al., 2010). These latter simulations were all performed at a strain rate of 1/s and the calculated stress-strain curves are shown in Fig. 9.20(a). An attempt to simulate an additional curve at 225 GPa and 300 K was made, but due to the extremely large activation enthalpy and low temperature at these conditions, the DD time-step required was too small to obtain meaningful results. However, the resolved yield stress for these conditions was estimated in the following manner. Using the standard strain-rate equation $\dot{\varepsilon} = \rho b v_s$ and a constant strain rate of 1/s together with Eq. (9.24) for the screw dislocation velocity v_s, the simulated dislocation density ρ at 225 GPa and 600 K could be used to estimate its value at 300 K. Then combining these data with the flow-stress data in Fig. 9.20(a), one could obtain an estimate of the resolved yield stress at 225 GPa and 300 K. This result, together with the resolved yield stresses obtained directly from the simulated stress-strain curves shown in Fig. 9.20(a), are given in Fig. 9.20(b). Again, one sees a strong dependence on pressure and temperature in the yield-stress results. For comparison, also shown in Fig. 9.20(b) are ambient pressure and temperature experimental data for the resolved yield stress in Mo from the Seeger group in Germany (Hollang et al., 1997; Seeger and Hollang, 2000) and from the Aono group in Japan (Aono et al., 1982). One should note, however, that these data were measured

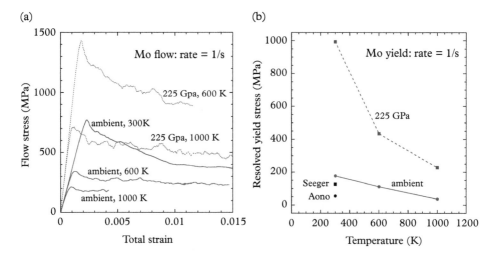

Fig. 9.20 *Simulated stress-strain relations and resolved yield stress values for single-crystal Mo at a strain rate of 1/s, as obtained with the ParaDiS DD code. (a) Stress-strain curves; (b) corresponding resolved yield stress values, with the point at 225 GPa and 300 K estimated only, as described in the text. References to the experimental points labeled Seeger and Aono, data obtained at lower quasistatic strain rates and shown only for comparison, are also given in the text.*

at quasistatic strain rates as opposed to the 1/s strain rate in the DD simulations. Since the resolved shear stress value is expected to decrease with decreasing strain rate, the measured data should be lower in magnitude than the corresponding simulated result, and they are. However, also note that the two experimental data points themselves have a rather large discrepancy. This shows the sensitivity of such experimental measurements to actual sample conditions.

9.5 Grain-boundary atomic structure in bcc transition metals

Understanding the fundamental properties of grain boundaries (GBs) and their interaction with dislocations and point defects are important to the multiscale modeling of strength and failure in polycrystalline materials. The prediction of GB atomic structure also represents an important validation test for QBIPs because this is an example of an extended defect where detailed qualitative and quantitative comparison with experiment is possible. Historically, this test was especially important to MGPT development for first demonstrating the essential nature of angular-force multi-ion potentials to defect structure, in addition to bulk structure, for central transition metals. Experimentally, the important enabling capability in this regard is high-resolution transmission electron microscopy (HRTEM). This technique has allowed accurate measurements at ambient conditions on high-quality bicrystals of symmetric tilt grain boundaries (STGBs), carefully fabricated by ultrahigh-vacuum diffusion bonding in prototype bcc transition metals. In particular, with such bicrystals, pioneering HRTEM measurements have been made on the $\Sigma 5$ (310)/[001] boundary for Nb, Mo and Ta by Campbell et al. (1993, 1999, 2000, 2002). In the initial successful HRTEM experiments on Nb, Campbell et al. (1993) observed a mirror-symmetric GB structure for the $\Sigma 5$ (310)/[001] boundary, with no atomic position shifts across the boundary plane. They further showed in this study, based on MC simulations, that the mirror GB structure was correctly predicted by the first-generation Nb1 MGPT multi-ion potentials (Moriarty, 1990b), but could not be accounted for with empirical radial-force EAM or FS potentials. The calculated MGPT $\Sigma 5$ (310)/[001] mirror GB structure for Nb is illustrated in the left panel of Fig. 9.21.

In subsequent joint experimental and theoretical STGB studies on Mo (Campbell et al., 1999) and Ta (Campbell et al., 2000), similar HRTEM measurements yielded a qualitatively different broken mirror-symmetry $\Sigma 5$ (310)/[001] structure in both metals. As in the case of Nb, the observed structures in Mo and Ta were correctly predicted by MGPT simulations, using the early-generation Mo2 and Ta3 multi-ion potentials, respectively. In both Mo and Ta, the broken-symmetry STGB structure exhibits a [001] displacement of atomic positions across the boundary plane, as shown in the center and right panels of Fig. 9.21. The magnitude of the [001] atomic displacements is material dependent, and in both Mo and Ta is well calculated by the MGPT potentials and in agreement with HRTEM measurements, as shown in Table 9.6.

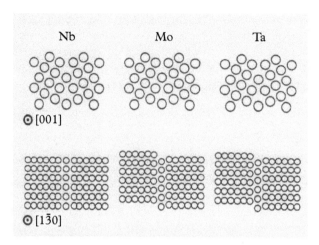

Fig. 9.21 *Atomic structure of the $\Sigma 5$ (310)/[001] grain boundary in the prototype bcc transition metals Nb, Mo and Ta, as obtained from MGPT simulations, using early-generation Nb1, Mo2 and Ta3 multi-ion potentials, respectively. Left panel: Nb; central panel Mo; right panel: Ta.*

The MGPT simulations on the Mo $\Sigma 5$ (310)/[001] STGB structure (Campbell et al., 1999) were performed as follows. First, an ideal mirror-symmetric structure was constructed based on the coincident site lattice (CSL), as defined by Bollmann (1970). To accommodate the full range of the MGPT potentials, a large computational cell of 960 atoms was defined with PBCs applied parallel to the GB plane and, at a sufficient distance, FBCs applied normal to the GB plane. The GB structure was then fully relaxed using MGPT-MD simulations together with a standard simulated annealing technique (Allen and Tidesley, 1987). The convergence of the results was confirmed by doubling the size of the computational cell to 1920 atoms with no change in the calculated GB structure. Later, it was discovered that the MGPT GB results on Mo could also be reproduced with the minimal 20-atom periodic cell defined by Ochs et al. (2000a) in their DFT studies of the $\Sigma 5$ (310)/[001] STGB structure for Nb, Mo, Ta and W, where the use of a small computational cell was essential to facilitate the self-consistent LDA electronic-structure calculations employed in these studies. The same 20-atom periodic cell was then also used in the MGPT-MD simulations of the Ta $\Sigma 5$ (310)/[001] STGB structure (Campbell et al., 2000). While the MGPT calculations correctly predicted a broken mirror-symmetry GB structure for Ta, the DFT-LDA Ta calculations predicted a mirror-symmetric structure instead, as indicated in Table 9.6.

In the cases of Nb and Mo, a comparative study of MGPT, BOP and DFT calculations on the $\Sigma 5$ (310)/[001] STGB was also done by Ochs et al. (2000b). The calculations were all performed with the smaller 20-atom periodic GB cell and included the predicted ground-state structure, higher-energy metastable structures and corresponding grain-boundary energies. The central results obtained from this study are included in Table 9.6. The BOP GB calculations were based on the early-generation

9.5 Grain-boundary atomic structure in bcc transition metals   421

Table 9.6 *MGPT, BOP and DFT calculations of ground-state (GS) and higher-energy metastable-state (HEMS) properties of Σ5 (310)/[001] STGBs in bcc Nb, Mo and Ta. Here "Mirror Yes" denotes a mirror-symmetric GB structure, while "[001]" denotes (mirror-breaking) atomic displacements in the [001] direction in units of the bulk bcc equilibrium lattice constant a_0. The corresponding GB energy E_{GB} is given in units of mJ/m^2.*

Metal	Method	Mirror	GS [001]	E_{GB}	Mirror	HEMS [001]	E_{GB}
Nb:							
	MGPT: Nb1[a,b,c]	Yes	0.0	1159	—	—	—
	Numerical BOP[d,c]	Yes	0.0	1396	—	—	—
	DFT: MBPP[e]	No	0.08	1288	Yes	0.0	1296
	Experiment: HRTEM[b]	Yes	0.0	—	—	—	—
Mo:							
	MGPT: Mo2[f,g,c]	No	0.19	1904	Yes	0.0	2124
	Numerical BOP[d,c]	No	0.22	1274	Yes	0.0	1365
	DFT: MBPP[e]	No	0.20	1702	Yes	0.0	1808
	Experiment: HRTEM[g]	No	0.25±0.05	—	—	—	—
Ta:							
	MGPT: Ta3[h,i,c]	No	0.17	—	—	—	—
	DFT: MBPP[e]	Yes	0.0	1544	—	—	—
	Experiment: HRTEM[i]	No	0.17±0.05	—	—	—	—

[a]Moriarty (1990b).
[b]Campbell et al. (1993).
[c]Ochs et al. (2000b).
[d]Mrovec et al. (2000).
[e]Ochs et al. (2000a).
[f]Moriarty (1994).
[g]Campbell et al. (1999).
[h]Moriarty et al. (1999).
[i]Campbell et al. (2000).

Nb and Mo numerical BOP potentials of Mrovec et al. (2000), while the DFT calculations were those from the LDA treatment of Ochs et al. (2000a). For Nb, the BOP calculations produced a mirror-symmetric GB structure, in agreement with MGPT and experiment. The DFT calculations, on the other hand, yielded a broken mirror-symmetry GB structure as the Nb ground state, but, at the same time, also produced a metastable mirror-symmetric structure that was only slightly higher in energy. For Mo, the BOP and DFT calculations both predicted broken mirror-symmetry structures as the ground state, in good agreement with MGPT and experiment; all three calculations also yielded a higher-energy metastable mirror-symmetric structure. However, while the calculated MGPT and DFT grain-boundary energies for Mo were found to be in reasonable 10–15% agreement, the corresponding BOP GB energies were calculated to be 25–35% lower in value, as shown in Table 9.6.

9.6 Defect properties in fcc transition metals

In the final section of this chapter, we turn briefly to the late-series fcc transition metals, whose physical properties, including basic point- and line-defect energies, are impacted by the strong sp-d hybridization of the nearly filled d bands with the NFE sp valence bands in these materials. In particular, we focus on the prototype case of Ni, for which the construction of corresponding MGPT transition-metal potentials at the effective pair-potential level was elaborated in Chapter 5. Here we consider calculations of a few important defect energies in nickel near ambient pressure conditions with the early-generation Ni2 MGPT potentials that were discussed in Sec. 5.2.4 and are illustrated in Fig. 5.22.

In sharp contrast to bcc metals, an important distinguishing characteristic feature of fcc and hcp metals is the existence of stable stacking faults in the observed sequence of close-packed planes. A useful theoretical tool with which to investigate such stacking faults is the GSFE or γ surface construction that we considered in Sec. 9.2.3 for the {110} and {211} slip planes of bcc metals. For fcc metals, the interesting γ surface to consider is that for the close-packed {111} plane. Yang (2009) has calculated this surface in the case of nickel near ambient pressure, both with the Ni2 MGPT potentials and with the first-principles DFT PP method for direct comparison. In Fig. 9.22 we plot the corresponding MGPT and DFT profiles of the {111} γ surface for Ni as a function of relative displacement d/d_0 along the <112> symmetry direction. The agreement between the MGPT and DFT calculations is clearly excellent, and the continuous MGPT curve reveals the two most important features of the γ surface. First, the maximum GSFE occurs at $d/d_0 = 1/3$, and in analogy with the bcc case, is often referred to as the unstable stacking-fault energy γ_{usf}. Second, the local minimum GSFE at $d/d_0 = 2/3$ is the observed intrinsic stacking-fault energy, which is usually denoted as γ_{isf}.

Beginning in the 1950s, experimental measurement of γ_{isf} in Ni has been attempted by many researchers using a variety of different techniques, but its value is still subject to a large quantitative uncertainty, as indicated in Fig. 9.22, as well as in Table 9.7. The experimental value of γ_{isf} most often cited in the literature is 125 mJ/m^2, and due to Carter and Holmes (1977), but note that this value is at the lower end of the experimental range plotted in Fig. 9.22. On the other hand, the calculated MGPT and DFT values of γ_{isf} are substantially higher in magnitude at 213 and 195 mJ/m^2, respectively. These latter values are more in the middle of the experimental range and are also supported by an earlier DFT calculation by Crampin et al. (1990), as indicated in Table 9.7. From the MGPT perspective, the large theoretical value of γ_{isf} in Ni reflects both the substantial contribution of sp-d hybridization to the result and the long-range nature of the effective pair potentials. In sharp contrast, early attempts to predict γ_{isf} with short-range

9.6 Defect properties in fcc transition metals

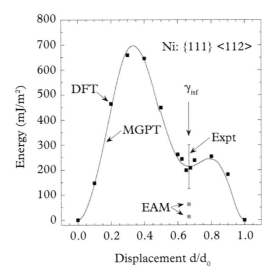

Fig. 9.22 Calculated MGPT and DFT generalized stacking-fault energies in the {111} plane along the <112> symmetry direction for fcc Ni near ambient pressure (Yang, 2009). Captured along this path is the observed intrinsic stacking fault at a relative displacement of $d/d_0 = 2/3$, and whose energy γ_{isf} is compared with both experiment and historical EAM calculations, as discussed in the text.

Table 9.7 Calculated MGPT and DFT defect properties for fcc Ni near ambient pressure conditions, including the relaxed vacancy formation energy E_{vac}^f, the intrinsic stacking-fault energy γ_{isf} and the $\Sigma 5$ (210) STGB energy E_{GB}.

Method	E_{vac}^f (eV)	γ_{isf} (mJ/m^2)	E_{GB} (mJ/m^2)
MGPT: Ni2[a]	1.79	213	1390
DFT: PP[a]	1.95	195	1090
DFT: FP-LMTO[b]	1.77	—	—
DFT: LKKR[c]	—	180	—
DFT: PAW[d]	—	—	1190
Experiment	1.79±0.05[e]	125[f], 79–415[g]	—

[a] Yang (2009).
[b] Korhonen et al. (1995).
[c] Crampin et al. (1990).
[d] Yamaguchi et al. (2005).
[e] Ehrhart (1991).
[f] Carter and Holmes (1977).
[g] Post-1970 range of γ_{isf} values from references cited in Carter and Holmes (1977).

empirical EAM potentials by Voter and Chen (1987) and Oh and Johnson (1988) produced very low values of 62 and 13 mJ/m^2, respectively, values that are also shown in Fig. 9.22. Partly as a consequence, subsequent empirical EAM potentials developed for Ni have been constrained to fit to the measured Carter and Holmes (1977) value of γ_{isf} (Zimmerman et al., 2000), thus still placing them at the lower end of the experimental range in Fig. 9.22.

Yang (2009) also calculated corresponding MGPT and DFT PP results for the relaxed vacancy formation energy $E_{\mathrm{vac}}^{\mathrm{f}}$ and the Σ5 (210) STGB energy E_{GB} in Ni. These results are listed in Table 9.7 together with additional available DFT calculations of the same quantities. In the case of $E_{\mathrm{vac}}^{\mathrm{f}}$ for Ni, the MGPT and DFT PP calculations were carried out only at a constant volume $\Omega = \Omega_0 = 73.82$ a.u. in Eq. (9.1), as opposed to constant pressure, so as to match the earlier DFT FP-LMTO treatment by Korhonen et al. (1995), but thus forgoing a calculation of $\Omega_{\mathrm{vac}}^{\mathrm{f}}$. There is good agreement found among MGPT, DFT and experiment for $E_{\mathrm{vac}}^{\mathrm{f}}$. The MGPT and DFT PP calculations of the Σ5 (210) STGB energy E_{GB} were performed at a slightly expanded volume of $\Omega = 76.55$ a.u., to match the DFT PAW calculation of Yamaguchi et al. (2005). There is similar good agreement among MGPT and DFT values of E_{GB}.

10
Alloys and Intermetallic Compounds

In this chapter, we consider extension of the QBIP principles and applications developed in the preceding chapters for bulk elemental metals to the much larger domain of alloys and intermetallic compounds. We will primarily concentrate on binary AB systems, but further extension to important ternary and quaternary systems, including high-entropy alloys (HEAs), will also be discussed. For treatment with volume-dependent QBIPs, such as those obtained from the GPT or MGPT, it is assumed that each alloy component is metallic in nature and belongs to one of two categories. In category 1 are nontransition metals, to be treated in the SM, EDB or FDB limits of the GPT, and with fixed values of sp valence and d-state occupation. In category 2 are transition metals, to be treated in the PFDB limit of either the GPT or MGPT, as well as by corresponding BOPs, and with either self-consistent or chosen values of sp valence and d-state occupation. Such multi-component potentials are applicable to both static and dynamic simulations on a wide variety of materials problems for alloys and intermetallic compounds, but our primary focus here will be on the basic issues of cohesion and structure. To date, $3d$ transition-metal aluminides have been the main prototype systems so investigated in this regard, including the development of first-principles GPT potentials across the series (Moriarty and Widom, 1997), together with significant applications to their complex binary and ternary phase diagrams (Widom and Moriarty, 1998; Widom et al., 2000), as well as a predicted quasicrystal structure (Mihalkovic et al., 2002). BOP investigations of $3d$ transition-metal aluminides have focused on potential development for TiAl compounds (Znam et al., 2003), with applications to dislocation core structure and mobility (Katzarov et al., 2007; Katzarov and Paxton, 2009). The still developing path forward to pure transition-metal alloys, via the simplified and efficient MGPT, will also be discussed.

10.1 General constrains with composition as an independent environmental variable

We begin by considering a general AB binary alloy or intermetallic compound described by volume-dependent QBIPs, with concentrations $c_A = N_A/N$ of A metal atoms and $c_B = N_B/N$ of B metal atoms. One can immediately reduce the composition dependence of the electron density and the total energy of this system to the single variable

$$x \equiv c_B = 1 - c_A . \tag{10.1}$$

From general considerations, one can expect that both the volume term and interatomic potentials in the elemental-metal total energy, Eq. (7.18), will become concentration dependent, as well as volume dependent, in the binary system. Thus, for example,

$$E_{\text{vol}}(\Omega) \to E_{\text{vol}}^{AB}(\Omega, x) \tag{10.2}$$

for binary alloys and intermetallic compounds, with Ω and x as independent variables. The necessary multiplicity of interatomic potentials appearing in the total energy of the binary system must also be accommodated, such that v_2 is replaced by three independent two-ion pair potentials,

$$v_2 \to v_2^{AA}, v_2^{AB}, v_2^{BB} ; \tag{10.3}$$

v_3 is replaced by four independent three-ion triplet potentials,

$$v_3 \to v_3^{AAA}, v_3^{AAB}, v_3^{ABB}, v_3^{BBB} ; \tag{10.4}$$

v_4 is replaced by five independent four-ion quadruplet potentials,

$$v_4 \to v_4^{AAAA}, v_4^{AAAB}, v_4^{AABB}, v_4^{ABBB}, v_4^{BBBB} ; \tag{10.5}$$

and so on. Each n-ion binary potential $v_n^{\alpha\beta\cdots}$ for component $\alpha\beta\cdots$ is assumed to be volume and concentration dependent in the form $v_n^{\alpha\beta\cdots}(ij\cdots, \Omega, x)$, so that the elemental-metal total-energy functional $E_{\text{tot}}(\mathbf{R}, \Omega)$, as given by Eq. (7.18), is generalized to

$$E_{\text{tot}}^{AB}(\mathbf{R}, \Omega, x) = N E_{\text{vol}}^{AB}(\Omega, x) + \frac{1}{2} \sum_{\alpha,\beta=A,B} \sum_{i,j}{}' v_2^{\alpha\beta}(ij, \Omega, x) + \frac{1}{6} \sum_{\alpha,\beta,\gamma=A,B} \sum_{i,j,k}{}' v_3^{\alpha\beta\gamma}(ijk, \Omega, x)$$

$$+ \frac{1}{24} \sum_{\alpha,\beta,\gamma,\delta=A,B} \sum_{i,j,k,l}{}' v_4^{\alpha\beta\gamma\delta}(ijkl, \Omega, x) + \cdots \tag{10.6}$$

for an AB binary alloy or intermetallic compound.

In addition, self-consistent electron-density constraints must be satisfied for AB binary intermetallics, constraints that directly link the sp valences Z_A and Z_B with the respective d-band occupation numbers Z_d^A and Z_d^B as a function of both volume and concentration. For chosen values of the average atomic volume Ω and concentration

x, the zero-order sp electron density of the binary system is Z/Ω, where Z is now the concentration-weighted average

$$Z = c_A Z_A + c_B Z_B = (1-x)Z_A + xZ_B. \tag{10.7}$$

It is also useful to define individual atomic volumes Ω_A and Ω_B such that

$$\Omega = c_A \Omega_A + c_B \Omega_B = (1-x)\Omega_A + x\Omega_B \tag{10.8}$$

and

$$Z/\Omega = Z_A/\Omega_A = Z_B/\Omega_B. \tag{10.9}$$

Then, in analogy with the elemental GPT, the parameter pairs Z_A, Ω_A and Z_B, Ω_B define zero-order pseudoatoms for the A and B components, respectively. These pseudoatoms are self-consistent A- and B-metal ions that have been properly neutralized by the common uniform electron-gas density Z/Ω and correctly establish shifted inner-core and d-state energy levels and corresponding basis wavefunctions that are needed in the full GPT formalism. Physically, one may think of the A- and B-metal pseudoatoms as being expanded or contracted from their elemental bulk sizes to ensure that Eq. (10.9) is satisfied. For category-1 nontransition-metal binary systems, where both Z_A and Z_B are fixed known quantities, it is a simple matter to use Eqs. (10.7) and (10.9) to calculate Z, Ω_A and Ω_B for any chosen values of Ω and x.

In transition-metal systems, on the other hand, where Z_A and/or Z_B are also volume and concentration dependent, the situation is more complex. With the transition-metal aluminides in mind, consider the case where component A is a category-1 nontransition metal with fixed Z_A and component B is a category-2 transition metal with variable Z_B. For such a system, one must simultaneously satisfy the equilibrium conditions between the sp valence Z_B and the d-electron occupation number Z_d^B for the B-metal component, which for the GPT is given from Eq. (2.169) of Chapter 2 in terms of the $\ell = 2$ phase shift δ_2^B of the B-metal pseudoatom potential:

$$Z_d^B = \frac{10}{\pi}\delta_2^B(\varepsilon_F), \tag{10.10}$$

with ε_F the free-electron Fermi energy of the binary system

$$\varepsilon_F = \frac{\hbar^2}{2m}\left(\frac{3\pi^2 Z}{\Omega}\right)^{2/3}. \tag{10.11}$$

In addition, one has the basic constraint

$$Z_B + Z_d^B = Z_a^B - Z_c^B, \tag{10.12}$$

where Z_a^B is the atomic number and Z_c^B is the number of inner-core electrons of the B-metal pseudoatom. Because the phase shift δ_2^B and the Fermi energy ε_F both depend on the intermetallic environment, Z_B and Z_d^B will be shifted away from their bulk elemental-metal values. In this regard, Eqs. (10.7), (10.9) and (10.10)–(10.12) represent six equations in six unknowns and must be iterated numerically, via the A- and B-metal GPT pseudoatoms, to achieve a self-consistent solution for Z, ε_F, Z_B, Z_d^B, Ω_A and Ω_B. An efficient strategy to accomplish this was proposed by Moriarty and Widom (1997) and is discussed in Sec. A2.8 of Appendix A2, together with technical details on how the GPT pseudoatom calculation is modified in the case of binary intermetallics. The primary modification concerns the common location of the zero of energy at the valence-band minimum, which is controlled through the constant V_0'' introduced in Chapters 3 and 4 for elemental metals, and a quantity that becomes concentration dependent in the binary system.

Figure 10.1 illustrates the calculated changes in value of Z_B and Ω_B for the 3d TM aluminides in going from the bulk transition-metal $x = 1$ limit to the aluminum-rich dilute-alloy $x = 0$ limit. These results have been obtained from self-consistent GPT pseudoatom calculations at the observed equilibrium volumes of the 3d transition metals for $x = 1$ and of aluminum for $x = 0$, using a TM canonical d-band boundary condition of $D_2^* = -3$, as is done for all TM-Al calculations discussed in this chapter. Quantitative values of Z_B and Ω_B are listed in Table I of Moriarty and Widom (1997).

As we previously discussed in connection with Fig. 5.3(a) for bulk transition metals, one has $Z_B \approx 1.4 - 1.6$ across the entire 3d series of elements. Bulk Al, on the other hand, with $Z_A = 3$ has a considerably higher sp valence electron density than any of

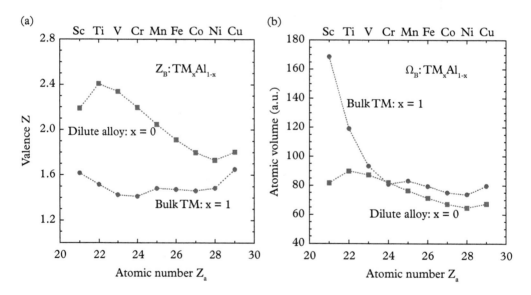

Fig. 10.1 *Transition-metal valence Z_B [panel (a)] and atomic volume Ω_B [panel (b)] for the 3d transition-metal aluminides TM_xAl_{1-x}, as calculated from self-consistent GPT pseudoatoms for the $x = 1$ and $x = 0$ concentration limits.*
From Moriarty and Widom (1997), with publisher (APS) permission.

the transition metals. Thus, when a transition metal is added to aluminum, one expects Z_B to rise in value and/or Ω_B to decrease in value in order to create a higher electron density. Inspection of Fig. 10.1 shows that both of these changes actually occur in the dilute $x = 0$ alloy, with the biggest quantitative impact on the LHS of the $3d$ series, where $Z_B > 2$ for Sc through Mn, and Ω_B is dramatically reduced for Sc and Ti. At the same time, one should note that the general increase in Z_B for TM_xAl_{1-x} alloys and compounds is balanced by a corresponding decrease in Z_d^B, which in turn is provided by the TM d bands rising in energy, broadening and partially unfilling in the alloy.

While the volume Ω and the concentration x are independent variables, there is usually a high correlation between the observed *equilibrium* volume $\Omega = \Omega_0$ and x in specific alloys and compounds. Approximate relations linking Ω_0 and x are often useful for preliminary calculations or in cases where data on particular phases of a material do not exist. In the case of the transition-metal aluminides, such a relationship can be easily derived from the above equations for aluminum-rich materials where x is small. Making simple Taylor expansions of $\Omega_A(x)$ and $\Omega_B(x)$ in Eq. (10.8), noting that $d\Omega_A/dx \ll \Omega_A$ and then using Eq. (10.9) in the $x = 0$ limit, one has to first order in x

$$\Omega_0 = [1 - (1 - Z_B^0/3)x]\Omega_0^{Al}, \qquad (10.13)$$

where Z_B^0 is Z_B evaluated at $x = 0$ and $\Omega_0^{Al} = 112.0$ a.u., the observed equilibrium volume of aluminum. Apart from the replacement of Z_B with Z_B^0, Eq. (10.13) is equivalent to the condition of constant electron density Z/Ω, which was suggested by Phillips et al. (1994) as an appropriate condition for treating the aluminum-rich phases of Co_xAl_{1-x}. Equation (10.13) can be applied more generally, however, and is useful in the range $x < 0.5$, as we demonstrate in Sec. 10.3.

10.2 Nontransition-metal binary alloys and compounds

The treatment of category-1 binary alloys and compounds as simple metals at the pair-potential level in the context of second-order PP perturbation theory has a substantial early history, which is well documented in the book by Hafner (1987a). This book also serves as a good introduction to the basic physics issues confronted in binary metallic systems, including cohesion, ordering and structure in solid, liquid and amorphous glass phases, and temperature-composition phase diagrams. Hafner and co-workers' specific treatments of binary intermetallics involving nontransition metals are all based on the original optimized nonlocal PP prescription of Harrison (1966) for NFE simple metals discussed in Chapter 3. This work includes, for example, calculations on Li-Mg alloys and alkali Laves-phase compounds (Hafner, 1976, 1977), as well as Ca-Mg, Ca-Al and Mg-Zn metallic glasses and liquid alloys (Hafner, 1980, 1983, 1988; Hafner et al., 1987b), with Ca and Zn treated in SM limit. A few model-potential applications to binary alloys using the DRT and RMP methods also appeared in this era, including Beauchamp et al. (1975) on Li-Mg alloys and Lam et al. (1980, 1981) on dilute Al alloys with Li, Na, K, Be, Mg, Ca and Zn. In these applications, Ca and Zn were treated with RMPs to allow for the incorporation of resonant d-state effects, as described in Chapter 4, while the remaining elements were treated with the DRT simple-metal potentials, whose parameters are listed in Table 3.1.

The GPT treatment of nontransition elemental metals in the SM, EDB or FDB limits, as described in Chapters 3 and 4, is readily extended to binary alloys and compounds. The total-energy functional given by Eq. (10.6) is then normally truncated at the pair-potential level:

$$E_{\text{tot}}^{\text{AB}}(\mathbf{R}, \Omega, x) = NE_{\text{vol}}^{\text{AB}}(\Omega, x) + \frac{1}{2} \sum_{\alpha,\beta=\text{A,B}} \sum_{i,j}{}' v_2^{\alpha\beta}(ij, \Omega, x). \qquad (10.14)$$

From Eqs. (3.58) and (3.59) for elemental metals in the GPT SM limit, one immediately derives the corresponding binary pair-potential

$$v_2^{\alpha\beta}(r, \Omega, x) = \frac{Z_\alpha^* Z_\beta^* e^2}{r} \left[1 - \frac{2}{\pi} \int_0^\infty F_N^{\alpha\beta}(q, \Omega, x) \frac{\sin(qr)}{q} dq \right], \qquad (10.15)$$

where the normalized energy-wavenumber characteristic is

$$F_N^{\alpha\beta}(q, \Omega, x) = \frac{q^2 \Omega^2}{4\pi Z_\alpha^* Z_\beta^* e^2} \left[-\frac{4}{(2\pi)^3} \int \frac{w^\alpha(\mathbf{k}, \mathbf{q}) w^\beta(\mathbf{k}, \mathbf{q})}{\varepsilon_\mathbf{k} - \varepsilon_{\mathbf{k+q}}} \Theta^<(\mathbf{k}) d\mathbf{k} \right.$$
$$\left. + \frac{4\pi e^2}{q^2} \left\{ [1 - G(q)] n_{\text{scr}}^\alpha(q) n_{\text{scr}}^\beta(q) + G(q) n_{\text{oh}}^\alpha(q) n_{\text{oh}}^\beta(q) \right\} \right] \qquad (10.16)$$

with

$$w^\alpha(\mathbf{k}, \mathbf{q}) \equiv \langle \mathbf{k} + \mathbf{q} | w^\alpha | \mathbf{k} \rangle. \qquad (10.17)$$

The quantities w^α, n_{scr}^α and n_{oh}^α for $\alpha = $ A or B are constructed from the wavefunctions and energies of the appropriate A or B metal pseudoatoms, with self-consistent screening performed at the alloy electron density Z/Ω, as given by Eqs. (10.7)–(10.9), and with $n_{\text{scr}}^\alpha(q)$ and $n_{\text{oh}}^\alpha(q)$ obtained from Eqs. (3.65) and (3.39), respectively.

The corresponding volume term $E_{\text{vol}}^{\text{AB}}$ in Eq. (10.14) can be readily expressed as a series of terms that are explicitly linear or quadratic in the concentration variables c_A and c_B:

$$E_{\text{vol}}^{\text{AB}} = c_\text{A} E_1^\text{A} + c_\text{B} E_1^\text{B} + c_\text{A}^2 E_2^{\text{AA}} + c_\text{A} c_\text{B} E_2^{\text{AB}} + c_\text{B}^2 E_2^{\text{BB}}$$
$$= (1-x) E_1^\text{A} + x E_1^\text{B} + (1-x)^2 E_2^{\text{AA}} + (1-x) x F_2^{\text{AB}} + x^2 F_2^{\text{BB}}, \qquad (10.18)$$

where all five energies E_1^A, E_1^B, E_2^{AA}, E_2^{AB} and E_2^{BB} retain a volume dependence on Ω and an additional implicit concentration dependence on x. In the SM limit, one finds

$$E_1^\alpha = E_{\text{fe}}^{0,\alpha} + \frac{2\Omega}{(2\pi)^3} \int w_{\text{pa}}^\alpha(\mathbf{k}) \Theta^<(\mathbf{k}) d\mathbf{k} + \delta E_{\text{oh}}^\alpha$$

$$+ \frac{1}{2}(Z_\alpha^\star e)^2 \left[\frac{1.8}{R_{\text{WS}}^\alpha} - \frac{2}{\pi} \int_0^\infty F_N^{\alpha\alpha}(q, \Omega, x) dq \right] - E_{\text{bind}}^{\text{atom},\alpha}, \quad (10.19)$$

$$E_2^{\alpha\alpha} = \frac{2\Omega}{(2\pi)^3} \int w_{\text{pa}}^\alpha(\mathbf{k}) p_c^\alpha(\mathbf{k}) \Theta^<(\mathbf{k}) d\mathbf{k} + \frac{\pi(Z_\alpha^\star e)^2}{\Omega} \frac{\partial^2 F_N^{\alpha\alpha}(0, \Omega, x)}{\partial q^2}, \quad (10.20)$$

for $\alpha =$ A or B, and

$$E_2^{AB} = \frac{2\Omega}{(2\pi)^3} \int [w_{\text{pa}}^A(\mathbf{k}) p_c^B(\mathbf{k}) + p_c^A(\mathbf{k}) w_{\text{pa}}^B(\mathbf{k})] \Theta^<(\mathbf{k}) d\mathbf{k} + (Z_A^\star Z_B^\star e^2) \frac{2\pi}{\Omega} \frac{\partial^2 F_N^{AB}(0, \Omega, x)}{\partial q^2}. \quad (10.21)$$

Equations (10.18)–(10.21) represent the generalization of Eq. (3.60) for an elemental simple metal to a binary alloy or compound of simple metals A and B. Here one has

$$E_{\text{fe}}^{0,\alpha} \equiv \frac{3}{5} Z_\alpha \varepsilon_F + Z_\alpha \varepsilon_{\text{xc}}(n_{\text{unif}}) - \frac{3}{5}(Z_\alpha e)^2 / R_{\text{WS}}^\alpha + Z_\alpha V_0', \quad (10.22)$$

with R_{WS}^α the Wigner–Seitz radius defined by atomic volume $\Omega_\alpha = 4\pi(R_{\text{WS}}^\alpha)^3/3$, in place of Eq. (3.37). Also, the quantity p_c^α is the inner-core projection operator for the atoms of type α, such that $p_c^\alpha(\mathbf{k}) \equiv \langle \mathbf{k} | p_c^\alpha | \mathbf{k} \rangle$ and the corresponding effective valence $Z_\alpha^\star$ is

$$Z_\alpha^\star = Z_\alpha + \frac{2\Omega}{(2\pi)^3} \int p_c^\alpha(\mathbf{k}) \Theta^<(\mathbf{k}) d\mathbf{k} \quad (10.23)$$

in place of Eq. (3.38), while

$$w_{\text{pa}}^\alpha(\mathbf{k}) \equiv \langle \mathbf{k} | w_{\text{pa}}^\alpha | \mathbf{k} \rangle = -\frac{3}{10}(Z_\alpha e)^2 / R_{\text{WS}}^\alpha + \langle \mathbf{k} | w_{\text{core}}^\alpha | \mathbf{k} \rangle - V_0' \quad (10.24)$$

in place of Eq. (3.35). As was done in deriving Eq. (3.95) for the volume term $E_{\text{vol}}(\Omega)$ of an elemental simple metal, the second derivatives of $F_N^{\alpha\beta}$ in Eqs. (10.20) and (10.21) can also be evaluated analytically for the optimum AHS form of the PP.

To illustrate a challenging contemporary application of the above GPT binary simple-metal formalism, we consider the Mg-Al alloy system, and more specifically the ordered compound $Mg_{17}Al_{12}$, which is a material of current scientific and technological interest, with at least five DFT studies in recent years on its structural, thermodynamic and elastic properties (Wang et al., 2008; Zhang et al., 2010, Hu et al., 2015; Zhuang et al., 2016;

Zhou et al., 2019). Illustrated in Fig. 10.2 are calculated first-principles Mg-Mg, Mg-Al and Al-Al GPT potentials in the SM limit both for the dilute $Mg_{1-x}Al_x$ alloy and for $Mg_{17}Al_{12}$. These results show an expected strong dependence of the potentials on the global environmental factors of alloy concentration x and volume Ω, resulting from the large mismatch in the equilibrium electron densities of elemental Mg and Al. At the assumed conditions of $x = 0$ and $\Omega = \Omega_A = 156.8$ a.u. in Fig. 10.2(a), the equilibrium conditions for elemental Mg, one has $\Omega_B = 235.2$ a.u for Al, which represents more than a two-fold expansion of the metal. At the conditions $x = 0.4138$ and $\Omega = 136.2$ a.u. used in Fig. 10.2(b) for the observed equilibrium conditions of $Mg_{17}Al_{12}$, one has $\Omega_A = 112.8$ a.u. for Mg, representing a 28% *compression* of the metal, while $\Omega_B = 169.3$ a.u. for Al, a 51% *expansion* of that metal.

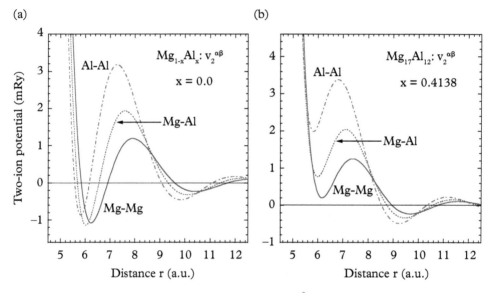

Fig. 10.2 *First-principles GPT binary alloy pair potentials $v_2^{\alpha\beta}$ calculated in the SM limit with IU xc for dilute $Mg_{1-x}Al_x$ and $Mg_{17}Al_{12}$. (a) Mg-Mg, Mg-Al and Al-Al potentials at $x = 0$ and $\Omega = 156.8$ a.u.; (b) Mg-Mg, Mg-Al and Al-Al potentials at $x = 0.4138$ and $\Omega = 136.2$ a.u., correponding to the observed ordered compound $Mg_{17}Al_{12}$.*

For comparison with the DFT calculations, as well as with experiment, we have used the $Mg_{17}Al_{12}$ GPT potentials displayed in Fig. 10.2(b) in calculations of the relaxed crystal structure, the corresponding cohesive energy $E_{\text{coh}}^{\text{AB}} = E_{\text{tot}}^{\text{AB}}/N$, and the formation enthalpy or heat of formation $\Delta H = E_{\text{coh}}^{\text{AB}} - E_{\text{coh}}^0$, where E_{coh}^0 is the concentration weighted average of the Mg and Al elemental-metal cohesive energies:

$$E_{\text{coh}}^0 \equiv c_A E_{\text{coh}}^{\text{Mg}} + c_B E_{\text{coh}}^{\text{Al}} = (1-x)E_{\text{coh}}^{\text{Mg}} + xE_{\text{coh}}^{\text{Al}}. \qquad (10.25)$$

The present calculated GPT results for these quantities are compared with DFT and experiment in Table 10.1.

Table 10.1 Comparison of first-principles GPT and DFT calculations of the structure, cohesive energy and heat of formation for $Mg_{17}Al_{12}$. The Wycoff structural parameters ($\alpha_1^{Mg2}, \alpha_1^{Mg3}, \alpha_2^{Mg3}, \alpha_1^{Al}, \alpha_2^{Al}$) are given in units of the cubic lattice constant a, while the energies E_{coh}^{AB}, E_{coh}^0 and ΔH are given in eV/atom.

Method	α_1^{Mg2}	α_1^{Mg3}	α_2^{Mg3}	α_1^{Al}	α_2^{Al}	E_{coh}^{AB}	E_{coh}^0	ΔH
GPT[a]	0.3297	0.3618	0.0495	0.0932	0.2834	−2.266	−2.488	0.222
						−2.266	−2.288	0.022
DFT: PAW[b]	0.3275	0.3567	0.0406	0.0904	0.2760	−2.208	−2.160	−0.048
DFT: PP[c]	—	—	—	—	—	−2.398	−2.377	−0.021
DFT: PP[d]	—	—	—	—	—	−2.385	−2.332	−0.053
Experiment[e]	0.3240	0.3582	0.0393	0.0954	0.2725	—	−2.288[f]	−0.039[g]–0.010[h]

[a]Present calculations in the SM limit, with IU xc.
[b]Wang et al. (2008). An apparent misprint in the value given for E_{coh}^{AB} has been corrected.
[c]Hu et al. (2015).
[d]Zhou et al. (2019).
[e]Zhang and Kelly (2005).
[f]Kittel (1976).
[g]Brown and Pratt (1970).
[h]Predel and Hulse (1978).

The observed $Mg_{17}Al_{12}$ compound has a complex cubic α-Mn crystal structure, with bcc primitive lattice vectors, but 29 basis atoms distributed across four inequivalent atomic sites, three of which are occupied by Mg atoms and denoted as Mg1, Mg2 and Mg3 sites. The $Mg_{17}Al_{12}$ structure is usually displayed in a larger 58-atom cubic unit cell of lattice constant a (e.g., see Fig. 1 of Zhou et al., 2019), where the Mg1 sites are coincident with the bcc positions defined by the primitive lattice vectors. The remaining Mg2, Mg3 and Al sites within the unit cell are specified by five variable Wycoff structural parameters, α_1^{Mg2}, α_1^{Mg3}, α_2^{Mg3}, α_1^{Al}, α_2^{Al}, whose magnitudes depend on the interatomic forces present and are only determined by fully relaxing the crystal structure. In all of the DFT calculations, the relaxation has been carried out at zero-temperature and -pressure conditions, $T = P_0 = 0$, while in the GPT calculations at $T = 0$ and constant volume at the observed room-temperature lattice constant $a_{300} = 10.54$ Å $= 19.92$ a.u. and corresponding atomic volume $\Omega = 136.2$ a.u. As shown in Table 10.1, the present GPT calculations and the DFT calculations of Wang et al. (2008) on the variable Wycoff structural parameters of $Mg_{17}Al_{12}$ are both in generally good agreement with the experimental measurements of Zhang and Kelley (2005). The calculated GPT and DFT values of the cohesive energy E_{coh}^{AB} for $Mg_{17}Al_{12}$ are also in reasonable accord and in the relatively narrow range from -2.2 to -2.4 eV, while the corresponding reference energies E_{coh}^0 are in the wider range -2.1 to -2.5 eV. In the case of the GPT results, the same overarching qualitative features that we noted for elemental metals in Chapter 6 still apply for binary systems. Namely, at constant volume, structure is totally determined by the interatomic potentials, while the cohesive energy is dominated by the volume term E_{vol}^{AB}. For $Mg_{17}Al_{12}$ we find $E_{vol}^{AB} = -2.391$ eV, while $E_{struc}^{AB} = 0.125$ eV.

Finally, there is the demanding calculation of the heat of formation ΔH for a binary compound, which in practice involves obtaining this small energy from the subtraction of the two much larger energies $E_{\text{coh}}^{\text{AB}}$ and E_{coh}^{0}, together with the requirement that one must have $\Delta H < 0$ if the compound is to be stable. As shown in Table 10.1, and confirmed in similar calculations by Zhang et al. (2010) and Zhuang et al. (2016), DFT predictions of ΔH for $Mg_{17}Al_{12}$ are all in the approximate range from -0.02 to -0.05 eV. This range overlaps the experimental range of -0.01 to -0.04 eV established by the two existing, but substantially differing, measurements of ΔH cited in Table 10.1. At the same time, it should be noted that the success of DFT in this regard owes much to an apparent good cancelation of errors between $E_{\text{coh}}^{\text{AB}}$ and E_{coh}^{0}, since it can also be seen from Table 10.1 that E_{coh}^{0} is calculated with an error of 0.05 eV or greater in each case. Unfortunately, such a favorable cancelation of errors does not occur in the SM GPT treatment of ΔH. The relatively poor calculated value of ΔH in the GPT case is largely due to a 9% overestimate in the magnitude of E_{coh}^{0}, amounting to 0.2 eV. As shown in Table 10.1, replacing the calculated E_{coh}^{0} with its experimental value in the GPT calculation of ΔH brings the result down to a reasonable magnitude of 0.022 eV, although still with the wrong sign. In this regard, the GPT cohesive energy $E_{\text{coh}}^{\text{AB}}$ appears to be well converged with respect to the sharp pair-potential cutoff of v_2^{AA}, v_2^{AB} and v_2^{BB} at $8.25R_{\text{WS}}$ that was used in this calculation. At the same time, however, the GPT value of $E_{\text{coh}}^{\text{AB}}$ has not been minimized with respect to volume here, as has been done in the DFT calculations.

In principle, the GPT calculation of the heat of formation for $Mg_{17}Al_{12}$ can also be improved by treating the Al component in the EDB limit of the GPT rather than in the SM limit, as we have discussed for elemental Al in Chapter 3. Although such a calculation has yet to be attempted for $Mg_{17}Al_{12}$, the necessary generalizations of the SM binary-alloy formalism given by Eqs. (10.15)–(10.24) can be readily derived from the EDB elemental-metal formalism described in Chapter 4.

More generally, we can elaborate the path forward to the case where metal A is treated in the SM limit, but metal B is treated in either the EDB or FDB limit. The total-energy functional $E_{\text{tot}}^{\text{AB}}$ remains of the general form expressed by Eq. (10.14) with the potentials v_2^{AA} and v_2^{AB} also still of the form (10.15). The v_2^{BB} potential is generalized to

$$v_2^{\text{BB}}(r,\Omega,x) = \frac{(Z_B^* e)^2}{r}\left[1 - \frac{2}{\pi}\int_0^\infty F_N^{\text{BB}}(q,\Omega,x)\frac{\sin(qr)}{q}dq\right] + v_{\text{ol}}^{\text{BB}}(r,\Omega,x), \quad (10.26)$$

where the additional d-state overlap potential $v_{\text{ol}}^{\text{BB}}$ is given by Eq. (4.55), noting that in the binary system the d-state basis states and matrix elements are implicitly impacted by the concentration variable x. With $\alpha = \beta = A$, the normalized energy-wavenumber characteristic F_N^{AA} is given by Eq. (10.16), while F_N^{AB} and F_N^{BB} are now both directly impacted by sp-d hybridization. With $w_0^B(\mathbf{k},\mathbf{q})$ and $h_1^B(\mathbf{k},\mathbf{q})$ defined by Eqs. (4.33) and (4.34), respectively, one has

10.2 Nontransition-metal binary alloys and compounds

$$F_N^{AB}(q, \Omega, x) = \frac{q^2\Omega^2}{4\pi Z_A^\star Z_B^\star e^2}\left[-\frac{4}{(2\pi)^3}\left\{\int \frac{w^A(\mathbf{k,q})w_0^B(\mathbf{k,q})}{\varepsilon_\mathbf{k}-\varepsilon_\mathbf{k+q}}\Theta^<(\mathbf{k})d\mathbf{k}\right.\right.$$
$$\left.+\int\frac{w^A(\mathbf{k,q})h_1^B(\mathbf{k,q})}{\varepsilon_\mathbf{k}-\varepsilon_\mathbf{k+q}}\Theta_>^<(\mathbf{k})d\mathbf{k}\right\}+\frac{4\pi e^2}{q^2}\left\{[1-G(q)]n_{\text{scr}}^A(q)n_{\text{scr}}^B(q)\right.$$
$$\left.\left.+G(q)n_{\text{oh}}^A(q)n_{\text{oh}}^B(q)\right\}\right] \tag{10.27}$$

and

$$F_N^{BB}(q,\Omega,x) = \frac{q^2\Omega^2}{4\pi(Z_B^\star e)^2}\left[-\frac{4}{(2\pi)^3}\left\{\int\frac{[w_0^B(\mathbf{k,q})]^2}{\varepsilon_\mathbf{k}-\varepsilon_\mathbf{k+q}}\Theta^<(\mathbf{k})d\mathbf{k}\right.\right.$$
$$\left.+\int\frac{2w_0^B(\mathbf{k,q})h_1^B(\mathbf{k,q})}{\varepsilon_\mathbf{k}-\varepsilon_\mathbf{k+q}}\Theta_>^<(\mathbf{k})d\mathbf{k}+\int\frac{[h_1^B(\mathbf{k,q})]^2}{\varepsilon_\mathbf{k}-\varepsilon_\mathbf{k+q}}\Theta_>^<(\mathbf{k})d\mathbf{k}\right\}$$
$$\left.+\frac{4\pi e^2}{q^2}\left\{[1-G(q)][n_{\text{scr}}^B(q)]^2+G(q)[n_{\text{oh}}^B(q)]^2\right\}\right], \tag{10.28}$$

where $n_{\text{scr}}^B(q)$ and $n_{\text{oh}}^B(q)$ are to be calculated from Eqs. (4.32) and (4.39), respectively.

The volume term E_{vol}^{AB} still has the general form expressed by Eq. (10.18), with E_1^A given by Eq. (10.19) and E_2^{AA} given by Eq. (10.20). The remaining energy components of E_{vol}^{AB} are generalized as follows:

$$E_1^B = E_{\text{fe}}^{0,B} + \frac{2\Omega}{(2\pi)^3}\left[\int w_0^{\text{pa},B}(\mathbf{k})\Theta^<(\mathbf{k})d\mathbf{k}+\int h_1^B(\mathbf{k})\Theta_>^<(\mathbf{k})d\mathbf{k}\right]+\delta E_{\text{oh}}^B$$
$$+\frac{1}{2}(Z_B^\star e)^2\left[\frac{1.8}{R_{\text{WS}}^B}-\frac{2}{\pi}\int_0^\infty F_N^{BB}(q,\Omega,x)dq\right]-E_{\text{bind}}^{\text{atom, B}}, \tag{10.29}$$

$$E_2^{BB} = \frac{2\Omega}{(2\pi)^3}\left\{\int w_0^{\text{pa},B}(\mathbf{k})p^B(\mathbf{k})\Theta^<(\mathbf{k})d\mathbf{k}+\int h_1^B(\mathbf{k})\left[p^B(\mathbf{k})-\frac{w_0^{\text{pa},B}(\mathbf{k})}{\varepsilon_\mathbf{k}-E_d^{\text{vol}}}\right.\right.$$
$$\left.\left.-\frac{h_1^B(\mathbf{k})}{\varepsilon_\mathbf{k}-E_d^{\text{vol}}}\right]\Theta_>^<(\mathbf{k})d\mathbf{k}\right\}+\frac{\pi(Z_B^\star e)^2}{\Omega}\frac{\partial^2 F_N^{BB}(0,\Omega,x)}{\partial q^2} \tag{10.30}$$

and

$$E_2^{AB} = \frac{2\Omega}{(2\pi)^3}\left[\int[w_{\text{pa}}^A(\mathbf{k})p^B(\mathbf{k})+p_c^A(\mathbf{k})w_0^{\text{pa},B}(\mathbf{k})]\Theta^<(\mathbf{k})d\mathbf{k}\right.$$
$$\left.+\int p_c^A(\mathbf{k})h_1^B(\mathbf{k})\Theta_>^<(\mathbf{k})d\mathbf{k}\right]+(Z_A^\star Z_B^\star e^2)\frac{2\pi}{\Omega}\frac{\partial^2 F_N^{AB}(0,\Omega,x)}{\partial q^2}. \tag{10.31}$$

In these equations, $w_0^{pa,B}(\mathbf{k})$ is given by Eq. (4.52); $h_1^B(\mathbf{k}) \equiv h_1^B(\mathbf{k}, 0)$; $p^B(\mathbf{k}) \equiv \langle \mathbf{k}| p^B |\mathbf{k}\rangle$, with p^B being the inner-core plus d-state projection operator on atom B; and Z_B^* is given by Eq. (4.37). Also, as above in the SM binary-alloy case, the second derivatives of F_N^{BB} in Eq. (10.30) and F_N^{AB} in Eq. (10.31) can be evaluated analytically. In addition to possible applications of this formalism to Mg-Al alloys and compounds, with Al treated in the EDB limit, another interesting prototype application would be to Mg-Zn binary systems, with Zn treated in the FDB limit. In systems like Ca-Al and Ca-Mg, on the other hand, Ca is under strong compression and exhibits transition-metal behavior, which can be treated in the PFDB limit with the additional SM-TM binary alloy formalism discussed in the next section.

10.3 Transition-metal aluminides and their phase diagrams

In this section, we move on to the important case of the transition-metal aluminides, as first treated in the context of the GPT by Moriarty and Widom (1997) across the $3d$ series of elements. For AB binary alloys and compounds of the form TM_xAl_{1-x}, with the metal-A atom as Al treated in the SM limit through two-ion interactions and metal-B atom as a TM element treated in the PFDB limit up to and including four four-ion interactions, the general total-energy functional expressed by Eq. (10.6) reduces to

$$E_{tot}^{AB}(\mathbf{R}, \Omega, x) = NE_{vol}^{AB}(\Omega, x) + \frac{1}{2} \sum_{\alpha,\beta=A,B} \sum_{i,j}{}' v_2^{\alpha\beta}(ij, \Omega, x) + \frac{1}{6} \sum_{i,j,k}{}' v_3^{BBB}(ijk, \Omega, x)$$

$$+ \frac{1}{24} \sum_{i,j,k,l}{}' v_4^{BBBB}(ijkl, \Omega, x). \qquad (10.32)$$

While this total-energy functional can be applied across the entire Al concentration range in TM_xAl_{1-x} alloys and compounds, the special leverage of the GPT treatment given by Eq. (10.32) comes in the Al-rich regime of $x < 0.5$. In this concentration range, the average interatomic spacing between transition-metal B atoms becomes larger and, as a result, the contributions of the multi-ion potentials v_3^{BBB} and v_4^{BBBB} in Eq. (10.32) diminish and can sometimes be neglected entirely. This is true, for example, in late transition-metal binary systems such as Co_xAl_{1-x} and Ni_xAl_{1-x}, where the pair potentials alone can explain the observed structural phase stability for $x \leq 0.25$. In the early transition-metal binary systems such as Ti_xAl_{1-x} and V_xAl_{1-x}, on the other hand, the three-ion TM potential contributions of v_3^{BBB} are significant even at $x = 0.25$, and they affect the structural phase stability of the trialuminide compounds Al_3Ti and Al_3V.

10.3.1 First-principles GPT interatomic potentials

For the TM_xAl_{1-x} alloys considered here, Al is still treated in the SM limit, so v_2^{AA} and F_N^{AA} are still given by Eqs. (10.15) and (10.16), respectively. The v_2^{AB} TM-Al potential

10.3.1 First-principles GPT interatomic potentials

is also still of the general form given by Eq. (10.15), but the corresponding normalized energy-wavenumber characteristic F_N^{AB} is now strongly impacted by both partial d-band filling and sp-d hybridization, and generalized to the form

$$F_N^{AB}(q,\Omega,x) = \frac{q^2\Omega^2}{4\pi Z_A^* Z_B^* e^2}\left[-\frac{4}{(2\pi)^3}\left\{\int \frac{w^A(\mathbf{k},\mathbf{q})w^B(\mathbf{k},\mathbf{q})}{\varepsilon_\mathbf{k}-\varepsilon_\mathbf{k+q}}\Theta^<(\mathbf{k})d\mathbf{k}\right.\right.$$
$$\left.+\int \frac{w^A(\mathbf{k},\mathbf{q})h_1^{tm,B}(\mathbf{k},\mathbf{q})}{\varepsilon_\mathbf{k}-\varepsilon_\mathbf{k+q}}d\mathbf{k}\right\} + \frac{4\pi e^2}{q^2}\left\{[1-G(q)]n_{scr}^A(q)n_{scr}^B(q)\right.$$
$$\left.\left.+G(q)n_{oh}^A(q)n_{oh}^B(q)\right\}\right], \tag{10.33}$$

where the TM energy function $h_1^{tm,B}(\mathbf{k},\mathbf{q})$ is defined by Eq. (5.52) and the quantities Z_B^*, n_{oh}^B and n_{scr}^B for the transition-metal B component are now defined by Eqs. (5.54)–(5.61).

The TM-TM potential v_2^{BB} is still of the general form expressed by Eq. (10.26), but with F_N^{BB} now generalized to

$$F_N^{BB}(q,\Omega,x) = \frac{q^2\Omega^2}{4\pi(Z_B^* e)^2}\left[-\frac{4}{(2\pi)^3}\left\{\int \frac{[w^B(\mathbf{k},\mathbf{q})]^2}{\varepsilon_\mathbf{k}-\varepsilon_\mathbf{k+q}}\Theta^<(\mathbf{k})d\mathbf{k} + \int \frac{2w^B(\mathbf{k},\mathbf{q})h_1^{tm,B}(\mathbf{k},\mathbf{q})}{\varepsilon_\mathbf{k}-\varepsilon_\mathbf{k+q}}d\mathbf{k}\right.\right.$$
$$\left.\left.+\int \frac{h_{21}^{tm,B}(\mathbf{k},\mathbf{q})}{\varepsilon_\mathbf{k}-\varepsilon_\mathbf{k+q}}d\mathbf{k}\right\} + \frac{4\pi e^2}{q^2}\left\{[1-G(q)][n_{scr}^B(q)]^2 + G(q)[n_{oh}^B(q)]^2\right\}\right], \tag{10.34}$$

where the additional TM function $h_{21}^{tm,B}(\mathbf{k},\mathbf{q})$ is defined by Eq. (5.74). In Eq. (10.26) for v_2^{BB}, the TM d-state overlap component v_{ol}^{BB} is obtained by first removing, and then transferring to F_N^{BB}, the sp-d hybridization contribution of the d-band-structure potential

$$v_2^{d,BB}(ij,\Omega,x) = \frac{2}{\pi}\mathrm{Im}\int_0^{\varepsilon_F} \ln\left[\det\left(I - T_{ij}^B T_{ji}^B\right)\right]dE$$
$$= -\frac{2}{\pi}\mathrm{Im}\int_0^{\varepsilon_F} \mathrm{Tr}\left[T_{ij}^B T_{ji}^B + \frac{1}{2}(T_{ij}^B T_{ji}^B)^2 + \cdots\right]dE, \tag{10.35}$$

where T_{ij}^B is the energy-dependent 5×5 d-state matrix that couples sites i and j, as given by Eq. (5.6). For elemental transition metals, this transfer was elaborated in detail in Chapter 5, beginning with Eq. (5.71). The resulting TM overlap potential v_{ol}^{BB} then consists of only short-ranged d-state interactions and is analogous in its form to that obtained in the EDB and FDB limits of the GPT formalism.

Using the above TM$_x$Al$_{1-x}$ pair-potential formalism, we first illustrate in Fig. 10.3 the TM concentration dependence of v_2^{AA}, v_2^{AB} and v_2^{BB} for the prototype binary system

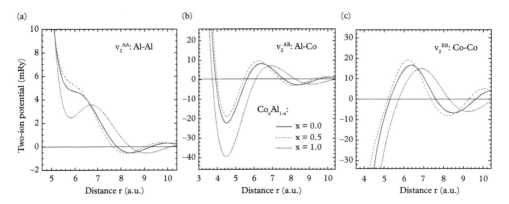

Fig. 10.3 *Concentration dependence of first-principles GPT binary alloy pair potentials $v_2^{\alpha\beta}$ for Co_xAl_{1-x}. (a) Al-Al; (b) Al-Co; (c) Co-Co.*
From Moriarty and Widom (1997), with publisher (APS) permission.

Co_xAl_{1-x} in equilibrium, as obtained by Moriarty and Widom (1997) with IU exchange-correlation.

At each concentration x considered, the potentials have been evaluated at either the observed or estimated [via Eq. (10.13)] equilibrium volume Ω_0 of the alloy, with the Co valence Z_B calculated self consistently at that volume. In Al-rich regime from $x = 0$ to $x = 0.5$, one sees that all three potentials change only very modestly in both magnitude and shape, whereas in the Co-rich regime from $x = 0.5$ to $x = 1.0$, the potentials vary significantly and become more attractive at short range below about 7 a.u. (3.7 Å). The nearly constant behavior of the three potentials in the Al-rich regime reflects the near constancy of the electron density Z/Ω_0 in that regime. Of course, as in elemental central transition metals, the substantial bonding contribution arising from partially filled d bands through $v_2^{d,BB}$ leads to a strongly attractive v_2^{BB} pair potential at short range, as displayed here for Co_xAl_{1-x} in Fig. 10.3(c). Although a deep potential well develops in v_2^{BB} below 4 a.u. (2.1 Å) for Co_xAl_{1-x}, this well is not physically accessible. At small x, TM near-neighbor distances are sufficiently large so as to avoid the short-ranged part of v_2^{BB} entirely, while at larger x, the attractive well of v_2^{BB} is balanced by repulsive multi-ion interactions at short range arising from v_3^{BBB} and v_4^{BBBB}. One can also see in Fig. 10.3 that for longer range, v_2^{BB} develops a qualitatively similar oscillatory structure to v_2^{AB}, but with larger-amplitude oscillations.

The aluminum-rich v_2^{AA} Al-Al, v_2^{AB} Al-Co and v_2^{BB} Co-Co potentials are more directly compared at $x = 0.25$ and $x = 0$ in Fig. 10.4(a). The systematic increase in energy scale in going from v_2^{AA} to v_2^{AB} to v_2^{BB} is typical of the central transition metals in general. The variation of v_2^{BB} with atomic number at $x = 1$, and thus corresponding to elemental $3d$ transition-series metals, is illustrated in Fig. 5.9. At the beginning and end of the $3d$ series in Sc, Ni and Cu, the energy scale of v_2^{BB} in the AB alloy becomes comparable to that of v_2^{AB} and v_2^{AA}, while the first minimum in v_2^{BB} moves outward to the vicinity of a close-packed nearest-neighbor distance. In the case of Cu_xAl_{1-x}, the first minimum in both v_2^{AB} and v_2^{BB} is also raised to positive energy, as shown at $x = 0$ in Fig. 10.4(b).

10.3.1 First-principles GPT interatomic potentials

The variation of v_2^{AB} across the 3d aluminides from Sc_xAl_{1-x} to Ni_xAl_{1-x} is illustrated in Fig. 10.5 for $x = 0$. Note that the position of the first minimum in the v_2^{AB} potential steadily decreases across the series from ~ 5.8 a.u. (3.1 Å) in Sc_xAl_{1-x} to ~ 4.5 a.u. (2.4 Å) in Ni_xAl_{1-x}. This trend has important structural consequences for Al-rich compounds, as is discussed in Sec. 10.3.2. In cases where the three- and four-ion contributions are needed for applications of the total-energy functional E_{tot}^{AB} given by Eq. (10.32), the corresponding GPT potentials are directly provided by the d band-structure components $v_3^{BBB} \equiv v_3^{d,BBB}$ and $v_4^{BBBB} \equiv v_4^{d,BBBB}$, respectively, which are the multi-ion analogs of $v_2^{d,BB}$ in Eq. (10.35). In applications to transition-metal aluminides, both short-ranged dd' and long-ranged sp-d hybridization contributions to the multi-ion potentials are retained, but the series expansions that define the potentials are truncated at fourth order in the T_{ij}^B:

$$v_3^{BBB}(ijk,\Omega,x) = -\frac{2}{\pi}\text{Im}\int_0^{\varepsilon_F}\text{Tr}\left[-2\left(T_{ij}^B T_{jk}^B T_{ki}^B\right) + \left(T_{ij}^B T_{ji}^B T_{ik}^B T_{ki}^B + T_{jk}^B T_{kj}^B T_{ji}^B T_{ij}^B\right.\right.$$
$$\left.\left. + T_{ki}^B T_{ik}^B T_{kj}^B T_{jk}^B\right)\right]dE \qquad (10.36)$$

and

$$v_4^{BBBB}(ijkl,\Omega,x) = -\frac{2}{\pi}\text{Im}\int_0^{\varepsilon_F}\text{Tr}\left[2\left(T_{ij}^B T_{jk}^B T_{kl}^B T_{li}^B + T_{ik}^B T_{kl}^B T_{lj}^B T_{ji}^B + T_{il}^B T_{lj}^B T_{jk}^B T_{ki}^B\right)\right]dE. \qquad (10.37)$$

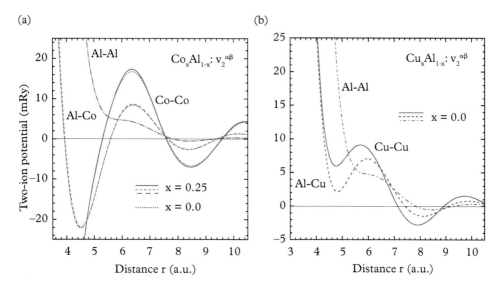

Fig. 10.4 *Dilute GPT binary alloy pair potentials $v_2^{\alpha\beta}$ for Co_xAl_{1-x} and Cu_xAl_{1-x}. (a) Al-Al, Al-Co and Co-Co potentials at $x = 0$ and $x = 0.25$ concentrations; (b) Al-Al, Al-Cu and Cu-Cu potentials at $x = 0$.*
From Moriarty and Widom (1997), with publisher (APS) permission.

Note that these equations are formally similar to Eqs. (5.92) and (5.93) used in constructing the MGPT multi-ion potentials, except that here with sp-d hybridization contributions retained, the three- and four-ion potentials exhibit more complex, long-ranged real-space behavior, such as that illustrated in Figs. 5.10 and 5.11 for elemental transition metals.

10.3.2 Basic trends in cohesion and structure

In this section, we consider a few illustrative applications of the TM$_x$Al$_{1-x}$ GPT potentials considered above to the basic issues of cohesion and structure in the $3d$ transition-metal aluminides. To proceed to the cohesive energy, however, we first need to complete the formal calculation of the remaining volume term E_{vol}^{AB} in Eq. (10.32) for the TM-Al total-energy functional. Once again, E_{vol}^{AB} can be expressed in the general form of Eq. (10.18), here with E_1^A and E_2^{AA} representing Al and obtained from Eqs. (10.19) and (10.20), respectively, for $\alpha = $ A. The corresponding TM energies E_1^B and E_2^{BB} are now generalized to the forms

$$E_1^B = E_{\text{fe}}^{0,B} + \frac{2\Omega}{(2\pi)^3} \int w_{\text{pa}}^B(\mathbf{k})\Theta^<(\mathbf{k})d\mathbf{k} + Z_d^B(\varepsilon_F - E_d^{\text{vol},B}) - \frac{10}{\pi}\int_0^{\varepsilon_F} \left[\delta_2^B(E) + \delta h_2^B(E)\right] dE$$

$$+ \delta E_{\text{oh}}^B + \frac{1}{2}(Z_B^\star e)^2 \left[\frac{1.8}{R_{\text{WS}}^B} - \frac{2}{\pi}\int_0^\infty F_N^{BB}(q,\Omega,x)dq\right] - E_{\text{bind}}^{\text{atom, B}}\left[Z_B, Z_d^B\right] + E_{\text{prep}}^B$$

(10.38)

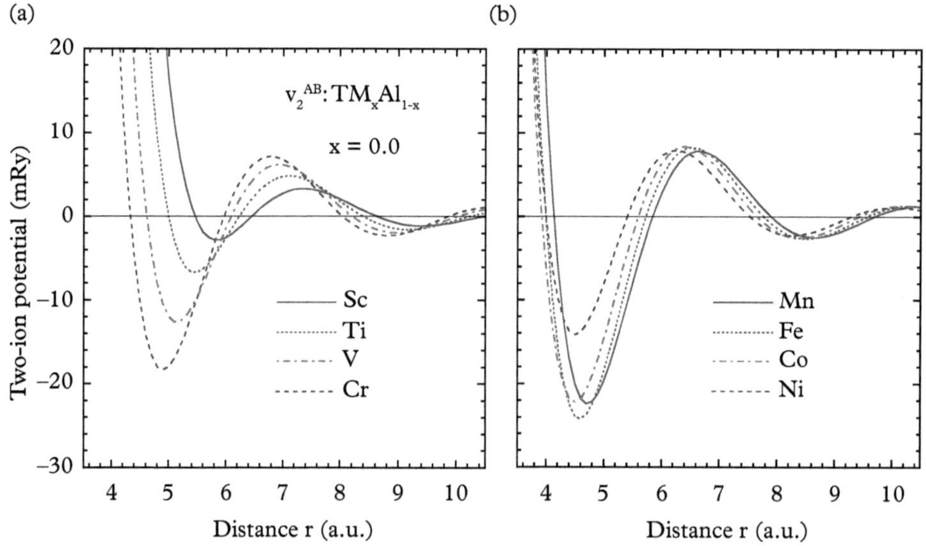

Fig. 10.5 The GPT TM$_x$Al$_{1-x}$ potential v_2^{AB} across the $3d$ series in the $x = 0$ limit. (a) Sc, Ti, V and Cr; (b) Mn, Fe, Co and Ni.
From Moriarty and Widom (1997), with publisher (APS) permission.

and

$$E_2^{BB} = \frac{2\Omega}{(2\pi)^3} \left\{ \int w_{pa}^B(\mathbf{k}) p_c^B(\mathbf{k}) \Theta^<(\mathbf{k}) d\mathbf{k} + \int \left[h_1^{tm,\,B}(\mathbf{k}) p_c^B(\mathbf{k}) + h_2^{tm,\,B}(\mathbf{k}) w_{pa}^B(\mathbf{k}) \right. \right.$$
$$\left. \left. + h_{22}^{tm,\,B}(\mathbf{k}) \right] d\mathbf{k} \right\} + \frac{\pi (Z_B^\star e)^2}{\Omega} \frac{\partial^2 F_N^{BB}(0, \Omega, x)}{\partial q^2}, \quad (10.39)$$

respectively. The final Al-TM energy E_2^{AB} has the form

$$E_2^{AB} = \frac{2\Omega}{(2\pi)^3} \left\{ \int \left[w_{pa}^A(\mathbf{k}) p_c^B(\mathbf{k}) + p_c^A(\mathbf{k}) w_{pa}^B(\mathbf{k}) \right] \Theta^<(\mathbf{k}) d\mathbf{k} \right.$$
$$\left. + \int \left[p_c^A(\mathbf{k}) h_1^{tm,\,B}(\mathbf{k}) + w_{pa}^A(\mathbf{k}) h_2^{tm,\,B}(\mathbf{k}) \right] d\mathbf{k} \right\} + (Z_A^\star Z_B^\star e^2) \frac{2\pi}{\Omega} \frac{\partial^2 F_N^{AB}(0, \Omega, x)}{\partial q^2}.$$
$$(10.40)$$

In Eqs. (10.38)–(10.40), $\delta h_2^B(E)$ is given by Eq. (5.83), $h_1^{tm,\,B}(\mathbf{k}) \equiv h_1^{tm,\,B}(\mathbf{k}, 0)$, $h_2^{tm,\,B}(\mathbf{k})$ is given by Eq. (5.56) and $h_{22}^{tm,\,B}(\mathbf{k})$ is given by Eq. (5.84). Again, the second derivatives of F_N^{BB} in Eq. (10.39) and F_N^{AB} in Eq. (10.40) can be evaluated analytically.

As first demonstrated by Moriarty and Widom (1997), the present theory implemented at the pair-potential level, i.e., omitting contributions from v_3^{BBB} and v_4^{BBBB} in the total-energy functional (10.32), can provide reasonable estimates of cohesive and structural properties for late-series transition-metal aluminides such as Co_xAl_{1-x} and Ni_xAl_{1-x} in the concentration range $x \leq 0.25$. These GPT capabilities are well illustrated by considering the cohesion curve and heat of formation of the complex aluminide compound Al_9Co_2 in its observed monoclinic phase, with 22 atoms per primitive cell and five inequivalent Al sites, and corresponding to a Co concentration of $x = 0.1818$. In Fig. 10.6(a) we have plotted the calculated GPT volume dependence of the cohesive energy, obtained both in unrelaxed form, from its observed structural parameters (Pearson, 1967), and in fully relaxed form, using a standard conjugate-gradient method to minimize the structural component of the energy at each volume considered. The quantitative effect of the relaxation is to lower the cohesion curve by 0.01 to 0.02 eV/atom in the vicinity of equilibrium. In calculating the pair-potential contributions to $E_{coh}^{AB} = E_{tot}^{AB}(\mathbf{R}, \Omega, x)/N$, a sharp cutoff of v_2^{AA}, v_2^{AB} and v_2^{BB} at $8.25 R_{WS}$ has been imposed in both calculations, which here provides smooth cohesion curves and adequate convergence of the results.

The calculated GPT equilibrium cohesive properties of Al_9Co_2 are listed in Table 10.2 and compared there with available DFT results and experimental data. The relaxed GPT equilibrium volume Ω_0 is within 5% of experiment and agrees well with the DFT results of Trambly de Laissardière et al. (1995), as does the corresponding calculation of the bulk modulus B_0. The GPT calculation of the heat of formation has been made by the direct subtraction of the compound and weighted elemental-metal cohesive energies, $\Delta H = E_{coh}^{AB} - E_{coh}^0$, where now E_{coh}^0 is given by

$$E_{coh}^0 = x E_{coh}^{Co} + (1 - x) E_{coh}^{Al}. \quad (10.41)$$

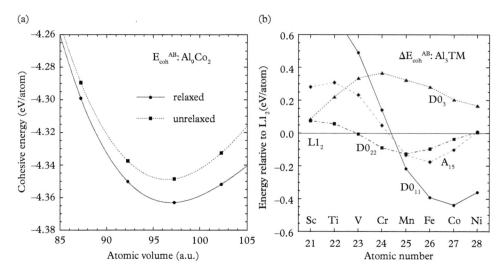

Fig. 10.6 *Cohesion and structural trends for Al-rich 3d transition-metal aluminides, as obtained in the GPT pair-potential limit discussed in the text. (a) E_{coh}^{AB} vs. volume for Al_9Co_2, where the solid symbols represent calculated values, while the solid and dashed lines are analytic fits to these results; (b) Relative cohesive energy ΔE_{coh}^{AB} for candidate $L1_2$, $D0_{22}$, $D0_3$, $A15$ and $D0_{11}$ structures of Al_3TM compounds, where the solid symbols are the calculated values.*
From Moriarty and Widom (1997), with publisher (APS) permission.

To optimize the cancellation of errors in ΔH, the cobalt cohesive energy E_{coh}^{Co} has been calculated at the same pair-potential level used to evaluate E_{coh}^{AB} for Al_9Co_2. As shown in Table 10.2, both the relaxed and unrelaxed GPT results for ΔH are then in line with the DFT calculations and in close agreement with experiment, although the latter agreement must be regarded as rather fortuitous, given the substantial absolute errors of ~ 0.5 eV in both E_{coh}^{AB} and E_{coh}^{0}.

Moriarty and Widom (1997) also examined the basic structural trends across the $3d$ series in the special case of transition-metal trialuminide compounds Al_3TM, corresponding to $x = 0.25$. Working as above in the GPT pair-potential limit with a sharp potential cutoff at $8.25R_{WS}$, five candidate structures were considered for each compound at its observed or estimated equilibrium volume. These structures are: cubic $L1_2$, with four atoms per primitive cell; tetragonal $D0_{22}$, with four atoms per primitive cell and an ideal c/a ratio of 2.0; cubic $D0_3$, with four atoms per primitive cell; cubic $A15$, with eight atoms per primitive cell; and the observed orthorhombic $D0_{11}$ structure in Al_3Ni, with 16 atoms per primitive cell. The relative cohesive energies of these five structures for Al_3TM compounds across the $3d$ series are plotted in Fig. 10.6(b). Among these five structures, the predicted structural sequence across the series is $L1_2 \to D0_{22} \to D0_{11}$. This predicted sequence is generally commensurate with what is observed for Al_3TM compounds: Al_3Sc is $L1_2$; Al_3Ti and Al_3V are $D0_{22}$, although with nonideal c/a ratios; and Al_3Ni is $D0_{11}$. In the cases of Al_3Cr and Al_3Mn, trialuminides do not form, while in

10.3.2 Basic trends in cohesion and structure

Table 10.2 *Comparison of first-principles GPT and DFT calculations of cohesion and the heat of formation for Al_9Co_2 at equilibrium. The equilibrium volume Ω_0 and bulk modulus B_0 are given in a.u. and GPa, respectively, while the energies E_{coh}^{AB}, E_{coh}^0 and ΔH are given in eV/atom.*

Method	Structural relaxation	E_{coh}^{AB}	Ω_0	B_0	E_{coh}^0	ΔH
GPT[a]	No	−4.35	96.5	119	−4.04	−0.31
GPT[a]	Yes	−4.36	96.9	105	−4.04	−0.32
DFT: LMTO-ASA[b]	No	—	96.2	108	—	−0.46
DFT: PAW[c]	Yes	—	—	—	—	−0.345
Experiment	—	—	−3.88[d]	102.2[e]	—	−3.57[f]	−0.31[g]

[a] Moriarty and Widom (1997), calculated in the pair-potential limit discussed in the text.
[b] Trambly de Laissardière et al. (1995).
[c] Mihalkovic and Widom (2007).
[d] Calculated from the measured values of E_{coh}^0 and ΔH.
[e] Villars and Calvert (1991).
[f] Kittel (1976).
[g] de Boer et al. (1988).

the cases of Al_3Fe and Al_3Co, more complex, nonstoichiometric structures are observed near $x = 0.25$. As we discuss in Sec. 10.3.3, $D0_{11}$ is indeed energetically competitive with these latter structures.

The close competition between the $L1_2$ and $D0_{22}$ structures for the early trialuminides has been investigated in more detail for Al_3Sc, Al_3Ti and Al_3V by Carlsson and Meschter (1989), using DFT total-energy calculations, and by Moriarty and Widom (1997), using the above GPT formalism, but adding the important contribution of the three-ion TM potential v_3^{BBB} to the total-energy functional, while also extending the real-space range of all potential interactions from $8.25R_{WS}$ to $10.30R_{WS}$. In both studies, however, only fixed values of the $D0_{22}$ c/a axial ratio have been treated: the ideal value of 2.0 and the observed value of 2.23 for Al_3Ti and Al_3V. For $c/a = 2.0$, the DFT and GPT predictions are in general agreement, as shown in Fig. 10.7(b), with the $L1_2$ structure favored in Al_3Sc and Al_3Ti, while the $D0_{22}$ structure is favored in Al_3V. For $c/a = 2.23$, the DFT calculated $D0_{22}$–$L1_2$ energy is substantially lowered in each case, such that the $D0_{22}$ structure is then favored in both Al_3Ti and Al_3V, as is observed. This result is only reproduced by the GPT treatment, however, if one considers only the TM potentials, $v_2^{BB} + v_3^{BBB}$, and, in particular, discards the unfavorable Al-TM pair potential v_2^{AB}, underscoring the importance of the direct d-state interactions to $D0_{22}$ structural phase stability, and as is illustrated in Fig. 14 of Moriarty and Widom (1997).

In an attempt to gain some further insight into this result, we have extended the GPT study by considering the actual minimization of the $D0_{22}$ total energy with respect to its c/a ratio. In the pair-potential limit, without v_3^{BBB} included in the total-energy functional

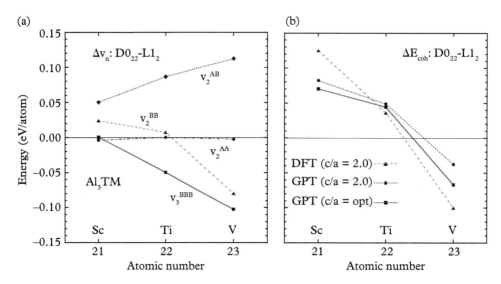

Fig. 10.7 GPT potential contributions and calculated $D0_{22}$–$L1_2$ cohesive-energy differences for the early 3d TM trialuminides. (a) The v_2^{AA} Al-Al, v_2^{AB} Al-TM, v_2^{BB} TM-TM and v_3^{BBB} TM-TM-TM potential contributions to $D0_{22}$–$L1_2$ energies, for optimum (opt) $D0_{22}$ c/a values of 1.85 in Al_3Sc, 2.06 in Al_3Ti and 2.10 in Al_3V; (b) Full DFT (Carlsson and Meschter, 1989) and GPT (Moriarty and Widom, 1997) cohesive-energy differences ΔE_{coh}, for ideal $D0_{22}$ with $c/a = 2.0$ and in the GPT case also with $c/a = $ opt.

E_{tot}^{AB}, the calculated optimum values of c/a are all at or below 2.0. These values are 1.81, 1.93 and 2.00 for Al_3Sc, Al_3Ti and Al_3V, respectively. When v_3^{BBB} is then restored in E_{tot}^{AB}, the optimum (opt) c/a all rise in value to 1.85, 2.06 and 2.10, respectively, which is indeed the desired trend. The corresponding two- and three-ion potential contributions to the respective $D0_{22}$–$L1_2$ energy differences are illustrated in Fig. 10.7(a). As found by Moriarty and Widom (1997) for $c/a = 2.0$, both v_2^{BB} and v_3^{BBB} increasingly favor $D0_{22}$ with increasing TM atomic number for $c/a = $ opt, but now with v_3^{BBB} making a larger contribution in that direction for both Al_3Ti and Al_3V. The corresponding $c/a = $ opt cohesive-energy differences are plotted in Fig. 10.7(b), and for each trialuminide compound, the $D0_{22}$–$L1_2$ energy is indeed lowered as desired, but the improvement for Al_3Ti is too small to account for its observed $D0_{22}$ phase stability.

Further progress on understanding phase stability in the early TM trialuminide compounds will require incorporating additional physics into the GPT treatment. As pointed out by Moriarty and Widom (1997), it may be necessary to include three-ion v_3^{AAB} and v_3^{ABB} potentials in the total-energy functional to compensate for the unfavorable impact of v_2^{AB} on $D0_{22}$ phase stability, but this has yet to be investigated. We have, however, investigated the inclusion of the four-ion TM potential v_4^{BBBB} in the total-energy functional, but interestingly, we have found this added contribution does not change the calculated optimum c/a values in either Al_3Ti or Al_3V, nor does it lower the $D0_{22}$–$L1_2$

energies of these compounds. A more promising path forward may be to go beyond the simple canonical d-band description that has been used heretofore to establish v_2^{BB} and v_3^{BBB}, and to optimize the TM d-basis states, while simultaneously minimizing the total energy to achieve optimum c/a values. As discussed in Chapter 5, this is readily feasible, but has also yet to be considered in this context.

10.3.3 Al-Co and Al-Ni binary phase diagrams

In this section, we turn to the aluminum-rich GPT phase diagrams of Al-Co and Al-Ni calculated for TM concentrations $x < 0.3$ by Widom and Moriarty (1998). In this work, great advantage was made of the very weak TM concentration dependence of the pair potentials at small x, as shown above for Co_xAl_{1-x} in Fig. 10.4(a). This simplification allowed the detailed consideration of 22 candidate structures for the Al-Co phase diagram with the $x = 0$ GPT pair potentials, including five nonstoichiometric structures with partial Al occupancy, where extensive MC simulations were necessary to establish the actual site occupations. Guided by these considerations, both $x = 0$ GPT and full concentration-dependent GPT phase diagrams for Al-Co and Al-Ni were constructed in the form of structural energy (ΔE_{coh}^{AB}) vs. concentration scatter plots. The full-GPT phase diagrams retained 14 of the most significant structures considered and are displayed in Fig. 10.8, where the predicted stable structures for Al-Co and Al-Ni are those connected along the convex hull of each material. The multiple thermodynamically stable structures so predicted at the GPT pair-potential level include the observed phases of fcc Al, Al_9Co_2 and $Al_{13}Co_4$ in Al-Co, while in Al-Ni they include the observed phases of fcc Ni and DO_{11}. Also in Al-Ni, the Al_9Co_2 structure properly lies just above the convex hull. The most noteworthy error incurred at this level of treatment is the prediction of the Al_5Fe_2 structure in favor of the observed Al_5Co_2 phase in Al-Co, an error that can be readily corrected by the inclusion of multi-ion GPT Co potentials, as discussed below.

The phase diagrams and structure predictions by the full GPT in Al-Co and Al-Ni also maintain very good qualitative and quantitative agreement with the corresponding $x = 0$ results. A few selected Al and Al-Co structural energies calculated with both $x = 0$ and full GPT potentials are compared with each other and with available DFT calculations in Table 10.3. As expected, the $x = 0$ and full GPT results are in close agreement. The agreement between GPT and DFT is also generally good, although for the Al_3Co compounds at $x = 0.25$, the comparison suffers somewhat from the fact that different equilibrium atomic volumes were assumed in the GPT and DFT calculations. In this regard, the assumed GPT atomic volume is derived from Eq. (10.13) with a self-consistent Co valence of 1.8, while the DFT atomic volume of Ögüt and Rabe (1994) is based on an assumed constant valence electron density and a Co valence of 1.5.

The five competitive nonstoichiometric structures considered by Widom and Moriarty (1998) derive from reported complex monoclinic and orthorhombic phases in the vicinity of $x = 0.25$ for Al-Co or Al-Fe. These structures include three variants of $Al_{13}Co_4$, all with large 102-atom unit cells: monoclinic M-$Al_{13}Co_4$ assigned with either the $C2/m$ space group (Freiburg et al., 1996) or the Cm space group (Hudd and Taylor,

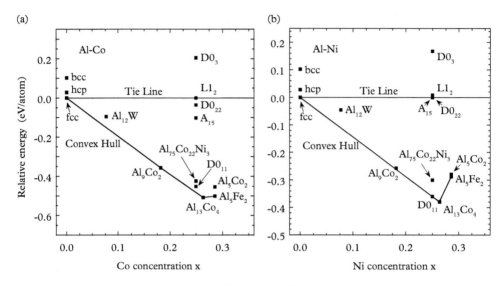

Fig. 10.8 *Aluminum-rich Al-Co and Al-Ni binary phase diagrams, here in the form of plotted $T = 0$ structural energies relative to the fcc–$L1_2$ tie line vs. concentration, as calculated from first-principles, full-concentration GPT pair potentials to $x = 0.3$. (a) Al-Co phase diagram; (b) Al-Ni phase diagram. From Widom and Moriarty (1998), with publisher (APS) permission.*

Table 10.3 *Calculated GPT and DFT structural energies, ΔE_{coh} or ΔE_{coh}^{AB}, for Al and Al-Co binary compounds at their observed or estimated equilibrium volumes, as discussed in the text. Energies are relative to the fcc–$L1_2$ tie line in units of $eV/atom$.*

Structure	x	DFT: PP[a]	DFT: PP[b]	$x = 0$ GPT	full GPT
hcp ($c/a = 1.63$)	0.0000	0.033	0.057	0.028	0.028
bcc	0.0000	0.102	0.103	0.102	0.102
$Al_{12}W$[c]	0.0769	—	−0.044	−0.066	−0.052
$D0_3$	0.2500	—	0.128	0.205	0.205
$D0_{22}$ ($c/a = 2.0$)	0.2500	—	−0.010	−0.035	−0.036
A15	0.2500	—	−0.083	−0.095	−0.102

[a]Lam and Cohen (1981).
[b]Ögüt and Rabe (1994).
[c]Unrelaxed observed structure.

1962), and orthorhombic O-$Al_{13}Co_4$ assigned with the $Pmn2_1$ space group (Grin et al., 1994). The remaining structures are monoclinic $Al_{75}Co_{22}Ni_3$, also with a $C2/m$ space group (Zhang et al., 1995), and orthorhombic Al_5Fe_2, with a $Cmcm$ space group (Villars and Calvert, 1991). Some of the important features that these layered structures have

10.3.3 Al-Co and Al-Ni binary phase diagrams

in common are very short Al-Al bond lengths (2.2–2.5 Å). The strongly repulsive Al-Al pair potential at such short distances ($r < 5$ a.u.), as can be seen in Fig. 10.3(a), produces effective negative vacancy formation energies in these compounds, so it becomes energetically favorable to create some number of vacancies within each unit cell and allow partial occupancy of the Al sites. The statistics of the Al partial occupation were studied in each structure using a fixed-site MC method (Cockayne and Widom, 1998), typically applied at temperatures up to 1000 K. These simulations started with the reported unit-cell atomic positions and Al composition, together with the assumption of full Co occupancy, and then allowed the Al atoms to hop freely from occupied to unoccupied sites. For M-$Al_{13}Co_4$ with the space group $C2/m$, and for Al_5Fe_2, complete agreement with experiment was found in the assignment of partially occupied sites, while substantial, but less than perfect, agreement was found for M-$Al_{13}Co_4$ with the space group Cm. In the case of O-$Al_{13}Co_4$, on the other hand, full Al occupancy was reported, but the expected partial occupancy was possibly masked by the Debye-Waller corrections used in the structure determination. In this case, it was decided to perform the MC simulations as a function of Al composition, from full occupancy, corresponding to 78 Al atoms per unit cell, down to the observed composition of M-$Al_{13}Co_4$, corresponding to 67 Al atoms per unit cell. At each composition treated, the structure was annealed down to $T = 0$ and then fully relaxed, searching for the lowest structural-energy assignment of the Al occupancy. This minimum structural energy occurs for the 67 Al-atom unit cell and corresponds to a Co concentration of $x = 0.2637$. This is the point where the $Al_{13}Co_4$ structural energy falls on the convex hull in the Al-Co phase diagram, as is illustrated in Fig. 10.8(a). The same is true for Ni-Al, as shown in Fig. 10.8(b).

Finally, we return to the issue noted in Fig. 10.8(a) that at the GPT pair-potential level of treatment, the Al_5Fe_2 structure is favored over the observed Al_5Co_2 structure in the predicted Al-Co phase diagram. We first note that although the cohesive energies for these two structures were only compared in Fig. 10.8(a) at the observed equilibrium volume for Al_5Co_2, the relative-ordering problem persists when the volume dependence of the cohesive energy of each structure is taken into account, as shown in Fig. 10.9(a). However, it is reasonable to suppose that neglected Co multi-ion interactions are more important in this case because of the larger TM concentration $x = 0.2857$. In particular, the observed Al_5Co_2 structure possesses near-neighbor equilateral triangles of Co atoms, each with a sizeable three-ion potential contribution: $v_3^{BBB} \cong -0.14$ eV. In contrast, the Al_5Fe_2 structure here has corresponding isosceles triangles of Co atoms, but each produces a much smaller three-ion potential contribution. When v_3^{BBB} and v_4^{BBBB} are summed over the relaxed pair-potential ion configurations of the two structures, as a function of atomic volume and to a radial cutoff of $8.25R_{WS}$, then the Al_5Co_2 cohesion curve is lowered, while the Al_5Fe_2 cohesion curve is raised, such that the Al_5Co_2 structure is now favored, as is shown in Fig. 10.9(b). Moreover, the calculated heat of formation ΔH for Al_5Co_2 is then improved from a value of -0.37 eV with the pair-potential treatment to a value of -0.45 eV with the multi-ion potential treatment, in close agreement with the experimentally measured value of -0.43 eV (deBoer et al., 1988).

10.3.4 Extension to ternary phase diagrams and quasicrystals

The leverage provided by the $x = 0$ GPT pair-potential treatment of the aluminum-rich Al-Co and Al-Ni binary phase diagrams can be readily extended to related ternary phase diagrams, including the interesting cases of Al-Co-Ni and Al-Co-Cu, where in both systems there exists a stable decagonal quasicrystal phase. Widom et al. (2000) so addressed the Al-Co-Cu ternary system in depth, considering a large number of observed and hypothetical candidate structures, some 30 for the Al-Cu binary phase diagram and an additional 20 for the full Al-Co-Cu phase diagram. They also made specific comparisons with the Al-Co-Ni phase diagram. While Al-Co-Cu and Al-Co-Ni can be treated directly as independent ternary systems, the calculated $x = 0$ TM valences of the Periodic-Table neighbors Co, Ni and Cu are so close in value (see Fig. 10.1) that the $x = 0$ GPT pair potentials from the binary-system treatments of Al-Co, Al-Ni and Al-Cu can be used instead with little additional error. Only needed then are TM potentials for Co-Ni and Co-Cu interactions, which Widom et al. (2000) approximated as an average of the Co-Co and Ni-Ni binary pair potentials for Co-Ni, and just the Ni-Ni pair potential for Co-Cu. The $x = 0$ Al-Cu and Cu-Cu GPT pair potentials are displayed in Fig. 10.4(b), where one sees that they have similar short-range magnitudes and character, with the Cu-Cu potential repulsive below about 7 a.u. (3.7 Å). Consequently,

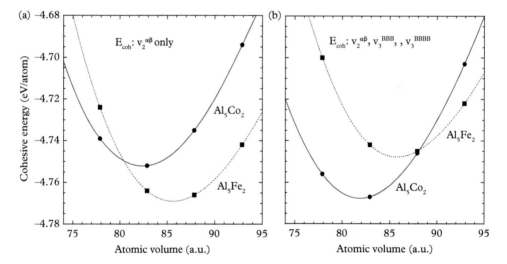

Fig. 10.9 *Impact of GPT multi-ion Co potentials on the energy ordering of the Al_5Co_2 and Al_5Fe_2 structures in the calculated Al-Co phase diagram. (a) Fully relaxed E_{coh}^{AB} vs. atomic volume cohesion curves obtained at the pair-potential level of treatment, where the solid symbols represent calculated values, while the solid and dashed lines are analytic fits to these results; (b) Corresponding E_{coh}^{AB} vs. atomic volume curves including the three- and four-ion Co potentials v_3^{BBB} and v_4^{BBBB}, respectively, but summed in the relaxed ion configurations of the pair-potential treatment (a).*
From Widom and Moriarty (1998), with publisher (APS) permission.

these pair potentials can be safely applied to compounds with Cu at any concentration, and without the need for additional multi-ion potentials. Cobalt and nickel concentrations, on the other hand, are again limited to values less than about 0.25 before multi-ion corrections are needed.

The observed experimental phase-diagram of Al-Co-Cu for Co concentrations $x \leq 0.5$ and Cu concentrations $y \leq 0.5$ is illustrated in Fig. 10.10. The LHS of the triangle effectively represents the binary Al-Co convex hull for $y = 0$, with x ranging from zero for fcc Al to 0.5 for the compound AlCo, which has a CsCl or B2 crystal structure. Likewise, the RHS of the triangle effectively represents the convex hull of binary Al-Cu for $x = 0$, with y ranging from zero for fcc Al to 0.5 for the compound AlCu, a vacancy-ordered phase based on the B2 structure, with both complex low-temperature monoclinic and high-temperature orthorhombic variants. On the Al-Cu convex hull, Widom et al. (2000) successfully calculated both the observed tetragonal $C16$ Al_2Cu phase at $y = 1/3$ and the low-temperature AlCu phase, but also found that an additional τ_5 phase touched the hull at $y = 0.375$ as well. The latter is one of a special family of vacancy-ordered phases of the form $Al_m Cu_n$, also derived from the B2 structure and following a Fibonacci sequence, with the whole family collectively known as the τ phases. These phases are metastable in the binary Al-Cu phase diagram (Sastry et al., 1980), where τ_3 is Al_3Cu_2 and τ_5 is Al_5Cu_3. In the full ternary Al-Co-Cu phase diagram, however, with Co partially substituted for Cu, some of the τ phases can be stabilized thermodynamically as well as mechanically (He et al., 1988), including τ_3, which then becomes Al_6CoCu_3, as shown in Fig. 10.10.

Widom et al. (2000) explored the interior of the Al-Co-Cu phase diagram by calculating relative cohesive energies of observed and hypothetical phases along several chosen xy lines of special interest. These lines were: $x = y$, $x + y = 1/4$ and $x + y = 1/3$. In each case, the calculated convex hull along these lines was found to include only observed phases, or combinations of observed phases, with all other hypothetical phases calculated at higher energies above the convex hull. Thus, for the most part and within the expected Co concentration limit of $x \leq 0.25$, the $x = 0$ GPT pair potential treatment of Al-Co-Cu is consistent with the experimental phase diagram depicted in Fig. 10.10. Also, in this regard, the structural tiling model used by Widom et al. (2000) for the Al-Co-Cu decagonal quasicrystal, denoted as "D" in Fig. 10.10, is that due to Cockayne and Widom (CW, 1998). In the CW quasicrystal model for Al-Co-Cu, there are no near-neighbor Co atoms, so the use of multi-ion Co potentials is not essential to obtain a viable description of the cohesive energy, which is, in fact, still provided by the $x = 0$ GPT pair-potential treatment. As revealed in the calculated Al-Co pair distribution function, all of the Al-Co near-neighbor separations of the quasicrystal lie within the first deep minimum of the Al-Co pair potential at 4.5 a.u. (2.4 Å), as displayed in Fig. 10.4(a). Moreover, the first two calculated peaks in the Co-Co pair correlation function for the quasicrystal also align perfectly with the second and third minima of the GPT Co-Co pair potential at 8.4 a.u. (4.4 Å) and 12.4 a.u. (6.4 Å).

In subsequent work, attention turned to the calculation of quasicrystal structures in Al-Co-Ni. To maintain the fast total-energy calculations that the above $x = 0$ GPT pair-potential treatment provides, while enabling full structural relaxation and atomistic

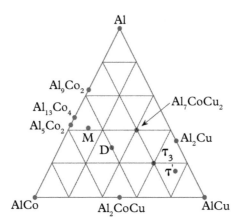

Fig. 10.10 *The Al-Co-Cu ternary phase diagram, indicating experimentally observed stable phases, with Co concentrations $x \leq 0.5$ and Cu concentrations $y \leq 0.5$. Here M denotes M - $(Al,Cu)_{13}Co_4$, D denotes a decagonal quasicrystal, τ_3 denotes Al_6CoCu_3 and τ' denotes $Al_{36}Co_3Cu_{24}$.*

simulations with TM-TM nearest-neighbor bonds, Al-Lehyani et al. (2001) developed an approximate way to incorporate the necessary balancing repulsive contribution of the multi-ion potentials to the strong short-range attraction of the TM-TM pair interactions for Co and Ni. Inspired by the MGPT canonical-d-band treatment of three- and four-ion potentials in central transition metals (Moriarty, 1990a), they argued that at sufficiently short range there should be a leading average repulsive term arising from the fourth-order d-band contribution to the three-ion potential v_3^{BBB} that varies as $\sim r^{-20}$. They proposed, therefore, adding to the GPT pair potentials for Co-Co, Co-Ni and Ni-Ni an effective short-range repulsive potential of the form $\delta v_2^{BB}(r) = a(r_0/r)^b$ to mimic the multi-ion potentials. To determine the parameters a, r_0 and b for the three TM-TM interactions, they introduced a convenient Al-Co-Ni quasicrystal approximant compound $Al_{34}Co_{10}Ni_6$. This is a layered periodic structure with an orthorhombic unit cell of 50 atoms and nearest-neighbor TM-TM bonds that can be treated by the modified GPT potentials, as well as by first-principles DFT calculations. They then focused on the comparative energetics of selected bond swaps between Co-Ni bonds and Co-Co plus Ni-Ni bonds to establish an optimum choice of parameters for $\delta v_2^{BB}(r)$. This procedure put r_0 at 2.55 Å (4.82 a.u.), with a values of 0.319, 0.237 and 0.140 eV, and b values of 16.6, 19.3 and 21.3, respectively, for the Co-Co, Co-Ni and Ni-Ni potentials. Note that the b values are indeed of the large magnitude expected, which effectively limits the range of $\delta v_2^{BB}(r)$ to about 3 Å (5.7 a.u.).

Using the modified GPT pair potentials for the TM-TM interactions of Co and Ni atoms together with the existing first-principles GPT pair potentials for Al-Al, Al-Co and Al-Ni interactions, Mihalkovic et al. (2002) have predicted the detailed atomic structure of an Al-Ni-Co quasicrystal from extensive MC simulations. The MC simulations used only minimal experimental constraints concerning the composition, lattice constants and geometry of the quasicrystal. In particular, no predetermined structural tiling model, such as the CW model mentioned above, was assumed. Rather, the MC simulations started with the chosen concentration $Al_{34}Ni_{12}Co_4$, a fixed unit cell size, and with all atoms randomly distributed at a high temperature of 23,208 K. The system was slowly cooled and annealed to a final temperature of 332 K, where a unique minimum-energy configuration of atoms was achieved. Regarding the predicted TM-TM bonds, the final atomic configuration had only Ni-Ni nearest neighbors, no doubt reflecting the small initial concentration of Co. The final atomic structure could then be described by a derived and highly constrained quasicrystal tiling model, obeying specific rules, where the atomic decoration of the tiles was virtually unique.

10.4 The special case of Ca-Mg

In addition to the transition-metal aluminides, the first-principles GPT SM-TM binary alloy formalism can be readily applied to other important physical systems. One such system of current interest is the binary alloy Ca_xMg_{1-x}, where for small concentrations x calcium is expected to exhibit transition-metal behavior. In this regard, we recall from Chapter 4 that elemental fcc Ca at ambient pressure behaves as an EDB metal through strong sp-d hybridization with the empty $3d$ bands just above the Fermi level. Under high pressure, however, the bottom of the $3d$ bands drops below E_F, driving the observed fcc $\rightarrow$ bcc phase transition near 20 GPa at room temperature (Olijnyk and Holzapfel, 1984; Anzellini et al., 2018). This behavior supports our more general conclusion reached in Chapter 7 that the appearance of the bcc structure in the heavy alkaline-earth metals at low temperature is a fingerprint for the presence of TM behavior. In the case of Ca, accompanying the pressure-induced transition to the bcc structure is a large volume compression of $\Omega_T/\Omega_0^{Ca} = 0.59$ at the transition pressure. In the dilute Ca_xMg_{1-x} alloy at ambient pressure, Ca undergoes a similar but even larger volume compression due to the mismatch of atomic sizes and equilibrium volumes with its Mg host: $\Omega_0^{Ca} = 294.5$ a.u. vs. $\Omega_0^{Mg} = 156.8$ a.u. In addition, the partial filling of the $3d$ bands coupled with the self-consistency requirements on the valence electron density, Eqs. (10.7)–(10.9), leads to TM values of $Z_B = 1.626$ and $\Omega_B = 127.5$ a.u. for Ca at $x = 0$, corresponding to an effective volume compression of $\Omega_B/\Omega_0^{Ca} = 0.43$. Calculated first-principles GPT pair potentials for Ca-Ca, Mg-Ca and Mg-Mg interactions at these conditions, with Ca treated as a $3d$ transition metal, are illustrated in Fig. 10.11.

The main current research interest in dilute Ca_xMg_{1-x} alloys arises from the technology-driven need to enhance and control complex hcp Mg mechanical properties, including dislocation mobility, anisotropic plasticity and ductility. At the impurity concentration level in Mg, this problem has been addressed historically though the use of rare-earth solute elements, but with the future cost and availability of these elements uncertain, research attention has turned to possible alternative solutes, including Li, Ca and Zn. At the same time, a more speculative and far-reaching idea is to use alloying with Ca in an attempt to extend the observed mechanical and thermodynamic stability of the high-pressure bcc Mg phase back to ambient pressure, and thus capitalize on the intrinsically more favorable mechanical properties of the bcc structure. The GPT Ca_xMg_{1-x} potentials illustrated in Fig. 10.11 can be usefully applied to both problems, but the latter one in particular has attracted our recent attention.

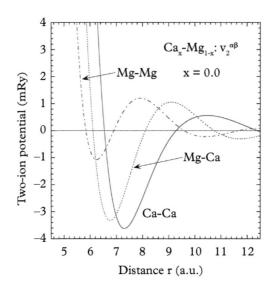

Fig. 10.11 *First-principles GPT binary alloy pair potentials for Ca_xMg_{1-x}, as calculated by Moriarty (2017) in the $x = 0$ magnesium-rich limit at $\Omega = 156.8$ a.u., treating calcium as a 3d transition metal.*

Using $x = 0$ GPT potentials and treating Ca_xMg_{1-x} as a bcc random substitutional alloy (RSA) at ambient pressure, Skinner and Moriarty (2017) and Skinner (2019) have studied the elastic constants and vibrational spectrum of the bcc alloy for Ca concentrations up to $x = 0.25$. Recall from Fig. 7.7 that in the elemental metal, bcc Mg is mechanically unstable at low pressure, with $C' < 0$ and corresponding imaginary phonon frequencies, for volumes $\Omega > 0.9\Omega_0^{Mg} \cong 141$ a.u. In the bcc Ca_xMg_{1-x} alloy, however, the mechanical instability is removed entirely for $x \geq 0.05$, with $C' > 0$ and no imaginary phonon frequencies calculated. Thus, the presence of Ca in the alloy indeed promotes bcc mechanical stability, even at ambient pressure.

These findings beg the question as to whether or not thermodynamic stability as well can be achieved for bcc Ca_xMg_{1-x}. The preliminary $T = 0$ calculations of Skinner (2019) on the heat of formation $\Delta H(x)$ at ambient pressure predicted a small positive value of 0.035 eV at $x = 0.05$, and $\Delta H(x)$ values that rise in magnitude in an approximate linear fashion with increasing Ca concentration above $x = 0.05$. These results are subject to the usual quantitative uncertainties, of course, but it is reasonable to assume that they are at least qualitatively correct. At the same time, one would expect that under pressure and temperature $\Delta H(x, P, T)$ would decrease with increasing either P or T at small fixed x and eventually become negative, since one knows that for $x = 0$ in pure Mg that a thermodynamically stable bcc phase exists above the hcp-bcc phase line (see Fig. 7.11). That phase line extends to about 50 GPa at low temperature and perhaps to 1200–1700 K at low pressure—boundaries that are well within the reach of modern DAC experimental capabilities. Thus, perhaps the best chance of actually realizing bcc $\text{Ca}_x\text{Mg}_{1-x}$ might be to create this alloy in the heated DAC at sufficiently high pressure and temperature, and then attempt to bring the alloy back very slowly to ambient pressure and temperature as a metastable phase. In this regard, it would be useful to calculate the pressure and temperature dependence of $\Delta H(x, P, T)$ for small fixed values of x with full volume- and concentration-dependent GPT potentials, to provide more concrete motivation and guidance for such experiments.

10.5 BOP treatment of transition-metal aluminides: TiAl

So far in this chapter, we have mainly discussed GPT treatments of AB binary compounds and alloys, with an emphasis on the $3d$ transition-metal aluminides. Correspondingly BOP research on the TM aluminides has focused more narrowly on important $\text{Ti}_x\text{Al}_{1-x}$ compounds and their mechanical properties in the titanium-rich concentration range $0.5 \leq x \leq 1.0$. This research specifically targets γ-TiAl, a well-known $x = 0.5$ compound with a tetragonal L1_0 structure, and α_2-Ti_3Al, an $x = 0.75$ compound with a hexagonal D0_{19} structure. Both of these compounds have direct practical relevance to high-temperature applications in the aerospace industry. Also considered in connection with the BOP research effort has been pure hcp Ti at $x = 1$.

Just as the BOP treatment of elemental transition metals is built on an orthogonal d-state TB model, the BOP treatment of $\text{Ti}_x\text{Al}_{1-x}$ is built on a corresponding orthogonal pd-state TB model, with Al represented quantum mechanically by p-state orbitals. The bond-order potentials for $\text{Ti}_x\text{Al}_{1-x}$ developed by Znam et al. (2003) generalize the numerical BOP framework for elemental transition metals discussed in Chapter 5 to treat SM-TM AB binary compounds. The BOP total-energy functional for such an AB intermetallic compound has the general form

$$E_{\text{tot}}^{\text{AB}}(\mathbf{R}) = E_{\text{bond}}^{\text{AB}}(\mathbf{R}) + E_{\text{rep}}^{\text{AB}}(\mathbf{R}), \tag{10.42}$$

where $E_{\text{bond}}^{\text{AB}}$ is the attractive bond energy captured by the AA pp', AB pd and BB dd' bonds and $E_{\text{rep}}^{\text{AB}}$ is the compensating repulsive energy associated with the effective compression of sp valence electrons and localized p- and d- orbital overlap in the presence of the strong directional bonds. Expressed in the usual way in terms of Slater-Koster matrices $h_{ij}^{\alpha\beta}$ for orbitals α and β centered on ion sites i and j, respectively, and corresponding bond-order functions $\Theta_{ji}^{\beta\alpha}$, the bond energy can be written as

$$E_{\text{bond}}^{\text{AB}}(\mathbf{R}) = \sum_{i,j}{}' \left\{ \sum_{p,p'} h_{ij}^{pp'} \Theta_{ji}^{p'p} + \sum_{p,d} h_{ij}^{pd} \Theta_{ji}^{dp} + \sum_{d,d'} h_{ij}^{dd'} \Theta_{ji}^{d'd} \right\}. \tag{10.43}$$

In general, the bond integrals entering the $h_{ij}^{\alpha\beta}$ in Eq. (10.43) can be modeled in the analytic form given by Eq. (5.182). For application to $\text{Ti}_x\text{Al}_{1-x}$, Znam et al. used this form with the adjustable parameter n_b set to zero, resulting in the following three-parameter power-law expressions:

$$pp\beta_m(R_{ij}) = pp\beta_m(R_{nn}^{AA})\left(\frac{R_{nn}^{AA}}{R_{ij}}\right)^{n_{AA}}$$

$$pd\beta_m(R_{ij}) = pd\beta_m(R_{nn}^{AB})\left(\frac{R_{nn}^{AB}}{R_{ij}}\right)^{n_{AB}}$$

$$dd\beta_m(R_{ij}) = dd\beta_m(R_{nn}^{BB})\left(\frac{R_{nn}^{BB}}{R_{ij}}\right)^{n_{BB}}, \qquad (10.44)$$

where $\beta_m = \sigma$ for $m = 0$, $\beta_m = \pi$ for $m = 1$, and in the case of the dd integral, $\beta_m = \delta$ for $m = 2$. Here R_{nn}^{AA} and R_{nn}^{AB} have been fixed at the observed nearest-neighbor distances for Al-Al and Al-Ti bonds, respectively, in γ-TiAl, while R_{nn}^{BB} was fixed at the observed Ti-Ti nearest-neighbor distance in hcp Ti. Also, the parameter n_{BB} was constrained by the hcp-fcc energy difference in pure Ti, as obtained from FPLAPW calculations. Subject to these constraints, the five remaining parameters in the bond integrals given by Eq. (10.44) were determined by fitting the DFT electronic structure of γ-TiAl, as obtained from self-consistent TB-LMTO calculations. In this procedure, as well as in subsequent applications, the bond integrals were smoothly truncated between the second and third nearest-neighbor distances. Znam et al. also showed that the bond integrals so determined demonstrate good transferability among γ-TiAl, α_2-Ti$_3$Al and hcp Ti.

Znam et al.'s treatment of the remaining repulsive energy in Eq. (10.42) follows the empirical approach used for elemental transition metals, as described in Chapter 5, with $E_{rep}^{AB}(\mathbf{R})$ containing nonlinear environmental and linear pair-potential contributions:

$$E_{rep}^{AB}(\mathbf{R}) = E_{env}^{AB}(\mathbf{R}) + E_{pair}^{AB}(\mathbf{R}). \qquad (10.45)$$

Here one has $E_{env}^{AB}(\mathbf{R}) = NE_{env}(\mathbf{R})$, with $E_{env}(\mathbf{R})$ modeled as the pair functional given by Eqs. (5.185) and (5.186), while $E_{pair}^{AB}(\mathbf{R}) = NE_{pair}(\mathbf{R})$, with $E_{pair}(\mathbf{R})$ expressed as a sum of cubic splines as in Eq. (5.187). However, in an AB compound one must account for AA, AB and BB interactions, so there are multiple chemical-specific parameters to be determined for both E_{env}^{AB} and E_{pair}^{AB}, in contrast to the case of an elemental transition metal. In their application to TiAl compounds, Znam et al. developed a detailed procedure to determine these multiple parameters, in which the main strategy was to fit E_{env}^{AB} to the observed negative Cauchy pressures $C_{13}-C_{44}$ and $C_{12}-C_{66}$ for both hcp Ti and L1$_0$ γ-TiAl, while fitting E_{pair}^{AB} to the observed cohesive energy, lattice constants and remaining elastic moduli for γ-TiAl.

Znam et al. tested the BOP potentials so determined against DFT calculations of various structural and defect properties in γ-TiAl, including bulk structural energy differences relative to L1$_0$, the {111} γ surface, and multiple stacking-fault-like and point-defect energies. Good qualitative agreement with DFT was found on the magnitude and ordering of all energies. They also tested the transferability of the BOP potentials to the cohesive, structural and elastic properties of D0$_{19}$ α_2-Ti$_3$Al, with somewhat more mixed results. The observed cohesive energy and lattice constants are well predicted, and the small, positive L1$_2$–D0$_{19}$ structural energy difference obtained with the BOP potentials qualitatively agrees with DFT calculations. At the same time, the BOP elastic constants calculated for α_2-Ti$_3$Al are considerably less accurate than for γ-TiAl, with quantitative errors of 20–60% for all moduli in comparison to experiment.

Independent computer simulations using the TiAl BOP potentials to investigate screw dislocation behavior in single-phase and lamellar γ-TiAl have been carried out by Katzarov et al. (2007) and by Katzarov and Paxton (2009). These studies were motivated in part by the industrial importance of a closely related two-phase alloy with a microstructure consisting of layers of both γ-TiAl and α_2-Ti$_3$Al, which is obtained by special heat treatments that improve the overall strength, ductility and toughness of the material. It is believed that the vast majority of the deformation in the two-phase alloy is still carried in the γ-TiAl phase, which has deformation modes of both slip and twinning operating on {111} planes. Slip can occur either by ordinary (1/2)<110] dislocations or by <101] super dislocations, so identified here with the special notation of Hug et al. (1988) for tetragonal TiAl. In single crystals of γ-TiAl, super dislocations dominate deformation at low temperature, while ordinary dislocations and twining dominate at high temperature. In lamellar γ-TiAl, on the other hand, twinning is the most important deformation mechanism at room temperature, although high density and glide of ordinary dislocations is also observed, while super dislocations only become significant at high temperature.

Katzarov et al. (2007) studied the core structure and glide of the (1/2)<110] screw dislocations in both single-phase and laminar γ-TiAl. In the ideal single-phase material, a nonplanar core structure is predicted, in which the dislocation is distributed symmetrically on two {111} planes. As a shear stress is applied, the dislocation begins to distort asymmetrically along the {111} planes, with the calculated value of the CRSS when the dislocation begins to move at $0.015 C_{44} = 1.64$ GPa. The resulting dislocation glide is characterized by zigzag motion on the two {111} planes. Dislocation behavior in laminar γ-TiAl was modeled in two lamellae forming a twin-γ/γ interface. The nonplanar core of the ideal screw dislocation is distorted asymmetrically when its elastic center is close to the interface, but the dislocation still glides on a {111} plane. The ordinary dislocation becomes an ordinary interfacial dislocation when it reaches the interface. With increasing applied stress, it can glide into the adjacent lamella, leaving no remnant interfacial dislocation.

Katzarorv and Paxton (2009) later investigated the corresponding core structure and glide of the <101] super dislocations in single-phase γ-TiAl, including mechanisms by which these dislocations can dissociate or become sessile by self-locking. The core structure was examined by starting with initially unrelaxed configurations corresponding to previously suggested super dislocation dissociations. A rich and complex set of possible planar (glissile) and nonplanar (sessile) core structures was found, and the response of each to an applied shear stress was studied.

10.6 Treating pure transition-metal alloys with the MGPT

In this section, we turn to the important problem of simulating pure transition-metal alloys with QBIPs, including treatment of both standard binary AB alloys, such as Ta_xW_{1-x}, and equiatomic, multi-component ABCD... alloys of current interest, such as NbMoTaW, with a concentration of 0.25 for each element. The latter systems are known as high-entropy alloys (HEAs) because of the high entropy of mixing of the alloying elements (Steurer, 2020). In both cases, we focus here on single-phase bcc alloys composed of central transition elements from groups VB and VIB (V, Nb, Ta, Cr, Mo, W). In terms of fabricating promising HEAs for high-temperature applications, these transition elements are also sometimes combined with the group-IV elements (Ti, Zr and Hf), which are bcc metals at high temperature and likewise refractory materials. HEAs made from any of these nine elements with concentrations in the range of 0.05 to 0.35 are often referred to as refractory high-entropy alloys (RHEAs) (Senkov et al., 2018).

Much of the current interest in RHEAs concerns their potential as high-temperature alloys with superior mechanical properties. Early fabrication and experimental study of the prototype systems NbMoTaW and VNbMoTaW showed that these systems exhibit exceptional microhardness, together with excellent high-temperature yield strength and good ductility, suggesting that an underlying solid-solution strengthening mechanism is at work here (Senkov et al., 2010). Thermodynamically, it is assumed that at sufficiently high temperature RHEAs are RSAs that form regular solid solutions, but at lower temperatures chemical short-range order (SRO) is expected, and phase transitions to more ordered structures are possible. Recent DFT and MLP studies have addressed some of these basic structural and mechanical issues in NbMoTaW. Kostiuchenco et al. (2019) used the lattice-based ML low-rank-potential (LRP) model of Shapeev (2017) in MC simulations to demonstrate the large impact of local atomic relaxation on ordering phase transitions. Using the simplified SNAP ML model, Li et al. (2020) calculated the dislocation core structure and CRSS for screw and edge dislocations in single crystals, and studied chemical SRO and segregation to grain boundaries in polycrystals. Yin et al. (2020) used DFT calculations to investigate the distribution of dislocation core properties and how they are influenced by chemical SRO. In addition, there has been dislocation-based model building at higher length scales aimed at explaining RHEA solid-solution strengthening. Zhou et al. (2021) have performed atomistically informed kinetic Monte Carlo (KMC) simulations on a prototype VNbTa alloy, which suggest that cross kinks control dislocation motion and material strength over a wide temperature range. Rao et al. (2021) have developed an analytic model of solid-solution strengthening directed at Ti-based RHEAs and built on the mechanisms of dipole dragging at low temperature and jog dragging at high temperature.

The simplified MGPT multi-ion potentials that we have successfully developed and applied to elemental transition metals are also highly compatible with binary TM alloys and RHEAs. For a binary AB alloy, the total-energy functional $E_{\text{tot}}^{\text{AB}}$ given by Eq. (10.6) still applies with components A and B now both transition metals. The form of the MGPT pair potential given by Eq. (5.99) for an elemental TM is generalized to

$$v_2^{\alpha\beta}(r,\Omega,x) = v_2^{\alpha\beta,\,sp}(r,\Omega,x) + v_2^{\alpha\beta,\,hc}(r,\Omega,x) + v_2^{\alpha\beta,\,d}(r,\Omega,x), \quad (10.46)$$

where $\alpha\beta$ = AA, AB or BB. The simple-metal component $v_2^{\alpha\beta,\,sp}$ is given by Eq. (10.15) and can be evaluated from first principles, as in the full GPT. Likewise, the hard-core-repulsion component $v_2^{\alpha\beta,\,hc}$ can be evaluated directly and is given by Eq. (A2.52) in Appendix A2. The remaining d-state component $v_2^{\alpha\beta,\,d}$ is generalized from Eq. (5.95) to

$$v_2^{\alpha\beta,\,d}(r,\Omega,x) = v_a^{\alpha\beta}(\Omega,x)[f_{\alpha\beta}(r)]^4 - v_b^{\alpha\beta}(\Omega,x)[f_{\alpha\beta}(r)]^2, \quad (10.47)$$

where $v_a^{\alpha\beta}$ and $v_b^{\alpha\beta}$ are volume- and concentration-dependent parameters, while the radial function $f_{\alpha\beta}$ is of the form

$$f_{\alpha\beta}(r) = (R_0/r)^{p_{\alpha\beta}}, \quad (10.48)$$

with $R_0 \equiv 1.8 R_{\text{WS}}$ and $p_{\alpha\beta}$ an additional variable parameter.

Assuming that canonical d bands are otherwise maintained in the AB alloy, the form of the three-ion triplet potential in $E_{\text{tot}}^{\text{AB}}$ is generalized from Eq. (5.100) to

$$\begin{aligned}v_3^{\alpha\beta\gamma}(r_1,r_2,r_3,\Omega,x) = \; & v_c^{\alpha\beta\gamma}(\Omega,x) f_{\alpha b}(r_1) f_{\beta\gamma}(r_2) f_{\gamma a}(r_3) L(\theta_1,\theta_2,\theta_3) \\ & + v_d^{\alpha\beta\gamma}(\Omega,x) \Big\{ [f_{\alpha\beta}(r_1) f_{\beta\gamma}(r_2)]^2 P(\theta_3) \\ & + [f_{\beta\gamma}(r_2) f_{\gamma\alpha}(r_3)]^2 P(\theta_1) + [f_{\gamma\alpha}(r_3) f_{\alpha\beta}(r_1)]^2 P(\theta_2) \Big\}, \quad (10.49)\end{aligned}$$

where $\alpha\beta\gamma$ = AAA, AAB, ABB or BBB, and the quantities $v_c^{\alpha\beta\gamma}$ and $v_d^{\alpha\beta\gamma}$ are additional volume- and concentration-dependent parameters for that mix of elements. The angular functions L and P for canonical d-bands in the AB alloy are the same as for an elemental transition metal and are given in analytic form by Eqs. (5.101) and (5.102), respectively, with the three-ion geometry assumed in Eq. (10.49) illustrated in Fig. 5.15(a).

The form of the corresponding four-ion quadruplet potential in $E_{\text{tot}}^{\text{AB}}$ is generalized from Eq. (5.103) to

$$\begin{aligned}v_4^{\alpha\beta\gamma\delta}(r_1,r_2,r_3,r_4,r_5,r_6,\Omega,x) = \; & v_e^{\alpha\beta\gamma\delta}(\Omega,x) \, [f_{\alpha\beta}(r_1) f_{\beta\gamma}(r_2) f_{\gamma\delta}(r_4) f_{\delta\alpha}(r_5) \\ & M(\theta_1,\theta_2,\theta_3,\theta_4,\theta_5,\theta_6) + f_{\alpha\gamma}(r_3) f_{\beta\gamma}(r_2) f_{\beta\delta}(r_6) f_{\delta\alpha}(r_5) \\ & M(\theta_7,\theta_8,\theta_9,\theta_{10},\theta_5,\theta_{12}) + f_{\alpha\beta}(r_1) f_{\beta\delta}(r_6) f_{\gamma\delta}(r_4) f_{\alpha\beta}(r_3) \\ & M(\theta_{11},\theta_{12},\theta_5,\theta_6,\theta_3,\theta_4)\,], \quad (10.50)\end{aligned}$$

where $\alpha\beta\gamma\delta$ = AAAA, AAAB, AABB, ABBB or BBBB, while $v_e^{\alpha\beta\gamma\delta}$ is an additional volume- and concentration-dependent parameter for that mix of elements. The angular function M for canonical d-bands is also the same as for an elemental transition metal, with the full, but algebraically complex, analytic result given in Appendix B of Moriarty (1990a), and the more computationally efficient matrix form given by Eq. (5.114). The specific four-ion geometry used in Eq. (10.50) is that depicted in Fig. 5.16.

For an elemental transition metal at a given volume Ω there are just seven parameters to be determined in an MGPT canonical-d-band treatment: E_{vol}, v_a, v_b, v_c, v_d, v_e and p. We have discussed in Chapter 5 how this can be accomplished in an optimum way for prototype metals such as Ta over a wide range of volumes, using a mix of experimental data and DFT calculations. For the binary AB alloy at volume Ω and concentration x, as considered above in the extended MGPT formalism given by Eqs. (10.46)–(10.50), the number of parameters to be determined increases significantly to 23. At the same time, because A and B are now similar chemical elements, one expects the concentration dependence of $E_{\text{tot}}^{\text{AB}}$ and the multi-ion potentials which define them to be weak. If so, then an obvious path forward to determine these parameters in an efficient way immediately presents itself, and is best illustrated by a specific example. Consider the case of the AB alloy Ta$_x$W$_{1-x}$, which forms a regular bcc substitutional solid solution across all values of concentration x. Then, focus on the 50–50 alloy for $x = 0.5$ at its equilibrium volume Ω_0^{TaW}, which can be treated accurately by DFT calculations, assuming only that the alloy becomes an ordered B2 compound at $T = 0$. With Ta as the A atom, all six of the pure Ta potential parameters, v_a^{AA}, v_b^{AA}, v_c^{AAA}, v_d^{AAA}, v_e^{AAAA} and p_{AA}, can be directly obtained from the elemental Ta MGPT potentials evaluated at the alloy volume $\Omega = \Omega_0^{\text{TaW}}$. Similarly, the six pure tungsten B-atom parameters, v_a^{BB}, v_b^{BB}, v_c^{BBB}, v_d^{BBB}, v_e^{BBBB} and p_{BB}, can be obtained from elemental tungsten MGPT potentials evaluated at $\Omega = \Omega_0^{\text{TaW}}$. The 11 remaining $x = 0.5$ alloy parameters in $E_{\text{tot}}^{\text{AB}}$, which are $E_{\text{vol}}^{\text{AB}}$, p_{AB}, v_a^{AB}, v_b^{AB}, v_c^{AAB}, v_c^{ABB}, v_d^{AAB}, v_d^{ABB}, v_e^{AAAB}, v_e^{AABB} and v_e^{ABBB}, can then be fitted to DFT ML data calculated for the B2 compound, including the cohesive energy, elastic constants, phonons frequencies and point defect energies. The whole procedure can be readily extended to other concentrations x by simply making the MGPT evaluations and DFT calculations at the appropriate equilibrium volume $\Omega_0^{\text{Ta}_x\text{W}_{1-x}}$. This volume can be obtained most easily by simple quadratic interpolation using

$$\Omega_0^{\text{Ta}_x\text{W}_{1-x}} = (1-x)(1-2x)\Omega_0^{\text{Ta}} + 4x(1-x)\Omega_0^{\text{TaW}} + x(2x-1)\Omega_0^{\text{W}}. \quad (10.51)$$

where Ω_0^{Ta} and Ω_0^{W} are the equilibrium volumes of elemental Ta and W, respectively.

As can be done for elemental transition metals, in cases where additional accuracy is needed in the description of the AB alloy, one can improve upon the above canonical-d-band treatment with a corresponding noncanonical-d-band treatment. In the alloy, the angular functions in the three- and four-ion potentials then effectively become material dependent, such that $L \to L_{\alpha\beta\gamma}$, $P \to P_{\alpha\beta\gamma}$ and $M \to M_{\alpha\beta\gamma\delta}$ in Eqs. (10.49) and (10.50). As we discussed in Chapter 5 for elemental transition metals, these functions must also now be obtained by matrix multiplication in terms of appropriately normalized Slater-Koster matrices. In an elemental TM at a given volume, one needs only a single such matrix H_{SK} defined by the bond-integral ratios $dd\sigma/dd\delta$ and $dd\pi/dd\delta$, resulting in two additional MGPT parameters to be determined. In an AB alloy at a given volume, three Slater-Koster matrices are needed: H_{SK}^{AA}, H_{SK}^{AB} and H_{SK}^{BB}, resulting in six additional noncanonical-d-band parameters, and a total of 29 parameters total.

The above formalism and strategy for treating binary transition metal alloys with MGPT multi-ion potentials can be readily extended to ternary, quaternary and quintenary RHEAs as well. The exact growth in both the numbers of two-, three- and four-ion potentials, and the number of canonical- or noncanonical-d-band parameters needed for such a treatment is summarized in Table 10.4. Beyond the binary AB alloy, the number of parameters that need to be determined at a given alloy volume roughly doubles with each additional TM element added, so that over 100 are needed to treat NbMoTaW, while over 200 are needed for VNbMoTaW. While these are large numbers of parameters, they are small compared to the number of parameters typically required to establish MLPs for RHEAs. For example, even using the simplified LRP model, Kostiuchenko et al. (2019) needed some 500 adjustable parameters to treat NbMoTaW. Moreover, if the MGPT parameterization strategy outlined above for a binary AB alloy is applied to RHEAs, the number of alloy parameters needed is roughly half of the totals listed in Table 10.4, and so is only about 50 or so for NbMoTaW, which is then an order of magnitude less than for even the most efficient MLPs. The relative simplicity of the MGPT treatment also bodes well for extending large-scale MGPT-MD simulations from elemental transition metals to binary alloys and RHEAs.

Table 10.4 *Number of MGPT multi-ion potentials and parameters needed to treat bcc elemental metals, binary alloys and refractory high-entropy alloys containing equiatomic concentrations x of group-VB and -VIB transition-metal elements. MGPT treatments are with either canonical d-band parameters (CDB) or noncanonical d-band parameters (NCDB).*

System type	Prototype	x	v_2	v_3	v_4	CDB	NCDB
Elemental: A	Ta	1.00	1	1	1	7	9
Binary: AB	TaW	0.50	3	4	5	23	29
Ternary: ABC	MoTaW	0.33	6	10	15	54	66
Quaternary: ABCD	NbMoTaW	0.25	10	20	35	106	126
Quintenary: ABCDE	VNbMoTaW	0.20	15	35	70	186	216

11
Local Volume Effects on Defects and Free Surfaces

In the final three chapters of this book, we consider several important extensions of the volume-dependent QBIP treatments of bulk nontransition and transition metals that we have developed and applied in the preceding chapters. In this chapter, we transform the bulk GPT total-energy functional $E_{\text{tot}}(\mathbf{R}, \Omega)$ for elemental metals, as given by Eq. (7.18), from a volume representation, based on the *global* atomic volume Ω, to an electron-density representation, based on the *local* valence electron density n_{val}. This local electron density (LED) representation of the GPT, which was first proposed by Moriarty and Phillips (1991), allows one to treat surface energies directly, as well as local volume effects on defect energies, which, as we indicated in Chapter 9, are essential for an accurate calculation of the vacancy formation energy in nontransition metals. Also possible in the case of transition metals is an alternate hybrid method proposed by Moriarty and Phillips (1991), which retains the LED framework for the non-d electrons, but for the d electrons joins the MGPT with a compatible local density-of-states (LDOS) representation of the d-state energy contributions. This hybrid method has shown promise for treating surface relaxation and reconstruction in central transition metals. More recently, Skinner et al. (2019) have refined and extended the LED version of GPT to include a first-principles calculation of forces and stresses at the pair-potential level, in what we now call the *adaptive* GPT or aGPT. This advance has paved the way for fully relaxed defect and surface calculations, as well as dynamic aGPT-MD simulations, in nontransition metals and series-end transition metals such as Ni and Cu. Further extension of the aGPT to include the multi-ion forces and stresses needed for central transition metals should be possible and is on the horizon for future development.

11.1 Local-density representations of the GPT and their application

11.1.1 Electron-density modulation

We first discuss the LED representation of the GPT obtained through the process of electron-density modulation. For simplicity and clarity, we begin at the pair-potential

11.1.1 Electron-density modulation

level relevant to nontransition metals in the SM, EDB and FDB limits of GPT, where the total electron density consists of a *volume-independent* inner-core plus occupied d-state density localized about each site, corresponding to $Z_d = 0$ or 10 d electrons per atom, and a *volume-dependent* valence electron density n_{val}, corresponding to the remaining Z electrons per atom. For any perfect single crystal with N equivalent ion positions, the bulk volume-dependent total-energy functional

$$E_{\text{tot}}(\mathbf{R}, \Omega) = NE_{\text{vol}}(\Omega) + \frac{1}{2} {\sum_{i,j}}' v_2(R_{ij}, \Omega), \qquad (11.1)$$

can then be transformed exactly to an electron-density-dependent functional of the form

$$E_{\text{tot}}(\mathbf{R}, n_{\text{val}}) = \sum_i E_{\text{vol}}(\bar{n}_i) + \frac{1}{2} {\sum_{i,j}}' v_2(R_{ij}, \bar{n}_{ij}), \qquad (11.2)$$

where $\bar{n}_i$ is an average value of the valence electron density n_{val} on the site i, and $\bar{n}_{ij}$ is the arithmetic average density

$$\bar{n}_{ij} = \frac{1}{2}(\bar{n}_i + \bar{n}_j). \qquad (11.3)$$

Thus, in the total-energy functional the global atomic volume Ω is replaced by the local electron densities $\bar{n}_i$ and $\bar{n}_{ij}$, although for the perfect crystal this is only a redefinition of variables with E_{tot} unchanged and all quantities still determined from first principles. The step forward achieved by Moriarty and Phillips (1991) was then, as an *ansatz*, to extend the application of Eq. (11.2) to all ion configurations in the metal, including free surfaces and defects. Calculating the formation energy for a free surface, or even a bulk defect that comes with a significant free volume, such as a vacancy, from Eq. (11.1) at constant Ω produces no contribution from the volume term $E_{\text{vol}}(\Omega)$, and consequently, a significant underestimate of the result. In contrast, as one can qualitatively appreciate from Fig. 11.1(a) for the case of Mg, a large contribution to the missing positive formation energy for surfaces and vacancies can be obtained from $E_{\text{vol}}(\bar{n}_i)$ in Eq. (11.2), because $\bar{n}_i$ is lower near a surface or vacancy site than at a bulk ion site. More generally in Eq. (11.2), the contributions of the volume term and the pair potential to E_{tot} are *modulated* by the local electron density. Qualitatively in this regard, the volume term then plays an analogous role in the GPT to the embedding term in the empirical EAM pair functional given by Eq. (1.23), as was pointed out in Chapter 1, and as is illustrated explicitly in Fig. 1 of Moriarty and Phillips (1991).

To implement the local-density GPT in real materials, Moriarty and Phillips (1991) developed a robust approach to calculate $\bar{n}_i$ from first principles. From the underlying volume-dependent GPT representation, one first recalls that the valence electron density n_{val} consists of the familiar components discussed in Chapters 3 and 4, namely, a uniform

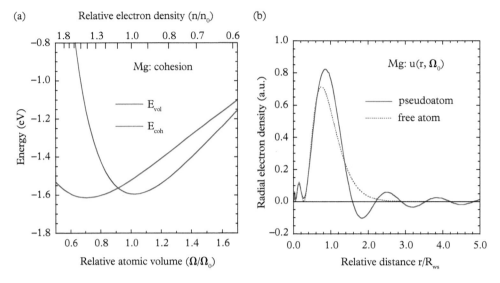

Fig. 11.1 *Key functionals in the LED representation of GPT, as illustrated for Mg, where $\Omega_0 = 156.8$ a.u. (a) Cohesion curve $E_{coh} = E_{tot}/N$ and volume term E_{vol} relevant to Eqs. (11.1) and (11.2), with $n_0 = Z/\Omega_0$ the average valence electron density for the bulk metal. (b) Radial valence electron density, $u(r, \Omega_0) = 4\pi r^2 n(r, \Omega_0)$, calculated for the self-consistent pseudoatom in the bulk metal ($n = n_{pa}$) and compared with that for the corresponding free atom ($n = n_{fa}$).*

free-electron density $n_{unif} = Z/\Omega$ plus small oscillatory and charge-neutral screening and orthogonalization-hole components, δn_{scr} and δn_{oh}:

$$n_{val}(\mathbf{r}) = n_{unif} + \delta n_{scr}(\mathbf{r}) + \delta n_{oh}(\mathbf{r})$$
$$= (Z^*/Z)n_{unif} + \sum_{\mathbf{q}}{}' S(\mathbf{q})n_{scr}(q)\exp(i\mathbf{q}\cdot\mathbf{r}) + \sum_i n_{oh}(\mathbf{r} - \mathbf{R}_i), \quad (11.4)$$

where we have used Eqs. (3.3), (3.5), (3.24) and (3.65). For application to the local-density total-energy given by Eq. (11.2), one then notes that without approximation one can rewrite n_{val} as a superposition of self-consistently screened pseudoatom densities:

$$n_{val}(\mathbf{r}) = \sum_i n_{pa}(\mathbf{r} - \mathbf{R}_i), \quad (11.5)$$

a result first recognized for simple metals by Ziman (1964). The exact form of the GPT pseudoatom density $n_{pa}(\mathbf{r})$ can be readily derived from Eq. (11.4). To do this, one inserts the explicit form of the structure factor

$$S(\mathbf{q}) = \frac{1}{N}\sum_i \exp(-i\mathbf{q}\cdot\mathbf{R}_i) \quad (11.6)$$

into Eq. (11.4) and then adds the $\mathbf{q} = 0$ term to the summation over $\mathbf{q}$ to account for the net uniform density $Z^* n_{unif}/Z$. Finally, converting the summation to an integral over $\mathbf{q}$, one infers that the single-site pseudoatom density is given by

$$n_{\text{pa}}(\mathbf{r}) = \frac{\Omega}{(2\pi)^3} \int n_{\text{scr}}(q) \exp(i\mathbf{q} \cdot \mathbf{r}) d\mathbf{q} + n_{\text{oh}}(\mathbf{r}), \tag{11.7}$$

with $n_{\text{scr}}(q)$ and $n_{\text{oh}}(\mathbf{r})$ to be obtained from Eqs. (3.65) and (3.39), respectively, in the SM limit, and from Eqs. (4.47) and (4.39), respectively, in the EDB or FDB limits. The calculated first-principles GPT pseudoatom density $n_{\text{pa}}(\mathbf{r})$ for Mg treated as a simple metal at equilibrium conditions is illustrated in Fig. 11.1(b) and compared there with the corresponding free-atom density $n_{\text{fa}}(\mathbf{r})$ for the valence $3s$ and $3p$ electrons. In the inner-core region, the pseudoatom well reproduces the density and core oscillations of the free atom. Beyond that point, the pseudoatom density is pushed outward relative to the free atom density, and $n_{\text{pa}}(\mathbf{r})$ has the familiar long-range screening oscillations.

To apply $n_{\text{pa}}(\mathbf{r})$ to the LED total-energy functional $E_{\text{tot}}(\mathbf{R}, n_{\text{val}})$ one needs to define $\bar{n}_i$ more explicitly in Eq. (11.2). To do this, we first combine Eqs. (11.4) and (11.5), and then spatially average the result around a particular ion site i with a localized weighting function $f_{\text{w}}(\mathbf{r} - \mathbf{R}_i)$ to obtain a basic constraining equation that both defines $\bar{n}_i$ and preserves the bulk properties in the perfect crystal:

$$n_{\text{unif}} = \sum_j \bar{n}_{\text{pa}}^{ij} - \delta\bar{n}_{\text{scr}}^i - \delta\bar{n}_{\text{oh}}^i \equiv \bar{n}_i, \tag{11.8}$$

where

$$\bar{n}_{\text{pa}}^{ij} = \int f_{\text{w}}(\mathbf{r} - \mathbf{R}_i) n_{\text{pa}}(\mathbf{r} - \mathbf{R}_j) d\mathbf{r}, \tag{11.9}$$

$$\delta\bar{n}_{\text{scr}}^i = \int f_{\text{w}}(\mathbf{r} - \mathbf{R}_i) \delta n_{\text{scr}}(\mathbf{r}) d\mathbf{r} \tag{11.10}$$

and

$$\delta\bar{n}_{\text{oh}}^i = \int f_{\text{w}}(\mathbf{r} - \mathbf{R}_i) \delta n_{\text{oh}}(\mathbf{r}) d\mathbf{r}. \tag{11.11}$$

Here the localized weighting function $f_{\text{w}}(\mathbf{r} - \mathbf{R}_i)$ is assumed to be properly normalized,

$$\int f_{\text{w}}(\mathbf{r} - \mathbf{R}_i) d\mathbf{r} = 1, \tag{11.12}$$

but is otherwise arbitrary and free to be optimized, if desired.

To develop a practical scheme that could be usefully extended to surfaces and defects, Moriarty and Phillips (1991) made a few additional assumptions and approximations. For the weighting function $f_{\text{w}}(\mathbf{r})$, they considered only the simplest possible form corresponding to a spherical average:

$$f_w(r) = \begin{cases} 1/V_w & r < R_a \\ 0 & r > R_a \end{cases}, \tag{11.13}$$

where V_w is the volume of the averaging sphere of radius R_a. Thus, the weighting function is reduced to the single input parameter R_a. They then broke the average density $\bar{n}_i$ into an effective on-site component $\bar{n}_a^i$ and an off-site or background component $\bar{n}_b^i$:

$$\bar{n}_i = \bar{n}_a^i + \bar{n}_b^i, \tag{11.14}$$

where

$$\bar{n}_a^i = \bar{n}_{pa}^{ii} - \delta\bar{n}_{scr}^i - \delta\bar{n}_{oh}^i \tag{11.15}$$

and

$$\bar{n}_b^i = \sum_{j \neq i} \bar{n}_{pa}^{ij}. \tag{11.16}$$

Moriarty and Phillips (1991) further assumed that $\bar{n}_a^i$ acts as a constant and only directly evaluated $\bar{n}_b^i$ as the independent variable, with $\bar{n}_a^i$ fixed by the bulk constraint given by Eq. (11.8). For vacancy and surface calculations at normal-density, they also neglected the volume dependence of the pseudoatom density $n_{pa}(\mathbf{r})$ and used only the $\Omega = \Omega_0$ form, which is plotted in Fig. 11.1(b) for the case of Mg. In addition to $E_{vol}(\Omega)$ and $v_2(r, \Omega)$, the only input quantities for a given material are then $n_{pa}(\mathbf{r})$ evaluated at $\Omega = \Omega_0$ and the weighting-sphere radius R_a.

In this local-density GPT scheme, the background density function $\bar{n}_{pa}^{ij} \equiv n_b(R_{ij}, \Omega_0)$ depends strongly on the choice of R_a, but at the same time, the fine details of $n_{pa}(\mathbf{r})$ are largely washed out for any value of R_a. This is demonstrated for Mg in Fig. 11.2 by comparing $n_b(R_{ij}, \Omega_0)$ calculated with $n_{pa}(\mathbf{r})$ and with the corresponding free-atom density $n_{fa}(\mathbf{r})$ for two widely different values of R_a. For intuitively plausible values of R_a in the range $R_a/R_{WS} = 1 - 2$, $n_b(R_{ij}, \Omega_0)$ looks much like the Gaussian function assumed in the empirical local-density models of Rosenfeld and Stott (1987) and Finnis et al. (1998), as shown in Fig. 11.2(a) for $R_a/R_{WS} = 1.8$. For larger values of R_a in the range $R_a/R_{WS} = 2 - 4$, $n_b(R_{ij}, \Omega_0)$ is nearly constant across the first two near-neighbor shells before dropping off rapidly, as shown in Fig. 11.2(c) for $R_a/R_{WS} = 3.4$. This latter behavior is nominally more consistent with the assumptions made above, because $\delta\bar{n}_{scr}^i$ and $\delta\bar{n}_{oh}^i$ will vanish in the limit of large R_a. In practice, however, most calculated physical properties are found to be only weakly dependent on the chosen value of R_a within the range $1.3 \leq R_a/R_{WS} \leq 3.4$. Within this range, R_a can be optimized, if desired.

The above local-density formalism can be readily extended to transition metals with a few straightforward modifications. First, one must recognize that in the bulk TM

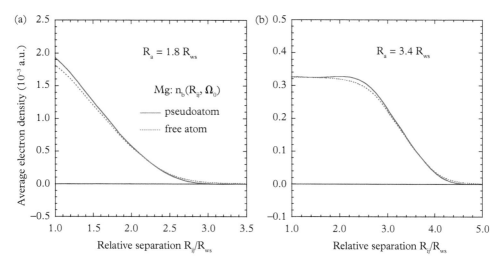

Fig. 11.2 *Background density function $n_b(R_{ij}, \Omega_0)$ calculated for Mg based on self-consistent pseudoatom and free-atom electron densities, with two different choices of weighting sphere radius R_a. (a) $R_a = 1.8 R_{WS}$; (b) $R_a = 3.4 R_{WS}$.*

GPT the sp valence and d-state occupation number are now volume-dependent quantities, $Z(\Omega)$ and $Z_d(\Omega)$, where electron transfer between the sp and d bands can take place via sp-d hybridization. In the TM local-density formalism, one thus has the replacements $Z(\Omega) \to Z(\bar{n}_i)$ and $Z_d(\Omega) \to Z_d(\bar{n}_i)$, where electron transfer can then vary site by site, provided such transfer is accommodated in the definition of $\bar{n}_i$. This accommodation can be made by scaling $\bar{n}_a^i$ and $\bar{n}_b^i$ by $Z(\bar{n}_i)/Z_0$ in Eq. (11.14) such that now

$$\bar{n}_i = \frac{Z(\bar{n}_i)}{Z_0} \left(\bar{n}_a^i + \bar{n}_b^i \equiv \bar{n}_i^0 \right), \qquad (11.17)$$

where $Z_0 = Z(n_0)$, with $n_0 = Z(\Omega_0)/\Omega_0$. Ideally, Eq. (11.17) should be solved self-consistently for $\bar{n}_i$, but in practice it is usually adequate to replace $\bar{n}_i$ with $\bar{n}_i^0$ in $Z(\bar{n}_i)$, so that Eq. (11.17) becomes

$$\bar{n}_i \cong \frac{Z(\bar{n}_i^0)}{Z_0} \bar{n}_i^0, \qquad (11.18)$$

which can be directly evaluated.

In the case of central transition metals, one must also account for the presence of angular-force multi-ion potentials in the local-density total-energy functional. Considering contributions though four-ion interactions, Eq. (11.2) is then replaced by

$$E_{\text{tot}}(\mathbf{R}, n_{\text{val}}) = \sum_i E_{\text{vol}}(\bar{n}_i) + \frac{1}{2} {\sum_{i,j}}' v_2(ij, \bar{n}_{ij}) + \frac{1}{6} {\sum_{i,j,k}}' v_3(ijk, \bar{n}_{ijk})$$
$$+ \frac{1}{24} {\sum_{i,j,k,l}}' v_4(ijkl, \bar{n}_{ijkl}), \qquad (11.19)$$

where

$$\bar{n}_{ijk} = \frac{1}{3}(\bar{n}_i + \bar{n}_j + \bar{n}_k) \qquad (11.20)$$

and

$$\bar{n}_{ijkl} = \frac{1}{4}(\bar{n}_i + \bar{n}_j + \bar{n}_k + \bar{n}_l). \qquad (11.21)$$

As in the case of bulk TM applications, local-density calculations on central transition metals are greatly facilitated with the neglect of sp-d hybridization and the use of the MGPT to establish the potentials v_2, v_3 and v_4. Without sp-d hybridization included, it is also appropriate to treat the valance Z as a constant, removing the added scaling factor in Eqs. (11.17) and (11.18), and thus reverting back to Eq. (11.14) to obtain $\bar{n}_i$.

Illustrative applications of the above LED formalism to the calculation of the unrelaxed vacancy formation energy $E_{\text{vac}}^{\text{f}}$ and surface energies $\gamma_{\text{surf}}^{nlm}$ for Mg, Cu and Mo are given in Table 11.1. Some important technical details on the respective local-density treatments of these metals are as follows:

(1) hcp Mg, treated at the GPT pair-potential level ($n = 2$) in the SM limit with $Z = 2$, using Eqs. (11.2) and (11.14) with $c/a = 1.62$ and $R_a/R_{\text{WS}} = 1.8$.

(2) fcc Cu, treated at the GPT pair potential level ($n = 2$) in the TM limit, with $D_2^* = -3.0$, $Z_0 = 1.651$ and full sp-d hybridization, using Eqs. (11.2) and (11.18) with $R_a/R_{\text{WS}} = 1.3$.

(3) bcc Mo, treated at the level of two-, three- and four-ion TM interactions ($n = 2, 3, 4$) obtained from first-generation MGPT potentials based on scheme 1 of Moriarty (1990a), with canonical d bands, $D_2^* = -2.15$, $Z = 1.402$, $p = 4.0$ and neglecting sp-d hybridization, using Eqs. (11.19) and (11.14) with $R_a/R_{\text{WS}} = 1.55$.

The LED calculations of $E_{\text{vac}}^{\text{f}}$ listed in Table 11.1 are based on the use of Eq. (9.1) at the observed equilibrium volume of each metal, with large computational cells of 250 atoms or more. These results are in excellent agreement with DFT calculations as well as with experiment for all three metals. In this regard, note that the bulk component of $E_{\text{vac}}^{\text{f}}$, which is given by Eq. (9.8), and in Table 11.1 is listed under the column labeled $E_{\text{bulk}} : v_n$, is only 29% and 61% of the total vacancy formation energy for Cu and Mg, respectively, while it is a full 94% of the total in Mo. As expected, the LED contribution to $E_{\text{vac}}^{\text{f}}$ from the volume term (E_{vol}) is large and positive for each of the metals. The additional LED contribution from the density dependence of the potentials, coming though the structural part of the energy ($E_{\text{struc}} : \delta v_n$), produces either a partially compensating negative contribution, as in Mg and Cu, or an almost completely cancelling negative contribution, as in Mo. The latter Mo result confirms the validity of the bulk MGPT treatment of point defects in central transition metals that was considered in Chapter 9. Physically, this simplification reflects the fact that defect energetics in central transition metals are dominated by d-electron physics, whose local character is already accounted for though the localized d basis states used to describe these metals in the bulk GPT and MGPT.

Table 11.1 *Unrelaxed vacancy-formation and surface energies, and their components, for hcp Mg, fcc Cu and bcc Mo calculated with the LED method, as discussed in the text. Here $E_{\text{vac}}^{\text{f}}$ and its components are given in eV, while $\gamma_{\text{surf}}^{hkl}$ and its components are given in J/m^2. Relaxed DFT and experimental results are shown for comparison.*

Metal	Energy	$E_{\text{bulk}} : v_n$	E_{vol}	$E_{\text{struc}} : \delta v_n$	E_{total}	DFT	Expt.
Mg[a]:							
	$E_{\text{vac}}^{\text{f}}$	0.44	0.47	−0.19	0.72	0.74[c]	0.79[d]
Cu[b]:							
	$E_{\text{vac}}^{\text{f}}$	0.4	1.9	−0.9	1.4	1.33[e]	1.28±0.05[d]
	$\gamma_{\text{surf}}^{100}$	0.35	1.50	−0.77	1.08	1.44[f]	−
	$\gamma_{\text{surf}}^{110}$	0.09	1.64	−0.41	1.32	−	<1.70>[g]
	$\gamma_{\text{surf}}^{111}$	0.03	1.43	−0.56	0.90	1.30[f]	−
Mo[b]:							
	$E_{\text{vac}}^{\text{f}}$	3.1	4.8	−4.6	3.3	2.8–3.0[h]	3.0–3.2[i]
	$\gamma_{\text{surf}}^{100}$	2.04	3.64	−2.26	3.42	3.15[f]	−
	$\gamma_{\text{surf}}^{110}$	1.66	3.11	−2.33	2.44	2.73[f]	<3.00>[j]

[a]LED results from Skinner et al. (2019).
[b]LED results from Moriarty and Phillips (1991).
[c]Uesugi et al. (2003) calculated with the PP method.
[d]Ehrhart (1991).
[e]Korzhavyi et al. (1999) calculated with the LSGF method.
[f]Lee et al. (2018) calculated with the PAW method.
[g]Mills and Su (2006) surface energy for an average face.
[h]See Table 9.2.
[i]Schultz (1991).
[j]de Boer et al. (1988) surface energy for an average face.

The LED calculations of surface energies γ_surf^{hkl} given in Table 11.1 for Cu and Mo were obtained with a semi-infinite slab geometry, where for an unrelaxed fcc or bcc structure at the observed equilibrium volume one has simply

$$\gamma_\text{surf}^{hkl} = \frac{1}{A_{hkl}} \sum_{s=1}^{s_\text{max}} [E_{hkl}(s) - E_\text{coh}], \qquad (11.22)$$

with the summation extending over single atomic layers, beginning with and parallel to the free surface, and then proceeding into the bulk crystal. Here, $E_{hkl}(s)$ is the total energy per atom in layer s, E_coh is the bulk equilibrium cohesive energy and A_{hkl} is the surface area per atom. The maximum surface layer one needs to calculate, s_max, is where $E_{hkl}(s_\text{max}) = E_\text{coh}$. Typically, for low-index, high-symmetry surfaces of central transition metals like Mo, s_max is in the range 5–9. Generally higher values of s_max are needed for series-end metals like Cu due to long-range sp-d hybridization contributions.

As indicated in Table 11.1, the LED γ_surf^{hkl} surface-energy results so calculated for Cu and Mo are in reasonable agreement with both recent DFT calculations (Lee et al., 2018) and with experimental estimates based on the extrapolation of liquid surface tension measurements. As expected, the bulk component of γ_surf^{hkl} is generally a smaller fraction of the total than is the case for E_vac^f: only 32%, 7% and 3% for the (100), (110) and (111) surfaces, respectively, in Cu, while 60% and 68% for the (100) and (110) surfaces in Mo. It is again the volume term that makes the largest positive contribution to γ_surf^{hkl} in all cases, with a partially compensating negative contribution from the density dependence of the potentials.

11.1.2 Local density-of-states modulation

In addition to the above LED approach to local volume effects in metals, Moriarty and Phillips (1991) proposed a second hybrid method for transition metals based on the local density-of-states modulation of d-state energy contributions to the total energy. This LDOS method is especially compatible with the simplified MGPT description of the d bands in central transition metals like Mo. In this formulation, one relates atomic volume Ω to the local second moment of the d-band density of states μ_2^i through the d-band width W_d. As we discussed in Chapter 5 and illustrated in Fig. 5.5, W_d typically varies with volume as $\Omega^{-p/3}$ for transition metals, where $p = 5$ for ideal canonical d bands, but more generally, p is a material parameter with $p = 4$ for Mo. At the same time, a TB representation of the d-band electronic structure shows that $W_d \propto (\mu_2^i)^{1/2}$, as discussed in Sec. 2.6.3 of Chapter 2. This result suggests a general transformation for the pure d-state energy components in the total energy E_tot of the form

$$(\Omega_0/\Omega)^{p/3} \to (\mu_2^i/\mu_2^{i0})^{1/2}, \qquad (11.23)$$

where μ_2^{i0} is the second moment of the equilibrium bulk crystal structure.

Proceeding further, one can partition the volume term E_vol in the total energy into an sp or non-d component E_vol^{sp} and a pure d component E_vol^d. Assuming that $E_\text{vol}^d \propto W_d$, one can write $E_{\text{vol}}^d(\Omega) = E_{\text{vol}}^d(\Omega_0)(\Omega_0/\Omega)^{p/3}$, which thus transforms as

$$NE_{\text{vol}}^d(\Omega) \to E_{\text{vol}}^d(\Omega_0) \sum_i (\mu_2^i/\mu_2^{i0})^{1/2} \equiv \sum_i E_{\text{vol}}^d(\mu_2^i). \quad (11.24)$$

One can then calculate and transform E_{vol}^{sp} as

$$NE_{\text{vol}}^{sp}(\Omega) = N\left[E_{\text{vol}}(\Omega) - E_{\text{vol}}^d(\Omega_0)(\Omega_0/\Omega)^{p/3}\right] \to \sum_i E_{\text{vol}}^{sp}(\bar{n}_i). \quad (11.25)$$

The quantity $E_{\text{vol}}^d(\Omega_0)$ acts a splitting parameter in the theory between the LED non-d and LDOS d contributions to E_{vol}, and in practice can be used as a variable quantity that can be optimized to advantage. For their application to Mo, Moriarty and Phillips (1991) chose this parameter so that the unrelaxed vacancy formation energy E_{vac}^f calculated with the hybrid LDOS approach would equal that calculated with the LED method, as given in Table 11.1. This procedure gave $E_{\text{vol}}^d(\Omega_0) = -2.7$ eV.

Noting the general form of the multi-ion potentials in the bulk MGPT: $v_2 = v_2^{sp} + v_2^{hc} + v_2^d$ from Eq. (5.99), $v_3 = v_3^d$ and $v_4 = v_4^d$, one then expects the hybrid LDOS total-energy functional to be in the form

$$E_{\text{tot}}(\mathbf{R}, \bar{n}_i, \mu_2^i) = \sum_i \left[E_{\text{vol}}^{sp}(\bar{n}_i) + E_{\text{vol}}^d(\mu_2^i)\right] + \frac{1}{2}\sum_{i,j}{}' \left[v_2^{sp}(ij, \bar{n}_{ij}) + v_2^{hc}(ij, \bar{n}_{ij}) + v_2^d(ij, \mu_2^{ij})\right]$$

$$+ \frac{1}{6}\sum_{i,j,k}{}' v_3^d(ijk, \mu_2^{ijk}) + \frac{1}{24}\sum_{i,j,k,l}{}' v_4^d(ijkl, \mu_2^{ijkl}), \quad (11.26)$$

where in analogy with the LED multi-ion total-energy functional given by Eq. (11.19), $\mu_2^{ij\cdots}$ is the arithmetic average $(\mu_2^i + \mu_2^j + \cdots)/n$ for the potential v_n^d with $n = 2$, 3 or 4. The specific dependence of the potentials v_n^d on these averages can be determined from their bulk analytic form, as given by Eq. (5.95) for v_2^d, Eq. (5.100) for v_3^d and Eq. (5.103) for v_4^d. To extract the full volume dependence from each potential and transform it a local moment dependence, we assume that the five volume-dependent coefficients $v_\alpha(\Omega)$ entering the potentials are all proportional to the d-band width, $v_\alpha \propto W_d$, where $\alpha = a, b, c, d$ or e. In addition, multiplying each coefficient $v_\alpha(\Omega)$ are n volume-dependent radial functions $f(r) = (1.8R_{\text{WS}}/r)^p$, so the total volume dependence can be extracted in the form

$$v_\alpha(\Omega)[f(r)]^n = v_\alpha(\Omega_0)[f_0(r)]^n (\Omega/\Omega_0)^{(n-1)p/3}, \quad (11.27)$$

where $f_0(r)$ is $f(r)$ evaluated at $\Omega = \Omega_0$. Transforming the two-ion potential v_2^d, with $n = 4$ for $\alpha = a$ and $n = 2$ for $\alpha = b$, one thus has

$$v_a(\Omega_0)(\Omega/\Omega_0)^p \to v_a(\Omega_0)(\mu_2^0/\mu_2^{ij})^{3/2} \quad (11.28a)$$

and
$$v_b(\Omega_0)(\Omega/\Omega_0)^{p/3} \to v_b(\Omega_0)(\mu_2^0/\mu_2^{ij})^{1/2}, \qquad (11.28b)$$

respectively. Transforming the three-ion potential v_3^d, with $n=3$ for $\alpha = c$ and $n=4$ for $\alpha = d$, one has

$$v_c(\Omega_0)(\Omega/\Omega_0)^{2p/3} \to v_c(\Omega_0)(\mu_2^0/\mu_2^{ijk}) \qquad (11.28c)$$

and

$$v_d(\Omega_0)(\Omega/\Omega_0)^p \to v_d(\Omega_0)(\mu_2^0/\mu_2^{ijk})^{3/2}, \qquad (11.28d)$$

respectively. Finally, transforming the four-ion potential v_4^d, with $n=4$ for $\alpha = e$, gives

$$v_e(\Omega_0)(\Omega/\Omega_0)^p \to v_e(\Omega_0)(\mu_2^0/\mu_2^{ijkl})^{3/2}. \qquad (11.28e)$$

Note that all of the second-moment modulation factors entering Eqs. (11.28a)–(11.28e) can be evaluated in terms of internal MGPT quantities alone. That is, the same d-state matrix elements that define the bulk MGPT potentials also establish the second moment μ_2^i. Thus, no additional parameters beyond $E_{\text{vol}}^d(\Omega_0)$ are necessary to implement the LDOS method in practice.

In Table 11.2 we compare LDOS calculations of the (100) and (110) unrelaxed surface energies for bcc Mo, as obtained by Moriarty and Phillips (1991), with the corresponding LED results that are given in Table 11.1. Excellent agreement between the two methods, as well as with DFT calculations and experimental data, is demonstrated, confirming the validity of both approaches. For central transition metals such as Mo, however, the LDOS method potentially provides enhanced sensitivity to the surface environment by greatly reducing the large cancellation between the modulated volume and potential contributions that is inherent in the LED method, as shown in Table 11.1.

The special advantage of the LDOS scheme in this regard is further demonstrated in Table 11.2, where the more subtle question of surface relaxation is also addressed. For the normal close-packed (110) surface of Mo, both the LED and LDOS treatments can explain the observed small inward relaxation of the first layer of the surface, as is measured in the low energy electron diffraction (LEED) experiments of Morales de la Garza and Clarke (1981), and is also well predicted by the DFT calculations of Lee et al. (2018). For the more anomalous open-packed (100) surface of Mo, however, the large observed relaxation of the first surface layer via LEED by Clarke (1980), and also earlier by Ignatiev et al. (1975), is well captured by the LDOS method, but not at all by the LED approach. Moreover, as shown in Table 11.2, the calculated magnitude of the relaxation in the LDOS treatment is in good agreement with both DFT calculations and experiment.

The (100) surface of Mo, as well as that of its group-VIB neighbor W, is also known to reconstruct. In the ideal case of W, one has a commensurate 2×2 reconstruction with an alternating $\langle 011 \rangle$ lateral shift of surface atoms. This 2×2 reconstruction model then serves as a prototype that is believed to underpin the actual observed incommensurate

Table 11.2 *Unrelaxed surface energies ($\gamma_{\text{surf}}^{hkl}$) and surface relaxations ($\Delta_{mn}, \Delta E_{hkl}$) for bcc Mo, as calculated with the LED and LDOS methods discussed in the text. Units: $\gamma_{\text{surf}}^{hkl}$ in J/m^2; Δ_{mn} in percentage change of spacing between layers m and n; E_{hkl} in mRy per surface atom. Relaxed DFT and experimental results are shown for comparison.*

	Quantity	LED[a]	LDOS[a]	DFT: MBPP[b]	DFT: PAW[c]	Expt.
Energies:						
	$\gamma_{\text{surf}}^{100}$	3.42	3.31	3.53	3.15	–
	$\gamma_{\text{surf}}^{110}$	2.44	2.20	–	2.73	<3.00>[d]
Relaxations:						
(100):	Δ_{12}	+0.4	−10.2	−10.7	−13.05	−9.5±2[e]
	Δ_{23}	−0.1	+1.3	+2.7	+4.20	–
	ΔE_{100}	−0.01	−8.1	−8	–	–
(110):	Δ_{12}	−1.7	−5.2	–	−4.74	−1.6±2[f]
	Δ_{23}	+1.5	+1.8	–	+0.73	–
	ΔE_{110}	−0.7	−3.7	–	–	–

[a]Moriarty and Phillips (1991).
[b]Wang et al. (1988).
[c]Lee et al. (2018).
[d]de Boer et al. (1988) surface energy for an average face.
[e]Clarke (1980) LEED measurement on reconstructed surface.
[f]Morales de la Garza and Clarke (1981) LEED measurement.

2.2 × 2.2 reconstruction of the Mo(100) surface. Neither the LED or LDOS treatments used in Table 11.2 stabilize the 2 × 2 reconstruction for Mo, but the LDOS treatment comes the closest, and the result is sensitive to the value of splitting parameter $E_{\text{vol}}^d(\Omega_0)$ used in Eq. (11.26) for the total-energy functional. Moriarty and Phillips (1991) found that if the magnitude of $E_{\text{vol}}^d(\Omega_0)$ is continuously increased, the unrelaxed (100) surface of Mo would reconstruct for $E_{\text{vol}}^d(\Omega_0) < -3.4$ eV. Moreover, as shown in Fig. 2 of their paper, for $E_{\text{vol}}^d(\Omega_0) = -4.1$ eV quite realistic reconstruction energetics are obtained, with an energy lowering of $\Delta E_{100} \simeq -0.7$ mRy per surface atom below the unreconstructed surface and a relative lateral shift of the surface atoms of $\delta/d_L \simeq 0.1$. This outcome suggests that with a fully optimized LDOS treatment one may be able to simultaneously describe both the relaxation and the reconstruction of the Mo(100) surface.

11.2 First-principles forces and stresses: the aGPT

The adaptive GPT or aGPT of Skinner et al. (2019) is a refinement and major extension of the LED representation of GPT for nontransition metals due to Moriarty and Phillips

(1991), as elaborated in Sec. 11.1.1. Skinner et al. (2019) began with the basic defining equations and assumptions of the LED formalism, as expressed in Eqs. (11.1)–(11.16), and then added one important refinement. Specifically, they noted that the weighting function $f_w(r)$, defined by Eq. (11.13) and used in the LED formalism to calculate the needed electron-density averages, is discontinuous at the averaging-sphere radius R_a. While this discontinuity only negligibly impacts the calculation of defect and surface energies in the LED and LDOS methods, it will lead to discontinuous radial derivatives of the total energy and hence severely limit one's ability to calculate accurate forces and stresses. Skinner et al. (2019) solved this problem by replacing Eq. (11.13) by the closely related, but smooth and continuous function

$$f_w(r) = \frac{1}{N_w} \begin{cases} \left[1 + \alpha(r/R_a - 1)^2\right] \exp\left[-\alpha(r/R_a - 1)^2\right] & r < R_a \\ 1 & r > R_a \end{cases}, \quad (11.29)$$

where N_w is the effective normalization volume

$$N_w = V_w + \left[\frac{8\pi}{\alpha} + \frac{5\pi^{3/2}}{2\alpha^{3/2}} + \frac{3\pi^{3/2}}{\alpha^{1/2}}\right] R_a^3, \quad (11.30)$$

which exactly preserves Eq. (11.12) for any value of the additional constant α. Note that the functional form of $f_w(r)$ in Eq. (11.29) is the same as that for the cutoff function $f_{cut}(r)$ defined in Eq. (3.112) and used to terminate smoothly GPT pair potentials at long range. Formally, one has $f_w(r) = (1/N_w) f_{cut}(r)$ with R_0 replaced by R_a, although in practice a smaller value of α is appropriate for the intended applications of Eq. (11.29). Skinner et al. (2019) recommended a value of $\alpha = 25$ for use in the aGPT, and showed in Fig. 3 of their paper that with this value $\bar{n}_{pa}(r)$ and its first two radial derivatives are indeed smooth and continuous functions of r as required.

11.2.1 Force equations and the stress tensor

For nontransition metals treated at constant volume, one can obtain the Cartesian force component μ on the ion i by differentiating the LED total-energy functional given by Eq. (11.2) and expressing the result in terms of three main contributions:

$$F_{i\mu} = -\frac{\partial E_{tot}(\mathbf{R}, n_{val})}{\partial r_{i\mu}}$$

$$= F_{i\mu}^{bulk} + F_{i\mu}^{evol} + F_{i\mu}^{v2}. \quad (11.31)$$

Here $F_{i\mu}^{bulk}$ is of the same general form as the bulk force component in the GPT, but is now derived from the radial derivative of $v_2(R_{ij}, \bar{n}_{ij})$, while $F_{i\mu}^{evol}$ and $F_{i\mu}^{v2}$ are local force components derived from the density derivatives of $E_{vol}(\bar{n}_i)$ and $v_2(R_{ij}, \bar{n}_{ij})$, respectively.

11.2.1 Force equations and the stress tensor

Using Eqs. (7.21) and (7.22), one has for $F_{i\mu}^{\text{bulk}}$ the explicit result

$$F_{i\mu}^{\text{bulk}} = -\sum_{j \neq i} \frac{\partial v_2(R_{ij}, \bar{n}_{ij})}{\partial R_{ij}} \frac{r_{ij\mu}}{R_{ij}}. \tag{11.32}$$

To obtain a comparable form for the second force component $F_{i\mu}^{\text{evol}}$, we first write

$$F_{i\mu}^{\text{evol}} = -\frac{\partial}{\partial r_{i\mu}} \sum_j E_{\text{vol}}(\bar{n}_j) = -\frac{\partial E_{\text{vol}}(\bar{n}_i)}{\partial \bar{n}_i} \frac{\partial \bar{n}_i}{\partial r_{i\mu}} - \sum_{j \neq i} \frac{\partial E_{\text{vol}}(\bar{n}_j)}{\partial \bar{n}_j} \frac{\partial \bar{n}_j}{\partial r_{i\mu}}. \tag{11.33}$$

Then, combining Eqs. (11.14) and (11.16) to obtain

$$\bar{n}_i = \bar{n}_a^i + \sum_{j \neq i} \bar{n}_{\text{pa}}^{ij}, \tag{11.34}$$

and noting that at fixed Ω with $\bar{n}_a^i$ taken as a constant and $\bar{n}_{\text{pa}}^{ij} = \bar{n}_{\text{pa}}^{ij}(R_{ij})$, one has

$$\frac{\partial \bar{n}_i}{\partial r_{i u}} = \sum_{j \neq i} \frac{\partial \bar{n}_{\text{pa}}^{ij}}{\partial r_{i u}} = \sum_{j \neq i} \frac{\partial \bar{n}_{\text{pa}}^{ij}}{\partial R_{ij}} \frac{r_{ij\mu}}{R_{ij}} \tag{11.35}$$

and

$$\frac{\partial \bar{n}_j}{\partial r_{i u}} = \frac{\partial \bar{n}_{\text{pa}}^{ij}}{\partial R_{ij}} \frac{r_{ij\mu}}{R_{ij}}, \tag{11.36}$$

where, consistent with our notation in Eq. (7.22), $r_{ij\mu} = r_{i\mu} - r_{j\mu}$. Finally, substituting Eqs. (11.35) and (11.36) back into Eq. (11.33) gives the end result

$$F_{i\mu}^{\text{evol}} = -\sum_{j \neq i} \left[\frac{\partial E_{\text{vol}}(\bar{n}_i)}{\partial \bar{n}_i} + \frac{\partial E_{\text{vol}}(\bar{n}_j)}{\partial \bar{n}_j} \right] \frac{\partial \bar{n}_{\text{pa}}^{ij}}{\partial R_{ij}} \frac{r_{ij\mu}}{R_{ij}}. \tag{11.37}$$

The third force component $F_{i\mu}^{\text{v2}}$ can be processed in a similar manner but involves extra ion-site summations. Using Eq. (11.3), one first has

$$F_{i\mu}^{\text{v2}} = -\frac{1}{2}\frac{\partial}{\partial r_{i\mu}} {\sum_{i,j}}' v_2(R_{ij}, \bar{n}_{ij}) = -\frac{1}{2}\sum_{j \neq i} \frac{\partial v_2(R_{ij}, \bar{n}_{ij})}{\partial \bar{n}_{ij}} \left(\frac{\partial \bar{n}_i}{\partial r_{i\mu}} + \frac{\partial \bar{n}_j}{\partial r_{i\mu}} \right)$$
$$- \frac{1}{4}\sum_{j \neq i}\sum_{k \neq j \neq i} \frac{\partial v_2(R_{jk}, \bar{n}_{jk})}{\partial \bar{n}_{jk}} \left(\frac{\partial \bar{n}_j}{\partial r_{i\mu}} + \frac{\partial \bar{n}_k}{\partial r_{i\mu}} \right). \tag{11.38}$$

Then, substituting Eqs. (11.35) and (11.36), plus the cyclic $j \to k$ permutation of the latter equation,

$$\frac{\partial \bar{n}_k}{\partial r_{iu}} = \frac{\partial \bar{n}_{\text{pa}}^{ik}}{\partial R_{ik}} \frac{r_{ik\mu}}{R_{ik}}, \tag{11.39}$$

all back into Eq. (11.38) yields the final result

$$F_{i\mu}^{\text{v2}} = -\sum_{j \neq i} \frac{\partial v_2(R_{ij}, \bar{n}_{ij})}{\partial \bar{n}_{ij}} \left(\frac{\partial \bar{n}_{\text{pa}}^{ij}}{\partial R_{ij}} \frac{r_{ij\mu}}{R_{ij}} + \frac{1}{2} \sum_{k \neq j \neq i} \frac{\partial \bar{n}_{\text{pa}}^{ik}}{\partial R_{ik}} \frac{r_{ik\mu}}{R_{ik}} \right)$$
$$- \frac{1}{4} \sum_{j \neq i} \sum_{k \neq j \neq i} \frac{\partial v_2(R_{jk}, \bar{n}_{jk})}{\partial \bar{n}_{jk}} \left(\frac{\partial \bar{n}_{\text{pa}}^{ij}}{\partial R_{ij}} \frac{r_{ij\mu}}{R_{ij}} + \frac{\partial \bar{n}_{\text{pa}}^{ik}}{\partial R_{ik}} \frac{r_{ik\mu}}{R_{ik}} \right). \tag{11.40}$$

The above formal aGPT force results, summarized by Eqs. (11.31), (11.32), (11.37) and (11.40), have been implemented by Skinner et al. (2019) to allow fully relaxed defect calculations as well as aGPT-MD simulations at constant volume. Some of their initial applications in this regard are discussed in Sec. 11.2.2.

In addition, as we did in Chapter 7 for the bulk GPT to calculate phonons at constant volume, one may go beyond the above formalism for individual ion forces to develop the full force-constant matrix $K_{\mu\nu}$ for the aGPT. The bulk GPT starting point for $K_{\mu\nu}$ given by Eq. (7.5) is then replaced in the aGPT by

$$K_{\mu\nu}(\mathbf{R}_i - \mathbf{R}_j, n_{\text{val}}) = -\left. \frac{\partial^2 E_{\text{tot}}(\mathbf{R}, n_{\text{val}})}{\partial r_{i\mu} \partial r_{j\nu}} \right|_{\mathbf{R}=\mathbf{R}_0}. \tag{11.41}$$

In this regard, Skinner et al. (2019) considered only the test case of the bulk phonon spectrum, where, of course, the bulk contribution to $K_{\mu\nu}$ dominates the result, and the density-derivative contributions effectively provide only small higher-order corrections. Specifically, they found that the GPT and aGPT phonon spectra so calculated for bcc and hcp Mg, respectively, are in excellent agreement, as is shown in Figs. 4 and 5 of their paper. At the same time, and potentially more interesting, is the modification of the bulk phonon spectrum in the vicinity of defects and surfaces, where localized modes and other phenomena are possible. The aGPT formalism and applications for such cases remain to be developed.

Furthermore, in addition to individual ion forces at constant volume, one can develop the aGPT stress tensor from Eq. (11.2) for an infinitesimal strain $\varepsilon_{\alpha\beta}$ as

$$\sigma_{\alpha\beta} = \frac{1}{N\Omega} \left. \frac{\partial E_{\text{tot}}(\mathbf{R}, n_{\text{val}})}{\partial \varepsilon_{\alpha\beta}} \right|_{\varepsilon_{\alpha\beta}=0}$$
$$= \sigma_{\alpha\beta}^{\text{bulk}} + \sigma_{\alpha\beta}^{\text{evol}} + \sigma_{\alpha\beta}^{\text{v2}}. \tag{11.42}$$

The bulk component of the stress tensor $\sigma_{\alpha\beta}^{\text{bulk}}$, aka the virial component, is just

$$\sigma_{\alpha\beta}^{\text{bulk}} = \frac{1}{2N\Omega} \sum_{i,j}{}' \frac{\partial v_2(R_{ij}, \bar{n}_{ij})}{\partial R_{ij}} \frac{r_{ij\alpha} r_{ij\beta}}{R_{ij}}, \tag{11.43}$$

where we have noted that in the limit $\varepsilon_{\alpha\beta} \to 0$

$$\frac{\partial R_{ij}}{\partial \varepsilon_{\alpha\beta}} \to \frac{r_{ij\alpha} r_{ij\beta}}{R_{ij}}, \tag{11.44}$$

as shown by Skinner et al. (2019). For the first density-derivative component of the stress tensor, $\sigma_{\alpha\beta}^{\text{evol}}$, one first has

$$\sigma_{\alpha\beta}^{\text{evol}} = \frac{1}{N\Omega} \sum_i \frac{\partial E_{\text{vol}}(\bar{n}_i)}{\partial \bar{n}_i} \frac{\partial \bar{n}_i}{\partial \varepsilon_{\alpha\beta}}. \tag{11.45}$$

Here we evaluate the strain derivative of $\bar{n}_i$ allowing for the fact that its contributions from $\bar{n}_a$ and $\bar{n}_{\text{pa}}^{ij}$ can be volume dependent. That is, we now assume $\bar{n}_a = \bar{n}_a(\Omega)$ and $\bar{n}_{\text{pa}}^{ij} = \bar{n}_{\text{pa}}^{ij}(R_{ij}, \Omega)$. Using the following identity obtained in the limit $\varepsilon_{\alpha\beta} \to 0$:

$$\frac{\partial \Omega}{\partial \varepsilon_{\alpha\beta}} \to \Omega \delta_{\alpha\beta}, \tag{11.46}$$

together with Eq. (11.44), one has

$$\frac{\partial \bar{n}_i}{\partial \varepsilon_{\alpha\beta}} = \Omega \left[\frac{dn_a(\Omega)}{d\Omega} + \sum_{j \neq i} \frac{\partial \bar{n}_{\text{pa}}^{ij}(R_{ij}, \Omega)}{\partial \Omega} \right] \delta_{\alpha\beta} + \sum_{j \neq i} \frac{\partial \bar{n}_{\text{pa}}^{ij}(R_{ij}, \Omega)}{\partial R_{ij}} \frac{r_{ij\alpha} r_{ij\beta}}{R_{ij}}. \tag{11.47}$$

Then, using Eq. (11.47) in Eq. (11.45), one has the final result

$$\sigma_{\alpha\beta}^{\text{evol}} = \frac{1}{N} \sum_i \frac{\partial E_{\text{vol}}(\bar{n}_i)}{\partial \bar{n}_i} \left[\frac{dn_a(\Omega)}{d\Omega} + \sum_{j \neq i} \frac{\partial \bar{n}_{\text{pa}}^{ij}(R_{ij}, \Omega)}{\partial \Omega} \right] \delta_{\alpha\beta}$$

$$+ \frac{1}{N\Omega} \sum_{i,j \neq i} \frac{\partial E_{\text{vol}}(\bar{n}_i)}{\partial \bar{n}_i} \frac{\partial \bar{n}_{\text{pa}}^{ij}(R_{ij}, \Omega)}{\partial R_{ij}} \frac{r_{ij\alpha} r_{ij\beta}}{R_{ij}}. \tag{11.48}$$

For the second density-derivative component of the stress tensor, $\sigma_{\alpha\beta}^{\text{v2}}$, one has

$$\sigma_{\alpha\beta}^{\text{v2}} = \frac{1}{4N\Omega} \sum_{i,j}{}' \frac{\partial v_2(R_{ij}, \bar{n}_{ij})}{\partial \bar{n}_{ij}} \left[\frac{\partial \bar{n}_i}{\partial \varepsilon_{\alpha\beta}} + \frac{\partial \bar{n}_j}{\partial \varepsilon_{\alpha\beta}} \right]. \tag{11.49}$$

Then, using Eq. (11.47) to evaluate the derivatives of $\bar{n}_i$ and $\bar{n}_j$ in Eq. (11.49), one arrives at the final result

$$\sigma_{\alpha\beta}^{v2} = \frac{1}{4N} \sum_{i,j}{}' \frac{\partial v_2(R_{ij}, \bar{n}_{ij})}{\partial \bar{n}_{ij}} \left\{ \left[2\frac{dn_a(\Omega)}{d\Omega} + \sum_{k \neq i} \frac{\partial n_{pa}^{ik}(R_{ik}, \Omega)}{\partial \Omega} + \sum_{k \neq j} \frac{\partial n_{pa}^{jk}(R_{jk}, \Omega)}{\partial \Omega} \right] \delta_{\alpha\beta} \right.$$
$$\left. + \frac{1}{\Omega} \left[\sum_{k \neq i} \frac{\partial n_{pa}^{ik}(R_{ik}, \Omega)}{\partial R_{ik}} \frac{r_{ik\alpha} r_{ik\beta}}{R_{ik}} + \sum_{k \neq j} \frac{\partial n_{pa}^{jk}(R_{jk}, \Omega)}{\partial R_{jk}} \frac{r_{jk\alpha} r_{jk\beta}}{R_{jk}} \right] \right\}. \quad (11.50)$$

The formal results for the aGPT stress tensor at the pair-potential level are thus summarized by Eqs. (11.41), (11.42), (11.48) and (11.50).

It is also instructive and useful to consider the corresponding GPT stress tensor

$$\sigma_{\alpha\beta}^{\text{GPT}} = \frac{1}{N\Omega} \left. \frac{\partial E_{\text{tot}}(\mathbf{R}, \Omega)}{\partial \varepsilon_{\alpha\beta}} \right|_{\varepsilon_{\alpha\beta}=0} \quad (11.51)$$

based on Eq. (11.1) for E_{tot}. Using only Eqs. (11.44) and (11.46), one readily obtains

$$\sigma_{\alpha\beta}^{\text{GPT}} = \frac{1}{2N\Omega} \sum_{i,j}{}' \frac{\partial v_2(R_{ij}, \Omega)}{\partial R_{ij}} \frac{r_{ij\alpha} r_{ij\beta}}{R_{ij}} + \left[\frac{dE_{\text{vol}}(\Omega)}{d\Omega} + \frac{1}{2N} \sum_{i,j}{}' \frac{\partial v_2(R_{ij}, \Omega)}{\partial \Omega} \right] \delta_{\alpha\beta}, \quad (11.52)$$

and for the diagonal $\alpha = \beta$ components, the expected and familiar result

$$\sigma_{\alpha\alpha}^{\text{GPT}} = \frac{1}{N} \frac{dE_{\text{tot}}(\mathbf{R}, \Omega)}{d\Omega} = -P_{\text{tot}}(\Omega), \quad (11.53)$$

consistent with Eq. (7.108) for the total $T = 0$ pressure P_{tot}. For the perfect crystal, the GPT and aGPT stress tensors are equivalent to within any small errors resulting from the additional approximations made in the aGPT. This equivalence can be directly tested through the calculation of the single-crystal elastic moduli, an application that is considered in the next section.

11.2.2 Initial applications of the aGPT

In this final section, we summarize and discuss some of the significant initial applications of the aGPT formalism made by Skinner et al. (2019) on hcp Mg. We begin with the elastic moduli $C_{\alpha\beta\gamma\delta}$, which, in keeping with the definition of the stress tensor $\sigma_{\alpha\beta}$ in Eqs. (11.42) and (11.51), can be defined by the corresponding strain derivative

$$C_{\alpha\beta\gamma\delta} = \left.\frac{\partial \sigma_{\alpha\beta}}{\partial \varepsilon_{\gamma\delta}}\right|_{\varepsilon_{\gamma\delta}=0}. \tag{11.54}$$

Skinner et al. (2019) implemented Eq. (11.54) numerically by choosing small strain parameters $\varepsilon_{\gamma\delta}$ and using a central difference method to evaluate the $C_{\alpha\beta\gamma\delta}$, in which the errors are only fourth order in the $\varepsilon_{\gamma\delta}$. This approach was applied to the GPT as well as aGPT stress tensors developed in Sec. 11.2.1. The GPT stress-strain elastic constants C_{ij} so determined for hcp Mg were previously given in Table 7.6, where it was shown that the values obtained are in good agreement with experiment, and for the non-shear constants, significantly improved over the virial-component values C_{ij}^{vir}. This latter improvement directly reflects the inclusion of the volume derivatives in the GPT stress tensor given by Eq. (11.52). The corresponding calculated aGPT elastic moduli for hcp Mg, obtained with weighting-sphere radii of $R_a = 1.8R_{\text{WS}}$ and $R_a = 3.4R_{\text{WS}}$, are compared with each other and with the GPT results in Table 11.3. Both sets of aGPT elastic moduli are seen to be in good agreement with the GPT results, demonstrating the accuracy of the aGPT approach, as well as its relative insensitivity to the actual value of the parameter R_a used in the calculations.

The aGPT ion forces developed in Sec. 11.2.1 have allowed Skinner et al. (2019) to make a number of noteworthy calculations of defect and thermal properties for hcp Mg, including the *relaxed* formation energy for monovacancies and binding energies of divacancies, γ surface profiles on the important hcp crystallographic planes, and thermal expansion of the bulk crystal via aGPT-MD simulations. Relaxed GPT and aGPT calculations of $E_{\text{vac}}^{\text{f}}$ at constant volume are compared in Table 11.4. While the GPT result is too small by nearly a factor of two, as expected, the aGPT values of $E_{\text{vac}}^{\text{f}}$ with $R_a = 1.8R_{\text{WS}}$ and $R_a = 3.4R_{\text{WS}}$ are in good agreement with both DFT and experiment, as well as with each other. Although the quantitative impact of relaxation on $E_{\text{vac}}^{\text{f}}$ is small, at just—0.01 eV in all three cases, this value agrees with the DFT results of Uesugi et al.

Table 11.3 *Comparison of calculated GPT and aGPT elastic constants (in GPa) for hcp Mg at its observed equilibrium volume $\Omega_0 = 156.8$ a.u. and axial ratio $c/a = 1.62$, with basal relaxation included in the calculations, as obtained via Eq. (11.54) by Skinner et al. (2019). The constant C_{hex} is defined by Eq. (7.113), while $B_0 = B_{\text{tot}}$ is the bulk modulus defined by Eq. (7.114).*

Method	C_{11}	C_{33}	C_{12}	C_{13}	C_{44}	C_{66}	C_{hex}	B_0
GPT	63.9	62.6	25.2	21.1	19.5	19.4	21.65	36.1
aGPT: $R_a = 1.8R_{\text{WS}}$	63.5	62.7	25.5	20.6	19.5	19.0	22.1	35.9
aGPT: $R_a = 3.4R_{\text{WS}}$	63.3	61.6	25.3	21.0	18.9	19.0	21.3	35.9

Table 11.4 *Relaxed vacancy-formation energy and its components (in eV) for hcp Mg, as calculated at a constant volume of $\Omega_0 = 156.8$ a.u. and $c/a = 1.62$, using the bulk GPT and local-density aGPT methods, by Skinner et al. (2019).*

Method	v_2	E_{vol}	δv_2	$\delta E_{\text{vac}}^{\text{relax}}$	$E_{\text{vac}}^{\text{f}}$	$\Omega_{\text{vac}}^{\text{f}}/\Omega_0$
GPT	0.44	–	–	–0.01	0.43	0.71
aGPT: $R_a = 1.8 R_{\text{WS}}$	0.44	0.47	–0.19	–0.01	0.71	0.65
aGPT: $R_a = 3.4 R_{\text{WS}}$	0.44	0.50	–0.23	–0.01	0.70	0.59
DFT[a]	–	–	–	–0.01	0.74	0.69
Expt.[b]	–	–	–	–	0.79	–

[a]Uesugi et al. (2003), calculated with the PP method.
[b]Erhart (1991).

(2003). Also considered in Table 11.4 is the corresponding vacancy formation volume $\Omega_{\text{vac}}^{\text{f}}$, obtained in the GPT and aGPT cases from the Harder and Bacon (1986) formula

$$\frac{\Omega_{\text{vac}}^{\text{f}}}{\Omega_0} = -\frac{1}{B_0} \frac{dE_{\text{vac}}^{\text{f}}(\Omega_0)}{d\Omega}, \qquad (11.55)$$

where B_0 is the equilibrium bulk modulus from Table 11.3. The GPT and aGPT calculations of $\Omega_{\text{vac}}^{\text{f}}/\Omega_0$ are also in general agreement with DFT, but in the case of the aGPT results, somewhat better for $R_a = 1.8 R_{\text{WS}}$ than for $R_a = 3.4 R_{\text{WS}}$.

Skinner et al. (2019) went on to consider GPT and aGPT divacancy binding energies as well, quantities that had been treated earlier in Mg with DFT by Uesugi et al. (2003). As done in the DFT calculations, the GPT and aGPT energies were obtained for vacancies separated as first nearest neighbors up to fifth nearest neighbors in the hcp structure. In all three treatments, the calculated binding energies were found to be small in magnitude, no more than 0.07 eV, but otherwise with variable signs and magnitudes. However, all results indicated a positive binding energy, and hence an unstable divacancy, for first and second nearest-neighbor separations, but a negative binding energy, and hence a stable divacancy, for a third nearest-neighbor separation.

Another significant application performed by Skinner et al. (2019) was the calculation of γ surface profiles along the [1$\bar{1}$00], [$\bar{1}$2$\bar{1}$0], [$\bar{1}$102] and [$\bar{2}$113] directions in the major Basal, Prism I, Pyramidal I and Pyramidal II crystallographic planes of hcp Mg, respectively. The latter four planes are depicted for the hcp unit cell in Fig. 1 of Yin et al. (2017), who had provided baseline DFT γ surface calculations on Mg for comparison. In this regard, the GPT and aGPT γ surface profiles are compared with the corresponding DFT results in Fig. 6 of Skinner et al. (2019), where the aGPT calculations were performed for $R_a = 1.8 R_{\text{WS}}$. Overall, there is good agreement among the GPT, aGPT and DFT profiles. The agreement is especially close for the predicted magnitudes of the stable stacking fault energies in the Basal and Pyramidal I planes.

Finally, Skinner et al. (2019) demonstrated the feasibility of performing accurate aGPT-MD simulations. The specific problem they addressed with these simulations was that of thermal expansion in hcp Mg at high temperature up to 700 K. Previously, Althoff et al. (1993) showed that ad hoc anharmonic corrections to first-principles GPT-QHLD calculations were necessary to explain the observed behavior of the thermal expansion coefficient β, which tends to a near-constant value of 7.5×10^{-5} K^{-1} in the 300–700 K temperature regime. Using an *NPT* ensemble, 512-atom aGPT-MD simulations in this temperature regime yielded a value of $\beta = 7.17 \times 10^{-5}$ K^{-1} at 500 K. Corresponding, but larger and presumably more accurate, 2000-atom aGPT-MD simulations produced a value of $\beta = 7.56 \times 10^{-5}$ K^{-1} at 500 K.

12
Extension to *f*-Band Actinide Metals and *p*-Band Simple Metals

Up until this point, we have developed DFT-based QBIPs for metals only in terms of their most common quantum-mechanical basis representations: plane waves for simple metals via the GPT and DRT pseudopotential approaches; localized d states for pure transition metals via the MGPT and BOP d-band treatments; and more generally, a combination of plane waves and localized d states for the extended transition-metal series via the GPT, encompassing metals with empty, filled and partially filled d bands alike. At the same time, however, the underlying pseudo-Green's function formalism, from which the GPT was developed in Sec. 2.7.2 in a mixed $|\mathbf{k}\rangle, |\phi_d\rangle$ basis set, does not actually impose any symmetry requirement on the character of the localized basis states. Thus, the GPT formalism set forth in Chapter 2 can be immediately extended to include, in addition to plane waves, any or all localized basis states that might be of physical interest in describing a specific metal or alloy. In particular, this extension includes localized p and f states, so schematically, one has quite generally

$$|\mathbf{k}\rangle, |\phi_d\rangle \to |\mathbf{k}\rangle, |\phi_p\rangle, |\phi_d\rangle, |\phi_f\rangle \cdots . \tag{12.1}$$

In practice, such an extension is most easily accomplished in the context of the simplified MGPT, with the canonical d-band approach developed in Chapter 5 acting as a prototype for corresponding canonical p-band or f-band treatments, as we discuss here in Sec. 12.1.

Perhaps the most obvious application of the MGPT in this regard is to the f-electron lanthanide and actinide metals, with the localized d states replaced by, or in some cases supplemented by, localized f states. At the same time, an adequate treatment for many of these metals is complicated by the presence of strong f-electron correlation beyond the scope of DFT, a trend that increases with atomic number in both series. Strong f-electron correlation is the case for the $4f$ lanthanide or rare-earth metals beginning with the low-density phases of Ce, and for the $5f$ actinides beginning with the low-density phases of Pu. In this challenging environment, two interesting applications made to actinide metals are discussed in Sec. 12.2. The first is to the α (orthorhombic), β (tetragonal) and γ (bcc) phases of uranium, all of which can be well treated within

DFT (Söderlind et al., 1995; Söderlind, 1998, 2002; Hood et al., 2008). The second application is to the low-density δ (fcc) phase of Pu, where a simple, novel treatment of the strong f-electron correlation (Moriarty et al., 2006) has been found to be remarkably effective, and may possibly be a stepping stone to a more rigorous dynamical mean field theory (DMFT) approach.

12.1 Localized *p* and *f* basis states in the GPT

Generally speaking, the p bands in metals are much broader than the d or f bands, and indeed, in good simple metals they are already well accounted for in a plane-wave basis representation, leading to accurate energies and interatomic forces within second-order PP perturbation theory, as we discussed at length in Chapter 3. We also pointed out there that in difficult cases involving larger pseudopotentials, such as the second-period metals Li and Be, with no p states in the ion core, or in heavy polyvalent metals such as Pb or Bi, adding localized p basis states to the plane-wave basis set might be a useful alternative to the traditional, but generally less tractable, higher-order PP perturbation theory that we also discussed in Chapter 3. In the context of the first-principles GPT, we have explored the possibility of treating Li in an empty-p-band representation, using the empty-d-band formalism discussed in Chapter 4, with the localized d basis states replaced by localized p states. However, the lowest-energy localized p states available correspond to spatially extended $2p$ states in the free atom or ion, and obtaining satisfactory basis states through the zero-order pseuodoatom construction, as we did in Chapter 4 for the d states, is computationally challenging in this case and still a work in progress. Likewise, a GPT treatment of f-band lanthanide and actinide metals brings with it the same complications discussed in Chapter 5 for partially filled d-band transition metals, plus the need for a fully relativistic treatment in the case of the actinides. In addition, there is the matter of treating strong electron correlation in many of these metals. These practical realities make a simplified MGPT representation incorporating canonical p or f bands an interesting alternative approach to consider for strong-PP simple metals and for the f-electron lanthanide and actinide metals, respectively.

12.1.1 Canonical *p* bands for strong-PP simple metals

Canonical p bands in the context of the MGPT can be formally described by TB-like matrix elements $\Delta_{pp'}^{ij}$ coupling ions i and j of the form

$$\Delta_{pp'}^{ij} \equiv \Delta_{pp'}(R_{ij}) = \alpha_m (R_{\mathrm{WS}}/R_{ij})^3 \delta_{mm'}, \qquad (12.2)$$

where on the RHS of Eq. (12.2) it is assumed that ions i and j are aligned along the z axis, and the power-law exponent of 3 derives from $2\ell + 1$ for $\ell = 1$. This result is analogous to Eq. (5.94) for MGPT canonical d bands with a power-law exponent of $p = 5$. Here the index m has values of 0 or ± 1 with $-\Delta_{pp'}^{ij} = pp\sigma$ and $pp\pi$, respectively, following the usual GPT sign convention for TB-like matrix elements. The corresponding coefficients α_m are likewise assumed to be independent of atomic structure and in a fixed ratio such that $\alpha_0 : \alpha_1$ is $2 : (-1)$.

The form of the MGPT multi-ion potentials v_2, v_3 and v_4 for canonical p bands can be readily worked out following the corresponding canonical d-band formalism elaborated in Chapter 5. In analogy with Eq. (5.97), the p-band contribution to the two-ion pair potential is of the form

$$v_2^p(r,\Omega) = v_a(\Omega)(R_0/r)^{12} - v_b(\Omega)(R_0/r)^6, \qquad (12.3)$$

where, as in the prototype d-band case, v_a and v_b are volume-dependent parameters to be optimized, and $R_0 = 1.8 R_{\mathrm{WS}}$. Note that for $v_a = v_b$, the canonical p-band potential v_2^p is exactly in the form of the classic Lennard-Jones (1924) potential, which was originally developed for insulating gases, with the r^{-6} term representing the attractive van der Waals interaction between atoms. In metals, on the other hand, one expects $v_b \gg v_a$, reflecting the stronger metallic bonding, as we indeed found to be the case for d-band transition metals in Chapter 5.

In analogy with Eq. (5.99) for d-band metals, the total pair potential for the p-band metal is then of the general form

$$v_2(r,\Omega) = v_2^{sp}(r,\Omega) + v_2^{hc}(r,\Omega) + v_2^p(r,\Omega). \qquad (12.4)$$

Here v_2^{sp} is usual the first-principles GPT simple-metal pair potential, as developed in Chapter 3, while v_2^{hc} is the repulsive hard-core overlap potential described in Chapter 5 and Sec. A2.7 of Appendix A2. In the case of the second-period metals Li and Be, the latter potential can probably be dropped entirely because the cores of these metals contain only highly localized $1s$ electrons. For heavy polyvalent metals like Pb and Bi, on the other hand, one would expect v_2^{hc} to be more significant because of the large cores of these elements.

The canonical p-band three-ion MGPT potential $v_3 = v_3^p$ has the general mathematical structure of Eq. (5.100) for canonical d bands, and can be expressed in terms of a p-state radial function $f(r) = (R_0/r)^3$ for three radial distances r_1, r_2 and r_3; angular functions L and P that depend on p symmetry; and two additional volume-dependent coefficients $v_c(\Omega)$ and $v_d(\Omega)$. The functions L and P can be calculated by matrix multiplication in terms of normalized 3×3 Slater-Koster matrices $H_{\mathrm{SK}}^{pp}(i,j)$ for the p-state TB interactions between sites i and j. To establish $H_{\mathrm{SK}}^{pp}(i,j)$ for canonical p bands, one needs to make two parameter substitutions in the standard p-state SK matrix (Slater and Koster, 1954; Harrison, 1980): $pp\sigma \to -S = -2$ and $pp\pi \to -P = 1$. Also note that these substitutions allow for the MGPT sign convention established in Eq. (12.2) for $\Delta_{pp'}^{ij}$. With the three-ion angles θ_1, θ_2 and θ_3 defined as in Fig. 5.15(a), the angular functions L and P for canonical p bands are given respectively by

$$L(x_1, x_2, x_3) = \frac{1}{A_3^p}\mathrm{Tr}\left[H_{\mathrm{SK}}^{pp}(1,2) H_{\mathrm{SK}}^{pp}(2,3) H_{\mathrm{SK}}^{pp}(3,1)\right], \qquad (12.5)$$

where $x_n = \cos\theta_n$ and $A_3^p = S^3 + 2P^3 = 6$, and by

$$P(x_2) = \frac{1}{A_4^p}\mathrm{Tr}\left[H_{SK}^{pp}(1,2)H_{SK}^{pp}(2,1)H_{SK}^{pp}(1,3)H_{SK}^{pp}(3,1)\right], \quad (12.6)$$

where $A_4^p = S^4 + 2P^4 = 18$, with $P(x_1)$ and $P(x_3)$ given by analogous equations. The characteristic angular behavior of the L and P functions for canonical p bands is illustrated in Fig. 12.1(a). These results can be compared with the corresponding behavior of L and P for canonical d bands displayed in Fig. 5.15(a). While the respective P functions behave similarly, the L functions show a strong dependence on the d versus p symmetry, with an additional strong oscillation for canonical d bands.

The more compact and symmetric character of p-state interactions also allows the angular functions L and P to be reduced to very simple analytic expressions. One finds

$$L(x_1, x_2, x_3) = -\frac{1}{2}\left(1 + 3x_1 x_2 x_3\right) \quad (12.7)$$

and

$$P(x_n) = \frac{1}{2}\left(1 + x_n^2\right), \quad (12.8)$$

where $x_n = x_1, x_2$ or x_3. Thus, the entire three-ion MGPT potential v_3 for canonical p bands can be obtained in a relatively transparent analytic form:

$$v_3(r_1, r_2, r_3, \Omega) = -\frac{1}{2}v_c(\Omega)\frac{R_0^9(1 + 3x_1 x_2 x_3)}{r_1^3 r_2^3 r_3^3}$$
$$+ \frac{1}{2}v_d(\Omega)\left\{\frac{R_0^6(1 + x_3^2)}{r_1^3 r_2^3} + \frac{R_0^6(1 + x_1^2)}{r_2^3 r_3^3} + \frac{R_0^6(1 + x_2^2)}{r_3^3 r_1^3}\right\}. \quad (12.9)$$

Note that the first term on the RHS of Eq. (12.9) is of the same form as the classic Axilrod-Teller (1943) potential, which was originally obtained as the lowest-order three-body correction to the van der Waals pair interaction.

The four-ion MGPT potential $v_4 = v_4^p$ for canonical p bands has the same mathematical structure as Eq. (5.99) for canonical d bands, and can be expressed in terms of the p-state radial function $f(r)$ for six radial distances r_1, r_2, r_3, r_4, r_5 and r_6; a third angular function M that depends on p symmetry and applies to the three possible four-ion configurations; and one final volume-dependent coefficient $v_e(\Omega)$. The canonical p-band function M corresponding to configuration (a) of Fig. 5.16 can be calculated as

$$M(x_1, x_2, x_3, x_4, x_5, x_6) = \frac{1}{A_4^p}\mathrm{Tr}\left[H_{SK}^{pp}(1,2)H_{SK}^{pp}(2,3)H_{SK}^{pp}(3,4)H_{SK}^{pp}(4,1)\right], \quad (12.10)$$

and its angular behavior is illustrated in Fig. 12.1(b). Like the three-ion function L, the four-ion function M depends strongly on d versus p symmetry, with the rapid

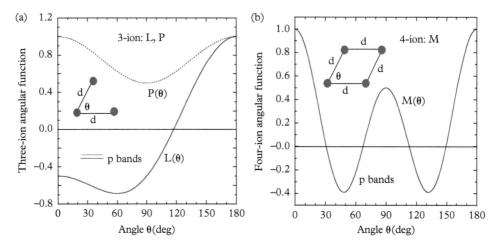

Fig. 12.1 *Canonical p-band MGPT multi-ion angular functions L, P and M. (a) $L(\theta)$ and $P(\theta)$ with $\theta = \theta_3$ and $\theta_1 = \theta_2 = (\pi - \theta)/2$ in the geometry of Fig. 5.15(a). (b) $M(\theta)$, with $\theta = \theta_3 = \theta_1$ and $\theta_2 = \theta_4$ in the geometry of configuration (a) in Fig. 5.16.*

four-minima oscillations for canonical d bands shown in Fig. 5.17(a) becoming just two-minima oscillations for canonical p bands in Fig. 12.1(b). The calculation of the canonical p-band M functions for configurations (b) and (c) of Fig. 5.16 can be done similarly. While, in principle, the expressions for M in all three configurations can be reduced to analytic forms, the results are not expected to be useful enough to justify the additional effort needed to obtain them, so this reduction has not been attempted.

The application of the above canonical p-band MGPT formalism to real materials is still on the horizon. Perhaps the test case of highest immediate priority and interest would be beryllium. As we discussed in Chapter 3, early local PP calculations to third order have established the importance of angular forces to the hcp elastic constants and phonons in Be, and these forces are conveniently provided in the MGPT framework, with five variable coefficients that can be adjusted to describe such basic properties. If additional flexibility is required, a more general *noncanonical p-band* approach is easily developed by allowing the $pp\sigma/pp\pi$ ratio and the radial function power law, $f(r) \to (R_0/r)^p$, to become variable as well.

12.1.2 Canonical f bands for lanthanide and actinide metals

To incorporate narrow canonical f bands described by localized f-state matrix elements $\Delta_{ff'}^{ij}$ into the MGPT, we again follow the approach established in Chapter 5 for canonical d bands, and generalize Eq. (5.94) for ions i and j aligned along the z axis in the form

$$\Delta_{ff'}^{ij} \equiv \Delta_{ff'}(R_{ij}) = \alpha_m (R_{\text{WS}}/R_{ij})^p \delta_{mm'} . \tag{12.11}$$

12.1.2 Canonical f bands for lanthanide and actinide metals

Here, as in the canonical d-band case, the quantity p is treated from the outset as a material-dependent parameter to be optimized. The index m in Eq. (12.11) now assumes values 0, ± 1, ± 2 or ± 3 such that $-\Delta_{ff}^{ij} = ff\sigma, ff\pi, ff\delta$ and $ff\phi$, respectively, retaining the standard GPT sign convention on Δ_{ff}^{ij}. The coefficients α_m are independent of atomic structure with fixed ratios between the m components such that $\alpha_0 : \alpha_1 : \alpha_2 : \alpha_3$ is $20 : (-15) : 6 : (-1)$, as is the case for pure canonical f bands obtained with $p = 7 = 2\ell + 1$ for $\ell = 3$.

As done in our canonical d-band treatment of central transition metals, we then seek to construct a cohesive-energy functional $E_{\text{coh}}(\mathbf{R}, \Omega)$ in terms of a volume term E_{vol} and multi-ion potentials v_2, v_3 and v_4. To do this in practice, we first need to assign physically meaningful values to the valence Z and to the f-electron occupation number Z_f. This is inherently a more difficult task for the actinides than for the transition metals because it now involves a rather complex set of contributions from $7s$, $7p$, $6d$ and $5f$ electrons. Here we only attempt the simplest reasonable assignment of values consistent with the MGPT formalism. For the f-band electrons in early actinide metals, we use the free-atom occupation number of $Z_f = 2, 3, 4$ and 6 for Pa, U, Np and Pu, respectively, while we combine the contributions from the remaining valence s, p and d electrons together to arrive at $Z = 3$ for Pa, U and Np, but $Z = 2$ for Pu. Then using Eqs. (5.95) and (5.99) for transition metals as a template, the MGPT pair potential v_2 is readily generalized to the early actinides in the form

$$v_2(r, \Omega) = v_2^{spd}(r, \Omega) + v_2^{\text{hc}}(r, \Omega) + v_a(\Omega)(R_0/r)^{4p} - v_b(\Omega)(R_0/r)^{2p}, \qquad (12.12)$$

where v_2^{spd} is a simple-metal GPT pair potential based on $Z=3$ or $Z=2$, as appropriate, arising from the spd valence electron density and v_2^{hc} is a corresponding repulsive hard-core potential, while v_a and v_b are variable f-band coefficients to be optimized.

The form of the three-ion potential v_3 is given directly by Eq. (5.100) with additional variable f-band coefficients $v_c(\Omega)$ and $v_d(\Omega)$, plus universal angular functions L and P that now depend on f symmetry instead of d symmetry. As is the case of d-band transition metals, these latter functions are most efficiently calculated by matrix multiplication, here in terms of appropriately normalized 7×7 Slater-Koster matrices $H_{\text{SK}}^{ff}(i,j)$ for the f-state coupling between sites i and j. To obtain $H_{\text{SK}}^{ff}(i,j)$ for canonical f bands, one makes the following parameter substitutions in the conventional f-state SK matrix, as defined by Sharma (1979) and by McMahan (1998): $ff\sigma \to -S = -20$, $ff\pi \to -P = 15$, $ff\delta \to -D = -6$ and $ff\phi \to -F = 1$, allowing for the overall minus sign included in the definition of Δ_{ff}^{ij}. Retaining the three-ion geometry defined in the inset of Fig. 5.15(a), the functions L and P are then given by

$$L(x_1, x_2, x_3) = \frac{1}{A_3^f} \text{Tr}\left[H_{\text{SK}}^{ff}(1,2) H_{\text{SK}}^{ff}(2,3) H_{\text{SK}}^{ff}(3,1) \right], \qquad (12.13)$$

where $x_n = \cos\theta_n$ and $A_3^f = S^3 + 2P^3 + 2D^3 + 2F^3 = 1680$, and

$$P(x_2) = \frac{1}{A_4^f} \text{Tr}\left[H_{SK}^{ff}(1,2)H_{SK}^{ff}(2,1)H_{SK}^{ff}(1,3)H_{SK}^{ff}(3,1)\right], \quad (12.14)$$

where $A_4^f = S^4 + 2P^4 + 2D^4 + 2F^4 = 263{,}844$, with $P(x_1)$ and $P(x_3)$ given similarly. The behavior of the functions L and P for canonical f bands is illustrated in Fig. 12.2, and compared there with the corresponding behavior of these functions for canonical d bands.

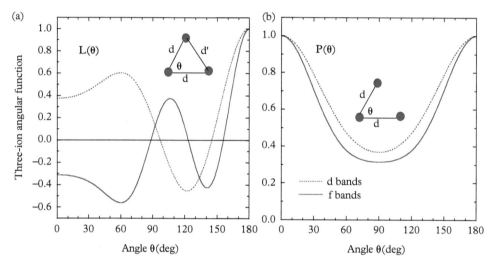

Fig. 12.2 *Three-ion MGPT angular functions L and P for canonical f bands compared with the corresponding functions for canonical d bands. (a) $L(\theta)$, as given by Eq. (12.13) for f bands and Eq. (5.108) for d bands, with $\theta = \theta_3$ and $\theta_1 = \theta_2 = (\pi - \theta)/2$ in the geometry of Fig. 5.15(a). (b) $P(\theta)$, as given by Eq. (12.14) for f bands and Eq. (5.109) for d bands.*

The general form of the additional four-ion potential v_4 is given by Eq. (5.103), which provides the fifth variable f-band coefficient $v_e(\Omega)$ and the third angular function M that contributes to the three possible four-ion configurations depicted in Figs. 5.16(a), (b) and (c). Using the four-ion geometry defined in Fig. 5.16(a) and proceeding clockwise starting from the lower left-side ion, taken as ion 1, one has for that configuration

$$M(x_1, x_2, x_3, x_4, x_5, x_6) = \frac{1}{A_4^f} \text{Tr}\left[H_{SK}^{ff}(1,2)H_{SK}^{ff}(2,3)H_{SK}^{ff}(3,4)H_{SK}^{ff}(4,1)\right], \quad (12.15)$$

where $x_n = \cos\theta_n$, and with similar expressions for M with configurations (b) and (c). The in-plane behavior of M for configuration (a) with canonical f bands is illustrated in Fig. 12.3, and compared with the corresponding behavior for canonical d bands.

12.2.1 Weak electron correlation: a canonical f-band treatment of uranium

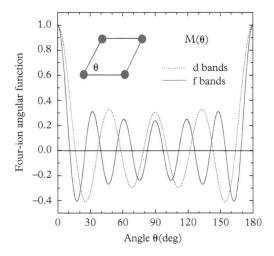

Fig. 12.3 *Four-ion MGPT angular function M for canonical f bands, as given by Eq. (12.15), compared with the corresponding function for canonical d bands, as given by Eq. (5.110), with* $\theta = \theta_3 = \theta_1$ *and.* $\theta_2 = \theta_4$ *in the geometry of Fig. 5.16(a).*

Note that the f-band $M(\theta)$ function depicted there has two extra oscillations and six total minima, which are displaced from the four minima of the d-band function. This behavior has important implications for structural phase stability in the central actinide metals versus the central transition metals. As discussed in Chapter 1 in connection with the four-ion transition-metal potentials shown in Fig. 1.4, the d-band angular function $M(\theta)$ favors the observed bcc structure over fcc for metals like V, Mo and Ta, while the f-band function $M(\theta)$ tends to favor more complex structures like the orthorhombic structure of α-U over simpler structures like bcc and fcc.

12.2 MGPT representations of the early actinides U and Pu

12.2.1 Weak electron correlation: a canonical *f*-band treatment of uranium

Just as the filling of canonical d bands can explain the ordering of the hcp, bcc and fcc structures of the nonmagnetic transition metals (e.g., see Fig. 1.6), the filling of canonical f bands can explain the ordering of the complex structures of the light actinide metals α-U, α-Np and α-Pu, as has been demonstrated by Söderlind et al. (1998). It is reasonable, therefore, to attempt a full canonical-f-band MGPT treatment of the α(orthorhombic), β (nominally bct) and γ (bcc) phases of uranium.

The above MGPT canonical f-band formalism has been applied to uranium over an extended volume range and pressures up to 100 GPa (Moriarty et al., 2006;

Moriarty, 2021). For convenience and to leverage our efficient MGPT scheme for bcc transition metals discussed in Chapter 5, we fitted E_{vol} and the f-band potential parameters v_a, v_b, v_c, v_d and v_e as a function of volume to first-principles FP-LMTO bcc uranium data on the cohesive energy E_{coh}, the bulk modulus B_{tot}, the C' and C_{44} shear elastic constants, the unrelaxed vacancy formation energy E_{vac}^0 and the average zone-boundary phonon frequency $\bar{\nu}_q$, as calculated by Söderlind (2003). Figure 12.4 tests the transferability of the uranium multi-ion potentials so determined by comparing the predicted MGPT structural energy differences for fcc, bct and α-U phases relative to bcc with first-principles FP-LMTO calculations. Qualitatively, the MGPT energy ordering of the structures is correctly predicted with the observed orthorhombic α-U structure of lowest energy and the observed high-temperature bct and bcc structures within close range. In this regard, the only additional constraint imposed in the MGPT treatment was to adjust the magnitude of the canonical f-band radial exponent parameter p below its ideal value of 7. An equilibrium value of $p = 4.2$ at $\Omega = \Omega_0$ was set, similar to the equilibrium value of 4.0 used in the MGPT treatments of Ta (see Table 5.3). This choice was then combined with an assumed weak volume dependence for $p(\Omega)$ to maintain realistic energy separations among the bcc, bct and α-U structures.

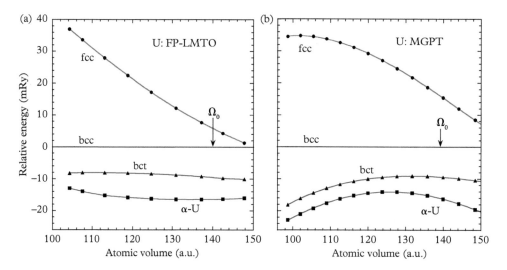

Fig. 12.4 *Predicted MGPT canonical f-band structural energy differences in uranium for the orthorhombic α-U, bct and fcc structures relative to bcc (Moriarty et al., 2006), as compared to the first-principles FP-LMTO results of Söderlind et al. (1995) and Söderlind (1998, 2002).*

The only major disappointment in this canonical f-band MGPT treatment of uranium is that the calculated α-U structure is found to be mechanically unstable. Although all nine of its orthorhombic elastic constants are calculated to be positive, as is observed, some of its acoustic phonon branches have imaginary frequencies, and this problem is not easily fixed within the present framework. The underlying problem is likely related to the long reach from the simple bcc structure, where the important multi-ion potential

12.2.1 Weak electron correlation: a canonical f-band treatment of uranium

constraints are imposed, to the complex α-U orthorhombic structure, whose MGPT description is only constrained through the value of the parameter $p(\Omega_0)$. While MGPT potential transferability from bcc to various different orthorhombic structures, including α-U, seems to be well-handled in the case of tantalum within a canonical d-band framework (Haskins and Moriarty, 2018), the canonical f-band treatment appears to be less robust in this regard for the case of uranium. A logical next step would be a noncanonical f-band treatment, with three additional parameters: $ff\sigma/ff\phi$, $ff\pi/ff\phi$ and $ff\delta/ff\phi$. The resulting ten available parameters in such a treatment, coupled with DFT input data on both the α-U and bcc structures, and implemented with a machine-learning strategy, might produce more accurate and transferable MGPT potentials for uranium. On the other hand, other missing physics such as five- and six-body interactions may also be needed, as well as spd-f hybridization, since the f bands in uranium are only about one-quarter filled, as opposed to the ideal one-half filling implicitly assumed in the MGPT approach.

At the same time, even within the MGPT canonical f-band framework for uranium, the theoretical situation looks more favorable to explain high-temperature bct and bcc solid phases, as well as the liquid. The PT boundaries of these phases are partially established by the measured portions of the phase diagram, which below 5 GPa is discussed by Young (1991). At ambient pressure, the orthorhombic α phase transforms to the tetragonal β phase at 940 K. While the β phase is commonly assumed to be bct, the actual space group of its tetragonal structure remains uncertain. In any case, however, β then transforms to the bcc γ phase at 1050 K, and uranium melts out of bcc at 1405 K. Under applied pressure, the β phase extends to only about 3 GPa, where it ends in a measured α–β–γ triple point, with the remaining α–γ phase line then extending to higher pressure, but measured only to 4 GPa. In high-pressure laser-heated DAC experiments by Yoo et al. (1993, 1998), the uranium melt curve was measured to 100 GPa, and the melting solid phase was confirmed to be bcc up 60 GPa.

On the theoretical side, the bcc structure in uranium is known to be mechanically unstable at $T = 0$, as is revealed in the input FP-LMTO data to the MGPT discussed above, where the C' elastic constant is negative over the entire volume range considered in Fig. 12.4. As illustrated in Fig. 12.5(a), the MGPT canonical f-band QH bcc phonon spectrum reflects this instability in the form of a calculated imaginary $T_1[110]$ branch of the spectrum. Using finite-temperature self-consistent phonon calculations, based on the FP-LMTO method with partial inclusion of anharmonic effects, Söderlind et al. (2012) showed that at ambient pressure the bcc structure for uranium has all real phonon frequencies, and hence mechanical stability at 1123 K, consistent with the observed appearance of the γ phase at 1050 K. At higher pressure near 50 GPa in uranium, high-temperature MGPT-MD simulations with the present canonical f-band potentials (Moriarty, 2021), which capture full anharmonic vibrational contributions beyond QHLD, predict that the bcc structure is mechanically stable at 1000 K. This result is illustrated in terms of the calculated pair correlation function in Fig. 12.5(b). The 1000 K temperature at 50 GPa is actually well below the assumed α–γ phase line placed in the Yoo et al. (1998) phase diagram, as shown in their Fig. 5. The actual α–γ phase line, however, wherever it is located in PT space, is *not* in any case a lower-bound temperature

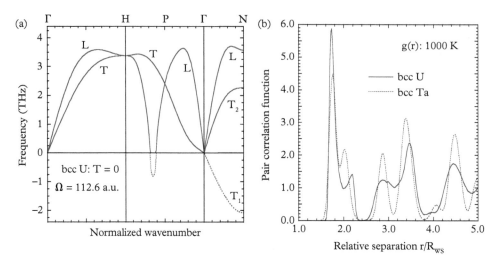

Fig. 12.5 *High-temperature stabilization of the bcc structure in uranium, as calculated near 50 GPa at a compressed volume of $\Omega = 112.6$ a.u. with MGPT canonical f-band multi-ion potentials. (a) Phonon spectrum of bcc U at $T = 0$, showing imaginary (negative) frequencies and hence mechanical instability. (b) Pair correlation function for bcc U, and also for a refence bcc metal Ta for comparison, at 1000 K, as obtained from MGPT-MD simulations (Moriarty, 2021).*

constraint on bcc mechanical stability, so the present results are consistent with current experimental knowledge. It should also be noted that it may be possible to calculate a realistic melt curve out of the bcc γ structure for uranium with the MGPT canonical f-band potentials for direct comparison to the measured Yoo et al. (1993, 1998) result, but such a calculation has yet to be attempted. In this regard, a logical starting point for a melting investigation in uranium would be to assume weak coupling between the ion- and electron-thermal degrees of freedom and apply the two-phase melting approach discussed in Chapter 8.

12.2.2 Strong electron correlation: a novel canonical *d*-band treatment of δ-Pu

Plutonium is one of the most complex metals in the Periodic Table with six observed phases at ambient pressure (Young, 1991). The α-Pu ground state has a high density of 19.86 g/cc and a complicated monoclinic crystal structure with 16 atoms per primitive cell and 8 inequivalent atomic sites. It is widely believed that the f electrons in α-Pu are itinerant and strongly bonding, so that this phase can be well described within DFT (e.g., Söderlind and Klepeis, 2009). In contrast, the five remaining high-temperature phases of plutonium are all of significantly lower density, and as such are prime candidates for strong f-electron correlation. The least dense of these phases is fcc δ-Pu with a density that is 20% less that of α-Pu. In pure plutonium, δ-Pu is the

12.2.2 Strong electron correlation: a novel canonical d-band treatment of δ-Pu

thermodynamically stable phase between 592 K and 724 K with an observed density of 15.92 g/cc. Of particular interest here is the fact that that δ-Pu can also be isolated at room temperature by alloying with a small amount of a group-IIIA metal such as Al or Ga. This alloying maintains the fcc crystal structure of the metal and can also be accomplished with little impact on either its density or its basic physical properties. From a theoretical point of view, Al- or Ga-stabilized δ-Pu can thus be treated as a room temperature elemental metal to a good approximation, and that is the perspective we take here.

To emphasize the importance of strong f-electron correlation in δ-Pu, Moriarty et al. (2006) first considered an alternative fully bonding canonical f-band treatment of the metal similar to that discussed above for uranium. In this treatment, MGPT potentials were only constructed for the 15.76 g/cc observed density of a typical Ga-stabilized metal, with 0.6 weight % of Ga, and corresponding to an equilibrium volume of $\Omega_0 = 167.7$ a.u. The assigned f-band occupation number was again taken as the free-atom value, giving $Z_f = 6$, and leaving an spd valence of $Z = 2$. The general forms of the v_2, v_3 and v_4 multi-ion potentials were otherwise the same as for uranium, and with a similar parameterization of the volume term and the five variable potential coefficients, but one based on experimental data for δ-Pu and not DFT calculations. This data included ultrasonic measurements of the fcc elastic constants by Ledbetter and Moment (1976) and inelastic X-ray scattering measurements of the fcc phonon spectrum by Wong et al. (2003, 2004). Force-constant fits to the latter phonon frequencies led to elastic constants in generally good agreement with the ultrasonic data, as well as a Debye temperature of $\Theta_D = 114.7$ K. The quantities used in a self-consistent fit of the volume term and potential coefficients for δ-Pu were then: E_{coh}, B_{tot}, C', C_{44}, Θ_D and an estimate of $E_{\text{vac}}^0 = 1.0$ eV. In this parameterization, it turned out that the fit could only be accomplished mathematically for large values of the radial exponent parameter p. To satisfy that requirement, a pure canonical f-band value of $p = 7$ was used.

The problematic nature of the MGPT canonical f-band treatment for δ-Pu becomes immediately apparent when the corresponding multi-ion potentials are applied to the fundamental question of structural phase stability and the prediction of the lowest energy crystal structure. This problem is illustrated in Fig. 12.6(a), where the f-bonding cohesive energy relative to that of the observed fcc structure is plotted along the Bain path, yielding a much lower-energy bct structure. Interestingly, this problem can be corrected with the introduction of a very simple model of strong f-electron correlation, in which each f-bonding potential contribution in the MGPT total energy is turned off and replaced by a corresponding d-bonding potential contribution. The d-bonding volume term and potential coefficients are then re-established by self-consistently fitting the same experimental input data used above for the f-bonding treatment. The MGPT d-bonding cohesive energy functional then correctly predicts fcc as the lowest energy structure for δ-Pu, as is also illustrated along the Bain path in Fig. 12.6(a).

In addition, the MGPT d-bonding potentials for δ-Pu give a very good account of the measured fcc phonon spectrum measured by Wong et al. (2003), as is demonstrated

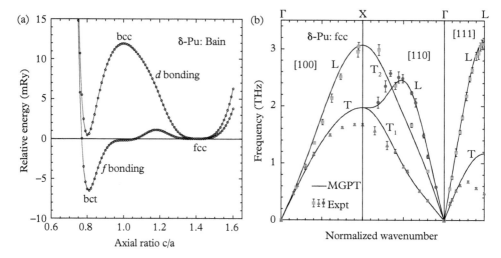

Fig. 12.6 *Structural phase stability and phonon spectra in δ–Pu at equilibrium ($\Omega = 167.7$ a.u.), as calculated with MGPT f- and d-bonding canonical-band potentials. (a) Structural energy relative to fcc along the Bain path, showing that, in contrast to the f-bonding treatment, which favors a bct structure, the d-bonding treatment favors the observed fcc structure. (b) Fcc phonons obtained from the canonical d-band potentials, compared against the experimental measurements of Wong et al. (2003). From Moriarty et al. (2006), with publisher (Springer) permission.*

in Fig. 12.6(b). In particular, note that the longitudinal (L) branches along the [100] and [111] directions, and the longitudinal L and transverse T_2 branches along the [110] direction are calculated accurately relative to experiment. It should further be noted that a more rigorous, first-principles treatment of f-electron correlation in δ-Pu, together with a calculation of the fcc phonon spectrum, has been obtained by Dai et al. (2003), using dynamical mean field theory. The calculated DMFT and MGPT phonon spectra are generally in good agreement, and, interestingly, both results overestimate the lower transverse frequencies near the X and L point zone boundaries is a similar way. In this regard, the main unexplained feature of the experimental phonon spectrum for δ-Pu is the very low measured transverse phonon frequency at the L point. This is possibly an alloy effect, which neither theoretical treatment has addressed.

The above success on δ-Pu also begs the more general question as to whether MGPT could be joined with a rigorous DMFT-like treatment of f-electron correlation. In this regard, the same pseudo-Green's function approach that underlies the GPT and MGPT methods was also the approach originally used by Anderson (1961) to treat strong electron correlation in the classic local magnetic-moment problem, as was briefly mentioned in Chapter 2. These connections appear to be worthy of further exploration.

13
Interatomic Potentials with Electron Temperature

In Chapters 7 and 8 of this book, we have treated the high temperature properties of metals in the conventional weak-coupling limit, where $T = 0$ QBIPs are applied to obtain both the cold and ion-thermal components of thermodynamic quantities in the solid as well as the liquid, supplemented by independent electron-thermal contributions calculated from the underlying electronic structure. In this final chapter, we consider the alternative and complementary strong-coupling limit, where the ion temperature T_{ion} and the electron temperature T_{el} are treated on an equal footing, leading to explicit T_{el}-dependent interatomic potentials and forces. Such a strong-coupling treatment can be important for narrow-band $3d$ and $4d$ transition metals near and above melt due to their high density of electronic states and the corresponding large observed specific heats (e.g., Shaner et al., 1977) for these materials. The development of robust T_{el}-dependent QBIPs could also allow much wider investigations of structural, thermodynamic, defect and mechanical properties at high temperature in such diverse application areas as alloy phase stability and materials design (e.g., Asker et al., 2008; Ozolins, 2009); embrittlement and stress corrosion cracking (e.g., Vashishta et al., 2008; Chen et al., 2010); and femotosecond laser heating (e.g., Recoules et al., 2006).

The fundamental Born-Oppenheimer approximation, which separates ion and electron motion, and was discussed near the outset of this book in Sec. 2.1 of Chapter 2, remains valid at high temperature and allows one to isolate T_{el} and equilibrate the ions and electrons separately. One can thereby define three main electron-temperature regimes of interest in the development and application of QBIPs:

(i) $T_{\text{el}} = 0$, with $T = T_{\text{ion}}$, as is assumed in the standard weak-coupling treatments of thermodynamic properties discussed in Chapters 7 and 8.

(ii) $T_{\text{el}} = T_{\text{ion}} = T$, as is normally assumed in high-T DFT calculations and quantum simulations such as QMD. We define this as the strong-coupling regime.

(iii) $T_{\text{el}} \gg T_{\text{ion}}$, with $T \cong T_{\text{el}}$, as occurs in ultrafast laser experiments, where the electrons are rapidly heated before the ions can move.

In case (iii) melt depends almost exclusively on T_{el}, but even in case (ii), melt in 3d and 4d metals like Mo can be strongly impacted by electron temperature. Following Moriarty et al. (2012), we use the Mermin (1965) formulation of temperature-dependent DFT to extend the first-principles GPT to finite T_{el}. We then develop and apply T_{el}-dependent MGPT potentials to high-pressure melting in Mo.

13.1   Some perspective on the importance of T_{el} in transition-metal melting

The special importance of T_{el} in transition-metal melting for a narrow-band 4d metal was suggested three decades ago in the first quantum-based calculation of the high-pressure melt curve for Mo (Moriarty, 1994). This calculation was performed in the weak-coupling limit, where the total free energy $A_{tot}(\Omega, T)$ in both the bcc solid and the liquid was obtained according to Eq. (7.115), as a sum of cold (E_0), ion-thermal (A_{ion}) and electron-thermal (A_{el}) contributions. The cold and ion-thermal free energies were calculated using the available Mo2 MGPT multi-ion potentials, which are based on canonical d bands, as we discussed in Chapter 5. In this regard, A_{ion} for the bcc solid was obtained via standard QHLD plus MGPT-MD simulations of anharmonic contributions, while the entire contribution to A_{ion} for the liquid was obtained from MGPT-MD simulations. Regarding the additional electron-thermal contribution, it was assumed that $A_{el} \propto \rho(E_F) T^2$, where $\rho(E_F)$ is a phase-dependent, $T_{el} = 0$ electronic DOS at the Fermi level E_F, with separate solid and liquid values. The most striking result arising from this weak-coupling treatment was that in the absence of A_{el}, the calculated Mo melt temperatures were found to be too large by a factor of two, and this was in spite of the fact that the calculated equilibrium Debye temperature for bcc Mo was overestimated by just 7%. Furthermore, to lower the calculated zero-pressure melt point to its observed value required an artificially large—and in retrospect, probably somewhat unphysical—increase in $\rho(E_F)$ in going from the bcc solid to the liquid, demonstrating that A_{el} is not a small correction to A_{ion}, as is implicitly assumed in the weak-coupling model.

The melting situation in Mo can be contrasted with that in the wider 5d-band metal Ta, where a similar weak-coupling treatment of the high-pressure melt curve generally works well using MGPT multi-ion potentials based on canonical d bands, and with A_{el} making only a modest 5% correction to A_{tot} in Eq. (7.115) for both the bcc solid and liquid phases. In addition, the quantitative accuracy of the Ta melt curve has improved with successive improvements in the canonical d-band MGPT potentials, from Ta3 (Moriarty et al., 2002a) to Ta4 (Moriarty et al., 2006) to Ta6.8x (Moriarty and Haskins, 2014), as we discussed in Chapter 5. The most recent 2014 Ta melt curve is illustrated in Fig 8.13 and shows good agreement with both DFT-TPT and QMD simulations, as well as with dynamic isobaric and shock data.

The weak-coupling treatment of melting fails in Mo, however, because T_{el} significantly alters the physics in this case due to the presence of relatively narrow 4d bands. This conclusion is directly supported by high-temperature DFT simulations on fundamental Mo properties, two examples of which are illustrated in Fig. 13.1. First, in

Fig. 13.1(a) one sees that the electronic DOS in the bcc solid is strongly temperature dependent and near melt is poorly represented by the $T = 0$ result that one would use to establish cold and electron-thermal energies in the weak-coupling model. Second, in Fig. 13.1(b) one sees that the elastic shear modulus C', a benchmark quantity used to establish MGPT potentials, is significantly dependent on the electron temperature T_{el}. In particular, the value of C' near melt conditions is lowered by 15–35% due to T_{el} alone, with $T_{\mathrm{ion}} = 0$ in these calculations. One can expect that the remaining elastic moduli, as well as the lattice vibrations that contribute to the ion thermal free energy, are similarly affected at high temperature.

At the same time, there is the simultaneous and related issue of developing an accurate *noncanonical* d-band representation of MGPT multi-ion potentials for Mo. In tandem with the Ta MGPT multi-ion potential improvements noted above, the canonical d-band representation for Mo has also been improved to version Mo5.2 (Moriarty et al., 2002a, 2006), as discussed in Chapter 5. The Mo5.2 MGPT multi-ion potentials do give a good description of low-temperature structural phase stability, as we discussed in Chapter 6, and of defects and mechanical properties, as we discussed in Chapter 9. They are not adequate to calculate a good phonon spectrum for Mo, however, as we pointed out in Chapter 7 in connection with Fig. 7.4. Thus, to treat Mo melting accurately, one needs *both* T_{el}-dependent and noncanonical d-band MGPT potentials, the development of which we discuss in Sec. 13.3. We first consider the formal development of the underlying T_{el}-dependent GPT in Sec. 13.2.

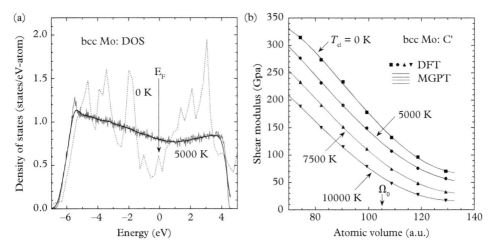

Fig. 13.1 *First-principles DFT calculations of two fundamental material properties as a function of electron temperature T_{el} in bcc Mo. (a) Electronic DOS at a 11.5% compressed volume of $\Omega = 92.99$ a.u. with $T_{el} = T_{ion}$, where the result at 5000 K represents conditions near melt. (b) Elastic shear modulus $C'(\Omega, T_{el})$ calculated in the static bcc lattice with $T_{ion} = 0$, and used to constrain T_{el}– dependent MGPT potentials in Sec. 13.3.*
From Moriarty et al. (2012), with publisher (APS) permission.

13.2 Extending the first-principles GPT to finite electron temperature

To go beyond the weak-coupling model and develop T_{el}-dependent GPT and MGPT interatomic potentials for transition metals in the strong-coupling limit, our starting point is the finite-temperature Mermin (1965) formulation of DFT. In this formalism, the focus is on the grand potential of the electrons Q_{el} in the presence of fixed nuclear potentials, which replaces the total energy E_{tot} given by Eq. (2.12) as the variational functional of the electron density $n(\mathbf{r})$ in the grand canonical ensemble (constant volume Ω, electron temperature T_{el} and chemical potential μ):

$$Q_{el} = E_{tot} - T_{el}S_{el} - \mu N_{el}, \tag{13.1}$$

where S_{el} is the electronic entropy and N_{el} is the number of electrons in the system. In practice, we are interested in the thermal excitation of the valence s, p and d electrons out of the ground state of the metal, with the ion cores fixed at their $T_{el} = 0$ values. Then, in direct analogy with Eqs. (2.28)–(2.36), which were derived at $T_{el} = 0$, one can extract an effective valence binding energy per ion E_{bind} from E_{tot}, so that Eq. (13.1) effectively becomes

$$Q_{el} = NE_{bind} - T_{el}S_{el} - \mu N_{el}. \tag{13.2}$$

In addition, for temperatures up to a few times the melt temperature, it is reasonable to assume that the sp valence Z and zero-order d-state occupation number Z_d remain close to their $T_{el} = 0$ values, and always with $Z + Z_d$ a fixed constant. One can thereby treat N_{el} as a constant and work in an equivalent canonical ensemble (constant volume Ω, T_{el} and N_{el}) with Q_{el} replaced by the free energy

$$F_{el} = NE_{bind} - T_{el}S_{el}. \tag{13.3}$$

Next, the binding energy E_{bind} can be extended to finite electron temperature starting from Eq. (2.36) and using the mixed-basis pseudo-Green's function formalism applied to a d-band metal given in Secs. 2.7.4 and 2.7.5. In particular, the dominant valence band-structure contribution E_{band}^{val} to NE_{bind} can be generalized to the form

$$E_{band}^{val} = \int_0^\infty f_{FD}(E) E \rho(E) dE - NZ_d E_d, \tag{13.4}$$

which replaces Eq. (2.177), where f_{FD} is the Fermi-Dirac distribution function:

$$f_{FD}(E) = \frac{1}{1 + \exp\left[(E - E_F)/k_B T_{el}\right]}. \tag{13.5}$$

As done in Chapter 2 with Eq. (2.177), it is useful integrate the first term on the RHS of Eq. (13.4) by parts so as to replace the DOS $\rho(E)$ with the integrated DOS $N(E)$. Approximating the first derivative of $f_{\text{FD}}(E)$ as $f'_{\text{FD}}(E) \cong -\delta(E - E_{\text{F}})$, one obtains the finite-temperature generalization of Eq. (2.178):

$$E_{\text{band}}^{\text{val}} = NZE_{\text{F}} + NZ_d(E_{\text{F}} - E_d) - \int_0^\infty f_{\text{FD}}(E) N(E) dE. \quad (13.6)$$

Expressing $N(E)$ as a sum of its four components N_0, N_d, δN_{sp} and δN_d, then using Eqs. (2.168)–(2.170) while making the well-tested approximation $E_{\text{F}} \cong \varepsilon_{\text{F}}$ throughout, Eq. (13.6) becomes

$$E_{\text{band}}^{\text{val}} = NZ\varepsilon_{\text{F}} + NZ_d(\varepsilon_{\text{F}} - E_d) - \int_0^\infty f_{\text{FD}}(E) \left[NZ \frac{E^{3/2}}{\varepsilon_{\text{F}}^{3/2}} + NZ_d \frac{\delta_2(E)}{\delta_2(\varepsilon_{\text{F}})} \right.$$
$$\left. + \delta N_{sp}(E) + \delta N_d(E) \right] dE \quad (13.7)$$

in place of Eqs. (2.188) and (2.189).

The corresponding entropy contribution to the free energy F_{el} is given quite generally by

$$S_{\text{el}} = -k_{\text{B}} \int_0^\infty \{f_{\text{FD}}(E) \ln f_{\text{FD}}(E) + [1 - f_{\text{FD}}(E)] \ln [1 - f_{\text{FD}}(E)]\} \rho(E) dE. \quad (13.8)$$

For the temperatures of interest here, the factor involving f_{FD} in curly brackets entering Eq. (13.8) will lead to contributions on the order of $k_{\text{B}} T_{\text{el}}$, making $T_{\text{el}} S_{\text{el}}$ on the order of $(k_{\text{B}} T_{\text{el}})^2$ in magnitude and generally small in comparison to $E_{\text{band}}^{\text{val}}$. Thus, in practice it is only necessary to approximate the DOS $\rho(E)$ in Eq. (13.8) to lowest order as just $\rho_0(E) + \rho_d(E)$. Using Eqs. (2.161) and (2.168), one can thus evaluate Eq. (13.8) as

$$S_{\text{el}} = -Nk_{\text{B}} \int_0^\infty \{f_{\text{FD}}(E) \ln f_{\text{FD}}(E) + [1 - f_{\text{FD}}(E)] \ln [1 - f_{\text{FD}}(E)]\} \left[a_{\text{fe}} E^{1/2} + a_d \delta'_d(E) \right] dE, \quad (13.9)$$

where $a_{\text{fe}} \equiv (2m/\hbar^2)^{3/2} (\Omega/2\pi^2)$, $a_d \equiv 10/\pi$ and $\delta'_d(E) \equiv d\delta_2(E)/dE$. The entropy S_{el} so evaluated is independent of atomic structure and the quantity $-T_{\text{el}} S_{\text{el}}$ can be absorbed into a generalized volume term $F_{\text{vol}}(\Omega, T_{\text{el}})$ in the free-energy functional $F_{\text{el}}(\mathbf{R}, \Omega, T_{\text{el}})$. In that form, S_{el} can still make a substantial contribution to the specific heat of the metal, but it makes no contribution to the interatomic potentials or forces.

Returning to Eq. (13.7) for $E_{\text{band}}^{\text{val}}$, the four terms directly proportional to Z and Z_d will also contribute to the temperature-dependent volume term in the free energy F_{el}, while δN_{sp} and δN_d will lead directly to temperature-dependent interatomic potentials. In this regard, $\delta N_{sp}(E)$ is still defined by the pseudopotential expansion given by Eq. (2.181) for the valence s and p electrons, with the second-order term in $W_{\mathbf{kk}'}$ producing a two-ion pair-potential contribution. The corresponding d-electron component $\delta N_d(E)$ is defined by Eq. (2.187), but is further simplified via Eq. (5.5) in terms of the normalized d-state interaction matrix T_{ij}, with a leading volume component $T_{dd'}^{\text{vol}}$ given by Eq. (5.6). Starting from Eqs. (5.5) and (5.7), the integrated structural-energy contribution to δN_d can then be developed into a multi-ion potential series analogous to that for $T_{\text{el}} = 0$:

$$NE_{\text{struc}}^d = \frac{2}{\pi} \text{Im} \int_0^\infty f_{\text{FD}}(E) \ln \left[\det\left(T_{dd'}^{\text{vol}}\right)\right] dE$$

$$= \frac{1}{2} \sum_{i,j}' v_2^d(ij) + \frac{1}{6} \sum_{i,j,k}' v_3^d(ijk) + \frac{1}{24} \sum_{i,j,k,l}' v_4^d(ijkl) + \cdots \quad (13.10)$$

As is true at $T_{\text{el}} = 0$, the multi-ion d-state potentials v_n^d in Eq. (13.10) have exact representations in terms of the T_{ij}, as well as useful expansions needed for the companion MGPT treatment. In this regard, the two-ion potential v_2^d is

$$v_2^d(ij) = \frac{2}{\pi} \text{Im} \int_0^\infty f_{\text{FD}}(E) L_{ij}(E) dE$$

$$= -\frac{2}{\pi} \text{Im} \int_0^\infty f_{\text{FD}}(E) \text{Tr}\left[T_{ij}T_{ji} + \frac{1}{2}(T_{ij}T_{ji})^2 + \cdots\right] dE, \quad (13.11)$$

with L_{ij} given by Eq. (5.15). The corresponding three-ion potential v_3^d is

$$v_3^d(ijk) = \frac{2}{\pi} \text{Im} \int_0^\infty f_{\text{FD}}(E) \left\{L_{ijk}(E) - [L_{ij}(E) + L_{jk}(E) + L_{ki}(E)]\right\} dE$$

$$= -\frac{2}{\pi} \text{Im} \int_0^\infty f_{\text{FD}}(E) \text{Tr}\left[-2T_{ij}T_{jk}T_{ki} + T_{ij}T_{ji}T_{ik}T_{ki} + T_{jk}T_{kj}T_{ji}T_{ij}\right.$$

$$\left. + T_{ki}T_{ik}T_{kj}T_{jk} + \cdots\right] dE, \quad (13.12)$$

13.2 Extending the first-principles GPT to finite electron temperature

with L_{ijk} given by Eq. (5.17). Finally, the four-ion potential v_4^d is

$$v_4^d(ijkl) = \frac{2}{\pi} \mathrm{Im} \int_0^\infty f_{\mathrm{FD}}(E) \{L_{ijkl}(E) - [L_{ijk}(E) + L_{jkl}(E) + L_{kli}(E) + L_{lij}(E)]$$

$$+ [L_{ij}(E) + L_{jk}(E) + L_{kl}(E) + L_{li}(E) + L_{ki}(E) + L_{lj}(E)]\} \, dE \quad (13.13)$$

$$= -\frac{2}{\pi} \mathrm{Im} \int_0^\infty f_{\mathrm{FD}}(E) \mathrm{Tr} \left[2 \left(T_{ij} T_{jk} T_{kl} T_{li} + T_{ik} T_{kl} T_{lj} T_{ji} + T_{il} T_{lj} T_{jk} T_{ki} \right) + \cdots \right] dE,$$

with L_{ijkl} given by Eq. (5.20). Note that the lower expansion forms in Eqs. (13.11)–(13.13) are carried out only to fourth order in the T_{ij} for MGPT applications.

To complete the T_{el}-dependent valence binding-energy contribution to F_{el} in Eq. (13.3), we proceed in the manner established by the $T_{\mathrm{el}} = 0$ formalism of Chapter 5. One first takes the above d-state multi-ion potentials, v_3^d as given by Eq. (13.12) and v_4^d as given by Eq. (13.13), to be the full three- and four-ion potentials: $v_3 = v_3^d$ and $v_4 = v_4^d$. The remaining contributions to a T_{el}-dependent volume term F_{vol} and two-ion pair potential v_2 can be derived starting from the general form of the full binding energy E_{bind} given by Eq. (2.36), including double-counting and exchange-correlation terms involving the valence electron density n_{val}. In particular, the screening and d-state orthogonalization hole components of n_{val}, i.e., δn_{scr} from Eq. (2.196) and δn_{oh}^d from Eq. (2.197), are then respectively generalized to

$$\delta n_{\mathrm{scr}}(\mathbf{r}) = -\frac{2}{\pi} \mathrm{Im} \int_0^\infty f_{\mathrm{FD}}(E) \sum_{k,k'}{}' \left[\frac{\langle \mathbf{r} | \mathbf{k}' \rangle W_{k'k} \langle \mathbf{k} | \mathbf{r} \rangle}{(E - \varepsilon_k)(E - \varepsilon_{k'})} + \sum_d \frac{\langle \mathbf{r} | \mathbf{k}' \rangle v'_{k'd} v'_{dk} \langle \mathbf{k} | \mathbf{r} \rangle}{(E - \varepsilon_k)(E - \varepsilon_{k'})(E - E_r)} \right] dE$$

(13.14)

and

$$\delta n_{\mathrm{oh}}^d(\mathbf{r}) = -\frac{2}{\pi} \mathrm{Im} \int_0^\infty f_{\mathrm{FD}}(E) \sum_k \sum_d \left[\frac{\langle \mathbf{r} | \mathbf{k} \rangle v'_{kd} v'_{dk} \langle \mathbf{k} | \mathbf{r} \rangle - \langle \mathbf{r} | \phi_d \rangle v'_{dk} v'_{kd} \langle \phi_d | \mathbf{r} \rangle}{(E - \varepsilon_k)^2 (E - E_r)} \right.$$

$$\left. + \frac{\langle \mathbf{r} | \phi_d \rangle v'_{dk} (\langle \mathbf{k} | \mathbf{r} \rangle - S_{kd} \langle \phi_d | \mathbf{r} \rangle) + \mathrm{c.c.}}{(E - \varepsilon_k)(E - E_r)} \right] dE, \quad (13.15)$$

using the definition of E_r given by Eq. (5.53). In the usual way, recasting the screening density $\delta n_{\mathrm{scr}}(\mathbf{r})$ in terms of its one-ion Fourier component $n_{\mathrm{scr}}(q)$, one has

$$n_{\mathrm{scr}}(q) = \frac{4}{(2\pi)^3} \left[\int \frac{f_{\mathrm{FD}}(\varepsilon_k) w(\mathbf{k}, \mathbf{q}) + h_1^{\mathrm{tm}}(\mathbf{k}, \mathbf{q})}{\varepsilon_k - \varepsilon_{k+q}} d\mathbf{k} \right], \quad (13.16)$$

in place of Eq. (5.51), with

$$h_1^{tm}(\mathbf{k},\mathbf{q}) = -\frac{1}{\pi}\text{Im}\int_0^\infty f_{FD}(E)\sum_d \frac{v'_{\mathbf{k}+\mathbf{q}d}v'_{d\mathbf{k}}}{(E-E_r)(E-\varepsilon_\mathbf{k})}dE, \quad (13.17)$$

in place of Eq. (5.52). The d-state orthogonalization-hole component $\delta n_{oh}^d(\mathbf{r})$ impacts both the effective valence Z^* given by Eq. (5.54), with the quantity h_2^{tm} in that equation now of the form

$$h_2^{tm}(\mathbf{k}) = -\frac{1}{\pi}\text{Im}\int_0^\infty f_{FD}(E)\sum_d \frac{v'_{\mathbf{k}d}v'_{d\mathbf{k}}}{(E-E_r)(E-\varepsilon_\mathbf{k})^2}dE, \quad (13.18)$$

and the full orthogonalization-hole density $n_{oh}(\mathbf{r})$ given by Eq. (5.57), with $\tilde{h}_2^{tm}$ in that equation now expressed by

$$\tilde{h}_2^{tm}(\mathbf{k},\mathbf{r}) = -\frac{1}{\pi}\text{Im}\int_0^\infty f_{FD}(E)\sum_d \left[\frac{\langle \mathbf{r}|\phi_d\rangle v'_{d\mathbf{k}}v'_{\mathbf{k}d}\langle\phi_d|\mathbf{r}\rangle}{(E-E_r)(E-\varepsilon_\mathbf{k})^2}\right.$$

$$\left. - \frac{\langle \mathbf{r}|\phi_d\rangle v'_{d\mathbf{k}}(\langle \mathbf{k}|\mathbf{r}\rangle - S_{\mathbf{k}d}\langle\phi_d|\mathbf{r}\rangle) + \text{c.c.}}{(E-E_r)(E-\varepsilon_\mathbf{k})}\right]dE. \quad (13.19)$$

The completion of the GPT free-energy functional $F_{el}(\mathbf{R},\Omega,T_{el})$ then largely parallels the completion of the $T_{el}=0$ cohesive-energy functional $E_{coh}(\mathbf{R},\Omega)$ starting from Eq. (5.62) and extending through Eq. (5.86). First, using Eq. (2.181) to express $\delta N_{sp}(E)$ to second order in $W_{\mathbf{kk'}}$, together with Eq. (5.5) to express $\delta N_d(E)$, one obtains from Eq. (13.7) and Eqs. (13.10)–(13.13), the full valence band-structure energy through four-ion interactions:

$$E_{band}^{val} = NZ\varepsilon_F + NZ_d(\varepsilon_F - E_d^{vol}) - \int_0^\infty f_{FD}(E)\left[NZ\frac{E^{3/2}}{\varepsilon_F^{3/2}} + NZ_d\frac{\delta_2(E)}{\delta_2(\varepsilon_F)}\right]dE$$

$$-\frac{2}{\pi}\text{Im}\int_0^\infty f_{FD}(E)\left[\sum_{\mathbf{k}}\frac{W_{\mathbf{kk}}}{E-\varepsilon_\mathbf{k}} + \frac{1}{2}\sum_{\mathbf{k},\mathbf{k'}}\frac{W_{\mathbf{kk'}}W_{\mathbf{k'k}}}{(E-\varepsilon_\mathbf{k})(E-\varepsilon_{\mathbf{k'}})}\right]dE$$

$$-\frac{2}{\pi}\text{Im}\int_0^\infty f_{FD}(E)\sum_d \left[\frac{E_d^{struc}(d\Gamma_{dd}^{vol}/dE) + \Gamma_{dd}^{struc} + \Lambda_{dd}^{vol}}{E-E_r}\right]dE$$

$$+\frac{1}{2}\sum_{i,j}{'}v_2^d(ij) + \frac{1}{6}\sum_{i,j,k}{'}v_3(ijk) + \frac{1}{24}\sum_{i,j,k,l}{'}v_4(ijkl). \quad (13.20)$$

Then, using Eq. (13.20) for E_{band}^{val} in Eq. (2.36) for NE_{bind} together with Eq. (13.9) for S_{el}, the final expression for the free-energy functional $F_{el}(\mathbf{R},\Omega,T_{el})$ can be fully elaborated. In the desired real-space representation, one has

$$F_{\text{el}}(\mathbf{R}, \Omega, T_{\text{el}}) = NF_{\text{vol}}(\Omega, T_{\text{el}}) + \frac{1}{2}\sum_{i,j}{}' v_2(ij, \Omega, T_{\text{el}}) + \frac{1}{6}\sum_{i,j,k}{}' v_3(ijk, \Omega, T_{\text{el}})$$

$$+ \frac{1}{24}\sum_{i,j,k,l}{}' v_4(ijkl, \Omega, T_{\text{el}}) . \qquad (13.21)$$

Here the full pair potential v_2 is of the form

$$v_2(r, \Omega, T_{\text{el}}) = \frac{(Z^*e)^2}{r}\left[1 - \frac{2}{\pi}\int_0^\infty F_N(q, \Omega, T_{\text{el}})\frac{\sin(qr)}{q}dq\right] + v_{\text{ol}}(r, \Omega, T_{\text{el}}) , \qquad (13.22)$$

where

$$F_N(q, \Omega, T_{\text{el}}) = -\frac{q^2\Omega}{2\pi(Z^*e)^2}F(q, \Omega, T_{\text{el}}) , \qquad (13.23)$$

with

$$F(q, \Omega, T_{\text{el}}) = \frac{2\Omega}{(2\pi)^3}\int \frac{f_{\text{FD}}(\varepsilon_\mathbf{k})[w(\mathbf{k},\mathbf{q})]^2 + 2w(\mathbf{k},\mathbf{q})h_1^{\text{tm}}(\mathbf{k},\mathbf{q}) + h_{21}^{\text{tm}}(\mathbf{k},\mathbf{q})}{\varepsilon_\mathbf{k} - \varepsilon_{\mathbf{k}+\mathbf{q}}}d\mathbf{k}$$

$$- \frac{2\pi e^2\Omega}{q^2}\left\{[1 - G(q)][n_{\text{scr}}(q)]^2 + G(q)[n_{\text{oh}}(q)]^2\right\} . \qquad (13.24)$$

In Eq. (13.24), we note that $h_1^{\text{tm}}(\mathbf{k}, \mathbf{q})$ is given by Eq. (13.17) and that

$$h_{21}^{\text{tm}}(\mathbf{k}, \mathbf{q}) = -\frac{1}{\pi}\text{Im}\int_0^\infty f_{\text{FD}}(E)\frac{\left(\sum_d v'_{\mathbf{k}+\mathbf{q}d}v'_{d\mathbf{k}}\right)^2}{(E - E_r)^2(E - \varepsilon_\mathbf{k})}dE , \qquad (13.25)$$

in place of Eq. (5.74). Additionally for the full pair potential v_2, one also has the important short-range d-state overlap contribution

$$v_{\text{ol}}(r, \Omega, T_{\text{el}}) = v_2^d(r, \Omega, T_{\text{el}}) - v_2^{\text{hyb}}(r, \Omega, T_{\text{el}}) + v_{\text{ol}}^0(r, \Omega) , \qquad (13.26)$$

in place of Eq. (5.71), with v_2^d now taken from Eq. (13.11) and

$$v_2^{\text{hyb}}(r, \Omega, T_{\text{el}}) = -\frac{2}{\pi}\text{Im}\int_0^\infty f_{\text{FD}}(E)\sum_{d,d'}\frac{\Gamma_{dd'}^{\text{vol}}(r, E)\Gamma_{d'd}^{\text{vol}}(r, E)}{(E - E_r)^2}dE , \qquad (13.27)$$

while v_{ol}^0 is still as calculated from Eq. (5.69).

Finally, the generalized volume term F_{vol} in Eq. (13.21) is

$$F_{\text{vol}}(\Omega, T_{\text{el}}) = E_{\text{fe}}^0(\Omega) - \frac{1}{2}\Omega B_{\text{eg}}(\Omega) + \frac{2\Omega}{(2\pi)^3}\int f_{FD}(\varepsilon_{\mathbf{k}})w_{\text{pa}}(\mathbf{k})\left[1 + p_c(\mathbf{k})\right]d\mathbf{k}$$

$$+ Z\left[\frac{2}{5}\varepsilon_F - \int_0^\infty f_{FD}(E)\frac{E^{3/2}}{\varepsilon_F^{3/2}}dE\right] + Z_d(\varepsilon_F - E_d^{\text{vol}}) - \frac{10}{\pi}\int_0^\infty f_{FD}(E)\left[\delta_2(E)\right.$$

$$\left. + \delta h_2(E)\right]dE + \frac{2\Omega}{(2\pi)^3}\int\left[h_1^{\text{tm}}(\mathbf{k})p_c(\mathbf{k}) + h_2^{\text{tm}}(\mathbf{k})w_{\text{pa}}(\mathbf{k}) + h_{22}^{\text{tm}}(\mathbf{k})\right]d\mathbf{k}$$

$$- Z^*\left[\bar{w}_{\text{core}}(0) + \bar{h}_1^{\text{tm}}(0)\Pi_{10}(0)\right] + \frac{3}{4}\frac{Z}{\varepsilon_F}\left[\overline{\delta w_{\text{core}}^2}(0) + \overline{\delta w_{\text{core}}h_1^{\text{tm}}}(0)\Pi_{10}(0)\right]$$

$$+ \frac{9}{10}\frac{(Z^*e)^2}{R_{\text{WS}}} - \frac{(Z^*e)^2}{\pi}\int_0^\infty F_N(q,\Omega,T_{\text{el}})dq + \delta E_{\text{oh}}^* - T_{\text{el}}S_{\text{el}}, \quad (13.28)$$

where $h_1^{\text{tm}}(\mathbf{k}) = h_1^{\text{tm}}(\mathbf{k}, 0)$, as given by Eq. (13.17), $h_2^{\text{tm}}(\mathbf{k})$ is given by Eq. (13.18) and

$$h_{22}^{\text{tm}}(\mathbf{k}) = -\frac{1}{\pi}\text{Im}\int_0^\infty f_{FD}(E)\frac{\left(\sum_d v'_{kd}v'_{dk}\right)^2}{(E - E_{\text{r}})^2(E - \varepsilon_{\mathbf{k}})^2}dE, \quad (13.29)$$

in place of Eq. (5.84).

13.3 Temperature-dependent MGPT potentials and the simulation of melt for Mo

The canonical ensemble used to establish the GPT free-energy functional $F_{\text{el}}(\mathbf{R}, \Omega, T_{\text{el}})$ in Eq. (13.21), coupled with additional simplifications provided by a corresponding MGPT treatment, permits robust MD simulations at constant volume and temperature on central transition metals like Mo. In particular, the T_{el}-dependent volume term F_{vol} and multi-ion potentials v_2, v_3 and v_4 retain their fundamental properties of being structure-independent and transferable to all bulk ion configurations. In an MD simulation, if the ions are equilibrated at a constant volume Ω and a temperature $T = T_{\text{ion}} = T_{\text{el}}$, then the total free energy of the strong-coupling system is

$$A_{\text{tot}}(\Omega, T) = \langle F_{\text{el}}\rangle - TS_{\text{ion}}, \quad (13.30)$$

where $\langle F_{\text{el}}\rangle$ is the configuration average of F_{el}, and S_{ion} is the remaining ionic entropy. This result then replaces the weak-coupling free energy given by Eq. (7.115) in terms of cold, ion-thermal and electron-thermal components.

13.3 Temperature-dependent MGPT potentials and the simulation of melt for Mo

We can then elaborate the generalizations needed for a T_{el}-dependent MGPT formalism in the case of noncanonical d bands, where v_3 and v_4 are altered in form with the addition of variable parameters $c_0 = dd\sigma/dd\delta$ and $c_1 = dd\pi/dd\delta$, which in general can be both volume and temperature dependent. In establishing the T_{el}-dependent d-state potentials v_n^d from Eqs. (13.11)–(13.13), the volume component of the normalized d-state interaction $T_{dd'}^{\text{vol}}$ is still of the form given by Eq. (5.90), with its energy dependence isolated from its dependence on interatomic separation. Thus, in performing the energy integrals in Eqs. (13.11)–(13.13), the explicit electron-temperature dependence of the potentials is thereby confined to the five basic d-state potential parameters v_a, v_b, v_c, v_d and v_e established in Chapter 5, while the radial and angular dependence of the potentials is formally the same as for noncanonical d bands at $T_{\text{el}} = 0$.

The full two-ion MGPT potential is taken in the form

$$v_2(r, \Omega, T_{\text{el}}) = v_2^{sp}(r, \Omega) + v_2^{hc}(r, \Omega) + v_a(\Omega, T_{\text{el}})(R_0/r)^{4p} - v_b(\Omega, T_{\text{el}})(R_0/r)^{2p}, \tag{13.31}$$

in place of Eq. (5.99), where $R_0 = 1.8R_{\text{WS}}$ and the radial exponent p remains a variable parameter to be optimized. In practice, we allow p to have a small implicit dependence on temperature as well as volume in this process. Also, note in Eq. (13.31) that we have neglected any weak dependence of v_2^{sp} and v_2^{hc} on electron temperature.

Retaining the geometry of Fig. 5.15(a), the corresponding three-ion MGPT potential $v_3 = v_3^d$ is of the form,

$$v_3(r_1, r_2, r_3, \Omega, T_{\text{el}}) = v_c(\Omega, T_{\text{el}}) f(r_1) f(r_2) f(r_3) L(\theta_1, \theta_2, \theta_3) + v_d(\Omega, T_{\text{el}})$$
$$\times \left\{ [f(r_1)f(r_2)]^2 P(\theta_3) + [f(r_2)f(r_3)]^2 P(\theta_1) + [f(r_3)f(r_1)]^2 P(\theta_2) \right\}, \tag{13.32}$$

in place of Eq. (5.100), with $f(r) = (R_0/r)^p$. For noncanonical d bands, the angular functions L and P must be defined in terms of an appropriately normalized SK matrix H_{SK}^{dd} with $S = c_0$, $P = c_1$ and $D = 1$. One then has

$$L(x_1, x_2, x_3) = \frac{1}{N_3^d} \text{Tr}\left[H_{\text{SK}}^{dd}(1,2) H_{\text{SK}}^{dd}(2,3) H_{\text{SK}}^{dd}(3,1) \right], \tag{13.33}$$

where $N_3^d = c_0^3 + 2c_1^3 + 2$, and

$$P(x_2) = \frac{1}{N_4^d} \text{Tr}\left[H_{\text{SK}}^{dd}(1,2) H_{\text{SK}}^{dd}(2,1) H_{\text{SK}}^{dd}(1,3) H_{\text{SK}}^{dd}(3,1) \right], \tag{13.34}$$

where $N_4^d = c_0^4 + 2c_1^4 + 2$, with $P(x_1)$ and $P(x_3)$ given by similar expressions.

The remaining four-ion MGPT potential $v_4 = v_4^d$ is of the form

$$v_4(r_1, r_2, r_3, r_4, r_5, r_6, \Omega, T_{el}) = v_e(\Omega, T_{el}) [f(r_1)f(r_2)f(r_4)f(r_5)M(\theta_1, \theta_2, \theta_3, \theta_4, \theta_5, \theta_6)$$
$$+ f(r_3)f(r_2)f(r_6)f(r_5)M(\theta_7, \theta_8, \theta_9, \theta_{10}, \theta_5, \theta_{12})$$
$$+ f(r_1)f(r_6)f(r_4)f(r_3)M(\theta_{11}, \theta_{12}, \theta_5, \theta_6, \theta_3, \theta_4)], \quad (13.35)$$

in place of Eq. (5.103), where for configuration (a) of Fig. 5.16,

$$M(x_1, x_2, x_3, x_4, x_5, x_6) = \frac{1}{N_4^d} \text{Tr} \left[H_{SK}^{dd}(1,2) H_{SK}^{dd}(2,3) H_{SK}^{dd}(3,4) H_{SK}^{dd}(4,1) \right], \quad (13.36)$$

with similar expressions for configurations (b) and (c).

Using the above formalism, a full T_{el}-dependent MGPT parameterization for Mo was carried out by Moriarty et al. (2012) over a wide volume range of $1.08 \leq \Omega/\Omega_0 \leq 0.593$, with $\Omega_0 = 105.1$ a.u., and an electron-temperature range of $0 \leq T_{el} \leq 10\,000$ K. This range of Ω and T_{el} covers the best estimate for the location of the high-pressure bcc melt curve in Mo up to 400 GPa. Within this range, the five MGPT d-state potential parameters (v_a, v_b, v_c, v_d, v_e), the noncanonical d-band parameters c_0 and c_1, and the volume term F_{vol} were established by fitting Ω- and T_{el}-dependent DFT data calculated for the bcc solid at $T_{ion} = 0$ on the equation of state (E_{tot} and B_{tot}), shear elastic moduli (C' and C_{44}), unrelaxed vacancy formation energy (E_{vac}^u), and zone-boundary phonon frequencies at the H point (ν_H) and the N point ($\nu_{N\text{-}L}, \nu_{N\text{-}T_1}, \nu_{N\text{-}T_2}$). The DFT data were calculated with the PP method in the LDA limit, as discussed by Moriarty et al. (2012).

The actual fitting process for Mo was significantly simplified by the discovery of some rather remarkable transferability properties for the noncanonical d-band parameters c_0 and c_1. Namely, if at a given temperature T_{el} and the normal equilibrium volume Ω_0, the eight MGPT parameters were fitted to corresponding DFT data on E_{tot}, B_{tot}, C', C_{44}, E_{vac}^u, ν_H, $\nu_{N\text{-}L}$ and $\nu_{N\text{-}T_2}$, then the values of c_0 and c_1 thereby obtained apply with equal accuracy to all other volumes at that temperature as well. In other words, c_0 and c_1 so obtained for Mo were found to be independent of the atomic volume Ω. The high accuracy of this result, as validated by parallel DFT calculations (Moriarty et al., 2012), is illustrated in Fig. 13.2. Here, four noncanonical MGPT zone-boundary phonon frequencies for bcc Mo have been calculated as a function of volume at $T_{el} = 5000$ K with $c_0 = 5.29$ and $c_1 = -0.81$ for all volumes considered. In almost complete contrast, a corresponding canonical d-band MGPT calculation of these frequencies, with $c_0 = 6$ and $c_1 = -4$, displays relatively poor agreement with the benchmark DFT data, as is also shown in Fig. 13.2. In both the noncanonical and canonical d-band treatments, the volume dependence of each of the remaining six parameters ($v_a, v_b, v_c, v_d, v_e, F_{vol}$) was obtained by fitting to calculated DFT data on E_{tot}, B_{tot}, C', C_{44}, E_{vac}^u and the average zone-boundary phonon frequency

$$\bar{\nu}_{zb} = \frac{1}{4} \left(\nu_H + \nu_{N\text{-}L} + \nu_{N\text{-}T_1} + \nu_{N\text{-}T_2} \right), \quad (13.37)$$

as a function of volume with the established values of c_0 and c_1.

13.3 Temperature-dependent MGPT potentials and the simulation of melt for Mo

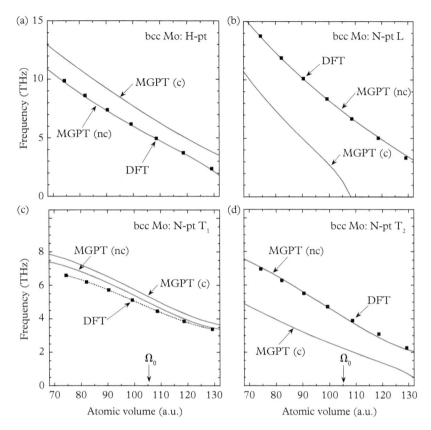

Fig. 13.2 *High-symmetry zone-boundary phonon frequencies in bcc Mo as a function of atomic volume, as calculated at an electron temperature of 5000 K with T_{el} – dependent canonical (c) and noncanonical (nc) MGPT multi-ion potentials and compared against corresponding DFT calculations. (a) H-point phonons; (b) N-point longitudinal L phonons; (c) N-point transverse T_1 phonons; (d) N-point transverse T_2 phonons.*
Part (a) from Moriarty et al. (2012) with publisher (APS) permission.

The entire MGPT noncanonical d-band fitting process for Mo was performed for electron temperatures $T_{el} = 0$, 5000, 7500 and 10 000 K. Across this temperature range, both c_0 and c_1 were found to vary smoothly, from values of $c_0 = 3.96$ and $c_1 = -0.81$ at $T_{el} = 0$ to values of $c_0 = 5.92$ and $c_1 = -0.19$ at $T_{el} = 10\,000$ K. Also, it should be mentioned that in the MGPT fitting process, the radial exponent p in the function $f(r)$ entering v_2, v_3 and v_4 has been very slowly varied with volume and temperature in the following manner. The volume variation at $T_{el} = 0$ was chosen to be similar to that of Mo5.2, as given in Table 5.3, with p allowed to increase with increasing volume in the range 4.6 – 5.2 and $p = 4.8$ at $\Omega = \Omega_0$. The temperature variation at each volume was chosen to be linear in T_{el}, with p *decreasing* by 0.3 for a temperature increase of 5000 K, such that, for example, $p = 4.2$ at $\Omega = \Omega_0$ and $T_{el} = 10\,000$ K. The

MGPT Mo free energy and interatomic potentials so established have been collectively designated as Mo12t, although the specific noncanonical d-band potentials calculated at $T_{\rm el} = 0$ have been denoted as Mo12.t0, as was done in Fig. 7.4(b) for the Mo bcc phonon spectrum and in Table 9.2 for point defect energies and volumes.

The net effect of electron temperature and noncanonical d bands on the multi-ion MGPT Mo12t potentials for representative conditions relevant to high-pressure melt is illustrated in Fig. 13.3. The two-ion pair potential v_2 has the familiar attractive well in the vicinity of near-neighbor distances ranging from about $r = 1.8 R_{\rm WS}$ to $2.5 R_{\rm WS}$. The depth of the well is strongly temperature dependent and increases in magnitude by a factor of two in going from $T_{\rm el} = 0$ to 10 000 K. The three-ion potential v_3 develops a strong positive barrier for angles below 90°, but one that rapidly decreases in magnitude with increasing temperature. For angles above 90°, v_3 becomes rather flat, except for an attractive dip in the potential above 150° at $T_{\rm el} = 10\,000$ K. The four-ion potential v_4 has an enhanced repulsive behavior near 90° at all temperatures and corresponding suppressed oscillations below 70° and above 110°. The rather skewed behavior of v_3 and v_4 seen here should be contrasted with that of the $T_{\rm el} = 0$ canonical d-band treatment of these potentials in the case of Mo5.2, which is shown in Fig. 1.4.

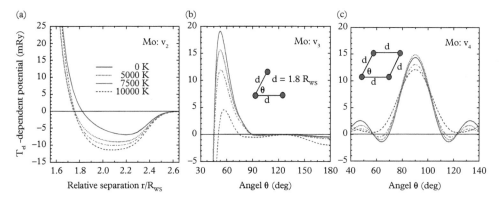

Fig. 13.3 T_{el} – dependent noncanonical d-band MGPT potentials for Mo at a representative compressed atomic volume of 92.99 a.u., or $\Omega/\Omega_0 = 0.885$, which corresponds to melting conditions of about 67 GPa and 5000 K (Moriarty et al., 2012). (a) Two-ion potential v_2; (b) Three-ion potential v_3; (c) Four-ion potential v_4.

We finally discuss the successful simulation of the high-pressure bcc melt curve for molybdenum with the $T_{\rm el}$-dependent Mo12t MGPT potentials, as obtained by Moriarty et al. (2012). The central MGPT-MD results in this regard are plotted in Fig. 13.4 together with validating QMD simulations and experimental data. The MGPT-MD simulations, together with confirming QMD simulations, were performed by Moriarty et al. using the robust two-phase melt method discussed in Chapter 8. A total of four QMD melt points up to 100 GPa were so obtained with a 256-atom, solid plus liquid, computational cell. In contrast, cell size was not a limitation in the MGPT-MD simulations

13.3 Temperature-dependent MGPT potentials and the simulation of melt for Mo 507

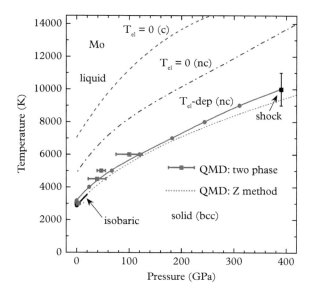

Fig. 13.4 *High-pressure bcc metal curve for Mo, as obtained from two-phase MD simulations with $T_{el} = 0$ and T_{el} − dependent MGPT potentials, using canonical (c) and noncanonical (nc) d-band treatments, as well as from two-phase and Z-method QMD simulations and experimental isobaric and shock data. The Z-method QMD melt results are from Belonoshko et al. (2008), the isobaric data from Shaner et al. (1977) and the shock data from Hixson et al. (1989).*
From Moriarty et al. (2012), with publisher (APS) permission.

and, although no melt size effects beyond 256 atoms were found, the final T_{el}-dependent MGPT melt curve was obtained with 87 808 atoms in the two-phase computational cell to minimize any statistical scatter among the eight melt points plotted in Fig. 13.4 up to 400 GPa. As is also indicated in Fig. 13.4, additional quantitative support for the T_{el}-dependent MGPT melt result is provided by the complementary Z-method QMD bcc melt simulations of Belonoshko et al. (2008), the isobaric measurements of Shaner et al. (1977), which provide the initial low-pressure melting slope, and the shock measurements of Hixson et al. (1989), which provide the observed shock melting point near 400 GPa.

Finally, the critical importance of electron temperature to the Mo12t MGPT melt result in Fig. 13.4 was made completely clear by performing parallel MGPT-MD two-phase melt calculations with $T_{el} = 0$ canonical and noncanonical d-band potentials, respectively. In the former canonical d-band case, the resulting melt temperatures are overestimated by a factor of two, confirming the original weak-coupling result of Moriarty (1994) discussed in Sec. 13.1. In the latter noncanonical d-band case, the melt temperatures are reduced significantly but are still 50% too high. Consequently, only when electron temperature is fully incorporated into the noncanonical d-band treatment is a quantitatively accurate MGPT melt curve achieved for Mo.

Appendix A1
Units, Conversion Factors and Useful Physical Data

Below is a brief summary of the main units, physical constants and conversion factors, to and from Rydberg atomic units (Ry-a.u.), as used in this book. For more extensive lists of physical constants and conversion factors, see http://physics.nist.gov/constants.

Length:

$$1 \text{ a.u.} = a_0 \equiv 0.529177 \text{ Å}$$
$$1\text{Å} = 10^{-8} \text{cm} = 0.1 \text{ nm}$$

Energy/Temperature:

$$E(\text{eV}) = 13.6057 \, E(\text{Ry})$$
$$E(\text{Ry}) = 0.0734986 \, E(\text{eV})$$
$$1\text{eV} = 1.60218 \times 10^{-19} \text{J}$$
$$k_B = 1.38065 \times 10^{-23} \text{ J/K}$$
$$= 8.61732 \times 10^{-5} \text{ eV/K}$$
$$= 6.33361 \times 10^{-6} \text{ Ry/K}$$

Mass/Density/Frequency:

$$1 \text{ amu} = 1.66054 \times 10^{-24} \text{g}$$
$$\rho(\text{g/cc}) = 11.2059 \, M(\text{amu}) / \Omega \text{ (a.u.)}$$
$$E(\text{Ry}) = 1312.77 \, E(\text{GPa cc/g})$$
$$\hbar = 1.05457 \times 10^{-34} \text{ J-s}$$
$$= 6.58212 \times 10^{-16} \text{eV-s}$$
$$= 4.83776 \times 10^{-17} \text{ Ry-s}$$

Volume:

$$1 \text{ a.u.} = a_0^3$$
$$\Omega\left(\text{Å}^3\right) = 0.148185 \, \Omega \, (\text{a.u.})$$
$$\Omega \, (\text{a.u.}) = 6.74834 \, \Omega\left(\text{Å}^3\right)$$

Pressure/Elastic constants:

$$P(\text{a.u.}) = dE(\text{Ry})/d\Omega \, (\text{a.u.})$$
$$P(\text{GPa}) = 14710.8 \, P(\text{a.u.})$$
$$C_{ij}(\text{GPa}) = C_{ij}(10^{10} \text{dynes}/\text{cm}^2)$$
$$1 \text{Mbar} = 100 \text{ GPa}$$
$$1 \text{GPa} = 10 \text{ kbar}$$

Area:

$$1 \text{ a.u.} = a_0^2 = 0.280028 \, \text{Å}^2$$

Surface/GB energy per unit area:

$$E(\text{J/m}^2) = 16.0218 \, E(\text{eV}/\text{Å}^2)$$
$$= 788.450 \, E(\text{Ry}/\text{a.u.})$$

In the following seven reference tables, we list fundamental observed and calculated materials properties of both nontransition and transition metals, which collectively have been the main prototype systems for QBIP development discussed in this book. These properties include: basic physical data in Tables A1.1 and A1.2; observed cohesive energies, bulk moduli and Debye and melt temperatures in Tables A1.3 and A1.4; observed fcc, bcc and hcp elastic constants in Tables A1.5 and A1.6; and observed crystal structures for elemental metals and intermetallic compounds in Table A1.7.

In Table A1.1, the electron radius r_s defined by Eq. (2.13) is calculated for the uniform electron density $n = Z/\Omega$ of nontransition metals in the form

$$r_s = R_{\text{WS}}/Z^{1/3}. \tag{A1.1}$$

In Table A1.2, the central theoretical quantities Z, Z_d, ε_F and $\varepsilon_F - E_d^{\text{vol}}$ for transition metals have been calculated directly from the GPT zero-order pseudoatom formalism discussed in Sec. 4.2 of Chapter 4 and Sec. 5.1.2.1 of Chapter 5.

Table A1.1 *Basic physical data on the nominal nontransition metals. The reference volume Ω_0 is that observed at the temperature T_0 and ambient pressure, after Pearson (1967). All quantities except T_0 (in K) and the observed atomic mass M_a (in amu) are given in Rydberg a.u. Values of M_a are from Hampel (1977).*

Metal	Z_a	T_0	Ω_0	R_{WS}	Z	r_s	ε_F	k_F	M_a
Li	3	78	144.0	3.25156	1	3.252	0.3483	0.5902	6.939
Na	11	5	255.2	3.93487	1	3.935	0.2379	0.4877	22.9898
K	19	5	481.3	4.86156	1	4.862	0.1558	0.3948	39.102
Rb	37	5	587.8	5.19653	1	5.197	0.1364	0.3693	85.47
Cs	55	5	745.4	5.62471	1	5.625	0.1164	0.3412	132.905
Be	4	293	54.73	2.35530	2	1.869	1.0539	1.0266	9.0122
Mg	12	298	156.8	3.34518	2	2.655	0.5224	0.7228	24.312
Ca	20	299	294.5	4.12730	2	3.276	0.3433	0.5859	40.08
Sr	38	298	380.1	4.49369	2	3.567	0.2895	0.5381	87.62
Ba	56	5	421.8	4.65235	2	3.693	0.2701	0.5197	137.34
Cu	29	293	79.68	2.66944	1	2.669	0.5169	0.7189	63.54
Ag	47	291	115.1	3.01760	1	3.018	0.4045	0.6360	107.870
Au	79	291	114.4	3.01147	1	3.011	0.4061	0.6372	196.967
Zn	30	298	102.7	2.90510	2	2.305	0.6929	0.8324	65.37
Cd	48	294	145.6	3.26355	2	2.591	0.5488	0.7408	112.40
Hg	80	5	155.1	3.33304	2	2.645	0.5263	0.7255	200.59
Al	13	298	112.0	2.99027	3	2.074	0.8566	0.9255	26.9815
Ga	31	4	130.5	3.14659	3	2.182	0.7738	0.8797	69.72
In	49	4	172.8	3.45529	3	2.396	0.6417	0.8011	114.82
Tl	81	291	192.9	3.58438	3	2.485	0.5963	0.7722	204.37
Si[a]	14	298	135.1	3.18314	4	2.005	0.9159	0.9570	28.086
Ge[a]	32	298	152.8	3.31649	4	2.089	0.8440	0.9187	72.59
Sn	50	110	181.1	3.50975	4	2.211	0.7534	0.8680	118.69
Pb	82	298	204.7	3.65603	4	2.303	0.6944	0.8333	207.19
Bi	83	298	239.0	3.84978	5	2.251	0.7267	0.8524	208.980

[a] data are for semiconducting volume of material

Units, Conversion Factors and Useful Physical Data 511

Table A1.2 *Basic physical data on transition metals. The reference volume Ω_0 is that observed at the temperature T_0 and ambient pressure, after Pearson (1967). The quantities Z, Z_d, ε_F and $\varepsilon_F - E_d^{vol}$ are representative values, calculated from self-consistent GPT zero-order pseudoatoms, assuming a canonical d-band boundary condition $D_2^* = -3$. All quantities except T_0 (in K) and M_a (in amu) are given in Rydberg a.u. The observed values of M_a are from Hampel (1977).*

Metal	Z_a	T_0	Ω_0	R_{WS}	Z	Z_d	ε_F	$\varepsilon_F - E_d^{vol}$	M_a
Ca	20	299	294.5	4.12730	1.568	0.432	0.2918	−0.2277	40.08
Sr	38	298	380.1	4.49369	1.508	0.492	0.2399	−0.2337	87.62
Ba	56	5	421.8	4.65235	1.285	0.715	0.2011	−0.1992	137.34
Sc	21	300	168.7	3.42775	1.615	1.385	0.4315	−0.1705	44.956
Y	39	300	222.8	3.76076	1.545	1.455	0.3481	−0.1935	88.905
La	57	300	250.6	3.91109	1.168	1.832	0.2671	−0.1856	138.91
Ti	22	298	119.2	3.05302	1.514	2.486	0.5210	−0.1385	47.90
Zr	40	298	157.1	3.34731	1.381	2.619	0.4076	−0.1652	91.22
Hf	72	297	150.6	3.30049	2.052	1.948	0.5459	−0.2244	178.49
V	23	300	93.23	2.81291	1.422	3.578	0.5886	−0.1135	50.942
Nb	41	293	122.0	3.07674	1.209	3.791	0.4415	−0.1409	92.906
Ta	73	300	121.6	3.07337	2.044	2.956	0.6280	−0.1823	180.948
Cr	24	293	80.94	2.68344	1.408	4.592	0.6425	−0.0818	51.996
Mo	42	293	105.1	2.92755	1.139	4.861	0.4686	−0.1080	95.94
W	74	298	107.0	2.94509	2.067	3.933	0.6891	−0.1283	183.85
Mn	25	293	83.16	2.70775	1.480	5.520	0.6525	−0.0366	54.9380
Tc	43	300	95.92	2.83971	1.122	5.878	0.4931	−0.0682	99.
Re	75	300	99.24	2.87210	2.107	4.893	0.7339	−0.0663	186.2
Fe	26	293	79.47	2.66709	1.470	6.530	0.6694	−0.0066	55.847
Ru	44	298	91.61	2.79652	1.129	6.871	0.5105	−0.0212	101.07
Os	76	293	94.43	2.82493	2.138	5.862	0.7661	−0.0007	190.2
Co	27	300	75.10	2.61728	1.459	7.541	0.6917	0.0265	58.9332
Rh	45	293	92.90	2.80959	1.142	7.858	0.5098	0.0339	102.905
Ir	77	293	95.45	2.83506	2.143	6.857	0.7617	0.0676	192.2
Ni	28	300	73.82	2.60233	1.482	8.518	0.7071	0.0808	58.71

Continued

Table A1.2 *Continued*

Metal	Z_a	T_0	Ω_0	R_{WS}	Z	Z_d	ε_F	$\varepsilon_F - E_d^{vol}$	M_a
Pd	46	295	99.37	2.87335	1.173	8.827	0.4962	0.1136	106.4
Pt	78	293	101.9	2.89753	2.081	7.919	0.7151	0.1344	195.09
Cu	29	293	79.68	2.66944	1.651	9.349	0.7219	0.2258	63.54
Ag	47	291	115.1	3.01760	1.615	9.385	0.5568	0.3763	107.870
Au	79	291	114.4	3.01147	2.104	8.896	0.6669	0.2515	196.967

In Tables A1.3–A1.6, it should be noted that observed values for the bulk modulus B_0 can depend sensitively on both the origin of the data and, for noncubic hcp metals, on any secondary assumptions made in calculating B_0 from that data. Nonetheless, the room-temperature values of B_0 quoted from Kittel (1976) in Tables A1.3 and Tables A1.4 mostly agree to better than 10% with the values of B_0 inferred from the elastic constants listed in Tables A1.5 and A1.6. Outliers for the cubic metals are bcc Cr (14%) and fcc Ca (41%), while outliers for the hcp metals are Y (13%), Be (14%), Zr (15%), Cd (23%), Zn (26%) and Sc (28%). Most of these discrepancies are readily explained.

Table A1.3 *Selected observed properties of nontransition metals at ambient pressure. These properties are the $T = 0$ cohesive energy E_{coh}, the $T = 300K$ bulk modulus B_0, the Debye temperatures Θ_D^{LT}, Θ_D^{MT} and Θ_D defined in Chapter 7 and the melting temperature T_m.*

Metal	$E_{coh}(eV)^a$	$B_0(GPa)^a$	$\Theta_D^{LT}(K)^a$	$\Theta_D^{MT}(K)^b$	$\Theta_D(K)^c$	$T_m(K)^b$
Li	1.63	11.6	344	400	–	453
Na	1.113	6.8	158	150	163[d]	371
K	0.934	3.2	91	100	100[e]	337
Rb	0.852	3.1	56	–	62[f]	312
Cs	0.804	2.0	38, 40[b]	–	43[g]	302
Be	3.32	100.3	1440	1000	–	1550
Mg	1.51	35.4	400	318	–	922
Ca	1.84	15.2	230	230	215[h], 211[i]	1111
Sr	1.72	11.6	147	–	134[j]	1043
Ba	1.90	10.3	110	–	100[j], 95[k]	998
Cu	3.49	137.	343	315	313[l]	1356
Ag	2.95	100.7	225	215	211[m]	1234
Au	3.81	173.2	165	170	180[n]	1337

Zn	1.35	59.8	327	234	–	693
Cd	1.16	46.7	209	120	–	594
Hg	0.67	38.2	72	100	–	234.3
Al	3.39	72.2	428	394	391º	933
Ga	2.81	56.9	320	240	–	303
In	2.52	41.1	108	129	–	429.8
Tl	1.88	35.9	78	96	–	577
Sn	3.14	111.	200	170	–	505
Pb	2.03	43.0	105	88	–	601
Bi	2.18	31.5	119	120	–	544.5

[a] Kittel (1976).
[b] Ashcroft and Mermin (1976).
[c] Present calculation using Eq. (7.48) and observed phonon spectra from listed references.
[d] Woods et al. (1962) for Na phonons at 90 K.
[e] Cowley et al. (1966) for K phonons at 9 K.
[f] Copley and Brockhouse (1973) for Rb phonons at 9 K.
[g] Nücker and Buchenau (1985) for Cs phonons at 50 K.
[h] Heiroth et al. (1986) for Ca phonons at 293 K.
[i] Stassis et al. (1983) for Ca phonons at 295 K.
[j] Buchenau et al. (1984) for Sr, Ba phonons at 293 K.
[k] Mizuki et al. (1985) for Ba phonons at 295 K.
[l] Nicklow et al. (1967) for Cu phonons at 298 K.
[m] Kamitakahara and Brockhouse (1969) for Ag phonons at 296 K.
[n] Lynn et al. (1973) for Au phonons at 296 K.
[o] Gilat and Nicklow (1966) for Al phonons at 300 K.

Table A1.4 *Selected observed properties of transition metals at ambient pressure. These properties are the $T = 0$ cohesive energy E_{coh}, the $T = 300\,\text{K}$ bulk modulus B_0, the Debye temperatures, Θ_D^{LT}, Θ_D^{MT} and Θ_D defined in Chapter 7 and the melting temperature T_m.*

Metal	$E_{\text{coh}}(\text{eV})$[a]	$B_0(\text{GPa})$[a]	$\Theta_D^{\text{LT}}(\text{K})$[a]	$\Theta_D^{\text{MT}}(\text{K})$[b]	$\Theta_D(\text{K})$[c]	$T_m(\text{K})$[b]
Sc	3.90	43.5	360, 359[b]	–	–	1812
Y	4.37	36.6	280, 256[b]	–	–	1796
La	4.47	24.3	142	132	–	1193
Ti	4.85	105.1	420	380	–	1933
Zr	6.25	83.3	291	250	–	2125

Continued

Table A1.4 *Continued*

Metal	$E_{coh}(eV)^a$	$B_0(GPa)^a$	$\Theta_D^{LT}(K)^a$	$\Theta_D^{MT}(K)^b$	$\Theta_D(K)^c$	$T_m(K)^b$
Hf	6.44	109	252	–	–	2495
V	5.31	169.9	380	390	350[d]	2163
Nb	7.57	170.2	275	275	279[e]	2741
Ta	8.10	200	240	225	226[f]	3269[j]
Cr	4.10	190.1	630	460	464[g]	2130
Mo	6.82	272.5	450	380	376[e]	2890
W	8.90	323.2	400	310	309[h]	3683
Mn	2.92	59.6	410	400	–	1518
Tc	6.85	–	–	–	–	2445
Re	8.03	372	430, 416[b]	–	–	3453
Fe	4.28	168.3	470	420	–	1808
Ru	6.74	320.8	600, 382[b]	–	–	2583
Os	8.17	–	500, 400[b]	–	–	3318
Co	4.39	191.4	445	385	–	1768
Rh	5.75	270.4	480, 350[b]	–	–	2239
Ir	6.94	355	420	430	–	2683
Ni	4.44	186	450	375	384[i]	1726
Pd	3.89	180.8	274	275	–	1825
Pt	5.84	278.3	240	230	–	2045

[a] Kittel (1976).
[b] Ashcroft and Mermin (1976).
[c] Present calculation using Eq. (7.48) and observed phonon spectra from listed references.
[d] Colella and Batterman (1970) for V phonons at 300 K.
[e] Powell et al. (1977) for Nb, Mo phonons at 296 K.
[f] Woods (1964) for Ta phonons at 296 K.
[g] Shaw and Muhlestein (1971) for Cr phonons at 300 K.
[h] Larose and Brockhouse (1976) for W phonons at 295 K.
[i] Birgeneau et al. (1964) for Ni phonons at 296 K.
[j] Hampel (1977).

In this regard, the elastic constants given in Tables A1.5 and A1.6 have been taken directly from accurate ultrasonic data, except in the case of the alkaline-earth metals Ca, Sr and Ba, where reliable ultrasonic data were not found. For those metals, we used instead elastic constants derived from force-constant fits to the corresponding phonon spectrum, a generally less reliable procedure.

Units, Conversion Factors and Useful Physical Data   515

This probably explains the overestimate of B_0 for Ca in Table A1.5. For the hcp metals listed in Table A1.6, B_0 was calculated from Eq. (7.114), which implicitly assumes that the c/a ratio is independent of strain. As pointed out in Chapter 7, this is not true for Zn and Cd, thus explaining their too-large values of B_0. This explanation likely applies to the Y, Be, Zr and Sc outliers as well.

Table A1.5 *Observed elastic constants (in GPa) for selected bcc and fcc metals at ambient pressure. Here* $C' = (C_{11} - C_{12})/2$, $A = C_{44}/C'$, $P_C = C_{12} - C_{44}$ *and* $B_0 = (C_{11} + 2C_{12})/3$.

Metal	$T_{\text{expt}}(K)$	Phase	C_{11}	C_{12}	C_{44}	C'	A	P_C	B_0
Li	78[a]	bcc	14.8	12.5	10.8	1.16	9.3	1.70	13.3
Na	4.2[b]	bcc	8.53	7.09	6.27	0.72	8.7	0.82	7.6
K	4.2[c]	bcc	4.16	3.41	2.86	0.38	7.5	0.55	3.7
Rb	4.2[d]	bcc	3.42	2.88	2.21	0.27	8.2	0.67	3.1
Cs	4.2[e]	bcc	2.60	2.16	1.60	0.22	7.3	0.56	2.3
Ca	295[f]	fcc	27.8	18.2	16.3	4.80	3.4	1.9	21.4
Sr	293[g]	fcc	15.3	10.3	9.9	2.48	4.0	0.4	12.0
Ba	293[g]	bcc	12.6	8.0	9.5	2.29	4.1	−1.5	9.5
V	298[h]	bcc	230.9	120.0	43.4	55.4	0.78	76.7	157.0
Nb	298[h]	bcc	246.2	132.9	28.7	56.6	0.51	104.2	170.7
Ta	298[h]	bcc	266.0	160.9	82.5	52.5	1.57	78.5	196.0
Cr	300[i]	bcc	348.4	70.2	100.7	139.1	0.72	−30.5	162.9
Mo	298[h]	bcc	464.8	161.8	108.9	151.6	0.72	52.6	262.6
W	298[h]	bcc	522.7	204.6	160.6	159.1	1.01	44.0	310.6
Fe	300[j]	bcc	230.4	134.1	115.9	48.2	2.41	18.2	166.2
Rh	300[k]	fcc	413.0	194.0	184.0	109.5	1.68	10.0	267.0
Ir	300[l]	fcc	580.0	242.0	256.0	169.0	1.51	−14.0	355.0
Ni	300[m]	fcc	251.6	154.4	122.0	48.6	2.51	32.4	186.8
Pd	300[n]	fcc	227.0	176.0	72.0	25.5	2.82	104.0	193.0
Pt	300[o]	fcc	346.7	250.7	76.5	48.0	1.59	174.2	282.7
Cu	300[p]	fcc	168.4	121.4	75.4	23.5	3.21	46.0	137.1
Ag	300[q]	fcc	124.0	93.7	46.1	15.2	3.04	47.6	103.8

Continued

Table A1.5 *Continued*

Metal	T_{expt}(K)	Phase	C_{11}	C_{12}	C_{44}	C'	A	P_C	B_0
Au	300[q]	fcc	192.3	163.1	42.0	14.6	2.88	121.1	172.9
Al	298[r]	fcc	106.8	60.7	28.2	23.1	1.22	32.5	76.1
Pb	296[s]	fcc	49.7	42.3	15.0	3.68	4.08	27.3	44.8

[a] Nash and Smith (1959).
[b] Diederich and Trivisonno (1966), as extrapolated to 4.2 K from data in range 78–195 K.
[c] Marquardt and Trivisonno (1965).
[d] Gutman and Trivisonno (1967), as extrapolated to 4.2 K from data in range 78–170 K.
[e] Kollarits and Trivisonno (1968), with C' extrapolated to 4.2 K from data in range 63–78 K.
[f] Stassis et al. (1983).
[g] Buchenau et al. (1984).
[h] Katahara et al. (1979).
[i] Palmer and Lee (1971).
[j] Adams et al. (2006).
[k] Simmons and Wang (1971).
[l] MacFarlane et al. (1966).
[m] Salama and Alers (1977).
[n] Rayne (1960).
[o] MacFarlane et al. (1965).
[p] Overton and Gaffney (1955).
[q] Neighbours and Alers (1958).
[r] Kamm and Alers (1964).
[s] Miller and Schuele (1969).

Table A1.6 *Observed elastic constants (in GPa) for selected hcp metals at ambient pressure. Here $C_{66} = C' = (C_{11} - C_{12})/2$, $C_{\text{hex}} = (C_{11} + 2C_{33} + C_{12} - 4C_{13})/6$ and $B_0 = (2C_{11} + C_{33} + 2C_{12} + 4C_{13})/9$. Values of the hcp c/a ratio are from Pearson (1967).*

Metal	T_{expt}(K)	c/a	C_{11}	C_{33}	C_{12}	C_{13}	C_{44}	C_{66}	C_{hex}	B_0
Be	300[a]	1.568	292.3	336.4	26.7	14.0	162.5	132.8	156.0	114.5
Mg	298[b]	1.624	59.4	61.6	25.6	21.4	16.4	16.9	20.4	35.2
Sc	300[c]	1.594	99.3	107.0	39.7	29.4	27.7	29.8	39.2	55.8
Y	300[d]	1.571	77.9	76.9	28.5	21.0	24.3	24.7	29.4	41.5
Ti	298[e]	1.587	162.4	180.7	92.0	69.0	46.7	35.2	56.6	107.3
Zr	298[e]	1.593	143.4	164.8	72.8	65.3	32.0	35.3	47.4	95.4
Hf	298[e]	1.581	181.1	196.9	77.2	66.1	55.7	52.0	64.6	108.7
Re	298[f]	1.615	612.6	682.7	270.0	206.0	162.5	171.4	237.3	363.5
Ru	300[c]	1.582	563.0	624.0	188.0	168.0	181.0	187.5	221.2	310.9
Co	298[g]	1.623	307.1	358.1	165.0	102.7	75.5	71.0	129.6	190.3

Zn	295[h]	1.856	163.7	63.5	36.4	53.0	38.8	63.6	19.2	75.1
Cd	293[i]	1.886	114.4	50.8	39.3	40.1	20.1	37.6	15.8	57.6
Tl	300[j]	1.598	40.8	52.8	35.4	29.0	7.3	2.7	11.0	35.7

[a] Smith and Arbogast (1960).
[b] Slutsky and Garland (1957).
[c] Brandes and Brook (1992).
[d] Smith and Gjevre (1960).
[e] Fisher and Renken (1964).
[f] Shepard and Smith (1965).
[g] McSkimin (1955).
[h] Alers and Neighbours (1958).
[i] Garland and Silverman (1960).
[j] Ferris et al. (1963).

Table A1.7 *Selected crystal structures of the elemental metals and intermetallic compounds discussed in this book.*

Type	Strukturbericht[a]	Pearson symbol[b]	Space group	Examples
fcc	$A1$	$cF4$	$Fm\bar{3}m$	Al, Ca, Cu, Pb, Ni, δ-Pu
bcc	$A2$	$cI2$	$Im\bar{3}m$	Na, K, Ba, V, Mo, Ta, Fe
hcp	$A3$	$hP2$	$P6_3/mmc$	Mg, Zn, δ-Hg, Tl, Ti, Co
dhcp	–	$hP4$	$P6_3/mmc$	La, Pr
9R	–	$hR9$	$P6_3/mmc$	Sm, Li, Na
bct	$A5$	$tI4$	$I4_1/amd$	β-Sn
bct	$A6$	$tI2$	$I4/mmm$	In, β-Hg
rhom	$A10$	$hR1$	$R\bar{3}m$	α-Hg, V at 69 GPa
ortho	$A11$	$oC8$	$Cmca$	α-Ga
cubic	$A12$	$cI58$	$I\bar{4}3m$	α-Mn, $Mg_{17}Al_{12}$
cubic	$A15$	$cP8$	$Pm\bar{3}n$	β-W
ortho	$A20$	$oC4$	$Cmcm$	α-U
hex–ω	–	$hP3$	$P6/mmm$	Ti, Zr, Hf
ortho	–	$oP4$	$Pmma$	MgCd
ortho	–	$oP4$	$Pnma$	Ca-VI, Am-IV

Continued

Table A1.7 *Continued*

Type	Strukturbericht[a]	Pearson symbol[b]	Space group	Examples
ortho	–	$oF2$	$Fddd$	γ-Pu, Am-III
cubic	B2	$cP2$	$Fm\bar{3}m$	CsCl, CoAl
tetrag	C16	$tI12$	$I4/mcm$	Al_2Cu
cubic	$D0_3$	$cF16$	$Fm\bar{3}m$	Fe_3Al
ortho	$D0_{11}$	$oP16$	$Pnma$	Al_3Ni
tetrag	$D0_{22}$	$tI8$	$I4/mmm$	Al_3Ti, Al_3V
hex	$D5_{13}$	$hP5$	$P\bar{3}m1$	Al_3Cu_2, Al_6CoCu_3
hex	$D8_{11}$	$hP28$	$P6_3/mmc$	Al_5Co_2
ortho	–	$oC16$	$Cmcm$	Al_5Fe_2
cubic	$L1_2$	$cP4$	$Pm\bar{3}m$	Al_3Sc
mono	–	$mP22$	$P2_1/a$	Al_9Co_2
cubic	–	$cI26$	$Im3$	$Al_{12}W$
mono	–	$mC34-1.8$	$C2/m$	$Al_{75}Co_{22}Ni_3$
ortho	–	$oP102$	$Pmn2_1$	O - $Al_{13}Co_4$
mono	–	$mC102$	$C2/m$	M - $Al_{13}Co_4$
tetrag	–	$tP40$	$P4/mnc$	Al_7CoCu_2
mono	–	$mC20$	$C2/m$	AlCu

[a] See http://wikipedia.org/wiki/Strukturbericht.
[b] Pearson (1967).

Appendix A2
Additional Elements of Generalized Pseudopotential Theory

A2.1 Analytic forms for the LDA correlation energy ε_c

The HL, VWN and PZ parameterizations of the LDA correlation energy for the uniform electron gas, as discussed in Sec. 2.2.1 of Chapter 2, are given as a function of the electron radius r_s defined by Eq. (2.13). The analytic HL correlation energy (Hedin and Lundqvist, 1971) is

$$\varepsilon_c = -C\left[(1+x^3)\ln\left(1+\frac{1}{x}\right) + \frac{1}{2}x - x^2 - \frac{1}{3}\right], \qquad (A2.1)$$

where $C = 0.045$ Ry, $x = r_s/r_0$ and $r_0 = 21$ a.u. The corresponding HL correlation potential is given by

$$\mu_c = \varepsilon_c - \frac{r_s}{3}\frac{d\varepsilon_c}{dr_s} = -C\ln\left(1+\frac{1}{x}\right). \qquad (A2.2)$$

The VWN correlation energy (Vosko et al., 1980) is

$$\varepsilon_c = A\left\{\ln\frac{x^2}{X(x)} + \frac{2b}{Q}\tan^{-1}\frac{Q}{2x+b} - \frac{bx_0}{X(x_0)}\left[\ln\frac{(x-x_0)^2}{X(x)} + \frac{2(b+2x_0)}{Q}\tan^{-1}\frac{Q}{2x+b}\right]\right\}, \qquad (A2.3)$$

where $A = 0.0621814$ Ry, $x = r_s^{1/2}$, $X(x) = x^2 + bx + c$ and $Q = (4c - b^2)^{1/2}$, with constants $x_0 = -0.10498$, $b = 3.72744$ and $c = 12.9352$ for r_s in a.u. The additional PZ correlation energy (Perdew and Zunger, 1981) is

$$\varepsilon_c = \gamma/(1 + \beta_1 x + \beta_2 x^2), \qquad (A2.4)$$

where $\gamma = -0.2846$ Ry and $x = r_s^{1/2}$, with $\beta_1 = 1.0529$ and $\beta_2 = 0.3334$ for r_s in a.u. For the latter VWN and PZ parameterizations of ε_c, corresponding analytic forms for the correlation potential μ_c can be readily obtained using the first equality in Eq. (A2.2).

A2.2 Free-atom ionization energy $E_{\text{bind}}^{\text{atom}}$ and preparation energy E_{prep}

The free-atom ionization energy needed in the simple-metal cohesive energy expression given by Eq. (2.31) can be calculated in the LDA of DFT as the binding energy

$$E_{\text{bind}}^{\text{atom}}(Z) = \sum_{m=\text{val}} \varepsilon_m - \frac{1}{2} n_{\text{val}}^{\text{at}} V_{\text{val}}^{\text{at}} + n_{\text{val}}^{\text{at}}[\varepsilon_{\text{xc}}(n_{\text{val}}^{\text{at}}) - \mu_{\text{xc}}(n_{\text{val}}^{\text{at}})], \quad (A2.5)$$

which is the analog of the metal binding energy $E_{\text{bind}}(Z)$ in Eq. (2.29). Here ε_m are the one-electron orbital energies for the free-atom valence s and p states (e.g., $3s$ and $3p$ for Al), and $n_{\text{val}}^{\text{at}}$ is the corresponding electron density.

In the case of transition metals, there are two additional complicating factors that must be taken into account in similarly calculating the cohesive energy in terms of valence binding energies, as envisaged in the second equality of Eq. (2.37). First, the self-consistent valence Z and corresponding d-state occupation number Z_d in the metal are generally different than the values of Z_0 and Z_d^0 for the ground state of the free atom. Second, because of the incomplete d shell in the free atom, there is a substantial spin-polarization energy contribution to its total energy, which requires replacing the usual LDA treatment with a local spin density (LSD) treatment, such as the spin density functional theory (SDFT) developed by von Barth and Hedin (1972).

Noting that Moruzzi et al. (1978) had previously calculated LSD free-atom values of Z_0, Z_d^0 and $E_{\text{tot}}^{\text{atom}}$ for the $3d$ and $4d$ transition series elements, Moriarty (1988a) developed the following approximate but efficient scheme to deal with the above issues. He first defined a spin-polarization energy correction

$$\delta E_{\text{sp}} \equiv E_{\text{tot}}^{\text{atom}}(Z_0, Z_d^0; \text{LDA}) - E_{\text{tot}}^{\text{atom}}(Z_0, Z_d^0; \text{LSD}), \quad (A2.6)$$

which is a positive constant and can be evaluated once and for all for each element, with the remaining promotion and binding-energy considerations to be performed in the LDA. To accomplish the needed s to d or d to s electron transfer to obtain Z valence s electrons and Z_d valence d electrons in the free atom, he proposed and validated the simple but accurate promotion-energy formula

$$E_{\text{pro}} = \frac{1}{2} \sum_{m=s,d} (\alpha_m - \alpha_m^0)(\varepsilon_m + \varepsilon_m^0), \quad (A2.7)$$

where α_m and α_m^0 are the occupation numbers and ε_m and ε_m^0 the one-electron orbital energies of the promoted and unpromoted atoms, respectively. The total preparation energy of the promoted atom E_{prep} entering Eq. (2.37) is then

$$E_{\text{prep}} = E_{\text{pro}} + \delta E_{\text{sp}}. \quad (A2.8)$$

The binding energy of the promoted atom $E_{\text{bind}}^{\text{atom}}(Z, Z_d)$ in Eq. (2.37) can then be calculated with Eq. (A2.5), noting only that the role of Z_d is now indirect and the sum on the RHS of that equation is over just the valence s orbitals corresponding to Z. Representative calculated values of δE_{sp}, E_{pro}, E_{prep} and $E_{\text{bind}}^{\text{atom}}(Z, Z_d) - E_{\text{prep}}$ for 20 $3d$ and $4d$ transition-series metals are given in Table 1 of Moriarty (1988a).

A2.3 Exchange-correlation correction terms $\delta\mu^*_{xc}(n_i, n_j)$ and $\delta\varepsilon^*_{xc}(n_i, n_j)$

As discussed by Moriarty (1988a), the two-ion exchange-correlation potential correction term $\delta\mu^*_{xc}(n_i, n_j)$ arises from a cluster-like expansion of the total exchange potential $\mu_{xc}(n)$ in terms of overlapping localized core plus d state electron densities $n_i, n_j \ldots$:

$$\mu_{xc}(n) = \mu_{xc}(n_{val}) + \sum_i \mu^*_{xc}(n_i) + \frac{1}{2}\sum_{i,j}{}' \delta\mu^*_{xc}(n_i, n_j) + \cdots, \quad (A2.9)$$

where the functional $\mu^*_{xc}(n_i)$ is defined by Eq. (2.26) and

$$\delta\mu^*_{xc}(n_i, n_j) \equiv \mu^*_{xc}(n_i + n_j) - \mu^*_{xc}(n_i) - \mu^*_{xc}(n_j), \quad (A2.10)$$

For use in calculating the d-state overlap matrix elements $\Delta^{ij}_{dd'}$ [see Eq. (4.92) and surrounding discussion in Sec. 4.2 of Chapter 4], the lowest-order approximation to $\delta\mu^*_{xc}(n_i, n_j)$ is written as $\delta v_{xc}(\mathbf{r} - \mathbf{R}_i, \mathbf{r} - \mathbf{R}_j)$ and is obtained by replacing n_{val} with n_{unif} in each functional $\delta\mu^*_{xc}$ on the RHS of Eq. (A2.10).

As also discussed by Moriarty (1988a), the two-ion xc energy correction term $\delta\varepsilon^*_{xc}(n_i, n_j)$ arises from a similar expansion of the total exchange-correlation energy $E_{xc} = n\varepsilon_{xc}(n)$. The last three two-ion xc terms in Eq. (2.36) for the binding energy of a d-band metal arise from the latter expansion, with

$$\delta\varepsilon^*_{xc}(n_i, n_j) \equiv (n_{val} + n_i + n_j)[\varepsilon^*_{xc}(n_i + n_j) - \varepsilon^*_{xc}(n_i) - \varepsilon^*_{xc}(n_j)]$$
$$+ n_i[\varepsilon^*_{xc}(n_j) - 0.5\mu^*_{xc}(n_j)] + n_j[\varepsilon^*_{xc}(n_i) - 0.5\mu^*_{xc}(n_i)], \quad (A2.11)$$

where, in analogy with Eq. (2.26) defining $\mu^*_{xc}(n_i)$, one has

$$\varepsilon^*_{xc}(n_i) \equiv \varepsilon_{xc}(n_{val} + n_i) - \varepsilon_{xc}(n_{val}). \quad (A2.12)$$

A2.4 The orthogonalization hole for an AHS pseudopotential

For an AHS pseudopotential (Austin et al., 1962) of the baseline form (2.54) used in the GPT, with plane-wave PP matrix elements given by Eq. (2.56), it is appropriate to calculate the corresponding simple-metal orthogonalization-hole density δn^c_{oh} to first order in the one-OPW limit. In this limit, the true wavefunction $\langle \mathbf{r} | \psi_\mathbf{k} \rangle$ is calculated from Eq. (2.62) with a single plane wave $\langle \mathbf{r} | \mathbf{k} \rangle$, initially normalized in the volume of the metal $N\Omega$ and substituted for the pseudowavefunction $\langle \mathbf{r} | \phi_\mathbf{k} \rangle$:

$$\langle \mathbf{r} | \psi_\mathbf{k} \rangle = C_\mathbf{k} \langle \mathbf{r} | (1 - P_c) | \mathbf{k} \rangle, \quad (A2.13)$$

where we recall that P_c is the *total* core-state projection operator given by Eq. (2.55) and where $C_\mathbf{k}$ is the normalization constant

$$C_\mathbf{k} = \frac{1}{[1 - \langle \mathbf{k} | P_c | \mathbf{k} \rangle]^{1/2}}. \quad (A2.14)$$

The charge-neutral orthogonalization-hole density δn_{oh}^c is then defined as

$$\delta n_{\text{oh}}^c(\mathbf{r}) \equiv 2 \sum_{\mathbf{k}} \left[\langle \mathbf{r} | \psi_{\mathbf{k}} \rangle \langle \psi_{\mathbf{k}} | \mathbf{r} \rangle - \langle \mathbf{r} | \phi_{\mathbf{k}} \rangle \langle \phi_{\mathbf{k}} | \mathbf{r} \rangle \right] \Theta^<(\mathbf{k})$$

$$= 2 \sum_{\mathbf{k}} \left[\left(\frac{\langle \mathbf{r} | \mathbf{k} \rangle \langle \mathbf{k} | \mathbf{r} \rangle - \langle \mathbf{r} | \mathbf{k} \rangle \langle \mathbf{k} | P_c | \mathbf{r} \rangle - \text{c.c.} + \langle \mathbf{r} | P_c | \mathbf{k} \rangle \langle \mathbf{k} | P_c | \mathbf{r} \rangle}{1 - \langle \mathbf{k} | P_c | \mathbf{k} \rangle} \right) - \langle \mathbf{r} | \mathbf{k} \rangle \langle \mathbf{k} | \mathbf{r} \rangle \right] \Theta^<(\mathbf{k})$$

$$= \frac{2\Omega}{(2\pi)^3} \int \left[\frac{\langle \mathbf{k} | p_c | \mathbf{k} \rangle}{1 - \langle \mathbf{k} | p_c | \mathbf{k} \rangle} - \frac{\sum_i \left(\langle \mathbf{r} | \mathbf{k} \rangle \langle \mathbf{k} | p_c^i | \mathbf{r} \rangle + \text{c.c.} - \langle \mathbf{r} | p_c^i | \mathbf{k} \rangle \langle \mathbf{k} | p_c^i | \mathbf{r} \rangle \right)}{1 - \langle \mathbf{k} | p_c | \mathbf{k} \rangle} \right] \Theta^<(\mathbf{k}) d\mathbf{k}.$$

(A2.15)

On the RHS of the third line of Eq. (A2.15), we have renormalized the plane waves in the single-site volume Ω, rather than the metal volume $N\Omega$, so that $\langle \mathbf{k} | P_c | \mathbf{k} \rangle \to \langle \mathbf{k} | p_c | \mathbf{k} \rangle$, with p_c the single-site core projection operator. Also, $p_c^i = p_c$ for the single specified site i.

Comparing Eq. (A2.15) with Eq. (3.5), one sees that the magnitude of the effective valence Z^* is established by the first term on the RHS of the third line of Eq. (A2.15):

$$Z^* = Z + \frac{2\Omega}{(2\pi)^3} \int \left[\frac{\langle \mathbf{k} | p_c | \mathbf{k} \rangle}{1 - \langle \mathbf{k} | p_c | \mathbf{k} \rangle} = \langle \mathbf{k} | p_c | \mathbf{k} \rangle \left(1 + \langle \mathbf{k} | p_c | \mathbf{k} \rangle + \cdots \right) \right] \Theta^<(\mathbf{k}) d\mathbf{k}. \quad (A2.16)$$

Retaining the full normalization factor $[1 - \langle \mathbf{k} | p_c | \mathbf{k} \rangle]^{-1}$ is appropriate for the Harrison AHS PP given by Eq. (3.102). However, since $\langle \mathbf{k} | p_c | \mathbf{k} \rangle$ is a first-order quantity, then expanding out the normalization factor produces only a second-order correction, which is appropriate to discard in the baseline AHS PP given by Eq. (2.56), as used in the GPT. Thus, in the GPT one obtains an effective valence given by Eq. (3.38). Similar reasoning applies to the density of the individual orthogonalization holes centered on each ion site, as collectively represented by the second group of terms on the RHS of the third line of Eq. (A2.15). Again, retaining the normalization factor is appropriate for the Harrison PP, but this factor is to be discarded for the GPT baseline PP, which leads to Eq. (3.39) for the single-site orthogonalization hole density $n_{\text{oh}}(\mathbf{r}) = n_{\text{oh}}^c(\mathbf{r})$ for a simple metal.

Finally, we discuss the two small orthogonalization-hole contributions δE_{oh} and γ_{oh} to the cohesive energy of a simple metal that were introduced in Chapter 3. The self-energy correction δE_{oh} associated with the finite size of $n_{\text{oh}}(\mathbf{r})$ enters the reciprocal-space volume term $E_{\text{vol}}^q(\Omega)$ in Eq. (3.43) and the real-space volume term $E_{\text{vol}}(\Omega)$ in Eqs. (3.95) and (3.96). Noting that $n_{\text{oh}}(\mathbf{r})$ is spherically symmetric about the site on which it is centered, we have evaluated δE_{oh} as

$$\delta E_{\text{oh}}(\Omega) = -\frac{9}{10}(Z^* - Z)^2 e^2 / R_{\text{WS}} + \frac{1}{2} \int_0^\infty u_{\text{oh}}(r) v_{\text{oh}}(r) dr + \int_0^\infty u_{\text{oh}}(r) \left[v_{\text{unif}}^{\text{oh}}(r) - v_{\text{unif}}^{\text{oh}}(0) \right] dr,$$

(A2.17)

where $u_{\text{oh}}(r) = 4\pi r^2 n_{\text{oh}}(r)$, $v_{\text{oh}}(r)$ is the Coulomb potential arising from $n_{\text{oh}}(r)$ and

$$v_{\text{unif}}^{\text{oh}}(r) = \frac{3}{2}(Z^* - Z)e^2 / R_{\text{WS}} - \frac{1}{2}(Z^* - Z)e^2 r^2 / R_{\text{WS}}^3. \quad (A2.18)$$

The dimensionless constant γ_{oh} can be calculated by noting that because of the spherical symmetry of $n_{oh}(\mathbf{r})$, one has

$$n_{oh}(q) = \frac{1}{\Omega} \int_0^\infty u_{oh}(r) j_0(qr) dr, \qquad (A2.19)$$

where $j_0(qr)$ is the $\ell = 0$ spherical Bessel function, which has a small q expansion of the general form

$$j_0(qr) = 1 + \frac{(qr)^2}{6} + \cdots. \qquad (A2.20)$$

Inserting Eq. (A2.20) in Eq. (A2.19) and comparing with the definition of γ_{oh} established by Eq. (3.94), it follows that

$$\gamma_{oh} = -\frac{k_F^2}{6} \int_0^\infty r^2 u_{oh}(r) dr / (Z^* - Z). \qquad (A2.21)$$

Also, it should be noted that Eq. (A2.17) for δE_{oh} and Eq. (A2.21) for γ_{oh} can be used more generally for d-band metals with appropriate values of Z^* and $n_{oh}(\mathbf{r})$, as given by Eqs. (4.37) and (4.39) for EDB and FDB metals, respectively, and by Eqs. (5.54) and (5.57) for transition metals.

A2.5 Band-structure energies for fcc metals at high-symmetry BZ points

We summarize here some useful analytic formulas for fcc band energies $E(\mathbf{k})$ at the high-symmetry Γ, X and L points in the BZ, as derived from the basic H-NFE-TB secular determinant given by Eq. (2.139) for d-band metals (Moriarty, 1982). For the $\mathbf{k} = 0$ Γ point, the use of a single $\mathbf{G} = 0$ reciprocal lattice vector is adequate, and puts the bottom of the conduction band at

$$\Gamma_1 = \langle 0 | w_0 | 0 \rangle, \qquad (A2.22)$$

while the d-band $\Gamma_{25'}$ and Γ_{12} states have energies

$$\Gamma_{25'} = E_d + 3dd\sigma + 4dd\pi + 5dd\delta \qquad (A2.23)$$

and

$$\Gamma_{12} = E_d + \frac{3}{2} dd\sigma + 6dd\pi + \frac{9}{2} dd\delta. \qquad (A2.24)$$

In the usual way, $dd\sigma$, $dd\pi$ and $dd\delta$ represent the $m = 0$, 1 and 2 components of the d-state overlap matrix element $-\Delta_{dd}^{ij} = -\Delta_{dd}(R_{ij}) \equiv -\langle \phi_d^i | \Delta | \phi_d^j \rangle$, here evaluated at the nearest-neighbor distance $R_{ij} = \sqrt{2}a/2$ in the fcc structure, with $\Omega = a^3/4$.

For the X point at $\mathbf{k}_0 = (2\pi/a)\hat{\mathbf{x}}$, two reciprocal lattice vectors are needed: $\mathbf{G} = 0$ and $\mathbf{G} = (4\pi/a)\hat{\mathbf{x}}$. With $k_0 = 2\pi/a$, the X-point energies encompassing the d bands are

$$X_1 = \frac{1}{2}[A(k_0) + E_d + D] \pm \left\{\frac{1}{4}[A(k_0) - E_d - D]^2 + 40\pi\Delta^2(k_0)\right\}^{1/2}, \quad (A2.25)$$

$$X_2 = E_d - \frac{3}{2}dd\sigma + 2dd\pi - \frac{9}{2}dd\delta, \quad (A2.26)$$

$$X_3 = E_d + 3dd\sigma - 4dd\pi - 3dd\delta, \quad (A2.27)$$

$$X_{4'} = B(k_0) \quad (A2.28)$$

and

$$X_5 = E_d - 3dd\sigma - dd\delta, \quad (A2.29)$$

where we have defined the quantities

$$D \equiv \frac{1}{2}dd\sigma - 6dd\pi + \frac{3}{2}dd\delta, \quad (A2.30)$$

$$A(k) \equiv \varepsilon_{\mathbf{k}} + w_0(\mathbf{k}, 0) + w_0(\mathbf{k}, -2\mathbf{k}) \quad (A2.31)$$

and

$$B(k) \equiv \varepsilon_{\mathbf{k}} + w_0(\mathbf{k}, 0) - w_0(\mathbf{k}, -2\mathbf{k}), \quad (A2.32)$$

with $w_0(\mathbf{k}, \mathbf{q}) = \langle \mathbf{k}+\mathbf{q}| w_0 |\mathbf{k}\rangle$ and $\langle \mathbf{k}| \Delta |\phi_d\rangle = -4\pi\Delta(k)Y_{2m}(\mathbf{k})$.

For band energies well removed from the center of the d bands [i.e., $E(\mathbf{k}) \gg E_d$ or $E_d \gg E(\mathbf{k})$], one can fold down the secular determinant (2.139) into the effective plane-wave form

$$\left|[\varepsilon_{\mathbf{k}-\mathbf{G}} - E(\mathbf{k})]\delta_{\mathbf{GG'}} + \langle \mathbf{k}-\mathbf{G'}| w_0 |\mathbf{k}-\mathbf{G}\rangle + \sum_d \frac{\langle \mathbf{k}-\mathbf{G'}| \Delta |\phi_d\rangle \langle \phi_d| \Delta |\mathbf{k}-\mathbf{G}\rangle}{E(\mathbf{k}) - E_d}\right| = 0, \quad (A2.33)$$

where only the reciprocal-lattice vectors $\mathbf{G}$ and $\mathbf{G'}$ are spanned. Equation (A2.33) can be used to calculate additional band energies at the $\mathbf{k}_1 = (\pi/a)(\hat{\mathbf{x}} + \hat{\mathbf{y}} + \hat{\mathbf{z}})$ L point for fcc EDB and FDB metals. Using reciprocal-lattice vectors $\mathbf{G} = 0$ and $\mathbf{G} = (2\pi/a)(\hat{\mathbf{x}} + \hat{\mathbf{y}} + \hat{\mathbf{z}})$, one finds L-point energies

$$L_1 = \frac{1}{2}[A(k_1) + E_d] \pm \left\{\frac{1}{4}[A(k_1) - E_d]^2 + 40\pi\Delta^2(k_1)\right\}^{1/2} \quad (A2.34)$$

and

$$L_{2'} = B(k_1), \quad (A2.35)$$

where $k_1 = \sqrt{3}\pi/a$. In Eq. (A2.34), the plus sign is for $\varepsilon_F > E_d$ and is appropriate for the upper L_1 state in FDB metals, while the minus sign is for $\varepsilon_F < E_d$ and is appropriate for the lower L_1 state in EDB metals.

A2.6 Solution of the pseudo-Green's function equations for a *d*-band metal

Following Moriarty (1972a), the solution of the full pseudo-Green's function equations (2.147)–(2.150) can be extended from the single transition-metal ion treated in Sec. 2.7.3 of Chapter 2 to the bulk *d*-band transition-series metal. To proceed with that task, one first solves Eq. (2.148) for G_{kd} in terms of $G_{d'd}$, using a perturbation expansion in powers of the pseudopotential $W_{kk'}$ that creates the series $\Sigma_{kk'} = W_{kk'} + \cdots$ given by Eq. (2.173) and uncouples $W_{kk'}$ and $G_{k'd}$ in the second term on the RHS of Eq. (2.148). This perturbation treatment yields

$$(E - \varepsilon_k)G_{kd} = S_{kd} + \sum_{d'} V'_{kd'} G_{d'd} + \sum_{k'} \frac{\Sigma_{kk'} S_{k'd}}{E - \varepsilon_{k'}} + \sum_{k'}\sum_{d'} \frac{\Sigma_{kk'} V'_{k'd'} G_{d'd}}{E - \varepsilon_{k'}}. \quad (A2.36)$$

Then, using Eq. (A2.36) in Eq. (2.148), one obtains the following coupled equations for $G_{dd'}$:

$$(E - E_d - \Gamma_{dd} - \Lambda_{dd}) G_{dd'} = S_{dd'} + \sum_k \frac{V'_{dk} S_{kd'}}{E - \varepsilon_k} + \sum_{k,k'} \frac{V'_{dk} \Sigma_{kk'} S_{k'd'}}{(E - \varepsilon_k)(E - \varepsilon_{k'})}$$

$$+ {\sum_{d''}}' (V'_{dd''} + \Gamma_{dd''} + \Lambda_{dd''}) G_{d''d'}, \quad (A2.37)$$

where $\Gamma_{dd'}$ and $\Lambda_{dd'}$ are given by Eqs. (2.175) and (2.176), respectively. The 5*N* coupled equations represented by Eq. (A2.37) are most readily solved by determinants, beginning with the $5N \times 5N$ determinant $D(E)$ given by Eq. (2.174). We further define the determinant $A(x_{d\alpha})$ as that formed by replacing the *d*th column in $D(E)$ by the column $x_{d\alpha}$. Returning to Eq. (A2.37) and combining the LHS of that equation with the final term on the RHS, the solution for $G_{dd'}$ is just

$$G_{dd'} = A(f_{dd'})/D(E), \quad (A2.38)$$

where

$$f_{dd'} = S_{dd'} + \sum_k \frac{V'_{dk} S_{kd'}}{E - \varepsilon_k} + \sum_{k,k'} \frac{V'_{dk} \Sigma_{kk'} S_{k'd'}}{(E - \varepsilon_k)(E - \varepsilon_{k'})}. \quad (A2.39)$$

Next, one can solve Eq. (2.149) for $G_{kk'}$ in terms of G_{dk} using the same perturbation expansion in $W_{kk'}$ as above:

$$(E - \varepsilon_k) G_{kk'} = \delta_{kk'} + \frac{\Sigma_{kk'}}{E - \varepsilon_{k'}} + \sum_d V'_{kd} G_{dk'} + \sum_{k''}\sum_d \frac{\Sigma_{kk''} V'_{k''d} G_{dk'}}{E - \varepsilon_{k''}}. \quad (A2.40)$$

Then, using Eq. (A2.40) in Eq. (2.150), one obtains a set of coupled equations for the last pseudo-Green's function component G_{dk}:

$$(E - E_d - \Gamma_{dd} - \Lambda_{dd}) G_{dk} = S_{dk} + \frac{V'_{dk}}{E - \varepsilon_k} + \sum_{k'} \frac{V'_{dk'} \Sigma_{k'k}}{(E - \varepsilon_{k'})(E - \varepsilon_k)}$$

$$+ {\sum_{d'}}' (V'_{dd'} + \Gamma_{dd'} + \Lambda_{dd'}) G_{d'k}. \quad (A2.41)$$

Once again, the $5N$ coupled equations represented in Eq. (A2.41) can be solved for G_{dk} by determinants:

$$G_{dk} = A(g_{dk})/D(E), \tag{A2.42}$$

where

$$g_{dk} = S_{dk} + \frac{V'_{dk}}{E - \varepsilon_k} + \sum_{k'} \frac{V'_{dk'} \Sigma_{k'k}}{(E - \varepsilon_{k'})(E - \varepsilon_k)}. \tag{A2.43}$$

Finally, one can insert Eq. (A2.42) into Eq. (A2.40) to obtain

$$(E - \varepsilon_k) G_{kk'} = \delta_{kk'} + \frac{\Sigma_{kk'}}{E - \varepsilon_{k'}} + \frac{1}{D(E)} \left[\sum_d V'_{kd} A(g_{dk'}) + \sum_{k''} \sum_d \frac{\Sigma_{kk''} V'_{k''d} A(g_{dk'})}{E - \varepsilon_{k''}} \right]. \tag{A2.44}$$

Using Eq. (A2.44) for $k = k'$ and Eq. (A2.38) for $d = d'$, one can assemble the density of states $\rho(E)$ for a d-band metal in the form

$$\rho(E) = -\frac{2}{\pi} \text{Im} \left[\sum_k G_{kk} + \sum_d G_{dd} \right]$$

$$= -\frac{2}{\pi} \text{Im} \left\{ \sum_k \frac{1}{E - \varepsilon_k} + \frac{\Sigma_{kk}}{(E - \varepsilon_k)^2} \right.$$

$$\left. + \frac{1}{D(E)} \left[\sum_d A(f_{dd}) + \sum_k \sum_d \frac{V'_{kd} A(g_{dk})}{E - \varepsilon_k} + \sum_{k,k'} \sum_d \frac{\Sigma_{kk'} V'_{k'd} A(g_{dk})}{(E - \varepsilon_k)(E - \varepsilon_{k'})} \right] \right\}. \tag{A2.45}$$

The first term on the RHS of the second line of Eq. (A2.45) is just the free-electron density of states $\rho_0(E)$, as given by Eq. (2.162). As shown by Moriarty (1972a), the final three terms in square brackets on the bottom line of Eq. (A2.45) are equal to $dD(E)/dE$, so $\rho(E)$ can be reduced to the final compact form given by Eq. (2.172).

A2.7 Approximate calculation of the *d*-state overlap kinetic energy δv_{ke}^{ij}

In this section, we consider the approximate calculation of the d-state overlap kinetic energy correction term $\delta v_{ke}^{ij} \equiv \delta v_{ke}(R_{ij})$ to the overlap potential $v_{ol}(R_{ij}, \Omega)$ that is currently used in GPT and MGPT applications. For EDB and FDB metals, δv_{ke}^{ij} appears explicitly in Eq. (4.55) for v_{ol}. In the GPT treatment of transition metals, δv_{ke}^{ij} appears in Eq. (5.69) for v_{ol}^0, while for the corresponding MGPT treatment, δv_{ke}^{ij} becomes the kinetic-energy component of the hard-core potential v_2^{hc} entering the full pair potential in Eq. (5.99).

Following Moriarty (1988a), the kinetic energy overlap correction term δv_{ke}^{ij} is calculated in the spirit of the xc overlap correction terms $\delta \mu_{xc}^*(n_i, n_j)$ and $\delta \varepsilon_{xc}^*(n_i, n_j)$ discussed in Sec. A2.3.

A2.7 Approximate calculation of the d-state overlap kinetic energy δv_{ke}^{ij}

Recall from Chapter 4 that the energy $\delta\varepsilon_{xc}^*$ adds directly to the overlap-potential contribution $\delta v_{es\text{-}xc}^{ij}$ defined by Eq. (4.59). For δv_{ke}^{ij}, one introduces the analog kinetic energy functional

$$\varepsilon_{ke}(n) \equiv \frac{3}{5}\frac{\hbar^2}{2m}(3\pi^2)^{2/3}n^{5/3} \qquad (A2.46)$$

and the companion localized functional

$$\varepsilon_{ke}^*(n_i) \equiv \varepsilon_{ke}(n_{val} + n_i) - \varepsilon_{ke}(n_{val})\,. \qquad (A2.47)$$

Then the two-ion overlap contribution to the kinetic energy is

$$\delta\varepsilon_{ke}^*(n_i, n_j) \equiv \varepsilon_{ke}^*(n_i + n_j) - \varepsilon_{ke}^*(n_i) - \varepsilon_{ke}^*(n_j)\,. \qquad (A2.48)$$

In practice, all the usual assumptions made about $\delta\mu_{xc}^*$ and $\delta\varepsilon_{xc}^*$ apply to $\delta\varepsilon_{ke}^*$ as well: n_{val} is replaced by n_{unif} in Eq. (A2.47); $n_i \equiv n_{core}^i = n_{core}(\mathbf{r} - \mathbf{R}_i)$ is the inner-core plus occupied d-state density on the ion site i; and on the RHS of Eq. (A2.48) there is an implied integration over all space.

From $\delta\varepsilon_{ke}^*(n_i, n_j) \equiv \delta\varepsilon_{ke}^*(R_{ij})$ in Eq. (A2.48), one must subtract the kinetic-energy contribution associated with d-state overlap alone, which is already contained in the d-state potential $v_2^d(R_{ij})$, defined most generally by Eq. (5.14). This contribution is given to a good approximation by

$$\delta\varepsilon_{ke}^0(R_{ij}) = 4\left[\frac{Z_d}{10}\right]\sum_{d,d'} S_{dd'}(R_{ij})\Delta_{d'd}(R_{ij})\,. \qquad (A2.49)$$

Noting that $S_{dd'}^{ij} \equiv S_{dd'}(R_{ij})$ and $\Delta_{d'd}^{ji} \equiv \Delta_{d'd}(R_{ij})$, the RHS of Eq. (A2.49) is the leading term of the overlap potential $v_{ol}(R_{ij}, \Omega)$ for EDB and FDB metals, as given by Eq. (4.55), with $Z_d = 0$ and $Z_d = 10$, respectively. Here we assume that Eq. (A2.49) is also adequate for transition metals as well, where $0 < Z_d < 10$.

The final form of the kinetic-energy overlap correction is then taken to be

$$\delta v_{ke}(R_{ij}) = \delta\varepsilon_{ke}^*(R_{ij}) - \delta\varepsilon_{ke}^0(R_{ij})\,. \qquad (A2.50)$$

At sufficiently large separation R_{ij}, where the overlap is confined entirely to the d states, $\delta\varepsilon_{ke}^*(R_{ij})$ does indeed approach $\delta\varepsilon_{ke}^0(R_{ij})$, as shown in Fig. 13 of Moriarty (1988a) for copper; in that figure, the notation $\delta E_{ke} \equiv \delta v_{ke}$, $E_{ke} \equiv \delta\varepsilon_{ke}^*$ and $E_{ke}^0 \equiv \delta\varepsilon_{ke}^0$ is used. At smaller separations, $\delta v_{ke}(R_{ij}) > 0$ as expected, and in the case of Cu, $\delta v_{ke}(R_{nn})$ amounts to 50% of $\delta\varepsilon_{ke}^*(R_{nn})$ at the fcc nearest-neighbor distance. However, the percentage contribution of $\delta v_{ke}(R_{nn})$ relative to $\delta\varepsilon_{ke}^*(R_{nn})$ is generally expected to drop as one moves to the left in the Periodic Table, away from the noble metals.

For the hard-core potential v_2^{hc} in the MGPT transition-metal pair potential v_2 expressed by Eq. (5.99), one has

$$v_2^{hc}(R_{ij}, \Omega) = \delta\varepsilon_{ke}^*(R_{ij}) + \delta v_{es\text{-}xc}(R_{ij})\,, \qquad (A2.51)$$

where $\delta v_{es\text{-}xc}(R_{ij}) \equiv \delta v_{es\text{-}xc}^{ij}$, as given by Eq. (4.59), and the volume dependence in the two terms on the RHS of Eq. (A2.51) is implicit. Equation (A2.51) can also be generalized to the

MGPT treatment of the AB components of transition-metal alloys, as is discussed in Sec. 10.6 of Chapter 10. The hard-core potential $v_2^{\alpha\beta,\,\text{hc}}$ entering the alloy pair potential $v_2^{\alpha\beta}$, expressed by Eq. (10.46), is then generalized to the form

$$v_2^{\alpha\beta\text{hc}}(R_{ij},\Omega,x) = \delta\overset{*}{\varepsilon}_{\text{ke}}(n_{i\alpha},n_{j\beta}) + \delta v_{\text{es-xc}}^{i\alpha,j\beta}, \qquad (A2.52)$$

where for $\alpha = A$ or B and $\beta = A$ or B

$$\delta\overset{*}{\varepsilon}_{\text{ke}}(n_{i\alpha},n_{j\beta}) = \overset{*}{\varepsilon}_{\text{ke}}(n_{i\alpha}+n_{j\beta}) - \overset{*}{\varepsilon}_{\text{ke}}(n_{i\alpha}) - \overset{*}{\varepsilon}_{\text{ke}}(n_{j\beta}), \qquad (A2.53)$$

with $n_{i\alpha} \equiv n_{\text{core}}^{i\alpha}$ and $n_{j\beta} \equiv n_{\text{core}}^{j\beta}$, generalizing Eq. (A2.48), and where

$$\delta v_{\text{es-xc}}^{i\alpha,j\beta} = n_{\text{core}}^{i\alpha}\left[\left(\frac{Z_a-Z_\beta}{Z_a}\right)v_{\text{nuc}}^{j\beta} + v_{\text{core}}^{j\beta} + v_{\text{xc}}^{j\beta}\right]$$
$$+ \left(\frac{Z_a+Z_\alpha}{Z_a}\right)n_{\text{nuc}}^{i\alpha}\left[\left(\frac{Z_a-Z_\beta}{Z_a}\right)v_{\text{nuc}}^{j\beta} + v_{\text{core}}^{j\beta}\right] + \delta\overset{*}{\varepsilon}_{\text{xc}}(n_{\text{core}}^{i\alpha},n_{\text{core}}^{j\beta}), \qquad (A2.54)$$

generalizing Eq. (4.59). One has the usual implied integration over all space on the RHS of Eqs. (A2.53) and (A2.54).

A2.8 Self-consistent GPT pseudoatoms and the zero of energy for metals and alloys

In this final section, we discuss some additional important results concerning the GPT zero-order pseudoatoms defined in Chapters 4 and 5 for elemental metals and in Chapter 10 for alloys and intermetallic compounds. The first such result comes from the use of the pseudoatom formalism to calculate the zero-of-energy constant $V_0' = V_0 - \mu_{\text{xc}}(n_{\text{unif}})$ in an elemental metal at a given atomic volume Ω. As discussed in Chapter 3, this constant is determined directly from the defined pseudoatom PP w_{pa} and the $\mathbf{k} = 0$ plane-wave condition $\langle 0|\,w_{\text{pa}}\,|0\rangle = 0$, which physically places the zero of energy at the bottom of the valence bands in the metal. Combining the definition of w_{pa} given by Eq. (3.34) in terms of the optimum ionic PP component w_{ion}, which is established by Eq. (3.103), together with the definition of the pseudoatom potential v_{pa} in Eq. (4.78), one has quite generally

$$\langle\mathbf{k}|w_{\text{pa}}|\mathbf{k}\rangle = \langle\mathbf{k}|v_{\text{pa}}|\mathbf{k}\rangle - V_0' + \sum_c [\varepsilon_\mathbf{k} - E_c^{\text{vol}}]\,\langle\mathbf{k}|\phi_c\rangle\,\langle\phi_c|\mathbf{k}\rangle, \qquad (A2.55)$$

where the volume component of the core energy E_c^{vol} in the metal can be expressed in terms of V_0' and the corresponding pseudoatom core energy E_c^{pa} as

$$E_c^{\text{vol}} = E_c^{\text{pa}} - \langle\phi_c|\,\delta V_{\text{unif}}\,|\phi_c\rangle - V_0', \qquad (A2.56)$$

with δV_{unif} given by Eq. (4.85). Taking the limit $\mathbf{k} \to 0$ in Eq. (A2.55) and setting the RHS of that equation to zero, one can then solve uniquely for V_0':

$$V_0' = \frac{\langle 0|v_{\text{pa}}|0\rangle - \sum_c (E_c^{\text{pa}} - \langle\phi_c|\delta V_{\text{unif}}|\phi_c\rangle)\,\langle 0|\phi_c\rangle\,\langle\phi_c|0\rangle}{1 - \langle 0|p_c|0\rangle}, \qquad (A2.57)$$

where p_c is the one-ion core projection operator. In practice, the magnitude of V'_0 is typically rather small near equilibrium conditions. For example, Moriarty (1988a) calculated $|V'_0| < 0.2$ Ry for all 20 $3d$ and $4d$ transition metals at normal density.

The above result for V'_0 in an elemental metal can be readily generalized to an AB alloy or intermetallic compound with concentrations c_A and c_B. In this case, we define an average pseudoatom PP w_{pa} at volume Ω with plane-wave matrix elements

$$\langle \mathbf{k} | w_{pa} | \mathbf{k} \rangle = c_A \langle \mathbf{k} | w_{pa}^A | \mathbf{k} \rangle + c_B \langle \mathbf{k} | w_{pa}^B | \mathbf{k} \rangle, \tag{A2.58}$$

where in analogy with Eqs. (A2.55) and (A2.56), one has

$$\langle \mathbf{k} | w_{pa}^\alpha | \mathbf{k} \rangle = \langle \mathbf{k} | v_{pa}^\alpha | \mathbf{k} \rangle - V'_0 + \sum_c [\varepsilon_\mathbf{k} - E_c^{vol,\alpha}] \langle \mathbf{k} | \phi_c \rangle \langle \phi_c | \mathbf{k} \rangle \tag{A2.59}$$

and

$$E_c^{vol,\alpha} = E_c^{pa,\alpha} - \langle \phi_c^\alpha | \delta V_{unif} | \phi_c^\alpha \rangle - V'_0, \tag{A2.60}$$

for $\alpha = $ A or B. As in the elemental-metal case, taking the limit $\mathbf{k} \to 0$ of Eq. (A2.58) and setting the RHS of that equation to zero, one can then solve for V'_0. The result may be put in the useful symmetric form

$$V'_0 = \frac{c_A \langle 0 | w_{pa}^A | 0 \rangle_0 + c_B \langle 0 | w_{pa}^B | 0 \rangle_0}{1 - c_A \langle 0 | p_c^A | 0 \rangle - c_B \langle 0 | p_c^B | 0 \rangle}, \tag{A2.61}$$

where we have defined

$$\langle 0 | w_{pa}^\alpha | 0 \rangle_0 \equiv \langle 0 | w_{pa}^\alpha | 0 \rangle + V'_0 (1 - \langle 0 | p_c^\alpha | 0 \rangle), \tag{A2.62}$$

which removes the dependence on V'_0 from $\langle 0 | w_{pa}^\alpha | 0 \rangle$.

The calculation of V'_0 is always necessary when either component A or B of the AB alloy or intermetallic compound is a transition metal, and in particular for the major case treated in Chapter 10 where A is a simple metal and B is a transition metal. In that case, the simple-metal A pseudoatom is coupled to the transition-metal B pseudoatom only indirectly through its volume Ω_A and does not depend on V'_0. The transition-metal B pseudoatom, on the other hand, is coupled to the A pseudoatom through both its volume Ω_B and its valence Z_B, which depends directly on V'_0. The latter dependence comes through the corresponding d-electron occupation number Z_d^B, which from Eqs. (10.10) and (5.45) is given by

$$Z_d^B = \frac{10}{\pi} \left\{ \delta_2^B (\varepsilon_F) = -\tan^{-1} \left(\frac{-\text{Im}\left[\Gamma_{dd}^{vol,B}(\varepsilon_F)\right]}{\varepsilon_F - E_d^{vol,B} - \text{Re}\left[\Gamma_{dd}^{vol,B}(\varepsilon_F)\right]} \right) \right\}, \tag{A2.63}$$

where

$$E_d^{vol,B} = E_d^{pa,B} - \langle \phi_d^B | \delta V_{unif} | \phi_d^B \rangle - V'_0. \tag{A2.64}$$

Within these constraints, Moriarty and Widom (1997) developed an efficient strategy to calculate the A and B pseudoatoms self-consistently for any desired binary-system with volume Ω and concentration $x = c_B$. This strategy is implemented in three steps:

(i) For the fixed value of the valence Z_A, choose a trial value for the volume Ω_A and calculate the A pseudoatom. This step provides values for $\langle 0|w_{pa}^{\alpha}|0\rangle_0$ and $\langle 0|p_c^{\alpha}|0\rangle$ entering Eq. (A2.61) for V_0'.

(ii) Evaluate the volume Ω_B from Eq. (10.8) and then calculate the B pseudoatom, assuming from Eq. (10.9) a valence electron density $Z/\Omega = Z_B/\Omega_B$. This step provides values for V_0', Z_d^B, Z_B and ε_F.

(iii) Iterate steps (i) and (ii) until the condition $Z_A/\Omega_A = Z_B/\Omega_B$ is satisfied. When this condition is satisfied, the condition assumed in step (ii), as well as Eq. (10.9), are also exactly true.

Glossary of Acronyms and Abbreviations

1D	one-dimensional
2D	two-dimensional
3D	three-dimensional
9R	nine hexagonal-layer close-packed rhombohedral structure of Sm metal
ACE	Atomic Cluster Expansion
ADP	Angular-Dependent Potential
aGPT	adaptive Generalized Pseudopotential Theory
AHS	Austin-Heine-Sham
aka	also known as
APD	Anti-Phase Defect
APS	American Physical Society
APW	Augmented Plane Wave
AS	Adiabatic Switching
ASA	Atomic-Sphere Approximation
BC	Bridgman Cell
bcc	body-centered cubic
bct	body-centered tetragonal
BOP	Bond-Order Potential
BP	Behler-Parrinello
BZ	Brillouin Zone
cp	close-packed
CRSS	Critical Resolved Shear Stress
CSL	Coincident Site Lattice
CW	Cockayne-Widom
DAC	Diamond-Anvil Cell
DD	Dislocation Dynamics
DFT	Density Functional Theory
dhcp	double hexagonal close-packed
DMFT	Dynamical Mean Field Theory
DOS	Density of States
DRC	Dynamic Ramp Compression

DRT	Dagens-Rasolt-Taylor
EAM	Embedded-Atom Method
EDB	Empty d Band
EMT	Effective Medium Theory
EOS	Equation of State
FBCs	Fixed Boundary Conditions
fcc	face-centered cubic
fct	face-centered tetragonal
FDB	Filled d Band
FPLAPW	Full-Potential Linear Augmented Plane Wave
FP-LMTO	Full-Potential–Linear Muffin-Tin Orbital
FS	Finnis-Sinclair
GAP	Gaussian Approximation Potential
GB	Grain Boundary
GF	Green's Function
GFBC	Green's Function Boundary Condition
GGA	Generalized Gradient Approximation
GPT	Generalized Pseudopotential Theory
GS	Ground State
GSFE	Generalized Stacking Fault Energy
GT	Geldart-Taylor
GTI	Group Theoretical Invariants
hcp	hexagonal close-packed
HEA	High-Entropy Alloy
HEMS	Higher-Energy Metastable State
hex	simple hexagonal
hex-ω	hexagonal omega
HF	Hartree-Fock
HL	Hedin-Lundqvist
H-NFE-TB	Hybrid–Nearly-Free-Electron–Tight-Binding
HPC	High Performance Computing
HRIXS	High-Resolution Inelastic X-ray Scattering
HRTEM	High-Resolution Transmission Electron Microscopy
ISLS	Impulsive Stimulated Light Scattering
IU	Ishimaru-Utsumi
KKR	Korringa-Kohn-Rostoker
KMC	Kinetic Monte Carlo

LANL	Los Alamos National Laboratory
lb	lower bound
LCGTO	Linear Combination of Gaussian-Type Orbitals
LDA	Local Density Approximation
LDOS	Local Density of States
LED	Local Electron Density
LEED	Low Energy Electron Diffraction
LHS	Left-Hand Side
LJ	Lennard-Jones
LKKR	Layer-Korringa-Kohn-Rostoker
LLNL	Lawrence Livermore National Laboratory
LMTO	Linear Muffin-Tin Orbital
LMTO-ASA	Linear Muffin-Tin Orbital–Atomic-Sphere Approximation
LRP	Low-Rank Potential
LSD	Local Spin Density
LSGF	Locally Self-consistent Green's Function
M	Million
MBPP	Mixed-Basis Pseudopotential
MC	Monte Carlo
MD	Molecular Dynamics
MEAM	Modified Embedded-Atom Method
MEMT	Modified Effective Medium Theory
MGPT	Model Generalized Pseudopotential Theory
ML	Machine Learning
MLP	Machine-Learning Potential
mono	monoclinic
MP	Model Potential
MS	Molecular Statics
MTP	Moment Tensor Potentials
NFE	Nearly Free Electron
NIM	Nuclear Impedance Match
NNP	Neural-Network Potential
OMP	Optimized Model Potential
OPW	Orthogonalized Plane Wave
ortho	orthorhombic
PAW	Projector Augmented Wave
PBCs	Periodic Boundary Conditions

PBE	Perdew-Burke-Ernzerhof
PES	Potential Energy Surface
PFDB	Partially Filled d Band
PLATO	Package of Linear Atomic-Type Orbitals
PP	Pseudopotential
PT	Pressure-Temperature
PW	Pettifor-Ward
PZ	Perdew-Zunger
QBIP	Quantum-Based Interatomic Potential
QH	Quasiharmonic
QHLD	Quasiharmonic Lattice Dynamics
QMC	Quantum Monte Carlo
QMD	Quantum Molecular Dynamics
RHEA	Refractory High-Entropy Alloy
rhom	rhombohedral
RHS	Right-Hand Side
RMP	Resonant Model Potential
RPA	Random Phase Approximation
RR	Rapid Resolidification
RSA	Random Substitutional Alloy
RSMD	Reversible Scaling Molecular Dynamics
SAX	Stress and Angle-resolved X-ray diffraction
sc	simple cubic
SDFT	Spin-Density Functional Theory
SIA	Self-Interstitial Atom
SK	Slater-Koster
SM	Simple Metal
SNAP	Spectral Neighbor Analysis Potential
SOAP	Smooth Overlap of Atomic Positions
SRO	Short-Range Order
STGB	Symmetric Tilt Grain Boundary
SW	Shock Wave
TB	Tight-Binding
TB-LMTO	Tight-Binding–Linear Muffin-Tin Orbital
tetrag	tetragonal
TF	Thomas-Fermi
thcp	triple hexagonal close-packed

TI	Thermodynamic Integration
TM	Transition Metal
TPT	Thermodynamic Perturbation Theory
ub	upper bound
UEOS	Universal Equation of State
VPT	Variational Perturbation Theory
VWN	Vosko-Wilk-Nusiar
WS	Wigner-Seitz
xc	exchange-correlation
XRD	X-Ray Diffraction

Bibliography

Ackland, G.J. and Thetford, R. An improved N-body semi-empirical model for body-centered cubic transition metals. *Philos. Mag. A* **56**, 15–30 (1987).
Ackland, G.J. and Vitek, V. Many-body potentials and atomic-scale relaxations in noble-metal alloys. *Phys. Rev. B* **41**, 10324–10333 (1990).
Adams, J.J., Agosta, D.S., Leisure, R.G. and Ledbetter, H. Elastic constants of monocrystal iron from 3 to 500 K. *J. Appl. Phys.* **100**, 113530 (2006).
Ahuja, R., Dubrovinsky, L., Dubrovinskaia, N., Guillen, J.M.O., Mattesini, M., Johannsson, B. and Bihan, T.L. Titanium metal at high pressure: synchrotron experiments and *ab initio* calculations. *Phys. Rev. B* **69**, 184102 (2004).
Akahama, Y., Kawaguchi, S., Hirao, N. and Ohishi, Y. High-pressure stability of bcc-vanadium and phase transition to a rhombohedral structure at 200 GPa. *J. Appl. Phys.* **129**, 135902 (2021).
Akahama, Y., Nishimura, M., Kinoshita, K. and Kawamura, H. Evidence of a fcc-hcp transition in aluminum at multimegabar pressure. *Phys. Rev. Lett.* **96**, 045505 (2006).
Albers, R.C., McMahan, A.K. and Müller, J.E. Electronic and X-ray-absorption structure in compressed copper. *Phys. Rev. B* **31**, 3435–3450 (1985).
Alers, G.A. and Neighbours, J.R. The elastic constants of zinc between 4.2 and 670 K. *J. Phys. Chem. Solids* **7**, 58–64 (1958).
Alfè, D. PHON: a program to calculate phonons using the small displacement method. *Comput. Phys. Comm.* **180**, 2622–2633 (2009).
Alfè, D., Gillan, M.J. and Price, G.D. Complementary approaches to the *ab initio* calculation of melting properties. *J. Chem. Phys.* **116**, 6170–6177 (2002).
Al-Lehyani, I., Widom, M., Wang, Y., Moghadam, N., Stocks, G.M. and Moriarty, J.A. Transition-metal interactions in aluminum-rich intermetallics. *Phys. Rev. B* **64**, 075109 (2001).
Allen, M.P. and Tildesley, D.J. *Computer Simulation of Liquids* (Oxford University Press, Oxford, 1987).
Almqvist, L. and Stedman, R. Phonons in zinc at 80 K. *J. Phys. F: Metal Phys.* **1**, 785–790 (1971).
Althoff, J.D., Allen, P.B., Wentzcovitch, R.M. and Moriarty, J.A. Phase diagram and thermodynamic properties of solid magnesium in the quasiharmonic approximation. *Phys. Rev. B* **48**, 13253–13260 (1993).
Andersen, O.K. Simple approach to the band-structure problem. *Solid State Commun.* **13**, 133–136 (1973).
Andersen, O.K. Linear methods in band theory. *Phys. Rev. B* **12**, 3060–3083 (1975).
Andersen, O.K., Arcangeli, A., Tank, R.W., Saha-Dasgusta, T., Krier, G., Jepsen, O. and Dasgusta, I. Third-generation TB-LMTO. In L. Colombo, A. Gonis and P. Turchi (eds.), *Tight-Binding Approach to Computational Materials Science*, MRS Symposium Proceedings, Vol. 491, pp. 3–34 (Materials Research Society, Pittsburgh, 1998).
Andersen, O.K. and Jepsen, O. Advances in the theory of one-electron energy states. *Physica* **91B**, 317–328 (1975).

Andersen, O.K. and Jepsen, O. Explicit, first-principles tight-binding theory. *Phys. Rev. Lett.* **53**, 2571–2574 (1984).

Anderson, P.W. Localized magnetic states in metals. *Phys. Rev.* **124**, 41–53 (1961).

Anderson, P.W. and McMillan, W.L. Multiple-scattering theory and resonances in transition metals. In *Rendiconti della Scuola Internazionale di Fisica "Enrico Fermi," XXXVII Corso*, pp. 50–86 (Academic, New York, 1967).

Anzellini, S., Errandonea, D., MacLeod, S.G., Botella, P., Daisenberger, D., De'Ath, J.M., Gonzalez-Platas, J., Ibáñez, J., McMahon, M.I., Munro, K.A., Popescu, C., Ruiz-Fuertes, J. and Wilson, C.W. Phase diagram of calcium at high pressure and high temperature. *Phys. Rev. Mater.* **2**, 083608 (2018).

Aoki, M. Rapidly convergent bond order expansion for atomistic simulations. *Phys. Rev. Lett.* **71**, 3842–3845 (1993).

Aoki, M., Nguyen-Manh, D., Pettifor, D.G. and Vitek, V. Atom-based bond-order potentials for modeling mechanical properties of metals. *Prog. Mater. Sci.* **52**, 154–195 (2007).

Aono, Y., Kuramoto, E., Kitajima, K. Fundamental plastic behaviors in high-purity bcc metals (Nb, Mo, Fe). In R.C. Gifkins (ed.), *Strength of Metals and Alloys*, Vol. 1, pp. 9–14 (Pergamon Press, Oxford, 1982).

Arsenlis, A., Cai, W. Tang, M., Rhee, M., Oppelstrup, T., Hommes, G., Pierce, T.G. and Bulatov, V.V. Enabling strain hardening simulation with dislocation dynamics. *Modell. Simul. Mater. Sci. Eng.* **15**, 553–595 (2007).

Ashcroft, N.W. Electron-ion pseudopotentials in metals. *Phys. Lett.* **23**, 48–50 (1966).

Ashcroft, N.W. and Mermin, N.D. *Solid State Physics* (Holt, Rinehart and Winston, Philadelphia, 1976).

Ashcroft, N.W. and Stroud, D. Theory of the thermodynamics of simple liquid metals. In H. Ehrenreich, F. Seitz and D. Turnbull (eds.), *Solid State Physics*, Vol. 33, pp. 1–81 (Academic, New York, 1978) and references therein.

Asker, C., Belonoshko, A.B., Mikhaylushkin, A.S. and Abrikosov, I.A. First-principles solution to the problem of Mo lattice stability. *Phys. Rev. B* **77**, 220102 (2008).

Austin, B.J., Heine, V. and Sham, L.J. General theory of pseudopotentials. *Phys. Rev.* **127**, 276–282 (1962).

Axilrod, B.M. and Teller, E. Interaction of the van der Waals type between three atoms. *J. Chem. Phys.* **11**, 299–300 (1943).

Bain, E.C. The nature of martensite. *Trans. AIME* **70**, 25–35 (1924).

Bartók, A., Kondor, R. and Csányi, G. On representing chemical environments. *Phys. Rev. B* **87**, 184115 (2013); Errata *Phys. Rev. B* **87**, 219902 (2013).

Bartók, A., Payne, M.C., Risi, K. and Csányi, G. Gaussian approximation potentials: the accuracy of quantum mechanics, without the electrons. *Phys. Rev. Lett.* **104**, 136403 (2010).

Basinski, Z.S., Duesbery, M.S., Pogany, A.P., Taylor, R. and Varshni, Y.P. An effective ion-ion potential for sodium. *Can. J. Phys.* **48**, 1480–1489 (1970).

Baskes, M.I. Modified embedded atom potentials for cubic materials and impurities. *Phys. Rev. B* **46**, 2727–2742 (1992).

Baskes, M.I. Atomistic model of plutonium. *Phys. Rev. B* **62**, 15532–15537 (2000).

Baskes, M.I. and Johnson, R.A. Modified embedded atom potentials for hcp metals. *Modell. Simul. Mater. Sci. Eng.* **2**, 147–163 (1994).

Beason, M.T., Mandal, A. and Jensen, B.J. Direct observation of the hcp-bcc phase transition and melting along the principal Hugoniot of Mg. *Phys. Rev. B* **101**, 024110 (2020).

Beauchamp, P., Taylor, R. and Vitek, V. Interatomic potentials for body centered cubic lithium-magnesium alloys. *J. Phys. F: Metal Phys.* **5**, 2017–2025 (1975).

Behler, J. and Parrinello, M. Generalized neural-network representation of high-dimensional potential-energy surfaces. *Phys. Rev. Lett.* **98**, 146401 (2007).

Belonoshko, A.B., Burakovsky, L., Chen, S.P., Johansson, B., Mikhaylushkin, A.S., Preston, D.L., Simak, S.I. and Swift, D.C. Molybdenum at high pressure and temperature: melting from another solid phase. *Phys. Rev. Lett.* **100**, 135701 (2008).

Belonoshko, A.B., Skorodumova, N.V., Rosengren, A. and Johansson, B. Melting and critical superheating. *Phys. Rev. B* **73**, 012201 (2006).

Benedict, U., Grosshans, W.A. and Holzapfel, W.B., Systematics of f electron delocalization in lanthanide and actinide elements under pressure. *Physica* **144B**, 14–18 (1986).

Beroni, S., de Gironcoli, S. and DalCorso, A. Phonons and related crystal properties from density-functional perturbation theory. *Rev. Mod. Phys.* **73**, 515–562 (2001).

Berthault, A., Arles, L. and Matricon, J. High-pressure, high-temperature thermophysical measurements on tantalum and tungsten. *Int. J. Thermophys.* **7**, 167–179 (1986).

Bertoni, C.M., Bortolani, V., Calandra, C. and Nizzoli, F. Three-body forces in the lattice dynamics of beryllium. *Phys. Rev. Lett.* **31**, 1466–1469 (1973).

Bertoni, C.M., Bortolani, V., Calandra, C. and Nizzoli, F. Third order perturbation theory and lattice dynamics of simple metals. *J. Phys. F: Metal Phys.* **4**, 19–38 (1974).

Birgeneau, R.J., Cordes, J., Dolling, G. and Woods, A.D.B. Normal modes of vibration in nickel. *Phys. Rev.* **136**, A1359–A1365 (1964).

Boercker, D.B. and Young, D.A. Variational limits on the Helmholtz free energy of simple fluids. *Phys. Rev. A* **40**, 6379–6383 (1989).

Boettger, J.C. and Trickey, S.B. High-precision calculation of crystallographic phase-transition pressures for aluminum. *Phys. Rev. B* **51**, 15623–15625 (1995).

Boettger, J.C. and Trickey, S.B. High-precision calculation of the equation of state and crystallographic phase stability for aluminum. *Phys. Rev. B* **53**, 3007–3012 (1996).

Bolef, D.I. and de Klerk, J. Anomalies in the elastic constants and thermal expansion of chromium single crystals. *Phys. Rev.* **129**, 1063–1067 (1963).

Bollmann, W. *Crystal Defects and Crystalline Interfaces* (Springer-Verlag, Berlin, 1970).

Born, M. and Huang, K. *Dynamical Theory of Crystal Lattices* (Clarendon Press, Oxford, 1954).

Born, M. and Oppenheimer, R. Zur quantentheorie der moleleln. *Ann. Physik* **389**, 457–484 (1927).

Bosak, A., Hoesch, M., Antonangeli, D., Farber, D.L., Fischer, I. and Krisch, M. Lattice dynamics of vanadium: inelastic X-ray scattering measurements. *Phys. Rev. B* **78**, 020301 (R) (2008).

Bosio, L. and Defrain, A. Structure cristalline du gallium β. *Acta Cryst. B* **25**, 995 (1969).

Brandes, E.A. and Brook, G.B., eds. *Smithells Metals Reference Book*, 7th edition, Table 15.3 (Butterworth-Heinemann, Oxford, 1992).

Briggs, R., Coppari, F., Gorman, M.G., Smith, R.F., Tracy, S.J., Coleman, A.L., Fernandez-Pañella, A., Millot, M. Eggert, J.H. and Fratanduono, D.E. Measurement of body-centered cubic gold and melting under shock compression. *Phys. Rev. Lett.* **123**, 045701 (2019).

Brockhouse, B.N., Abou-Helal, H.E. and Hallman, E.D. Lattice vibrations in iron at 296 K. *Solid State Commun.* **5**, 211–214 (1967).

Brovman, E.G. and Kagan, Yu. Long-wave phonons in metals. *Zh. Eksp. Teor. Fiz.* **57**, 1329–1341 (1969) [*Sov. Phys. JETP* **30**, 721–727 (1970)].

Brovman, E.G., Kagan, Yu. and Holas, A. An analysis of the static and dynamic properties of metals and in particular of magnesium (the role of many-ion interaction). *Zh. Eksp. Teor. Fiz.* **61**, 737–752 (1971) [Sov. *Phys. JETP* **34**, 394–402 (1972)].

Brovman, E.G. and Kagan, Yu.M. Phonons in nontransition metals. *Usp. Phys. Nauk.* **112**, 369–426 (1974) *[Sov. Phys. Usp.* **17**, 125–152 (1974).

Brown, J.A. and Pratt, J.N. The thermodynamic properties of solid Al-Mg alloys. *Metall. Trans.* **1**, 2743–2750 (1970).

Brown, J.M. and Shaner, J.W. Rarefaction velocities in shocked tantalum and the high pressure melting point. In R.A. Graham and G.K. Straub (eds.), *Shock Waves in Condensed Matter—1983*, pp. 91–94 (Elsevier, Amsterdam, 1984).

Buchenau, U., Heiroth, M., Schober, H.R., Evers, J. and Oehlinger, G. Lattice dynamics of strontium and barium. *Phys. Rev. B* **30**, 3502–3505 (1984).

Bulatov, V.V. and Cai, W. *Computer Simulations of Dislocations* (Oxford University Press, Oxford, 2006).

Burakovsky, L., Chen, S.P., Preston, D.L., Belonoshko, A.B., Rosengren, A. Mikhaylushkin, A.S., Simak, S.I. and Moriarty, J.A. High-pressure–high-temperature polymorphism in Ta: resolving an ongoing experimental controversy. *Phys. Rev. Lett.* **104**, 255702 (2010).

Burakovsky, L., Chen, S.P., Preston, D.L. and Sheppard, D.G. Z methodology for phase diagram studies: platinum and tantalum as examples. *J. Phys.: Conf. Series* **500**, 162001 (2014).

Burgers, W.G. On the process of transition of the cubic-centered modification into the hexagonal-close-packed modification of zirconium. *Physica* **1**, 561–586 (1934).

Byggmästar, J., Nordlund, K. and Djurabekova, F. Gaussian approximation potentials for body-centered-cubic transition metals. *Phys. Rev. Mater.* **4**, 093802 (2020).

Cák, M., Hammerschmidt, T., Rogal, J., Vitek, V. and Drautz, R. Analytic bond-order potentials for the bcc refractory metals Nb, Ta, Mo and W. *J. Phys.: Condens. Matter* **26**, 195501 (2014).

Campbell, G.H., Belak, J. and Moriarty, J.A. Atomic structure of the $\sum 5$ (310)/[001] symmetric tilt grain boundary in molybdenum. *Acta Mater.* **47**, 3977–3985 (1999).

Campbell, G.H., Belak, J. and Moriarty, J.A. Atomic structure of the $\sum 5$ (310)/[001] symmetric tilt grain boundary in tantalum. *Scripta Mater.* **43**, 659–664 (2000).

Campbell, G.H., Foiles, S.M., Gumbsch, P., Rühle, M. and King, W.E. Atomic structure of the (310) twin in niobium: experimental determination and comparison with theoretical predictions. *Phys. Rev. Lett.* **70**, 449–452 (1993).

Campbell, G.H., Kumar, M., King, W.E., Belak, J., Moriarty, J.A. and Foiles, S.M. The rigid displacement observed at the $\sum = 5$ (310)–[001] symmetric tilt grain boundary in central transition bcc metals. *Philos. Mag. A* **82**, 1573–1594 (2002).

Carlsson, A.E. Beyond pair potentials in elemental transition metals and semiconductors. In H. Ehrenreich, F. Seitz and D. Turnbull (eds.), *Solid State Physics*, Vol. 43, pp. 1–91 (Academic, New York, 1990).

Carlsson, A.E. Angular forces in group-VI transition metals: application to W (100). *Phys. Rev. B* **44**, 6590–6597 (1991).

Carlsson, A.E., Gelatt, Jr., C.D. and Ehrenreich, H. An *ab initio* pair potential applied to metals. *Philos. Mag. A* **41**, 241–250 (1980).

Carlsson, A.E. and Meschter, J.P. Relative stability of $L1_2$, $D0_{22}$ and $D0_{23}$ structures in MAl_3 compounds. *J. Mater. Res.* **4**, 1060–1063 (1989).

Carter, C.B. and Holmes, S.M. The stacking-fault energy of nickel. *Philos. Mag.* **35**, 1161–1172 (1977).

Cawkwell, M.J., Nguyen-Manh, D., Pettifor, D.G. and Vitek, V. Construction, assessment, and application of a bond-order potential for iridium. *Phys. Rev. B* **73**, 064104 (2006).

Ceperley, D.M. and Alder, B.J. Ground state of the electron gas by a stochastic method. *Phys. Rev. Lett.* **45**, 566–569 (1980).

Chang, K.J. and Cohen, M.L. Solid-solid phase transitions and soft phonon modes in highly condensed Si. *Phys. Rev. B* **31**, 7819–7826 (1985).

Che, J.G., Chan, C.T., Jian, W.-E. and Leung, T.C. Surface atomic structures, surface energies, and equilibrium crystal shape of molybdenum. *Phys. Rev. B* **57**, 1875–1880 (1998).

Chen, C., Deng, Z., Tran, R., Tang, H., Chu, I-H. and Ong, S.P. Accurate force field for molybdenum by machine learning large materials data. *Phys. Rev. Mater.* **1**, 043603 (2017).

Chen, H.-P., Kalia, R.K., Kaxiras, E., Lu, G., Nakano, A., Normura, K., van Duin, A.C.T., Vashishta, P. and Yuan, Z. Embrittlement of metal by solute segregation-induced amorphization. *Phys. Rev. Lett.* **104**, 155502 (2010).

Chen, S.H. and Brockhouse, B.N. Lattice vibrations of tungsten. *Solid State Commun.* **2**, 73–77 (1964).

Chen, Y., Ho, K.-M., Harmon, B.N. and Stassis, C. Anomalously low [100] longitudinal phonon branch in Ba: the role of the d hybridization. *Phys. Rev.* **33**, 3684–3687 (1986).

Chernyshov, A.A., Sukhoparov, V.A. and Sadykov, R.A. Effect of pressure on the phase transitions in Li and Na. *JETP Lett.* **37**, 405–409 (1983).

Christensen, N.E., Satpathy, S. and Pawlowska, Z. First-principles theory of tetrahedral bonding and crystal structure of lead. *Phys. Rev. B* **34**, 5977–5980 (1986).

Christiansen, N.E., Ruoff, A.L. and Rodriguez, C.O. Pressure strengthening: a way to multi-megabar static pressures. *Phys. Rev. B* **52**, 9121–9124 (1995).

Clarke, L.J. LEED analysis of the surface structure of Mo(001). *Surf. Sci.* **91**, 131–152 (1980).

Cockayne, E. and Widom, M. Structure and phason energetics of Al-Co decagonal phases. *Philos. Mag. A* **77**, 593–619 (1998).

Cockayne, E. and Widom, M. Ternary model of an Al-Cu-Co decagonal quasicrystal. *Phys. Rev. Lett.* **81**, 598–601 (1998).

Cohen, S.S., Klein, M.L., Duesbery, M.S. and Taylor, R. Correction to a previous pseudopotential calculation of the elastic constants of sodium. *J. Phys. F: Metal Phys.* **6**, L271–273 (1976).

Colella, R. and Batterman, B.W. X-ray determination of phonon dispersion in vanadium. *Phys. Rev. B* **1**, 3913–3921 (1970).

Colella, R. and Merlini, A. A study of the (222) "forbidden" reflection in germanium and silicon. *Phys. Status Solidi* **18**, 157–166 (1966).

Coleridge, P.T. and Templeton, I.M. High precision de Haas-van Alphen measurements in the noble metals. *J. Phys. F: Metal Phys.* **2**, 643–656 (1972).

Copley, J.R.D. and Brockhouse, B.N. Crystal dynamics of rubidium. I. Measurements and harmonic analysis. *Can. J. Phys.* **51**, 657–675 (1973).

Cortella, L., Vinet, B., Desre, P.J., Pasturel, A., Paxton, A.T. and van Schilfgaarde, M. Evidences of transitory metastable phases in refractory metals solidified from highly undercooled liquids in a drop tube. *Phys. Rev. Lett.* **70**, 1469–1472 (1993).

Cousins, C.S.C. The calculation of the elastic shear constants of hexagonal metals using the optimized model potential. *J. Phys. C: Solid State Phys.* **3**, 1677–1692 (1970).

Cowley, R.A., Woods, A.D.B. and Dolling, G. Crystal dynamics of potassium: I. Pseudopotentials analysis of phonon dispersion curves at 9 K. *Phys. Rev.* **150**, 487–494 (1966).

Crampin, S., Hampel, K., Vvedensky, D.D. and MacLaren, J.M. The calculation of stacking fault energies in close-packed metals. *J. Mater. Res.* **5**, 2107–2119 (1990).

Crowhurst, J.C., Abramson, E.H., Slutsky, L.J., Brown, J.M., Zaug, J.M. and Harrell, M.D. Surface acoustic waves in the diamond anvil cell: an application of impulsive stimulated light scattering. *Phys. Rev. B* **64**, 100103 (2001).

Cutler, P.H., Day, R. and King III, W.F. On the role of the orthogonalization hole potential in Harrison's first principles pseudopotential theory. *J. Phys. F: Metal Phys.* **5**, 1801–1816 (1975).

Cynn, H., Evans, W., Yoo, C.S., Ohishi, Y., Sata, N. and Shimormura, O. Behavior of magnesium at high pressures and high temperatures. *Bull. Am. Phys. Soc., APS March Meeting 2004, Sec. W20.009* (2004).

Cynn, H. and Yoo, C.-S. Equation of state of tantalum to 174 GPa. *Phys. Rev. B* **59**, 8526–8529 (1999).

Cynn, H. and Yoo, C.-S. Elasticity of tantalum to 105 GPa using a stress and angle-resolved X-ray diffraction. In M.H. Manghnani, W.J. Nellis and M.F. Nicol (eds.), *Science and Technology of High Pressure*, Vol. 1, Proceedings of AIRAPT-17, pp. 432–435 (Universities Press, Hyderabad, India, 2000).

Cynn, H. and Yoo, C.-S. Elastic constants of tantalum to high pressure using a DAC-SAX technique (unpublished and private communication, 2001).

Dacorogna, M.M. and Cohen, M.L. First-principles study of the structural properties of alkali metals. *Phys. Rev. B* **34**, 4996–5002 (1986).

Dagens, L. The resonant model potential form factor: general theory and application to copper, silver and calcium. *J. Phys. F: Metal Phys.* **6**, 1801–1817 (1976).

Dagens, L. The resonant model potential: II Total energy: theory and application to copper, silver, gold and calcium. *J. Phys. F: Metal Phys.* **7**, 1167–1191 (1977a).

Dagens, L. Calculation of a resonant model potential for copper, silver and gold. *Phys. Status Solidi B* **84**, 311–324 (1977b).

Dagens, L., Rasolt, M. and Taylor, R. Charge densities and interionic potentials in simple metals: nonlinear effects. II. *Phys. Rev. B* **11**, 2726–2734 (1975).

Dai, C., Hu, J. and Tan, H. Hugoniot temperatures and melting of tantalum under shock compression determined by optical pyrometry. *J. Appl. Phys.* **106**, 043519 (2009).

Dai, X., Savrasov, S.Y., Kotliar, G., Migliori, A., Ledbetter, H. and Abrahams, E. Calculated phonon spectra of plutonium at high temperatures. *Science* **300**, 953–955 (2003).

Daw, M.S. and Baskes, M.I. Semiempirical, quantum mechanical calculation of hydrogen embrittlement in metals. *Phys. Rev. Lett.* **50**, 1285–1288 (1983).

Daw, M.S. and Baskes, M.I. Embedded atom method: derivation and application to impurities, surfaces and other defects in metals. *Phys. Rev. B* **29**, 6443–6453 (1984).

Daw, M.S., Foiles, S.M. and Baskes, M.I. The embedded-atom method: a review of theory and applications. *Mat. Sci. Reports* **9**, 251–310 (1993).

de Boer, F.R., Boom, R., Mattens, W.C.M., Miedema, A.R. and Niessen, A.K. *Cohesion in Metals Transition Metal Alloys* (Elsevier, Amsterdam, 1988).

de Koning, M., Antonelli, A. and Yip, S. Optimized free-energy evaluation using a single reversible-scaling simulation. *Phys. Rev. Lett.* **83**, 3973–3977 (1999).

Degtyareva, O. Crystal structure of simple metals at high pressures. *High Pressure Res.* **30**, 343–371 (2010).

Derlet, P.M., Nguyen-Manh, D. and Dudarev, S.L. Multiscale modeling of crowdion and vacancy defects in body-centered-cubic transition metals. *Phys. Rev. B* **76**, 054107 (2007).

Dewaele, A., Loubeyre, P. and Mezouar, M. Equations of state of six metals above 94 GPa. *Phys. Rev. B* **70**, 094112 (2004).

Dewaele, A., Mezouar, M., Guignot, N. and Loubeyre, P. High melting points of tantalum in a laser-heated diamond anvil cell. *Phys. Rev. Lett.* **104**, 255701 (2010).

Dezerald, L., Ventelon, L., Clouet, E., Denoual, C., Rodney, D. and Willaime, F. *Ab initio* modeling of the two-dimensional energy landscape of screw dislocations in bcc transition metals. *Phys. Rev. B* **89**, 024104 (2014).

Dharma-wardana, M.W.C. and Aers, G.C. Determination of the pair potential and the ion-electron pseudopotential for aluminum from experimental structure-factor data for liquid aluminum. *Phys. Rev. B* **28**, 1701–1710 (1983).

Diederich, M.E. and Trivisonno, J. Temperature dependence of the elastic constants of sodium. *J. Phys. Chem. Solids* **27**, 637–642 (1966).

Ding, Y., Ahuja, R., Shu, J., Chow, P., Luo, W. and Mao, H.-K. Structural phase transition of vanadium at 69 GPa. *Phys. Rev. Lett.* **98**, 085502 (2007).

Drain, J.F., Drautz, R. and Pettifor, D.G. Magnetic analytic bond-order potential for modeling the different phases of Mn at zero Kelvin. *Phys. Rev. B* **89**, 134102 (2014).

Drautz, R. Atomic cluster expansion for accurate and transferable interatomic potentials. *Phys. Rev. B* **99**, 014104 (2019).

Drautz, R., Hammerschmidt, T., Cák, M. and Pettifor, D.G. Bond-order potentials: derivation and parameterization for refractory elements. *Modell. Simul. Mater. Sci. Eng.* **23**, 074004 (2015).

Drautz, R. and Pettifor, D.G. Valence-dependent analytic bond-order potential for transition metals. *Phys. Rev. B* **74**, 174117 (2006).

Drautz, R. and Pettifor, D.G. Valence-dependent analytic bond-order potential for magnetic transition metals. *Phys. Rev. B* **84**, 214114 (2011).

Ducastelle, F. *Order and Phase Stability in Alloys* (Elsevier, Amsterdam, 1991).

Duclos, S.J., Vohra, Y.K. and Ruoff, A.L. hcp to fcc transition in silicon at 78 GPa and studies to 100 GPa. *Phys. Rev. Lett.* **58**, 775–777 (1987).

Duclos, S.J., Vohra, Y.K. and Ruoff, A.L. Experimental study of the crystal stability and equation of state of Si to 248 GPa. *Phys. Rev. B* **41**, 12021–12028 (1990).

Duesbery, M.S. On kinked screw dislocations in the b.c.c. lattice–I. The structure and Peierls stress of isolated kinks. *Acta Metall.* **31**, 1747–1758 (1983a).

Duesbery, M.S. On kinked screw dislocations in the b.c.c. lattice–II. Kink energies and double kinks. *Acta Metall.* **31**, 1759–1770 (1983b).

Duesbery, M.S. On non-glide stresses and their influence on the screw dislocation core in body-centred cubic metals. I. The Peierls stress. *Proc. R. Soc. London, Ser. A* **392**, 145–173 (1984a).

Duesbery, M.S. On non-glide stresses and their influence on the screw dislocation core in body-centred cubic metals. II. The core structure. *Proc. R. Soc. London, Ser. A* **392**, 175–197 (1984b).

Duesbery, M.S. Kink-pair formation on {211} planes in the bcc transition-metal lattice. (unpublished and private communication, 1999).

Duesbery, M.S., Jacucci, G. and Taylor, R. The use of long-range metallic pair potentials in computer simulations. *J. Phys. F: Metal Phys.* **9**, 413–424 (1979).

Duesbery, M.S. and Taylor, R. The exact summation of asymptotic ion-pair potentials in simple metallic structures. *J. Phys. F: Metal Phys.* **7**, 47–60 (1977).

Duesbery, M.S. and Taylor, R. Some properties of ion-implanted Li in Be. *J. Phys. F: Metal Phys.* **9**, L19–22 (1979).

Duesbery, M.S. and Vitek, V. Plastic anisotropy in b.c.c. transition metals. *Acta Mater.* **46**, 1481–1492 (1998).

Duesbery, M.S., Vitek, V. and Bowen, D. The effect of shear stress on the screw dislocation core structure in body-centred cubic lattices. *Proc. R. Soc. London, Ser. A* **332**, 85–111 (1973).

Duthie, J.C. and Pettifor, D.G. Correlation between d-band occupancy and crystal structure in the rare earths. *Phys. Rev. Lett.* **38**, 564–567 (1977).

Ecolessi, F. and Adams, J.B. Interatomic potentials from first-principles calculations: the force-matching method. *Europhys. Lett.* **26**, 583–588 (1994).

Ecolessi, F., Parrinello, M. and Tosatti, E. Simulation of gold in the glue model. *Philos. Mag. A* **58**, 213–226 (1988).

Ehrhart, P. The configuration of atomic defects as determined from scattering studies. *J. Nucl. Mater.* **69–70**, 200–214 (1978).

Ehrhart, P. Defect properties of fcc and hcp metals. In H. Ullmaier (ed.), *Atomic Defects in Metals, Landolt-Börnstein-Group III Condensed Matter*, Vol. 25 (Springer, Berlin, 1991).

Errandonea, D. The melting curve of ten metals up to 12 GPa and 1600 K. *J. Appl. Phys.* **108**, 033517 (2010).

Errandonea, D., Boehler, R. and Ross, M. Melting of the alkaline-earth metals to 80 GPa. *Phys. Rev. B* **65**, 012108 (2001b).

Errandonea, D., Burakovsky, L., Preston, D.L., MacLeod, S.G., Santamaría-Perez, D., Chen, S., Cynn, H., Simak, S.I., McMahon, M.I., Proctor, J.E. and Mezouar, M. Experimental and theoretical confirmation of an orthorhombic phase transition in niobium at high pressure and temperature. *Commun. Mater.* **1**, 60 (2020).

Errandonea, D., Meng, Y., Häusermann, D. and Uchida, T. Study of the phase transformations and equation of state of magnesium by synchrotron X-ray diffraction. *J. Phys.: Condens. Matter* **15**, 1277–1289 (2003a).

Errandonea, D., Schwager, B., Ditz, R., Gessmann, C., Boehler, R. and Ross, M. Systematics of transition-metal melting. *Phys. Rev. B* **63**, 132104 (2001a).

Errandonea, D., Somayazulu, M., Häusermann, D. and Mao, H.K. Melting of tantalum at high pressure determined by angle dispersive X-ray diffraction in a double-sided laser-heated diamond anvil cell. *J. Phys.: Condens. Matter* **15**, 7635–7649 (2003b).

Evans, D.J. Computer "experiment" for nonlinear thermodynamics of Couette flow. *J. Chem. Phys.* **78**, 3297–3302 (1983).

Evans, D.J. and Holian, B.L. The Nose-Hoover thermostat. *J. Chem. Phys.* **83**, 4069–4074 (1985).

Farber, D.L., Krisch, M., Antonangeli, D., Beraud, A., Badro, J., Occelli, F. and Orlikowski, D. Lattice dynamics of molybdenum at high pressure. *Phys. Rev. Lett.* **96**, 115502 (2006).

Featherston, F.H. and Neighbours, J.R. Elastic constants of tantalum, tungsten, and molybdenum. *Phys. Rev.* **130**, 1324–1332 (1963).

Ferris, R.W., Shepard, M.L. and Smith, J.F. Elastic constants of thallium single crystals in the temperature range 4.2–400 K. *J. Appl. Phys.* **34**, 768–770 (1963).

Finnis, M.W. *Interatomic Forces in Condensed Matter* (Oxford University Press, Oxford, 2003).

Finnis, M.W., Kear, K.L. and Pettifor, D.G. Interatomic forces and phonon anomalies in bcc $3d$ transition metals. *Phys. Rev. Lett.* **52**, 291–294 (1984).

Finnis, M.W. and Sinclair, J.E. A simple empirical N-body potential for transition metals. *Philos. Mag. A* **50**, 45–55 (1984). Errata *Philos. Mag.* A**53**, 161 (1986).

Finnis, M.W., Walker, A.B. and Gumbsch, P. Representations of the local atomic density. *J. Phys.: Condens. Matter* **10**, 7983–7993 (1998).

Fiquet, G., Narayana, C., Bellin, C., Shukla, A., Estève, I., Ruoff, A.L., Garbarino, G. and Mezouar, M. Structural phase transitions in aluminum above 320 GPa. *Comptes Rendus Geoscience* **351**, 243–252 (2019).

Fisher, E.S. and Renken, C.J. Crystal elastic moduli and the hcp → bcc transformation in Ti, Zr, and Hf. *Phys. Rev.* **135**, A482–A494 (1964).

Foiles, S.M. Interatomic interactions for Mo and W based on the low-order moments of the density of states. *Phys. Rev. B* **48**, 4287–4298 (1993).

Frank, W., Elsässer, C. and Fähnle, M. *Ab initio* force-constant method for phonon dispersion in alkali metals. *Phys. Rev. Lett.* **74**, 1791–1794 (1995).

Frederiksen, S.L. and Jacobsen, K.W. Density functional theory studies of screw dislocation core structures in bcc metals. *Philos. Mag.* **83**, 365–375 (2003).

Frederiksen, S.L., Jacobsen, K.W., Brown, K.S. and Sethna, J.P. Bayesian ensemble approach to error estimation of interatomic potentials. *Phys. Rev. Lett.* **93**,165501 (2004).

Freiburg, C., Grushko, B., Wittenberg, R. and Reichert, W. Once more about monoclinic $Al_{13}Co_4$. *Mater. Sci. Forum* **228–231**, 583–586 (1996).

Frenkel, D. and Smit, B. *Understanding Molecular Simulation: From Algorithms to Applications* (Academic Press, San Diego, 2002).

Friedel, J. Transition metals. Electronic structure of the d-band. Its role in the crystalline and magnetic structures. In J. Ziman (ed.), *The Physics of Metals, Vol 1.—Electrons*, pp. 340–408 (Cambridge University Press, Cambridge, 1969).

Fuchs, K. A quantum mechanical investigation of the cohesive forces of metallic copper. *Proc. Roy. Soc. London* **A151**, 585–602 (1935).

Gal, J. The influence of liquid-liquid phase transition in molten argon on the controversial melting results of Mo, Ta and W. *Physica B* **619**, 413082 (2021).

Garland, C.W. and Silverman, J. Elastic constants of cadmium from 4.2 K to 300 K. *Phys. Rev.* **119**, 1218–1222 (1960). Errata *Phys. Rev.* **127**, 2287 (1962).

Gaspar, R. *Acta Phys. Acad. Sci. Hung.* **3**, 263 (1954).

Gathers, G.R. Correction of specific heat in isobaric expansion data. *Int. J. Thermophys.* **4**, 149–157 (1983). This paper corrects pyrometry calibration errors on the liquid temperatures reported by Shaner et al. (1977).

Gehlen, P.C., Beeler, Jr., J.R. and Jaffee, R.I. (eds.) *Interatomic Potentials and Simulation of Lattice Defects* (Plenum, New York, 1972).

Geldart, D.J.W. and Taylor, R. Wave-number dependence of the static screening function of an interacting electron gas. II. Higher-order exchange and correlation effects. *Can. J. Phys.* **48**, 167–181 (1970).

Gilat, G. and Nicklow, R.M. Normal vibrations in aluminum and derived thermodynamic properties. *Phys. Rev.* **143**, 487–494 (1966).

Girshick, A., Bratkovsky, A.M., Pettifor, D.G. and Vitek, V. Atomistic simulation of titanium I. A bond-order potential. *Philos. Mag. A* **77**, 981–997 (1998).

Glosli, J.N. Fast algorithm for calculating multi-ion MGPT forces in molecular dynamics simulations. (unpublished and private communication, 2001).

Glosli, J.N. and Streitz, F.H. ddcMD: an internal LLNL domain-decomposition molecular dynamics code. (unpublished and private communication, 2002).

Gong, X.G., Chiarotti, G.L., Parrinello, M. and Tosatti, E. α-gallium: a metallic molecular crystal. *Phys. Rev. B* **43**, 14277–14280 (1991).

Goodwin, L., Skinner, A.J. and Pettifor, D.J. Generating transferable tight-binding parameters: application to silicon. *Europhys. Lett.* **9**, 701–706 (1989).

Greeff, C.W. and Moriarty, J.A. *Ab initio* thermoelasticity of magnesium. *Phys. Rev. B* **59**, 3427–3433 (1999).

Grin, J., Burkhardt, U. and Ellner, M. Crystal structure of orthorhombic Co_4Al_{13}. *J. Alloys Compd.* **206**, 243–247 (1994).

Gschneidner, Jr., K.A. Physical properties and interrelationships of metallic and semimetallic elements. In F. Seitz and D. Turnbull (eds.), *Solid State Physics*, Vol. 16, pp. 275–426 (Academic, New York, 1964).

Gutman, E.J. and Trivisonno, J. Temperature dependence of the elastic constants of rubidium. *J. Phys. Chem. Solids* **28**, 805–809 (1967).

Hafner, J. Structural, thermochemical and thermomechanical properties of binary alloys. *J. Phys. F: Metal Phys.* **6**, 1243–1257 (1976).

Hafner, J. Structure, bonding, and stability of topologically close-packed intermetallic compounds. *Phys. Rev. B* **15**, 617–630 (1977).

Hafner, J. Theory of the formation of metallic glasses. *Phys. Rev. B* **21**, 406–426 (1980).

Hafner, J. Structure and vibrational dynamics of the metallic glass $Ca_{70}Mg_{30}$. *Phys. Rev. B* **27**, 678–695 (1983).

Hafner, J. *From Hamiltonians to Phase Diagrams* (Springer-Verlag, Berlin, 1987a).

Hafner, J. Inherent structure theory of local order in liquid and amorphous alloys: I. The nearly-free-electron case. *J. Phys. F: Metal Phys.* **18**, 153–181 (1988).

Hafner, J., Egami, T., Aur, S. and Giessen, B.C. The structure of calcium-aluminium glasses: X-ray diffraction and computer simulation studies. *J. Phys. F: Metal Phys.* **17**, 1807–1815 (1987b).

Hamann, D.R., Schlüter, M. and Chiang, C. Norm-conserving pseudopotentials. *Phys. Rev. Lett.* **43**, 1494–1497 (1979).

Hammerschmidt, T., Madsen, G.K.H., Rogal, J. and Drautz, R. From electrons to materials. *Phys. Status Solidi B* **248**, 2213–2221 (2011).

Hampel, C.A. *Periodic Table of Properties of the Elements* (Materials Research Corporation, 1977).

Harder, J.M. and Bacon, D.J. Point-defect and stacking-fault properties in body-centred-cubic metals with n-body interatomic potentials. *Philos. Mag. A* **54**, 651–661 (1986).

Harrison, W.A. *Pseudopotentials in the Theory of Metals* (Benjamin, New York, 1966).

Harrison, W.A. Transition metal pseudopotentials. *Phys. Rev.* **181**, 1036–1053 (1969).

Harrison, W.A. Multi-ion interactions and structures in simple metals. *Phys. Rev. B* **7**, 2408–2415 (1973).

Harrison, W.A. *Electronic Structure and the Properties of Solids* (Freeman, San Francisco, 1980).

Hasegawa, H., Finnis, M.W. and Pettifor, D.G. Phonon softening in ferromagnetic bcc iron. *J. Phys. F: Metal Phys.* **17**, 2049–2055 (1987).

Hasegawa, M. Third-order perturbation theory and structures of liquid metals: sodium and potassium. *J. Phys. F: Metal Phys.* **6**, 649–626 (1976).

Haskins, J.B. and Moriarty, J.A. Polymorphism and melt in high-pressure tantalum. II. Orthorhombic phases. *Phys. Rev. B* **98**, 144107 (2018).

Haskins, J.B., Moriarty, J.A. and Hood, R.Q. Polymorphism and melt in high-pressure tantalum. *Phys. Rev. B* **86**, 224104 (2012).

Haydock, R., The recursive solution of the Schrödinger equation. In H. Ehrenreich, F. Seitz and D. Turnbull (eds.), *Solid State Physics*, Vol. 35, pp. 215–294 (Academic, New York, 1980).

Haydock, R., Heine, V. and Kelly, M.J. Electronic structure based on the local atomic environment for tight-binding bands. *J. Phys. C: Solid State Phys.* **5**, 2845–2858 (1972).

He, L.X., Li, X.Z., Zhang, Z. and Kuo, K.H. One dimensional quasicrystal in rapidly solidified alloys. *Phys. Rev. Lett.* **61**, 1116–1118 (1988).

Hedin, L. and Lundqvist, B.I. Explicit local exchange-correlation potentials. *J. Phys. C: Solid State Phys.* **4**, 2064–2083 (1971).

Heine, V. s-d interaction in transition metals. *Phys. Rev.* **153**, 673–682 (1967).

Heine, V. Crystal structure of gallium metal. *J. Phys. C (Proc. Phys. Soc.)* **1**, 222–231 (1968).

Heine, V. and Abarenkov, I. A new method for the electronic structure of metals. *Philos. Mag.* **9**, 451–465 (1964).

Heine, V. and Weaire, D. Pseudopotential theory of cohesion and structure. In H. Ehrenreich, F. Seitz and D. Turnbull (eds.), *Solid State Physics*, Vol. 24, pp. 249–463 (Academic, New York, 1970).

Heiroth, M., Buchenau, U., Schober, H.R. and Evers, J. Lattice dynamics of fcc and bcc calcium. *Phys. Rev. B* **34**, 6681–6689 (1986).

Herring, C. A new method for calculating wave functions in crystals. *Phys. Rev.* **57**, 1169–1177 (1940).

Hirth, J.P. Dislocation-pressure interactions. In S. Yip (ed.), *Handbook of Materials Modeling*, Vol. 1, pp. 2879–2882 (Springer, Amsterdam, 2005).

Hirth, J.P. and Lothe, J. *Theory of Dislocations* (Wiley-Interscience, New York, 1982).

Hixson, R.S., Boness, D.A., Shaner, J.W. and Moriarty, J.A. Acoustic velocities and phase transitions in molybdenum under strong shock compression. *Phys. Rev. Lett.* **62**, 637–640 (1989).

Hodges, L., Ehrenreich, H. and Lang, N.D. Interpolation scheme for band structure of noble and transition metals: ferromagnetism and neutron diffraction in Ni. *Phys. Rev.* **152**, 505–526 (1966).

Hohenberg, P. and Kohn, W. Inhomogeneous electron gas. *Phys. Rev.* **136**, B864–B871 (1964).

Hollang, L., Hommel, M. and Seeger, A. The flow stress of ultra-high purity molybdenum single crystals. *Phys. Status Solidi A* **160**, 329–359 (1997).

Holmes, N.C., Moriarty, J.A., Gathers, G.R. and Nellis, W.J. The equation of state of platinum to 660 GPa (6.6 Mbar). *J. Appl. Phys.* **66**, 2962–2967 (1989).

Hood, R.Q. Two-phase melting algorithm developed for QMD application in the Yang (2000) DFT PP code and for QBIP potentials in the LLNL ddcMD code (unpublished and private communication, 2012).

Hood, R.Q., Yang, L.H. and Moriarty, J.A. Quantum molecular dynamics simulations of uranium at high pressure and temperature. *Phys. Rev. B* **78**, 024116 (2008).

Hoover, W.G. Canonical dynamics: equilibrium phase-space distributions. *Phys. Rev. A* **31**, 1695–1697 (1985).

Hoover, W.G. *Computational Statistical Mechanics* (Elsevier, Amsterdam, 1991).

Hoover, W.G., Ladd, A.J.C. and Moran, B. High-strain-rate plastic flow via non-equilibrium molecular dynamics. *Phys. Rev. Lett.* **48**, 1818–1820 (1982).

Horsfield, A.P., Bratkovsky, A.M., Fern, M., Pettifor, D.G. and Aoki, M. Bond-order potentials: theory and implementation. *Phys. Rev. B* **53**, 12694–12712 (1996b).

Horsfield, A.P., Bratkovsky, A.M., Pettifor, D.G. and Aoki, M. Bond-order potential and cluster recursion for the description of chemical bonds: efficient real-space methods for tight-binding molecular dynamics. *Phys. Rev. B* **53**, 1656–1666 (1996a).

Hu, W.-C., Liu, Y., Hu, X.-W., Li, D.-J., Zeng. X.-Q., Yang, X., Xu, Y.-X., Zeng, X., Wang, K.-G. and Huang, B. Predictions of mechanical and thermodynamic properties of $Mg_{17}Al_{12}$ and Mg_2Sn from first-principles calculations. *Philos. Mag.* **95**, 1626–1645 (2015).

Hubbard, J. The approximate calculation of electronic band structure. *Proc. Phys. Soc. (London)* **92**, 921–937 (1967).

Hubbard, J., The approximate calculation of electronic band structures III. *J. Phys. C: Solid State Phys.* **2**, 1222–1229 (1969).

Hubbard, J. and Dalton, N.W. The approximate calculation of electronic band structures II. Application to copper and iron. *J. Phys. C: Solid State Phys.* **1**, 1637–1649 (1968).

Hudd, R.C. and Taylor, W.H. The structure of Co_4Al_{13}. *Acta Crystallogr.* **15**, 441–442 (1962).

Hug, G., Loiseau, A. and Veyssière, P. Weak-beam observation of a dissociation transition in TiAl. *Philos. Mag. A* **57**, 499–523 (1988).

Hultgren, R., Desai, P.D., Hawkins, D.T., Gleiser, M., Kelley, K.K. and Wagman, D.D. *Selected Values of the Thermodynamical Properties of the Elements*, pp. 490–495 (American Society for Metals, Metals Park, Ohio, 1973).

Ignatiev, A., Jona, F., Shih, H.D., Jepsen, D.W. and Marcus, P.M. The structure of the clean Mo[001] surface. *Phys. Rev. B* **11**, 4787–4794 (1975).

Inglesfield, J.E. The structure and phase changes of gallium. *J. Phys. C (Proc. Phys. Soc.)* **1**, 1337–1346 (1968).

Ishimaru, S. and Utsumi, K. Analytic expression for the dielectric screening function of strongly coupled electron liquids at metallic and lower densities. *Phys. Rev. B* **24**, 7385–7388 (1981).

Ismail-Beigi, S. and Arias, T.A. *Ab initio* study of screw dislocations in Mo and Ta: a new picture of plasticity in bcc transition metals. *Phys. Rev. Lett.* **84**, 1499–1502 (2000).

Ito, K. and Vitek, V. Atomistic study of non-Schmid effects in the plastic yielding of bcc metals. *Philos. Mag. A* **81**, 1387–1407 (2001).

Jacobs, R.L. The theory of transition metal band structures. *J. Phys. C: Solid State Phys.* **1**, 492–506 (1968).

Jacobsen, K.W., Nørskov, J.K. and Puska, M.J. Interatomic interactions in the effective medium theory. *Phys. Rev. B* **35**, 7423–7442 (1987).

Jacucci, G. and Taylor, R. The calculation of vacancy formation energies in the alkali metals Li, Na and K. *J. Phys. F: Metal Phys.* **9**, 1489–1501 (1979).

Jacucci, G., Taylor, R., Tenenbaum, A. and van Doan, N. The use of a pair potential for the study of defects and disorder in aluminium. *J. Phys. F: Metal Phys.* **11**, 793–804 (1981).

Johnson, R.A. Interstitials and vacancies in α iron. *Phys. Rev.* **134**, A1329–A1336 (1964).

Johnson, R.A. Empirical potentials and their use in the calculation of energies of point defects in metals. *J. Phys. F: Metal Phys.* **3**, 295–321 (1973).

Jona, F. and Marcus, P.M. Lattice parameters of aluminium in the Mbar range by first-principles. *J. Phys.: Condens. Matter* **18**, 10881–10888 (2006).

Jones, H.D. Theory of the thermodynamic properties of liquid metals. *Phys. Rev. A* **8**, 3215–3226 (1973).

Jones, R.O. and Gunnarsson, O. The density functional formalism, its applications and prospects. *Rev. Mod. Phys.* **61**, 689–746 (1989).

Kamitakahara, W.A. and Brockhouse, B.N. Crystal dynamics of silver. *Phys. Lett.* **29A**, 639–640 (1969).

Kamm, G.N. and Alers, G.A. Low-temperature elastic moduli of aluminum. *J. Appl. Phys.* **35**, 327–330 (1964).

Kanamori, J., Terakura, K. and Yamada, K. The approximate expression of Green's function for the calculation of electronic structure in metals and alloys. *Progr. Theoret. Phys. (Kyoto)* **41**, 1426–1437 (1969).

Kanamori, J., Terakura, K. and Yamada, K. Resonance orbital, off-resonance orbital and pseudo-Greenian. *Progr. Theoret. Phys. (Kyoto) Suppl.* **46**, 221–243 (1970).

Karandikar, A. and Boehler, R. Flash melting of tantalum in a diamond cell to 85 GPa. *Phys. Rev. B* **93**, 054107 (2016).

Katahara, K.W., Manghnani, W.H. and Fisher, E.S. Pressure derivatives of the elastic moduli of bcc Ti-V-Cr, Nb-Mo and Ta-W alloys. *J. Phys. F: Metal Phys.* **9**, 773–790 (1979).

Katzarov, I.H., Cawkwell, M.J., Paxton, A.T. and Finnis, M.W. Atomistic study of ordinary <110]/2 screw dislocations in single-phase and lamellar γ-TiAl. *Philos. Mag.* **87**, 1795–1809 (2007).

Katzarov, I.H. and Paxton, A.T. Atomistic studies of <110] screw dislocation core structures and glide in γ-TiAl. *Philos. Mag.* **89**, 1731–1750 (2009).

King III, W.F. and Cutler, P.H. Lattice dynamics of magnesium from a first-principles nonlocal pseudopotential approach. *Phys. Rev. B* **3**, 2485–2496 (1971).

Kittel, C. *Quantum Theory of Solids* (Wiley, New York, 1963).

Kittel, C. *Introduction to Solid State Physics*, 5th edition (Wiley, New York, 1976).

Kohn, W. Image of the Fermi surface in the vibration spectrum of a metal. *Phys. Rev. Lett.* **2**, 393–394 (1959).

Kohn, W. and Rostoker, N. Solution of the Schrödinger equation in periodic lattices with an application to metallic lithium. *Phys. Rev.* **94**, 1111–1120 (1954).

Kohn, W. and Sham, L.J. Self-consistent equations including exchange and correlation effects. *Phys. Rev.* **140**, A1133–A1138 (1965).

Koizumi, H., Kirchner, H.O.K. and Suzuki, T. Nucleation of trapezoidal kink pairs on a Peierls potential. *Philos. Mag. A* **69**, 805–820 (1994).

Kollarits, F.J. and Trivisonno, J. Single-crystal elastic constants of cesium. *J. Phys. Chem. Solids* **29**, 2133–2139 (1968).

Korhonen, T., Puska, M.J. and Nieminen, R.M. Vacancy-formation energies for fcc and bcc transition metals. *Phys. Rev. B* **51**, 9526–9532 (1995).

Korringa, J. On the calculation of the energy of a Bloch wave in a metal. *Physica* **13**, 392–400 (1947).

Korzhavyi, P.A., Abrikosov, I.A., Johansson, B., Ruban, A.V. and Skriver, H.L. First-principles calculations of the vacancy formation energy in transition and noble metals. *Phys. Rev. B* **59**, 11693–11703 (1999).

Kostiuchenko, T., Körmann, F., Neugebauer, J. and Shapeev, A. Impact on phase transitions in a high-entropy alloy studied by machine-learning potentials. *npj Comput. Mater.* **5**, 55 (2019).

Kraus, R.G., Coppari, F., Fratanduono, D.E., Smith, R.F., Lazicki, A., Wehrenberg, C. Eggert, J.H., Rygg, J.R. and Collins, C.W. Melting of tantalum at multimegabar pressures on the nanosecond timescale. *Phys. Rev. Lett.* **126**, 255701 (2021).

Kresse, G., Furthmüller, J. and Hafner, J. *Ab initio* force constant approach to phonon dispersion relations of diamond and graphite. *Europhys. Lett.* **32**, 729–734 (1995).

Ladd, A.J.C. and Woodcock, L.V. Interfacial and co-existence properties of the Lennard-Jones system at the triple point. *Mol. Phys.* **36**, 611–619 (1978).

Lam, N.Q., Dagens, L. and Doan, N.V. Calculations of the properties of self-interstitials and vacancies in the face-centred cubic metals Cu, Ag and Au. *J. Phys. F: Metal Phys.* **13**, 2503–2516 (1983).

Lam, N.Q., Doan, N.V. and Adda, Y. Molecular dynamics study of interstitial-solute interactions in irradiated alloys: I. Configuration, binding and induced migration of mixed dumbbells in Al-Zn alloys. *J. Phys. F: Metal Phys.* **10**, 2359–2373 (1980).

Lam, N.Q., van Doan, N., Dagens, L. and Adda, Y. Molecular dynamics study of interstitial-solute interactions in irradiated alloys: II. Configurations and binding energies of interstitial-solute complexes in Al-Be, Al-Ca, Al-K, Al-Li and Al-Mg alloys. *J. Phys. F: Metal Phys.* **11**, 2231–2245 (1981).

Lam, P.K. and Cohen, M.L. *Ab initio* calculation of the static structural properties of Al. *Phys. Rev. B* **24**, 4224–4229 (1981).

Lam, P.K. and Cohen, M.L. Calculation of high-pressure phases of Al. *Phys. Rev. B* **27**, 5986–5991 (1983).

Landa, A., Klepeis, J., Söderlind, P., Naumov, I. Velikikhatnyi, O., Vitos, L. and Ruban, A. *Ab initio* calculations of elastic constants of the V–Nb system at high pressures. *J. Phys. Chem. Solids* **67**, 2056–2064 (2006a).

Landa, A., Klepeis, J., Söderlind, P., Naumov, I. Velikokhatnyi, O., Vitos, L. and Ruban, A. Fermi surface nesting and pre-martensitic softening in V and Nb at high pressures. *J. Phys.: Condens. Matter* **18**, 5079–5085 (2006b).

Landau, L.D. and Lifshitz, E.M. *Quantum Mechanics* (Pergamon, New York, 1965).

Langreth, D.C. and Mehl, M.J. Beyond the local-density approximation in calculations of ground-state electronic properties. *Phys. Rev. B* **28**, 1809–1834 (1983); Errata *Phys. Rev. B* **29**, 2310 (1984).

Langreth, D.C. and Perdew, J.P. Theory of nonuniform electronic systems. I. Analysis of the gradient approximation and a generalization that works. *Phys. Rev. B* **21**, 5469–5493 (1980).

Larose, A. and Brockhouse, B.N. Lattice vibrations in tungsten at 22° C studied by neutron scattering. *Can. J. Phys.* **54**, 1819–1823 (1976).

Ledbetter, H.M. Elastic properties of zinc: a compilation and review. *J. Phys. Chem. Ref. Data* **6**, 1181–1203 (1977).

Ledbetter, H.M. and Moment, R.L. Elastic properties of face-centered-cubic plutonium. *Acta Metall.* **24**, 891–899 (1976).

Lee, B., Rudd, R.E., Klepeis, J.E., Söderlind, P. and Landa, A. Theoretical confirmation of a high-pressure rhombohedral phase in vanadium metal. *Phys. Rev. B* **75**, 180101 (2007).

Lee, B-J. and Baskes, M.I. Second nearest-neighbor modified embedded-atom-method potential. *Phys. Rev. B* **62**, 8564–8567 (2000).

Lee, J.K. (ed.) *Interatomic Potentials and Crystal Defects* (Metallurgical Society of AIME, New York, 1981).

Lee, J.-Y., Punkkinen, M.P.J., Schönecker, S., Nabi, Z., Kádas, K., Zólyomi, V., Koo, Y.M., Hu, Q.-M., Ahuja, R., Johansson, B., Kóllar, J., Vitos, L. and Kwon, S.K. The surface energy and stress of metals. *Surf. Sci.* **674**, 51–68 (2018).

Lennard-Jones, J.E. On the determination of molecular fields. II. From the equation of state of a gas. *Proc. R. Soc. London, Ser. A* **106**, 463–477 (1924).

Li, P., Gao, G., Wang, Y. and Ma, Y. Crystal structures and exotic behavior of magnesium under pressure. *J. Phys. Chem. C* **114**, 21745–21749 (2010).

Li, X.-G., Chen, C., Zheng, H., Zuo, Y. and Ong, S.P. Complex strengthening mechanism in the NbMoTaW multi-principal element alloy. *npj Comput. Mater.* **6**, 70 (2020).

Li, X.-G., Hu, C., Chen, C., Deng, Z. Luo, J. and Ong, S.P. Quantum-accurate spectral neighbor analysis potential models for Ni-Mo binary alloys and fcc metals. *Phys. Rev. B* **98**, 094104 (2018).

Liberman, D.A. Self-consistent field model for condensed matter. *Phys. Rev. B* **20**, 4981–4989 (1979).

Lin, Y.-S., Mrovec, M. and Vitek, V. A new method for development of bond-order potentials for transition bcc metals. *Modell. Simul. Mater. Sci. Eng.* **22**, 034002 (2014).

Lin, Y.-S., Mrovec, M. and Vitek, V. Bond-order potential for magnetic body-centered-cubic iron and its transferability. *Phys. Rev. B* **93**, 214107 (2016).

Liu, G., Nguyen-Manh, D., Liu, B.-G. and Pettifor, D.G., Magnetic properties of point defects in iron with the tight-binding-bond Stoner model. *Phys. Rev. B* **71**, 174115 (2005).

Liu, Q., Fan, C. and Zhang, R. First-principles study of high-pressure structural phase transitions of magnesium. *J. Appl. Phys.* **105**, 123505 (2009).

Liu, X.-Y., Adams, J.B., Ercolessi, F. and Moriarty, J.A. EAM potential for magnesium from quantum mechanical forces. *Modell. Simul. Mater. Sci. Eng.* **4**, 293–303 (1996).

Liu, Z.-L., Cai, L.-C., Zhang, X.-L. and Xi, F. Predicted alternate structure for tantalum metal under high pressure and high temperature. *J. Appl. Phys.* **114**, 073520 (2013).

Lloyd, P. and Sholl, A. A structural expansion of the cohesive energy of simple metals in an effective Hamiltonian approximation. *J. Phys. C: Solid State Phys.* **1**, 1620–1632 (1968).

Löwdin, P.-O. Quantum theory of cohesive properties of solids. *Advanc. Phys.* **5**, 1–171 (1956).

Lynn, J.W., Smith, H.G. and Nicklow, R.M. Lattice dynamics of gold. *Phys. Rev. B* **8**, 3493–3499 (1973).

Lysogorskiy, Y., van der Oord, C., Bochkarev, A., Menon, S., Rinaldi, M., Hammerschmidt, T., Mrovec, M., Thompson, A., Csányi, G., Ortner, C. and Frautz, R. Performant implementation of the atomic cluster expansion (PACE) and application to copper and silicon. *npj Comput. Mater.* **7**, 97 (2021).

Ma, P.-W. and Dudarev, S.L. Effect of stress on vacancy formation and migration in body-centered cubic metals. *Phys. Rev. Mater.* **3**, 063601 (2019a).

Ma, P.-W. and Dudarev, S.L. Symmetry-broken self-interstitial defects in chromium, molybdenum, and tungsten. *Phys. Rev. Mater.* **4**, 043606 (2019b).

MacDonald, A.H., Daams, J.M., Vosko, S.H. and Koelling, D.D. Non-muffin-tin and relativistic interaction effects on the electronic structure of noble metals. *Phys. Rev. B* **25**, 713–725 (1982); Errata *Phys. Rev. B* **26**, 3473 (1982).

MacFarlane, R.E., Rayne, J.A. and Jones, C.K. Anomalous temperature dependence of shear modulus C_{44} for platinum. *Phys. Lett.* **18**, 91–92 (1965).

MacFarlane, R.E., Rayne, J.A. and Jones, C.K. Temperature dependence of elastic moduli of iridium. *Phys. Lett.* **20**, 234–235 (1966).

Mackintosh, A.R. and Andersen, O.K. The electronic structure of transition metals. In M. Springford (ed.), *Electrons at the Fermi Surface*, pp. 149–224 (Cambridge University, Cambridge, 1980).

Manninen, M., Jena, P., Nieminen, R.M. and Lee, J.K. *Ab initio* calculation of interatomic potentials and electronic properties of a simple metal—Al. *Phys. Rev. B* **24**, 7057–7070 (1981).

Marquardt, W.R. and Trivisonno, J. Low temperature elastic constants of potassium. *J. Phys. Chem. Solids* **26**, 273–278 (1965).

Martin, D.L. The specific heat of sodium from 20 to 300 K: the martensitic transformation. *Proc. R. Soc. London, Ser. A* **254**, 433–443 (1960).

Martin, R.M. *Electronic Structure Basic Theory and Practical Methods* (Cambridge University Press, Cambridge, 2004).

McMahan, A.K. Pressure-induced changes in the electronic structure of solids. *Physica* **139 & 140B**, 31–41 (1986).

McMahan, A.K. Two-center s–f Slater-Koster integrals. *Phys. Rev. B* **58**, 4293–4303 (1998).

McMahan, A.K. and Moriarty, J.A. Structural phase stability in third-period simple metals. *Phys. Rev. B* **27**, 3235–3251 (1983).

McSkimin, H.J. Measurement of the elastic constants of single crystal cobalt. *J. Appl. Phys.* **26**, 406–409 (1955).

Mehta, S., Price, G.D. and Alfè, D. *Ab initio* thermodynamics and phase diagram of solid magnesium: a comparison of the LDA and GGA. *J. Chem. Phys.* **125**, 194507 (2006).

Mermin, N.D. Thermal properties of the inhomogeneous electron gas. *Phys. Rev.* **137**, A1441–A1443 (1965).

Methfessel, M., van Schilfgaarde, M. and Casali, R.A. A full-potential LMTO method based on smooth Hankel functions. In H. Dreysse (ed.), *Electronic Structure and Physical Properties of Solids: The Uses of the LMTO Method*, pp. 114–147 (Springer, Heidelberg, 2000).

Mihalkovic, M., Al-Lehyani, I., Cockayne, E., Henley, C.L., Moghadam, N., Moriarty, J.A., Wang, Y. and Widom, M. Total-energy prediction of a quasicrystal structure. *Phys. Rev. B* **65** 104205 (2002).

Mihalkovic, M. and Widom, M. First-principles calculations of cohesive energies in the Al-Co binary alloy system. *Phys. Rev. B* **75**, 014207 (2007).

Miller, R.A. and Schuele, D.E. The pressure derivatives of the elastic constants of lead. *J. Phys. Chem. Solids* **30**, 589–600 (1969).

Mills, K.C. and Su, Y.C. Review of surface tension data for metallic elements and alloys part 1—pure metals. *Int. Mater. Rev.* **51**, 329–351 (2006).

Minkiewicz, V.J., Shirane, G. and Nathans, R. Phonon dispersion relation for iron. *Phys. Rev.* **162**, 528–531 (1967).

Mishin, Y. and Lozovoli, A.Y. Angular-dependent interatomic potential for tantalum. *Acta Mater.* **54**, 5013–5026 (2006).

Mishin, Y., Mehl, M.J. and Papconstantopoulos, D.A. Phase stability in the Fe-Ni system: investigation by first-principles calculations and atomistic simulations. *Acta Mater.* **53**, 4029–4041 (2005).

Mitchell, A.C. and Nellis, W.J. Shock compression of aluminum, copper, and tantalum. *J. Appl. Phys.* **52**, 3363–3374 (1981).

Mitchell, A.C., Nellis, W.J., Moriarty, J.A., Heinle, R.A., Holmes, N.C., Tipton, R.E. and Repp, G.W. Equation of state of Al, Cu, Mo, and Pb at shock pressures up to 2.4 TPa (24 Mbar) *J. Appl. Phys.* **69**, 2981–2986 (1991).

Mizuki, J., Chen, Y., Ho, K.-M. and Stassis, C. Phonon dispersion curves of bcc Ba. *Phys. Rev. B* **32**, 666–670 (1985).

Morales de la Garza, L. and Clarke, L.J. The surface structure of Mo(110) determined by LEED. *J. Phys. C: Solid State Phys.* **14**, 5391–5401 (1981).

Moriarty, J.A., Pseudopotential form factors for copper, silver and gold. *Phys. Rev. B* **1**, 1363–1370 (1970).

Moriarty, J.A. Pseudo Green's functions and the pseudopotential theory of d-band metals. *Phys. Rev. B* **5**, 2066–2080 (1972a).

Moriarty, J.A. Total energy of copper, silver and gold. *Phys. Rev. B* **6**, 1239–1252 (1972b).

Moriarty, J.A. Localized d states for pseudopotential calculations: application to the alkaline-earth metals. *Phys. Rev. B* **6**, 4445–4458 (1972c).

Moriarty, J.A. Hybridization and the fcc-bcc phase transitions in calcium and strontium. *Phys. Rev. B* **8**, 1338–1345 (1973).

Moriarty, J.A. Zero-order pseudoatoms and the generalized pseudopotential theory. *Phys. Rev. B* **10**, 3075–3091 (1974).

Moriarty, J.A. Equivalence of resonance and tight binding descriptions of the d band in transition metals. *J. Phys. F: Metal Phys.* **5**, 873–882 (1975).

Moriarty, J.A. Density-functional formulation of the generalized pseudopotential theory. *Phys. Rev. B* **16**, 2537–2554 (1977).

Moriarty, J.A. Simplified local-density theory of the cohesive energy of metals. *Phys. Rev. B* **19**, 609–619 (1979).

Moriarty, J.A. Density-functional formulation of the generalized pseudopotential theory. II. *Phys. Rev. B* **26**, 1754–1780 (1982).

Moriarty, J.A. First-principles phonon spectra in Ca and Sr. *Phys. Rev. B* **28**, 4818–4821 (1983).

Moriarty, J.A., First-principles interatomic potentials in transition metals. *Phys. Rev. Lett.* **55**, 1502–1505 (1985).

Moriarty, J.A. High-pressure ion-thermal properties of metals from *ab initio* interatomic potentials. In Y.M. Gupta (ed.), *Shock Waves in Condensed Matter*, Proceedings of the 1985 APS Topical Conference, pp. 101–106 (Plenum, New York, 1986a).

Moriarty, J.A. First-principles phonon spectrum in bcc Ba: three-ion forces and transition-metal behavior. *Phys. Rev. B* **34**, 6738–6745 (1986b).

Moriarty, J.A. Density-functional formulation of the generalized pseudopotential theory. III. Transition-metal interatomic potentials. *Phys. Rev. B* **38**, 3199–3231 (1988a).

Moriarty, J.A. High-pressure structural phase stability in Hg to 1 TPa (10 Mbar). *Phys. Lett. A* **131**, 41–46 (1988b).

Moriarty, J.A. Analytic representation of multi-ion interatomic potentials in transition metals. *Phys. Rev. B* **42**, 1609–1628 (1990a).

Moriarty, J.A. First-principles interatomic potentials in transition metals: multi-ion interactions and their analytic representation. In R.M. Nieminen, M.J. Puska and M. Manninen (eds.) *Many-Atom Interactions in Solids*, Springer Proceedings in Physics **48**, pp. 158–167 (Springer-Verlag, Berlin, 1990b).

Moriarty, J.A. Ultrahigh-pressure structural phase transitions in Cr, Mo, and W. *Phys. Rev. B* **45**, 2004–2014 (1992).

Moriarty, J.A. Angular forces and melting in bcc transition metals: a case study of molybdenum. *Phys. Rev. B* **49**, 12431–12445 (1994).

Moriarty, J.A. First-principles equations of state for Al, Cu, Mo and Pb to ultrahigh pressures. *High Pressure Res.* **13**, 343–365 (1995).

Moriarty, J.A. Performance MGPT-MD and GPT-MD simulations on liquid metals (unpublished, 2016).

Moriarty, J.A. First-principles GPT interatomic potentials for Ca_xMg_{1-x} (unpublished, 2017).

Moriarty, J.A. A canonical f-band treatment of the high-temperature bcc phase of uranium (unpublished, 2021).

Moriarty, J.A. and Althoff, J.D. First-principles temperature-pressure phase diagram of magnesium. *Phys. Rev. B* **51**, 5609–5616 (1995).

Moriarty, J.A., Belak, J.F., Rudd, R.E., Söderlind, P., Streitz, F.H. and Yang, L.H. Quantum-based atomistic simulation of materials properties in transition metals. *J. Phys.: Condens. Matter.* **14**, 2825–2858 (2002a).

Moriarty, J.A., Benedict, L.X., Glosli, J.N., Hood, R.Q., Orlikowski, D.A., Patel, M.V., Söderlind, P., Streitz, F.H., Tang, M. and Yang, L.H. Robust quantum-based interatomic potentials for multiscale modeling in transition metals. *J. Mater. Res.* **21**, 563–573 (2006).

Moriarty, J.A., Glosli, J.N., Hood, R.Q., Klepeis, J.E., Orlikowski, D.A., Söderlind, P. and Yang, L.H. Quantum-based atomistic simulations of metals at extreme conditions. In *TMS 2008 Annual Meeting Supplemental Proceedings Volume I: Materials Processing and Properties*, pp. 313–319 (TMS, Warrendale, 2008).

Moriarty, J.A. and Haskins, J.B. Efficient wide-range calculation of free energies in solids and liquids using reversible-scaling molecular dynamics. *Phys. Rev. B* **90**, 054113 (2014).

Moriarty, J.A., Hood, R.Q. and Yang, L.H. Quantum-mechanical interatomic potentials with electron temperature for strong-coupling transition metals. *Phys. Rev. Lett.* **108**, 036401 (2012).

Moriarty, J.A. and McMahan, A.K. High-pressure structural phase transitions in Na, Mg, and Al. *Phys. Rev. Lett.* **48**, 809–812 (1982).

Moriarty, J.A. and Phillips, R. First-principles interatomic potentials for transition-metal surfaces. *Phys. Rev. Lett.* **66**, 3036–3039 (1991).

Moriarty, J.A., Vitek, V., Bulatov, V.V. and Yip, S. Atomistic simulations of dislocations and defects. *J. Computer-Aided Mater. Design* **9**, 99–132 (2002b).

Moriarty, J.A. and Widom, M. First-principles interatomic potentials for transition-metal aluminides: theory and trends across the $3d$ series. *Phys. Rev. B* **56**, 7905–7917 (1997).

Moriarty, J.A., Xu, W., Söderlind, P., Belak, J., Yang, L.H. and Zhu, J. Atomistic simulations for multiscale modeling in bcc metals. *J. Eng. Mater. Tech.* **121**, 120–125 (1999).

Moriarty, J.A., Young, D.A. and Ross, M. Theoretical study of the aluminum melting curve to very high pressure. *Phys. Rev. B* **30**, 578–588 (1984).

Morris, J.R., Wang, K.M., Ho, K.M. and Chan, C.T. Melting line of aluminum from simulations of coexisting phases. *Phys. Rev. B* **49**, 3109–3115 (1994).

Morris, Jr., J.W., Krenn, C.R., Roundy, D. and Cohen, M.L. Elastic stability and the limit of strength. In P.E. Turchi and A. Gonis (eds.), *Phase Transformations and Evolution in Materials*, p. 187 (TMS, Warrendale, PA, 2000).

Morse, P.M. Diatomic molecules according to the wave mechanics. II. Vibrational levels. *Phys. Rev.* **34**, 57–64 (1929).

Moruzzi, V.L., Janak, J.F. and Williams, A.R. *Calculated Electronic Properties of Metals* (Pergamon, New York, 1978).

Mrovec, M., Gröger, R., Bailey, A.G. Nguyen-Manh, D., Elsässer, C. and Vitek, V. Bond-order potential for simulations of extended defects in tungsten. *Phys. Rev. B* **75**, 104119 (2007).

Mrovec, M., Nguyen-Manh, D., Elsässer, C. and Gumbsch, P. Magnetic bond-order potential for iron. *Phys. Rev. Lett.* **106**, 246402 (2011).

Mrovec, M., Nguyen-Manh, D., Pettifor, D.G. and Vitek, V. Bond-order potential for molybdenum: application to dislocation behavior. *Phys. Rev. B* **69**, 094115 (2004).

Mrovec, M., Vitek, V., Nguyen-Manh, D., Pettifor, D.G., Wang, L.G. and Sob, M. Bond-order potentials for molybdenum and niobium: an assessment of their quality. In V. Bulatov, T. Diaz de la Rubia, R. Phillips, T. Kaxiras and N. Ghoniem (eds.), *Multiscale Modeling of Materials, Materials Research Society Symposium Proceedings*, Vol. 538, pp. 529–534 (Materials Research Society, Pittsburgh, 2000).

Mueller, F.M. Combined interpolation scheme for transition and noble metals. *Phys. Rev.* **153**, 659–669 (1967).

Nash, H.C. and Smith, C.S. Single-crystal elastic constants of lithium. *J. Phys. Chem. Solids* **9**, 113–118 (1959).

Neaton, J.B. and Ashcroft, N.W. Pairing in dense lithium. *Nature* **400**, 141–144 (1999).

Neaton, J.B. and Ashcroft, N.W. On the constitution of sodium at higher densities. *Phys. Rev. Lett.* **86**, 2830–2833 (2001).

Neighbours, J.R. and Alers, G.A. Elastic constants of silver and gold. *Phys. Rev.* **111**, 707–712 (1958).

Nellis, W.J., Moriarty, J.A., Mitchell, A.C., Ross, M., Dandrea, R.G., Ashcroft, N.W., Holmes, N.C. and Gathers, G.R. Metals physics at ultrahigh pressure: aluminum, copper, and lead as prototypes. *Phys. Rev. Lett.* **60**, 1414–1417 (1988).

Nguyen, J.H., Akin, M.C., Chau, R., Fratanduono, D.E., Ambrose, W.P., Fat'yanov, O.V., Asimow, P.D. and Holmes, N.C. Molybdenum sound velocity and shear modulus under shock compression. *Phys. Rev. B* **89**, 174109 (2014).

Nguyen-Manh, D., Horsfield, A.P. and Dudarev, S.L. Self-interstitial atom defects in bcc transition metals: group-specific trends. *Phys. Rev. B* **73**, 020101 (2006).

Nguyen-Manh, D., Pettifor, D.G. and Vitek, V. Analytic environment-dependent tight-binding bond integrals: application to $MoSi_2$. *Phys. Rev. Lett.* **85**, 4136–4139 (2000).

Nicklow, R.M., Gilat, G., Smith, H.G., Raubenheimer, L.J. and Wilkinson, M.K. Phonon frequencies in copper at 49 and 298 K. *Phys. Rev.* **164**, 922–928 (1967).

Nosé, S. A molecular dynamics method for simulations in the canonical ensemble. *Mol. Phys.* **52**, 255–268 (1984).

Nücker, N. and Buchenau, U. Phonons in cesium. *Phys. Rev. B* **31**, 5479–5482 (1985).

Ochs, T., Beck, O., Elsässer, C. and Meyer, B. Symmetrical tilt grain boundaries in bcc transition metals: an *ab-initio* local-density-functional study. *Philos. Mag. A* **80**, 351–372 (2000a).

Ochs, T., Elsässer, C., Mrovec, M., Vitek, V., Belak, J. and Moriarty, J.A. Symmetrical tilt grain boundaries in bcc transition metals: comparison of semiempirical with *ab-initio* total-energy calculations. *Philos. Mag. A* **80**, 2405–2423 (2000b).

Ögüt, S. and Rabe, K.M. *Ab initio* pseudopotential calculations for aluminum-rich cobalt compounds. *Phys. Rev. B* **50**, 2075–2084 (1994).

Oh, D.J. and Johnson, R.A. Simple embedded atom method model for fcc and hcp metals. *J. Mater. Res.* **3**, 471–478 (1988).

Olijnyk, H. and Holzapfel, W.B. Phase transitions in alkaline earth metals under pressure. *Phys. Lett. A* **100**, 191–194 (1984).

Olijnyk, H. and Holzapfel, W.B. High-pressure structural phase transition in Mg. *Phys. Rev. B* **31**, 4682–4683 (1985).

Oppelstrup, T. Fast algorithm for implementing multi-ion MGPT potentials in molecular dynamics simulations (unpublished and private communication, 2015).

Oppelstrup, T. and Moriarty, J.A. LAMMPS implementation of quantum-based, multi-ion MGPT potentials for transition metals: application to Ta, Mo and V (unpublished, 2018).

Orlikowski, D. First-principles thermoelasticity of transition metals at high pressure: anharmonicity with tantalum as a prototype (unpublished and private communication, 2007).

Orlikowski, D., Söderlind, P. and Moriarty, J.A. First-principles thermoelasticity of transition metals at high pressure: tantalum prototype in the quasiharmonic limit. *Phys. Rev. B* **74**, 054109 (2006).

Overton, Jr., W.C. and Gaffney, J. Temperature variation of the elastic constants of cubic elements. I. Copper. *Phys. Rev.* **98**, 969–977 (1955).

Ozolins, V. First-principles calculations of free energies of unstable phases: the case of fcc W. *Phys. Rev. Lett.* **102**, 065702 (2009).

Paidar, V., Wang, L.G., Sob, M. and Vitek, V. A study of the applicability of many-body central force potentials in NiAl and TiAl. *Modell. Simul. Mater. Sci. Eng.* **7**, 369–381 (1999).

Palmer, S.B. and Lee, E.W. The elastic constants of chromium. *Philos. Mag.* **24**, 311–318 (1971).

Panitz, J., Cutler, P.H. and King III, W.F. A first-principles generalized pseudopotential calculation for the phonon spectra of zinc. *J. Phys. F: Metal Phys.* **4**, L106–L110 (1974).

Paxton, A.T., Gumbsch, P. and Methfessel, M. A quantum mechanical calculation of the theoretical strength of metals. *Philos. Mag. Lett.* **63**, 267–274 (1991).

Pearson, W.B. *Handbook of Lattice Spacings and Structures of Metals and Alloys, Vol. 2.* (Pergamon, Oxford, 1967).

Pélissier, J.L. and Wetta, N. A model-potential approach for bismuth (I). Densification and melting curve calculation. *Physica A* **289**, 459–478 (2001).

Perdew, J.P., Burke, K. and Ernzerhof, M. Generalized gradient approximation made simple. *Phys. Rev. Lett.* **77**, 3865–3868 (1996). Errata *Phys. Rev. Lett.* **78**, 1396 (1997).

Perdew, J.P. and Zunger, A. Self-interaction correction to density-functional approximations for many-electron systems. *Phys. Rev. B* **23**, 5048–5079 (1981).

Pettifor, D.G. An energy-independent method of band-structure calculation for transition metals. *J. Phys. C: Solid State Phys.* **2**, 1051–1069 (1969).

Pettifor, D.G. Accurate resonance parameter approach to transition-metal band structure. *Phys. Rev. B* **2**, 3031–3034 (1970a).

Pettifor, D.G. Theory of the crystal structures of transition metals. *J. Phys. C: Solid State Phys.* **3**, 367–377 (1970b).

Pettifor, D.G. First principle basis functions and matrix elements in the H-NFE-TB representation. *J. Phys. C: Solid State Phys.* **5**, 97–120 (1972).

Pettifor, D.G., New many-body potential for the bond order. *Phys. Rev. Lett.* **63**, 2480–2483 (1989).

Pettifor, D.G. *Bonding and Structure of Molecules and Solids* (Oxford University Press, Oxford, 1995).

Pettifor, D.G. and Ward, M.A. An analytic pair potential for simple metals. *Solid State Commun.* **49**, 291–294 (1984).

Phillips, J.C. and Kleinman, L. New method of calculating wave functions in crystals and molecules. *Phys. Rev.* **116**, 287–294 (1959).

Phillips, R., Zou, J., Carlsson, A.E. and Widom, M. Electronic-structure-based pair potentials for aluminum-rich cobalt compounds. *Phys. Rev. B* **49**, 9322–9330 (1994).

Pines, D. and Nozieres, P. *The Theory of Quantum Liquids* (Benjamin, New York, 1966).

Plimpton, S. Fast parallel algorithms for short-range molecular dynamics. *J. Comput. Phys.* **117**, 1 (1995). The website for the LAMMPS code is http://lammps.sandia.gov.

Polsin, D.N., Fratanduono, D.E., Rygg, J.R., Lazicki, A., Smith, R.F., Eggert, J.H., Gregor, M.C., Henderson, B.H., Delettrez, J.A., Kraus, R.G., Celliers, P.M., Coppari, F., Swift, D.C., McCoy, C.A., Seagle, C.T., Davis, J.-P., Burns, S.J., Collins, G.W. and Boehly, T.R. Measurement of body-centered-cubic aluminum at 475 GPa. *Phys. Rev. Lett.* **119**, 175702 (2017). Errata *Phys. Rev. Lett.* **120**, 029902 (2018).

Powell, B.M., Martel, P. and Woods, A.D.B. Phonon properties of niobium, molybdenum and their alloys. *Can. J. Phys.* **55**, 1601–1612 (1977).

Predel, B. and Hulse, K. Some thermodynamic properties of aluminium-magnesium alloys. *Z. Metallkd.* **69**, 661–666 (1978).

Pynn, R. and Squires, G.L. Measurements of the normal-mode frequencies of magnesium. *Proc. R. Soc. London, Ser. A* **326**, 347–360 (1972).

Rao, S., Hernandez, C., Simmons, J., Parthasarathy, T. and Woodward, C. Green's function boundary conditions in two-dimensional and three-dimensional atomistic simulations of dislocations. *Philos. Mag. A* **77**, 231–256 (1998).

Rao, S.I. and Woodward, C. Atomistic simulations of a/2<111> screw dislocations in bcc Mo using a modified generalized pseudopotential theory potential. *Philos. Mag. A* **81**, 1317–1327 (2001).

Rao, S.I., Woodward, C., Akdim, B., Senkov, O.N. and Miracle, D. Theory of solid solution strengthening of bcc chemically complex alloys. *Acta Mater.* **209**, 116758 (2021).

Rasolt, M. and Taylor, R. A study of energy dependence and non locality in metallic pseudopotentials. *J. Phys. F: Metal Phys.* **3**, 67–74 (1973).

Rasolt, M. and Taylor, R. Charge densities and interionic potentials in simple metals: nonlinear effects. I. *Phys. Rev. B* **11**, 2717–2725 (1975).

Rayne, J.A. Elastic constants of palladium from 4.2 to 300 K. *Phys. Rev.* **118**, 1545–1549 (1960).

Rayne, J.A. and Chandrasekhar, B.S. Elastic constants of iron from 4.2 to 300 K. *Phys. Rev.* **122**, 1714–1716 (1961).

Recoules, V., Clérouin, J., Zérah, G., Anglade, P.M. and Mazevet, S. Effect of intense laser irradiation on the lattice stability of semiconductors and metals. *Phys. Rev. Lett.* **96**, 055503 (2006).

Regnaut, C., Fusco, E. and Badiali, J.P. Structure and thermodynamics of the liquid noble metals from the generalized OPW and resonant model approaches. *Phys. Status Solidi B* **120**, 373–382 (1983).

Regnaut, C., Fusco, E. and Badiali, J.P. Parameterized version of the generalized pseudopotential theory for noble metals: application to copper. *Phys. Rev. B* **31**, 771–784 (1985).

Rodney, D., Ventelon, L., Clouet, E. Pizzagalli, L. and Willaime, F. *Ab initio* modeling of dislocation core properties in metals and semiconductors. *Acta Mater.* **124**, 633–659 (2017).

Rosenfield, A.M. and Stott, M.J. Density-dependent pair potentials and the compressibility problem. *J. Phys. F: Metal Phys.* **17**, 605–627 (1987).

Ruffino, M., Skinner, G.C.G., Andritos, E.I. and Paxton, A.T. Ising-like models for stacking faults in a free electron metal. *Proc. R. Soc. London, Ser. A* **476**, 20200319 (2020).

Ruoff, A.L., Rodriquez, C.O. and Christensen, N.E. Elastic moduli of tungsten to 15 Mbar, phase transition at 6.5 Mbar, and rheololgy to 6 Mbar. *Phys. Rev. B* **58**, 2998–3002 (1998).

Ruoff, A.L., Xia, H., Luo, H. and Vohra, Y.K. Miniaturization techniques for obtaining static pressures comparable to the pressure at the center of the earth: X-ray diffraction at 416 GPa. *Rev. Sci. Instrum.* **61**, 3830–3833 (1990).

Ruoff, A.L., Xia, H. and Xia, Q. The effect of a tapered aperture on X-ray diffraction from a sample with a pressure gradient: studies on three samples with a maximum pressure of 560 GPa. *Rev. Sci. Instrum.* **63**, 4342–4348 (1992).

Salama, K. and Alers, G.A. The composition dependence of the third-order elastic constants of the Cu-Ni system. *Phys. Status Solidi A* **41**, 241–247 (1977).

Santamaría-Perez, D., Ross, M., Errandonia, D., Mukherjee, G.D., Mezouar, M. and Boehler, R. X-ray diffraction measurements of Mo melting to 119 GPa and the high-pressure phase diagram. *J. Chem. Phys.* **130**, 124509 (2009).

Sastry, G.V.S., Bariar, K.K., Chattopadhyay, K. and Ramachandrarao, P. Vacancy-ordered phases in rapidly solidified aluminium-copper alloys. *Z. Metallkd.* **71**, 756–759 (1980).

Schiff, L.I. *Quantum Mechanics* (McGraw-Hill, New York, 1955).

Schmunk, R.E. and Smith, C.S. Pressure derivatives of the elastic constants of aluminum and magnesium. *J. Phys. Chem. Solids* **9**, 100–112 (1959).

Schulte, O. and Holzapfel, W.B. A new structure of mercury under pressure. *Phys. Lett. A* **131**, 38–40 (1988).

Schulte, O. and Holzapfel, W.B., Phase diagram for mercury up to 67 GPa and 500 K. *Phys. Rev. B* **48**, 14009–14012 (1993).

Schultz, H. Defect properties of bcc metals. In H. Ullmaier (ed.), *Atomic Defects in Metals, Landolt-Börnstein-Group III Condensed Matter*, Vol. 25 (Springer, Berlin, 1991).

Seeger, A. and Würthrich, C. Dislocation relaxation processes in body-centred cubic metals. *Nuovo Cimento B* **33**, 38–75 (1976).

Seeger, A. and Hollang, L. The flow stress asymmetry of ultra-pure molybdenum single crystals. *Mater. Trans. JIM* **41**, 141–151 (2000).

Seko, A., Togo, A. and Tanaka, I. Group-theoretical high-order rotational invariants for structural representations: application to linearized machine learning interatomic potentials. *Phys. Rev. B* **99**, 214108 (2019).

Senkov, O.N., Miracle, D.B., Chaput, K.J. and Couzinie, J.P. Development and exploration of refractory high-entropy alloys—a review. *J. Mater. Res.* **33**, 3092–3128 (2018).

Senkov, O.N., Wilkes, G.B., Miracle, D.B., Chuang, C.P. and Liaw, P.K. Refractory high-entropy alloys. *Intermetallics* **18**, 1758–1765 (2010).

Shaner, J.W., Gathers, G.R. and Minichino, C. Thermophysical properties of liquid tantalum and molybdenum. *High Temp. High Press.* **9**, 331–343 (1977).

Shapeev, A. Accurate representation of formation energies of crystalline alloys with many components. *Comput. Mater. Sci.* **139**, 26–30 (2017).

Shapeev, A.V. Moment tensor potentials: a class of systematically improvable interatomic potentials. *Mult. Model. Simul.* **14**, 1153–1173 (2016).

Sharma, R.R. General expressions for reducing the Slater-Koster linear combination of atomic orbitals integrals to the two-center approximation. *Phys. Rev. B* **19**, 2813–2822 (1979). Errata *Phys. Rev. B* **22**, 5015 (1980).

Sharma, S.M., Turneaure, S.J., Winey, J.M., Li, Y., Rigg, P., Schuman, A., Sinclair, N. Toyoda, Y., Wang, X., Weir, N., Zhang, J. and Gupta, Y.M. Structural transformation and melting in gold shock compressed to 355 GPa. *Phys. Rev. Lett.* **123**, 045702 (2019).

Shaw, R.W. Optimum form of a modified Heine-Abarenkov model potential for the theory of simple metals. *Phys. Rev.* **174**, 769–781 (1968).

Shaw, Jr., R.W. and Harrison, W.A. Reformulation of the screened Heine-Abarenkov model potential. *Phys. Rev.* **163**, 604–611 (1967).

Shaw, Jr., R.W. and Pynn, R. Optimized model potential: exchange and correlation corrections and calculation of magnesium phonon spectrum. *J. Phys. C: Solid State Phys.* **2**, 2071–2088 (1969).

Shaw, W.M. and Muhlestein, L.D. Investigation of the phonon dispersion relations of chromium by inelastic neutron scattering. *Phys. Rev. B* **4**, 969–973 (1971).

Shepard, M.L. and Smith, J.F. Elastic constants of rhenium single crystals in the temperature range 4.2–298 K. *J. Appl. Phys.* **36**, 1447–1450 (1965).

Simmons, G. and Wang, H. *Single Crystal Elastic Constants and Calculated Aggregate Properties: A Handbook*, 2nd edition (MIT Press, Cambridge, MA, 1971).

Sin'ko, G.V. and Smirnov, N.A. *Ab initio* calculations of elastic constants and thermodynamic properties of bcc, fcc, and hcp Al crystals under pressure. *J. Phys.: Condens. Matter* **14**, 6989–7005 (2002).

Singwi, K.L., Sjölander, A., Tosi, M.P. and Land, R.H. Electron correlations at metallic densities. IV. *Phys. Rev. B* **1**, 1044–1053 (1970).

Skinner, G.C.G. Novel quantum-based interatomic potentials applied to magnesium and its alloys. Ph.D. thesis, King's College London (2019).

Skinner, G.C.G. and Moriarty, J.A. Stability of the bcc structure in dilute Ca_xMg_{1-x} alloys. (unpublished, 2017).

Skinner, G.C.G., Paxton, A.T. and Moriarty, J.A. Local volume effects in the generalized pseudopotential theory. *Phys. Rev. B* **99**, 214107 (2019).

Skriver, H.L. *The LMTO Method* (Springer, Berlin, 1984).

Skriver, H.L. Crystal structure from one-electron theory. *Phys. Rev. B* **31**, 1909–1923 (1985).

Slater, J.C. A simplification of the Hartree-Fock method. *Phys. Rev.* **81**, 385–390 (1951).

Slater, J.C. *The Self-Consistent Field for Molecules and Solids: Quantum Theory of Molecules and Solids*, Vol. 4 (McGraw-Hill, New York, 1974).

Slater, J.C. and Koster, G.F. Simplified LCAO method for the periodic potential problem. *Phys. Rev.* **94**, 1498–1524 (1954).

Slutsky, L.J. and Garland, C.W. Elastic constants of magnesium from 4.2 K to 300 K. *Phys. Rev.* **107**, 972–976 (1957).

Smith, J.F. and Arbogast, C.L. Elastic constants of single crystal beryllium. *J. Appl. Phys.* **31**, 99–102 (1960).

Smith, J.F. and Gjevre, J.A. Elastic constants of yttrium single crystals in the temperature range 4.2–400 K. *J. Appl. Phys.* **31**, 645–647 (1960).

Söderlind, P. Theory of the crystal structures of cerium and the light actinides. *Advanc. Phys.* **47**, 959–998 (1998).

Söderlind, P. First-principles FP-LMTO calculation of the A15-bcc structural energy difference in Mo. (unpublished and private communication, 2001).

Söderlind, P. First-principles elastic and structural properties of uranium metal. *Phys. Rev. B* **66**, 085113 (2002).

Söderlind, P. First-principles FP-LMTO calculations of the equation of state, elastic moduli, phonons and unrelaxed vacancy formation energy as a function of volume for bcc uranium (unpublished and private communication, 2003).

Söderlind, P., Ahuja, R., Eriksson, O., Johansson, B. and Wills, J.M. Theoretical predictions of structural phase transitions in Cr, Mo, and W. *Phys. Rev. B* **49**, 9365–9371 (1994a).

Söderlind, P., Ahuja, R., Eriksson, O., Wills, J.M. and Johansson, B. Crystal structure and elastic-constant anomalies in the magnetic 3d transition metals. *Phys. Rev. B* **50**, 5918–5927 (1994b).

Söderlind, P., Eriksson, O., Johansson, B., Wills, J.M. and Boring, A.M., A unified picture of the crystal structures of metals. *Nature* **374**, 524–525 (1995).

Söderlind, P., Grabowski, B., Yang, L., Landa, A., Bjorkman, T., Sourvatzis, P. and Eriksson, O. High-temperature phonon stabilization of γ-uranium from relativistic first-principles theory. *Phys. Rev. B* **85**, 060301 (2012).

Söderlind, P. and Klepeis, J.E. First-principles elastic properties of α-Pu. *Phys. Rev. B* **79**, 104110 (2009).

Söderlind, P. and Moriarty, J.A. First-principles theory of Ta up to 10 Mbar pressure: structural and mechanical properties. *Phys. Rev. B* **57**, 10340–10350 (1998).

Söderlind, P., Wills, J.M. and Eriksson, O. Simple model for complex structures. *Phys. Rev. B* **57**, 1320–1323 (1998).

Söderlind, P., Yang, L.H., Moriarty, J.A. and Wills, J.M. First-principles formation energies of monovacancies in bcc transition metals. *Phys. Rev. B* **61**, 2579–2586 (2000).

Stassis, C., Zaretsky, J., Misemer, D.K., Skriver, H.L., Harmon, B.N. and Nicklow, R.M. Lattice dynamics of fcc Ca. *Phys. Rev. B* **27**, 3303–3307 (1983).

Steinhardt, P.J., Nelson, D.R. and Ronchetti, M. Icosahedral bond orientational order in super-cooled liquids. *Phys. Rev. Lett.* **47**, 1297–1300 (1981).

Steurer, W. Single-phase high-entropy alloys—a critical update. *Mater. Characterization* **162**, 110179 (2020).

Stinton, G.W., MacLeod, S.G., Cynn, H., Errandonea, D., Evans, W.J., Proctor, J.E., Meng, Y. and McMahon, M.I. Equation of state and high-pressure/high-temperature phase diagram of magnesium. *Phys. Rev. B* **90**, 134105 (2014).

Stoner, E.C. Collective electron ferromagnetism. *Proc. R. Soc. London, Ser. A* **165**, 372–414 (1938).

Stoner, E.C. Collective electron ferromagnetism II. Energy and specific heat. *Proc. R. Soc. London, Ser. A* **169**, 339–371 (1939).

Streitz, F.H., Glosli, J.N. and Patel, M.V. Beyond finite-size scaling in solidification simulations. *Phys. Rev. Lett.* **96**, 225701 (2006a).

Streitz, F.H., Glosli, J.N., Patel, M.V., Chan, B., Yates, R.K., deSupinski, B.R., Sexton, J. and Gunnels, J.A. Simulating solidification in metals at high pressure: the drive to petascale computing. *J. Phys.: Conf. Series* **46**, 254–267 (2006b).

Streitz, F.H. and Moriarty, J.A. Metastable A15 phase during rapid resolidification of tantalum. LLNL Report No. UCRL-JC-148950 (2002).

Stricker, M., Yin, B. Mak, E. and Curtin, W.A. Machine learning for metallurgy II. A neural-network potential for magnesium. *Phys. Rev. Mater.* **4**, 103602 (2020).

Stroh, A.N. Dislocations and cracks in anisotropic elasticity. *Philos. Mag.* **3**, 625–646 (1958).

Stroh, A.N. Steady state problems in anisotropic elasticity. *J. Math. and Phys.* **41**, 77–103 (1962).

Suzuki, N. and Otani, M. Theoretical study on the lattice dynamics and electron-phonon interaction of vanadium under high pressures. *J. Phys.: Condens. Matter* **14**, 10869–10872 (2002).

Suzuki,T., Kaminura, Y. and Kirchnner, H.O.K. Plastic homology of bcc metals. *Philos. Mag. A* **79**, 1629–1642 (1999).

Szlachta, W.J., Bartók, A.P. and Csányi, G. Accuracy and transferability of Gaussian approximation potential models for tungsten. *Phys. Rev. B* **90**, 104108 (2014).

Tadano, T., Gohda, Y. and Tsuneyuki, S. Anharmonic force constants extracted from first-principles molecular dynamics: applications to heat transfer simulations. *J. Phys.: Condens. Matter* **26**, 225402 (2014). The main website for the ALAMODE code is https://sourceforge.net/projects/alamode.

Tadmor, E.B. and Miller, R.E. *Modeling Materials: Continuum, Atomistic and Multiscale Techniques* (Cambridge University Press, Cambridge, UK, 2011).

Taioli, S., Carzorla, C., Gillan, M.J. and Alfè, D. Melting curve of tantalum from first principles. *Phys. Rev. B* **75**, 214103 (2007).

Takemura, K., Nakano, S., Ohishi, Y. Nakamoto, Y. and Fujihisa, H. High-pressure structural study of solid mercury up to 200 GPa. *Mater. Res. Express* **2**, 016502 (2015).

Takeuchi, S., Kuramoto, E. and Suzuki, T. Orientation dependence of slip in tantalum single crystals. *Acta Metall.* **20**, 909–915 (1972).

Tang, M. A lattice-based screw-edge dislocation dynamics simulation of body center cubic single crystals. In S. Yip (ed.), *Handbook of Materials Modeling*, Vol. 1, pp. 827–837 (Springer, Amsterdam, 2005).

Tang, M., Kubin, L.P. and Canova, G.R. Dislocation mobility and the mechanical response of b.c.c. single crystals: a mesoscopic approach. *Acta Mater.* **46**, 3221–3235 (1998).

Taylor, R. A critique of the practice of fitting pair potentials to experimental data. In J.K. Lee (ed.), *Interatomic Potentials and Crystal Defects*, pp. 71–83 (Metallurgical Society of AIME, New York, 1981).

Taylor, R. and MacDonald, A.H. Harmonic phonons and phonon-limited resistivities for Rb and Cs from first-principle pseudopotentials. *J. Phys. F: Metal Phys.* **10**, 2387–2394 (1980).

Thompson, A.P., Swiler, L.P., Trott, C.R., Foiles, S.M. and Tucker, G.J. Spectral neighbor analysis method for automated generation of quantum-accurate interatomic potentials. *J. Comp. Phys.* **285**, 316–330 (2015).

Torrens, I.M. *Interatomic Potentials* (Academic, New York, 1972).

Touloukian, Y.S., Kirby, R.K., Taylor, R.E. and Desai, P.D. *Thermophysical Properties of Matter (Thermal Expansion: Metallic Elements and Alloys)*, Vol. 12 (Plenum, New York, 1975).

Trambly de Laissardière, G., Mayou, D. and Nguyen Manh, D. Electronic structure of transition atoms in quasi-crystals and Hume-Rothery alloys. *Europhys. Lett.* **21**, 25–30 (1993).

Trambly de Laissardière, G., Nguyen Manh, D., Magaud, L., Julien, J.P., Cyrot-Lackmann, F. and Mayou, D. Electronic structure and hybridization effects in Hume-Rothery alloys containing transition elements. *Phys. Rev. B* **52**, 7920–7933 (1995).

Tran, R., Xu, Z., Radhakrishnan, B., Winston, D., Sun, W., Persson, K. and Ong, S.P. Surface energies of elemental crystals. *Sci. Data* **3**,160080 (2016).

Tsuppayakorn-aek, P., Zhang, J., Luo, W., Ding, Y., Ahuja, R. and Bovornratanaraks, T. Bain deformation mechanism and Lifshitz transition in magnesium under pressure. *Phys. Status Solidi B* **258**, 2000279 (2020).

Tyson, W.R. and Miller, W.A. Surface free energies of solid metals: estimation from liquid surface tension measurements. *Surf. Sci.* **62**, 267–276 (1977).

Uesugi, T., Kohyama, M. and Higashi, K. *Ab initio* study on divacancy binding energies in aluminum and magnesium. *Phys. Rev. B* **68**, 184103 (2003).

Upadhyaya, J.C. and Dagens, L. Resonant model potential and the phonon frequencies in copper. *J. Phys. F: Metal Phys.* **8**, L21–24 (1978).

Upadhyaya, J.C. and Dagens, L. Dispersion relations for noble metals in the resonant model potential. *J. Phys. F: Metal Phys.* **9**, 2177–2184 (1979).

Upadhyaya, J.C. and Dagens, L. Relativistic effects on the lattice dynamics of gold. *J. Phys. F: Metal Phys.* **12**, L137–139 (1982a).

Upadhyaya, J.C. and Dagens, L. Phonon spectra and crystal structure of Zn and Cd using the resonant model-potential approach. *Phys. Rev. B* **26**, 743–752 (1982b).

Vaks, V.G., Katsnelson, M.I., Koreshkov, V.G., Likhtenstein, A.I., Parfenov, O.E., Skok, V.F., Sukhoparov, V.A., Trefilov, A.V. and Chernyshov, A.A. An experimental and theoretical study of martensitic phase transitions in Li and Na under pressure. *J. Phys.: Condens. Matter* **1**, 5319–5335 (1989).

Vallera, A.M. High temperature phonons in iron. *J. Physique Coll.* **42**, C6 (suppl. 12), 398–400 (1981).

van der Oord, C., Dusson, G., Csányi, G. and Ortner, C. Regularised atomic body-ordered permutation-invariant polynomials for the construction of interatomic potentials. *Mach. Learn.: Sci. Technol.* **1**, 015004 (2020).

Varma, C.M. and Weber, W. Phonon dispersion in transition metals. *Phys. Rev. B* **19**, 6142–6154 (1979).

Vashishta, P., Kalia, R.K., Nakano, A., Kaxiras, E., Grama, A., Lu, G., Eidenbenz, S., Voter, A.F., Hood, R.G., Moriarty, J.A. and Yang, L.H. Hierarchical petascale simulation framework for stress corrosion cracking. *J. Phys. Conf. Ser.* **125**, 012060 (2008).

Vasiliu, E.V. Computational study of the effective three-ion interaction potentials in liquid metals with high density of electron gas. *J. Phys. Studies* **7**, 161–166 (2003).

Villars, P. and Calvert, L.D. *Pearson's Handbook of Crystallographic Data for Intermetallic Phases*, 2nd edition (ASM International, Materials Park, 1991).

Vinet, P., Smith, J.R., Ferrante, J. and Rose, J.H. Temperature effects on the universal equation of state of solids. *Phys. Rev. B* **35**, 1945–1953 (1987).

Vinet, P., Rose, J.H., Ferrante, J. and Smith, J.R. Universal features of the equation of state of solids. *J. Phys. Condens. Matter* **1**, 1941–1964 (1989).

Vitek, V. Theory of the core structures of dislocations in body-centered cubic metals. *Cryst. Lattice Defects* **5**, 1–34 (1974).

Vitos, L., Ruban, A.V., Skriver, H.L. and Kóllar, J. The surface energy of metals. *Surf. Sci.* **411**, 186–202 (1998).

von Barth, L. and Hedin, L. A local exchange-correlation potential for the spin polarized case: I. *J. Phys. C: Solid State Phys.* **5**, 1629–1642 (1972).

Vosko, S.H., Wilk, L. and Nusair, M. Accurate spin-dependent electron liquid correlation energies for local spin density calculations: a critical analysis. *Can. J. Phys.* **58**, 1200–1211 (1980).

Voter, A.F. and Chen, S.P. Accurate interatomic potentials for Ni, Al and Ni_3Al. In R.W. Siegel, J.R. Weertman and R. Sinclair (eds.), *Characterization of Defects in Materials, Materials Research Society Symposium Proceedings*, Vol. 82, pp. 175–180 (Materials Research Society, Pittsburgh, 1987).

Walker, A.B. and Taylor, R. Density-dependent potentials for simple metals. *J. Phys.: Condens. Matter* **2**, 9481–9499 (1990).

Walker, E. and Bujard, P. Anomalous temperature behavior of the shear elastic constant C_{44} in tantalum. *Solid State Commun.* **34**, 691–693 (1980).

Wallace, D.C. *Thermodynamics of Crystals* (Wiley, New York, 1972).

Wang, N., Yu, W.-Y., Tang, B.-Y., Peng, L.-M. and Ding, W.-J. Structural and mechanical properties of $Mg_{17}Al_{12}$ and $Mg_{24}Y_5$ from first-principles calculations. *J. Phys. D: Appl. Phys.* **41**, 195408 (2008).

Wang, X.W., Chan, C.T., Ho, K.M. and Weber, W. Role of surface-state nesting in the incommensurate reconstruction of Mo(001). *Phys. Rev. Lett.* **60**, 2066–2069 (1988).

Wasserbach, W. Plastic deformation and dislocation arrangement of Nb–34 at.% Ta alloy single crystals. *Philos. Mag. A* **53**, 335–356 (1986).

Watanabe, M. and Reinhardt, W.P. Direct dynamical calculation of entropy and free energy by adiabatic switching. *Phys. Rev. Lett.* **65**, 3301–3304 (1990).

Weinberger, C.R., Tucker, G.J. and Foiles, S.M. Peierls potential of screw dislocations in bcc transition metals: Predictions from density functional theory. *Phys. Rev. B* **87**, 054114 (2013).

Wentzcovitch, R.M. and Cohen, M.L. Theoretical model for the hcp-bcc transition in Mg. *Phys. Rev. B* **37**, 5571–5576 (1988).

Wetta, N. and Pélissier, J.L. A model-potential approach for bismuth (II). Behavior under shock loading. *Physica A* **289**, 479–497 (2001).

Widom, M., Al-Lehyani, I. and Moriarty, J.A. First-principles interatomic potentials for transition-metal aluminides: III. Extension to ternary phase diagrams. *Phys. Rev. B* **62**, 3648–3657 (2000).

Widom, M. and Moriarty, J.A. First-principles interatomic potentials for transition-metal aluminides: II. Application to Al-Co and Al-Ni phase diagrams. *Phys. Rev. B* **58**, 8967–8979 (1998).

Wills, J.M., Alouani, M., Andersson, P., Delin, A., Eriksson, O. and Grechnyev, O. *Full-Potential Electronic Structure Method* (Springer, Heidelberg, 2010).

Wills, J.M., Eriksson, O., Alouani, M. and Price, D.L. Full-potential LMTO total energy and force calculations. In H. Dreysse (ed.), *Electronic Structure and Physical Properties of Solids: The Uses of the LMTO Method*, pp. 148–167 (Springer, Heidelberg, 2000).

Wills, J.M., Eriksson, O., Söderlind, P. and Boring, A.M. Trends of the elastic constants of cubic transition metals. *Phys. Rev. Lett.* **18**, 2802–2805 (1992).

Wilson, B., Sonnad, V., Sterne, P. and Isaacs, W. Purgatorio—a new implementation of the Inferno algorithm. *JQSRT* **99**, 658–679 (2006).

Woll, Jr., E.J. and Kohn, W. Images of the Fermi surface in the phonon spectra of metals. *Phys. Rev.* **126**, 1693–1697 (1962).

Wong, J., Krisch, M., Farber, D.L., Occelli, F., Schwartz, A.J., Chiang, T-C., Wall, M., Boro, C. and Xu, R., Phonon dispersions of fcc δ-plutonium-gallium by inelastic X-ray scattering. *Science* **301**, 1078–1080 (2003).

Wong, J., Wall, M., Schwartz, A.J., Xu, R., Holt, M., Hong, H., Zschack, P. and Chiang, T-C. Imaging phonons in a fcc Pu-Ga alloy by thermal diffuse X-ray scattering. *Appl. Phys. Lett.* **84**, 3747–3749 (2004).

Woods, A.D.B. Lattice dynamics of tantalum. *Phys. Rev.* **136**, A781–783 (1964).

Woods, A.D.B., Brockhouse, B.N., March, R.H., Stewart, A.T. and Browers, R. Crystal dynamics of sodium at 90 K. *Phys. Rev.* **128**, 1112–1120 (1962).

Woodward, C. First-principles simulations of dislocation cores. *Mater. Sci. Eng. A* **400–401**, 59–67 (2005).

Woodward, C. and Rao, S.I. *Ab initio* simulation of isolated screw dislocations in bcc Mo and Ta. *Philos. Mag. A* **81**, 1305–1316 (2001).

Woodward, C. and Rao, S.I. Flexible *ab initio* boundary conditions: simulating isolated dislocations in bcc Mo and Ta. *Phys. Rev. Lett.* **88**, 216402 (2002).

Wu, C., Söderlind, P., Glosli, J.N. and Klepeis, J.E. Shear-induced anisotropic plastic flow from body-centred-cubic tantalum before melting. *Nature Mater.* **8**, 223–228 (2009).

Xia, H., Duclos, S.J., Ruoff, A.L. and Vohra, Y.K. New high-pressure phase transition in zirconium metal. *Phys. Rev. Lett.* **64**, 204–207 (1990a).

Xia, H., Parthasarathy, G., Luo, H. Vohra, Y.K. and Ruoff, A.L. Crystal structures of group IVA metals at ultrahigh pressures. *Phys. Rev. B* **42**, 6736–6738 (1990b).

Xu, W. and Moriarty, J.A. Atomistic simulation of ideal shear strength, point defects, and screw dislocations in bcc transition metals: Mo as a prototype. *Phys. Rev. B* **54**, 6941–6951 (1996).

Xu, W. and Moriarty, J.A. Accurate atomistic simulations of the Peierls barrier and kink-pair formation energy for <111> screw dislocations in bcc Mo. *Comput. Mater. Sci.* **9**, 348–356 (1998).

Yamaguchi, M., Shiga, M. and Kaburaki, H. Grain boundary decohesion by impurity segregation in a nickel-sulfur system. *Science* **307**, 393–397 (2005).

Yang, L.H. Advanced atomic-level materials design. In A.E. Koniges (ed.), *Industrial Strength Parallel Computing*, pp. 297–316 (Morgan Kaufmann, San Francisco, 2000).

Yang, L.H. First-principles DFT PP calculations of the volume dependence of high-symmetry phonon frequencies in fcc nickel. (unpublished and private communication, 2008).

Yang, L.H. DFT PP and MGPT calculations of selected defect properties in fcc nickel (unpublished and private communication, 2009).

Yang, L.H., Söderlind, P. and Moriarty, J.A. Accurate atomistic simulation of (a/2)<111> screw dislocations and other defects in bcc tantalum. *Philos. Mag. A* **81**, 1355–1385 (2001).

Yang, L.H., Tang, M. and Moriarty, J.A. Dislocations and plasticity in bcc transition metals at high pressure. In J.P. Hirth and L. Kubin (eds.), *Dislocations in Solids*, Vol. 16, Chap. 92, pp. 1–46 (Elsevier, Amsterdam, 2010).

Yao, Y. and Klug, D.D. Reconstructive structural phase transitions in dense Mg. *J. Phys.: Condens. Matter* **24**, 265401 (2012).

Yao, Y. and Klug, D.D. Stable structures of tantalum at high temperature and high pressure. *Phys. Rev. B* **88**, 054102 (2013).

Yin, B., Wu, Z. and Curtin, W.A. Comprehensive first-principles study of stable stacking faults in hcp metals. *Acta Mater.* **123**, 223–234 (2017).

Yin, S., Ding, J., Asta, M. and Ritchie, R.O. Ab initio modeling of the energy landscape for screw dislocations in body-centered cubic high-entropy alloys. *npj Comput. Mater.* **6**, 110 (2020).

Yoo, C-S., Akella, J. and Moriarty, J.A. High-pressure melting temperatures of uranium: laser-heating experiments and theoretical calculations. *Phys. Rev. B* **48**, 15529–15534 (1993).

Yoo, C-S., Cynn, H. and Söderlind, P. Phase diagram of uranium at high pressures and temperatures. *Phys. Rev. B* **57**, 10359–10362 (1998).

Yoo, C.S., Cynn, H., Söderlind, P. and Iota, V. New β(fcc)-cobalt to 210 GPa. *Phys. Rev. Lett.* **84**, 4132–4135 (2000).

Young, D.A., *Phase Diagrams of the Elements* (University of California, Berkeley, 1991).

Young, D.A. and Rogers, F.J. Variational fluid theory with inverse 12th power reference potential. *J. Chem. Phys.* **81**, 2789–2793 (1984).

Zaug, J.M., Abramson, E.H., Brown, J.M., Slutsky, L.J., Aracne-Ruddle, C.M. and Hansen, D.W. A study of the elasticity of Ta at high temperature and pressure. In M.H. Manghnani, W.J. Nellis and M.F. Nicol (eds.), *Science and Technology of High Pressure*, Vol. 1, Proceedings of AIRAPT-17, pp. 804–806 (Universities Press, Hyderabad, India, 2000).

Zaug, J.M. and Crowhurst, J.C. Elastic constants of Ta at high pressure using an ISLS technique. (unpublished and private communication, 2001).

Zhang, B., Gramlich, V. and Steurer, W. $Al_{13-x}(Co_{1-y}Ni_y)_4$, a new approximate of the decagonal quasicrystal in the Al-Co-Ni system. *Z. Kristallogr.* **210**, 498–503 (1995).

Zhang, H., Shang, S.L., Wang, Y., Saengdeejing, A., Chen, L.Q. and Liu, Z.K. First-principles calculations of the elastic, phonon and thermodynamic properties of $Al_{12}Mg_{17}$. *Acta Mater.* **58**, 4012–4018 (2010).

Zhang, M.-X. and Kelly, P.M. Edge to edge match and its applications part II. Application to Mg-Al, Mg-Y and Mg-Mn alloys. *Acta Mater.* **53**, 1085–1096 (2005).

Zhou, X., He, S. and Marian, J. Cross-kinks control screw dislocation strength in equiatomic bcc refractory alloys. *Acta Mater.* **211**, 116875 (2021).

Zhou, Y., Dang, M., Sun, L., Zhai, W., Dong, H., Gao, Q., Zhao, F. and Peng, J. First-principle studies on the electronic structural, thermodynamics and elastic properties of $Mg_{17}Al_{12}$ intermediate phase under high pressure. *Mater. Res. Express* **6**, 0865e1 (2019).

Zhuang, H. Chen, M. and Carter, E.A. Elastic and thermodynamic properties of complex Mg-Al intermetallic compounds via orbital-free density functional theory. *Phys. Rev. Appl.* **5**, 064021 (2016).

Ziman, J.M. The method of neutral pseudo-atoms in the theory of metals. *Advanc. Phys.* **13**, 89–138 (1964).

Ziman, J.M. The T matrix, K matrix, d bands and l-dependent pseudo-potentials in the theory of metals. *Proc. Phys. Soc. (London)* **86**, 337–353 (1965).

Zimmerman, J.A., Gao, H. and Abraham, F.F. Generalized stacking fault energies for embedded atom fcc metals. *Modell. Simul. Mater. Sci. Eng.* **8**, 103–115 (2000).

Znam, S. Nguyen-Manh, D., Pettifor, D.G. and Vitek, V. Atomistic modeling of TiAl I. Bond-order potentials with environmental dependence. *Philos. Mag.* **83**, 415–438 (2003).

Subject Index

Adaptive GPT (aGPT) for local treatment of energies and forces
  formalism 471–6
  initial tests and applications for hcp Mg
    elastic moduli 476–7
    relaxed vacancy formation and divacancy binding energies 477–8
    γ surface profiles 478
    thermal expansion 479
ALAMODE lattice-dynamics code 295–6
Alloys and intermetallic compounds
  GPT treatment of general AB binary alloys 425–7
  SM-SM systems 429–31
  SM-EDB or SM-FDB systems 434–6
  SM-TM systems 427–9, 436–40
  MGPT treatment of multi-component TM high-entropy alloys 456–9
  See also $Mg_{17}Al_{12}$ ordered compound and Transition-metal aluminides
Aluminum
  baseline physical data 510
  basic observed properties 513, 516–7
  DFT-LDA electronic structure
    KKR energy bands and density of states 49
    simplified NFE PP energy bands 61
  DFT-based SM pseudopotential perturbation theory

    electron screening functions 109
    GPT form factor 110
    GPT pair potential 162
    pseudopotential nonlocality 116
    exchange and correlation 42, 117
  DRT normalized energy-wavenumber characteristic 113
  pseudopotential strength 131
  DRT SM treatment
    Debye temperature 298
    elastic constants 318
    equilibrium crystal structure 264
    pair potential 13, 121
    phonons 298
  empirical Morse and LJ pair potentials 13
  force-matching EAM potential 29–30
  GPT EDB treatment
    Bain path 259
    cohesive energy 254–5
    Debye temperature 298
    elastic constants 318
    equilibrium crystal structure 264
    high-pressure phase transitions 275–8
    pair potential 121, 357
    phonons 298
Andersen force theorem 256
Angular correlation function defined 341
  See also Molecular dynamics simulation
Atomic units 10–1, 508–9

Bain transformation path 257–60
Barium

  baseline physical data 510–1
  basic observed properties 512, 515, 517
  GPT treatment
    EDB pair potential 232–3
    TM three-ion potential 272–3
    TM equilibrium crystal structure 264–5, 268
    SM, EDB and TM phonons 301–3
Beryllium
  baseline physical data 510
  basic observed properties 512, 516–7
  DFT-LDA density of states 50
  GPT SM treatment
    equilibrium crystal structure 264
    form factor and pseudopotential strength 131
Bismuth
  baseline physical data 510
  basic observed properties 513
  GPT SM treatment
    pseudopotential strength 131
Bond-order potentials for transition metals
  cohesive-energy TB functional in terms of the d-state bond order 23
  numerical BOPs: inter-site bond-order representation of the d-bond energy and forces 234–42
  analytic BOPs: simplified Chebyshev polynomial representation

Subject Index

of the d-bond energy
  and forces 242–5
parameterization of bond
  integrals and the
  remaining repulsive
  energy 246–9
BOP vs MGPT
  methods 240–1,
  246–7
Born-Oppenheimer
  approximation 35–6
Bulk modulus
  for free electron gas 112
  from stress-strain
    relations 311–2
  from measured EOS 322–4
  from QHLD for GPT
    pair-potentials
    virial component 314
    non-virial
      correction 315–6
  from QHLD for
    MGPT multi-ion
    potentials 319–22
  hcp metals at constant axial
    ratio 325
  isothermal modulus
    defined 330
  See also Compressibility
    sum rule
Burgers transformation
  path 350

Cadmium
  baseline physical data 510
  basic observed
    properties 513,
    517
  GPT FDB treatment
    equilibrium crystal
      structure 264–5
    pair potential 172–3
  RMP treatment
    equilibrium crystal
      structure 264–5
Calcium
  baseline physical data 510–1
  basic observed
    properties 512,
    515, 517
  DFT-LDA density of
    states 53
  GPT EDB treatment
    Bain path 257–8
    band energies 221–222

cohesive energy 254–5
Debye temperature 164,
  298
d-state hybridization and
  optimization 163–5
elastic constants 318
equilibrium crystal
  structure 264–5
force-constant
  functions 290
form factor vs SM
  treatment 149
orthogonalization hole vs
  SM treatment 146–7
overlap potential 153
pair potential vs SM
  treatment 156–7
  and vs Sr and Ba
  potentials 172–3
phonons 296–8
GPT TM treatment
  pair potential 202
RMP treatment
  elastic constants 318
  equilibrium crystal
    structure 264
Ca-Mg binary alloy
  calcium as a transition
    metal 451–2
  GPT Mg-rich Ca_xMg_{1-x} pair
    potentials 452
  mechanical stability of
    the bcc random
    substitutional
    alloy 452–3
Canonical TB energy bands
  Derived d-band bonds
    integrals 68–9
  Basic origin in KKR-ASA
    equations 69–71
  MGPT canonical energy
    bands and bond
    integrals
    p-band 481
    d-band 209
    f-band 484
  See also Tight-binding d
    bands and Slater-
    Koster matrix
    relations
Cauchy pressure 14, 16–7,
  319
Cesium
  baseline physical data 510

basic observed
  properties 512,
  515, 517
DRT SM treatment
  equilibrium crystal
    structure 264
  pair potential 124
GPT SM treatment
  equilibrium crystal
    structure 263–4
Chromium
  baseline physical data 511
  basic observed
    properties 514,
    515, 517
  BOP TM treatment
    equilibrium crystal
      structure 268
  GPT TM treatment
    multi-ion
      potentials 201–4
    screening of sp-d
      hybridization 205–7
Cobalt
  baseline physical data 511
  basic observed
    properties 514,
    516–7
  GPT TM treatment
    pair potential 202
Cohesion and structure 253–6
Compressibility sum rule
  basic form 219, 317
  consequences for multi-ion
    potentials 319–20
  incorporation into
    MGPT 320–2
  See also Bulk modulus and
    Elastic moduli
Copper
  baseline physical data 510,
    512
  basic observed
    properties 512,
    515, 517
  DFT-LDA electronic
    structure
    KKR energy bands and
      density of states 52
    simplified H-NFE-TB
      energy bands 79
  GPT modified FDB
    treatment
    cohesive energy 254–5
    Debye temperature 299

Copper (*Continued*)
  elastic constants 318
  equilibrium crystal
    structure 264
  form factor and pair
    potential 168–9
  phonons 299
  GPT TM treatment
    pair potential 202
  zero-order pseudoatom
    properties
      cohesive energy 193–4
      d-band width 190, 192
      self-consistent
        valence 190, 192
      sharp resonance
        behavior 83–4
      valence energy level
        evolution 191–3
  Regnaut
    parameterization 169
  RMP treatment
    elastic constants 318
    equilibrium crystal
      structure 264
    hybridization potential vs.
      GPT 170–1
    pair potential 171
    phonons 299
Crystal structures predicted by
  QBIPs
    nontransition metals 262–7
    transition metals 267–71

d-band width
  from Wigner-Seitz boundary
    conditions 71, 189
  from TB bond integrals 73,
    189
  relationship to d resonance
    width 66–9
  3d and 4d transition-metal
    values 190
  volume dependence 191–2
ddcMD code using
  MGPT multi-
    ion potentials 338,
    340, 352
  See also Molecular dynamics
    simulation
Debye temperature
  defined 396–7
Density functional theory
  local density approximation
    (LDA)

basic self-consistent-field
  equations 37–9
HL, VMN and PZ
  parameterizations
    of exchange and
    correlation 39–40,
    519
  total energy
    calculation 38–9
methods beyond the LDA
  generalized gradient
    approximation 40–1
  many-body perturbation
    theory: GT and IU
    methods for valence
    electrons 41–2
small-core approximation
  separation of valence and
    core electrons 41–3
  effective core xc
    potential 43
  starting LDA
    equations 42–4
valence binding energy and
  cohesive energy
    free-atom binding or
      ionization energy
      and preparation
      energy 520
    simple metals 44
    transition-series
      metals 44–6
with electron temperature
  grand potential
    defined 496
  reduction to an electronic
    free energy 496–7
  electronic entropy 497
Density of states
  DFT-calculated results
    for selected
    metals 49–53
  free-electron form 48, 82
  Friedel model for
    transition-metal
    d bands 51–3
  general form for valence
    energy bands 48
  GPT form for d-
    band transition
    metals 85–6
  integrated DOS 48, 51
  moments defined 53–4
  temperature depen-
    dence near melt

  in 4d transition
    metals 494–5
Dislocation dynamics
  simulation
    See Multiscale modeling of
      plasticity
Dislocations in bcc metals
  core structure and
    polarization 401–5
  critical resolved shear
    stress 407–9
  high temperature
    velocity 412–4
  kink-pair formation energy
    405-7 and activation
    enthalpy 409–12
  pressure dependence of the
    Peierls stress 408–10
DRT model potential
  See Pseudopotentials in
    metals
Dynamical matrix 381–3, 387

Effective valence
  See Generalized pseu-
    dopotential
    theory
Elastic moduli
  from stress-strain in linear
    elasticity
      general relations 309–10
      cubic elastic
        constants 310–4
      hcp elastic
        constants 323–6
  from QHLD for QBIP
    pair-potentials
      cubic nontransition
        metals 316–9
  from QHLD for MGPT
    multi-ion potentials
      cubic transition
        metals 319–21
  See also Bulk modulus,
    Shear moduli and
    Thermoelasticity
Elastic anisotropy ratio
  defined 317
Electron exchange interaction
  and energy
    Hartree-Fock method 37
    Hartree-Fock-Slater
      methods 37–8, 40

Subject Index 567

Kohn-Sham-Gaspar
  exchange
  potential 40
  See also Density functional
    theory
Electron radius 39, 682
Electron screening
  functions 105–110
  dielectric function 106, 109
  Lindhard function 107–8
  polarizability 106
  random phase
    approximation 106
  susceptibility 107, 109
  Thomas-Fermi
    wavenumber 107
Electron-thermal free
  energy 356–8
Electronic entropy
  See Density functional
    theory
Electrostatic energy 94–6, 99
Energy-wavenumber
  characteristic
  simple metals 97–104,
    110–3
  empty or filled d-band
    metals 151, 154–6
  transition metals 198–200,
    671
  See also Generalized
    pseudopotential
    theory
Enthalpy difference between
  two structures under
  pressure 260–2
Equation of state 477–84
Equilibrium crystal structures
  of the elements
  nontransition metals 262–7
  transition metals 267–71
Environmental dependence
  of interatomic
  potentials 8

Failure of pure pair potentials
  in metals 11–15
Fermi-Dirac distribution
  function 496
Form factor defined 59, 109
  See also Pseudopotentials in
    metals
Free surfaces
  See Local volume
    GPT/MGPT

representations for
  defects and surfaces
Friedel oscillations and
  screening
  asymptotic pair potential in
    metals 25
  exact calculation of
    static lattice
    properties 125–6
  multi-ion screening of sp-d
    hybridization 205–7
  phonon anomalies 25,
    403–7
  potential damping tech-
    niques for atomistic
    simulations 126–30

Gallium
  baseline physical data 510
  basic observed prop-
    erties 513,
    517
  GPT FDB treatment
    equilibrium crystal
      structure 264–6
    pair potential 358
Generalized pseudopotential
  theory (GPT)
  plane-wave basis: simple
    metals
    NFE band
      structure 58–60
    PP perturbation
      theory 60–2
    valence electron density
      orthogonalization
      hole and effective
        valence 99, 521–3
    self-consistent
      screening 105–7
    reciprocal-space
      cohesive-energy
      functional 93–100
    real-space cohesive-energy
      functional 91, 100–5
    pair potential 103–4,
      121
    volume term 102–5,
      114
  plane-wave, localized d-state
    basis: d-band metals
    atomic-like d-state
      orbitals 62–3

hybridization potential
  and associated matrix
  elements 75–7
H-NFE-TB band
  structure 77–9,
    523–4
pseudo-Green's functions
  and the density of
  states 79–80, 525–6
resonant transition-
  metal ion reference
  system 81–5
valence band-structure
  energy 85–8, 137–44
valence electron den-
  sity 88–90, 144–8,
    251–3, 260–3
  orthogonalization
    hole and effective
    valence 145–7,
      195–6
  self-consistent screen-
    ing 144–5, 147–8,
      195–6
real-space cohesive-energy
  functional for empty-
  and filled d-band
  metals 136, 150–7
  pair potential 153–4, 157,
    172–5
  volume term 154, 156–7
real-space cohesive-energy
  functional for tran-
  sition metals 176,
    178–88, 197–201
  multi-ion d-state
    potentials 180–3
  pair potential 198–200
  partial TB moments and
    d-band occupation
    functions 183–8
  volume term 200–1
special modified FDB
  treatment of the noble
  metals 166–9
temperature-dependent
  real-space free-
    energy functional
    for transition
    metals 501–2
  multi-ion d-state
    potentials 498–9
  pair potential 501
  volume term 502

568 *Subject Index*

Generalized pseudopotential theory (GPT) (*Continued*)
  See also Adaptive GPT and Model generalized pseudopotential theory
Generalized stacking fault energy (γ) surfaces in bcc metals 394–8
Gold
  baseline physical data 510, 512
  basic observed properties 512, 516–7
  GPT modified FDB treatment
    equilibrium crystal structure 264
    form factor and pair potential 168–9
  RMP treatment
    pair potential 171
Grain-boundary atomic structure in bcc metals 419–22
Green's function simulation method for dislocations 398–401
  See also Dislocations in bcc metals
Grüneisen parameter
  low-temperature value defined from QHLD 327–30
  high-temperature value 362–3
Guiding principles for QBIPs 15–34

Heat of formation 432–4, 441–3, 447, 452
High-entropy transition-metal alloys 456–9
High-pressure phase transitions
  sp-d electron transfer in metals 272–5
  GPT and MGPT confirmed predictions 275–81
Hybridization potential
  See Generalized pseudopotential theory

Ideal shear strength 390–4, 410
Indium
  baseline physical data 510
  basic observed properties 513, 517
  GPT FDB treatment
    equilibrium crystal structure 264, 266–7
    pair potential vs SM treatment 266
INFERNO high-temperature electron-thermal code 358, 361
Iridium
  baseline physical data 511
  basic observed properties 513, 515, 517
  BOP TM treatment
    equilibrium crystal structure 268
Iron
  baseline physical data 511
  basic observed properties 514, 515, 517
  BOP TM treatment
    equilibrium crystal structure 268
    magnetism 250–1
    phonons 308
  GPT TM treatment
    multi-ion potentials 202–4
    proposed treatment of magnetism 251–2

Lead
  baseline physical data 510
  basic observed properties 513, 516–7
  GPT SM treatment
    form factor and pseudopotential strength 131
LAMMPS MD code with USER-MGPT package 36, 340
  See also Molecular dynamics simulation
Lennard-Jones pair potential 11

Liquid structure factor
  defined 457
Lithium
  baseline physical data 510
  basic observed properties 512, 515, 517
  DFT-LDA density of states 68
  DRT SM treatment
    bcc mechanical instability 263
    pair potential 124
    equilibrium crystal structure 262–4
  GPT SM treatment
    bcc mechanical stability 263
    equilibrium crystal structure 262–4
    rare-earth structures 263
Local density approximation
  See Density functional theory
Local volume GPT/MGPT representations for defects and surfaces
  local electron density modulation 460–8
  local density of states modulation 468–71
  See also Adaptive GPT

Machine-learning potentials
  force-matching EAM method 29–30
  Gaussian approximation potentials 30–2
  neural-network potentials 30–1
  spectral neighbor analysis potentials 32
Magnesium
  baseline physical data 510
  basic observed properties 512, 516–7
  DRT SM treatment
    cohesive energy 123
    Debye temperature 122, 300
    equilibrium crystal structure 264
    pair potential 121

phonons 300–1
structural energies 122
volume term 123
GPT SM treatment
  Bain path 257–8
  bulk modulus 322–6
  cohesive energy 101, 123, 254–5
  Debye temperature 122, 300, 329
  elastic constants 323–5, 368–9
  equilibrium crystal structure 264
  form factor 110
  Grüneisen parameter 329, 362–3
  hcp-bcc phase line 334–5
  high-pressure phase transitions 275–8
  impact of long-range cutoff 130
  melt curve and phase diagram 371–4
  pair potential 121, 124, 128
  phonons 300–1
  pseudopotential strength 131
  shock Hugoniot 372
  stable hcp stacking faults 34
  structural energies 122
  volume term 101, 105, 123
GPT EDB treatment
  high-pressure phase transitions 275–8
See also Adaptive GPT
Magnetism in BOP and MGPT potentials 250–2
Manganese
  baseline physical data 511
  basic observed properties 514, 517
  BOP TM treatment
    equilibrium crystal structure 268
    magnetic behavior 252
  DFT-LDA density of states 53
  GPT TM treatment
    multi-ion potentials 202, 204

Melt curve calculation
  from free energies 370–1
  two-phase method 352–4
  Z-method 354–5
  See also Pressure-temperature phase diagrams
Mermin temperature-dependent DFT
  See Density functional theory
Mercury
  baseline physical data 510
  basic observed properties 513, 517
  GPT FDB treatment
    equilibrium crystal structure 264–5
    high-pressure phase transition 278–9
$Mg_{17}Al_{12}$ ordered compound
  GPT pair potentials 431–2
  structure, cohesion and heat of formation 432–4
Model generalized pseudopotential theory (MGPT)
  d-band bcc transition metals with multi-ion potentials
    a simplification from GPT 208–9
    analytic MGPT for canonical d bands 209–16
    general matrix MGPT, including non-canonical d bands 216–8
    optimization of parameters 219–24
    special extension to five- and six-ion potentials 227–34
    temperature-dependent multi-ion potentials near melt 502–6
  d-band fcc transition metals with sp-d hybridization 224–7

p-band simple metals
  with canonical-band multi-ion potentials 481–4
f-band lanthanide and actinide metals
  with canonical-band multi-ion potentials 484–7
See also Slater-Koster matrix relations
Molecular dynamic simulation
  QBIP pair and multi-ion forces 337–41
  pair and angular correlation functions 341–2
  reversible-scaling MD for free energies 342–9
  structure factor for liquids 342
  variable-friction thermostats
    Nose-Hoover 337–8
    Gaussian 338
    Langevin 338
  See also Melt curve calculation, and ddcMD and LAMMPS codes
Morse pair potential 15
Molybdenum
  baseline physical data 511
  basic observed properties 514, 515, 517
  BOP TM treatment
    critical resolved shear stress 408–9
    equilibrium crystal structure 268
    phonons 306–7
    structural energy differences 267–71
    vacancy and self-interstitial formation energies 388
  GAP TM treatment
    self-interstitial formation energies 388
    vacancy formation and migration energies and volume 388
  MGPT TM treatment
    Bain bath 258–9
    cohesive energy 254–5
    critical resolved shear stress 408–9

Molybdenum (*Continued*)
  Debye temperature 306
  dislocation core structure 402–5
  elastic moduli under pressure 389–91; at high electron temperature 495
  equilibrium crystal structure 268
  grain boundary structure and energy 419–22
  γ surfaces 395–8
  ideal shear strength 393–4, 410
  kink-pair formation energy 406–7 and activation enthalpy 411–2
  melt curve to high pressure at high electron temperature 506–7
  multi-ion potentials 20, 222; at high electron temperature 506
  phonons 304–7; at high electron temperature 504–5
  pressure dependence of the Peierls stress 410
  self-interstitial formation and migration energies and volumes 388
  stress-strain and yield stress 418–9
  structural energy differences 267–71
  vacancy formation and migration energies and volumes 388
Multi-ion potential geometry 7
Multiscale modeling of single-crystal plasticity in bcc metals
  analytic forms of the activation enthalpy and dislocation velocity 413–6
  dislocation dynamics simulations of stress-strain and yield stress 1, 416–9
  See also Dislocations in bcc metals, ParaDiS code

Nickel
  baseline physical data 511
  basic observed properties 514, 515, 517
  GPT TM treatment
    equilibrium crystal structure 268
    pair potential 202, 227
  MGPT TM treatment
    equilibrium crystal structure 268
    pair potential 225–7
    phonons 308
    stacking fault energy 422–4
    vacancy formation energy 423–4
Niobium
  baseline physical data 511
  basic observed properties 514, 515, 517
  BOP TM treatment
    equilibrium crystal structure 268
  MGPT TM treatment
    equilibrium crystal structure 268
    grain boundary structure and energy 419–22
Norm-conserving model potential defined 57
  See also Pseudopotentials in metals

Orthogonalization hole
  See Generalized pseudopotential theory
Overlap potential for d states in the total energy
  empty or filled d-band metals 151–3, 526–8
  transition metals 198–9, 526–8
  See also Generalized pseudopotential theory

Pair correlation function defined 341
  See also Molecular dynamics simulation

ParaDiS dislocation dynamics code 414
  See also Multiscale modeling of single-crystal plasticity in bcc metals
Peierls stress defined 408
  See also Dislocations in bcc metals
Periodic Table 47
Pettifor-Ward pair-potential damping 127–8
  See also Friedel oscillations and screening
PHON lattice-dynamics code 295, 334
Phonon spectra: calculated vs. experiment
  nontransition metals 296–301
  transition metals 301–8
Physical convergence 12
Plutonium
  MGPT novel strong electron-correlation canonical d-band treatment of fcc δ-Pu
  d- vs f-band parameterization and Bain-path structural phase stability 490–2
  phonon spectra validated by experiment 491–2
Point defect formation and migration energies and volumes
  vacancies 382–8, 466–7, 477–8
  self-interstitials 385–8
Polymorphism in transition metals 375–81
Potassium
  baseline physical data 510
  basic observed properties 512, 515, 517
  DRT SM treatment
    Debye temperature 298
    elastic constants 318
    equilibrium crystal structure 264
    pair potential 124
    phonons 298
  GPT SM treatment
    cohesive energy 254–5

Debye temperature 298, 329
elastic constants 318
equilibrium crystal structure 263–4
Grüneisen parameter 329
phonons 296–8
rare-earth structures 263
Pressure-temperature phase diagrams 331–5, 371–5, 506–7
See also Magnesium, Molybdenum and Tantalum
Pseudopotentials in metals
Austin-Heine-Sham non-local PPs 55–6, 57–8
Moriarty GPT PP for simple metals 56, 118
Harrison optimized PP for simple metals 118
Heine-Abarenkov-type model potentials 56–57
Ashcroft local empty-core MP 58
Dagens-Rasolt-Taylor nonlocal norm-conserving MP 114–5, 120
Shaw optimized nonlocal MP 57
transition-metal extensions
Harrison transition-metal PP 78
Moriarty GPT PP for EDB and FDB metals 145, 149, 168–9
Dagens resonant model potential 169–72
See also Generalized pseudopotential theory
PURGATORIO high-temperature electron-thermal code 358

Quasicrystals
See Transition-metal aluminides

Quasiharmonic lattice dynamics 282–94
See also Phonon spectra: calculated vs. experiment

Rapid resolidification 376–7
Reference atomic volume 10–1
Resonant model potential
See Pseudopotentials in metals
Reversible-scaling MD
See Molecular dynamics simulation
Rubidium
baseline physical data 510
basic observed properties 512, 515, 517
DRT SM treatment
Debye temperature 298
elastic constants 318
equilibrium crystal structure 264
pair potential 124
phonons 298
GPT SM treatment
Debye temperature 298
elastic constants 318
equilibrium crystal structure 263–4
phonons 298

s-d electron transfer
See High-pressure phase transitions
Scandium
baseline physical data 511
basic observed properties 513, 516–7
GPT TM treatment
equilibrium crystal structure 268
multi-ion potentials 202–3
Self-interstitial formation and migration
See Point defect formation and migration energies
Shear elastic moduli and their pressure dependence 389–91
See also Elastic moduli

Shock physics and the Hugoniot 359–61
Silicon
baseline physical data 510
GPT SM pair potential 121
GPT EDB high-pressure phase transition 275–8
Silver
baseline physical data 510, 512
basic observed properties 512, 515, 517
GPT modified FDB treatment
equilibrium crystal structure 264
form factor and pair potential 168–9
RMP treatment
equilibrium crystal structure 264
pair potential 171
Slater-Koster matrix relations
standard TB representations for s, p, d and f states 73–74, 485
normalized SK representations of MGPT angular functions L, P and M
p states 482–4
d states 291–2
f states 485–7
reduction of angular functions L and P to analytic forms
p states 483
d states 211
Sodium
baseline physical data 510
basic observed properties 512, 515, 517
DRT SM treatment
Debye temperature 298

572 Subject Index

Sodium (Continued)
  elastic constants 318
  equilibrium crystal
    structure 264
  pair potential 121, 124
  phonons 298
  GPT SM treatment
    cohesive energy 254–5
    Debye temperature 298
    elastic constants 318
    equilibrium crystal
      structure 263–4
    form factor 110
    high-pressure phase
      transitions 275–7
    pair potential 121
    rare-earth structures 263
    phonons 298
  GPT EDB treatment
    high-pressure phase
      transitions 275–7
Sound velocity 367–8
Specific heat defined 329–30
Stacking faults in close-packed
    metals
  {111} fault in fcc
    metals 422–4
  basal-plane faults in hcp
    metals 34, 641–2
  See also Generalized
    stacking fault energy
    (γ) surfaces in bcc
    metals
Strontium
  baseline physical data 510–1
  basic observed
    properties 512,
    515, 517
  GPT EDB treatment
    equilibrium crystal
      structure 264–5
    pair potential vs Ca and
      Ba 172–3
Structure factor defined 59,
    93
  See also Pseudopotentials in
    metals and Molecular
    dynamics simulation
Surface energies and relax-
    ations 466–8,
    470–1

Tangential and radial force
    constant functions
  pair potentials 385–9
  multi-ion potentials 388–94
Tantalum
  baseline physical data 511

  basic observed proper-
    ties 514, 515,
    517
  BOP TM treatment
    equilibrium crystal
      structure 268
    structural energy
      differences 270–1
    vacancy and self-
      interstitial formation
      energies 387
  GAP TM treatment
    self-interstitial formation
      energies 387
    vacancy formation and
      migration energies
      and volume 387
  MGPT TM treatment
    Bain path 258–9
    cohesive energy 254–5
    critical resolved shear
      stress 407–9
    Debye temperature 306,
      329
    dislocation core
      structure 402–5
    elastic moduli and
      their tempera-
      ture and pressure
      dependence 366,
      389–91
    equilibrium crystal
      structure 268
    grain boundary structure
      and energy 419–22
    γ surfaces 394–8
    Grüneisen parameter 329,
      362–3
    high-T correlation
      functions 341–2
    ideal shear
      strength 393–4,
      410
    kink-pair formation energy
      406–7 and activation
      enthalpy 411–2
    melt curve and phase
      diagram 370–1,
      373–5
    multi-ion potentials 20,
      220–4
    pressure dependence of
      the Peierls stress 410
    phonons 303–6
    polymorphism 375–81
    rapid resolidification 376–
      7
    self-interstitial formation
      and migration

      energies and
      volumes 387
    shock Hugoniot 359–60
    shock melting 374
    sound velocity 368
    specific heat 330–1,
      361–4
    stress-strain and yield
      stress 416–8
    structural energy
      differences 270–1
    thermal expansion
      coefficient 330–1,
      361–4
    vacancy formation and
      migration energies
      and volumes 384,
      387
Temperature-induced
    solid-solid phase
    transitions 331–5
Tetragonal transformation
    path 350
Thallium
  baseline physical data 510
  basic observed prop-
    erties 513,
    517
  GPT FDB treatment
    equilibrium crystal
      structure 264, 267
    pair potential 358
Thermal expansion coefficient
    defined 330
Thermodynamic properties
    in the QHLD
    limit 326–331
Thermoelasticity 363–9
Tight-binding d bands
  density of states moments
    and d-band
    width 71–4
  equivalence of resonance
    and TB d bands 63–8
  resonance position and
    width 63, 67
  bond integral defined 64
  nonorthogonality integral
    defined 66
  GPT and BOP bond
    integrals for cen-
    tral transition
    metals 217–8, 246–9
  See also Canonical TB
    energy bands and

Subject Index   573

Slater-Koster matrix
    relations
Titanium
    baseline physical data 511
    basic observed properties 513, 516, 693
    BOP TM treatment
        equilibrium crystal structure 268
    GPT TM treatment
        pair potential 202
Transferability of interatomic potentials 8–9
Transition-metal aluminides
    BOP treatment of Ti-Al compounds and their mechanical properties 453–5
    GPT treatment of TM_xAl_{1-x} compounds and alloys
        self-consistent TM sp valence and volume 427–9
        total-energy functional 436–41
        $3d$ pair potentials 437-40 and structural energies 442–5
        Al_9Co_2 cohesion and heat of formation 441–3
        Al-Co and Al-Ni binary phase diagrams 445–7
        Al-Co-Ni and Al-Co-Cu ternary phase diagrams 448–50
        Al-Ni-Co quasicrystal 449–51
Trigonal transformation path 260
Two-phase melting 349–350
    See Melt curve calculation

Universal equation of state 431–4
Uranium
    MGPT canonical f-band treatment
        basic potential parameters and their parameterization 482–8
        calculated structural energies vs. DFT 488
        mechanical stability of the observed orthorhombic α-U structure 488–9
        phonons and 1000-K pair correlation function of high-temperature bcc structure 489–90

Vanadium
    baseline physical data 511
    basic observed properties 514–5, 517
    BOP TM treatment
        vacancy formation energy 387
    MGPT TM treatment
        cohesive energy 254–5
        dislocation core structure 402–5
        elastic moduli and their pressure dependence 280, 389–91
        equilibrium crystal structure 268
        γ surfaces 395, 397–8
        high-pressure phase transition 279–81
        ideal shear strength 393–4, 410
        kink-pair formation energy 406–7
        multi-ion potentials 20, 201–5, 221–3
        pressure dependence of the Peierls stress 410
        self-interstitial formation energies and volume 387
        vacancy formation and migration energies and volume 387
Vacancy formation and migration
    See Point defect formation and migration energies
Variational perturbation theory 349–52
Volume term in the total energy functional
    simple metals 98–100, 102–4
    empty or filled d-band metals 150, 154–7
    transition metals 198–201
    See also Generalized pseudopotential theory and Model generalized pseudopotentital theory
Voigt notation in elasticity 414

Zero-order pseudoatoms
    core and d-state energies and orbitals 157–9
    d-state localization potential and boundary conditions 159–64
    metal zero of energy, PP and extension to alloys 98, 528–30
    transition-metal valence, d-band width and cohesive energy 188–94
Zinc
    baseline physical data 510
    basic observed properties 513, 517
    GPT FDB treatment
        Bain path 259
        band energies vs DFT 165–6
        cohesive energy 254–5
        Debye temperature 164, 301
        d-state hybridization and optimization 163–5
        equilibrium crystal structure 264–5
        force constant functions 290
        form factor vs SM treatment 149
        orthogonalization hole vs SM treatment 146–7
        overlap potential 153
        phonons 297, 300–1
        pair potential vs SM treatment 156–7
        and vs Cd and Hg potentials 172–3
    RMP treatment
        equilibrium crystal structure 264–5
        phonons 300–1
Z-method melting
    See Melt curve calculation